Flavins and Flavoproteins 1987

Flavins and Flavoproteins 1987

Proceedings of the
Ninth International Symposium
Atlanta, Georgia, USA, June 7-12, 1987

Editors
D. E. Edmondson · D. B. McCormick

W DE G

Walter de Gruyter · Berlin · New York 1987

Editors
Dale E. Edmondson, Ph.D.
Professor of Biochemistry
Department of Biochemistry

Donald B. McCormick, Ph.D.
Fuller E. Callaway Professor and Chairman
Department of Biochemistry
Executive Associate Dean for Science

Emory University, School of Medicine
Atlanta, GA 30322
U.S.A.

Library of Congress Cataloging in Publication Data

Flavins and flavoproteins, 1987.

"The Ninth International Symposium on Flavins and Flavoproteins"--Pref.
Includes bibliographies and indexes.
1. Flavins--Congresses. 2. Flavoproteins--Congresses.
I. Edmondson, D. E. II. McCormick, Donald B. (Donald Bruce)
III. International Symposium on Flavins and Flavoproteins
(9th : 1987 : Atlanta, Ga.)
QP671.F5F52 1987 574.19'258 87-30478
ISBN 0-89925-306-7 (U.S.)

CIP-Kurztitelaufnahme der Deutschen Bibliothek

Flavins and flavoproteins ... : proceedings of the ... internat. symposium. – Berlin ; New York : de Gruyter.
1987. Proceedings of the ninth international symposium, Atlanta, Georgia, USA, June 7 – 12, 1987. – 1987
ISBN 3-11-010950-6

 Printing: Gerike GmbH, Berlin. Binding: Lüderitz & Bauer GmbH, Berlin. – Printed in Germany.

Preface

The Ninth International Symposium on Flavins and Flavoproteins was held June 7-12, 1987 on the campus of Emory University, Atlanta GA. As in recent symposia held in Ann Arbor and in Brighton, the format of the meeting consisted of morning lectures, afternoon poster sessions, and evening discussions. This format again proved to be successful in stimulating interactions among the 150 participants in attendence from all over the world.

The Organizing Committee and participants are indebted to the following organizations for their financial support: the National Science Foundation, the International Union of Biochemistry (through the Interest Group on Oxidases and Redox Enzymes), the National Institutes of Health, the Coca-Cola Company, Emory University, and Beckmann Instruments. This support made it possible to subsidize the registration fees for the meeting and to support the local expenses for a number of younger scientists in the field.

The chairman and co-chairman of the symposium wish to thank the members of the organizing committee for their help and advice in the organization of the program. Special mention should be made of the tireless efforts of Ms. Rae Ellen Young (the Conference Secretary) in seeing to all of the innumerable details and tasks that are required for a successful meeting.

The time and effort contributed by Joan Edmondson in advising and working with us on the planning of the meeting is also acknowledged and is deeply appreciated. Finally, we wish to thank the staff of the Division of Continuing Medical Education at Emory and the students and postdoctorals in each of our respective laboratories for their assistance during the meeting.

The diversity in subject matter of the papers presented here reflects a "maturation" of the field. A broad approach to the fundamental question of how flavoenymes work is evident by the combined efforts of x-ray crystallography, NMR spectroscopy, site-directed mutagenesis, in conjunction with more "classical" kinetic and mechanistic experiments. These approaches, in combination, promise to provide new experimental insights into the ongoing dilemma of how binding of the flavin to its site in the enzyme can dramatically alter its chemistry and oxidation-reduction properties. Thus, the field of flavin enzymology is embarking on a new era where molecular insights into structure and function will be of enormous benefit for a wide variety of biological and biomedical disciplines; ranging from neurochemistry to environmental detoxification.

The editors accept full responsibility for the organization of the contributed papers into the categories outlined in this book. Such a task always reflects the personal bias of the editors in their view of the field. To ensure the rapid publication of the proceedings of this symposium, we were prompted by the success of the previous symposium to again use the camera-ready format as the basis for this book. While such a format leads to more typographical errors than would be encountered in a printed volume with copy-editing, the benefit of rapid publication of the contributed articles, in our opinion, outweighs this deficiency.

Finally, we and our colleagues look forward to further exciting advances that will evolve during the next three years and constitute the basis for what will be reported in the tenth symposium on this subject to be hosted by Professor Bruno Curti and his associates in Italy.

Donald B. McCormick

Dale E. Edmondson

August, 1987

ORGANIZING COMMITTEE

D.B. McCormick (Atlanta) Chairman

D.E. Edmondson (Atlanta) Co-Chairman

R.C. Bray (Sussex)

B. Curti (Milano)

S. Ghisla (Konstanz)

F. Lederer (Paris)

V. Massey (Ann Arbor)

S.G. Mayhew (Dublin)

Y. Miyake (Osaka)

F. Müller (Wageningen)

G. Schultz (Freiburg)

C. Walsh (Cambridge, MA)

K. Yagi (Nagoya)

ACKNOWLEDGEMENTS

The Symposium Organizing Committee gratefully acknowledges financial support received from the following:

International Union of Biochemistry

National Science Foundation

National Institutes of Health

The Coca-Cola Company

Emory University

Beckmann Instruments

CONTENTS

STRUCTURE AND MECHANISMS OF FLAVOENZYMES

FLAVOENZYME STRUCTURE AND ELECTRON TRANSFER MECHANISMS

FLAVIN BIOSYNTHESIS, METABOLISM, AND MEDICAL ASPECTS

FLAVOENZYME HYDROXYLASES

Flavin-Dependent Bioluminescence

STRUCTURE, MECHANISM, AND BIOSYNTHESIS OF FLAVOENZYMES CONTAINING COVALENTLY-BOUND FLAVINS

Structure and Mechanisms of Flavoenzymes

ABSOLUTE STEREOCHEMISTRY OF FLAVINS IN ENZYME-CATALYZED REACTIONS

Dietmar J. Manstein, Vincent Massey*, and Emil F. Pai

Department of Biophysics, Max-Planck-Institute for Medical Research, Jahnstrasse 29, D-6900 Heidelberg, FRG, and
*Department of Biological Chemistry, The University of Michigan, Ann Arbor, MI 48109, USA

Introduction

Enzymes, as a rule, show extraordinary chiral recognition and use complex steric control to achieve a very high degree of stereospecificity of the reaction they catalyze. Sophisticated methods have been developed to explore the detailed stereochemistry going on at an enzyme's active site (1). In the case of flavoenzymes this knowledge has been restricted to the substrates; the isoalloxazine ring, the reactive part of the flavin prosthetic group, is not easily probed. Upon reduction possible isotopic labels would be attached to N1 and N5 from where they would be immediately lost to solvent. Therefore, it is not practicable to use label transfer to and from common flavins in order to determine the stereospecificity of flavoenzymes.

C4α and N5 have been identified as the positions at which most substrate molecules interact with the isoalloxazine ring system (2-7). In a first attempt to find out which one of the two prochiral faces of the flavin ring is used in any particular reaction 5-deaza-5-carba-analogues of the coenzymes FMN and FAD were synthesized (8-10). Further work with these analogues which had their prochiral center made inert to solvent exchange conclusively showed that there is direct hydrogen transfer from substrates to 5-deazaflavin coenzymes bound to the active site of flavoenzymes (10-13). Rapid scrambling of label between the

Flavins and Flavoproteins 1987

oxidized and reduced forms of 5-deazaflavins, however, made these analogues unsuited for determining relative stereospecificity (14).

In a series of coupled redox reactions leading to methane as the final product methanogenic bacteria use a 7,8-didemethyl-8-hydroxy-5-deazaflavin derivative, named F_{420}, as cofactor (15, 16). Its hydrolysis products include the riboflavin and the FMN forms of this analogue. In their case the analogous exchange of label is much slower making it feasible to use them for the determination of relative (17) or absolute (18) stereospecificity of F_{420}-dependent enzymes. As neither the riboflavin nor the FMN form of this compound is converted to the FAD level by the established enzymatic procedures (19), analyses of FAD-dependent flavoenzymes would have required chemical adenylylation.

The first flavoenzyme for which the absolute stereochemistry of the complete catalytic reaction became known, was glutathione reductase from human erythrocytes. X-ray crystallography elucidated the threedimensional structure of several reaction intermediates thereby establishing that the nicotinamide substrate interacts with the re-face of the isoalloxazine ring (20). Another result crucial for the development of our method of determination of absolute stereospecificity of flavoenzyme prosthetic groups, was the finding that after general acyl-CoA-dehydrogenase reconstituted with 5-deaza-FAD had been treated with NaB^3H_4, about 90% of tritium label associated with the flavin had been incorporated on one side of the ring system (21).

Results and Discussion

Slightly modifying the method described by Ashton and Brown (22) we were able to obtain good yields of pure 8-OH-5-deaza-riboflavin. This analogue proved to be a good substrate for the FAD synthetase from B. ammoniagenes, when the pH of the reaction

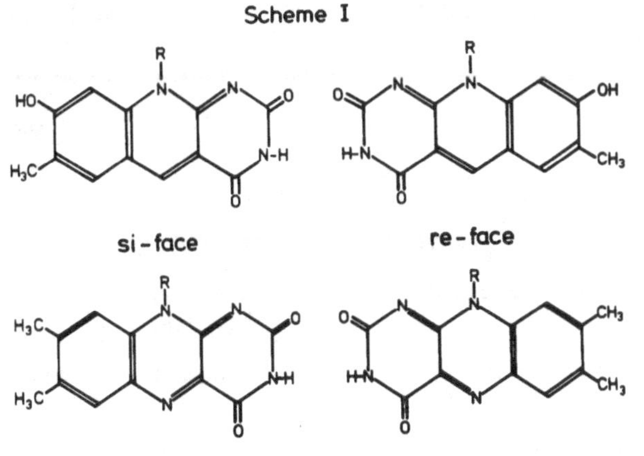

Scheme I: si- and re-face views of regular flavin and of the 8-OH-5-deaza-analogue.

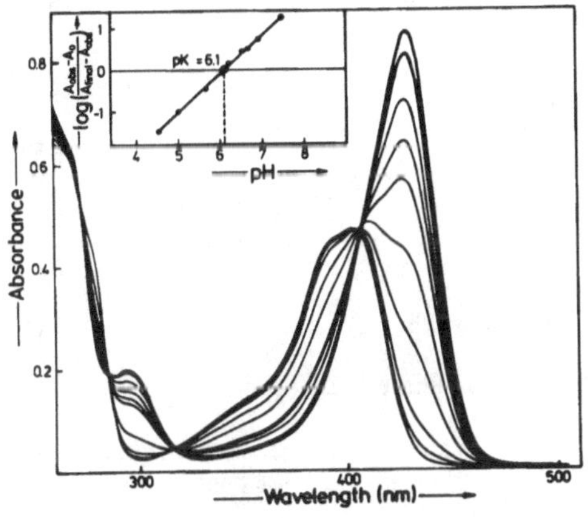

Fig. 1: UV-spectra of 8-OH-5-deaza-FAD at different pH-values. In a mixed buffer system NaOH was used to increase the pH from 3.9 via 4.6, 5.0, 5.7, 6.1, 6.5, 6.9, 7.6 to 10.5 which shows the highest absorption at 430 nm. Optical density at 430 nm (corrected for dilution) as a function of pH is plotted in the inset. As indicated, the resulting pK-value is 6.1.

Table I. Absorption Maxima and Extinction Coefficients of Proteins Reconstituted with 8-OH-5-Deaza-Flavins.[a]

Protein	Source	λ_{max} [nm]	ϵ_{max} [$cm^{-1}.M^{-1}$]
General Acyl-CoA Dehydrogenase	pig kidney	403	33,700
Glutathione Reductase	human erythrocytes	434	29,000
Mercuric Reductase	Pseudomonas aeruginosa	437	n.d.
Mercuric Reductase-Ser-135[b]	Pseudomonas aeruginosa	432	n.d.
Thioredoxin Reductase	E. coli	414	31,000
Glucose Oxidase	Aspergillus niger	413	32,000
p-Hydroxybenzoate Hydroxylase	Pseudomonas fluorescens	435	39,500
+ 50 µM p-hydroxybenzoate		443	40,000
Melilotate Hydroxylase	Pseudomonas sp.	433	36,600
+ 20 µM melilotate		433	41,000
Anthranilate Hydroxylase	Trichosporum cutaneum	440	40,200
+ 100 µM anthranylic acid		440	42,400
+ 100 µM salicylic acid		442	45,500
Ferredoxin:NADP$^+$-oxidoreductase	spinach	432	36,800
L-Lactate Oxidase	Mycobacterium smegmatis	446	n.d.
D-Lactate Dehydrogenase	Megasphera elsdenii	430	36,500
D-Aminoacid Oxidase	pig kidney	432	39,000
Flavodoxin	Megasphera elsdenii	430	29,600

[a]in buffer A as solvent; [b]single site mutant with Cys-135 replaced by Ser-135;

was kept at 5.9 pH units. This was rather crucial as the rate of conversion steadily increased when the pH was lowered, until at pH values between 5.8 and 5.6 the enzyme started to become unstable. It seems, as if the enzyme accepts only the neutral form of this anlogue as substrate. The reaction products were purified by HPLC. The FMN form was obtained by treating the corresponding FAD with Naja naja venom.

Most of the apoenzymes tested so far showed tight binding of the respective 8-OH-5-deaza-flavin, both in the oxidized and in the reduced state. When reconstituted enzymes were subjected to gel filtration no loss of label could be detected. The spectral features (Table I) remained constant. None of the 8-OH-5-deaza-flavin enzymes we have tested had its native activity. However, anyone of them showed at least partial reactiviy towards either substrates or products enabling us to determine its stereospecificity (26).

We used two methods to introduce the isotopic label, reduction of the analogue with NaB^3H_4 while it was bound to gen. acyl-CoA dehydrogenase or reaction of tritiated glucose with 8-OH-5-deaza-

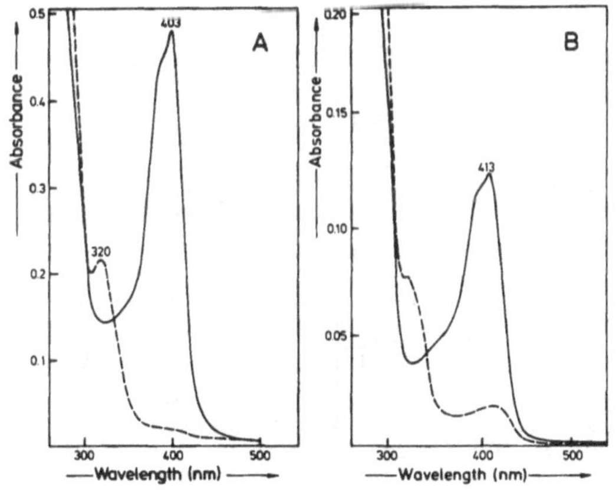

Fig. 2: Spectra of gen. acyl-CoA dehydrogenase (A) and glucose oxidase (B) complexed with oxidized (——) or reduced (---) analogue, respectively. The visible absorption band provides a convenient way of recording reduction or reoxidation rates.

General procedure of stereoanalysis

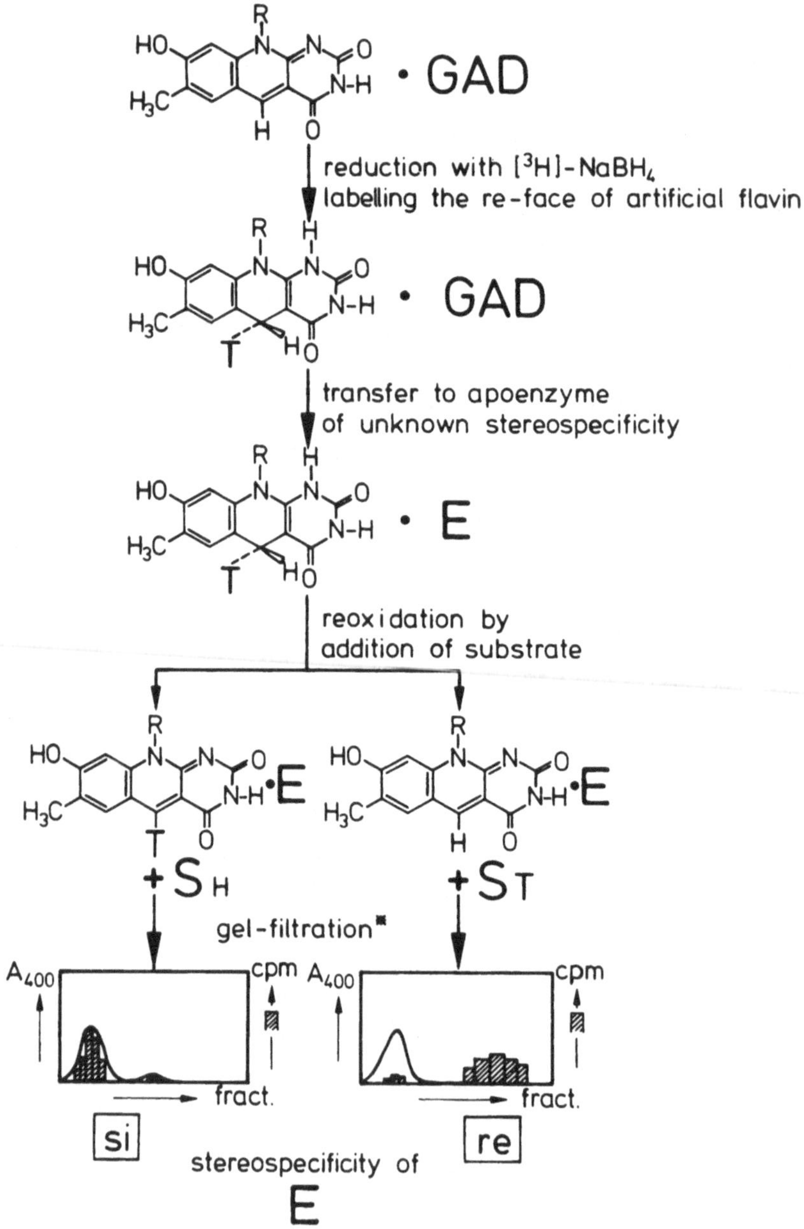

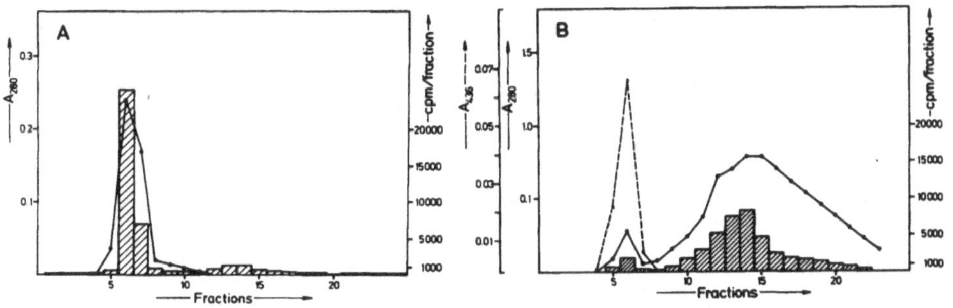

Fig. 3: Gel filtration profile of glutathione reductase reconstituted with reduced 8-OH-5-deaza-FAD which had been stereoselectively labeled in gen. acyl-CoA dehydrogenase (A). Gel filtration of the same sample after treating it with a large excess of NAD^+ over protein. After reoxidation the counts are found in the small molecule fraction of the eluate. This establishes that gen. acyl-CoA dehydrogenase and glutathione reductase have had the same side of the flavin ring interact with sodium borohydride and NAD^+, respectively.

FAD·glucose oxidase. Following the "General Procedure " shown on the last page, the now stereospecifically labeled analogue was set free by incubating the solution at 100°C for a short time. After centrifugation to remove denatured protein and reconstitution of a second apoenzyme with the reduced flavin, the analogue was reoxidized by adding the appropriate substrate. When the radioactivity was found in the peak of a gel filtration column which contained the small molecules, like nucleotides or amino acids, we concluded that the two flavoenzymes studied used the same side of the flavin ring to interact with their respective substrates. When, however, the radioactivity was found together with the protein opposite sides had been involved. Figure 3 gives an example for the couple gen. acyl-CoA dehydrogenase and glutathione reductase. Exact reaction conditions had to be established by preliminary tests and were slightly different for every pair of enzymes. In addition to gel filtration we recently used HPLC to analyze protein and substrate fractions. which helped in directly identifying the compound accepting the label from the flavin analogue.

DETERMINATION OF STEREOSPECIFICITY OF FLAVIN-SUBSTRATE INTERACTION

enzyme	source	substrate	distr. of label flavin	substr.	face of flavin
glutathione reductase	human red blood cells	$NADP^+$	9 %	91 %	re
mercuric reductase	Pseudomonas aeruginosa	$NADP^+$	16 %	84 %	re
(mutant enzyme: Cys-Ser-135)			7 %	93 %	
thioredoxin reductase	Escherichia coli	$NADP^+$	9 %	91 %	re
p-OH-benzoate hydroxylase	Pseudomonas fluorescens	$NADP^+$	5 %	95 %	re
melilotate hydroxylase	Pseudomonas sp.	NAD^+	9 %	91 %	re
anthranilate hydroxylase	Trichosporum cutaneum	$NADP^+$	24 %	76 %	re
MHPC-* oxygenase	Pseudomonas sp. MA-1	NAD^+	20 %	80 %	re
cyclohexanone oxygenase	Acinetobacter NCIB	$NADP^+$	5 %	95 %	re
D-amino acid oxidase	pig kidney	pyruvate + NH_4^+	7 %	93 %	re
L-lactate oxidase	Mycobacterium smegmatis	pyruvate	88 %	12 %	si
D-lactate dehydrogenase	Megasphera elsdenii	pyruvate	95 %	5 %	si
glucose oxidase	Aspergillus niger	glucose	7 %	93 %	re[A]
gen. acyl-CoA dehydrogenase	pig kidney	acyl-CoA $NaBH_4$	5 %	95 %	re[B]

* 2-methyl-3-hydroxypyridine-5-carboxylic acid oxygenase

[A] analogue was labelled in glucose oxidase and analyzed in mercuric reductase.

[B] Ghisla et al., Biochemistry 23, 3154-3161 (1984)
gen. acyl-CoA-dehydrogenase was always used to label the analogues except with glucose oxidase (see [A]). By linking it to glutathione reductase it was established that this enzyme uses the re-face.

It is not yet possible to correlate the type of enzyme and the stereospecificity of its prosthetic flavin group. There are however, some trends. All enzymes which use pyridine nucleotides have this substrate reacting at the re-face of the isoalloxazine ring. Sequence homology (glutathione reductase, mercuric reductase, thioredoxin reductase, and D-amino acid oxidase) as well as structure analysis by X-ray crystallography (glutathione reductase and p-hydroxybenzoate hydroxylase) (23, 24) suggest similar binding domains for the flavin prosthetic group of these enzymes. Similarly, inspection of the crystal structure of glycollate oxidase (25) reveals that this enzyme binds its substrate at the si-side of the flavin ring, as does L-lactate oxidase. It would be interesting to see a crystal structure of the latter protein to find out whether it is related to the ß-barrel type of glycollate oxidase which in turn has no similarity to the chain folds of the reductases mentioned above.

Acknowledgement

We thank our colleagues S. Ghisla, L.M. Schopfer, and C. Walsh for very helpful discussions. Many scientists contributed by providing samples of pure enzymes. We are grateful to all of them. This work was supported in part by Grant GM-11106 from the U.S. Public Health Service and Grant Ma-1081/1 from the DFG.

References

1. Retey, J., J.A. Robinson. 1982. Stereospecificity in Organic Chemistry and Enzymology. Verlag Chemie, Weinheim.

2. Jorns, M.S., L.B. Hersh. 1974. J.Am.Chem.Soc. 96, 4012.

3. Fisher, J., C. Walsh. 1974. J.Am.Chem.Soc. 96, 4345.

4. Hersh, L.B., C. Walsh. 1981. Meth.Enzymol. 66, 277.

5. Thorpe, C., C.H. Williams. 1976. J.Biol.Chem. 251, 7726.

6. Brüstlein, M., T.C. Bruice. 1972. J.Am.Chem.Soc. 94, 6548.

7. Loechler, E.L., T.C. Hollocher. 1980. J.Am.Chem.Soc. 102, 7312.

8. O'Brien, D., L. Weinstock, C. Cheng. 1970. J.Heterocycl. Chem. 7, 99.

9. Hersh, L.B., M.S. Jorns. 1975. J.Biol.Chem. 250, 8728.

10. Fisher, J., R. Spencer, C. Walsh. 1976. Biochem. 15, 1054.

11. Jorns, M.S., L.B. Hersh. 1975. J.Biol.Chem. 250, 3620.

12. Jorns, M.S., L.B. Hersh. 1976. J.Biol.Chem. 251, 4872.

13. Hersh, L.B., M.S. Jorns, J. Peterson, M. Currie. 1976. J.Am.Chem.Soc. 98, 865.

14. Spencer, R., J. Fisher, C. Walsh, 1976. Biochem. 15, 1043.

15. Zeikus, J.G., G. Fuchs, W. Kenealy, R.K. Thauer. 1977. J.Bacteriol. 132, 604.

16. Eirich, L.D., G.D. Vogels, R.S. Wolfe.1978. Biochem. 17, 4583.

17. Yamazaki, S., L. Tsai, T.C. Stadtman, F.S. Jacobson, C. Walsh. 1980. J.Biol.Chem. 255, 9025.

18. Yamazaki, S. L. Tsai, T.C. Stadtman, T. Teshima, A. Nakaji, T. Shiba. 1985. Proc.Natl.Acad.Sci. 82, 1364.

19. Walsh, C., J. Fisher, R. Spencer, R.W. Graham, W.T. Ashton, J.E. Brown, R.D. Brown, E.F. Rogers. 1978. Biochem. 17, 1942.

20. Pai, E.F., G.E. Schulz. 1983. J.Biol.Chem. 258, 1752.

21. Ghisla, S., C. Thorpe, V. Massey. 1984. Biochem. 23, 3154.

22. Ashton, W.T., R.D. Brown. 1980. J.Heterocycl.Chem. 17, 1709.

23. Rice, D.W., G.E. Schulz, J.R. Guest. 1984. J.Mol.Biol. 174,483

24. Wierenga, R.K., J. Drenth, G.E. Schulz. 1983. J.Mol.Biol. 167, 725.

25. Lyndqvist, Y., C.-I. Brändén. 1985. Proc.Natl.Acad.Sci. 82, 6855.

26. Manstein, D.J., E.F. Pai, L.M. Schopfer, V. Massey. 1986. Biochem. 25, 6807.

CONSERVED CYSTEINE PAIRS OF MERCURIC ION REDUCTASE: AN INVESTIGATION OF FUNCTION VIA SITE-DIRECTED MUTAGENESIS

Christopher T. Walsh, Melissa J. Moore, Mark D. Distefano
Departments of Biology and Chemistry, Massachusetts Institute of Technology, Cambridge, MA 02139

Introduction

Mercuric ion reductase (MR) catalyzes the reduction of mercuric ions to elemental mercury with concomitant oxidation of NADPH to $NADP^+$. This reaction constitutes the last step in bacterial mercury detoxification, a pathway that involves several proteins which carry out a vareity of regulatory, transport, and chemical processing functions (Figure 1) (See 1,2 for review).

MR contains both a flavin (FAD) (3) and an active site redox-active disulfide (4). These elements are characteristic of a family of flavoproteins, classified as the disulfide oxidoreductases, that includes the enzymes glutathione reductase (GR) (5), lipoamide dehydrogenase (LAD) (5), and trypanothione reductase (TR) (6). These enzymes possess a four electron redox capacity, two of which can be accommodated by the redox-active disulfide and the other two by the FAD cofactor. All of the above proteins catalyze the reduction of various disulfide substrates by thio/disulfide interchange reactions via the redox-active disulfide in the two electron reduced enzyme species (EH_2) (5). However, while GR, LAD and MR are all able to bind mercuric ions in their EH_2 forms (7,8,9), only MR is capable of catalyzing their subsequent reduction; GR and LAD remain potently inhibited.

Flavins and Flavoproteins 1987

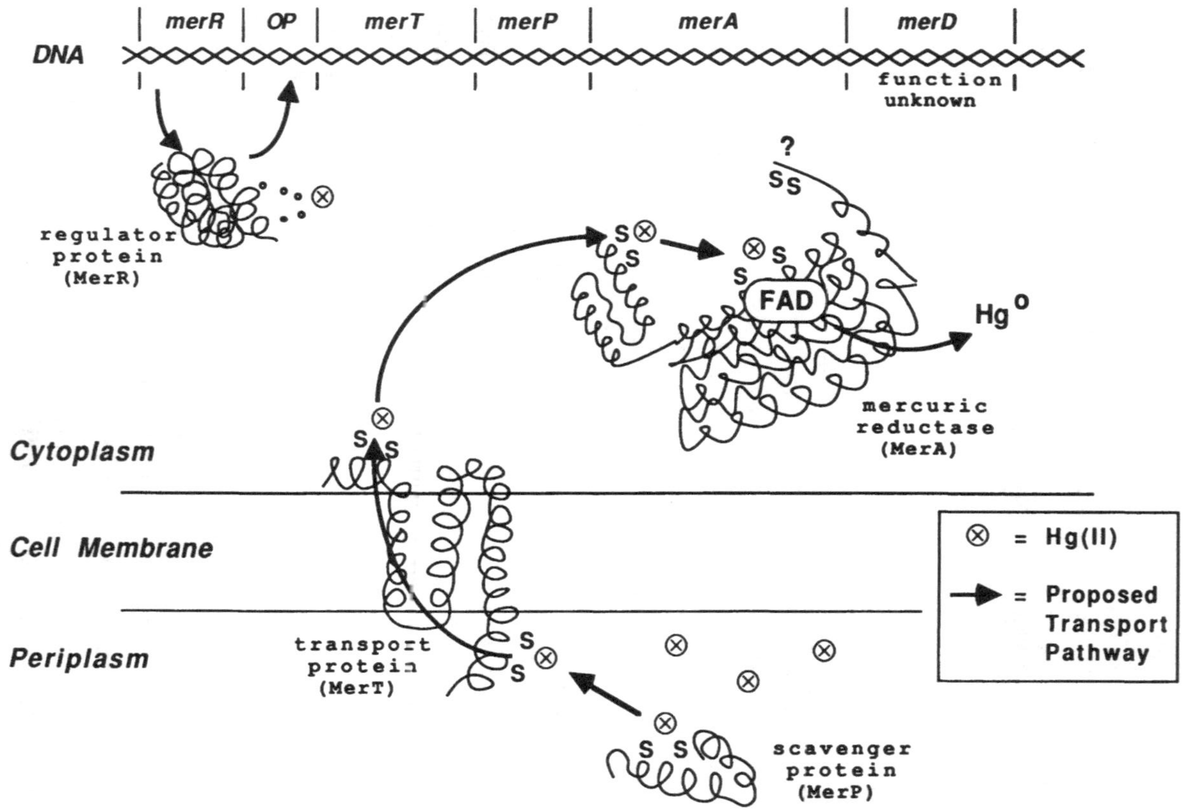

Figure 1: Proposed Hg(II) transport pathway and roles of proteins encoded by the mer operon.

Our research has been directed towards delineating the differences between MR and other disulfide oxidoreductases with an emphasis on understanding their mechanistic implications. To assess the function of potentially important residues we have used site-directed mutagenesis to alter them, followed by characterization of the resulting mutants by both in vivo and in vitro methods.

The means of choice for selecting appropriate amino acid residues as targets for mutagenesis has always been by examination of a well-defined protein crystal structure. Unfortunately, despite intensive efforts, no such structure exists for MR. However, examination of the primary sequences of MR from three sources (10,11,12) (Figure 2) clearly shows the absolute conservation of three cysteine pairs in each protein. One of these cysteine pairs forms the redox-active disulfide characteristic of this oxidoreductase family (13). These cysteines have been implicated as participants in

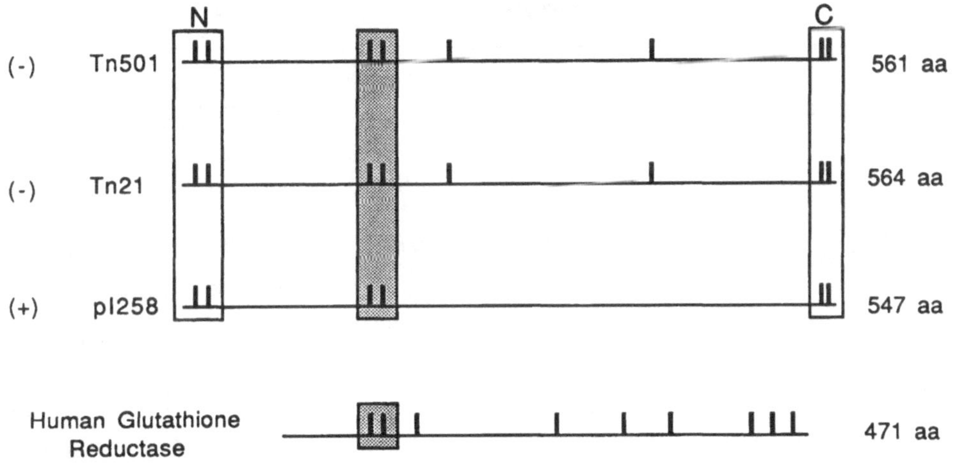

Figure 2. Location of cysteine residues in the primary sequences of MR from Gram-negative (-) and Gram-positive (+) sources. Also shown for comparison is the location of cysteines in human GR.

catalysis of iodoacetamide inactivation studies (13), mercury binding experiments (7) and early mutagenesis work (14). The second and third conserved cysteine pairs are located at the N- and C-terminii of the MR primary structure. These cysteine residues are present in regions which possess no sequence homology with the other proteins in this family and may thereby be likely to fulfill key roles in conferring upon MR its unique ability to reduce mercuric ions.

In this paper we describe the mutant MR enzymes shown in Table 1 and summarize their in vivo properties (22) as well as provide some information concerning the physical, spectral and catalytic properties of the purified proteins in vitro (22,23). Finally, we discuss the mechanistic implications of these data with a view toward understanding how MR binds and reduces mercuric ions.

Table I: Mutations of Cysteines in Mercuric Ion Reductase

Region	Native	Mutant
N-terminus	cys_{10} cys_{13}	1) cys_{10} ala_{13}
		2) ala_{10} ala_{13}
Redox active disulfide	cys_{135} cys_{140}	3) ala_{135} cys_{140}
		4) ser_{135} cys_{140}
		5) cys_{135} ala_{140}
		6) cys_{135} ser_{140}
		7) ala_{135} ala_{140}
C-terminus	cys_{558} cys_{559}	8) ala_{558} ala_{559}

Methods

In vivo characterization of mutants. Functionally, the structural genes of the Tn401 mer operon encode two distinct activities: 1) transport of mercurials and 2) Hg(II) reduction (Figure 1). Mutations causing disruption of either activity can be distinguished in vivo by observing the $HgCl_2$ resistance phenotype of cells harboring the mutation. Such an in vivo test of function is especially important for those mutations which affect transport, but not reduction, since no analogous in vitro assay exists for this multicomponent activity.

If one measures the growth of cells containing various mer alleles in liquid media containing differing amounts of $HgCl_2$ (15), three resistance phenotypes become readily apparent: resistant, sensitive, and supersensitive. Resistant cells are those which contain a functionally intact mer operon and can withstand $HgCl_2$ concentrations in excess of 50 mM $HgCl_2$. Sensitive cells either lack a working Hg(II) transport system or the entire mer operon; they are viable at low Hg(II) concentrations, but do not grow in 50 mM HgCl2. Conversely, supersensitive cells retain the transport activity, but not the ability to complete the pathway by reducing Hg(II) to Hg(0). These cells essentially commit suicide by concentrating mercurials in their cytoplasms and characteristically show no growth above 5 mM $HgCl_2$.

In vitro characterization. The mutant MR genes have been cloned into plasmids behind the tac promoter to yield good protein overproduction (ca. 5% of total soluble protein) (14). A one step purification using Orange A affinity chromatography (4) allows the isolation of homogeneous protein in 100 mg quantities. With the purified proteins in hand, we have characterized the mutant MR's using a variety of physical and enzymological methods.

In the absence of a crystal structure, physical methods are valuable tools for providing confirmation of tertiary structure integrity in the mutant enzymes, in assessing the specific effects of mutations and in complementing enzymological data. Thus, for several mutant MR's we have measured the native and subunit molecular weight, the number of titratable thiols by DTNB and their susceptibility to alkylation by iodoacetamide. We have also carried out detailed spectrophotometric measurements to determine flavin k_{max}, extinction coefficients, thiolate/flavin charge transfer absorbances and redox potentials. Lastly, fluorescence studies were performed to determine charge transfer fluorescence quenching and fluorescence intensity relative to free flavin.

MR has a variety of catalytic activities (14). As shown in Table 2, (see following page) catalytic mercury reduction can be assayed by monitoring NADPH consumption or $^{203}Hg(0)$ production. Single turnover events can be observed by titrating the enzymes to their reduced flavin state followed by addition of excess Hg(II), which causes flavin reoxidation in a pseudo-first order fashion.

As a member of the disulfide oxidoreductase family, MR can also reduce small disulfides, such as DTNB (16), by a process thought to involve the redox-active disulfide. Additionally, MR can catalyze the reduction of other electron acceptors, namely the $NADP^+$ analogue, $thioNADP^+$, and O_2. These latter activities do not require residues on the flavin si face (17), where Hg(II) and DTNB reduction occurs, thus allowing the distinction between those residues involved in Hg(II) binding/reduction and those involved in flavin reduction. (It should be pointed out that, where noted in Table 2, several enzyme activities were assayed anaerobically; this was done to eliminate any possible interference from the O_2 reductase activity or its product, H_2O_2).

TABLE II: Native and mutant mercuric ion reductase activity assays.

	Activity	Parameter Monitored	Reaction
Aerobic:	1) Hg(II) reduction	NADPH consumption (340 nm)	$Hg(SR)_2 + NADPH + H^+ \longrightarrow Hg(0) + NADP^+ + 2\ RSH$
	2) Transhydrogenation	NADPH consumption (340 nm or fluorescence)	$thioNADP^+ + NADPH \longrightarrow thioNADPH + NADP^+$
	3) O_2 reduction	O_2 or NADPH consumption (Oxygen electrode or 340 nm)	$O_2 + NADPH + H^+ \longrightarrow H_2O_2 + NADP^+$
Anaerobic:	4) DTNB reduction	TNB^- production (412 nm)	$RSSR + NADPH \longrightarrow 2RS^- + NADP^+ + H^+$
	5) Hg(II)- dependant $E\text{-}FADH_2$ reoxidation (stoichiometric)	reappearance of oxidized flavin spectrum (430 to 450 nm region)	$Enz\text{-}FADH_2 + Hg(II)[Ligand^*] \longrightarrow Enz\text{-}FAD + Hg(0) + [Ligand^*]$
	6) Hg(II) reduction (multiple turnovers)	NADPH consumption or $^{203}Hg(0)$ volatilization (340 nm or liquid scintillation counting)	$NADPH + Hg(II)[Ligand^*] \longrightarrow NADP^+ + Hg(0)\uparrow + [Ligand^*]$

* Ligand can be either EDTA or thiol $(SR)_2$

Results

<u>N-terminal mutants</u>. The most N-terminal cysteine pair of MR, cys_{10},cys_{13}, is found in an 85 amino acid domain that is totally absent in other members of the disulfide oxidoreductase family (Figure 3). While this domain is conserved in MR's from a variety of sources (10,11,12), previous in vitro proteolysis experiments have shown it to have no detectable function in catalysis (13). However, there is ca. 30% homology in this region with MerP, the perplasmic Hg(II) binding protein (1) (Figure 3). Thus Brown (18) has suggested a transport role for this domain, perhaps involving specific transfer of Hg(II) from MerT via the cys_{10},cys_{13} pair to the active site of MR (Figure 1). This scheme is particularly inviting because at no time would Hg(II) be free in the cytoplasm to react with other cellular thiols. To test the above hypothesis, we constructed two MR mutants, cys_{10},ala_{13} and ala_{10},ala_{13}, and examined their conferred

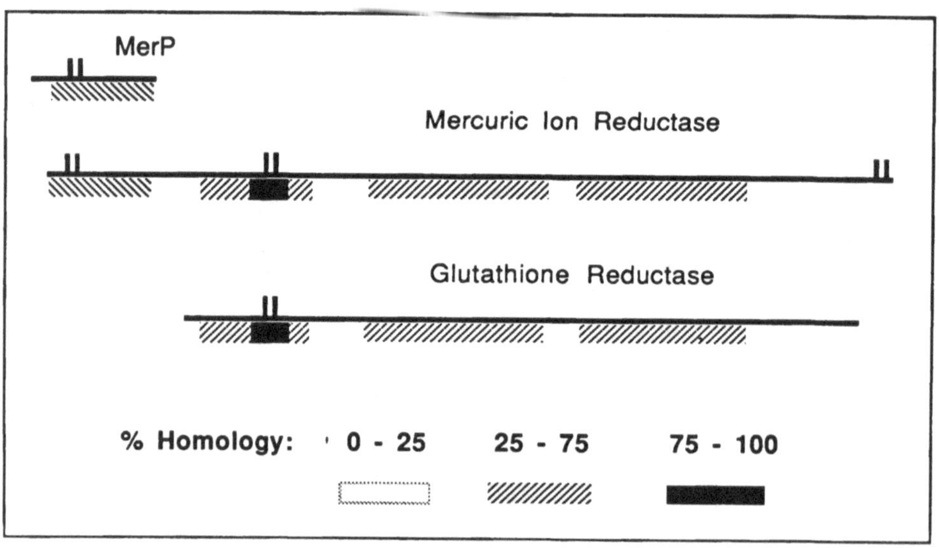

Figure 3: Areas of homology between the primary sequences of MerP, MR and GR with cysteine pairs shown.

$HgCl_2$ resistance phenotypes. However, when either mutation is present in E. coli on a multiple copy plasmid, the resultant phenotype is indistinguishable from that of the wild type, i.e. the cells are <u>fully $HgCl_2$ resistant</u>. To ask whether the entire N-terminal domain is necessary at all in vivo, we then constructed a shortened version of the merA gene that should encode for "clipped" MR (i.e. residues 86 - 561). This mutation results in a <u>supersensitive</u> phentotype, but further studies revealed that no "clipped" protein is detectable in the cell, suggesting no translation or a very short halflife for the polypeptide (e.g. due to a folding problem).

<u>Redox-active disulfide mutants</u>. As discussed briefly above, all members of the disulfide oxidoreductase family possess a redox active disulfide and FAD cofactor, resulting in a 4 electron redox capacity per monomer (Figure 4). Input of two electrons first reduces the redox-active disulfide to generate a species denoted EH_2; addition of two more electrons gives rise to the reduced flavin species, termed EH_4. In disulfide oxidoreductases, such as GR and LAD, which perform dithiol/disulfide redox chemistry, catalysis is thought to

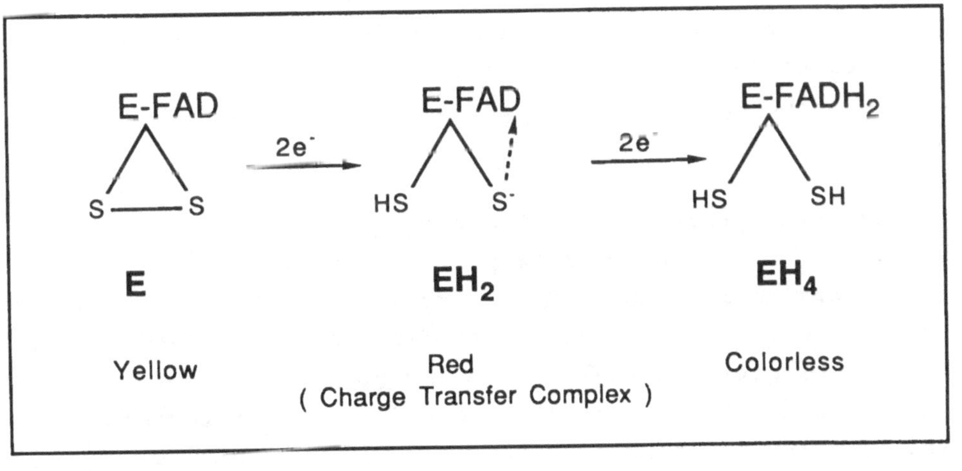

Figure 4: Redox states of MR

occur by enzymatic cycling between the E and EH_2 forms (5); thus the FAD moiety serves merely as an electron conduit between the disulfide and nicotinamide. However, in MR the roles of the redox active disulfide and FAD remain unclear. To assess the specific function of the cysteine residues, cys_{135} and cys_{140}, which form the active site disulfide, we have constructed cys to ser and cys to ala mutants at both positions and have studied their physical and enzymological properties in vitro.

Physical characterization of the mutant enzymes listed in Table 2 has revealed that their subunit and native molecular weights are identical to those of native MR. The mutant enzymes also titrate the predicted number of thiols in both their oxidized and reduced states (e.g. three in cys_{135}, ala_{140}). Furthermore, all the mutants possess only a 2 electron redox capacity per subunit, consistent with the removal of the redox active disulfide. While each mutant is distinct spectroscopically, the existence of charge transfer absorbance and dramatic fluorescence quenching in the $ser_{135}cys_{140}$ and $ala_{135}cys_{140}$ mutants represents the most significant departure from the native spectrum. However, this charge transfer absorbance mirrors that seen in the native EH_2 spectrum and is consistent with the localization of cys_{140} proximal to the flavin, with the cys_{140} thiolate anion as charge transfer donor and oxidized FAD the charge transfer acceptor.

The spectral properties of the $ser_{135}cys_{140}$ (red) and $ala_{135}cy_{s}140$ (green) mutants provide a useful optical window for observing binding of Hg(II) to cys_{140}. While addition of $HgCl_2$ does eliminate charge transfer absorbance (and converts the red or green enzyme back to the yellow color of unperturbed FAD), it can be regained by adding low amounts of thiols or EDTA (molar amounts $\geq 2[HgCl_2]$). This is in contrast to the results observed by Miller et al for native MR

(7) and clearly illustrates the decreased affinity for Hg(II) binding to cys_{140} in the absence of cys_{135}.

Catalytically, all the active site mutant MR's retain the oxidase and transhydrogenase activities. For all but one mutant, $ser_{135}cys_{140}$, these k_{cat} values are actually higher than those for native MR. Interestingly, a linear correlation between enzyme-bound flavin redox potential and k_{cat} is observed, implicating the flavin in a rate determining step in these two reactions. In aerobic NADPH consumption assays no mercuric reductase activity was detected for any of the active site mutants (?5% sensitivity limit). We then switched to a more sensitive single turnover assay which measured E-$FADH_2$ reoxidation by Hg(II) and found that <u>all</u> active site mutants are in fact reoxidized by Hg(II), but at rates of 10^2 to 10^4 times slower than native MR. To search for multiple turnover events a sensitive $^{203}Hg(II)$ to volatile $^{203}Hg(0)$ assay was employed. While the $cys_{135}ser_{140}$, $cys_{135}ala_{140}$ mutants were too slow to give reliable numbers, the $ser_{135}cys_{140}$ and $ala_{135}cys_{140}$ mutants carried out at least 30 turnovers at rates comparable to those observed in the single turnover experiments.

<u>C-terminus mutants</u>. Like cys_{10},cys_{13}, the C-terminal cysteine pair, cys_{558},cys_{559}, is located in an extension of 15 residues having no counterpart in the other members of this oxidoreductase family (Figure 3). By modelling this "tail" onto the crystal structure of human GR, Brown et al (18) have proposed that these C-terminal cysteines are capable of involvement at the active site of the opposite subunit, perhaps in a Hg(II) binding capacity. Thus as the third target in our mutagenesis program, we have constructed the MR mutant ala_{558},ala_{559}.

Bacteria containing the ala_{558},ala_{559} mutation exhibit a $HgCl_2$ supersensitive phenotype, suggesting an enzyme that is no longer catalytically competent. We have now overproduced and purified this mutant MR in good yield, and have shown that the enzyme is very simlar to native MR physically; in fact their visible spectra are indistinguishable. However, like the redox active disulfide mutants, ala_{558},ala_{559} MR shows no detectable mercuric reductase activity aerobically. Similarly, this mutant MR retains both the oxidase and transhydrogenase activities, with k_{cat}'s similar to native MR. (Interestingly, addition of $Hg(SR)_2$ increases the enzyme's oxidase activity in a saturable fashion). It is also capable of reducing DTNB at a rate comparable to the native enzyme demonstrating the intactness of the redox active disulfide. Further analysis of this mutant awaits more anaerobic assay data and redox potential determination.

Discussion

Bacterial mercury detoxification is a complex process and involves the interaction of several proteins. Using site-directed mutagenesis, we have begun to address some of the mechanistic details of both mercuric ion transport and its ultimate binding and reduction by MR.

Mutagenesis of the N-terminal cysteine pair of MR was undertaken to address whether these residues might have an in vivo effect, possibly by interacting with MerT in an Hg(II) transport role. Our results with the mutations on multicopy plasmids show no difference between the N-terminal mutants and wild type in conferred $HgCl_2$ resistance. However, such gene amplified systems are not necessarily good models for the natural single-copy phenomenon. The resulting overproduction of mer gene products may effectively swamp out some small pertubation in transport pathway. Nevertheless, we have shown that in contrast to the redox active and C-terminal cysteine pairs, the N-terminal pair is not absolutely essential for mercury detoxification.

Mutagenesis of the redox-active disulfide of MR has provided a means for understanding the role of these residues in Hg(II) reduction. First, our observation of catalytic turnover (albeit quite sluggish) by mutants lacking a redox active disulfide demonstrates that such a disulfide is not absolutely necessary for Hg(II) reduction; in these mutants this reaction must occur via a direct electron transfer from a reduced flavin intermediate. In fact, our lower estimate for the rate of Hg(II)-dependent reoxidation of the EH_4 native enzyme (containing $FADH_2$) is ca. 10% of that observed for multiple catalytic turnovers. This shows that E-$FADH_2$ mediated Hg(II) reduction can proceed at sufficiently high rates to account for the catalytic rates observed with native MR.

The reduction of Hg(II) by oxidative addition/elimination to olefinic organic compounds is well precedented (19). Furthermore, it has previously been shown that flavin semiquinone can reduce Hg(II) by an outer sphere electron transfer mechanism (20). We have extended these results by demonstrating that free $FADH_2$ can reduce a variety of Hg(II) salts at rates comparable to those observed with the mutant MR's (M. Distefano, unpublished data). Thus, the hypothesis that native MR may reduce Hg(II) via reduced flavin intermediate is chemically feasible as well.

All of the above results are in agreement with the elegant work of Miller et al who showed that the native MR EH_2 form (E-FAD, active site dithiol) is not capable of reducing Hg(II) without further electron input (7), as well as the work of Sahlman et al who clearly demonstrated the facile formation of the MR EH_2-NADPH complex, a possible progenitor of an EH_4-$NADP^+$ species (21). While those investigators did not observe EH_2-NADPH to EH_4-$NADP^+$ conversion, their experiments were performed in the absence of Hg(II). We know that in the presence of Hg(II), some quenching of the EH_2 charge transfer absorbance can occur (7). Since the thiolate-flavin charge transfer interaction is thought to be largely responsible for the low EH_2/EH_4 redox couple, abolition of this complex upon Hg(II) binding is likely to result in a substantial increase in flavin redox potential which would be conducive to transient EH_4 formation.

While the active site cysteines may not perform the actual redox chemistry of Hg(II) reduction, that these residues must have some role in catalysis is evidenced by the drastic decrease in catalytic rate observed upon their removal. Given the known tendency for Hg(II) to react with thiols with high affinity, it is tempting to speculate that cys_{135} and cys_{140} are involved in Hg(II) binding. Under in vitro assay conditions in the absence of pyridine nucleotide, it has previously been shown that the thiolate/flavin charge transfer absorbance is almost completely quenched upon addition of Hg(II) to native EH_2, suggesting that cys_{140} can bind Hg(II) (7). However, no such quenching was observed during Hg(II) titrations of mutant MR's lacking cys_{135}. This behavior may be attributable to either the participation of cys_{135} in a cys_{135}-Hg(II)-cys_{140} chelate or its requisite involvement in reduction of a second disulfide (see below) whose thiols may interact with cys_{140}-bound Hg(II). While thses cys_{140}-Hg(II) complexes are observable under nonturnover conditions, it should be pointed out that they are not visible in stopped-flow experiments performed under catalytic turnover conditions (24). Thus in the absence of further information, the roles of cys_{135} and cys_{140} remain unclear.

Considering the high degree of conservation, between MR and other disulfide oxidoreductases in the region around the redox active disulfide (Figure 3), perhaps it is not surprizing that mutagenesis in this area drastically changes activity. In contrast, it was not immediately obvious that removal of the C-terminal cysteine pair would yield essentially the same result. Since removal of the C-terminal cysteine pair has little effect on the visible absorbance spectrum of the enzyme, it is unlikely that this pair is as intimately associated with the flavin as the cys_{135},cys_{140} thiolate pair. However, recent fluorescence and thiol titration studies (S. Miller, unpublished data) have shown native MR to possess a second reducible disulfide. Additionally, iodoacetamide labelling studies (M. Moore, unpublished data) have demonstrated that this disulfide is reducible via thiol/disulfide interchange with the cys_{135},cys_{140} pair and that it involves at least one of the C-terminal cysteines. These data suggest that cys_{558} and cys_{559} are positioned at the active site and perhaps play a role in Hg(II) binding. Thus, with these findings, we have begun to delineate the structural features of MR that, by conferring upon it the unique ability to catalyze Hg(II) reduction, set it apart from the other disulfide oxidoreductases.

Acknowledgements. We would like to thank Drs. David Ballou, Vincent Massey, Sue Miller and Charles Williams, Jr., for many valuable discussions. This work is supported in part by NIH grant GM216439. M.J.M. was the recipient of a National Science Foundation predoctoral fellowship.

References

1. Summers, A.O. (1986) Ann. Rev. Microbiol. 40, 607.

2. Foster, T.J. (1983) Microbiol. Rev. 47, 361.

3. Schottel, J.L. (1978) J. Biol. Chem. 253, 4341.

4. Fox, B.S. and Walsh, C. (1982) J. Biol. Chem. 257, 2498.

5. Williams, C.H. (1976) In: The Enzymes (Boyer, P.D., ed) 3rd Ed., Academic Press, New York, 13, 89..

6. Shames, S.L., Fairlamb, A.H., Cerami, A. and Walsh, C.T. (1986) Biochem. 25, 3519.

7. Miller, S.M., Ballou, D.P., Massey, V., Williams, C.H., Jr. and Walsh, C.T. (1986) J. Biol. Chem. 263, 8081.

8. Massey, V. and Williams, C.H., Jr. (1965) J. Biol. Chem. 240, 4470.

9. Casola, L. and Massey, V. (1966) J. Biol. Chem. 241, 4985.

10. Brown, N.L., Ford, S.J., Pridmore, R.D. and Fritzinger, D.C. (1983) Biochem. 22, 4089.

11. Misra, T.K., Brown, N.L., Haberstroh, L., Schmidt, A., Goddette, D. and Silver, S. (1985) Gene 34, 253.

12. Laddaga, R.A., Chu, L., Misra, K.T. and Silver, S. (in press) PNAS USA.

13. Fox, B.S. and Walsh, C.T. (1985) Biochem 22, 4082.

14. Schultz, P.G., Au, K.A. and Walsh, C. (1985) Biochem. 24, 6840.

15. Nakahara, H., Silver, S., Miki, T. and Rownd, R.H. (1979) J. Bact. 140, 160.

16. Rinderle, S.J., Booth, J.E. and Williams, J.W. (1983) Biochem 22, 869.

17. Manstein, D.J., Pai, E.F., Schopfer, L.M. and Massey, V. (1986) Biochem. 25, 6807.

18. Brown, N.L. (1985) TIBS 10, 400.

19. Rappoport, A., Dyall, L.K., Winstein, S. and Young, W.G. (1970) Tet. Lett. 40, 3483.

20. Singh, A.N., Gerlerinter, E. and Gould, E.S. (1982) Inorg. Chem. 21, 1232.

21. Sahlman, L., Lambeir, A.-M., Lindskog, S. and Dunford, H.B. (1984) J. Biol. Chem. 259, 12403.

22. Moore, M.J. and Walsh, C.T. (manuscript in preparation).

23. Distefano, M.D., Au, K.A. and Walsh, C.T. (manuscript in preparation).

24. Sandstrom, A. and Lindskog, S., (this volume).

REDUCED FORMS OF NATIVE MERCURIC REDUCTASE AND THEIR REACTIONS WITH MERCURIC COMPOUNDS-ACTIVITY VS INHIBITION

S.M. Miller, V. Massey, D.P. Ballou, C.H. Williams, Jr., Department of Biological Chemistry, University of Michigan and M. Moore, C.T. Walsh, Chemistry Department, Massachusetts Institute of Technology.

Mercuric Reductase, which catalyzes the reduction of Hg^{2+} by NADPH, contains both FAD and a reducible active site disulfide (1). Addition of reducing equivalents produces an EH_2 species with a distinct spectrum (Fig. 2, dashed line spectrum), analogous to that seen with other flavin-disulfide oxidoreductases and attributed to a charge transfer interaction between the thiolate of Cys-140 and FAD (1). Low temperature (5 C) rapid reaction studies (2) coupled with thiol titrations on denatured E_{ox} and EH_2 enzyme forms (1) (all conducted aerobically) indicated that only 2 electron equivalents are required to form EH_2. In this and in an accompanying paper (3) we show that an additional thiol pair (most likely the C-terminal cysteines 558 & 559) forms a disulfide reducible via the active site. In the reduced form these thiols are involved in proper binding of Hg^{2+} for normal catalytic turnover.

REDUCTION OF EXTERNAL DISULFIDE Under rigorously anaerobic conditions we have found that E_{ox} mixed with 1eq NADPH initially gives rise to EH_2-$NADP^+$, but at 4 C this converts over a period of minutes to an E_{ox}-$NADP^+$ complex, Figure 1. A further 1eq NADPH is required to obtain a stable EH_2-$NADP^+$. Since only 2-4 thiols/monomer of the 8 cysteines in mercuric reductase are titratable in E_{ox}, it seemed likely that the extra reducing equivalents required to produce a stable EH_2-$NADP^+$ were being consumed by an additional disulfide in communication with the active site. As expected, thiol titrations on enzyme denatured under anaerobic conditions show that when native E_{ox} is reacted with excess NAD(P)H until no further changes occur at 340 nm, the thiol titer of EH_2 is increased by ca. 4 from that of E_{ox}. An activated E_{ox} containing 2 thiols/monomer more than unactivated E_{ox} (but spectrally identical) can be generated by treatment with either DTT or NADPH followed by gel filtration purification. This activated enzyme forms a stable EH_2-$NADP^+$ complex when mixed with 1eq NADPH and shows only 2 additional thiols/monomer in EH_2 vs E_{ox}.

Flavins and Flavoproteins 1987

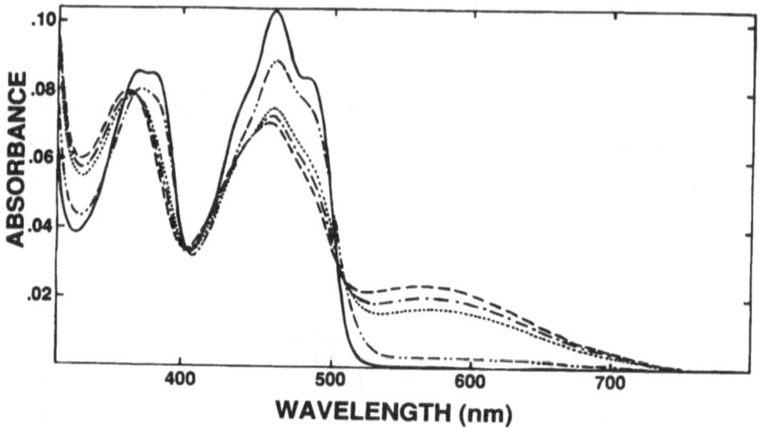

Figure1. Anaerobic reaction of 8.7μM native E_{ox} with 1eq NADPH at 4 C. (---) 1min after mixing; (-·-) 6min; (· · ·) 15min; (- · · -) after 1hr 20min at 25 C.

Brown & coworkers have proposed that the cysteine pair at residues 558 & 559 may be involved in Hg^{2+} binding at the active site of mercuric reductase (4). In agreement with this we found that the catalytically inactive Ala-558, Ala-559 mutant enzyme (5) forms a stable EH_2-$NADP^+$ complex (identical to that with native enzyme!) with only 1eq NADPH and shows only 2 additional thiols/monomer in EH_2 vs E_{ox}. Thus, eliminating cysteine residues 558 & 559 has removed the thiols which communicate with the active site and which appear to be involved in turnover.

<u>REACTIONS WITH $Hg(CN)_2$</u> Figure 2 illustrates the reaction of $4e^-$-reduced (activated) EH_2 with excess (4.4eq) $Hg(CN)_2$. 2.6eq NADH were used to titrate to EH_2 leaving a slight excess of NADH. Upon addition of $Hg(CN)_2$ a species forms which is spectrally similar to E_{ox} but is blue-shifted by 7 nm. Further addition of NADH to this species causes reduction of the FAD with no development of a charge transfer band. Since NADH is thermodynamically incapable of reducing the flavin of enzyme in the absence of Hg^{2+}, this new species must contain Hg^{2+} bound to the active site thiols. Thus, with the C-terminal thiols reduced, an excess of any mercuric compound with easily displaced, weaker binding ligands than thiols will readily form an inhibited complex with the reduced enzyme.

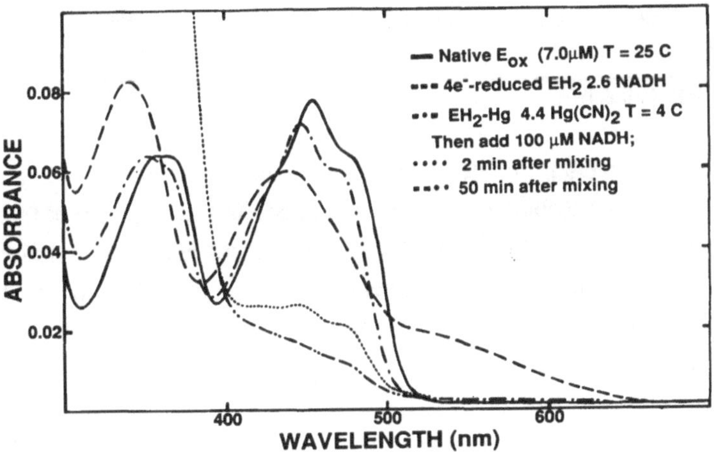

Figure 2. Reaction of 4e⁻-reduced native EH_2 with $Hg(CN)_2$ and subsequent reduction by NADH.

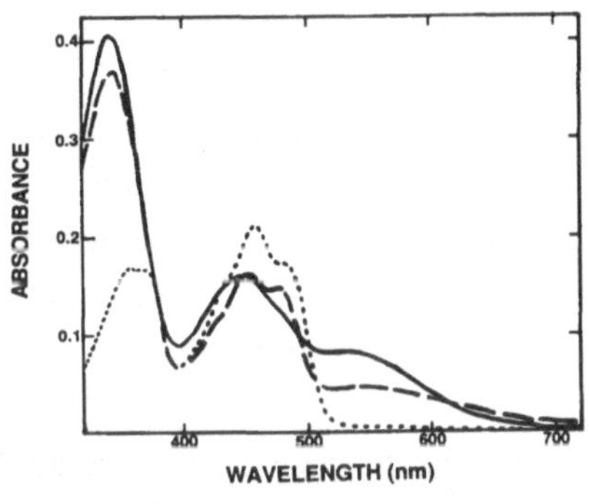

Figure 3. Stopped-flow reaction of 17.7μM unactivated E_{ox} with 3.6eq NADPH + 1eq $Hg(CN)_2$ (- - -) compared to that of 18.6μM activated E_{ox} with 4eq NADPH + 1eq $Hg(CN)_2$ (—). Spectra are taken within 7 sec after mixing. (· · ·) activated E_{ox} before reduction.

Figure 3 illustrates the reaction of unactivated native E_{ox} and activated native E_{ox} with a mixture of excess NADPH + 1eq $Hg(CN)_2$ at 4 C. With the unactivated enzyme reduction of the external disulfide takes several minutes at 4 C. Therefore, the rapid reaction with 1eq $Hg(CN)_2$ results in an inhibited species analogous to that seen with free EH_2 but which contains bound NADPH. At 340 nm we see only the kinetics for formation of 2e⁻-reduced EH_2. These results are practically identical to those obtained with the Ala-558 Ala-559 mutant enzyme (3). By contrast, activated E_{ox} (with reduced external thiols) reduces 1eq Hg^{2+} as indicated by the presence of an additional phase in the kinetics at 340 nm with an amplitude corresponding to uptake of 1eq NADPH. The final spectrum in this case is that of EH_2-NADPH (Fig. 3). With 2eq $Hg(CN)_2$, Hg^{2+} reduction also occurs, but with 4eq, the inhibited species forms with no Hg^{2+} reduction (data not shown).

From these studies it is clear that the C-terminal thiols are intimately involved in liganding Hg^{2+} to form a viable complex for reduction. In the absence of those thiols either by oxidation to the disulfide, overloading with excess Hg^{2+}, or mutation to alanine residues (3) proper binding and efficient reduction of Hg^{2+} cannot occur. A working model showing participation of these thiols in catalysis is shown in the accompanying paper (3).

In summary: 1) A disulfide external to the active site, apparently comprised of the C-terminal cysteine pair (558, 559), can communicate with the active site to be reduced. 2) Efficient reduction of Hg^{2+} requires that Cys-558 and Cys-559 be reduced; otherwise, an inhibited Hg^{2+} complex forms which can reduce Hg^{2+} but only very slowly. 3) In the inhibited complex Hg^{2+} binds at the active site, obliterating the charge transfer interaction between the Cys-140 thiolate and FAD. The visible spectrum is thereby blue-shifted ca. 7nm relative to that of E_{ox}. 4) The reduction potential for FAD in the inhibited complex is substantially raised as evidenced by the formation of fully reduced enzyme in the EH_2-Hg^{2+} complex under conditions where EH_2 is the stable form in the uncomplexed enzyme. Binding of NADPH to the EH_2-Hg^{2+} complex lowers the reduction potential of the FAD as evidenced by the incomplete formation of the fully reduced $FADH_2$ in that complex.

Acknowledgements. This work was supported by grants from the U.S. Public Health Service; GM 11106 (V.M.), GM 20877 (D.P.B.), GM 21444 (C.H.W.), and GM 21643 (C.T.W.).

REFERENCES

1. Fox, B.S. & C.T. Walsh. 1983. Biochemistry 22, 4082-4088.
2. Sahlman, L. & S. Lindskog. 1983. Biochem. Biophys. Res. Commun. 117, 231-237.
3. Moore, M., M. Distefano, C.T. Walsh, S.M. Miller, V. Massey, C.H. Williams, Jr., D.P. Ballou. 1987. This volume accompanying communication.
4. Brown, N.L., S.J. Ford, R.D. Pridmore & D.C. Fritzinger. 1983. Biochemistry 22, 4089-4095.
5. Walsh, C.T., M. Distefano & M. Moore. 1987. Protein Structure & Design, UCLA Symposia on Molecular & Cell Biology. (in press)

RAPID-SCAN STOPPED-FLOW STUDIES OF THE FLAVOENZYME MERCURIC REDUCTASE DURING CATALYTIC TURNOVER

A. Sandström, S. Lindskog

Department of Biochemistry, University of Umeå
S-901 87 Umeå, Sweden

Introduction

Detoxication of Hg^{2+} by mercury-resistant bacteria involves the catalytic action of the flavoenzyme mercuric reductase (1). The Hg^{2+} ions are reduced to metal by NADPH produced in normal metabolism: Hg^{2+} + NADPH → Hg^0 + $NADP^+$ + H^+.
To yield catalytic turnover in vitro, the Hg^{2+} must be complexed with some ligand, such as mercaptoethanol or cysteine (2-4). The enzyme molecule is dimeric and contains an FAD molecule and a redox-active disulfide group per subunit (3). Oxidized mercuric reductase, E, is rapidly reduced by NADPH to yield a two-electron-reduced enzyme, EH_2, where FAD is in the oxidized state and the disulfide group has been reduced. One of the active-site thiols in EH_2 interacts with FAD giving rise to a long-wavelength absorption band (3). The shape and intensity of this band are sensitive to ligands such as $NADP^+$ and NADPH (5). The slowest step (57 s^{-1} at 25 °C) in the reaction of E with an excess of NADPH is the formation of a stable EH_2-NADPH complex (5). Fully active mercuric reductase has a turnover number of about 13 s^{-1} per FAD at pH 7.3 and 25 °C (4). This suggests that the reduction of Hg^{2+} limits the rate of turnover and that, therefore, a two-electron-reduced form of the enzyme should dominate during the steady state at saturating substrate concentrations. To test this hypothesis we have performed rapid-scan stopped-flow measurements to observe mercuric reductase during the catalytic reaction.

Flavins and Flavoproteins 1987

Results and Discussion

Rapid-scan and single-wavelength stopped-flow experiments were performed aerobically with Tn501-encoded and cysteine-activated (4) mercuric reductase from *Pseudomonas aeruginosa* at 25 °C in 50 mM potassium phosphate buffer, pH 7.3, containing 1 mM L-cysteine. With 15 µM enzyme-bound FAD and an excess of NADPH (90 µM NADPH and 45 µM $HgCl_2$) there was a rapid (43 s^{-1}) formation of a pseudosteady-state spectrum quite similar to that previously assigned to an EH_2-NADPH complex (5). The spectrum changed only little during the remainder of the reaction which reached an apparent equilibrium after about 2 s. The results indicate that the rate of formation of EH_2-NADPH is only marginally affected by the presence of $HgCl_2$, that EH_2-NADPH is a true catalytic intermediate, and that the rate-limiting step is the oxidation of EH_2-NADPH by Hg^{2+} (presumably bound to an active-site thiol group).

With an excess of $HgCl_2$ over NADPH (conditions as above except that 45 µM NADPH and 90 µM $HgCl_2$ were used) three kinetic phases were observed as shown in Fig. 1. In a rapid presteady-state phase a spectrum similar to that of EH_2-NADPH was formed and this spectrum remained essentially unchanged during a short pseudosteady-state phase lasting about 60 ms. A spectrum recorded during the pseudosteady-state phase is shown in Fig. 2. The final poststeady-state phase did not result in the formation of fully oxidized enzyme as one might expect under conditions when the reducing substrate is limiting, but in a reduced species having a long-wavelength band centered near 580 nm as illustrated in Fig. 2. This spectrum is quite similar to that previously assigned essentially to an EH_2-$NADP^+$ complex (5). The most straightforward interpretation of this result is that only the EH_2-NADPH complex is kinetically competent and that the EH_2-$NADP^+$ complex cannot reduce the Hg^{2+} substrate. Thus, when NADPH is exhausted the catalytic

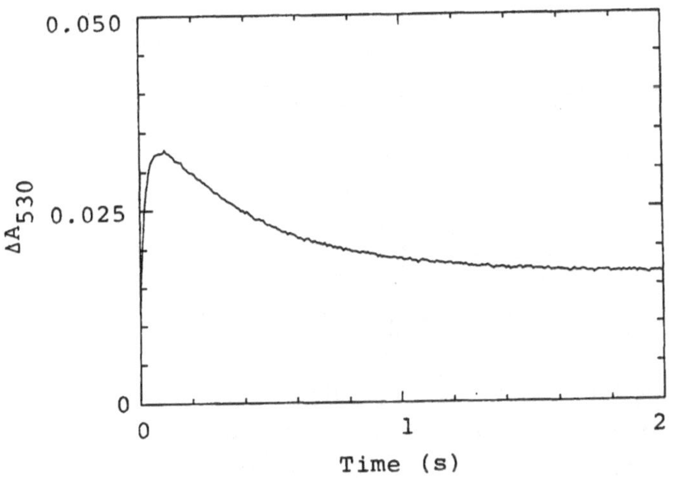

Fig. 1. Absorbance change at 530 nm during the reaction catalyzed by mercuric reductase with NADPH as the limiting substrate. Conditions: 15 μM enzyme-bound FAD in 50 mM potassium phosphate buffer, pH 7.3, containing 1 mM cysteine; 45 μM NADPH; 90 μM $HgCl_2$; 25 °C.

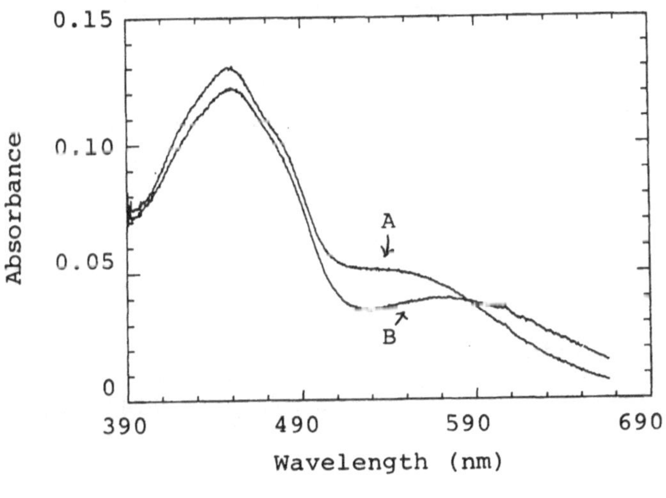

Fig. 2. Rapid-scan spectra of mercuric reductase during the catalyzed reaction with NADPH as the limiting substrate. The scanning time was 5 ms. Conditions as in Fig. 1. Spectrum A was recorded during the pseudosteady-state phase 98-103 ms after the stop of flow and spectrum B near the apparent end of the reaction 1513-1518 ms after the stop of flow.

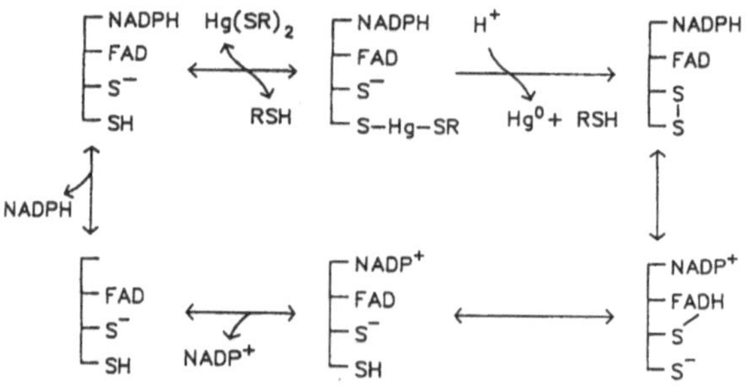

Fig. 3. Proposed schematic mechanism of mercuric reductase.

cycle stops and the rather stable EH_2-$NADP^+$ complex accumulates. Our results are in accordance with the schematic mechanism of Fig. 3 (cf. (6)). However, Fig. 3 is most likely an over-simplification. The results of Miller et al. (6) indicate that the Hg^{2+} substrate can bind to all reduced forms of the enzyme. Moreover, we have found that if the oxidized enzyme is incubated with an excess of radioactive Hg^{2+} and only one equiv. of NADPH, then about 0.8 equiv. Hg^0 is formed within 5 min whereas Fig. 3 predicts no Hg^0 formation. We have also found that under conditions as in Fig. 2 the EH_2-$NADP^+$ type spectrum undergoes further changes in the minutes time range. Our tentative interpretation is that EH_2-$NADP^+$ can reduce Hg^{2+} but not sufficiently fast to be kinetically important.

References

1. Summers, A.O. 1986. Ann. Rev. Microbiol. 40, 607-634.
2. Schottel, J.L. 1978. J. Biol. Chem. 257, 4341-4349.
3. Fox, B. and C.T. Walsh. 1982. J. Biol. Chem. 257, 2498-2503.
4. Sandström, A. and S. Lindskog. 1987. Eur. J. Biochem. 164, 243-249.
5. Sahlman, L., A.-M. Lambeir, S. Lindskog and H.B. Dunford. 1984. J. Biol. Chem. 259, 12403-12408.
6. Miller, S.M., D.P. Ballou, V. Massey, C.H. Williams, Jr. and C.T. Walsh. 1986. J. Biol. Chem. 261, 8081-8084.

THE PENULTIMATE CYSTEINES IN MERCURIC REDUCTASE AID IN THE REDUCTION OF MERCURY

Melissa Moore, Mark Distefano, Christopher Walsh
Dept. Chemistry., Massachusetts Institute of Technology, Cambridge, MA 02139

Susan Miller, Vincent Massey, Charles H. Williams, Jr., and David Ballou
Dept. Biological Chemistry, Univ of. Michigan, Ann Arbor, MI 48109

Introduction. Mercuric reductase (MR) catalyzes the reduction of Hg^{2+} by NADPH as part of the process of detoxification of mercury by microorganisms. MR is an FAD-disulfide oxidoreductase closely related to the glutathione reductase family of enzymes(1). It contains an active site disulfide involving residues Cys-135 and Cys-140. In the reduced form (EH_2) Cys-140 participates in charge-transfer interactions (C-T) with the FAD to give characteristic spectral properties. Another cysteine pair at residues 558 and 559 is also found in MR. Studies in the previous paper(2) have shown that these residues are in redox communication with the active site and that they may be involved in catalysis. Site-directed mutations to convert these residues to alanines(3) have enabled us to examine more carefully their role in catalysis.

MR with Ala-558, Ala-559 is inactive(3) in Hg^{2+} reduction. Moreover, cells carrying the Mer operon with this construction of MR are supersensitive to Hg^{2+}. This is due to the ability of the Mer operon gene products to transport Hg^{2+} into the cells without the requisite capacity to reduce the Hg^{2+}.

Methods and Conditions. All studies discussed here were with the mutant Ala558, Ala 559 MR and were carried out at 4° in 50 mM KPi, pH 7.3 anaerobically in a stopped-flow spectrophotometer. Spectra were recorded in a period of 7 sec with initiation at the times indicated after mixing.

Flavins and Flavoproteins 1987

Results. Reduction of MR (Ala-558, Ala-559) by NADPH follows closely the course of reduction of native MR.(4) This rapid process gives rise to EH_2-NADP ($k \sim 3 \times 10^7\ M^{-1}s^{-1}$) and is followed by an exchange reaction leading to EH_2-NADPH ($k \sim 5\ s^{-1}$) (See Fig. 1). This fact, coupled with the knowledge that spectra of this mutant and of native MR are identical, imply that no gross changes in the enzyme structure result from the mutations.

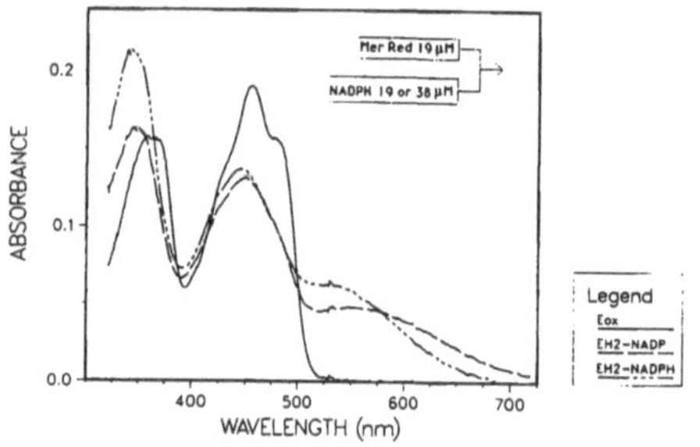

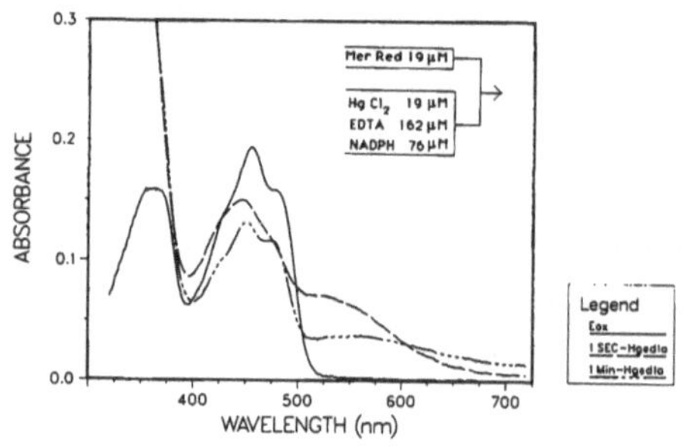

Figure 1 (left) MR (Ala 558, Ala 559) (19 uM) was mixed with an equal volume of a) buffer (———) b) NADPH (19 uM) (— — —) c) NADPH (38 uM). (— ··· —) All curves are final spectra. Figure 2 (right) MR (19 uM) was mixed with an equal volume of a) buffer (———) b) and c) a mixture of $HgCl_2$ (19 uM), EDTA (162 uM), NADPH (76 uM). Spectra recorded at b) 1 sec (— — —) and c) 1 min (— ··· —).

Reaction with 1 eq Hg(Cl_2-EDTA) and 4 eq NADPH (Fig. 2) involves a very rapid formation of EH_2-NADPH (as in Fig. 1) followed by a slower conversion to an inhibited enzyme form ($k \sim 2\ s^{-1}$). Fig. 2 shows that at 1 sec EH_2-NADPH is present, but that by 1 min the inhibited form is established. This species is stable and is essentially incompetent in reducing Hg^{2+}. The reaction with $Hg(CN)_2$ forms the same type of complex. However, in this case the complex is completely formed by 3 sec. Only one Hg^{2+} per flavin is required to form the complex, thus establishing that no reduction of Hg^{2+} occurs.

The spectrum of this inhibited form of MR Ala 558, Ala 559 is very similar to those of the NADPH complexes of two other mutant forms of MR: (Cys-135, Ser -140) and (Ala-135, Ala-140). This correlation is strong evidence that Hg^{2+} binds to Cys-140 in the inhibited complex. Thus in the presence of NADPH when Cys-140 is either liganded to Hg^{2+} or replaced by serine or alanine, a species with a strong C-T interaction between FAD and NADPH (eg. the inhibited enzyme) can form. By contrast, neither of the mutants, (Ser-135, Cys-140) nor (Ala-135, Cys-140) give rise to this C-T spectrum. Instead, these mutants exhibit "permanent" EH_2-like spectra with a Cys-140 C-T to FAD. That the C-T interaction of the inhibited species involves NADPH is also supported by the fact that the spectrum of the inhibited form does not appear when NADH (which binds only weakly to MR) is used as the reductant.

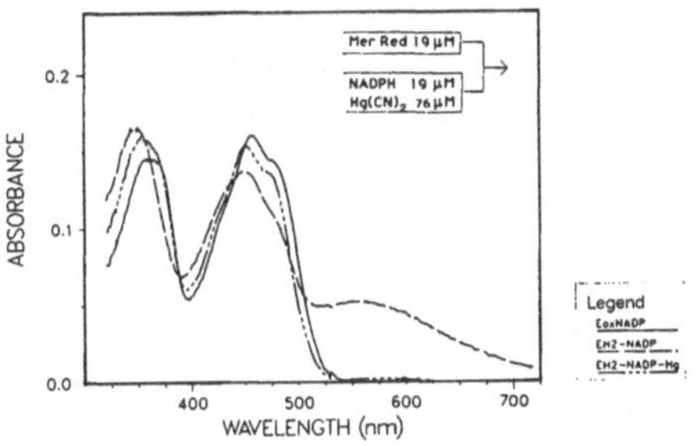

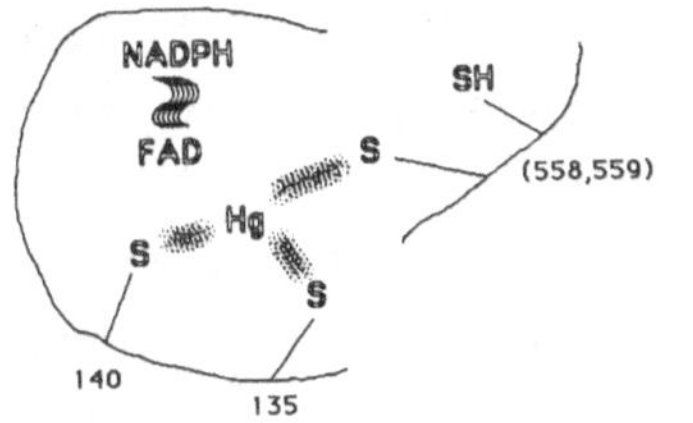

Catalytic Complex

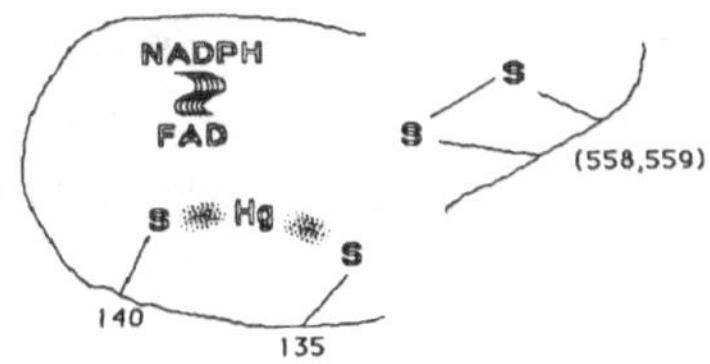

Inhibited Complex

Figure 3 (left) MR (19 uM) was mixed with an equal volume of a) NADP (19 uM) (——). b) a mixture of NADPH (19 uM) and $Hg(CN)_2$ (76 uM). Spectrum after initial reduction by NADPH (— — —). Spectrum recorded at 1 min after mixing (— ··· —).

Fig. 3 shows the reaction of MR (Ala 558,Ala 559) with 1 eq NADPH and excess $Hg(CN)_2$. EH_2-NADP forms rapidly, but is converted to a species that looks like E_{ox}-NADP. On careful inspection, it can be seen that the latter spectrum is perturbed from that of E_{ox}-NADP (Fig. 3), with all of the spectral features in the visible region being shifted ca 6-8 nm to the blue. Thus this spectrum must be of a species with Hg^{2+} bound to it, presumably to Cys-140 and possibly to Cys-135. This also confirms that the spectrum of the inhibited species results from C-T between FAD and NADPH, not NADP.

In summary, the present results implicate Cys-558 and/or Cys-559 to be involved in catalysis of Hg^{2+} reduction. A possible role is presented in the cartoon below. Hg^{2+} binds to thiols with a K_a of 10^{40} - 10^{45} M^{-2} (5). This has the effect of greatly lowering the redox potential for mercury. In fact, the Hg-S bond can be calculated to be 27 - 30 kcal in strength and therefore has substantial covalent character. Since Hg-SR exchange reactions are very rapid, they must occur by an associative rather than a dissociative mechanism. A dissociative process involving a complex of this stability would be extremely slow. The combination of a third and/or fourth thiol ligand to the complex must substantially change the nature of the Hg-S bonding, thus raising the redox potential of the Hg^{2+}, so that reduction may become more facile.

We propose the novel idea that the Cys-558 and/or Cys-559 thiols serve(s) this role of weakening the Hg-S bond so that NADPH (via the flavin) can reduce the Hg^{2+}. The cartoon shows the Catalytic Complex of EH_2 which involves a C-T interaction between NADPH and FAD (just before electron transfer) and a $Hg(SR)_3$ complex in which each Hg-S bond is weaker than those in a $Hg(SR)_2$ complex. An inhibited complex forms when Cys residues 558 and 559 are either oxidized or mutated to alanines. This complex involves a tight $Hg(SR)_2$ interaction which lowers the redox potential of the Hg^0/Hg^{2+} couple and prevents reduction. This inhibited complex exhibits a fully formed C-T interaction between NADPH and FAD.

Acknowledgement

This research was supported by NIH grant nos. GM11106(VM), GM20877(DB), and GM21444(CHW), and GM21643(CW) M.M. is a recipient of a NSF predoctoral fellowship.

References

1. Fox, B.S. and Walsh, C.T. 1983 Biochemistry 22, 4082-4088.

2. Miller, S. Ballou, D. Massey, V. Williams Jr., C.H., Moore, M., Walsh, C.T. 1987 (This volume)

3. Walsh, C.T. , Distefano, M., and Moore, M. 1987 Protein Structure & Design, UCLA Symposia on Molecular and Cell Biology, (in press).

4. Sahlman, L., Lambeir, A.M., Lindskog, S., and Dunford, H.B. 1984 J. Biol. Chem. 259, 12403.

5. Casas, J.J. and Jones, M.M. 1980 J. Inorg. Nucl. Chem. 42, 99-102.

STUDIES ON THE ACTIVE SITE OF MERCURIC REDUCTASE EMPLOYING SITE-DIRECTED MUTANTS AND THE THIOL-REACTIVE FLAVIN, 6-SCN-FAD.

V. Massey, S.M. Miller, D.P. Ballou, C.H. Williams, Jr., Department of Biological Chemistry, University of Michigan and M. Moore, M. Distefano, C.T. Walsh, Chemistry Department, Massachusetts Institute of Technology.

Extensive spectral and sequence homologies have been found to exist between glutathione reductase, lipoamide dehydrogenase and the TN501-encoded mercuric reductase grown in E. coli (1,2). These similarities permitted identification of an active site disulfide in mercuric reductase as comprising cysteine residues 135 and 140, and led to the production of a series of active site mutants where these residues were replaced by serine and alanine (3). In addition another pair of cysteines (Cys 558, Cys 559) appear to be essential for catalysis, since the mutant form (Ala 558, Ala 559) is inactive in catalyzing Hg^{2+} reduction (4).

The relative positioning of the active site disulfide and enzyme bound FAD has been probed by replacing the native FAD with 6-SCN-FAD. This derivative is readily converted to 6-mercapto FAD by reaction with thiols (5). When 6-SCN-FAD was bound to the apoprotein of native mercuric reductase, a stable spectrum of 6-SCN flavin in a hydrophobic environment was obtained. On addition of dithiothreitol (DTT), the 6-mercapto FAD enzyme was formed. This is shown in Fig. 1A for apoenzyme pre-treated with 0.5 mM methylmethane thiolsulfonate (MMTS) at 0° for 10 min before addition of 6-SCN FAD. This pretreatment of the apoenzyme is necessary to prevent the partial spontaneous conversion to 6-mercapto FAD, which is probably due to reaction of 6-SCN FAD with some of the apoenzyme thiols before complete formation of the holoenzyme. Such spontaneous conversion is variable, depending on the history of the apoenzyme, and can amount to as much as 25% formation of 6-mercapto FAD enzyme. On addition of DTT to the 6-SCN FAD enzyme there is a rapid development of long wavelength absorption and a partial bleaching and shift to shorter wavelengths of the 455 nm band. This spectrum has therefore some of the characteristics of the "EH_2" form

Flavins and Flavoproteins 1987

of the native enzyme, in which there is a charge transfer interaction between the thiolate of Cys 140 of the active site dithiol pair and enzyme-bound oxidized FAD (1,3). This is followed by the slower conversion to 6-mercapto FAD enzyme, with isosbestic points at 348 and 470 nm. The final spectrum is identical with that obtained by titrating 6-mercapto FAD with apoenzyme. That the conversion of the 6-SCN enzyme to the 6-mercapto FAD enzyme was due to reaction of the former with an active site thiol produced by reaction with DTT, was shown in various ways. First, the conversion is also brought about by the addition of NADH or NADPH; in the latter case the final spectrum is that of the 6-mercapto FAD enzyme-$NADP^+$ complex. Studies with active site mutants allow us to specify that it is, in fact, Cys 135 which reacts with the bound 6-SCN FAD. The Ala 558 Ala 559 mutant behaves identically to native enzyme, with the significant exception that no spontaneous conversion of 6-SCN FAD to 6-mercapto FAD occurs in the absence of added reducing agents. When the apoprotein of the Cys 135, Ser 140 mutant was used to titrate a solution of 6-SCN FAD, there was spontaneous conversion to the 6-mercapto FAD holoenzyme, without any addition of external reducing agent (Fig. 1B). In contrast, when 6-SCN FAD was mixed with the apoprotein of the Ser 135 Cys 140 mutant, the spectrum was not the same as that obtained on mixing 6-mercapto FAD and the apoenzyme (Fig. 1C). The spectrum of the 6-SCN FAD enzyme has marked similarities to that of the EH_2 form of native enzyme. It is also quite stable, and is not converted to the 6-mercapto FAD enzyme even on prolonged incubation with DTT. The apoenzyme of the Ala 135 Ala 140 mutant also forms a very stable 6-SCN holoenzyme, which again cannot be converted to the 6-mercapto FAD form on incubation with DTT. It is, however, readily reduced by NADPH or NADH, without any sign of an EH_2-type spectrum (Fig. 1D).

In all cases where there is no active site thiol present the spectrum of the 6-SCN holoenzyme shows a distinct peak in the 350 nm region (see Fig. 1D) compared with the shoulder in this region displayed by free 6-SCN FAD (see Fig. 1B). A similar spectral resolution is obtained when 6-SCN-3-methyl lumiflavin (a kind gift of Dr. S. Ghisla) is dissolved in non polar solvents. This suggests that in the oxidized native enzyme the

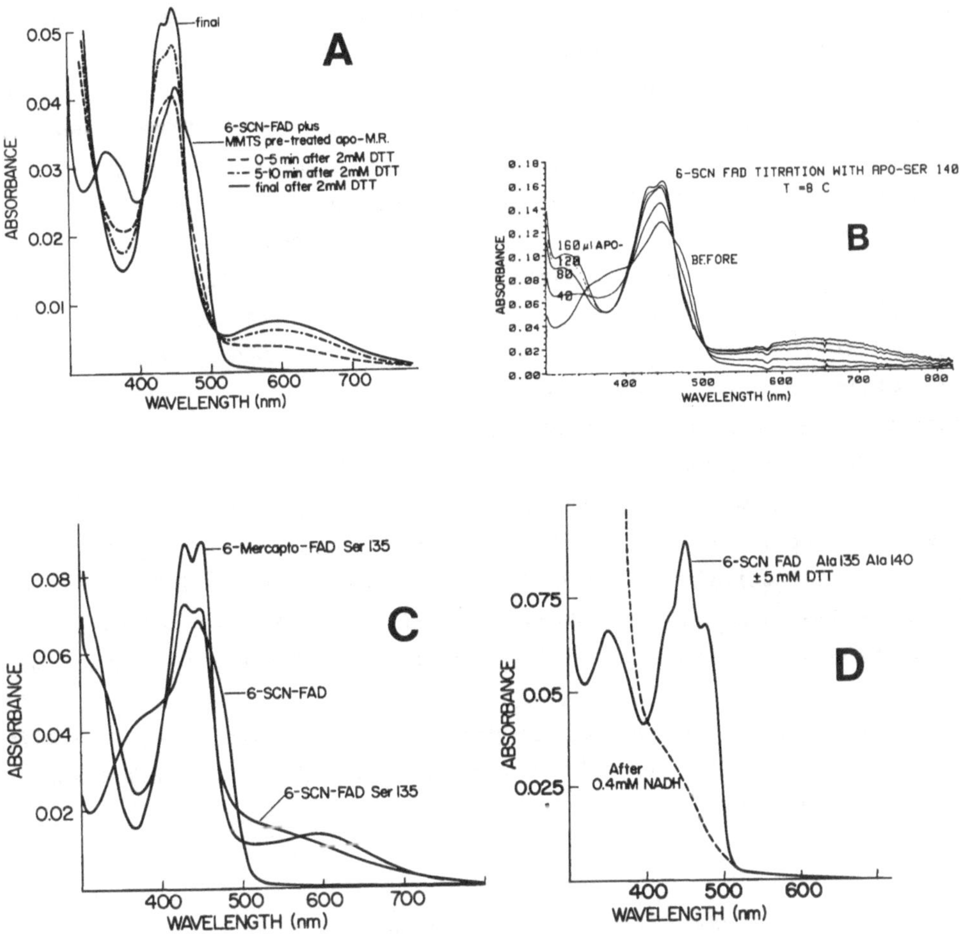

Figure 1. All reactions were carried out in 0.05 M phosphate, pH 7.0, at 20°, unless specified otherwise. (A) Effect of 2mM DTT on 6-SCN FAD native mercuric reductase prepared by brief pretreatment of apoprotein with 0.5 mM MMTS before addition of 6-SCN FAD. The resultant holoprotein was centrifuged twice through a Centricon 30 microconcentrator with an intermediate buffer wash before reaction with DTT. (B) Titration of 6-SCN FAD with the apoprotein of the Cys 135, Ser 140 mutant mercuric reductase, showing spontaneous conversion to the 6-mercapto FAD enzyme. Temperature, 8° (C) Spectra of free 6-SCN FAD, 6-SCN FAD Ser 135, Cys 140 mercuric reductase and 6-mercapto FAD enzyme (D) Spectrum of 6-SCN FAD Ala 135 Ala 140 mercuric reductase. The dashed line shows the spectrum directly after the addition of 0.4 mM NADH. After aerobic turnover of the NADH, the initial spectrum was regained completely. The spectrum is completely unaffected by 5 mM DTT for at least 90 min.

flavin is in a rather hydrophobic environment. A similar phenomenon is found with glutathione reductase (6). As with glutathione reductase (7), it is known that the pyridine nucleotide binding pocket of mercuric reductase is located on the _re_-side of the flavin (8). Mercuric reductase is probably similar in having the active site disulfide located on the opposite (_si_-) side of the flavin, so that electron transfer between pyridine nucleotide and active site disulfide (and hence the $Hg(SR)_2$ substrate) occurs via the sandwiched flavin. From the studies with 6-SCN FAD and the various forms of mercuric reductase we can conclude that the active site Cys 135 is located in such a position that it can act as a nucleophile in the displacement of CN^- from bound 6-SCN FAD. Dithiothreitol is incapable of reaching directly the 6-SCN linkage, but it is capable of reducing the disulfide of Cys 135 Cys 140. These findings are in contrast with those obtained with human erythrocyte glutathione reductase, where the active site disulfide is also reduced by DTT, but where neither active site thiol is able to react with the 6-SCN linkage, and where the conversion of bound 6-SCN FAD to 6-mercapto FAD by DTT is mediated by DTT gaining access via the pyridine nucleotide pocket (6).

Acknowledgements. This work was supported by grants from the U.S. Public Health Service; GM 11106 (V.M.), GM 20877 (D.P.B.) GM 21444 (C.H.W.) and GM 21643 (C.T.W.)

References

1. Fox, B.S. and Walsh, C.T. (1983) _Biochemistry_ _22_, 4082-4088.
2. Brown, N.L., Ford, S.J., Pridmore, R.D. and Frintzinger, D.C. (1983) _Chemistry_ _22_, 4089-4095.
3. Schultz, P.G., Au, K.G. and Walsh, C.T. (1985) _Biochemistry_ _24_, 6840-6848.
4. Walsh, C.T., Distefano, M. and Moore, M. (1987) Protein Structure & Design, UCLA Symposia on Molecular & Cell Biology, in press.
5. Massey, V., Ghisla, S. and Yagi, K. (1986) _Biochemistry_ _25_, 8103-8112.
6. Massey, V., Ghisla, S., Ermler, U. and Schulz, G.E. (1987) This volume accompanying communication.
7. Pai, E.F. and Schulz, G.E. (1983) _J. Biol. Chem._ _238_, 1752-1758.
8. Manstein, D.J., Pai, E.F., Schopfer, L.M. and Massey, V. (1986) _Biochemistry_ _25_, 6807-6816.

THE ACTIVE CENTER OF GLUTATHIONE REDUCTASE AT 1.54 Å RESOLUTION

P.A. Karplus and G.E. Schulz
Institut für Organische Chemie und Biochemie der Universität Freiburg, Albertstr. 21, 78oo Freiburg, F.R.Germany

Introduction

Glutathione reductase (EC 1.6.4.2) catalyses the NADPH dependent reduction of glutathione disulfide, serving to maintain a high ratio of reduced to oxidized glutathione in living cells (1,2). There are indications that other enzymes, such as trypanothione reductase (3,4), lipoamide dehydrogenase (5,6) and mercuric reductase (7) have the same basic structure. Therefore, results from the structural analysis of glutathione reductase will also further our understanding of these systems.

Human erythrocyte glutathione reductase, a homo-dimer of M_r = 104800, is structurally the best known member of this family. The amino acid sequence analysis (8) and crystal structure analysis at 2 Å resolution (9) have led to a three-dimensional model of the enzyme. As the crystals are catalytically competent, they have also been used for the structural analysis at 3 Å resolution of a number of intermediates of the catalytic cycle of the reaction (10). The results of these studies have already answered a number of questions regarding the mechanism and kinetics of this enzyme. However, these studies did not yield the level of accuracy necessary to address questions having to do with the details of the structure: Where are hydrogen-bonds and what is their geometry? What is the nature of the various enzyme bound solvent molecules? What is the geometry of the flavin moiety of FAD? In order to begin to answer these questions, we have refined the structure of the

Flavins and Flavoproteins 1987

oxidized enzyme using a restrained least-squares method (11) and extended the resolution of the analysis to 1.54 Å (12).

The refined protein model

At 1.54 Å resolution, the individual bonded atoms are not resolved, but the electron density is very straightforward to interpret. As an example, Figure 1 shows two different representations of the final electron density for the flavin. In Figure 1b, one can also note how at this resolution the oxygen atoms have higher density than the nitrogen atoms, which have higher density than the carbon atoms. The statistics of the refinement suggest that the overall accuracy of the co-ordinates is around 0.15 Å. The more rigid atoms in the structure have an accuracy better than this, while the more mobile atoms are less well determined.

The refinement has made significant changes in the model. On the average, the main chain and side chain atoms have moved

(a) (b)

Fig.1 Final 2Fo-Fc electron density. a) cage contours around FAD at 1.3 e/Å^3. b) in the plane of the flavin at 0.3, 0.6...2.7 e/Å^3; stippled lines indicate decreasing density.

0.6 Å and 1.4 Å, respectively. Some side chain atoms have moved more than 10 Å. Even atoms at the active center, where the 2 Å electron density map was fairly clear, have moved around 0.5 Å. Another way the model has improved is through the inclusion of a rather extensive model for the protein bound solvent. Aside from a phosphate atom observed to bind at the binding site for the 2'-phosphate of NADPH, all observed sites of protein bound solvent have been assigned to be water molecules. For identification purposes the water molecules have been numbered from Sol-1 (the strongest water site) through Sol-523 (the weakest water site). Thirdly, the refined model gives information about the dynamics of the structure. The refinement yields a so-called temperature factor for each atom, which has been shown to be a reasonable indicator of the atomic mobility.

The active center is highly immobile

The mobility of the main chain of glutathione reductase is shown in Figure 2. More mobile and less mobile stretches alternate along the chain. In general, the more mobile stretches are on the surface of the protein and the less mobile stretches are internal. Interestingly, a survey of the chain segments which directly take part in the active center

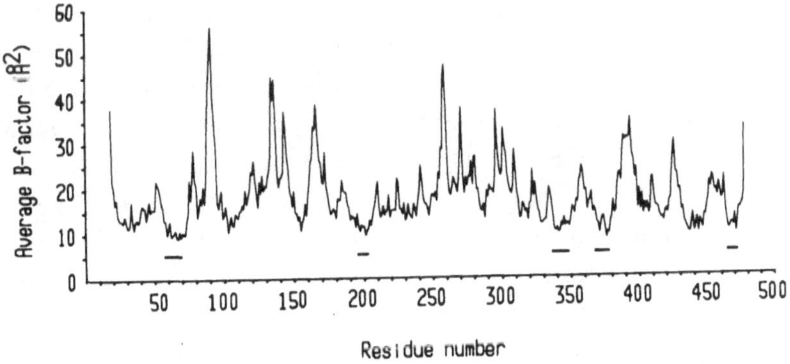

Fig.2 Average main chain temperature factors of glutathione reductase. Bars designate residues at the active center.

(see Fig. 2) shows that they have the lowest mobilities of the whole protein. Thus, the protein chain is very well fixed at the active site and shows increasing mobility as one moves farther away. The flavin is the most immobile part of the structure, with an average temperature factor of 8.7 $Å^2$.

There is a growing awareness that the mobility of proteins is of great importance to an understanding of their structure and function. Often for substrate binding to occur, movements of amino acid side chains and sometimes even of the main chain is necessary. This is, for instance, the case for the binding of NADPH by glutathione reductase (10). However, it seems that when catalysis requires a change in the electronic properties of a prosthetic group or amino acid side chain, that the groups involved tend to be very rigidly held in a particular position (13,14). Thus, the observed rigidity already suggests that the protein:flavin interactions are likely to modulate the flavin structure for fine tuning this particular reaction.

New aspects of the binding of FAD

A stereo view of FAD bound to glutathione reductase is shown in Figure 3. The FAD makes extensive van der Waals contacts with the protein and is held specifically by approximately 20 hydrogen-bonds. Most of these interactions were reported in the unrefined structure (15) and, although the hydrogen-bond distances have changed somewhat, they will not be discussed here. A detailed report of the refined interactions is given elsewhere (12). The important role of solvent molecules in FAD binding is emphasized by the observation that 25% of the hydrogen-bonds to FAD involve tightly bound water molecules, although tightly bound solvent makes up only about 2% of the mass of the protein (12).

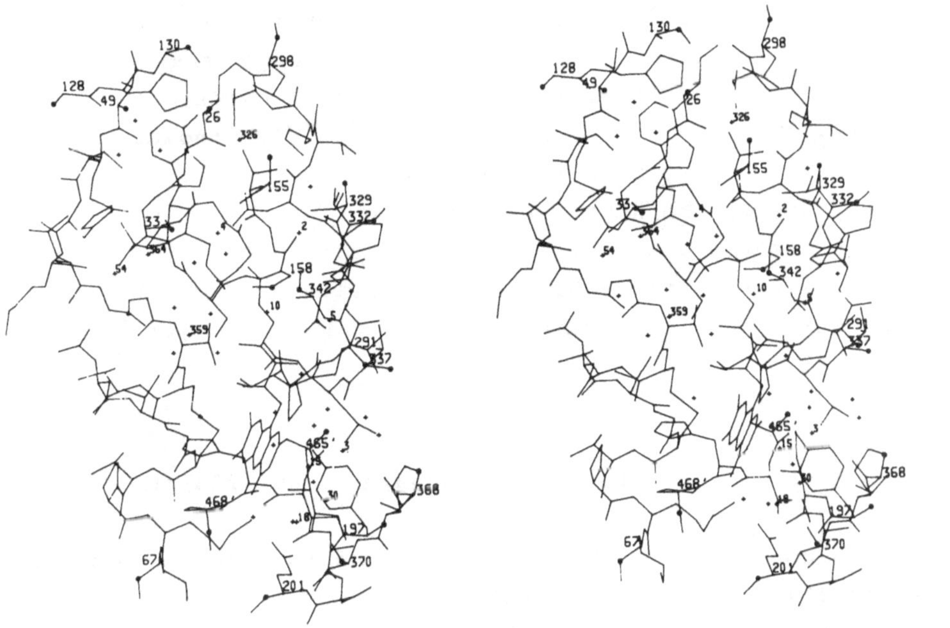

Fig.3 Stereo view of FAD bound to glutathione reductase. The residues at the ends of the chain segments (dots) are labelled and water molecules are indicated by plusses.

Two solvent molecules of particular interest are those which interact with the buried pyrophosphate of FAD. Density for these solvent molecules was already visible at 2 Å resolution, but it was not possible to identify their chemical nature. Since no positively charged side chains were located close to the pyrophosphate, it was suggested that one of the two solvent molecules could be a cation which would provide charge compensation for this negatively charged group (15). Based on the refined structure, two different lines of reasoning lead to the conclusion that both of these molecules are neutral water molecules: each site accepts and donates two hydrogen-bonds in a tetrahedral arrangement, and the density at each site is appropriate for an oxygen atom. Stabilization of this buried negative charge probably comes to a large extent from the helix beginning at residue 29 (15,16). In this case, the oxygen atom

OF1 which is located directly at the N-terminus of this helix would be expected to experience greater charge compensation than OF2. Interestingly, the refined bond lengths also show an asymmetry (PF-OF1=1.56 Å, PF-OF2=1.39 Å) which indicates that the PF-OF1 bond has more single bond character and thus OF1 carries more of the negative charge. A similar asymmetry is observed for the other half of the pyrophosphate where OA1, which receives partial charge compensation through an indirect contact with the side chain of Arg-291, has a longer bond length (12).

Environment and structure of the enzyme bound flavin

Hydrogen-bonding between the protein and the flavin is probably one important mechanism by which the redox properties of the flavin are tuned for a particular reaction (17). In the following, these interactions will be discussed in some detail. As mentioned above, the flavin is held very tightly by the protein. Figure 4 shows the environment of the flavin. The pteridine moiety of the flavin makes extensive hydrogen-bonds and van der Waals contacts with the enzyme. The benzene ring, however, points into a solvent accessible region putting C7a and C8a into a partly aqueous environment. The closest contact made by these methyl groups to another protein atom is 3.51 Å from C8a to the guanidinium of Arg-291 (Fig. 4). The hydrogen-bonding potential of the flavin seems to be well satisfied by its protein environment. N3 donates a strong hydrogen-bond to the main chain carbonyl of His-467' from the other subunit. The main chain nitrogen of Thr-339, which is at the N-terminus of an α-helix, points towards both N1 and O2a. Although the potential hydrogen-bond to N1 is more linear (see Fig. 4), that to O2a is about 0.4 Å shorter, suggesting that it is the stronger interaction. The relevance of these interactions, which can serve to stabilize a negative charge at the N1/O2a locus, for catalysis has been discussed previously (15,18).

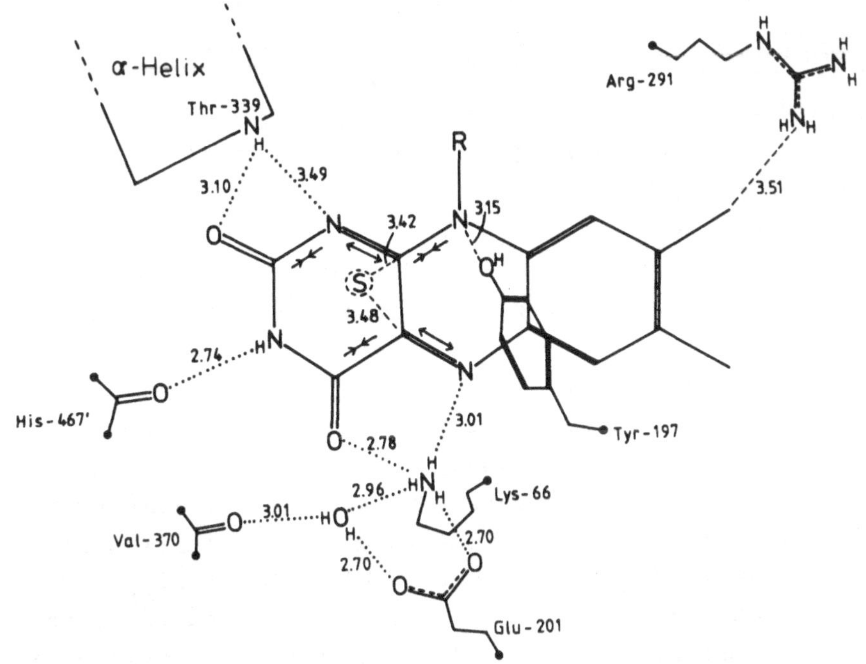

Fig.4 Sketch of the flavin environment. The view is approximately perpendicular to the flavin. Distances of approach and possible hydrogen bonds are indicated.

N10 is approached by the hydroxyl group of Tyr-197, but the geometry of the approach is such that no good hydrogen-bond can be formed (Fig. 4).

Near N5 and O4a, contact is made with the side chain of Lys-66 (Fig. 4). During refinement, the previously reported (15) distance of 3 Å from N5 to 66-CE has increased to 4.0 Å. This means that this atom does not cause steric hinderance to protonation at N5. The amino group of Lys-66 is within 3 Å of 4 possible hydrogen-bond acceptors. Since the X-ray data we have are not sufficient to give information on the position of the hydrogen atoms, we cannot say for certain which of the 4 possible hydrogen-bonding interactions occur. However since N5, Sol-18 and Glu-201-OE1 are situated approximately 120° from each other when viewed along the rotation axis of the ammonium rotor (Fig. 4), it seems very likely that these are the hydrogen-bonded partners. If this interpretation is correct, the strong interaction at N5 might have a large effect on the electronic properties of the flavin. The situation in the presence of NADPH may be different because Sol-18 is displaced upon NADPH binding.

If the protein environment of the flavin is such that certain resonance forms are stabilized relative to others, these effects may be visible in deviations from the expected flavin geometry. Although the accuracy of this structure does not justify a discussion of small deviations in bond lengths, we feel it is worth mentioning that there are 5 bonds of the flavin, all near the redox active part, which deviate by 0.03 Å to 0.05 Å from their ideal values (19). Interestingly, two deviations are a lengthening of the double bonds to N1 and N5. The directions of the 5 deviations are indicated by the double-headed arrows in Figure 4. However, we emphasize that these deviations are not significantly above the error level, and may be artefacts. The flavin as a whole deviates slightly from planarity. It is bent approximately 3° towards the redox-active disulfide around the N5...N10 virtual bond and there is a twist of nearly 2° along the long dimension of the flavin. In flavodoxin, the semiquinone structure showed a similar magnitude of "butterfly" bending, while the oxidized structure was planar (20).

The proximal sulfur (Cys-63-SG) sits behind the flavin plane of Figure 4. It is closest to atoms C10a and C4a, but the geometry is such that an interaction with C4a would be favored. An intermediate C4a adduct has been suggested based on spectroscopic data (21). In thinking about the structural features of such an intermediate, it is interesting to note that the protein atoms 339-N and 467'-O, which are hydrogen-bonded to N1/O2a and N3, respectively, are both around 0.8 Å behind the flavin plane in Figure 4. This means that these hydrogen-bonds would become stronger if the pyrimidine part of the flavin were to move back towards the proximal sulfur. Such a motion would occur during a possible "butterfly" bending of the N5-reduced flavin and would lead to a closer approach of C4a to 63-SG. By having the hydrogen-bonds set up to stabilize this hypothetical intermediate, the activation barrier could be reduced.

The redox-active disulfide loop is strained

The redox-active disulfide bridge in glutathione reductase connects Cys-58 and Cys-63. Surveys of known structures show that its conformation fits into none of the previously described classes (22,23). Although the disulfide bond length was not restrained during refinement, it turned out to be normal. One indication of strain in the bridge is the departure of two of the torsion angles by more than 30° from the expected minimum energy values. A closer inspection reveals that in addition to this, the bond angle 63-CB:63-SG:58-SG deviates by 12° from its ideal value (11). The third feature which indicates strain is that the peptide between residue 58 and 63 all show significant (ca. 5°) and approximately equal deviation from planarity. Nowhere else in the chain is such an extended segment of deviating residues seen. The strain in the disulfide should change its redox potential in favor of reduction.

Acknowledgements

This research was supported by grants from the Alexander von Humboldt-Stiftung and the Deutsche Forschungsgemeinschaft.

References

1. Williams,C.H.,Jr..1976. In: The Enzymes, Vol. 13 (P.D. Boyer, ed.). Academic Press, New York. p. 89.
2. Chance,B., H.Sies, A.Boveris. 1979. Physiol.Rev. 59, 527.
3. Shames,S.L., A.H.Fairlamb, A.Cerami, C.T.Walsh. 1986. Biochemistry 25, 3519.
4. Krauth-Siegel,R.L., B.Enders, G.B.Henderson, A.H.Fairlamb, R.H.Schirmer. 1987. Eur.J.Biochem. 164, 123.
5. Stephens,P.E., H.M.Lewis, M.G.Darlison, J.R.Guest. 1983. Eur.J.Biochem. 135, 519.
6. Rice,D.W., G.E.Schulz, J.R.Guest. 1984. J.Mol.Biol.174, 483.
7. Brown,N.L., S.J.Ford, R.D.Pridmore, D.C.Fritzinger. 1983. Biochemistry 22, 4o89.
8. Krauth-Siegel,R.L., R.Blatterspiel, M.Saleh, E.Schiltz, R.H.Schirmer, R.Untucht-Grau. 1982. Eur.J.Biochem. 121,259.
9. Thieme,R., E.F.Pai, R.H.Schirmer, G.E.Schulz. 1981. J.Mol. Biol. 152, 763.
1o. Pai,E.F., G.E.Schulz. 1983. J.Biol.Chem. 258, 1752.
11. Tronrud,D.E., L.F.Ten Eyck, B.W.Matthews. 1987. Acta Crystallogr. (in press).
12. Karplus,P.A., G.E.Schulz. 1987. J.Mol.Biol. 195,(in press).
13. Moult,J., F.Sussman, M.N.G.James. 1985. J.Mol.Biol. 182, 555.
14. James,M.N.G., A.R.Sielecki. 1983. J.Mol.Biol. 163, 299.
15. Schulz,G.E., R.H.Schirmer, E.F.Pai. 1982. J.Mol.Biol. 16o, 287.
16. Hol,W.G.J., P.T.van Duijnan, H.J.C.Berendsen. 1978. Nature 273, 443.
17. Massey,V., P.Hemmerich. 198o. Biochem.Soc.Trans. 8, 246.
18. Krauth-Siegel,R.L., R.H.Schirmer, S.Ghisla. 1985. Eur.J. Biochem. 148, 335.
19. Wang,M., C.J.Fritchie Jr..1973. Acta Crystallogr. B29,2o4o.
2o. Smith,W.W., R.M.Burnett, G.D.Darling, M.L.Ludwig. 1977. J.Mol.Biol. 117, 195.
21. Huber,P.W., K.G.Brandt. 198o. Biochemistry 19, 4568.
22. Thornton,J..1981. J.Mol.Biol. 151, 261.
23. Richardson,J..1981. Adv.Prot.Chem. 34, 167.

KINETIC ISOTOPE EFFECT STUDIES ON YEAST, SPINACH AND HUMAN ERYTHROCYTE GLUTATHIONE REDUCTASES

M. A. Vanoni
Department of General Physiology and Biochemistry, University of Milano, Milano, Italy

K. K. Wong, J. S. Blanchard
Department of Biochemistry, Albert Einstein College of Medicine, Bronx, New York, 10461

Introduction

Glutathione reductase (E.C.1.6.4.2.) catalyzes the reversible reduced pyridine nucleotide dependent reduction of oxidized glutathione to form two moles of the tripeptide, reduced glutathione. Yeast, spinach and human erythrocyte enzymes have been extensively studied (1), and the three-dimensional structure of the human erythrocyte enzyme has been determined (2). The enzymes are highly homologous, and this homology extends to other flavoprotein reductases which contain a redox-active disulfide. However, the active sites of the three glutathione reductases might differ due to specific amino acid substitutions which may affect the stabilization of the substrate and product binary complexes, E-NADPH and EH_2-NADP, as well as the transition state.

The analysis of steady-state kinetic isotope effects using deuterated and tritiated low-commitment pyridine nucleotide substrates has allowed us to investigate differences among the three enzymes.

Flavins and Flavoproteins 1987

Results and Discussion

Glutathione reductases from yeast, spinach and human erythrocyte differ in their specificity towards reduced pyridine nucleotides. The observed V_m at saturating [GSSG] using NADH were 3.8, 18.4, and 73 % of that obtained using NADPH for the yeast, spinach and human erythrocyte enzymes, respectively. In all cases, NADPH and phosphorylated nucleotides are characterized by low K_m, while nonphosphorylated nucleotides exhibit high K_m. Together with the low K_m determined for cyclic-NADPH, these results suggest that the negatively charged 2'-phosphate contributes substantially to tight binding, although there appears to be little difference between dianionic and monoanionic forms.

We determined steady-state kinetic isotope effects (KIE) using [4S-4-^{2}H]NADPH and [4S-4-^{2}H]NADH at several levels of GSSG with both the yeast and spinach enzyme. The observed kinetic isotope effects measured using either NADPH or NADH were independent of the [GSSG]. These data confirm the ping-pong kinetic mechanism proposed for glutathione reductase (1), and suggest that, if ternary complexes do form, substrate occupancy at either substrate binding site does not alter the rates of the two half-reactions.

Kinetic isotope effect data were analyzed according to the scheme shown below for which the following equations can be derived (4):

$$E_{ox} + NAD(P)H \underset{k_2}{\overset{k_1}{\rightleftharpoons}} E_{ox}\ NAD(P)H \underset{k_4^*}{\overset{k_3^*}{\rightleftharpoons}} E_{red}\text{-}NAD(P)^+ \xrightarrow{k_5} E_{red} + NAD(P)^+$$

$$E_{red} + GSSG \xrightarrow{k_7} E_{ox} + 2\ GSH$$

$${}^{D}V/K = \frac{{}^{D}k_3 + c_f + c_r\,{}^{D}K_{eq}}{1. + c_f + c_r} \quad \text{and} \quad {}^{D}V = \frac{{}^{D}k_3 + c_{Vf} + c_r\,{}^{D}K_{eq}}{1. + c_{Vf} + c_r}$$

where $c_f = (k_3/k_2)$, $c_r = (k_4/k_5)$, $c_{Vf} = (k_3/k_5 + k_3/k_7)$, and $^DK_{eq} = k_{3H}k_{4D}/k_{3D}k_{4H} = 0.98$.

At saturating levels of GSSG, no $^DV/K$ but ^DV's of 1.4, 1.5 and 1.1 for the spinach, yeast and human enzyme were observed using NADPH as reductant, indicating that NADPH exhibits a large commitment to catalysis as expected from its low K_m. $^DV/K$'s of approximately 4.0, 2.7 and 1.3 were observed with NADH for these enzymes, respectively, in agreement with its high K_m. In these cases, small ^DV's were observed, but their value is poorly determined.

The comparison of $^DV/K$ and $^TV/K$ exhibited by NADH using each enzyme has allowed us to examine the magnitude of the intrinsic deuterium kinetic isotope effect on the C-H bond cleavage step (k_3).

If the intrinsic kinetic isotope effect on NADH oxidation were fully expressed in $^DV/K$, then we would predict that the observed $^DV/K$'s and $^TV/K$'s would fulfill the Swain-Schaad (3) relationship ($^Dk^{1.442} = {}^Tk$).

Table 1: Comparison of Deuterium and Tritium KIE's.

	$^DV/K$ (exp)	$^TV/K$ (calc)	$^TV/K$ (exp)
Human	1.29 ± .04	1.44	1.40 ± .13
Yeast	2.73 ± .06	4.18	4.17 ± .28
Spinach	4.04 ± .25	7.48	7.40 ± .39

The good agreement between calculated and experimental values suggests that $^Dk_3$ is fully expressed in $^DV/K_{NADH}$ for the three enzymes. This finding is supported by the independent determination of a kinetic isotope effect of 2.7 on NADPH oxidation catalyzed by the yeast enzyme (4).

Steady-state kinetic isotope effect analysis shows that despite their high homology, the enzymes differ either in the magnitude of the intrinsic isotope effect or in the suppression of an identical large intrinsic isotope effect.

If $^{D}k_3$ were fully expressed in $^{D}V/K_{NADH}$, the comparison of $^{D}V/K_{NADH}$, measured for the three glutathione reductases, would suggest that their transition states differ significantly, although they catalyze the same reaction between the same redox partners. This is in concert with established theories on the enormous modulating ability of enzymes to stabilize or destabilize enzyme bound substrates, products or intermediates, and transition states.

Acknowledgement

This research was supported by NIH GM-33449 and the Sinsheimer Foundation.

References

1. C.H. Williams, Jr. 1976. In: The Enzymes, Vol. XIII (P.D.Boyer, ed.). Academic Press, New York. pp.89-173.

2. Thieme, R., E.F. Pai, R.H. Schirmer and G.E. Schulz. 1981. J. Mol. Biol. 152, 763.

3. Swain, C.G., E.C. Stivers, J.F. Reuwer and L.J. Schaad. 1958. J. Am. Chem. Soc. 80, 5885.

4. Huber, P.W. and K.G. Brandt. 1980. Biochemistry. 19, 4568.

EXCITED ELECTRONIC STATES OF FLAVIN-CONTAINING COENZYME MODELS

F. Gückel, D. Schweitzer
Abteilung Molekulare Physik, Max-Planck-Institut für Medizinische Forschung, Heidelberg

K. Becker, R.H. Schirmer
Institut für Biochemie II der Universität Heidelberg

M.F. Zipplies, H.A. Staab
Abteilung für Organische Chemie, Max-Planck-Institut für Medizinische Forschung, Heidelberg

Introduction

In order to gain further insight into the physical basis of flavin-catalyzed reactions, the interactions of a flavin with a second flavin and with other aromatic ring systems were studied. For this purpose compounds 2 through 5 (Fig.1) were synthesized (1,2). These compounds contain the interacting units in defined geometric orientation. A monomeric flavin (1) and glutathione reductase (6), a flavoenzyme of known active-site chemistry (3), were included as reference molecules. The present report deals with excited triplet states of the flavin compounds as studied by optical spectroscopy and optically detected magnetic resonances (ODMR) at 1.3K.

Results and Discussion

The flavin-containing compounds shown in Figure 1 were studied in polycrystalline form as well as at low concentrations in a 1:1 mixture of ethylene glycol/water which forms a glass matrix at 1.3K. Singulet and triplet states (via intersystem

Flavins and Flavoproteins 1987

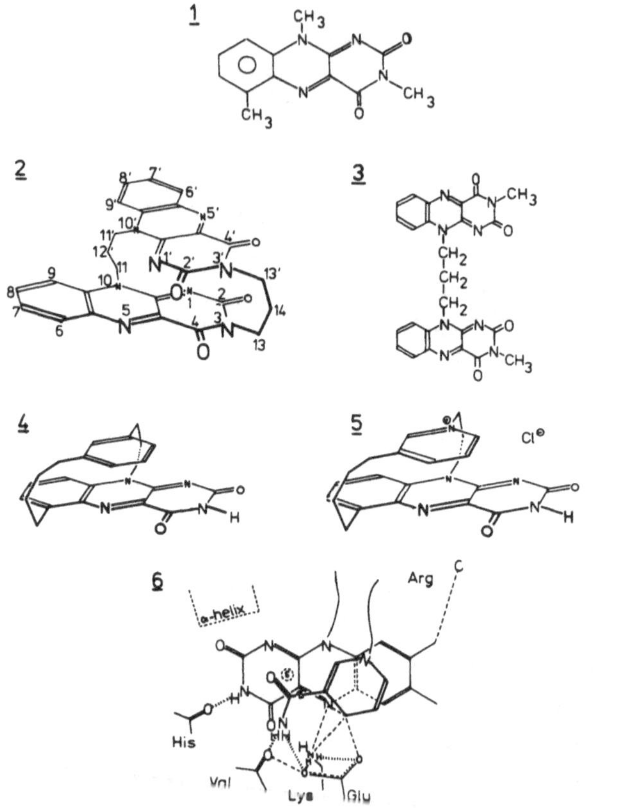

Figure 1. The Structures of the Flavin-containing Compounds 1-6.

1 = 3,6,10-Trimethylisoalloxazine,
2 = (3,3)(3,10)-Isoalloxazinophane,
3 = (10,10'-Trimethylenebis-(3,3'-dimethylisoalloxazine),
4 = (4)Metacyclo-(3)(10,6)-isoalloxazinophane,
5 = (4)(3,1)Pyridinio-(3)(10,6)-isoalloxazinophane chloride,
6 = sketch of one active site of glutathione reductase. The ring systems of flavin and nicotinamide are stacked. A number of amino acid residues located at the catalytic site are indicated.

crossing) were excited by using the 476.5 nm line of an argon ion laser. The characteristic emission spectra of flavins were obtained (Figs.2a and 2b, lower part)(4). In order to identify emissionswith longer lifetimes than 1 ms - phosphorescence (Ph) and delayed fluorescence (DF) - the fluorescence of the spectra was eliminated by using crossed rotors in the excitation and detection pathways. All compounds showed a typical flavin phosphorescence; in addition, the flavin dimers 2 and 3 exhibited a relatively strong delayed fluorescence (Ref.4 and Fig.2a, upper part). While other flavin compounds (1,4-6) did not show DF, in the case of the dimers 2 and 3 a weak delayed fluorescence was observed even at a concentration of <10 μmol per liter glass matrix (Fig.2b).

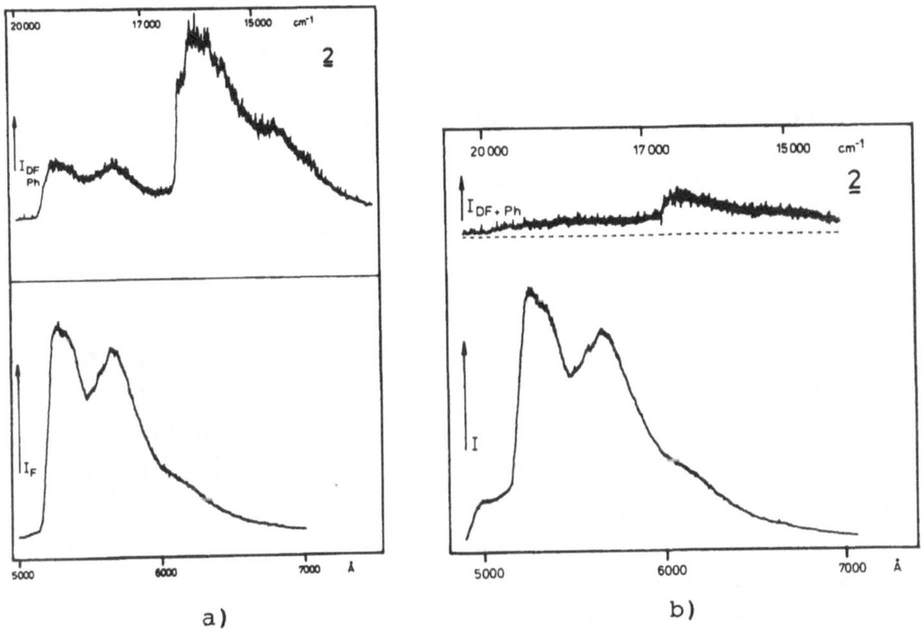

Figure 2. Emission Spectra of Compound 2 in Polycrystalline Form (a) and in 'Dissolved' Form (b). In each panel, the lower spectrum represents the total emission whereas the upper spectrum shows only the long-living emission contributed by delayed fluorescence (DF) and phosphorescence (Ph).

DF which was described so far only in crystals can be explained in the following way (5). The lowest excited triplet state is populated via intersystem crossing from the singulet manifold. In crystals, there are singulet and triplet exciton bands in which excitation energy can be transported over many lattice sites. A detailed description of such Frenkel excitons is given in ref. (6). The collision of two excitonic triplet states or one excitonic triplet state and a localized triplet trap state results in the formation of a pair state as shown in the following scheme.

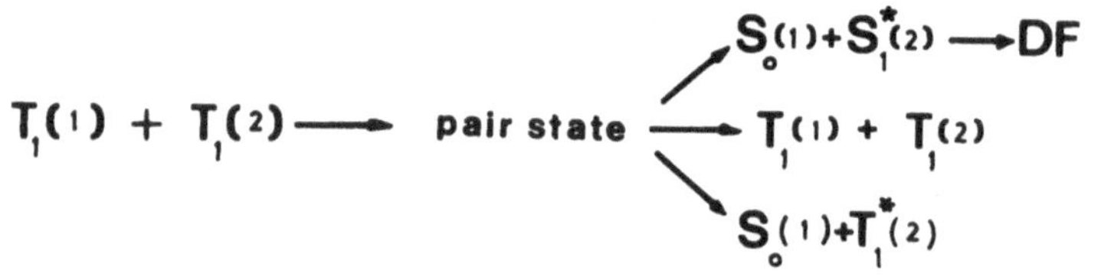

Scheme of a Triplet-Triplet Annihilation

The two interacting triplet states are able to add their energies and spins and then to decay according to group theoretical selection rules. One of the possible pathways leads to an excited singulet state. The deactivation of this state can be observed as a delayed fluorescence with a lifetime in the ms range. According to theory, this lifetime corresponds to the half-lifetime of a triplet exciton state (7). Within experimental error, our values of 23.5 ms and 8.5 ms meet this expectation (Table 1, lines 5 and 6).
As emphasized above, DF is based on an effective intersystem crossing process resulting in a high density of excitonic triplet states within crystals. The really surprising property of the flavin dimers is therefore the moderate DF observed in the case of isolated dimer molecules.

The pair formation of triplet states in an isolated dimer can occur only if both monomeric moieties are excited. Since the lifetime of triplet states is much longer than of the singulet state simultaneous excitation of triplet states of neighbouring units by singulet absorption seems possible provided that effective intersystem crossing takes place. In addition, the relatively short average distance between the two units allows an excitonic interaction leading to pair formation. In this context it should be emphasized that the interaction of the two flavin moieties in the excited triplet state must be very weak because otherwise symmetric charge transfer terms would prevent the two monomers from being excited simultaneously (8,9).

Table 1. Lifetimes of the Triplet States. Ph, phosphorescence; DF, delayed fluorescence

Compound	Emmission Wavelength nm	Lifetime at 1.3K ms
1	Ph, 630	53
2	Ph, 630	< 3
2	Ph, 700	< 3
2	DF, 550	< 3
3	Ph, 610	23.5
3	DF, 550	8.7
4	Ph, 630	62
4	Ph, 650	62
5	Ph, 650	27
6	Ph, 630	41.6
Riboflavin-	Ph, 605	170
tetrabutyrate (15)	Ph, 640	100

An alternative explanation, namely that the observed DF is due to an effective absorption of two quanta by the same monomer, is unsatisfactory for two reasons: (i) the relatively long lifetime of the DF (Table 1) cannot be explained sufficiently by a 2-quanta absorption; (ii) we were unable to obtain a DF by direct singulet-triplet excitation ($S_o \rightarrow T_1$) using a narrow banded dye laser. This indicates that in spite of the comparatively high quantum yield of the intersystem crossing process ($S_1 \rightarrow T_1$) (10) the $S_o \rightarrow T_1$ transition is strongly forbidden. In consequence, an effective 2-quanta absorption seems unfavourable in this case. In contrast, the process of pair formation is not limited by intercombination selection rules and can - especially under these conditions - be most effective. Thus the indirect decay of the triplet state via pair formation and delayed fluorescence is able to compete with the direct phosphorescence decay. On the other hand the occurrence of DF in the case of the dimeric flavins 2 and 3 is not associated with a fundamental change of the electronic properties with respect to monomeric flavins.

Evidence for this are other spectroscopic data, such as the energy levels of the singulet and triplet states as well as the zero field splitting parameters and the triplet lifetimes. These data are shown in Tables 1 - 3 together with literature values for other flavins (11 - 15). In particular, the comparison of the $|D|$ and $|E|$ parameters of compounds 2 to 6 with those of monomeric flavins indicates that no appreciable charge transfer occurs in the triplet state T_1; otherwise one would expect a strong reduction of the $|D|$ parameters in 2 to 6 (16).

Table 2. Energies of the Singulet and Triplet Levels of Compounds 1 to 6 Measured as Polycrystalline Material. Protein crystals contain approximately 50% solvent; consequently the spectrum of 6 represents a mixture of crystalline and solvated molecules.

Compound	S_1 (cm^{-1})	T_1 (cm^{-1})
1	19840	16750
2	19460	16370
3	20200	17390
4	20450	17290
5	20120	17180
6	20600	17250
N3-Undecyllumiflavin (15)	20956	-
8-Methylisoalloxazine (14)	20410	17545
Lumiflavin (14)	21275	-

Table 3. Fine Structure Parameters. Dashes(-) referring to 1 to 6 indicate that these parameters could not be determined experimentally.

Compound	|D+E| (MHz)	|D-E| (MHz)	|2E| (MHz)	|D| (cm^{-1})	|E| (cm^{-1})
1	2164	1086,3	-	0,0542	0,0179
2	1985	1008	-	0,0499	0,0163
3	-	-	-	-	-
4	2071	1040	-	0,05188	0,0172
5	-	-	-	-	-
6	-	-	-	-	-
FAD (12,13)	-	-	-	0,059	0,019
Riboflavin (13)	-	-	-	0,059	0,019
Riboflavin (11)	-	-	-	0,061	0,015
Alloxazine (13)	-	-	-	0,079	0,019
FMN (13)	-	-	-	0,058	0,019

Conclusion

For the two compounds 2 and 3 containing two flavin units we detected delayed fluorescence (DF) associated with a most efficient excitonic interaction. In this process, two low-energetic quanta are transformed into one high-energetic quantum with an unusually high quantum yield. An activation energy as high as 34000 cm^{-1} ≙ 4.2 eV ≙ 96.8 kcal per mol can be provided in this process. Therefore analogous molecular arrangements might be relevant for flavoenzyme-catalyzed reactions.

References

1. Zipplies, M.F., PhD thesis, Heidelberg University 1984.
2. Zipplies, M.F., H.A. Staab 1984. Tetrahedron Lett. 25, 1035.
3. Thieme, R., E.F. Pai, R.H. Schirmer, G.E. Schulz 1981. J. Mol. Biol. 152, 763.
4. Gückel, F., MD thesis Heidelberg University 1986.
5. Birks, J.B. 1970. Photophysics of Aromatic Molecules. Wiley Interscience, p. 142.
6. Davydov, A.S. 1971. Theory of Molecular Excitons. Plenum Press, p. 23.
7. Benz, K.W., PhD thesis, Stuttgart University 1970.
8. Schweitzer, D., K.H. Hausser, H. Vogler, H.A. Staab, V. Bockelheide 1986. Chem. Phys. 110, 33.
9. Gückel, F., D. Schweitzer, J.P. Collman, S. Bencosme, J. Sessler 1984. Chem. Phys. 86, 161.
10. Moore, W.M., I.C. McDaniels, J.A. Hen 1977. Photochem. Photobiol. 25, 505.
11. Lhoste, J.M., A. Haug, P. Hemmerich 1966. Biochemistry 5, 3290.
12. Moore, T.A., A.L.C. Kwiram 1976. In: Flavins and Flavoproteins (T.P. Singer ed.) Elsevier Scientific Publ., p. 782.
13. Moore, T.H., A.L. Kwiram 1974. Biochemistry 13, 5403.

14. Sun, M., T.A. Moore, P.S. Song 1972. J. Am. Chem. Soc. 94, 1730.

15. Song, P.S., T.A. Moore, W.E. Kurtin 1972. Z. Naturforsch. 27B, 1011.

16. Schweitzer, D. Habilitationsschrift Heidelberg University 1980.

TRYPANOTHIONE REDUCTASE FROM *TRYPANOSOMA CRUZI*

R.L. Krauth-Siegel, R.H. Schirmer, T. Schöllhammer
Institut für Biochemie II der Universität Heidelberg,
West Germany

Introduction

Trypanosomes are unicellular eukaryotes which belong to the family trypanosomatidae. Recently it was discovered that these protozoal parasites do not possess glutathione reductase - as their hosts do - but trypanothione reductase instead (1). Here we report on the properties of trypanothione reductase from *Trypanosoma cruzi* (2). The enzyme catalyzes the reduction of trypanothione disulfide to dihydrotrypanothione by NADPH:

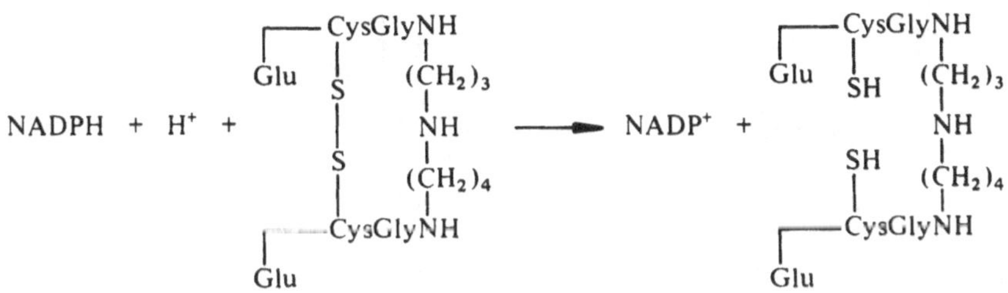

The structural differences between trypanothione reductase of *T. cruzi* and human glutathione reductase, an enzyme of known atomic structure, offer an opportunity for rational drug design against Chagas' disease in man (3).

Results and Discussion

From 33.4g epimastigotes of *T. cruzi* 628 enzyme units were extracted. The enzyme was purified 2400 fold in three steps with an overall yield of 45% and crystallized (2).

Flavins and Flavoproteins 1987

Table 1: Properties of *T.cruzi* Trypanothione Reductase and Human Glutathione Reductase (2)

	trypanothione reductase	glutathione reductase
subunit Mr	50000	52500
cofactor	FAD	FAD
oligomeric structure	dimer	dimer
intracellular concentration	1.25 µM	0.1 µM
substrate at the pyridine nucleotide site:		
NADPH	K_m = 5 µM	K_m = 8.5 µM
activity with NADH/ activity with NADPH	0.03	0.05
substrates at the disulfide site:		
glutathione disulfide	no substrate	K_m = 65 µM
trypanothione disulfide	K_m = 45 µM	no substrate
glutathionylspermidine disulfide	K_m = 275 µM	no substrate
specific activity	284 U/mg	240 U/mg
turnover number	14200 min^{-1}	12600 min^{-1}
inhibition by nifurtimox	negligible	yes
highly reactive cysteine residues per subunit in the native		
oxidized enzyme species E	0.0	0.3
reduced enzyme species EH_2	1.0	1.3
catalytic disulfide	yes	yes (Cys58-Cys63)
$\lambda_{max,vis}$	461 nm	463 nm
absorption at 530 nm in EH_2	yes	yes
A^{1mM}_{530nm} of the EH_2·NADPH complex	4.9	4.5

Trypanothione reductase is a member of a closely knit protein family which includes the FAD-cystine enzymes lipoamide dehydrogenase, glutathione reductase, mercuric ion reductase, asparagusate reductase and pantethinediphosphate reductase(4). Table 1 shows some properties of trypanothione reductase from *T. cruzi* in comparison with those of human glutathione reductase. As shown in Fig.1 the absorption spectrum of trypanothione reductase is nearly indistinguishable from that of glutathione reductase.

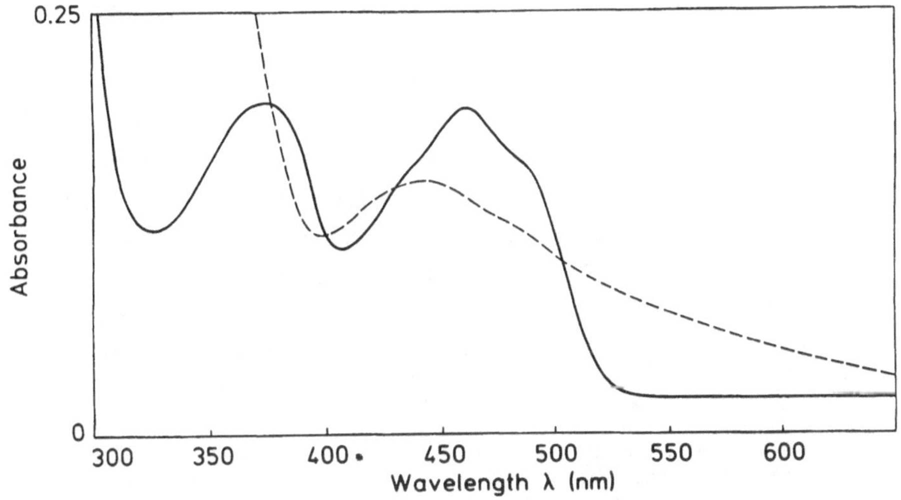

Fig.1: Absorption Spectra of Trypanothione Reductase in Oxidized Form E and after Reduction with NADPH.
17 µM trypanothione reductase in 50 mM potassium phosphate, 1 mM EDTA, pH 7.0 (——) was reduced with 90 µM NADPH (----); subsequent addition of trypanothione disulfide (17 µM) and GSSG (500 µM) restored the spectrum of the oxidized enzyme.

Reduction of the enzyme with excess NADPH leads to a long wavelength absorption around 550 nm with a concomitant decrease of the absorption at 461 nm. This probably reflects a charge transfer complex between a thiolate anion of the reduced enzyme and the flavin ring amplified by the presence of NADPH. Addition of trypanothione disulfide restores the original spectrum of the oxidized enzyme E. This behaviour shows that the catalytic cycle of trypanothione reductase can be subdivided into two half-reactions:

$$E + NADPH + H^+ \longrightarrow EH_2 + NADP^+ \quad \text{and}$$

$$EH_2 + \text{trypanothione disulfide} \longrightarrow E + \text{dihydrotrypanothione.}$$

Glutathione reductase and its relatives possess two catalytic Cys residues which form a disulfide bridge in the oxidized enzyme E. In the two electron-reduced species EH_2, a dithiol is formed and one of these thiols is easily alkylated by iodoacetamide. This is the case also in trypanothione reductase. The EH_2 form of the enzyme incorporates exactly one radioactive carboxamidomethyl group per subunit whereas the oxidized enzyme E does not react with iodoacetamide (2). The most obvious difference between trypanothione reductase and human glutathione reductase are the mutually exclusive specificities for their respective disulfide substrates. Beside 'its' substrate trypanothione reductase is able to reduce glutathionylspermidine disulfide which can be isolated from *Escherichia coli* (5). The relatively high K_m value of

Table 2: Isolation of Glutathionylspermidine and Trypanothione. Glutathionylspermidine from *Escherichia coli* was determined according to (5). *Crithidia fasciculata* and *Trypanosoma cruzi* were processed as described in (6). Two different cultures of *C. fasciculata* (A and B) were tested. The different glutathionylspermidine concentrations remained unexplained. The concentration of glutathione (GSH-equivalents) in *T. cruzi* epimastigotes was found to be 130 nmol/g cells.

Species	Final yield (nmol per g cells) Glutathionylspermidine	Trypanothione
Escherichia coli (stationary phase)	360	not determined
Crithidia fasciculata	60 (A) <10 (B)	210 (A) 340 (B)
Trypanosoma cruzi	< 2	80

this substrate measured in phosphate buffer pH 7.0 (Table 1) decreased to 60 µM, when assaying the enzyme in 50 mM N-ethylmorpholinium acetate, pH 7.7.

As a first step in the elucidation of the parasite's thiol metabolism we determined glutathione, trypanothione and glutathionylspermidine in extracts of *T. cruzi* epimastigotes (Table 2). Thiol redox cycles play a crucial role in protecting parasites from reactive oxygen species which are produced via the flavoenzyme NADPH oxidase of host cells (for a review see ref.3). As trypanothione reductase does not occur in mammalian cells this antioxidant enzyme is a promising target in the design of drugs against diseases caused by leishmanias and trypanosomes.

References

1. Fairlamb, A.H., P. Blackburn, P. Ulrich, B.T. Chait, A. Cerami 1985. Science 227, 1485-1487.

2. Krauth-Siegel, R.L., B. Enders, G.B. Henderson, A.H. Fairlamb, R.H. Schirmer 1987. Eur. J. Biochem. 164, 123-128.

3. Schirmer, R.H., T. Schöllhammer, G. Eisenbrand, R.L. Krauth-Siegel 1987. Free Radical Res. Comms. 3, 3-12.

4. Schirmer, R.H., G.E. Schulz 1987. In Coenzymes and Cofactors, Vol.2 (D. Dolphin, O. Avramovic eds.) Wiley Publ. p. 168-202.

5. Tabor, H., C.W. Tabor 1975. J. Biol. Chem. 250, 2648-2654.

6. Schöllhammer, T., MD Thesis, Universität Heidelberg 1987.

INACTIVATION OF TWO-ELECTRON REDUCED *ESCHERICHIA COLI* GLUTATHIONE REDUCTASE BY NADPH AND NADH

L. David Arscott, Donna M. Drake and Charles H. Williams, Jr.

Veterans Administration Medical Center and Department of Biological Chemistry, University of Michigan, Ann Arbor, MI 48105, USA

Introduction

Glutathione reductase from several sources is inactivated in the presence of the physiological substrate NADPH. Worthington and Rosemeyer (1) associated inactivation with aggregation at high concentrations of the human erythrocyte enzyme. Lopez-Barea and his colleagues studied low concentrations of mouse liver (2), yeast (3) and *E. coli* (4) enzymes and ascribed the inactivation to intramolecular thiol oxidation. They found that NADH, thiols and dithionite also inactivate and noted the dependence of the final level of inactivation on the enzyme concentration (4). They attributed reactivation by glutathione to reduction of the new disulfide and reactivation by oxidized glutathione to "disulfide exchange" (3).

We have studied the inactivation of *E. coli* glutathione reductase by both NADPH and NADH and also find that the extent to which the inactivation proceeds is inversely related to the enzyme concentration, suggesting to us that dimer dissociation is involved. Modeling the reaction in which NADH is used is consistent with an inactivation due to monomer formation in the two-electron reduced enzyme. The use of NADH rather than NADPH to model the inactivation process is necessitated by the fact that the two-electron reduced form of glutathione reductase forms tight complexes with NADPH and NADP and these complexes markedly affect the inactivation process (6). The reactivation promoted by glutathione adheres to second order kinetic dependence on the concentration of the putative monomer. Veeger and his associates were the first to propose a dimer-monomer equilibrium affecting the activity of the closely related enzyme, lipoamide dehydrogenase (7, 8).

Flavins and Flavoproteins 1987

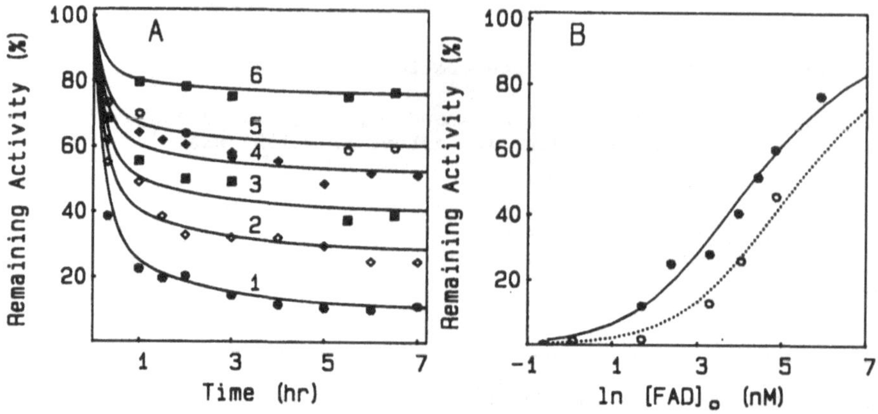

Figure 1. Panel A: Inactivation of various concentrations of *E. coli* glutathione reductase by incubation with 150 µM NADH. Enzyme at 5.4, 26.8, 53.4, 85.1, 127 and 372 nM (curves 1-6 respectively) is incubated at 25 °C in 0.05 M TRIS, 0.3 mM EDTA, pH 7.5. Activity is determined by adding aliquots of the incubation mixture to a NADPH/oxidized glutathione assay solution (5). The diluted enzyme concentration in the assay is ca. 1.8 nM. Panel B: Dependence of the remaining activity level on the ln of the initial enzyme concentration. o, The activity remaining after 5-7 h aerobically as a function of enzyme concentration. o, The anaerobic 5-7 h inactivation level (primary data not shown); conditions are as in panel A. Activity = $100*2[FAD_d]/[FAD]_o$, where $[FAD]_o = 2[FAD_d] + [FAD_m]$.

Results and Discussion

When glutathione reductase is incubated with 150 µM NADH and aliqouts assayed over time by the addition to saturating substrates (5), activity decreases and reaches a stable value after about 5-7 h of incubation. Figure 1A shows the per cent remaining activity at various concentrations of enzyme versus time incubated. If it is assumed that remaining catalytic activity reflects a dimer-monomer equilibrium at the 2-electron reduced level in which dimer is active (FAD_d) and monomer (FAD_m) is inactive, an equilibrium described by: $K_d = [FAD_m]^2/[FAD_d]$, can be used to express the data of panel A, and is shown in panel B. The solid line reflects an assumed K_d of 72.7 nM. Due to the inherent oxidase activity of glutathione reductase, this is an apparent K_d. Anaerobic inactivation is 10% to 20% greater at each enzyme concentration yielding a true K_d of 221 nM, dotted line panel B. The good fits of the data lend credence to the suggestion of a dimer-monomer equilibrium.

NADPH inactivates glutathione reductase much faster than NADH but only at concentrations which are 10 times lower than the K_d for the EH_2-NADPH complex of 12 µM (6). At 1 µM NADPH, 1 nM enzyme is inactivated with a half time of 1 minute. However, the presence of 0.5 µM to 20 µM NADP added either at the start of inactivation or up to 10 minutes later provides impressive protection and/or a modest reactivation. Thus, due to the oxidase activity in the presence of NADPH and to the protection against inactivation by low levels of NADP, the use of NADPH to study inactivation as a function of enzyme concentration is too complex.

Figure 2A shows reactivation of glutathione reductase by glutathione added to the inactivating enzyme at various times. The kinetics (Figure 2B) are second order in monomer concentration for a large part of the reactivation. The average slope of the four lines yields a rate constant of 1.3×10^7 $M^{-1}min^{-1}$, a rate well within experimentally determined rates from other protein-protein interactions (9, 10). The fits of the data suggest that reassociation of a putative monomer species leads to reactivation.

Oxidized glutathione spans the interface of the two monomers of the human erythrocyte enzyme, i.e. there are binding sites for each end of each half molecule in each subunit; moreover, each active site is composed of elements from each polypeptide chain (11). These facts, along with the high degree of homology between erythrocyte enzyme and _E. coli_ enzyme (12), give strong support to the idea that the monomer must be catalytically inactive. Dimer reformation and reactivation then, would be promoted by glutathione (Figure 2) or oxidized glutathione (3). A mechanism of inactivation involving thiol oxidation (2-4), for which an electron acceptor is not apparent especially under the conditions of inactivation, should not be dependent on enzyme concentration. Neither should reactivation promoted by glutathione follow the kinetic pattern shown in Figure 2. The implications for physiological control by low levels of substrate and enzyme have been noted (2-4). It is therefore important that the mechanism be understood. The dimer-monomer equilibrium indicated by the data presented here offers a simple explanation for the inactivation process.

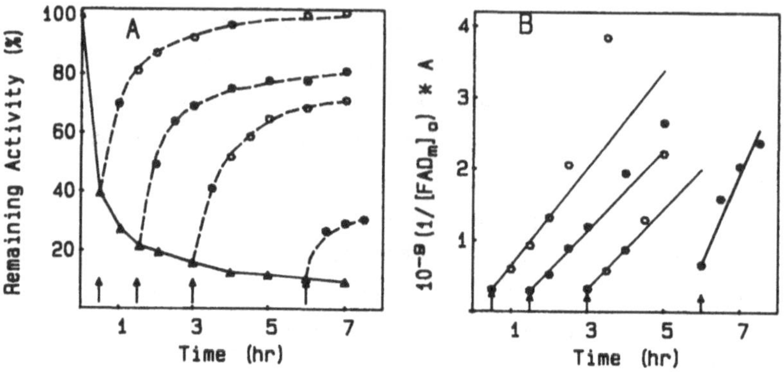

Figure 2. Panel A: Time course of inactivation of 5.4 nM E. coli glutathione reductase with 152 µM NADH and subsequent reactivation with glutathione (conditions and assays as in Fig 1). At the arrows, aliquots of inactivating enzyme were made 1 mM in glutathione. Panel B: Second order kinetic plot for reactivation from panel A. $A = 1/1-[A_x-A_i/A_a-A_i]$ where A_x, A_i and A_a are respectively, the % remaining activity at time x, initially and finally (estimated).

Acknowledgment: This work has been supported by the Medical Research Service, Veterans Administration and by Grant 21444 from the National Institute of General Medical Sciences.

References

1. Worthington, D.J., M.A. Rosemeyer. 1976, Eur. J. Biochem. 67, 231-238

2. Lopez-Barea, J., C. Lee. 1979, Eur. J. Biochem. 98, 487-499

3. Pinto, M.C., A.M. Mata, J. Lopez-Barea. 1985, Eur. J. Biochem. 151, 275-281 and 1984, Arch. Biochem. Biophys. 228, 1-12

4. Mata, A.M., M.C.Pinto, J.Lopez-Barea.1985, Mol. Cell.Biochem. 67, 65-76

5. Massey, V., C.H. Williams, Jr.. 1965, J. Biol. Chem. 240, 4470-4480

6. Williams, C.H. Jr.. 1976, in Flavins and Flavoproteins (Singer, T. P., ed.) pp. 455-463, Elsivier Scientific, Amsterdam

7. Kalse, J.F., C. Veeger. 1968, Biochim. Biophys. Acta 159, 244-256

8. Visser, J., C. Veeger. 1968, Biochim. Biophys. Acta 159, 265-275

9. Ip, S.H.C., G.K. Ackers. 1977, J. Biol. Chem. 252, 82-87

10. Parr, G.R., G.G. Hammes. 1976, Biochem. 15, 857-862

11. Pai, E.F., G.E. Schulz. 1983, J. Biol. Chem. 258, 1752-1757

12. Greer, S., R.N. Perham. 1986, Biochem. 25, 2736-2742

STUDIES ON THE ACTIVE SITE OF HUMAN ERYTHROCYTE GLUTATHIONE REDUCTASE USING 6-SCN-FAD AND 6-MERCAPTO-FAD

Vincent Massey*, Sandro Ghisla†, Ulrich Ermler**& Georg E. Schulz**.
*Department of Biological Chemistry, University of Michigan, †Fakultät Biologie, Universität Konstanz and **Institüt für Organishe Chemie und Biochemie, Albert-Ludwigs Universität, Freiburg.

Glutathione reductase (GR) is a member of a group of mechanistically similar flavoproteins, which all contain a redox-active disulfide in addition to FAD. Because its structure has been solved by X-ray diffraction analysis (1,10), GR from human erythrocytes has become the prototype of this group of enzymes, especially in view of the extensive spectral and sequence homologies found within this group (2,3). Such homologies have been used to identify successfully the catalytically essential disulfide of mercuric reductase and in the production of a series of active site mutants of that enzyme (4). Despite the impressive similarities, there do exist, however, significant structural differences in the relative orientation of the flavin and active site disulfide, as will be shown in this and an accompanying communication (5). This was made possible by replacing the native FAD with the chemically reactive flavins, 6-SCN-FAD and 6-mercapto-FAD (6).

Preparation of Modified Holoenzymes

The apoprotein of GR was prepared by the method of Fritsch as described in Manstein et al. (7); DTT, however, was not included in the final solution. The apoprotein was dissolved in 0.05 M phosphate, pH 7.0, containing 0.3 mM EDTA, and was added in slight excess to a solution of 6-SCN-FAD or 6-mercapto-FAD in the same buffer. In general the mixture was incubated in ice for 20 min followed by incubation at 20° for 10 min. The resulting holoenzyme was either used as such, or concentrated by centrifuging through an Amicon Centricon 30 Ultrafilter. In the case of preparation of the 6-mercapto FAD enzyme for crystal studies, 6-mercapto FAD was added in slight excess over apoprotein; after reconstitution the excess flavin was removed by Centricon ultrafiltration.

Flavins and Flavoproteins 1987

Conversion of 6-SCN-FAD Enzyme to 6-Mercapto-FAD Enzyme

6-Thiocyanato-FAD binds tightly to the apoprotein of GR and is very stable in the absence of added thiols. The holoenzyme has λ_{max} of 456 nm (compared to 446 nm for free 6-SCN-FAD) and a well resolved maximum at 380 nm (compared to a shoulder with the free flavin). On addition of DTT there was an immediate color change from yellow to red, with a spectral change (Figure 1, curve 1) typical of the EH_2-form of the native enzyme, i.e. a charge transfer complex involving Cys 63 of the reduced active site disulfide as donor and oxidized flavin as acceptor (8). This is followed by a secondary isosbestic reaction in which the flavin is converted to 6-mercapto-FAD. The final spectrum shown (curve 10) is identical to that obtained by titration of 6-mercapto-FAD (λ_{max} 443 nm, ε = 18900 M^{-1} cm^{-1}) with apo-GR (λ_{max} 450 nm, ε = 18,100 M^{-1} cm^{-1}).

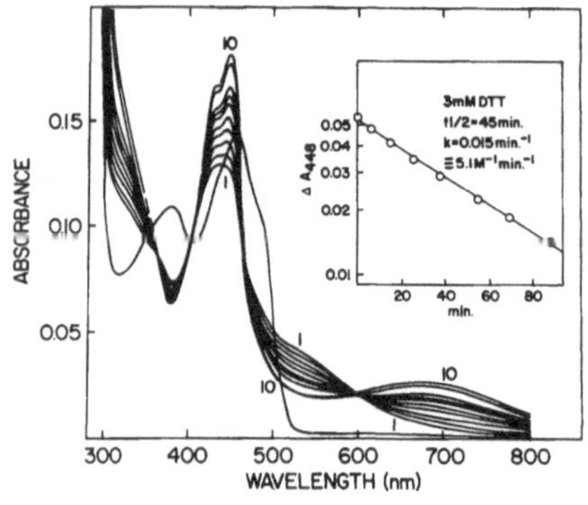

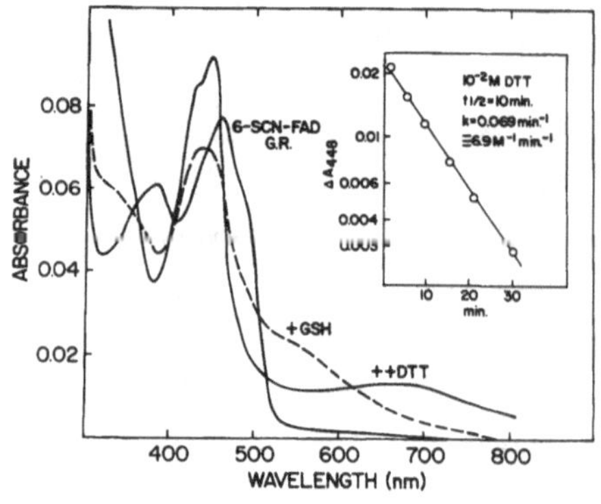

Figure 1. (left) Conversion of 6-SCN-FAD glutathione reductase to the 6-mercapto-FAD form by reaction with DTT. The spectrum of the 6-SCN-FAD enzyme is that with λ_{max} of 456 and 380 nm. Immediately after addition of 3 mM DTT the spectrum changed to that of curve 1. This was followed by the slow secondary change shown by curves 2-10. Conditions; 0.05 M phosphate + 0.3 mM EDTA, pH 7.0, 20°C.

Figure 2. (right) Effect of reduced glutathione followed by DTT on 6-SCN-FAD glutathione reductase. On addition of 2 mM GSH (K_d ~ 0.1 mM) the dashed-line spectrum was obtained and remained unchanged over a period of more than 1 hour. On the subsequent addition of 10 mM DTT, the 6-mercapto-FAD enzyme was formed with a $t_{1/2}$ of 10 min (see inset). Conditions, as in Fig. 1.

The rate of conversion of the 6-SCN-FAD enzyme to the 6-mercapto-FAD form is dependent on the concentration of DTT employed. The inset to Figure 1 shows a semi log plot of the absorbance increase at 448 nm; an average second order rate constant of 4.5 M^{-1} min^{-1} was obtained in several such experiments. These results indicate that the conversion is brought about by DTT itself, rather than by interaction of the enzyme active site dithiol with the flavin 6-SCN substituent. This conclusion was confirmed by the finding that reduced glutathione (GSH) produced the same EH_2 spectrum as found initally with DTT, but failed to bring about the secondary conversion, even on prolonged incubation. Several lines of evidence indicate that DTT gains success to the flavin 6-SCN substituent via the pyridine nucleotide pocket on the re-side (8) of the flavin. First, DTT added after saturating GSH brings about the conversion to 6-mercapto FAD, as shown in Figure 2. In fact the second order rate constant is slightly greater than that found in the absence of GSH, perhaps indicating some distortion of the active site to make the flavin re-face more accessible when GSH is bound (presumably as the mixed disulfide with Cys 58). Second, $NADP^+$ binds tightly to the EH_2 form produced by GSH, and blocks completely the subsequent conversion to the 6-mercapto-FAD form by DTT (results not shown).

The spectrum of the 6-mercapto-FAD enzyme shown in Figure 1 is that with the active site disulfide reduced. On removal of DTT and air reoxidation the long wavelength peak at 680 nm shifts to 720 nm with essentially the same extinction coefficient. This effect is readily reversed by DTT. The spectrum is that typical of the anionic species of 6-mercapto flavin (9) in which the negative charge is localized in the N(1)-C(2) locus of the flavin. When bound to GR the anionic species is highly stabilized, the spectrum being unchanged on dialysis of the 6-mercapto-FAD enzyme vs 0.1 M acetate, pH 5.0. Thus the pK of 5.9 with the free flavin (9) is lowered by at least 1.5 pH units on binding to the protein. This might be attributed to the stabilizing effect of a protein α-helix whose N-terminal end contributes a strong partial positive charge directed toward the N(1)-region of the flavin (8).

Crystal Structure of 6-mercapto-FAD Glutathione Reductase

Crystals of the 6-mercapto-FAD enzyme were grown by the hanging drop method using the same crystallization conditions as with native enzyme (1). The crystals, obtained within three weeks, diffracted to at least 2A° resolution. They are much better ordered than the crystals of apoprotein and appear to reach the quality of the native crystals. The space group and unit cell parameters are identical with those of the native enzyme. We collected 10,934 reflections up to 3A° resolution with an average internal R value of 8.3% using a 4-circle diffractometer (1).

For identification of the structural changes, the difference map between the electron density of the 6-mercapto-FAD enzyme and native enzyme was used. The structure factors of the model were taken from the 1.54A° crystal structure of glutathione reductase refined to an R-factor of 18.6% (10). The structure of the 6-mercapto-FAD enzyme was refined at 3A° resolution by a restrained least-squares procedure (11) to an R-factor of 13.7% with an overall root-mean-square deviation of 0.02A° for bond lengths and 2.1° for bond angles.

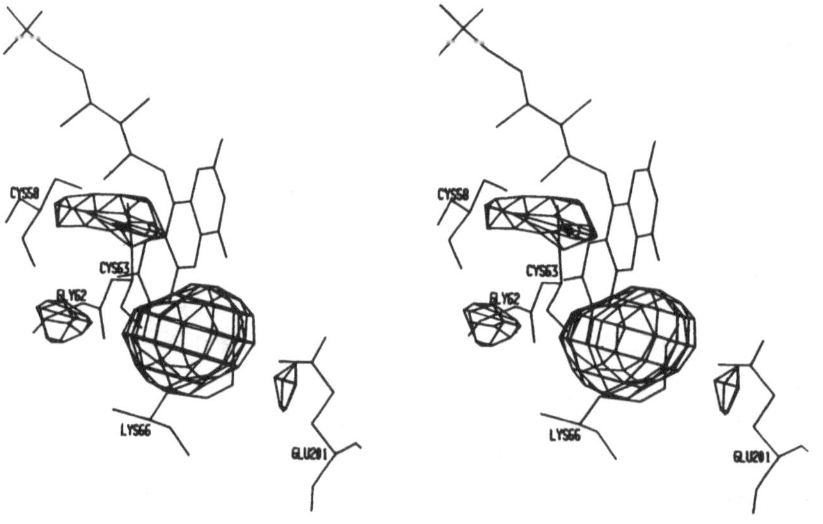

Figure 3. Positive difference-fourier map around the 6-mercapto group of 6-mercapto-FAD glutathione reductase, and native enzyme. The contours are drawn at 15% of the maximum of this map.

The highest peak in the difference fourier map (Fig. 3) was four times higher than the second highest peak, and corresponds unequivocally to the sulfur of 6-mercapto-FAD. The bond length between the S6- and C6-atoms is about 1.6A°, strongly supporting the double bond character of the flavin-sulfur linkage in the anionic 6-mercapto flavin (expected single bond length = 1.8A°, expected double bond length = 1.7A°).

The model of the native enzyme (10) shows free space within a radius of about 5A° around the C6-atom of the flavin. This space clearly facilitates the incorporation of 6-SCN-FAD and explains why there are only minor alterations of the enzyme structure and the FAD-binding position with 6-mercapto-FAD. The 6-mercapto group is surrounded directly by residues Gly 62, Cys 63, Lys 66 and Glu 201. The electron density map of Fig. 3 reveals that Lys 66 is shifted only a fraction of an Angström away from the flavin 6-S linkage. The carboxyl group of Glu 201, which forms a strong salt bridge to Lys 66, is moved about 0.2A° sideways, evading the sulfur of 6-mercapto-FAD. The carboxyl group of Gly 62, which has only little mobility in the native enzyme (10), is shifted about 0.2A° away from the sulfur. This effect, and the spatial proximity of Cys 63 to the 6S-atom of the FAD, induces a small movement of the cystine bridge Cys 58:Cys 63 in a direction parallel to the plane of the isoalloxazine moiety.

The above results confirm fully the conclusions reached from the study of the conversion of the 6-SCN-FAD enzyme to the 6-mercapto-FAD form by DTT. The 6-SCN residue can be readily accomodated by the flavin-binding pocket of the enzyme, and be available for attack by DTT from the nicotinamide side of the flavin. On the other hand, with a distance of about 6.3A°, the sulfur of the 6-SCN group would be too far away from the active site disulfide for the conversion to occur through reduction of the disulfide, and the 6-SCN group would not be accessible to DTT from the glutathione side of the flavin because of steric hindrance. These results are in sharp distinction to those obtained with mercuric reductase (5) and point to significant differences in the active sites of the two enzymes.

Acknowledgements. Supported by grants from the U.S. Public Health Service, GM11106 (V.M.), the Deutsche Forschungsgemeinschaft, Gh2/4-4 (S.G.) and Fsb60/D-14 (U.E. & G.E.S.)

References

1. Thieme, R., Pai, E.F., Schirmer, R.H. and Schulz, G.E. (1981) J. Mol. Biol. 152, 763-782
2. Williams, C.H., Arscott, L.D. and Schulz, G.E. (1982) Proc. Natl. Acad. Sci. U.S.A. 79, 2199-2201
3. Brown, N.L., Ford, S.J., Pridmore, R.D. and Frintzinger, D.C. (1983) Biochemistry 22, 4089-4095
4. Schultz, P.G., Au, K.G. and Walsh, C.T. (1985) Biochemistry 24, 6840-6848
5. Massey, V., Miller, S.M., Ballou, D.P., Williams, C.H., Moore, M., Distefano, M. and Walsh, C.T. (1987), this volume, accompanying communication
6. Massey, V., Ghisla, S. and Yagi, K. (1986) Biochemistry 25, 8103-8112
7. Manstein, D.J., Pai, E.F., Schopfer, L.M. and Massey, V. (1986) Biochemistry 25, 6807-6816
8. Pai, E.F. and Schulz, G.E. (1983) J. Biol. Chem. 238, 1752-1758
9. Ghisla, S., Massey, V. and Yagi, K. (1986) Biochemistry 25, 3282-3289
10. Karplus, P.A. and Schulz, G.E. (1987) J. Mol. Biol. 195, in press
11. Tronrud, D.E., Ten Eyck, L.E. and Matthews, B.W. (1987) Acta Crystallogr. Sect. A, in press

PROTEIN ENGINEERING OF GLUTATHIONE REDUCTASE FROM *ESCHERICHIA COLI*.

N.S.Scrutton, A.Berry and R.N.Perham

Department of Biochemistry, University of Cambridge, Tennis Court Road, Cambridge, UK., CB2 1QW.

Introduction

Glutathione reductase (EC 1.6.4.2) catalyses the reduction of oxidized glutathione by NADPH. It is a widespread member of an important family of flavoprotein oxidoreductases that includes dihydrolipoamide dehydrogenase (1,2), thioredoxin reductase (3) and mercuric reductase (4). All these enzymes are dimers with an M_r of about 105000 and all possess a redox-active disulphide bridge in each subunit.

Considerable light has been thrown on the reaction mechanism of human glutathione reductase from a knowledge of the amino acid sequence (5) and X-ray crystallographic analysis of the protein at 0.2nm resolution (6,7). The gene (*gor*) encoding the *Escherichia coli* glutathione reductase has been cloned and sequenced (8). The human and *E.coli* enzymes are highly homologous, making it certain that their 3-dimensional structures will also be similar. Attempts to crystallize *E.coli* glutathione reductase are in hand. Meanwhile, availability of an over-expressed *E.coli* *gor* gene and a rapid purification scheme for the enzyme (9) has made it possible to test some of the predictions made about the catalytic mechanism of the enzyme (7) by means of site-directed mutagenesis.

Results and Discussion

We report here on the properties of engineered glutathione reductases bearing mutations at residue Tyr-177. This residue is the counterpart of Tyr-197 in the human enzyme (8), which lies in the NADPH binding pocket and shields the isoalloxazine ring of the flavin in the absence of NADPH. On binding of NADPH, the

Flavins and Flavoproteins 1987

tyrosine moves to allow the nicotinamide ring of the NADPH to come close to the flavin for electron transfer (7) and it has been postulated (10) that this residue acts as a "lid" in the NADPH pocket to prevent loss of electrons from the flavin moiety. Two mutations at position 177 have been made, one (Y177F) of tyrosine to phenylalanine, and the other (Y177S) of tyrosine to serine.

After mutagenesis and confirmation by dideoxy sequencing (11), the mutated gene was subcloned into the expression vector pKK223-3 (9) and transformed into a gor deletion strain (SG5) of E.coli (8). The mutant and wild-type enzymes were purified (9). Y177F was almost as active as wild-type whereas Y177S showed only about 25% of the wild-type activity (Table 1).

Table 1. Kinetic Parameters of Wild-Type and Mutant Enzymes.

	Wild type	Tyr-177 to Phe-177	Tyr-177 to Ser-177
Specific Activities (under saturating assay conditions)			
Units/mg	251.8	246.1	72.3
Thermal Stability (T_{50} = Temp. at which 50% activity remained after 10 min.)			
T_{50} − NADPH	65°C	65°C	65°C
T_{50} + NADPH	42°C	25°C	32.5°C
Forward Reaction (GSSG before NADPH)			
Mechanism	Ping-Pong	Ping-Pong	Sequential
K_m NADPH	38.0 ± 4.1 µM	23.9 ± 4.5 µM	30.1 ± 8.0 µM
K_m GSSG	96.6 ± 11.7 µM	53.3 ± 8.8 µM	1.77 ± 1.54 µM
V_{max}	52.1 ± 3.8 nmol/min	42.8 ± 5.3 nmol/min	46.4 ± 7.9 nmol/min
k_{cat}	36059 ± 2631 min^{-1}	31436 ±3889 min^{-1}	8189 ± 1407 min^{-1}
Forward Reaction (NADPH before GSSG)			
Mechanism	Ping-Pong	Sequential	
K_m NADPH	57.9 ± 6.3 µM	45.5 ± 9.0 µM	
K_m GSSG	86.4 ± 10.7 µM	32.4 ± 10.4 µM	Not
V_{max}	41.5 ± 3.2 nmol/min	45.4 ± 4.3 nmol/min	Determined
k_{cat}	28766 ± 2243 min^{-1}	11123 ± 1067 min^{-1}	

The results of a detailed study of the forward reaction are shown in Table 1. The kinetic parameters were measured using a 5 x 5 matrix of substrate concentrations at 30°C and the results were analysed by using a least squares regression analysis (12). The wild-type enzyme showed predominantly ping-pong kinetics, independent of the order of substrate addition, consistent with the mechanism postulated for the human enzyme (7). The mechanism for mutant Y177F was also predominantly ping-pong if GSSG was added to the enzyme first and the reaction started by adding NADPH. However, the reaction appeared more sequential if the substrates were added in the opposite order, consistent with a hybrid ping-pong/ordered sequential mechanism (Fig 1) postulated for the yeast enzyme (13). The Y177S mutant differed more in that it displayed an ordered sequential mechanism even when GSSG was added before NADPH. Unexpectedly, since the mutation Y177S is not near the GSSG-binding site (6,7), the K_m for GSSG was about 50 times lower than that of the wild-type. This may account for the change in kinetics since high concentrations of GSSG favour the sequential pathway over the ping-pong pathway for the hybrid mechanism (13).

Figure 1. Reaction Scheme for the Branching Mechanism of Yeast Glutathione reductase (adapted from (13)).

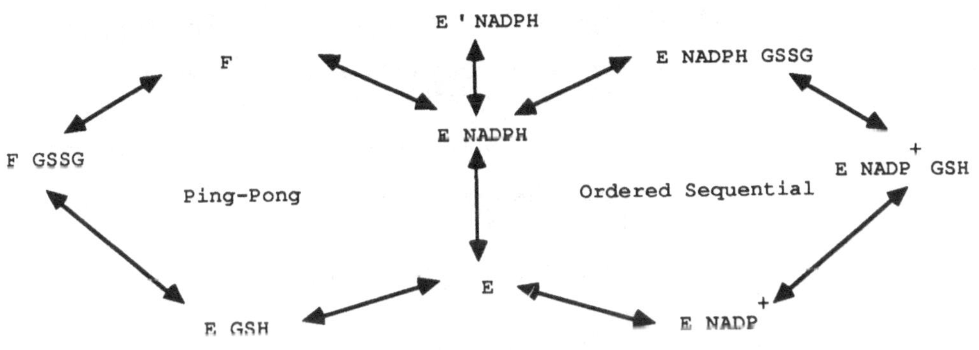

Further information came from a study of the thermal stability of the enzyme. In the absence of NADPH, the thermal stabilities of the mutant and wild-type enzymes were the same, but in the presence of NADPH, the apparent stability was markedly diminished by mutations Y177F and Y177S (Table 1). When the enzymes inactivated by exposure to NADPH were incubated with GSSG, however, activity was restored after about 1 h. Thus mutation of Tyr-177 appears to favour reversible formation of an inactive NADPH-enzyme complex, which is rescued by subsequent addition of GSSG. This too may be connected with the shift to a sequential mechanism by the Y177F mutant when NADPH is added before GSSG (Table 1).

For the *E.coli* glutathione reductase the hybrid kinetic mechanism proposed for the yeast enzyme (13) appears to operate. Our experiments demonstrate further that the balance of flux between the ping-pong and sequential ordered pathways can be crucially tipped by mutation of just one residue, Tyr-177.

Acknowledgements

We are grateful to the SERC for financial support (Grant GR/D 05790) and a Research Studentship (to N.S.S.) and to St. John's College, Cambridge for the award of a Benefactors' Studentship (to N.S.S.). We thank Dr. G. Hunter and Mr. C. Fuller for help with the synthesis of oligonucleotides, and Dr. P. Henderson for useful discussions.

References

1. Reed, L.J. (1974) Acc. Chem. Res. 1, 40-46.
2. Packman, L.C. & Perham, R.N. (1982) FEBS Letts. 139, 155-158.
3. Holmgren, A. (1980) in "Dehydrogenases Requiring Nicotinamide"(Jeffrey, J.(ed)). Experientia Suppl. 36, 149-180.
4. Schultz, P.G., Au, K.G. & Walsh, C.T. (1985) Biochem. 24, 6840-6848.
5. Krauth-Siegel, R.L., Blatterspiel, R., Saleh, M., Schlitz, E., Schirmer,R.H. & Untuch-Grau, R. (1982) Eur. J. Biochem. 121, 259-267.
6. Thieme, R., Pai, E.F., Schirmer, R.H. & Schulz, G.E. (1981) J. Mol. Biol. 152, 763-782.
7. Pai, E.F. & Schulz, G.E. (1983) J. Biol. Chem 258, 1752-1757.
8. Greer, S. & Perham, R.N. (1986) Biochem. 25, 2736-2742.
9. Scrutton, N.S., Berry, A. & Perham, R.N. (1987) Biochem. J. In Press.
10. Rice, D.W., Schulz, G.E. & Guest, J.R. (1984) J. Mol. Biol. 174, 483-496.
11. Carter, P., Bedouelle, M., Waye, M.M. & Winter, G. (1985). In: Oligonucleotide Site-Directed Mutagenesis in M13. A Laboratory Manual. Anglian Biotechnology.
12. Cleland, W.W. (1979). In: Methods in Enzymology, 63, Academic Press, New York and London.
13. Mannervik, B. (1973) Biochem. Biophys. Res. Comm. 53, 1151-1158.

STUDIES ON THE ACTIVE SITES OF LIPOAMIDE DEHYDROGENASE AND GLUTATHIONE REDUCTASE.

Lena Sahlman and Charles H. Williams, Jr.

Veterans Administration Medical Center and Department of Biological Chemistry, University of Michigan, Ann Arbor 48105, Michigan, USA

Introduction

The pyridine nucleotide-disulfide oxidoreductases form a group of flavoproteins which are all dimeric enzymes with a redox-active disulfide in each active site. The well characterized members of this group are lipoamide dehydrogenase, glutathione reductase, thioredoxin reductase (1), and mercuric reductase (not technically a disulfide reductase) (2); trypanothione reductase has only recently been discovered (3). Lipoamide dehydrogenase, glutathione reductase and mercuric reductase have a high degree of homology in amino acid sequence (4 - 7) especially impressive in the active site, the FAD-binding region, and the pyridine nucleotide binding region. Although these enzymes are theoretically capable of being reduced with four electrons per subunit, the catalytically active species is a two-electron reduced enzyme, EH_2. In this species the active-site disulfide is reduced, and the FAD is oxidized, but there is charge-transfer between one nascent thiol anion and the FAD, giving rise to a characteristic absorption band with a maximum at 530 - 540 nm (1, 2). The spectral properties of the FAD in EH_2 depend on its environment, and on the juxtaposition of the thiolate, which is the donor in the charge-transfer, and whatever influences the thiolate pK value.

Studies of the pH dependence of the rate of reduction of lipoamide dehydrogenase by dihydrolipoamide and of the redox potentials suggested the presence of a base and the mechanism shown in Scheme 1 in which species 1 is the oxidized enzyme with dithiol substrate bound and species 5 is EH_2 with disulfide substrate bound (8). The invariance of the rate of

Flavins and Flavoproteins 1987

Scheme 1

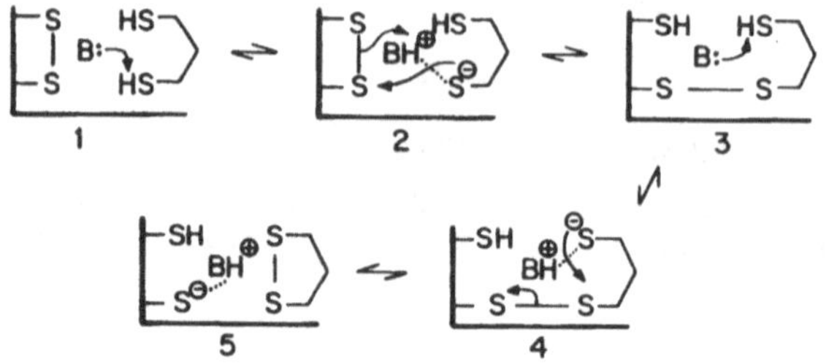

reduction at pH values above 5.5 indicated that the base has a pK_a below 5.5 in the oxidized enzyme. The pH dependence of the rate of alkylation of EH_2 reveals a pK of 7.9 which has been ascribed to the base (8). It has been shown that the nascent thiolate alkylated by iodoacetamide and interacting directly with the dithiol substrate (lower sulfur in Scheme 1, referred to as the interchange thiol) is that nearer the N-terminus of the protein in lipoamide dehydrogenase (9) and glutathione reductase (10) and that this is not the charge transfer donor thiolate (upper sulfur in Scheme 1). Reference to both thiols as thiolates in EH_2 is intended to imply that they share a single proton at neutral pH. The population of molecules with the upper thiol deprotonated can exhibit charge transfer while those molecules with the lower thiol deprotonated can be alkylated or interchange to initiate the second half-reaction in glutathione reductase catalysis.

The X-ray crystal structure of human erythrocyte glutathione reductase has been determined at high resolution (11, 12). Each active site is made up of elements from both polypeptide chains. The thiol nearer the C-terminus is within bonding distance of the FAD and the other thiol is pointing toward the substrate binding site. The base is His-467 with its imidazole roughly equidistant from the two thiols such that when protonated it might help stabilize either as an anion. An alpha-helix is positioned near the thiols in such a way as to allow the positive dipole to give additional stabilization to the thiolate, a feature also present in papain.

A study of the pH dependent changes in the spectrum of yeast glutathione reductase EH_2 has been used to determine macroscopic pK_a values reflecting the properties of the thiol anion in charge-transfer and of the histidine (10). By analogy with papain (13, 14), the pK_a observed at 4.8

Scheme 2 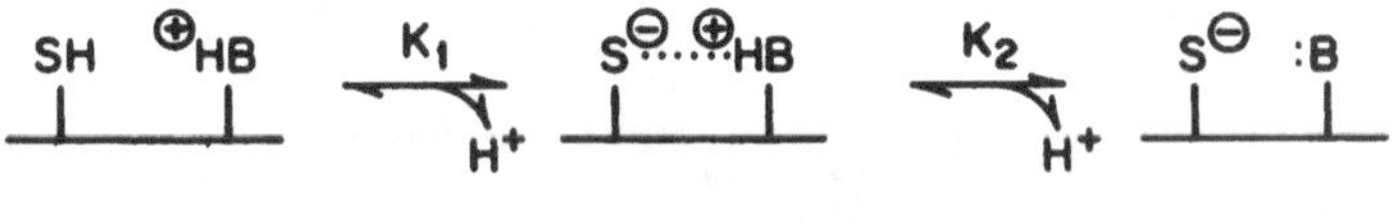

was assigned to the thiol and that at 7.4 to the His residue. We have refined these studies and extended them to pig heart lipoamide dehydrogenase and to alkylated glutathione reductase.

Alkylated glutathione reductase

The analogy between glutathione reductase EH_2 and papain is not exact due to the presence of the extra thiol as well as the FAD. We decided to improve the analogy by alkylation which eliminates the interchange thiol in EH_2, leaving the charge-transfer thiolate in interaction with the His residue. In papain (and glyceraldehyde-3-phosphate dehydrogenase) extensive experiments have shown that the ion pair predominates at neutral pH (Scheme 2) due presumably to the apolar environment. Thus, Cys-25 has a pK of 3.3 and His-159 has a pK_a of 8.5 (13, 14). The active sites of the disulfide reductases are apolar (1). The alkylated glutathione reductase fully retains its charge-transfer band (10).

EH_2 was formed by anaerobic titration with dithionite at neutral pH and 4° C. The pH was changed by mixing with appropriate buffer and the spectrum recorded. Representative spectra from the pH-jump are given in Fig. 1. It can be seen that at low pH the spectrum of EH_2 closely resembled that of oxidized enzyme, but as the pH increased, the 462 nm band of FAD decreased and blue shifted while the charge-transfer band increased. The spectra showed two distinct phases in the titration. In the early stage of the titration an isosbestic point was formed at 490 nm shifting to 508 in the later stage indicating an intermediate at neutral pH. The extinction coefficient of the charge-transfer band and of the flavin peak at 462 nm are plotted vs. pH in Fig. 2. There are two well separated pK_a values at pH 3.6 and pH 9.2 at both wavelengths roughly corresponding to the pK_a values of 3.3 and 8.5 for Cys-25 and His-159 in papain (14).

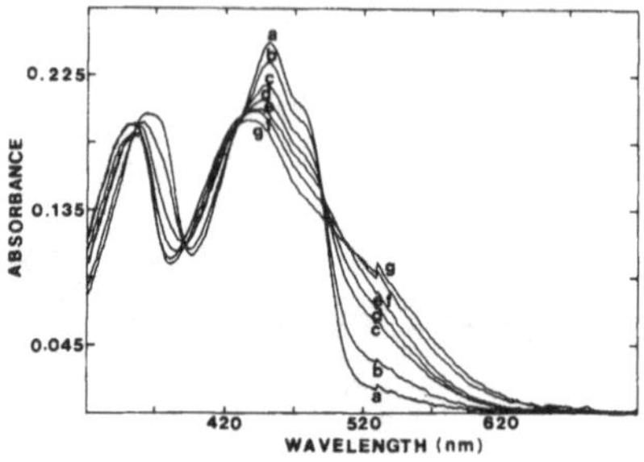

Figure 1. The spectra of aerobic, alkylated yeast glutathione reductase. a: pH 3.0, b: pH 3.5, c: pH 4.0, d: pH 4.4, e: 6.4, f: 9.2, g: 10.5. The spectra were recorded starting at 1 s and finishing at 9 s after mixing in a stopped-flow apparatus at 4 °C. The enzyme was in a buffer containing 10 mM phosphate, 50 mM NaCl, 0.3 mM EDTA at pH 6.9. It was mixed with citrate-phosphate buffer, TRIS-phosphate buffer, or glycine-NaOH buffer to give pH values 3.0 to 7.7, 7.6 to 9.9, and 9.2 to 10.3, respectively to give a final concentration of 11.8 µM enzyme in a cell with 2 cm pathlength. All buffers contained NaCl to a constant ionic strength of 0.4 M before mixing. All spectra are corrected for the baseline normalized to 0 at 720 nm.

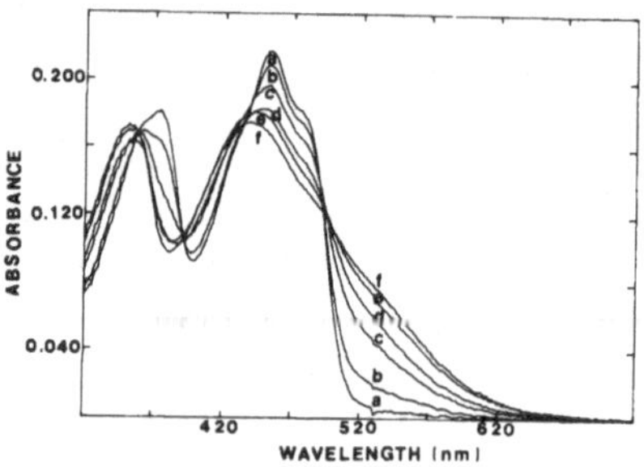

Figure 3. The spectra of anaerobic, two-electron reduced yeast glutathione reductase, 10.1 µM after mixing, other conditions as in fig. 1. a: pH 3.5, b: pH 4.5, c: pH 5.4, d: pH 6.5, e: pH 8.0, f: pH 9.9.

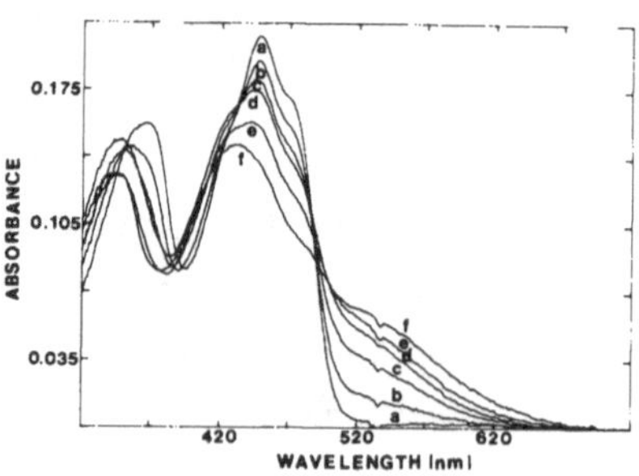

Figure 5. The spectra of anaerobic, two-electron reduced lipoamide dehydrogenase, 9.2 µM after mixing, other conditions as in fig. 1. a: pH 3.6, b: pH 4.0, c: pH 4.8, d: pH 6.5, e: pH 8.4, f: pH 10.3.

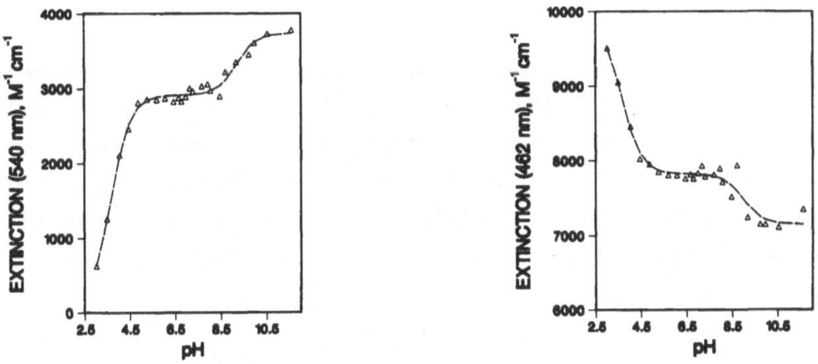

Figure 2. The extinction coefficients at 540 nm and 462 nm from spectra as in fig. 1. Each data point represents the average of values from two to four spectral values. For data analysis a non-linear curve fit was used optimizing the pK values for best fit to the data, using both two and three pK values for each data set. The curves are best fit for two pK values at 3.67 and 9.21, and at 3.65 and 8.99 for 540 nm and 462 nm, respectively.

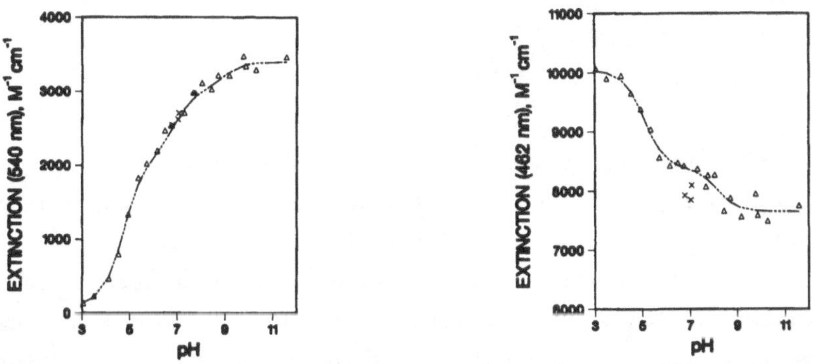

Figure 4. The extinction coefficients at 540 nm and at 462 nm from spectra as in fig. 2. (Δ): buffers with ionic strengths of 0.4 M (X): 10 mM phosphate buffer. The curves are the best fit using three pK values at 4.84, 7.11 and 9.22 for 540 nm, and two pK values at 5.08 and 8.16 for 462 nm, where (X) were not included in the curve fitting.

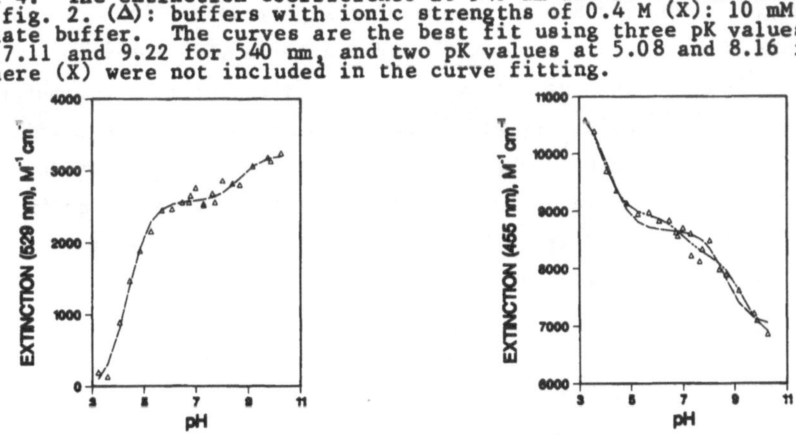

Figure 6. The extinction at 529 nm and 455 nm from spectra in fig. 3. The dashed curves are the best fit with two pK values at 4.35 and 8.67, and at 4.16 and 8.65 for 529 nm and 455 nm, respectively. The dashed-dotted line is the best fit with three pK values at 3.86, 6.98 and 9.32 for 455 nm.

Glutathione reductase

The previous pH-jump study of native two-electron reduced yeast glutathione reductase carried out in a static spectrophotometer was repeated in a rapid-scan stopped-flow spectrophotometer with additional points and extending the range to higher pH values. The spectra are shown in Fig. 3. An isosbestic point at 489 nm was associated with the titration at low pH while one at 495 appeared at high pH. When the extinction coefficient at 540 nm is plotted against pH as in Fig. 4, in contrast to the alkylated enzyme, a three pK_a assumption in the curve fit gives a better fit to the data, yielding pK_a values of 4.84, 7.11 and 9.22. The absorbance at 462 nm shows a pH dependence with only two pK_a value at 5.08 and 8.16.

Lipoamide dehydrogenase

Lipoamide dehydrogenase is not stable at pH values below 5.5, so the pH jump study in a rapid-scan stopped-flow spectrophotometer was essential. Some representative spectra are shown in Fig. 5. Isosbestic points at 432 and 486 are associated with the low pH spectra while an isosbestic at 499 is associated at high pH. The spectra in the 420-440 nm region show FAD reduction (d to e to f) as the pH was raised, while at the same time, the 530 nm region shows further formation of charge-transfer complex. In retrospect, this can be seen to a lesser degree in the glutathione reductase spectra. The extinction coefficients of the charge-transfer band at 529 nm and the flavin peak at 455 nm are plotted versus pH in Fig. 6. Although an assumption of 2 pK_a values gives a good fit for the data at 529 nm, a 3 pK_a assumption gives a better fit to the data at 455 nm. The pK_a values are presented in Table 1 together with those for native and alkylated glutathione reductase.

Table 1. Summary of the pK_a Values

Lipoamide dehydrogenase			
529 nm:	4.35		8.67
455 nm:	3.86	6.98	9.32
Glutathione reductase			
540 nm:	4.84	7.11	9.22
462 nm:	5.08		8.16
Alkylated glutathione reductase			
540 nm:	3.67		9.21
462 nm:	3.65		8.99

Discussion

We do not feel that all the data in Table 1 can be satisfactorily rationalized except in the case of alkylated glutathione reductase which is analogous to papain albeit with the imposition of the charge-transfer to the FAD. However, it seems safe to state, on the basis of the papain analogy and the charge-transfer absorbance, that the lowest macroscopic pK_a, in each case, reflects deprotonation of a thiol to an anion that can act as the donor in the charge-transfer complex. Since this would be an unusually low pK_a for a cysteine residue, it has been argued in the case of papain, that the protonated histidine stabilizes the unprotonated cysteine (8, 14). The crystal structure of human erythrocyte glutathione reductase shows that the His-467 residue is approximately equidistant from the two sulfur atoms. Lacking a crystal structure for lipoamide dehydrogenase and in view of the impressive homology in all domains between it and glutathione reductase (4, 5), we make the logical assumption that the folding pattern in the two enzymes is similar and that the His residue is positioned relative to the two sulfurs in a similar manner. This means that in the native EH_2 of these enzymes, there is an equilibrium between enzyme species with the charge-transfer thiol deprotonated or protonated, and that the extinction coefficients observed are not maximal for the

Scheme 3

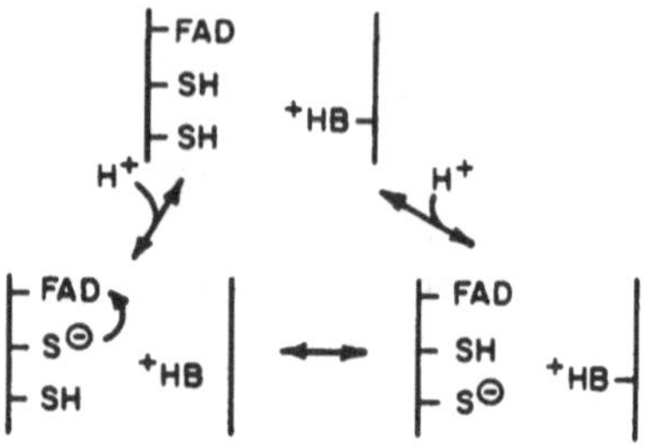

charge-transfer, since the charge-transfer thiol is not 100% ionized (Scheme 3 and Fig. 7 of Ref. 15). The equilibrium between the two forms at neutral pH must lie distinctly in favor of the charge-transfer species given that the extinction coefficients are high. This is in contrast to E. coli lipoamide dehydrogenase where the equilibrium only slightly favors the charge transfer species (15). Because iodoacetamide reacts only with a thiolate anion, it reacts with the species without charge-transfer absorption (the right-hand species in Scheme 3). It is known that EH_2 looses a proton upon alkylation and alkylated enzyme is properly represented as EHR (16). Thus, in EHR there is still an ion-pair between the unalkylated thiolate and the His residue, and charge transfer is still possible.

The highest macroscopic pK_a observed in these enzymes reflects the ionization of the His residue. This conclusion is clear for alkylated glutathione reductase where the analogy to papain can be made and we feel that it is logical also for the two native enzymes.

The putative pK_a values near pH 7 in the observations of native lipoamide dehydrogenase at 455 nm and of native glutathione reductase at 540 nm can not be assigned with any confidence as reflecting specific ionizations. Had they been noted in the charge-transfer bands of both enzymes, they could have been ascribed to the ionization of the His residue in the species having the charge transfer thiol ionized. The higher pK_a would then reflect the His in enzyme having the interchange thiol ionized. This seems unlikely however, in view of the substantial change in the extinction coefficient of the thiolate in papain upon deprotonation of His-159, and of the charge-transfer complex in alkylated glutathione reductase. Several factors may contribute to these pK_a values and to differences between pK_a values derived from the charge-transfer band and from the principal flavin band. The error bars are far too small to ascribe the latter differences to experimental error.

First, as the crystallographers have pointed out (12), the thiolate and the charge-transfer are stabilized not only by the imidazolium, but also by the dipole of the well oriented alpha-helix. Changes in this orientation, dependent on pH would be expected. The same is possible for other residues which might impact on the macroscopic pK_a values. These changes may variously affect the principal flavin band and the charge transfer band. Second, these enzymes are dimeric, and if it is assumed that the active site of lipoamide dehydrogenase consists of amino acids from both subunits, as is the case in glutathione reductase (12), it is possible that there is a cooperative effect between the subunits and that this is seen more clearly in the flavin absorbance of lipoamide dehydrogenase and in the charge-transfer absorbance of glutathione reductase.

The data presented here have provided insight into the interactions of the nascent thiols with the His residue in the active site of two-electron reduced lipoamide dehydrogenase and glutathione reductase. However, data derived exclusively from optical spectra will not give a complete correlation between macroscopic pK_a values and the groups being titrated. Efforts to apply NMR to this problem have not been successful.

Acknowledgment

This work was supported by the Medical Research Service of the Veterans Administration and by grant GM21444 from the National Institute of General Medical Sciences.

References

1. Williams, C.H., Jr. 1976. In: The Enzymes, 3rd Ed. 13, pp. 89-173.

2. Fox, B., C.T. Walsh. 1982. J. Biol. Chem. 257, 2498.

3. Shames, S.L., A.H. Fairlamb, A. Cerami, C.T. Walsh. 1986. Biochemistry 25, 3519.

4. Williams, C. H., Jr., L.D. Arscott, G.E. Schulz. 1982. Proc. Nat. Acad. Sci. USA 79, 2199.

5. Rice, D.W., G.E. Schulz, J.R. Guest. 1984. J. Mol. Biol. 174, 483.

6. Brown, N.L., S.J. Ford, R.D. Pridmore, D.C. Fritzinger. 1983. Biochemistry 22, 4089.

7. Williams, C. H., Jr., L.D. Arscott, R.P. Swenson. 1984. In: Flavins and Flavoproteins (Bray, R. C., Engel, P. C. and Mayhew, S. G., eds) Walter de Gruyter & Co., Berlin, pp. 95-109.

8. Matthews, R.G., D.P. Ballou, C. Thorpe, C.H. Williams, Jr. 1977. J. Biol. Chem. 252, 3199.

9. Thorpe, C., C.H. Williams, Jr. 1976. J. Biol. Chem. 251, 3553.

10. Arscott, L.D., C. Thorpe, C.H. Williams, Jr. 1981. Biochemistry 20, 1513.

11. Pai, E.F., G.E. Schulz. 1983. J. Biol. Chem. 258, 1752.

12. Schulz, G. E., R.H. Schirmer. 1987. In: Coenzymes and Cofactors, 2, (D. Dolphin, ed.). J. Wiley & Sons, New York, pp 168-200.

13. Polgar, L., P. Halasz. 1982. Biochem. J. 207, 1.

14. Lewis, S.D., F.A. Johnson, J.A. Shafer. 1976. Biochemistry 15, 5009.

15. Wilkinson, K. D., C.H. Williams, Jr. 1979. J. Biol. Chem. 254, 852.

16. Thorpe, C., C.H. Williams, Jr. 1981. Biochemistry 20, 1507.

CLONING, ORGANISATION AND NUCLEOTIDE SEQUENCE OF THE *AZOTOBACTER VINELANDII* GENE FOR LIPOAMIDE DEHYDROGENASE.

Arie de Kok and Adrie H.Westphal
Department of Biochemistry, Agricultural University, De Dreyen 11, 6703BC Wageningen, The Netherlands.

Lipoamide dehydrogenase (E3) is a component of the pyruvate and 2-oxoglutarate dehydrogenase multienzyme complexes. It catalyses the oxidation of dihydrolipoyl groups which are covalently attached to the lipoate acetyltransferase component (E2) of both complexes:

$$\text{E2-lip(SH)}_2 + \text{NAD}^+ \rightleftharpoons \text{E2-lipS}_2 + \text{NADH} + \text{H}^+$$

Lipoamide dehydrogenase from *E.coli* has been cloned and sequenced (1). It is part of the gene cluster which encodes the pyruvate dehydrogenase complex. Transcription analysis (2) has revealed that part of E3 is transcribed under influence of its own promotor and used for complementation of the 2-oxoglutarate dehydrogenase complex, while the other part is transcribed in a single transcription unit containing the three structural genes of the pyruvate dehydrogenase complex.

Lipoamide dehydrogenase from *A.vinelandii* has been studied extensively in this laboratory (3,4,5). The enzyme has been crystallized and the structure is determined at Groningen University (6). We report here in a brief way the cloning and nucleotide sequence determination.

At the start of this work *E.coli* probes, kindly provided by Dr J.R. Guest, were used to select the *A.vinelandii* E3 by hybridization. This approach failed, probably because the codon usage is quite different. Therefore we choose to clone the *A.vinelandii* DNA in the expression vector pUC9 (7) and select for E3 polypeptides by immunochemical methods. Antibodies against *A.vinelandii* E3 are not cross-reactive with *E.coli*

Flavins and Flavoproteins 1987

E3. 9-23 kb fragments obtained by partial digestion of Azotobacter DNA with Sau3A were ligated into the multiple cloning site of pUC9. E.coli TG2 cells were then transformed with this plasmid. Of 1600 colonies tested 2 were positive on plates and one prooved positive on nitrocellulose blots of cellfree extracts. The insert of 14.7 kb was subcloned in the same vector and a 3200 bp fragment was selected for the nucleotide sequence determination by the dideoxy chain termination method.

The gene organisation of the 3200 bp fragment and its orientation in pUC9 is given in fig 1. The nucleotide sequence revealed, in contrast to E.coli, that the E3 gene is part of the gene cluster which encodes the 2-oxoglutarate dehydrogenase complex. The transsuccinylase fragment has 71% homology with the E.coli sequence 122-410. Furthermore part of the internal aminoacid sequence of transacetylase has been determined by direct methods (8), indicating that this fragment does not code for transacetylase. Between the stopcodon of E2 and the startcodon of E3 a short intracistronic region of 85 nucleotides is found. This region lacks a consensus promotor sequence but a ribosome binding site is indicated 7-12 bases upstream of the start codon.

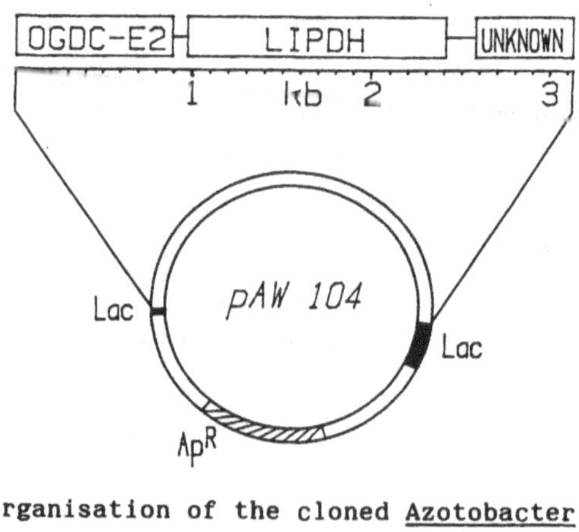

Figure 1. Gene organisation of the cloned Azotobacter DNA fragment and its orientation in pUC9. Ap^R = ampicilline resistence factor.

Nevertheless the expression is under control of this region and not by the lac promotor as indicated by the lack of induction by IPTG. Moreover inversion of the insert in the vector pUC8 results in the same high level of expression. Thus from a 24L culture of transformed E.coli cells, grown in the presence of ampicilline, 1100 mg of pure active lipoamide dehydrogenase was obtained.

The aminoacid sequence as deducted from the nucleotide sequence is given in figure 2. The molecular weight as calculated from the sequence is 50233, FAD included.The N-terminal sequence is identical with the first 10 residues determined by automated Edman degradation. Within experimental error the aminoacid composition as calculated from the sequence is identical with that reported previously (9). The homology with E.coli lipoamide dehydrogenase is evident. As expected the sequence

```
          1          11         21         31         41         51
A.v.       SQKFDVIVIG AGPGGYVAAI KSAQLGLKTA LIEKYKGKEG KTALGGTCLN VGCIPSKALL
             *  * * * *** ** **     * *** *     * *          *** *** **********
E.c.      STEIKTQVVVLG AGPAGYSAAF RCADLGLETV IVERYNT--- ---LGGVCLN VGCIPSKALL
           3          13         23         33                    47
          61         71         81         91         101        111
           DSSYKFHEAH ESFKLHGIST GEVAIDVPTM IARKDQIVRN LTGGVASLIK ANGVTLFEGH
              *  *         ***   **   *          *     **** *   *    *    *
           HVA-KVIEEA KALAEHGIVF GEPKTDIDKI RTWKEKVINQ LTGGLAGMAK GRKVKVVNGL
          57         66         76         86         96         106
          121        131        141        151        161        171
           GKLLAGKKVE VTAADGSSQV LDTENVILAS GSKPVEIPPA PVDQDVIVDS TGALDFQNVP
           **        * *    *   *      * * *  ** *    *    *     * ** * **     **
           GKFTGANTLE VEGENGKT-V INFDNAIIAA GSRPIQLPFI PHEDPRIWDS TDALELKEVP
          116        126        135        145        155        165
          181        191        201        211        221        231
           GKLGVIGAGV IGLELGSVWA RLGAEVTVLE AMDKFLPAVD EQVAKEAQKI LTKQGLKILL
             * * * *  **** * *    **    * *   *   ** *    *   *    *     *
           ERLLVMGGGI IGLEMGTVYH ALGSQIDVVE MFDQVIPAAD KDIVKVFTKR ISK-KFNLML
          175        185        195        205        215        225
          241        251        261        270        280        290
           GARVTGTEVK NKQVTVKFVD AEGE-KSQAF DKLIVAVGRR PVTTDLLAAD SGVTLDERGF
               ** * *         *             *    *   ** **   *     * *     **  * ***
           ETKVTAVEAK EDGIYVTMEG KKAPAEPQRY DAVLVAIGRV PNGKNLDAGK AGVEVDDRGF
          234        244        254        264        274        284
          300        310        320        330        340        350
           IYVDDYCATS VPGVYAIGDV VRGAMLAHKA SEEGVVVAER IAGHKAQMNY DLIPAVIYTH
           * **     *  **    ****  *   *****     ** * **  *** *          **    **
           IRVDKQLRTN VPHIFAIGDI VGQPMLAHKG VHEGHVAAEV IAGKKHYFDP KVIPSIAYTE
          294        304        314        324        334        344
          360        370        380        390        400        410
           PEIAGVGKTE QALKAEGVAI NVGVFPFAAS GRAMAANDTA GFVKVIADAK TDRVLGVHVI
           ** * ** **      *   *         ** *** *** *        *   * * *       ** *
           PEVAWVGLTE KEAKEKGISY ETATFPWAAS GRAIASDCAD GMTKLIFDKE SHRVIGGAIV
          354        364        374        384        394        404
          420        430        440        450        460        470
           GPSAAELVQQ GAIAMEFGTS AEDLGMMVFA HPALSEALHE AALAVSGHAI HVANRKK*
           *     **         * * *    ***         * ** * *     **     *          * *
           GTNGGELLGE IGLAIEMGCD AEDIALTIHA HPTLHESVGL AAEVFEGSIT DLPNPKAKKK*
          414        424        434        444        454        464
```

Figure 2. Aminoacid sequence of lipoamide dehydrogenase from Azotobacter vinelandii. The homology with the E.coli sequence is indicated with asteriks.

around the redox active cystine is well conserved. Of interest is the insertion of 6 aminoacids just before the active site sequence. This is lacking in all other redox active disulfide containing enzymes sequenced thus far. The sequence around the ADP binding sites of FAD and NAD is well conserved as are most of the isoalloxazine and interface contact residues (10). Further research will be devoted to 3 features in which the A.vinelandii enzyme differs from that of E.coli: the NADH reduced enzyme is stabilized in the 2 electron reduced state like the enzyme from pigheart, it binds and is reduced by NADPH and it dissociates readily in the reduced state(5).

References

1. Stephens, P.E., Lewis, H.M., Darlison, M.G. and Guest, J.R.: Eur.J.Biochem. 135, 519-527 (1983).
2. Spencer, M.E. and Guest, J.R.: Mol.Gen.Genet. 200, 145-154 (1985).
3. Veeger, C., Krul, J., Bresters, T.W., Haaker, H., Wassink, J.H., Santema, J.S. and De Kok, A.: in Enzymes: Structure and Function, Drenth, J., Oosterbaan, R.A. and Veeger, C., eds., Elsevier North Holland, Amsterdam p. 217-234 (1972).
4. De Kok, A., Visser, A.J.W.G. and De Graaf-Hess, A.C. in: Flavins and Flavo-proteins, Massey,V. and Williams, C.H., eds, Elsevier North Holland, Amsterdam, p. 61-65 (1982).
5. De Kok, A. and Visser, A.J.W.G. in: Flavins and Flavoproteins, Bray, R.C., Engel, P.C. and Mayhew, S.G., eds, De Gruyter, Berlin, p. 149-152 (1984).
6. Schierbeek, B., Groendijk, H., Van der Laan, J.H., Wierenga, R. and Drenth, J.: J.Mol.Biol. 165, 563-564 (1983).
7. Vieira, J. and Messing, J.: Gene 19, 259-268 (1982).
8. Hanemaaijer, J.R.O., De Kok, A. and Veeger, C.: Eur.J.Biochem., submitted.
9. Bosma, H.J., De Kok, A. and Veeger, C.: Eur.J.Biochem. 142, 541-549 (1984).
10. Williams, C.H., Arscott, L.D. and Schulz, G.E.: Proc.Natl.Acad.Sci. U.S.A. 79, 2199-2201 (1982).

REACTION OF PIG HEART LIPOAMIDE DEHYDROGENASE WITH MONOBROMOACETONE AND 1,3-DIBROMOACETONE

James S. Lee and Charles H. Williams Jr.

Veterans Administration Medical Center and Department of Biological Chemistry, University of Michigan, Ann Arbor, Michigan

Introduction

Lipoamide dehydrogenase from pig heart catalyzes the reversible reoxidation of dihydrolipoamide by NAD^+. The enzyme is a dimer of identical subunits with a 51,000 dalton monomer molecular weight. Each subunit contains an FAD and an oxidation-reduction active disulfide and shuttles between the oxidized and two electron reduced forms during catalysis; it can accept four electrons under various forcing conditions (1, 2). In the stable two electron reduced form (EH_2) produced by anaerobic addition of one equivalent of dithionite, the FAD is oxidized and the disulfide is reduced. One of the thiols exists as a thiolate forming an ion pair with a nearby protonated base between pH 5.5 and 7.0 (3). This thiolate is the charge transfer donor to the oxidized FAD, the spectrum of which has a diminished, blue shifted, flavin absorbance and a new band at 530 nm (1).

Incubation of oxidized lipoamide dehydrogenase with 1,3-dibromoacetone (DBA) or monobromoacetone (MBA) at pH 6.0 results in rapid irreversible inhibition of the enzyme (4, 5). This is somewhat surprising since oxidized enzyme is inactivated by 100 mM iodoacetamide only very slowly (24 h) due to modification of a single cysteine residue (6). Evidence for specific binding of MBA and DBA to lipoamide dehydrogenase, and alteration of the active site environment by the reaction of MBA with the enzyme are presented in this paper.

Results and Discussion

Initial studies of the inactivation of lipoamide dehydrogenase by DBA

Flavins and Flavoproteins 1987

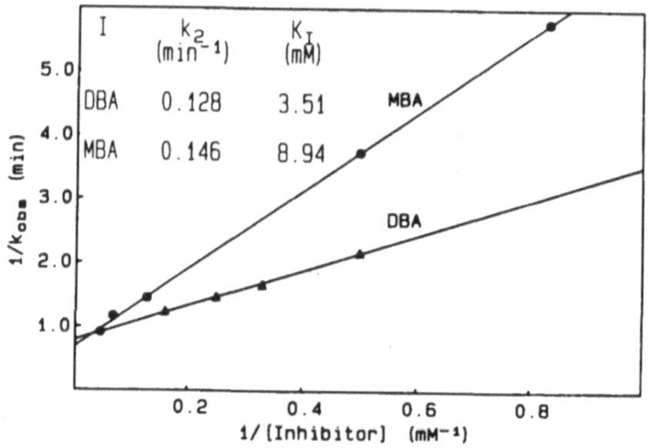

Figure 1. Double reciprocal plot of k_{obs} for the rate of inactivation of lipoamide dehydrogenase _versus_ [DBA] (▲) and [MBA] (●). Lipoamide dehydrogenase (10 µM) was inactivated by the indicated concentrations of DBA and MBA. The assay followed the reduction of 3-acetylpyridine adenine dinucleotide by dihydrolipoamide.

indicated that the reagent bound to the enzyme prior to inactivation (4). The reaction scheme can be summarized as follows,

$$E + I \xrightleftharpoons{K_I} [E \cdot I] \xrightarrow{k_2} E_{inactive}$$

If the equilibrium between enzyme and the complex is rapid with respect to the rate of inactivation, and [I]>>[E] then the following expression holds for the rate of inactivation (7),

$$1/k_{obs} = (K_I/k_2)(1/[I]) + 1/k_2$$

where k_{obs} is the observed rate of inactivation. A plot of $1/k_{obs}$ _vs._ 1/[I] should be linear with an ordinate intercept of $1/k_2$ and a slope of K_I/k_2. Figure 1 shows such a plot for DBA and MBA. In both instances the plot is linear and does not pass through the origin. Therefore, the data are consistent with the two step mechanism proposed. The results are shown in the inset to Figure 1. Assuming that both reagents are reacting with the same nucleophile(s) on the enzyme, the similarity of the rates of reaction (k_2) suggests that the major influence over the reactivity of the bromomethyl group toward the nucleophile on the enzyme is the adjacent carbonyl, with little effect coming from the substituent three carbons away. The difference in the reaction of DBA and MBA with the enzyme lies in the affinity of the enzyme for the inhibitor. The dipole moments of the bromo groups of DBA are aligned with that of the carbonyl. Hence DBA is less polar than MBA. As expected, DBA is eluted before MBA from a Waters µPorosil column with a mobile phase of 0.5% tetrahydrofuran in hexane. In this system the least polar compounds are eluted first. This suggests that the active site of lipoamide dehydrogenase is somewhat apolar in that it binds DBA tighter than MBA, albeit neither binds very tightly.

The effect of these modifications on the thiolate in two electron reduced enzyme has been investigated. DBA modified lipoamide dehydrogenase is very fragile while MBA modified enzyme is reasonably stable to anaerobic titration with dithionite. MBA inactivated enzyme was subjected to pH jump experiments using a stopped-flow spectrophotometer equipped for scanning. Figure 2a shows representative scans at various pH values. Note the oxidized character of the low pH spectrum and the two electron reduced appearance of the high pH scan. Figure 2b is a replot of the absorbance at 529 nm vs. pH. The curve through the points is the best fit to the data using a nonlinear least squares fitting program. The indicated pK is 4.72 ± 0.06, and is associated with the protonation of the thiolate donor in the charge transfer complex. The pK of the thiolate in native enzyme is 4.35 (8). Thus modification of lipoamide dehydrogenase by MBA has raised the pK of the thiolate in two electron reduced enzyme by 0.37 pH units and greatly diminished the intensity of the charge transfer complex. In papain, which also contains an ion pair, the pK of the thiolate changes from 3.3 to 7.6 upon disruption of the ion pair (9).

The kinetics of the reaction of MBA with the enzyme were not simple first order indicating that the modified enzyme may be heterogeneous and this is supported by our inability to repeatably isolate a modified peptide. A peptide isolated from a SAP digest of enzyme reacted with MBA had an amino acid analysis matching closely that of residues 345-375 (10). Structural studies on enzyme modified by DBA, and reduced by NADH (4), also showed heterogeneity. The data suggest initial reaction with unknown residues (perhaps one of the nucleophiles in the MBA-implicated peptide) as well as limited initial reaction with His, extensive hydrolysis and subsequent cross-linking of unhydrolyzed reagent with the nascent thiols.

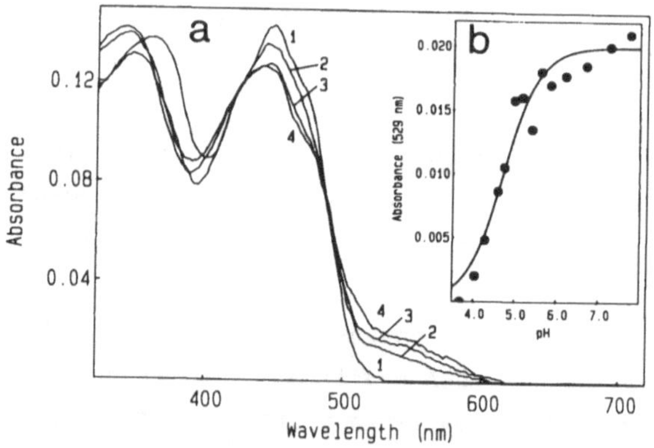

Figure 2. a) Representative spectra of two electron reduced MBA modified lipoamide dehydrogenase at pH values 3.68, 5.41, 6.26, and 7.76 (1-4 respectively). b) Replot of absorbance at 529 nanometers *versus* pH.

Acknowledgement This work has been supported by the Medical Research Service of the Veterans Administration and by Grant GM21444 from the National Institute of General Medical Sciences.

References

1. Williams, C. H., Jr. 1976. In: The Enzymes, Vol. 13 (P. D. Boyer, ed.). Academic Press, New York. pp. 89-173.

2. Massey, V., Q. H. Gibson, C. Veeger. 1960. Biochem. J. 77, 341-351.

3. Matthews, R. G., D. P. Ballou, C. Thorpe, C. H. Williams, Jr. 1977. J. Biol. Chem. 252, 3199-3207.

4. Lee, J. S., and C. H. Williams, Jr. 1984. In: Flavins and Flavoproteins (R. C. Bray, P. C. Engel, S. G. Mayhew, eds.). Walter de Gruyter & Co., Berlin. pp. 161-164.

5. Lee, J. S., and C. H. Williams, Jr. 1986. Fed. Proc. 45, 1618.

6. Matthews, R. G., and C. H. Williams, Jr. 1974. Biochim. Biophys. Acta 370, 26-38.

7. Kitz, R., and I. B. Wilson. 1962. J. Biol. Chem. 237, 3245-3249.

8. Sahlman, L., and C. H. Williams, Jr. In: (this volume)

9. Lewis, S. D., F. A. Johnson, J. A. Shafer. 1981. Biochem. 15, 5009-5017

10. Arscott, L. D., M. Bermann, C. H. Williams, Jr. 1984. In: Flavins and Flavoproteins (R. C. Bray, P. C. Engel, S. G. Mayhew, eds.). Walter de Gruyter & Co., Berlin. pp. 95-109.

SITE-DIRECTED MUTAGENESIS OF THIOREDOXIN REDUCTASE

Andrew J. Prongay, David R. Engelke, and Charles H. Williams, Jr.

Veterans Administration Medical Center, and The Department of Biological Chemistry, University of Michigan, Ann Arbor, Michigan, 48105 U.S.A.

Introduction

Thioredoxin reductase (E.C.1.6.4.5) from E. coli is a member of the pyridine dinucleotide-disulfide oxidoreductase family of flavoproteins which have a redox active disulfide. Unlike other members of this family, the functions of the nascent thiols in 2-electron reduced enzyme - interchange with thiol substrate and electron transfer from the FAD - have not been determined in thioredoxin reductase (1). Interchange occurs through the nascent thiol nearer the N-terminus while electron transfer is via the other thiol in lipoamide dehydrogenase and glutathione reductase.

Russel and Model (2) have created the constructs f1R366 (a derivative of filamentous phage f1) and pPMR14 (a plasmid expression vector), which contain the gene for thioredoxin reductase (trxB) from E. coli. These two constructs have been used in a site-directed mutagenesis of one of the half-cystines of the active-site disulfide.

Results and Discussion

OLAJP1 is a 24 base oligonucleotide whose sequence, with the exception of a 3 base pair mismatch, is complementary to the nucleotide sequence in the region of the active site disulfide on the noncoding strand of the trxB gene in ssf1R366 (Fig. 1). A site-directed mutagenesis that changed the active site cysteine (TGT) nearer the amino terminus of the protein to a serine (AGC) has been performed using OLAJP1 as the primer and ssf1R366 as the template DNA at a 34:1 molar ratio (3,4). Transfected E. coli cells were plated as a lawn of plaques, and potential mutant phage (ca 3%) were

Flavins and Flavoproteins 1987

	NH3-Val-Ser-Ala-CYS-Ala-Thr-Cys-Asp-COOH
pPMR14, f1R366	
	5'-GTT TCT GCT TGT GCA ACC TGC GAC-3'
	• • •
OLAJP1	3'-CAA AGA CGA TCG CGA TGG ACG CTG-5'
	• • •
	5'-GTT TCT GCT AGC GCT ACC TGC GAC-3'
pAJP1, f1AJP1	
	NH3-Val-Ser-Ala-SER-Ala-Thr-Cys-Asp-COOH

Figure 1. The amino acid and nucleotide sequences occurring in the mutated region of the trxB gene. The nucleotide sequences shown are those that hybridize to the sequence of the mutagenic oligonucleotide OLAJP1. The bases that were altered are indicated by the dots between the gene sequences and the oligonucleotide sequences. f1R366 and pPMR14 are the ~~vectors carrying the wild type trxB gene. f1AJP1 and pAJP1 are the vectors~~ carrying the mutated trxB gene.

isolated by plaque hybridization screening (5). The mutations created a new HaeII site (AGCGCT) in the trxB gene that changed a fragment from 1095 bp into two fragments of 935 bp and 160 bp, allowing further screening of the positive plaques by HaeII restriction endonuclease analysis and finally by dideoxynucleotide sequencing of the mutant viral ssDNA (6,7)(Fig. 2). Candidate (30 of 90) viral DNAs giving positive hybridization signals had the expected altered sequence.

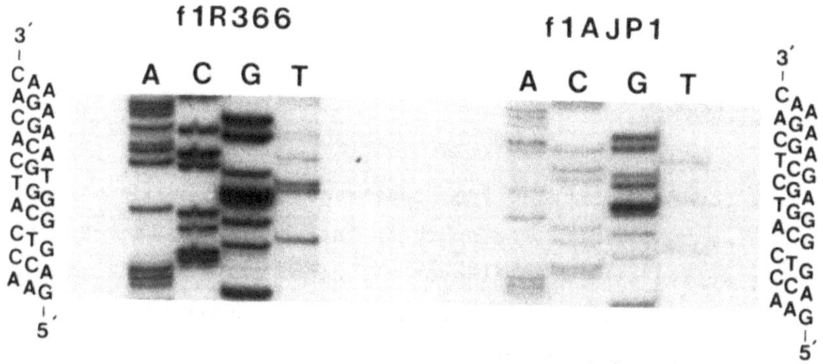

Figure 2. Autoradiograms of dideoxynucleotide sequences of f1R366 and f1AJP1 in the mutation region. The reactions were performed using 1x10^{5} cpm of 5'-^{32}P-SQAJP1, a 16 base oligonucleotide sequencing primer, and 100 ng ssf1R366 or ssf1AJP1 template DNA in 50 mM HEPES, pH 7.9, 40 mM KCl, 10 mM DTT, 60 mM $MgCl_2$, 0.6 mM non-limiting dNTPs, 0.06 mM limiting dNTP, 0.02 mM ddNTP and 11 units AMV reverse transcriptase at 50°C for 30 mins. The reactions were electrophoresed through an 8% polyacrylamide-25% formamide-7 M urea gel in a Tris-Borate-EDTA buffer.

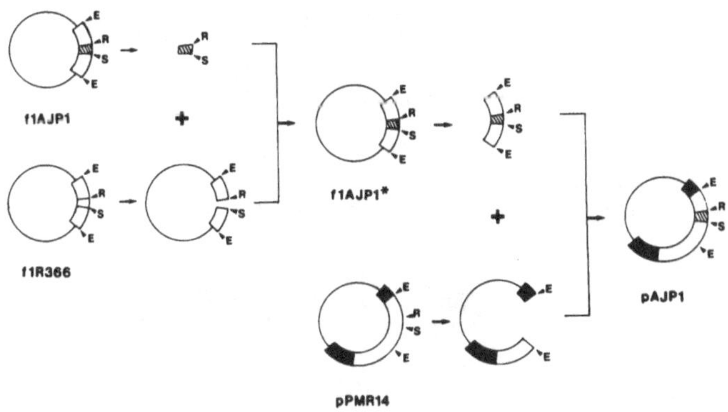

Scheme 1. Subcloning of the mutated trxB gene. See the text for details. R, RsrII; S, SacII; E, EcoRI. The solid blocks are lpp promoter and 3' flanking sequences. The open blocks are trxB gene sequences. The hatched blocks are the 290 bp fragments carrying the mutation.

Additional directed-mutageneses on the trxB gene are in progress. To avoid the need to resequence the entire gene to confirm that no second site mutations have occurred in any of the mutants created, a "cassette" method (Scheme I) has been designed for use with all of our mutants. A 350 base pair region of the trxB gene containing the mutation site was sequenced on ssf1R366 and ssf1AJP1, and with the exception of the directed base changes, no differences in the sequences were found. The restriction endonucleases RsrII (CGGTCCG) and SacII (CCGCGG) both have single sites occurring on the constructs f1R366 and f1AJP1 that are located within this 350 bp region. The dsf1AJP1 DNA was digested with these enzymes, creating a 290 base pair "cassette" containing the altered sequence. This cassette was subcloned as a replacement for the same 290 base pair region in f1R366, creating the hybrid f1AJP1*. The 1.8 kb EcoRI gene insert from f1AJP1* was isolated and used to replace the 1.8 kb EcoRI gene insert of pPMR14 in a ligation reaction that created pAJP1, a plasmid containing the sequences of the altered trxB gene. The presence of the mutation was confirmed by HaeII restriction endonuclease digestion and DNA:DNA blot analysis (Fig. 3). The pAJP1 has been isolated and introduced into E. coli A238, from which the altered protein will be isolated. A238 was created by insertion of the transposon Tn10 into the trxB gene, making the cell deficient in thioredoxin reductase activity (1).

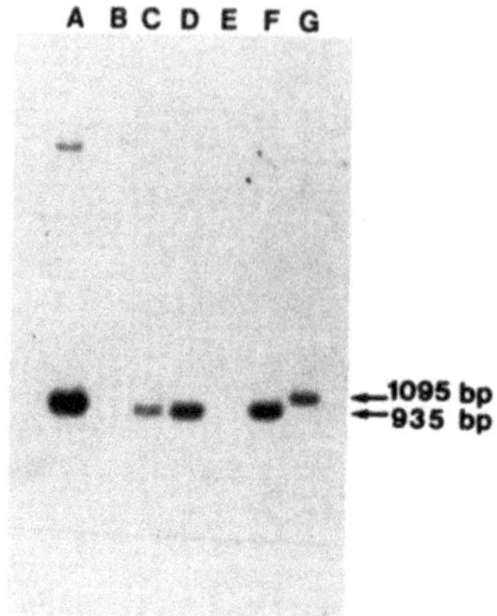

Figure 3. DNA:DNA Blot Analyses. Plasmid and viral DNA was digested with *Hae*II at 37°C for 15 hours and then electrophoresed through a 1% agarose gel. The gel was blotted with nitrocellulose paper. The blot was probed with 5'-^{32}P-SQAJP3, a 16 base oligonucleotide that binds 78 bases upstream of the mutation site. Only those fragments containing the sequences of the gene insert in the active site region will give a signal. Lanes: A. pPMR14, B. pAJP1-1, C. pAJP1-2, D. pAJP1-3, E. pAJP1-4, F. f1AJP1, G. f1R366. The wild type gene is represented in lanes A and G.

Acknowledgements - We are grateful to Drs. M. Russel and P. Model of Rockefeller University for their generous gifts of f1R366, pPMR14 and *E. coli* A238, which made this project possible, and for their advice. We also wish to thank J. Huibregtse and Dr. S. Freytag of The University of Michigan for their assistance. This work was supported by the Medical Research Service of the Veterans Administration and by Grant GM21444 from the National Institute of General Medical Sciences.

References

1. O'Donnell, M. E. and Williams, C. H., Jr.: Biochem. 24, 7617-7621.

2. Russel, M. and Model, P.: J. Bact. 163, 238-242 (1985).

3. Zoller, M. and Smith, M.: Methods Enzymol. 100, 468-501 (1983)

4. Fulford, W. and Model, P.: J. Mol. Biol. 178, 137-153 (1984)

5. Benton, W. and Davis, R.: Science 196, 180-182 (1977)

6. Sanger, F., Nicklen, S., and Coulson, A.: Proc. Natl. Acad. Sci. USA, 74, 5463-5467 (1977).

7. Huibregtse, J. and Engelke, D.: Gene 44, 151-158 (1986).

EFFECTS OF UREA ON REDOX PROPERTIES OF STREPTOCOCCAL NADH PEROXIDASE

Leslie B. Poole and Al Claiborne
Department of Biochemistry, Wake Forest University Medical Center, Winston-Salem, North Carolina 27103

Introduction

Chemical denaturants such as urea or guanidine•HCl can bind to proteins and cause localized perturbations at concentrations well below those required to completely disrupt protein structure (1). In some cases these localized effects include altered catalytic and spectroscopic properties. Lipoamide dehydrogenase from pig heart, for example, exhibits an altered absorbance spectrum, lowered NADH-lipoamide activity, and increased reactivity of a single thiol when studied at concentrations of guanidine•HCl below 1.0 M (2).

The streptococcal NADH peroxidase is a tetrameric flavoprotein which appears to possess a single redox-active disulfide per subunit, in addition to the bound FAD (3, 4). The two-electron reduced enzyme (EH_2) reacts with the sulfhydryl reagent phenylmercuric acetate (PMA), but not with iodoacetamide. The rate of the PMA reaction with EH_2 is, however, very slow in the absence of any denaturant and proceeds with a $t_{1/2}$ of 17 hr; the sluggishness of this reaction makes interpretation of the data difficult, as flavin dissociation becomes a problem over the course of the modification. Low concentrations of urea were found to have a dramatic effect on the rate of thiol modification. Additional studies have been pursued to further characterize the denaturant's effects on the structure and redox properties of this enzyme.

Flavins and Flavoproteins 1987

Results and Discussion

Structural properties in 1.3 M urea

Incubation of NADH peroxidase with 1.3 M urea over several hours at neutral pH has little effect on the catalytic activity of the enzyme (80-85% activity retained, as assayed at pH 5.4 in the presence or absence of 1.3 M urea) and no effect on the absorbance spectrum of the oxidized enzyme. However, the pseudo-first order rate of phenylmercuric acetate modification of the nascent sulfhydryl group(s) of EH_2 is increased 340-fold in the presence of this low urea concentration (Fig. 1). This increased thiol reactivity suggests that urea causes some structural perturbations at or near the active site, although the level of catalytic

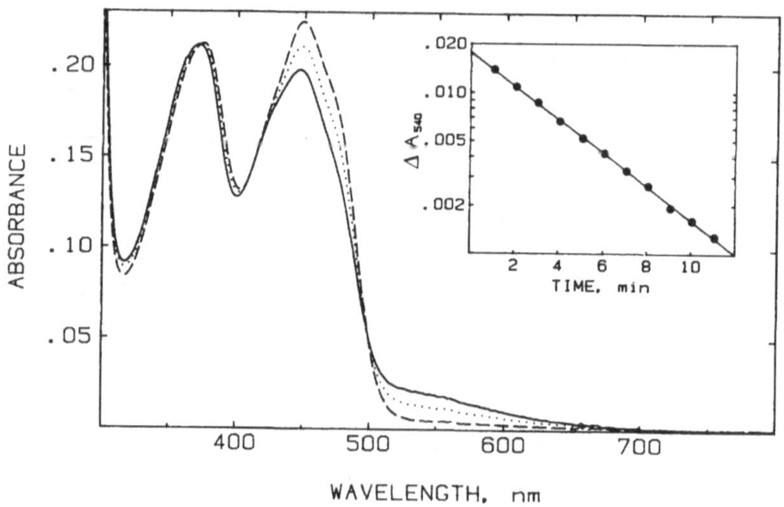

Figure 1. Phenylmercuric acetate (PMA) modification of the 2-electron reduced (EH_2) form of NADH peroxidase in the presence of 1.3 M urea. Enzyme after anaerobic addition of 1.06 eq of dithionite at pH 7 (——); same enzyme 3 min after addition of a 10-fold excess of PMA (····); final spectrum of PMA modification taken 20 min after tip to PMA (----). Inset: semilog plot of the change in A_{540} as a function of time after tip to PMA, $t_{1/2}$=3 min (25°C).

activity retained raises doubt that the structural perturbations extend to the overall conformation of the polypeptide.

As probes of both specific and generalized structural changes in the enzyme, tryptophan and FAD fluorescence measurements were made at increasing concentrations of urea (Fig. 2). These results indicate that, at concentrations below 2 M, urea has very little effect on the overall structure of the enzyme at neutral pH (FAD fluorescence), and possibly a small effect on the protein environment surrounding one or more of the six tryptophan residues. Circular dichroism spectra of the enzyme in the UV and visible regions show a small increase in the molar ellipticity of the flavin in the presence of 1.3 M urea, but little or no change in the conformation of the peptide backbone (data not shown).

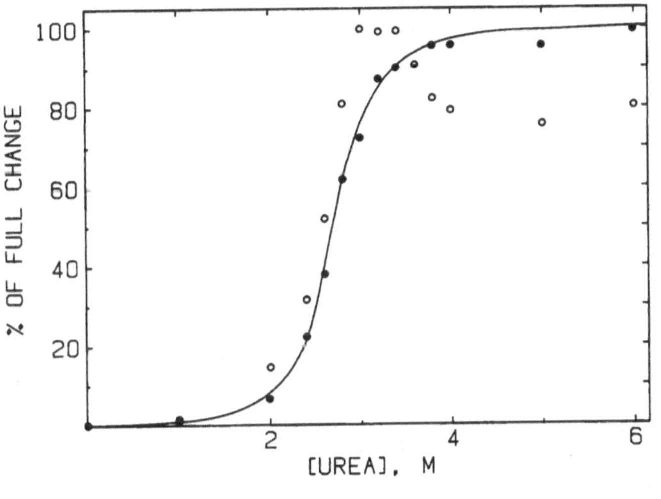

Figure 2. Effects of urea on FAD and tryptophan fluorescence of NADH peroxidase at pH 7. Titration of 4.4 µM enzyme with stock urea to a final enzyme concentration of 1.8 µM, at 25°C. FAD fluorescence intensity at 527 nm, excitation wavelength 450 nm, corrected for solvent effects of urea on free flavin (•); tryptophan fluorescence intensity at 347 nm, excitation wavelength 287 nm (o).

Preliminary sedimentation equilibrium studies indicate that the tetrameric nature of the protein is largely maintained under these conditions.

Redox properties in 1.3 M urea, pH 7

Although the gross structure and catalytic activity of the enzyme are only slightly affected by the presence of 1.3 M urea, the redox properties of the enzyme are significantly altered. Anaerobic addition of excess NADH to the enzyme, which in the absence of urea leads to formation of the stable $EH_2 \cdot NADH$ complex (3), completely reduces the enzyme to the EH_4 level in the presence of 1.3 M urea (Fig. 3). 3-Acetyl-pyridine adenine dinucleotide (AcPyADH), a pyridine nucleotide analogue with a higher redox potential than NADH

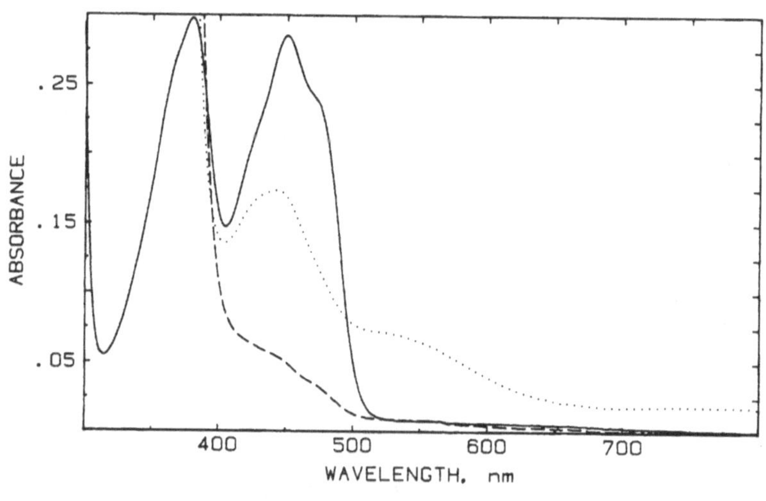

Figure 3. Anaerobic addition of excess NADH to NADH peroxidase in the presence and absence of 1.3 M urea, pH 7.0. Oxidized enzyme without NADH or urea (——); enzyme 30 min after addition of a 10-fold excess of NADH, in the absence of urea (····); enzyme 20 min after addition of a 20-fold excess of NADH, in 1.3 M urea (----). The same result was obtained with a 10-fold excess of NADH, although the spectral changes were slower.

(-248 mV versus -320 mV for NADH), does not, however, fully reduce the enzyme under these same denaturant conditions, even when added in a 10-fold excess. These results suggest that the reduction potential associated with the flavin center is increased in the presence of 1.3 M urea to allow full reduction by NADH, but not AcPyADH.

Reductive titration of NADH peroxidase in 1.3 M urea at pH 7 requires the expected 1.9-2.0 eq of dithionite for complete reduction, but generates an intermediate spectrum with decreased absorbance at both 450 and 540 nm in comparison with the EH_2 spectrum generated in the absence of urea (Fig. 4). The spectrum of the fully reduced enzyme is also altered in

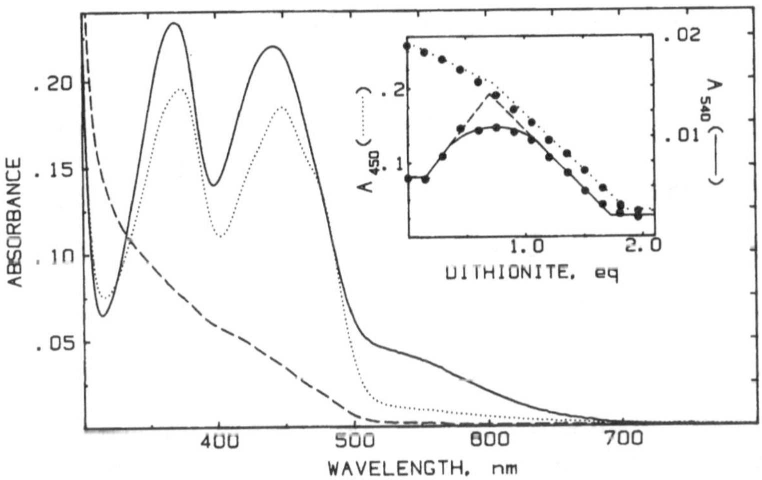

Figure 4. Two-electron reduced (EH_2) NADH peroxidase in the presence and absence of 1.3 M urea, pH 7.0. EH_2 spectrum of enzyme after anaerobic addition of 0.93 eq of dithionite in the absence of urea (———); EH_2 spectrum of enzyme after anaerobic addition of 0.75 eq of dithionite in the presence of 1.3 M urea (····);fully reduced (EH_4) spectrum of enzyme after anaerobic addition of 1.96 eq of dithionite in the presence of 1.3 M urea (----). Inset: A_{450} (●····●) and A_{540} (●———●) as a function of added dithionite in the presence of 1.3 M urea.

the presence of 1.3 M urea, with loss of the small absorption maximum around 358 nm that is seen in the native reduced enzyme (3), to give a spectrum more similar to that of a neutral reduced flavin (5).

The plot of the absorbance changes throughout the course of the titration demonstrates the enhanced disproportionation of EH_2 to E and EH_4 which takes place in urea (Fig. 4, inset). Although in the absence of urea the EH_2 spectrum seems to represent a single redox form of the enzyme possessing a charge-transfer thiol, the EH_2 spectrum in 1.3 M urea at pH 7 is best interpreted as a mixture of species. In addition to the increased contributions of E and EH_4 on 2-electron reduction, the presence of an EH_2 microform possessing a reduced flavin is also suggested (microform II):

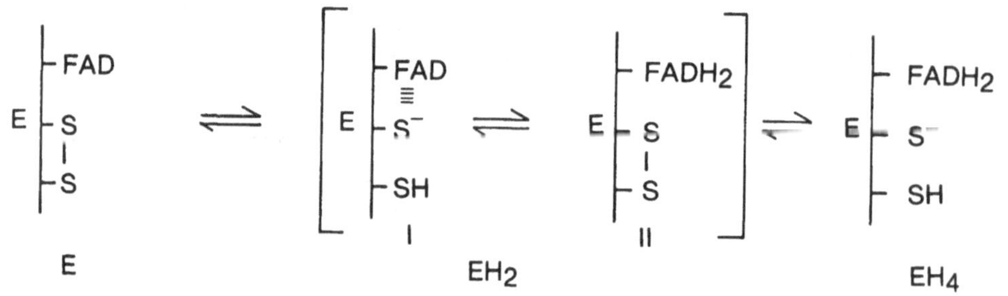

The contribution of microform II is clearly indicated since, in the absence of this species, the extent of disproportionation of EH_2 microform I (Fig. 4, inset) is insufficient to account for the rather reduced nature of the intermediate spectrum. By the same token, although the intensity of the charge-transfer absorbance band in urea is only 30% that of native EH_2, the fraction of enzyme actually present as EH_2 microform I must be significantly greater than 30%; the flavin peaks appear too oxidized to represent a combination of spectra including only 30% EH_2 microform I, with the remaining 70% contributed by E, EH_4, and EH_2 microform II (at half-reduction, E and EH_4 are present due to disproportionation only and should be present in the same molar concentrations). Although the true spectrum of EH_2 microform I cannot be determined from these data, it is likely that the molar extinction coefficient of the charge-transfer absorbance band is decreased in the presence of 1.3 M urea at pH 7.

When a single equivalent of NADH is added to native enzyme at pH 7, the $EH_2 \cdot NAD^+$ species generated is stable to autoxidation for more than 24 h. Anaerobic addition of one eq of NADH to the enzyme in 1.3 M urea at pH 7 also results in the generation of the EH_2 species expected from the dithionite titration described above (Fig. 4); under these conditions, however, the air-stability of the species generated is significantly reduced. The presence of a small amount of EH_4 in equilibrium with other species at 2-electron reduction could be at least partially responsible for the decreased air-stability observed, as EH_4 is readily autoxidizable. Furthermore, NAD^+ may not be as tightly bound to the EH_2 form(s) of the enzyme in 1.3 M urea at pH 7, a factor which could also increase oxygen-sensitivity at this redox level.

The alterations in redox properties which occur in the presence of 1.3 M urea at pH 7 support the conclusion that the reduction potential for the flavin center is increased relative to the potential associated with the disulfide center under these conditions.

Redox properties in 1.3 M urea, pH 8.5

A more pronounced shift in the relative redox potentials is exhibited by the enzyme in the presence of 1.3 M urea at pH 8.5. In the absence of urea, dithionite titration of the enzyme at pH 8.5 proceeds through an EH_2 intermediate nearly identical with that at pH 7 (3). In the presence of urea at the higher pH, however, the flavin is preferentially reduced, although full flavin reduction ultimately requires the addition of approximately 2 equivalents (1.7-1.9) of dithionite (Fig. 5). The flavin reduction potential is thus higher than the disulfide potential under these conditions, and EH_2 microform II is favored over EH_2 microform I at this higher pH. Although some disulfide as well as flavin is being reduced by dithionite throughout the titration at pH 8.5 in 1.3 M urea, no charge-transfer absorbance band is generated, even though the pH is presumably well above the pK_a for the charge-transfer thiolate.

That the titration requires approximately 2 eq of dithionite for complete reduction of the enzyme implies that the flavin and disulfide centers are in redox contact; the orientation of these centers may not, however, be optimal for the charge-

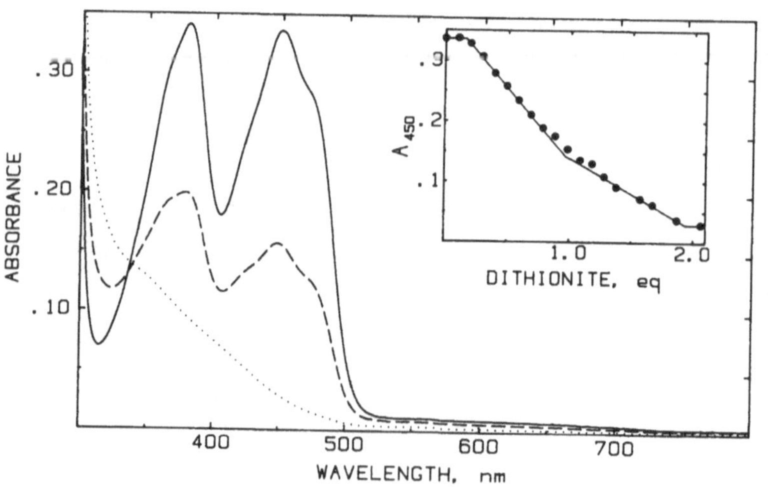

Figure 5. Anaerobic dithionite titration of NADH peroxidase in the presence of 1.3 M urea at pH 8.5. Oxidized enzyme (——); enzyme after addition of 0.98 eq of dithionite (----); enzyme after addition of 2.06 eq of dithionite (····). Inset: A_{450} as a function of added dithionite.

transfer interaction normally stabilized at pH 7 with 1.3 M urea or at pH 8.5 in the absence of urea.

Evidence for a significant alteration in the conformation of the enzyme as a result of both the higher pH and urea conditions is obtained by measurement of the tryptophan and FAD fluorescence intensities. The denaturation curve generated by titration of the enzyme with urea at pH 8.5 is similar to that at pH 7 (Fig. 2), but indicates that the changes associated with denaturation occur over a broader range of urea concentrations, beginning below 1 M urea; the FAD and tryptophan fluorescence values after 30 min in 1.3 M urea at pH 8.5 are 20 and 32%, respectively, of the full changes expected for the fully-denatured enzyme. Thus, the conformation of the enzyme may be altered in 1.3 M urea at pH 8.5 to an extent sufficient to change the relative redox potentials of the two centers as well as the charge-transfer interactions between thiolate and oxidized FAD.

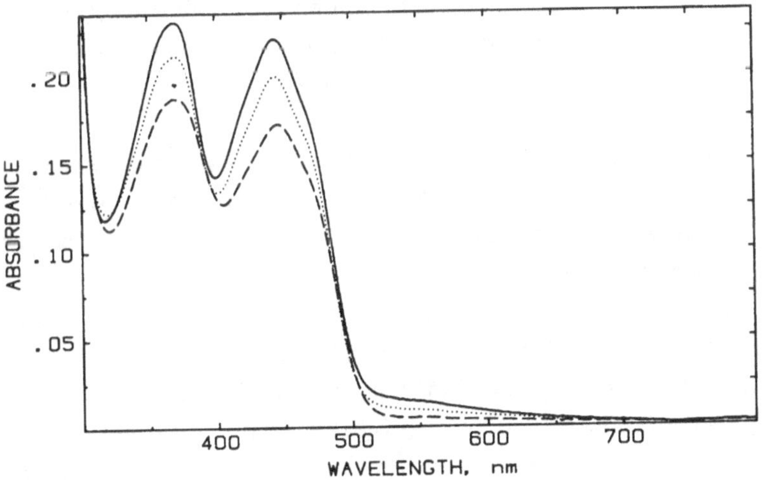

Figure 6. pH effect on the spectrum of EH_2 in 1.3 M urea. Peroxidase in 1.3 M urea at pH 8.5 after addition of 0.97 eq of dithionite (----); 1 min after tip to pH 7 buffer (••••); 3 1/2 hr after tip (——).

To ensure that the changes described above in 1.3 M urea at pH 8.5 are fully reversible, the EH_2 species of the enzyme under these conditions was tipped to buffer at pH 7. The spectral changes which ensued are shown in Fig. 6, and proceeded to a final spectrum nearly identical to that expected from the dithionite titration performed at pH 7 in the presence of 1.3 M urea; the charge-transfer absorbance band appears, and the equilibrium mixture of species at the EH_2 level in 1.3 M urea, pH 7, is established.

Summary

The effect of urea on NADH peroxidase at neutral pH is best described as a localized perturbation of the enzyme focused primarily at or near the active site. This perturbation results in a significant increase in the reactivity of one or both active site thiol(s) of EH_2, a small alteration in the CD signal of the flavin, and a shift in the relative redox potentials of the two centers. Overall conformation of the protein backbone is largely unchanged, as shown by the tryptophan and FAD fluorescence, the far-UV CD spectrum, and the significant retention of catalytic activity.

Further investigations into urea's effect on the enzyme were pursued at pH 8.5 to determine whether a more intense charge-transfer absorbance band would be generated at the higher pH, where there would potentially be more thiolate anion. In fact, an even larger structural perturbation was noted at this pH, and no charge-transfer absorbance band is generated. The potentials of the two redox centers are inverted with respect to one another, and both FAD and tryptophan fluorescence intensities are increased, although the enzyme appears to remain intact throughout reductive titrations.

Acknowledgment

This research was supported by NIH grant no. GM 35394.

References

1. Tanford, C. 1970. Protein denaturation. Adv. Protein Chem. 24, 1-95.

2. Thorpe, C. and C. H. Williams, Jr. 1974. Effects of low concentrations of guanidine hydrochloride on pig heart lipoamide dehydrogenase. Biochemistry 13, 3263-3268.

3. Poole, L. B. and A. Claiborne. 1986 Interactions of pyridine nucleotides with redox forms of the flavin-containing NADH peroxidase from *Streptococcus faecalis*. J. Biol. Chem. 261, 14525-14533.

4. Dolin, M. I. 1975. Reduced diphosphopyridine nucleotide peroxidase. J. Biol. Chem. 250, 310-317.

5. Ghisla, S., V. Massey, J.-M. Lhoste and S. G. Mayhew. 1974. Fluorescence and optical characteristics of reduced flavines and flavoproteins. Biochemistry 13, 589-597.

THREE DIMENSIONAL STRUCTURE OF BAKER'S YEAST FLAVOCYTOCHROME $\underline{b}_2$

F. S. Mathews and Z.-x. Xia

Departments of Cell Biology and Physiology and of Biological Chemistry, Washington University Medical School, St. Louis, MO 63110, USA

Introduction

Flavocytochrome $\underline{b}_2$ is a tetramer of M_r 230,000 located in the intermembrane space of yeast mitochondria. It catalyzes the oxidation of lactate to pyruvate with subsequent transfer of electrons to cytochrome c. The enzyme contains two cofactors, FMN and heme, both bound noncovalently (1). Lactate oxidation occurs via transfer of two electrons to FMN to form the flavin hydroquinone. The flavin is then reoxidized in a two step process, with formation of a semiquinone intermediate, by one-electron transfer to the heme group coupled to electron transfer to cytochrome c (2,3).

The amino acid sequence of the enzyme is known (4,5). The first 100 residues form a trypsin-resistant heme-binding domain which is homologous to the soluble fragment of microsomal cytochrome $\underline{b}_5$ (6). Crystals of flavocytochrome $\underline{b}_2$ contain two subunits per asymmetric unit and are about 68% solvent (7). They diffract to a maximum resolution of 2.4 Å.

Structure Analysis

The structure of flavocytochrome $\underline{b}_2$ was determined initially at 3.0 Å resolution with x-ray diffraction data collected from native and two heavy atom derivatives (7). The data were collected at the area detector facility at the University of California at San Diego (UCSD) (8). The crystals were transferred from the mother liquor containing 30% 2-methyl-

Flavins and Flavoproteins 1987

2,4-pentanediol (MPD) to 30% 2,5-hexanediol to facilitate heavy atom binding. The 3.0 Å map was improved by solvent leveling procedures (9) and by averaging the two images of the molecule contained in the asymmetric unit (10). After initially tracing the polypeptide chain in a minimap (9) the full sequence was fitted to the electron density on the MMS-X molecular graphics system using the program NEWNIP (11).

Data were collected to 2.4 Å resolution from native crystals in MPD by oscillation photography using synchrotron radiation at the Stanford Synchrotron Radiation Laboratory (SSRL). A total of 14 crystals were used to record 701,250 reflections on 296 film packs to yield 68,896 unique reflections. These data were combined with data recorded between 10.0 Å and 3.3 Å from a single crystal on the area detector at SSRL (12) using synchrotron radiation tuned to near the iron absorption edge.

The data were used to refine the model of flavocytochrome $\underline{b}_2$ using the Hendrickson-Konnert restrained refinement procedure (13,14). A total of 81 cycles of refinement were carried out, first with the 3.0 Å model using the UCSD data and then at 2.4 Å refinement using the SSRL data. Isotropic temperature factors were allowed to vary during the last 35 cycles. The two subunits were refined independently. The model was rebuilt a total of 5 times using maps calculated with $(2F_o-F_c)$ coefficients. A total of 283 water molecules were included in the final model. The final R factor was 0.185 and the rms deviation from ideal bond lengths was 0.034 Å.

Structure of the Molecule

The tetramer of flavocytochrome $\underline{b}_2$ is shown in figure 1. The four subunits are related to one another by a molecular four-fold axis of symmetry. Each subunit is composed of two domains, a cytochrome domain, residues 1-100, and a flavin-binding domain, residues 101-486. In addition, there is a C terminal tail, residues 487-511, which extends into the center of the tetramer, wraps around the four-fold axis and makes contact with each of the other 3 subunits.

Only two of the four cytochrome domains are visible in the electron density map. The other two domains are positionally disordered, meaning

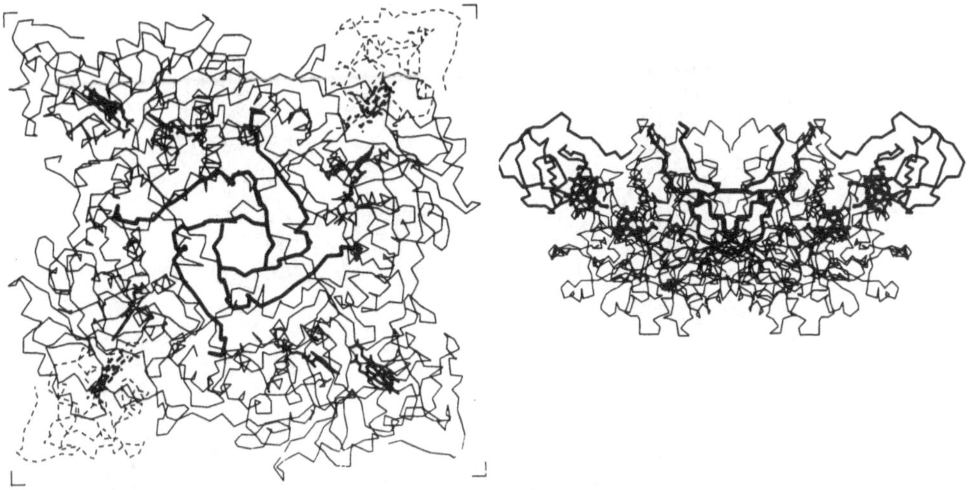

Fig. 1. C_α tracing of flavocytochrome $\underline{b}_2$ tetramer. Left: view along molecular 4-fold axis. Dashed lines are the positionally disordered cytochrome domains. The heme, flavin and C terminal tail are drawn with thick lines. Right: view with 4-fold axis vertical and the disordered cytochrome domains omitted.

that the folded chains probably occupy a number of discrete orientations in the crystal. This situation is possible because the 4-fold symmetry is local in nature and not imposed by the crystal symmetry. In fact, the local four-fold symmetry axis is coincident with the crystallographic two-fold axis which is exact and relates half the tetramer in the asymmetric unit to the other half.

The four flavin-binding domains pack together about the 4-fold axis to form an ellipsoidal disk about 100 Å in diameter and about 60 Å thick. The four cytochrome domains are located on the periphery of the disk and protrude from above the mid plane of the disk. A view of the tetramer normal to the four-fold axis is also shown in figure 1.

Heme and Flavin Binding

The ordered monomer of flavocytochrome $\underline{b}_2$ is shown in figure 2. The heme group is located at the interface between the cytochrome domain and the flavin-biding domain. It is tucked into a hydrophobic crevice in the

Fig. 2. Stereograph of flavocytochrome b_2. The ordered cytochrome domain is below and the flavin-binding domain is above. Both the heme and flavin (solid circles) are buried. Picture produced by a computer program written by A. M. Lesk and K. D. Hardman (20,21).

cytochrome domain formed by 2 pairs of antiparallel helices and a 6-stranded mixed β-sheet. The ligands of the heme iron atom are formed by the N_ε atoms of His 43 and His 66. The heme propionate groups are both extended away from the cytochrome domain toward the center of the molecule and the FMN cofactor. Each propionate is hydrogen bonded to a tyrosine

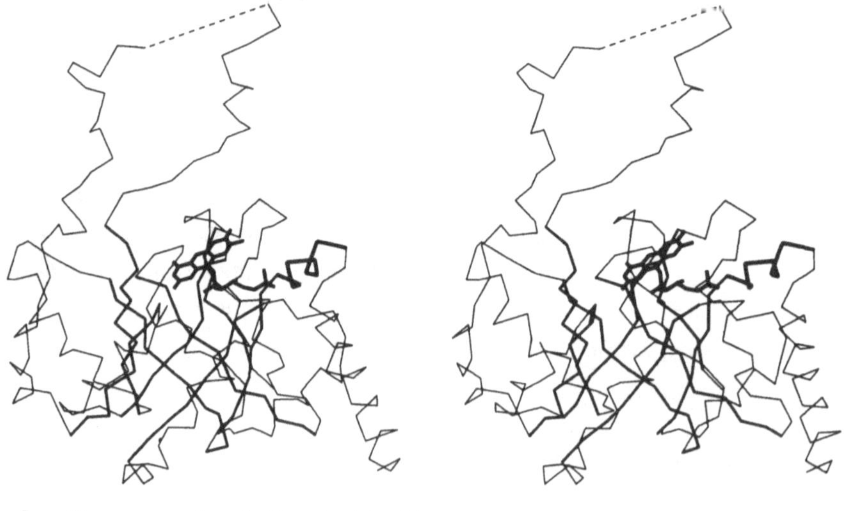

Fig. 3 Stereograph of flavin-binding domain. The central β-barrel and FMN are drawn with thick lines as is the helix after β-strand 8. The latter is oriented with its N terminus pointing toward the FMN phosphate group. The dashed lines represent a disordered segment, residues 300-315.

side chain, one to Tyr 97 of the cytochrome domain and the other to Tyr 143 of the flavin-binding domain.

The heme and flavin groups are nearly coplanar. The distance from the iron atom to the center of the flavin ring is about 16 Å. The minimum separation of the two planes, from N5 of the flavin ring to the nearest atom of the pyrrole ring, is about 10 Å.

The flavin-binding domain is composed of a parallel $\beta_8\alpha_8$ barrel motif. The FMN is located at the C terminal end of the central β-barrel (figure 3). The isoalloxazine ring is tilted so that its plane is roughly parallel to β-strand 1. The pyrimidine ring of the flavin lies close to the barrel axis while the benzenoid ring lies near the periphery of the β-barrel. The ribityl chain is extended and lies in a groove between β-strands 7 and 8. The phosphate group is located at the N terminus of a short helix at the end of β-strand 8.

The FMN makes contact with main and side chain atoms from residues located on six of the eight β-strands (figure 4). The main chain interactions are limited to strands β-1 and β-8. The phosphate group forms salt bridges to two basic residues, Arg 413 (slightly beyond the end of β-7) and Arg 433 (β-8) as well as hydrogen bonds to the peptide nitrogens of residues 432 and 433 (β-8). The ribityl oxygens O3* interacts with Asp 409 (β-7) while

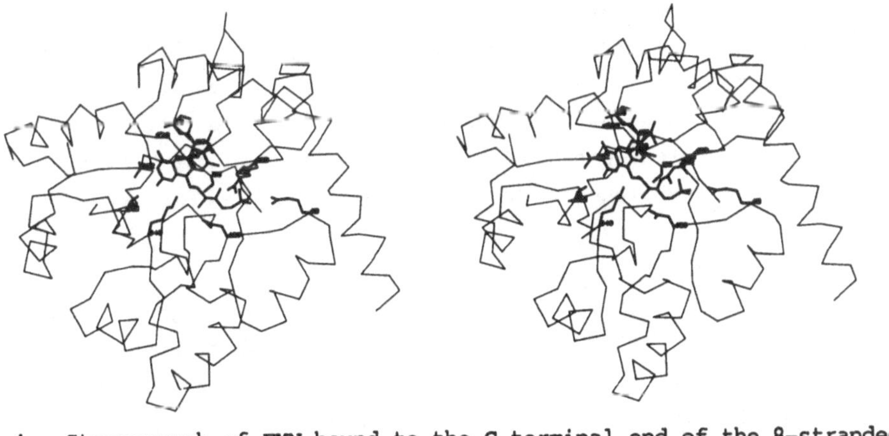

Fig. 4. Stereograph of FMN bound to the C terminal end of the 8-stranded β-barrel. Main and side chain residues in contact with the FMN are drawn in thick lines (see text).

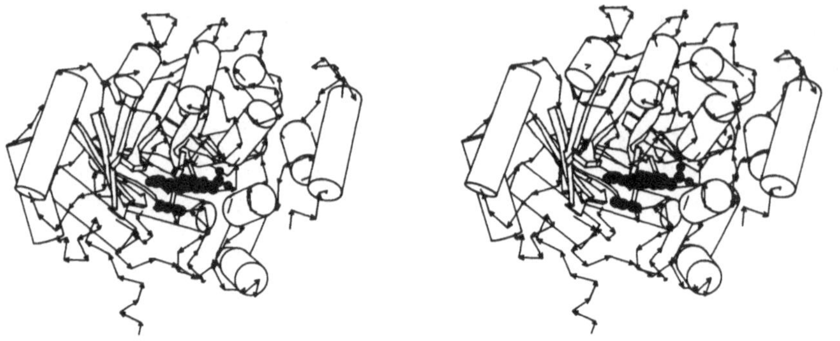

Fig. 5. Stereograph of subunit 2 (with the positionally disordered cytochrome domain) showing bound ligand. The FMN (above) and pyruvate (below) are drawn as solid circles. Picture produced by a computer program written by A. M. Lesk and K. D. Hardman (20,21).

O2* interacts with Ser 195 (β-1), Lys 349 (β-5) and the carbonyl oxygen of residue 196 (β-1). In the flavin ring, N1 and O2 both interact with Lys 349 (β-5, which also interacts with O2*) while the latter interacts with Thr 280 (β-4) as well. N3 interacts with Gln 252 (β-3) and O4 with Ser 228 (β-2). Finally, N5 interacts with the peptide nitrogen of residue 198 (β-1).

Ligand Binding

In one subunit, an isolated piece of density was found close to the FMN in the unaveraged 3.0 Å resolution electron density map. This subunit contains the disordered cytochrome domain. The density is suggestive of a bound ligand and is about the size of a molecule of lactate or pyruvate. Its center is located near the N-5 containing edge of the flavin ring about 4 Å below the plane of the ring (figure 5).

The ligand is likely to be pyruvate, which has a much higher affinity for the reduced enzyme than does lactate (15). Although pyruvate was not present when the crystals were prepared, it could arise through the slow turnover of the crystalline enzyme in the presence of excess lactate (present at 50mM concentration) and oxygen as the electron acceptor.

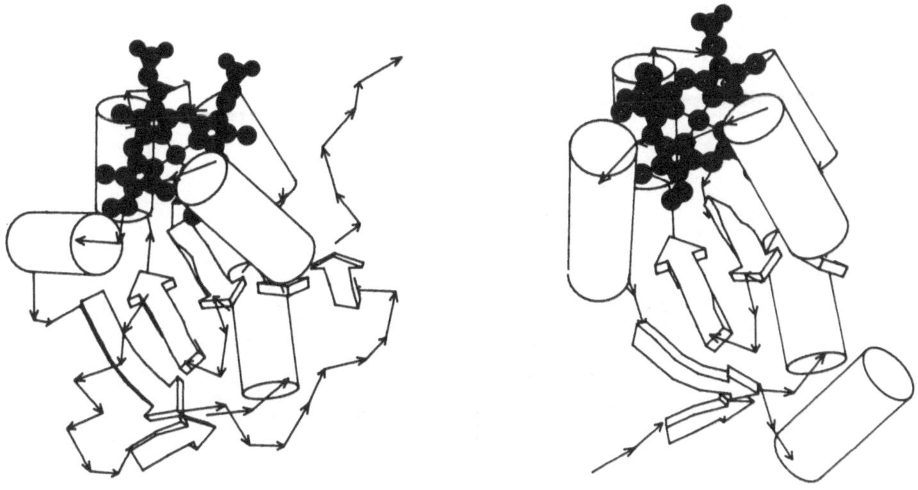

Fig. 6. Schematic diagram of cytochrome domain of flavocytochrome $\underline{b}_2$ (left) compared to cytochrome b_5 (right). Only one helix (left front in both diagrams) differs considerably in orientation in the two structures. Picture produced by a computer program written by A. M. Lesk and K. D. Hardman (20,21).

Since the electron density at 3.0 Å resolution was relatively featureless the orientation of the pyruvate could only be deduced from the chemical nature of the groups surrounding it. In its present refined position the pyruvate molecule is oriented so that its carboxyl group interacts with Arg 376 and Tyr 143 while its keto oxygen is located close to Tyr 254 and His 373. Its methyl group is located in a relatively uncrowded environment. In the other subunit, the one with the ordered cytochrome domain, the ligand binding site is occupied by a cluster of water molecules.

Comparison with Other Proteins

The cytochrome domain is very similar in conformation to microsomal cytochrome $\underline{b}_5$ (16). The two structures are compared in figure 6. The sequence identity is very high (about 37%) from near the N terminus of the $\underline{b}_2$ cytochrome domain to about position 63 (6). Beyond this point, one deletion and two insertions are needed in the $\underline{b}_2$ sequence to maintain

Fig. 7. C_α drawing of the flavin-binding domain of glycolate oxidase (dashed lines) superimposed on that of flavocytochrome $\underline{b}_2$ (solid).

maximal but reduced sequence similarity. The β-sheet and two of the four helices which form the heme crevice (in the back in figure 6) are oriented very similarly in the two structures. On the other hand, helix 4 and, to some extent helix 3 are oriented somewhat differently and reflect considerable difference in local conformation of the polypeptide chain. This results in the heme group of flavocytochrome $\underline{b}_2$ appearing to be more exposed than in cytochrome $\underline{b}_5$. The heme group is also oriented somewhat differently in the two structures.

The $\beta_8\alpha_8$ motif of the flavin binding domain has been observed in a number of proteins of diverse character (17). Two of these, trimethylamine dehydrogenase (TMADH) (18) and glycolate oxidase (19) , contain FMN. The FMN is bound to all three enzymes in a very similar fashion. When the overall $\beta_8\alpha_8$ barrel structure of TMADH is compared with flavocytochrome $\underline{b}_2$, the rms deviation of about 180 equivalent C_α atoms is approximately 2.7 Å, a little smaller than the deviation between C_α atoms in comparisons between functionally unrelated $\beta_8\alpha_8$-containing structures. However, glycolate oxidase is much closer in structure to flavocytochrome $\underline{b}_2$, with an rms deviation of about 1.3 Å for about 325 equivalent C_α atoms. The structures are compared in figure 7. The major difference between the structure of the $\underline{b}_2$ flavin-binding domain and glycolate oxidase is near the N-5 atom of the flavin ring. This area is covered by the cytochrome domain in flavocytochrome $\underline{b}_2$ but by an extensive loop between β-strand 4 and helix 4 in glycolate oxidase. The corresponding loop in flavocytochrome $\underline{b}_2$ is disordered.

Discussion

The crystal structure of flavocytochrome $\underline{b}_2$ displays the enzyme in two distinct states. In one of these the cytochrome domain is fixed with respect with the flavin-binding domain and appears to be poised to receive electrons. The electron density map of the refined structure shows a cluster of about 15 water molecules fixed at the interface between the two domains.

In the other state the cytochrome domain is mobile, suggesting that the binding affinity of the two domains is slight and that in solution the electron transfer complex is transitory in nature. In this state, the refinement shows much fewer water molecules are fixed at the active site.

The difference between the two subunits in crystals of flavocytochrome $\underline{b}_2$ may arise from crystal packing forces. Although the solvent content of the crystals is quite high, the molecules are quite tightly packed about the crystallographic 3-fold screw axes. The ordered cytochrome domain makes more contacts with neighboring molecules than the disordered domain and is probably held in place by these contacts. If the ordered cytochrome domain is firmly held in place by intermolecular contacts, there may not be enough room for pyruvate to diffuse into the active site region. This would account for the presence of the ligand only in the second subunit where the mobility of the cytochrome domain would permit access of substrates to the active site.

Acknowledgements

This research was supported by NIH grant no. GM-20530 and by NSF grant no. DMB-8518583. We wish to thank Drs. C.-I. Brändén and Dr. Y. Lindqvist for use of the unpublished coordinates of glycolate oxidase.

References

1. Appleby, C. A. and Morton, R. K. (1954) Nature (Lond) 173, 749-752.
2 Capeillère-Blandin, C., Bray, R. C., Iwatsubo, M. and Labeyrie, F. (1975) Eur. J. Biochem 54, 549-566.
3 Morton, R. K. and Sturtevant, J. M. (1964) J. Biol. Chem. 239, 1614-1624.
4. Lederer, F., Cortial, S., Bécam, A. M., Haumont, P. Y. and Perez, L. (1985) Eur. J. Biochem 152, 419-429.
5. Guiard, B. (1985) EMBO J. 4, 3265-3272.
6. Guiard, B., Groudinsky, O. and Lederer, F. (1974) Proc. Nat. Acad. Sci. U. S. A. 71, 2539-2543.
7. Xia, Z.-x., Shamala, N., Bethge, P. H., Lim, L. W., Bellamy, H. D., Xuong, Ng.-h., Lederer, F. and Mathews, F. S. (1987) Proc. Natl. Acad. Sci. USA 84, 2629-2633.
8. Xuong, Ng.h., Sullivan, D., Nielsen, C. and Hamlin, R. (1985) Acta Cryst. B41, 267-269.
9. Wang, B. C. (1985) Methods Enzymol. 115, 90-112.
10. Bricogne, G. (1976) Acta Crystallographica A32, 832-847.
11. Lederer, F., Glatigny, A., Bethge. P. H., Bellamy, H. D. and Mathews, F. S. (1981) J. Mol. Biol. 148, 427-448.
12. Phizackerley, R. P., Cork, C. W. and Merritt, E. A. (1986) Nucl. Inst. and Meth. A246, 579-595.
13. Hendrickson, W. A. (1985) Methods Enzymol. 115, 252-270.
14. Finzel, B. C. (1987) J. Appl. Cryst. 20, 53-55.
15. Urban, P., Alliel, P. M. and Lederer, F. (1983) Eur. J. Biochem. 134, 275-281.
16. Mathews, F. S., Argos, P. and Levine, M. (1972) Cold Spring Harbor Symp. Quant. Biol. 36, 387-395.
17. Muirhead, H. (1983) Trends Biochem. Sci. 8, 326-330.
18. Lim, L. W., Shamala, N., Mathews, F. S., Steenkamp, D. J., Hamlin, R., and Xuong, Ng.h. (1986) J. Biol. Chem., 261, 15140-15146.
19. Lindqvist, Y., and Brändén C.-I. (1985) Proc. Natl. Acad. Sci. USA 82, 6855-6859.
20 Lesk, A. M. and Hardman, K. D. (1982) Science 216, 539-540.
21. Lesk, A. M. and Hardman, K. D. (1985) Methods Enzymol. 115, 381-390.

MECHANISM OF L-LACTATE DEHYDROGENATION CATALYZED BY FLAVOCYTOCHROME b_2 FROM BAKER'S YEAST

F. Lederer

INSERM U 25, CNRS LA 122, Hôpital Necker, 75743 Paris Cedex 15, France

F.S. Mathews

Department of Cell Biology and Physiology, Washington University School of Medicine, St. Louis, MO 63110, USA

Introduction

The three-dimensional structure of flavocytochrome b_2 (1,2) gives a unique opportunity for examining the protein structure around a flavin in a dehydrogenase-electron transferase whose substrate is not a nicotinamide. One can start correlating the structural features with the results of studies carried out in solution and with current theories concerning structure-function relationships in flavoproteins.

The non-protein piece of electron density seen close to the flavin in the electron density map has been of great help in mechanistic interpretations. It has been ascribed to a three-carbon ligand, namely, pyruvate (2,3). But it cannot be totally excluded that it could be that of a semiquinone-pyruvate complex, since pyruvate has been shown to stabilize the semiquinone (4). EPR measurements on the crystals should settle the question. The following discussion is based on the hypothesis that the structure we observe is that of the reduced enzyme-pyruvate complex.

The Structure: Isoalloxazine Ring Environment

It is known that it is the protein which selects the function of a flavin cofactor. The first hypotheses concerning the role of flavin-protein interactions were put forth by Müller (5) and Hemmerich (6). They suggested that region-specific hydrogen bonds from the protein to the N(1)-O(2) and N(5) areas governed the oxidase and the electron transferase function,

Flavins and Flavoproteins 1987

respectively. A most recent version of these ideas was expressed in (7,8). It was noted that "flavoenzymes of the oxidase class...all stabilize the anionic semiquinone and the anionic fully reduced flavin and give substantial stabilization of the anionic forms of 6-hydroxy-, 6-mercapto- and 8-hydroxyflavins" (8); these proteins also stabilize the blue benzoquinoid form of 8-mercaptoflavins and form a reversible adduct with sulfite at the flavin N(5) position. All these properties were ascribed to "the stabilizing influence of a protein positively charged residue in the neighbourhood of the flavin N(1)-C(2)=O locus" (8).

In a number of respects, flavocytochrome b_2 resembles the oxidases, in particular lactate oxidase, even though its flavin is reoxidized in a monoelectronic fashion. It forms a tight complex with sulfite (9) and stabilizes the anionic semiquinone (10) - which is functional, contrary to that in oxidases - , as well as the benzoquinoid form of 8-mercaptoflavin (D. Pompon, unpublished experiments). It is therefore of considerable interest to find that its structure provides a beautiful verification of the postulated existence of a positive charge in the N(1)-C(2) = 0 area (Fig. 1). Lys 349 lies close to flavin N(1) and O(2) (2.80 and 2.81 Å respectively). Although the structure does not provide the information, it is reasonable to assume that the ε amino group is positively charged. The interaction between the flavin and Lys 349 might be one of the main factors responsible for the bound flavin high redox potential (-50 mV (10) versus - 220 mV for free flavin). Furthermore, the presence of a charged lysine in close contact with the N(1)-O(2) area could easily explain why flavocytochrome b_2 is unable to bind 1-deazaflavin (11).

There are several potential hydrogen bond donors around the pyrimidine ring (Fig. 1): Gln 252 (amide oxygen to N(3): 3.00 Å), Thr 280 (hydroxyl oxygen to flavin O(2): 2.77 Å) and Ser 228 (hydroxyl oxygen to flavin 0(4): 2.60 Å). For the time being we have no way of assessing the potential importance of interactions with residues other than Lys 349.

The benzene ring is shielded from solvent, as is most of the cofactor. Its closest neighbours are the aromatic rings of Tyr 143 and Tyr 144 and part of Leu 436 side chain. Within 3.5 to 5 Å of the benzenoid ring also lie the polar extremities of Arg 433, which makes one of the two electro-

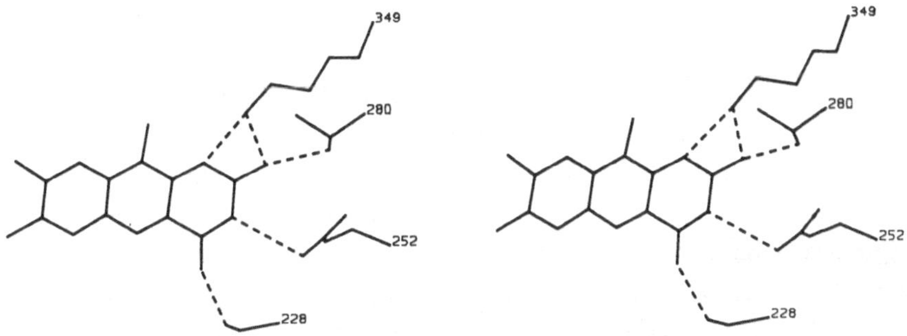

Fig. 1: Side chains interacting with the cofactor pyrimidine ring (stereoview): Ser 228, Gln 252, Thr 280, Lys 349. Dashed lines represent hydrogen bonds and number represent Cα positions.

static contacts with the cofactor phosphate group, and of Arg 376 which is one of the substrate-binding side chains, as is also Tyr 143.

Interestingly enough, part of the β barrel segment β1 runs more or less centrally across the cofactor (Fig. 2). Ala 196 amide oxygen forms a hydrogen bond with the ribityl O(2) (2.69 Å) and is not far from N(10), C(4a) and C(10a). Another close contact is that between Ala 198 backbone nitrogen and N(5) (3.01 Å).

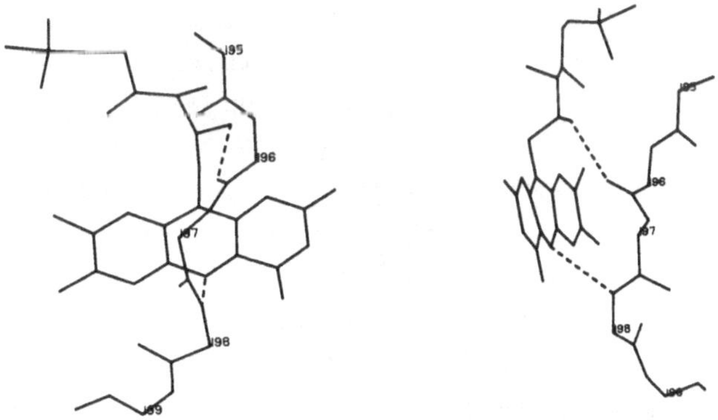

Fig. 2: Interaction between the cofactor and the backbone atoms 196 O and 198 N. Left: front view, with the flavin above. Right: side view.

Before the Structure: Mechanistic Hypotheses

For flavoproteins catalyzing the dehydrogenation of α-amino- and α-hydroxyacids, it was proposed that the first step in the reaction pathway was α-hydrogen abstraction (12,13). The topic was recently reviewed in (14). The main arguments rested on the observation that a number of flavodehydrogenases could catalyze halide ion elimination from β-halo-substrates (14) and on the inactivation of lactate oxidase by hydroxy-butynoate, a suicide substrate (15,16). As underlined by Ghisla (14), however, the existing evidence was not unambiguous.

With flavocytochrome b_2, the finding that the enzyme can catalyse both α-hydroxy acid oxidation and α-keto acid reduction (3) enabled us to study the dehydrohalogenation reaction (17-19), as well as inactivation by the β-acetylenic substrate (20), under both forward and reverse reaction conditions. The results were best interpreted in terms of a carbanion mechanism, but did not constitute an absolute proof. The study of intermolecular hydrogen transfer catalyzed by flavocytochrome b_2 brought the strongest arguments in favour of a carbanion mechanism (19).

The first argument derives from the measurement of the specific radioactivity of [2-^{3}H]bromolactate and [3-^{3}H]pyruvate when the enzyme is incubated in the presence of [2-^{3}H]lactate and bromopyruvate (19). Incorporation into bromolactate-derived pyruvate was significantly higher than into water and significantly lower than into bromolactate. These results are best explained as arising from ketonization of an intermediate enol pyruvate (formed in the dehydrohalogenation reaction) partially on the enzyme, partially in the solvent. A hydride or a radical mechanism cannot be reconciled with such a finding. The second argument comes from measurements of deuterium isotope effects in the reverse reaction when flavocytochrome b_2 is incubated in the presence of deuterolactate and bromopyruvate (17,19). The finding of an isotope effect on bromopyruvate disappearance but of an inverse isotope effect on pyruvate formation requires a branched pathway where a common reaction intermediate can lead to bromolactate in an isotope-sensitive step and to pyruvate in an isotope-insensitive step. Clearly the hydride mechanism does not account for these results. The epoxide mechanism proposed by Visser (21) for the dehydrohalogenation reaction is also ruled out by the specific radioactivity measurements and the isotope effects: this reaction pathway is expected to lead to the same results as the hydride mechanism.

The Structure: Active Site Description

Fig. 3 shows the residues that seem to be implicated in binding and catalysis. In front of the benzene ring lies the side chain of Arg 376, which is expected to bind the substrate carboxyl end. The ligand thus lies in the active site with the carboxylate closer to the cofactor benzene moiety. This orientation is in perfect agreement with that deduced from the structure of the hydroxybutynoate-flavin adduct (16,20). Arg 376 appears to be making a hydrogen bond through NE with one of the two carboxylate oxygens (3.05 Å). There must also be ion pairing between the two charged groups. Furthermore, Tyr 143 phenolic oxygen lies within hydrogen bonding distance of the other ligand carboxylate oxygen (2.81 Å).

His 373 imidazole ring lies above the flavin isoalloxazine moiety, with its plane roughly parallel to it. Asp 282 has one of its carboxyl oxygens oriented toward His 373, within hydrogen bonding distance of N(1) (2.31 Å). It thus appears that Asp 282 provides an anchor for His 373 positioning, as well as electrostatic stabilization for an imidazolium ion. The position of Asp 282 thus also defines His 373 N(3), which lies near the presumed ligand carbonyl oxygen (2.61 Å). In close proximity

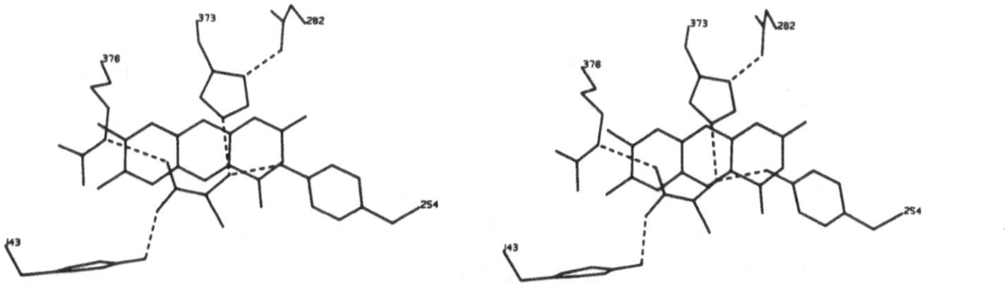

Fig. 3: Side chains interacting with pyruvate (stereoview): Tyr 143, Tyr 254, Asp 282, His 373, Arg 376.

to this same atom is another potential hydrogen bonding group, Tyr 254 phenolic oxygen (2.39 Å).

The ligand molecule defines a plane which is roughly parallel to that of the flavin. Its central carbon atom is about equally distant from His 373 N(3) (3.56 Å), flavin N(5) (3.74 Å) and Tyr 254 phenolic oxygen (3.65 Å). In order to form L-lactate upon reduction, the pyruvate carbonyl, as oriented, has to receive electrons from the flavin side. This orientation is that required both for a hydride mechanism and a carbanion mechanism. Another conclusion from the active site topography concerns the stereochemistry of flavin reduction in flavocytochrome b_2: reducing equivalents enter the flavin from the si side contrary to the situation in a number of nicotinamide-dependent flavoenzymes and in glucose oxidase and general acyl-CoA dehydrogenase (24).

The Structure: Mechanistic Interpretation

When one looks at Fig 3, what strikes immediately is the closeness of His 373 N(3) to the ligand oxygen. It looks as if His 373 were ready to give a proton to an incipient oxyanion arising as the result of hydride donation to substrate C(2) by the reduced flavin. It has to be noted that in nicotinamide-linked malate dehydrogenase, the substrate carboxyl is oriented by an arginine, and that a histidine, held in place at N(1) by an aspartate, is believed to play the role of a general acid base catalyst with respect to the substrate C(2) oxygen (23). These side chains can be aligned in space with identical residues in NAD^+-linked lactate dehydrogenase (23). The relative disposition of the catalytic side chains, however, is different from that in the flavoprotein. Therefore a mechanistic extrapolation is meaningless.

The first argument against a hydride mechanism for flavocytochrome b_2 is the result of the solution studies reported above. But the structure, upon closer analysis, also speaks against it. The hydride transfer mechanism does not provide a role for Tyr 254. Furthermore, we have to remark that the structure is that of a reduced enzyme poised for reduction of pyruvate, which is, in solution, a highly inefficient substrate (k_{cat} =

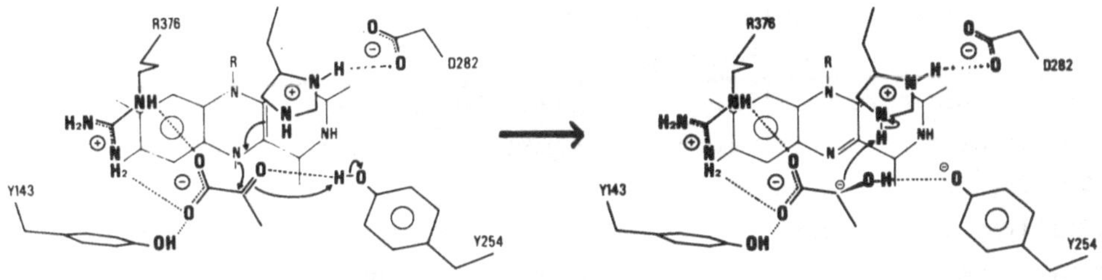

Fig. 4: Proposed role for the side chains in a carbanion mechanism.

0.08 s^{-1} versus k_{cat} = 120 s^{-1} for lactate oxidation; these figures essentially reflect the rate of hydrogen abstraction from and donation to C(2) (3,19)). In other words, a geometry which is at first sight not favorable for proton donation to C(2) is rather appropriate for an inefficient catalyst. After these considerations, the following proposal can be made for substrate reduction by reduced flavocytochrome b_2 (reverse reaction) (Fig. 4). In the complex with pyruvate, N(5) is proposed to be anionic; the charge may be stabilized by interaction with the positively polarized main chain 198 NH mentioned above and substrate C(2). At this point, it clearly cannot be decided if N(5) is also anionic in free reduced enzyme or if it is protonated and the proton is expelled upon keto acid binding. Electrons are transferred to the substrate C(2), which becomes tetrahedral. In the process, either C(2) may move closer to flavin N(5) or the carbonyl oxygen would be pushed out of the plane, away from His 373 N(3), and be protonated by Tyr 254, which would thus afford general acid catalysis. Electrostatic interactions and strong hydrogen bonding on the substrate carboxylate side may help stabilize the carbanion. In the final step, the latter would take the proton from His 373 N(3). The structure gives no indication as to how the electrons could be transferred from the flavin to pyruvate. The static picture of the active site probably does not rule out a covalent intermediate in the transition state (14).

Conclusion

The active site structure of flavocytochrome b_2 suggests that the catalytic elements for hydroxyacid oxidation (and ketoacid reduction) are:

-binding of the carboxylate group by Arg 376 and Tyr 143 (electrostatic interaction and hydrogen bonding)
- general acid base catalysis with respect to C(2) provided by His 373
- general acid base catalysis with respect to the central oxygen provided by Tyr 254
- facilitation of electron entry into the cofactor by Lys 349, whose positive charge stabilizes the anionic reduced flavin.

The intermolecular hydrogen transfer experiments (19) led to the finding that the active site base which abstracts the substrate α-proton is a monoprotonic one. This conclusion appears to be verified. The experiments also led the authors to calculate an approximate pK of 15 for the ionization of the protein-bound substrate α-proton in the unliganded reduced enzyme. This implies a pK shift from the normal of 8 to 9 pH units. His 57 in the serine proteases catalytic triad has been ascribed a pK around 7 (25-27) and His 181 in holomalate dehydrogenase a pK of 7.4 (28). It thus appears that electrostatic stabilization of an imidazolium ion by a carboxylate is insufficient to raise the pK of a histidine to 15. Urban and Lederer suggested hydrogen bonding, perhaps to a flavin N(5) anion, as a possible explanation for the rise in pK. The structure as we see it does not make highly probable a hydrogen bond between His 373 N(3)-H and flavin N(5) (4.34 Å between the two nitrogens). On the other hand, Tyr 254 might be a good candidate for accepting a hydrogen bond from His 373 N(3)-H.

Finally, one can try to make deductions, from what we see in the reduced enzyme-pyruvate complex, concerning a probable mechanism for the reaction with 5-deazaflavin. Pompon and Lederer (11) observed that 5 deazaflavocytochrome b_2, like other flavoenzymes reconstituted with the 5 deaza-analog (29), catalyzes transfer of the substrate α-hydrogen to and from the cofactor C(5), but is unable to achieve heme reduction. It has been suggested that 5-deazaflavins behave rather like "flavin-shaped nicotinamides" than like flavin analogs (30) and this has been considered as a "mechanistic dilemma" (8,14): can an enzyme change mechanism upon cofactor replacement? We propose that the answer is yes. There is only need for switching off the function of one of the two active site general acid base catalysts, possibly His 373. Tyr 254 could facilitate hydride transfer to and from the 5-deazaflavin, just as histidine does in nicotinamide-linked lactate dehydrogenase.

Acknowledgements

This work was supported by NIH grant No. GM-20530 and by NATO grant No. 174/87. One of the authors (F. L.) is indebted to Dr. J. P. Mornon (Université Paris VI) and to his collaborators for access to their molecular graphics system and for instruction in the use of their molecular graphics program MANOSK.

References

1. Xia, Z-x., N. Shamala, P.H. Bethge, L.W. Lim, H.D. Bellamy, N.H. Xuong, F. Lederer, F.S. Mathews. 1987. Proc.Natl.Acad.Sci.(USA) 84, 2629-2633.

2. Mathews, F.S. and Z.-x. Xia. 1978. This volume.

3. Urban, P., P.M. Alliel, F. Lederer. 1983. Eur.J.Biochem. 134, 275-281.

4. Tegoni, M., J.M. Janot, F. Labeyrie. 1986. Eur.J.Biochem. 155, 491-503.

5. Müller, F. 1972. Z. Naturforsch. 27b, 1023-1026.

6. Hemmerich, P. 1976. Fortsch.Chem.Org.Naturst. 33, 451-528.

7. Massey, V., P. Hemmerich. 1980. Biochem.Soc.Transactions 8, 246-257.

8. Ghisla, S., V. Massey. 1986. Biochem.J. 239, 1-12.

9. Lederer, F. 1978. Eur.J.Biochem. 88, 425-431.

10. Capeillère-Blandin, C., R.C. Bray, M. Iwatsubo, F. Labeyrie. 1975. Eur.J.Biochem. 54, 549-566.

11. Pompon, D., F. Lederer. 1979. Eur.J.Biochem. 96, 571-579.

12. Cornforth, J.W. 1959. J. Lipid. Res. 1, 3-28.

13. Hemmerich, P. 1964. In: Mechanismus Enzymatischer Reaktionen, Mosbach Symposium. Springer Verlag, Berlin, pp. 183-209.

14. Ghisla, S. 1982. In: Flavins and Flavoproteins (V. Massey and C.H. Williams, eds.). Elsevier North Holland, Inc., Amsterdam, pp. 133-142.

15. Ghisla, S., H. Ogata, V. Massey, A. Schonbrunn, R.H. Abeles, C.T. Walsh. 1976. Biochemistry 15, 1791-1797.

16. Schonbrunn, A., R.H. Abeles, C.T. Walsh, S. Ghisla, H. Ogata, V. Massey. 1976. Biochemistry 15: 1798-1807.

17. Lederer, F. 1984. In: Flavins and Flavoproteins (R.C. Bray, P.C. Engel and S.G. Mayhew, eds.) Walter de Gruyter and Co., Berlin, pp. 513-526.

18. Urban, P., F. Lederer. 1984. Eur.J.Biochem. 144, 345-351.

19. Urban, P., F. Lederer. 1985. J. Biol. Chem. 260, 11115-11122.

20. Pompon, F., F. Lederer. 1985. Eur. J. Biochem. 148, 145-154.

21. Visser, C.M. 1980. Naturwissenschaften 67, 549-555.

22. Kennard, O., J. Walker. 1963. J. Chem. Soc. 5513-5531.

23. Birktoft, J.J., L.J. Banaszak. 1983. J.Biol.Chem. 258, 472-482.

24. Manstein, D.J., E.F. Pai, L.M. Schopfer, V. Massey. 1986. Biochemistry 25, 6807-6816.

25. Robillard, G., R.G. Shulman. 1974. J. Mol. Biol. 86, 519-540.

26. Markley, J.L., Ibanez, I.B. 1978. Biochemistry 17, 4627-4640.

27. Bachovchin, W.W., J.D. Roberts. 1978. J. Am. Chem. Soc. 100, 8041-8047.

28. Lodola, A., D.M. Parker, R. Jack, J.J. Holbrook. 1978. Biochem. J. 173, 427-448.

29. Walsh, C. 1978. Annu. Rev. Biochem. 45, 881-931.

30. Hemmerich, P., V. Massey, H. Fenner. 1977. FEBS Lett. 84, 5-21.

INACTIVATION OF YEAST FLAVOCYTOCHROME b_2 WITH FLUOROPYRUVATE. MODIFICATION OF THE ACTIVE SITE HISTIDINE.

Philippe Urban and Florence Lederer
CNRS LA 122, INSERM U 25, Hôpital Necker, F-75730 Paris Cedex 15, France

Oxidized flavocytochrome b_2 (L-lactate : cytochrome c oxidoreductase), an FMN-containing hemoprotein, is inactivated by fluoropyruvate which is known to be a substrate for the reduced enzyme (1).

A model developed by Ray & Koshland (2) :

$$E \xrightarrow[k_1]{FP} \underset{\text{actif}}{E_1} \xrightarrow[k_2]{FP} \underset{\text{inactif}}{E_2}$$

satisfactorily fits the inactivation kinetics. A formal interpretation of this scheme is that a non-essential residue is modified during the first step yielding a still fully active enzyme. This modification affects the protein molecule so that a previously unreactive essential residue can be attacked by the reagent, with concomitant enzyme inactivation. The inactivation is not due to accumulation in solution of a transformation product of fluoropyruvate. Using a multi-linear optimization procedure, the rate constants for both steps are shown to be fluoropyruvate-concentration dependent, and lowered in the presence of a competitive inhibitor (D-lactate). The first step appears to correspond to a second order with reaction, while the second step shows a saturation effect.

The chemical event which occurs during the second step, is the N-alkylation of a histidine side chain. After reduction with $[^3H]NaBH_4$, the tritiated derivative is shown to belong to proteolytic fragment β, which possesses only one histidine residue, His 373. In the flavocytochrome b_2 three-dimensional structure (3), His 373 is seen to lie in the active site. This fact suggests that His 373 is the catalytic base which abstracts the substrate α-proton during the L-lactate oxidation reaction, and gives it back to pyruvate during the reverse reaction (4). It is appropriate that its alkylation renders the enzyme inactive.

On the other hand it has been impossible to demonstrate that a chemical reaction takes place in the first step. Possible interpretations will be discussed in the poster.

1. P. Urban & F. Lederer (1983) Eur. J. Biochem. 134, 275-281.
2. W.J. Ray, Jr. & D.E. Koshland, Jr. (1963) J. Biol. Chem. 236, 1973-1979.
3. Z.-X. Xiu, N. Shamala, P.H. Bethge, L.W. Lim, H.D. Bellamy, N.H. Xuong, F. Lederer & F.S. Mathews (1987) Proc. Nat. Acad. Sci. USA, in press.
4. P. Urban & F. Lederer (1985) J. Biol. Chem. 260, 11115-11122.

Flavins and Flavoproteins 1987

EVIDENCE FOR AN ALTERATION OF THE FLAVIN BEHAVIOR INDUCED BY 2-METHYL-2,4-PENTANDIOL IN S. cerevisiae FLAVOCYTOCHROME b_2.

M. Tegoni & F. Labeyrie
Laboratoire d'Enzymologie Physicochimique du Centre de Génétique Moléculaire du CNRS, 91190 Gif-sur-Yvette, France.

Introduction

S. cerevisiae flavocytochrome b_2 (L-lactate cytochrome c oxydoreductase) has been crystallized in its "intact" (non proteolyzed) form at low ionic strength in the presence of 2-methyl-2,4-pentandiol (MPD) (1). The **three dimensional structure** of this enzyme has recently been **solved at 2.4 Å** resolution (2).

By microspectrophotometry on single crystals, we have shown that **the molecules within the crystal keep their catalytic competence** towards specific electron donors (L-lactate) and acceptors (O_2 and cytochrome c); nevertheless, it has not been possible to determine the true reaction rates constants (3).

We have more recently analyzed **some thermodynamic and kinetic properties of S.cerevisiae flavocytochrome b_2 "intact" in solution in the presence of 2-methyl-2,4-pentandiol.**

Results

Thermodynamic parameters

The presence of MPD in solution does not sensibly alter the dichroic spectrum of flavocytochrome b_2 in the far U.V. region (190-270 nm) suggesting no marked alteration of the protein folding in this solvent. In contrast, **major changes in the ellipticity of the oxidized and reduced flavin** have been evidenced, proportional to the concentration of the solvent.

Flavins and Flavoproteins 1987

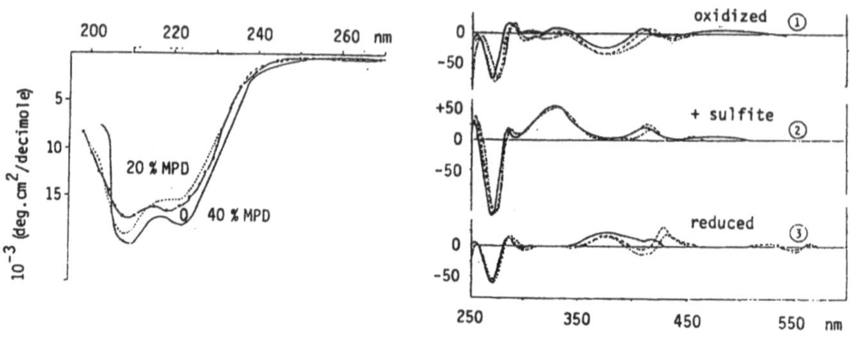

DICHROIC SPECTRA OF FLAVOCYTOCHROME b_2 AT VARIABLE CONCENTRATION OF MPD

The **dissociation of the prosthetic flavin** at low ionic strenght takes place much faster in the presence of the solvent. In fact, this reaction exhibits here a biphasic behavior, **2 classes of flavins** (half/half), dissociating approximately 50 and 10 times faster than in the absence of MPD, become apparent. Under these conditions, we have shown that the **binding of cytochrome c** at the stoichiometry of **one mole per tetrameric** flavocytochrome b_2 entirely restores the usual dissociation behavior.

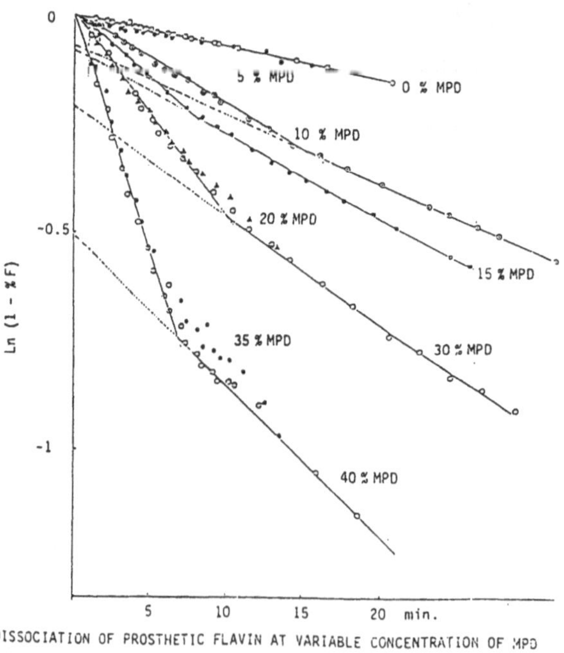

DISSOCIATION OF PROSTHETIC FLAVIN AT VARIABLE CONCENTRATION OF MPD

<u>Kinetic parameters</u>

The catalytic activity of flavocytochrome b_2 under steady state conditions has been tested at variable concentration of MPD. The inhibition exerted by MPD (fully reversible) is formally **uncompetitive** : K_M and V_{max} decrease in a parallel way and linearly with the square concentration of MPD. On the basis of the temperature dependence we can affirm that step 2 <u>ie</u> the electron transfer from fixed L-lactate to the prosthetic flavin (k_2) **is still limiting the overall reaction** in the presence of the solvent. We can calculate k_1, rate constant for the binding of L-lactate to the oxidized enzyme : indeed $V_{max}/K_M = (k_2 \times k_1)/(k_2 + k_{-1})$, being $k_{-1} << k_2$ in our case, the ratio V_{max}/K_M approximates to k_1. At variable concentration of MPD, **k_1 value does not sensibly differ** from that formerly calculated by Pompon et al (4) from kinetic experiments in identical conditions but in the absence of the solvent.

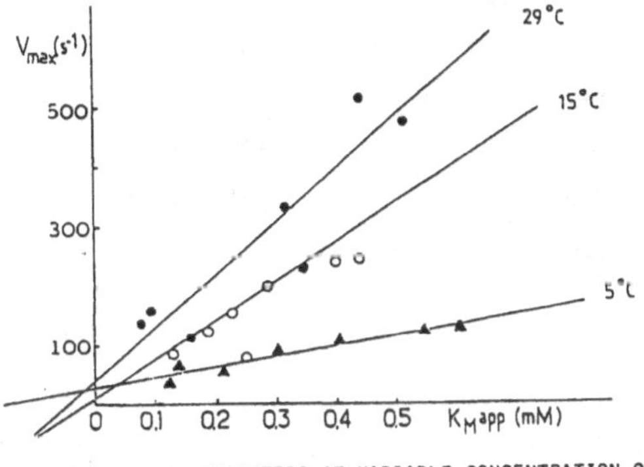

STEADY STATE KINETIC PARAMETERS AT VARIABLE CONCENTRATION OF MPD

The **reduction of flavocytochrome b_2 by L-lactate** in excess has been recorded by stopped-flow, under pre-steady state conditions at variable concentration of MPD. It appears that the apparent reduction rate constants of prosthetic flavin and heme are both decreased at half their normal values, at 30% MPD. In the sequential process (5), step 2 is the only common step a part L-lactate binding.

$$F_{ox}\text{-}H_{ox} + \text{L-lactate} \underset{k_{-1}}{\overset{k_1}{\rightleftharpoons}} \underset{\text{L-lactate}}{F_{ox}\text{-}H_{ox}} \xrightarrow{k_2} \underset{\text{Pyruvate}}{F_{red}\text{-}H_{ox}} \underset{k_{-3}}{\overset{k_3}{\rightleftharpoons}} \underset{\text{Pyruvate}}{F_{sq}\text{-}H_{red}}$$

In our interpretation, the presence of **MPD mainly affects step** 2 (*ie* reduction of flavin by bound L-lactate).

In summary, **all the changes** in equilibrium and kinetic parameters evidenced in the presence of MPD can be accounted for by **some alteration at the level of the flavin site.** Possible **specific interaction** between flavin and MPD (which structure is not so far from substrate of flavocytochrome b_2) to give a partially productive ternary complex responsible of the effects observed, should in our opinion be excluded on the basis of the very low affinity constant detected under steady state conditions ($K_I = K_{diss} = 1\ M^2$).

The classical formulation ($\ln k = \ln k_0 + (Z_A Z_B / D \times r) \times (e^2/kT)$) allows to calculate ($Z_A Z_B/r$) from experimental data at variable **dielectric constant** value. In our case, it yields a reasonable value for **r**= 1 Å, with $Z_A Z_B = -1$ **suggesting that the flavin reduction rate is controlled** by **the ionisation of a nearby group.**

It is interesting to note that the finding, described above, of two classes of flavins, each 50 % of total sites with different dissociation behaviors, at low ionic strength in the presence of MPD, is fully explained by the tridimensional structure of the tetramer showing actually that two flavins out of four have a different environment in the crystal (2).

1. Mathews, F.S. & F. Lederer. 1976. J. Mol. Biol. 102, 853.

2. Xia, Z., N. Shamala, P.H. Bethge, L.W. Lim, H.D. Bellamy, N. Xuong, F. Lederer, F.S. Mathews. 1987. Proc. Nat. Acad. Sci. (in press).

3. Tegoni, M., A. Mozzarelli, G.L. Rossi, F. Labeyrie. 1983. J. Biol. Chem. 258, 5424.

4. Pompon, D., M. Iwatsubo, F. Lederer. 1980. Eur. J. Biochem. 104, 479.

5. Capeillère-Blandin, C., R.C. Bray, M. Iwatsubo, F. Labeyrie. 1975. Eur. J. Biochem. 54, 549.

MEDIUM CHAIN ACYL-CoA DEHYDROGENASE FROM PIG KIDNEY: ASPECTS OF THE REDUCTIVE HALF-REACTION

C. Thorpe

Department of Chemistry and Biochemistry, University of Delaware, Newark, Delaware 19716

Introduction

The acyl-CoA dehydrogenases are tetrameric flavoproteins which catalyze the first oxidative step of mitochondrial β-oxidation yielding trans-2-enoyl-CoA derivatives (1). The reducing equivalents are then transferred to a second flavoprotein, electron transferring flavoprotein which mediates their entry into the respiratory chain at the level of ETF-CoQ oxidoreductase. Three categories of mammalian acyl-CoA dehydrogenases have been described with overlapping specificities for short, medium and long acyl chains (1). These enzymes appear mechanistically very similar, and thus studies performed with the readily available medium chain enzyme should be generally relevant to all of the dehydrogenases in this group.

Over the last few years there has been a marked resurgence of interest in the structure, mechanism of action and control of these important flavoproteins. These studies have gained added impetus with the recognition of a number of genetic diseases which stem from acyl-CoA dehydrogenase deficiencies. This chapter will deal with mechanistic aspects of the reductive half-reaction of the medium chain acyl-CoA dehydrogenase from pig kidney.

Reduction of the Dehydrogenase by Acyl-CoA Thioesters

Rapid reaction studies suggest that reduction of the medium chain enzyme by a variety of deuterated substrates is concerted, with multiplicative

Flavins and Flavoproteins 1987

primary kinetic isotope effects at α and β positions (2-4). Several lines of evidence suggest that the pro-R-α-hydrogen is removed as a proton by an active site base (5-7), whereas the pro-R-β-hydrogen is transferred to the flavin 5-position as a hydride (6,7).

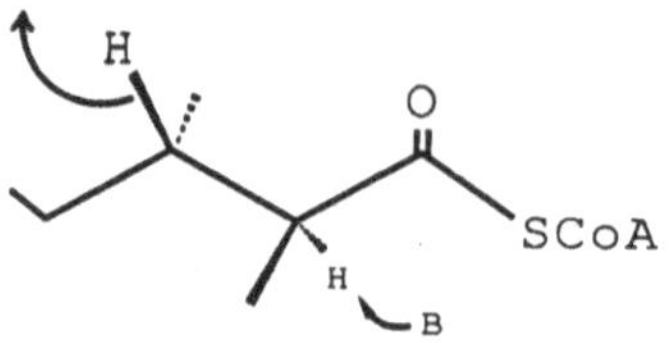

Although reduction of the native dehydrogenase by normal thioester substrates is probably concerted, we were interested in the possibility of uncoupling proton abstraction from hydride transfer via selection of suitable substrate and flavin analogs. In the extreme case, we wished to examine whether carbanion formation (and α-proton exchange) could occur when hydride transfer to the isoalloxazine ring was blocked. We hope that these studies will lead to a greater understanding of the reductive half reaction in the acyl-CoA dehydrogenases, and may aid in the design of more effective inhibitors of the enzyme.

Effects of Acyl-CoA Substrate Analogs Substituted at C-3

<u>Trans-3-octenoyl-CoA.</u> We have recently observed that this unsaturated derivative (Scheme 1, compound A) binds to the medium chain dehydrogenase

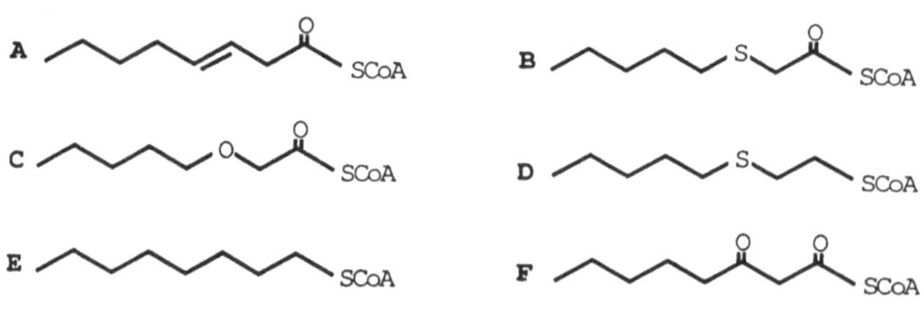

Scheme 1. CoA-derivatives used in this work

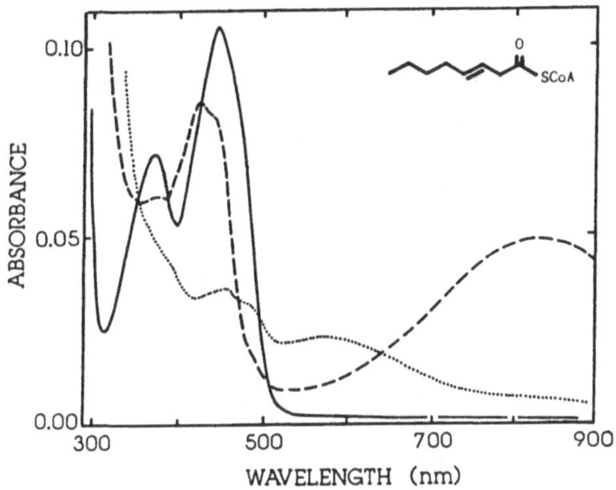

Figure 1. Interaction of trans-3-octenoyl-CoA with medium chain enzyme. Oxidized dehydrogenase (——); immediately (---), and 2 hours (...) after the addition of 1.3 equivalents of trans-3-octenoyl-CoA.

with the rapid formation of an intense long wavelength band at 820 nm (8; Figure 1). Trans-3-octenoyl-CoA binds tightly (with an apparent K_d of about 80 nM), and is an effective inhibitor of the dehydrogenation of octanoyl-CoA. The long wavelength band is sensitive to flavin substitution (8), and most reasonably represents a charge transfer complex (9) between an electron rich form of the thioester ligand as donor, and oxidized flavin as acceptor. NMR experiments show that the dehydrogenase catalyzes the rapid exchange of one α-proton in D_2O without significant isomerization to the corresponding trans-2-enoyl derivative (8). On a much slower time scale the long wavelength band decays with reduction of the enzyme (Figure 1).

3-Thia- and 3-oxa-octanoyl-CoA. Compounds B and C in Scheme 1 were chosen because they would not be expected to reduce the enzyme, although the lone pairs on the heteroatom might lead to interesting effects. These thioesters were synthesized from the corresponding acids by the mixed anhydride procedure and purified by HPLC (10). Figure 2 shows the intense stable charge transfer band (extinction coefficient of 8,700 M^{-1} cm^{-1} at

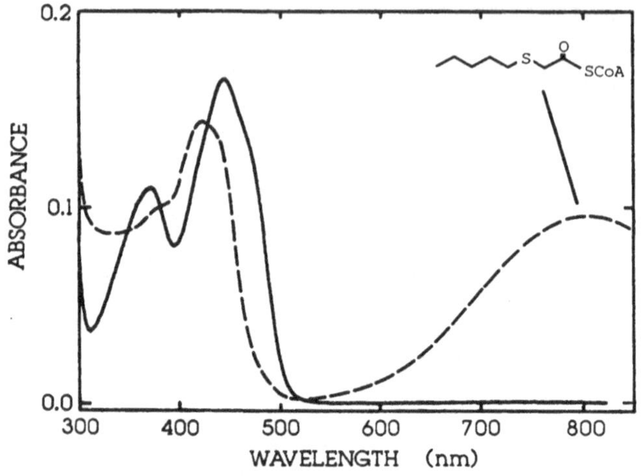

Figure. 2. Effect of 3-thia-octanoyl-CoA. The oxidized medium chain dehydrogenase (——) was treated with 6 equivalents of 3-thia-octanoyl-CoA (---).

804 nm) generated on the addition of the 3-thia-analog to the medium chain enzyme. Binding shows a stoichiometry of 1.1/FAD with a K_d of 0.4 µM. One might envision the polarizable thioether sulfur atom as the charge transfer donor in this complex, and thus we tested the corresponding oxa-analog (compound C) since it would have a considerably higher ionization potential at oxygen. Compound C gave a long wavelength band at only slightly shorter wavelengths (780 nm), making this original ascription highly unlikely.

Further evidence against the heteroatom as charge transfer donor is provided by compound D which retains an equivalently placed thioether sulfur atom as in B, but lacks the thioester carbonyl moiety. Previous work from our laboratory suggests that this carbonyl group contributes little to the binding of substrates in the Michaelis complex, and that alkyl-SCoA thioethers (e.g. compound E, Scheme 1) are valid substrate analogs in this regard (8). Binding of D perturbs the oxidized flavin chromophore, with a red shift and increased resolution of the main absorbance band similar to that encountered with octyl-SCoA (compound E). No detectable long wavelength band is observed.

Nature of the Charge Transfer Donor

These data suggest that the charge transfer bands observed with compounds A-C involve either the enolate or the enol generated on α-proton abstraction:

The enolate would represent the better charge transfer donor and is also more likely in view of the preferred binding of anionic CoA analog to the medium chain enzyme (11). In the case of the comparatively acidic 3-ketoacyl-CoA thioesters (pK 9), e.g. compound F, spectroscopic evidence for the bound enolate has been obtained (8, 12-14). However if enolates are invoked to account for the intense charge transfer bands seen in Figure 1 and 2, then we must explain how the protein stabilizes these species (whose free pK values are in the 18 - 20 range). Strategies for the stabilization of the bound enolate might include: raising the pK of the (carboxylate?; 5) base upon thioester binding and electrostatic and structural complementarity with the planar enolate unit (15). As discussed previously, such stabilization is necessary to ensure that abstraction of an α-proton is sufficiently rapid to sustain turnover (15).

A possible arrangement of substrate and flavin is sketched in Figure 3. The substrate is positioned to transfer the pro-R- β-hydride to the re face of the isoalloxazine at the N-5 position (6,7,16). During reduction with normal substrates, the developing enolate might be stabilized both by stacking over the flavin and by a positive charge on residue X. There is presumably no energy minimum corresponding to the carbanion along the reaction coordinate for these substrates, and the reaction will proceed with concerted elimination of the hydride (17). Hydride transfer might be facilitated by the presence of a positively charged group Y, since binding of both substrates and products stabilize negatively charged flavin species at the N-1 - C-2=O locus (11).

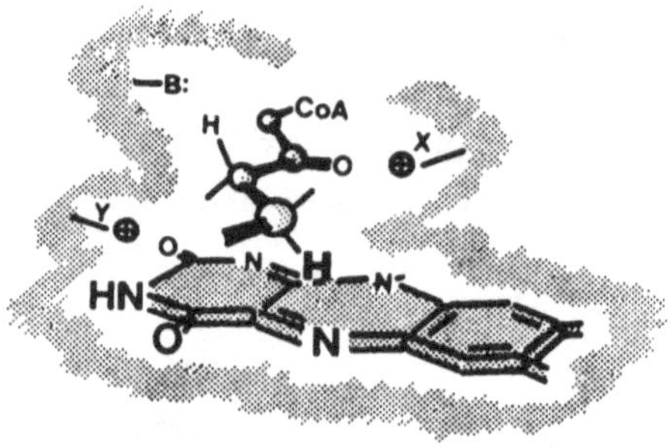

Figure 3. Hypothetical active site for the acyl-CoA dehydrogenase.

α-Proton Exchange Studies with the Medium Chain Dehydrogenase.

It has been known for a number of years that mammalian acyl-CoA dehydrogenases catalyze the exchange of the pro-R-α-proton in the absence of a suitable electron acceptor, whereas the β-protons do not exchange significantly (2,7,18,19). However the mechanism of this exchange reaction is controversial. Murfin (2) and Ghisla et al. (7) have suggested that exchange occurs from the reduced enzyme.enoyl-CoA complex, because dehydrogenation is concerted (see above). Retention of the α-hydrogen was proposed to reflect the fact that this atom, reversibly transferred to the flavin, is sequestered from solvent and non-exchangeable. In contrast, Ikeda et al. suggest that exchange occurs at the level of the oxidized enzyme, and that this partial reaction constitutes proof of a carbanion mechanism (19).

We have recently explored this issue (20) using dehydrogenase substituted with flavin analogs (such as 1- and 5-deaza-FAD and 6-OH-FAD) which are thermodynamically incapable of significant reduction by bound acyl-CoA thioesters. If proton abstraction and hydride elimination are obligatorily coupled, then no α-proton exchange would be observed in these instances. The initial results we obtained were puzzling. The 5-deaza-enzyme is incapable of catalyzing α-proton exchange, whereas the 1-deaza- and 6-OH-FAD enzymes catalyze exchange at about 10 and 100% of the rate of the native enzyme under standard conditions. Obviously exchange can occur in the absence of hydride transfer, and we have reason to believe that the behavior of the 5-deaza-FAD substituted enzyme is anomalous.

Further evidence for the uncoupling of proton and hydride transfer comes from the study of acyl-CoA thioesters which lack methylene hydrogens at the β-position (and which are therefore non-substrates of the enzyme). Exchange occurs rapidly with trans-3-octenoyl-CoA and with phenylacetyl-CoA for example (20). The medium chain dehydrogenase even catalyzes a slow exchange of the methyl protons of acetyl-CoA (20). These data suggest that the reduction of the dehydrogenase by normal substrates is an example of an "uncoupled-concerted" reaction (17).

Nature of Octanoyl-CoA Complexes of the Medium Chain Enzyme

A useful description of the reductive half-reaction involves the following internal redox step (21,22):

$$E.FAD_{ox}.SH_2 \overset{K_2}{\rightleftharpoons} E.FAD_{2e}.P$$

which is both dependent on chain length and sensitive to the redox potential of the flavin used (11,22). K_2 lies far to the right with octanoyl-CoA, leading to the characteristic purple complexes ascribed to a charge transfer interaction between reduced flavin as the donor and trans-2-octenoyl-CoA (P) as acceptor (9,23). The nature of these complexes has recently been questioned on the basis of chemical quenching experiments. The substrate reduced enzyme was quenched in 0.1 M KOH, and the liberated fatty acids were extracted and analyzed by gas chromatography (19). Only octanoate (and not octenoate derived from P) was detected, leading to the claim that "the hydride ion transfer from C-3 of the substrate to the enzyme-FAD is not yet complete in the charge transfer complex" (19). We note two problems with these experiments. First, quenching is likely to be too slow compared to equilibration via K_2. Second, our control experiments show that the analytical procedure gives very poor recoveries of octenoate (24). In brief, trans-2-octenoyl-CoA hydrolyzes relatively slowly and is subject to Michael addition from molecules of $CoAS^-$ released earlier during thioester hydrolysis (24). Accordingly, we have performed rapid quenching experiments in 1 M anaerobic HCl analyzing CoA-thioesters directly by HPLC: trans-2-octenoyl-CoA comprised greater than 80% of the CoA derivatives released from the enzyme (24). Thus these results are fully consistent with the original ascription that these purple species represent charge transfer complexes between reduced flavin and octenoyl-CoA.

Acknowledgements

The work performed in the author's laboratory was supported by NIH GM 26643. The author would like to thank Drs. Sandro Ghisla and Lawrence Schopfer for helpful discussions.

References

1. Beinert, H. 1963. The Enzymes, Vol. 7, 2nd Ed. (P.D. Boyer, H. Lardy, and K. Myrback, eds.) Academic Press, New York, pp. 447-466.

2. Murfin, W. W. 1974. Ph.D. Thesis, Washington University, St. Louis, MO.

3. Reinsch, J. Katz, A., Wean, J., Aprahamian, G., and McFarland, J. T. 1980. J. Biol. Chem. 255, 9093-9097.

4. Pohl, B., Raichle, T., and Ghisla, S. 1986. Eur. J. Biochem. 160, 109-115.

5. Fendrich, G., and Abeles, R. H. 1982. Biochemistry 21, 6685-6695.

6. Ghisla, S. 1984. In: Flavins and Flavproteins (R. C. Bray, P. C. Engel, and S. G. Mayhew, eds.). Walter de Gruyter, Berlin. pp. 385-401.

7. Ghisla, S., Thorpe, C., and Massey, V. 1984. Biochemistry 23, 3154-3160.

8. Powell, P. J., Lau, S-M., Killian, D., and Thorpe, C. 1987. Biochemistry (in press).

9. Massey, V. and Ghisla, S. 1974. Ann. N. Y. Acad. Sci. 227, 446-465.

10. Lau, S.-M., Brantley, R., and Thorpe, C., unpublished.

11. Thorpe, C., and Massey, V. 1983. Biochemistry 22, 2972-2978.

12. McKean, M. C., Frerman, F. E., Mielke, D. M. 1979. J. Biol. Chem. 254, 2730-2735.

13. Benecky, M., Li, T. Y., Schmidt, J., Frerman, F., Walters, K. L., and McFarland, T. 1979. Biochemistry, 18, 3471-3476.

14. Williamson, G., Engel, P. C., Nishina, Y., and Shiga, K. 1982. FEBS Lett. 138, 29-32.

15. Gilbert, H. F. 1981. Biochemistry 20, 5643-5649.

16. Manstein, D. J., Pai, E. F.,Schopfer, L. M., and Massey, V. 1986. Biochemistry 25, 6807-6816.

17. Jencks, W. P. 1981. Chem. Soc. Rev. 10, 345-375.

18. Biellmann, J. F., and Hirth, C. G. 1970. FEBS. Lett. 9, 55-56; 335-336.

19. Ikeda, Y., Hine, D. G., Okamura-Ikeda, K., and Tanaka, K. 1985. J. Biol. Chem. 260, 1326-1337.

20. Zhou, J. Z., Ghisla, S., and Thorpe, C., in preparation.

21. Beinert, H., and Page, E. 1957. J. Biol. Chem. 225, 479-497.

22. Thorpe, C., Matthews, R. G., and Williams, C. H. 1979. Biochemistry 18, 331-337.

23. Engel, P. C., and Massey, V. 1971. Biochem. J. 125, 889-902.

24. Lau, S. M., and Thorpe, C., in preparation.

CRYSTAL STRUCTURE OF THE GENERAL ACYL-CoA DEHYDROGENASE

J.J.P. Kim and J. Wu

Department of Biochemistry, Medical College of Wisconsin
Milwaukee, WI 53226, USA

The general acyl-CoA dehydrogenase is an FAD containing mitochondrial enzyme involved in the β-oxidation of fatty acids. The enzyme has a molecular weight of 180,000 and contains four identical subunits, each containing one equivalent of FAD. The three dimensional crystal structure has been determined at 5.5Å resolution by the X-ray diffraction method.

The enzyme was crystallized from polyethylene glycol solutions. Crystals have the space group $C222_1$ with a=128.2, b=136.1 and c=106.3Å. The asymmetric unit contains two monomers of the tetrameric dehydrogenase molecule. An electron density map was calculated with phases determined from three isomorphous derivatives. It showed clear contrast between solvent and protein regions and the local two fold axes calculated by the Rotation Function study were confirmed. The molecule has 222 symmetry and approximate dimensions of 64x64x75Å. The diffraction data at 2.4Å has been collected using the multiwire area detector system at Univ. of Calif., San Diego. Calculation of electron density map at this resolution is in progress.

This research was supported by NIH grant No. GM 29076.

Flavins and Flavoproteins 1987

INACTIVATION OF GENERAL ACYL-CoA DEHYDROGENASE FROM PIG KIDNEY BY THE SUICIDE SUBSTRATE METHYLENECYCLOPROPYLACETYL-CoA. STRUCTURE OF ONE OF THE COVALENT FLAVIN-INHIBITOR ADDUCTS.

Hans-Dieter Zeller and Sandro Ghisla
Fakultät Biologie der Universität Konstanz
D-7750 Konstanz, FRG

Introduction

General acyl-CoA dehydrogenase from pig kidney (GAD), a typical fatty acyl-CoA dehydrogenase, has previously been shown to be irreversibly inactivated by methylenecyclopropylacetyl-CoA (MCPA-CoA), a metabolite derived from the poisonous amino acid hypoglycine (1). The inactivation results from covalent addition of the inactivator molecule to the flavin coenzyme, FAD. Elucidation of the structure of the flavin inhibitor adduct(s) and of the detailed inactivation mechanism were hampered by the instability of the adducts. We have succeeded in synthesizing labelled MCPA-CoA (^{14}C at C(1) of MCPA). Inactivation with labelled inhibitor leads to incorporation of the label into the two major flavin adducts, and no reaction of the inactivator with the protein occurs. In this report we present data on the structure of the minor adduct II, which results from the attack of the inhibitor at C(6) position of the flavin.

Results and Discussion

General acyl-CoA dehydrogenase (GAD) was completely inactivated by three equivalents of the suicide substrate, MCPA-CoA, per mol of flavin. Optimization of the protein denaturation procedure (90% v/v methanol), and of the purification (HPLC) lead to the isolation of two adducts, which could be partially characterized by optical spectroscopy and ^{1}H-NMR. The major adduct, I, shows the properties of cyclic C(4a), N(5)-dihydro

Flavins and Flavoproteins 1987

flavin derivatives. This adduct is unstable and decays to oxidized flavin. Figure 1 shows the spectral properties of the second minor adduct, II, which is stable. There is a remarkable correspondence between its absorption spectrum and that of a C(6)-substituted flavin derived from an adduct formed during irreversible inactivation of D-lactate dehydrogenase from Megasphaera elsdenii by the suicide substrate D-2-hydroxy-3-butynoic acid (2,3).

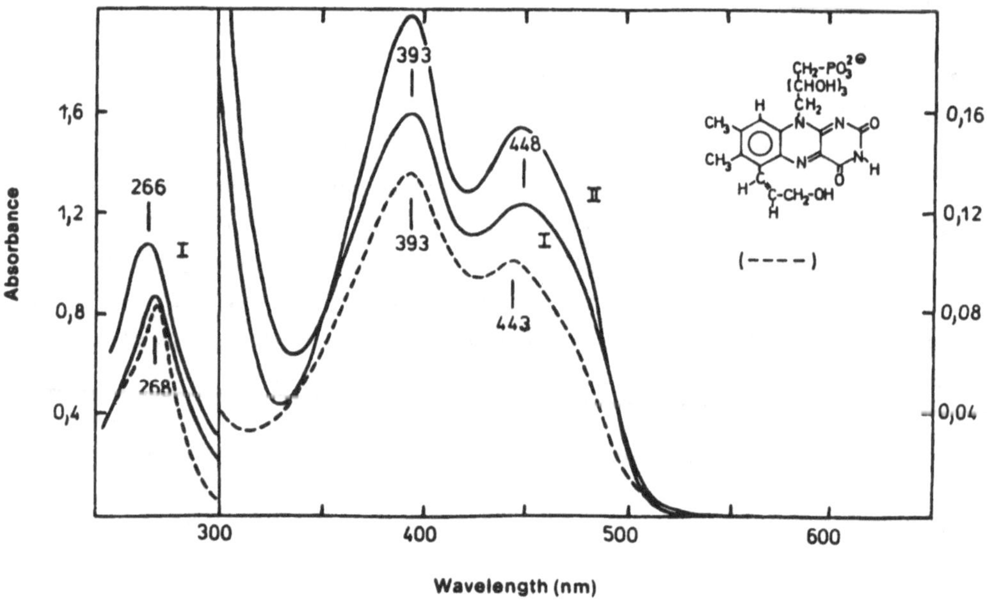

Fig. 1. Absorption spectrum of adduct II after HPLC purification
Curve I shows the spectrum of adduct II in potassium phosphate buffer pH 6.0. Curve II was obtained after treatment of adduct II with snake venom phosphodiesterase to generate the corresponding FMN analog. For comparison the spectrum of 6-(3-hydroxy-1-propenyl)-FMN is shown by curve (---).

Together with the data from the fluorescence emission spectrum of adduct II, which is compatible to that expected for a C(6)-substituted flavin, this spectral comparison suggests a covalent C(6)-flavin adduct structure. The ^{1}H-NMR-spectrum (Figure 2) is also compatible with substitution at C(6), particularly since the signal expected for the C(6) hydrogen is absent.

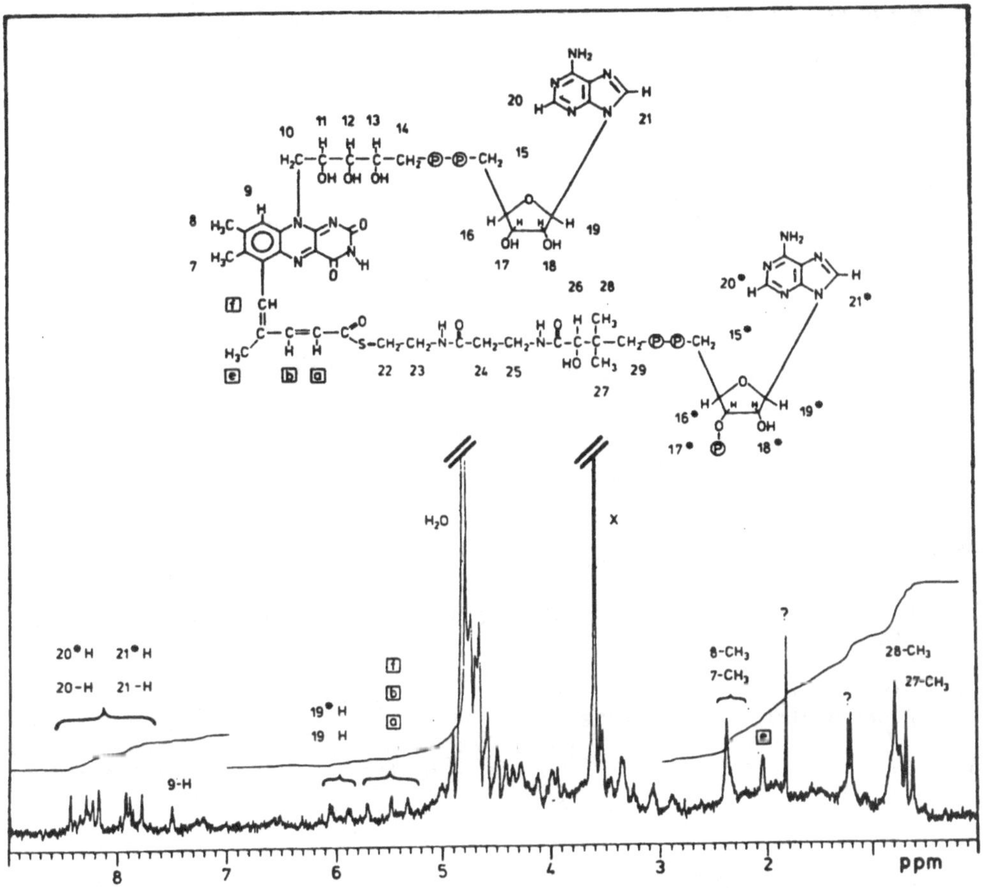

Fig. 2. ^{1}H-NMR-spectrum of adduct II and its proposed structure
The HPLC-purified chromophore was dissolved several times in D_2O and the NMR data aquired with a 250 MHz-FT-NMR spectrophotometer (1280 pulses). X and ? denote signals attributed to impurities.

Conclusions

The inactivation of GAD with radiolabelled MCPA-CoA (^{14}C at position C(1) of the MCPA-molecule) leads to the incorporation of the label into two major flavin adducts. No reaction of the inactivator with the protein occurs. The minor adduct II results from the attack of the inhibitor molecule at the C(6) position of the flavin. The structure proposed is the following:

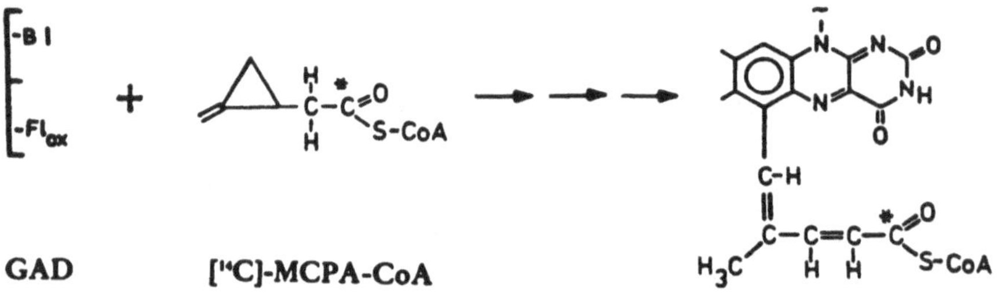

References

1. Wenz, A., C. Thorpe, S. Ghisla. 1981. J. Biol. Chem. 256, 9809-9812.

2. Olson, S.T., V. Massey, S. Ghisla, C.D. Whitfield. 1979. Biochemistry 18, 4724-4732.

3. Ghisla, S., S.T. Olson, V. Massey, J.M. Lhoste. 1979. Biochemistry 18, 4733-4742.

2-BROMO-OCTANOYL CoA: A PSEUDO-SUBSTRATE FOR GENERAL ACYL CoA DEHYDROGENASE.

Carole L. Hall
School of Applied Biology, Georgia Institute of Technology, Atlanta, Georgia 30332.

Introduction

2-bromo-octanoate has been reported to inhibit mitochondrial β-oxidation, and it was suggested that it did so as the 3-keto derivative, which then inhibited thiolase (1). These observations suggested that 2-bromo-octanoyl CoA would be an interesting compound with which to study the interaction of substrates with general acyl CoA dehydrogenase (G-AD).

Results

2-bromo-octanoyl CoA was synthesized via a variation of the mixed anhydride method. HPLC analysis (C18 reverse phase) showed one major peak (86%) and several small ones. Two of the small peaks were tentatively identified as n-octanoyl CoA (3%) and 2-Br-octenoyl CoA (5%). The 2-Br-octanoyl CoA was used without further purification.

As has been previously noted (2), when G-AD is titrated either aerobically or anaerobically with octanoyl CoA, the plots appear to saturate at 1 nmole octanoyl CoA per enzyme flavin, and attempts to ascertain binding constants are difficult by the usual graphical methods based on absorbance changes (3).

Flavins and Flavoproteins 1987

Addition of 2-Br-octanoyl CoA to G-AD resulted in spectral perturbations similar to those seen with butyryl- or hexanoyl-CoA, which are "less-favored" substrates for G-AD (3). The enzyme was titrated aerobically and anaerobically with 2-Br-octanoyl CoA. Some of the titration curves from the aerobic experiment are shown in Fig. 1 (left side).

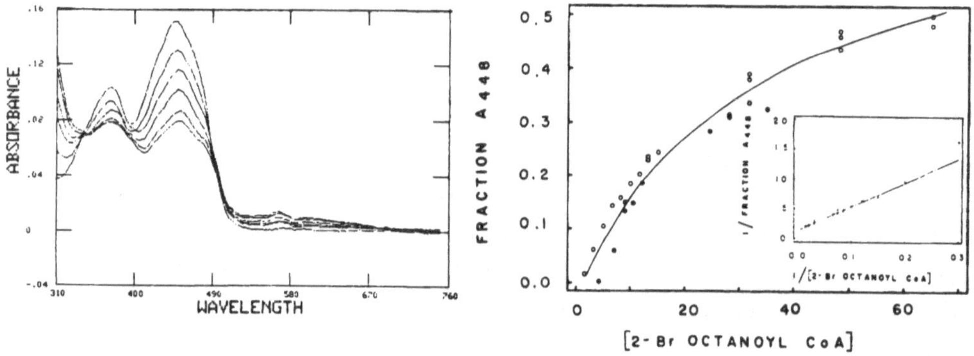

Figure 1. Aerobic and Anaerobic Titrations of G-AD with 2-Br-octanoyl CoA.

(LHS) G-AD (10.57 nmoles enzyme flavin) was diluted to 1.2 mL and the spectrum recorded. 200 data points per spectrum were collected using an OLIS 3820 data system interfaced to a Perkin-Elmer 559A spectrophotometer. The sample was titrated aerobically with 2-Br-octanoyl CoA. Spectra obtained after addition of 8,18,38,58, and 78 nmoles of 2-Br-octanoyl CoA are shown.

(RHS) Fractional absorbance changes at 445 nm vs. nmoles 2-Br-octanoyl CoA/nmole enzyme flavin are shown for the aerobic titration (●) and for a similar experiment using the same lot of enzyme performed anaerobically (○). The inset shows a double reciprocal plot of the data.

In contrast to the case with n-octanoyl CoA, the data fitted a straight line when plotted as 1/fractional absorbance vs. 1/[ligand, L]. This plot yielded a dissociation constant obtained from plots of 1/fractional absorbance vs. 1/[ligand] for the binding of 2-Br-octanoyl CoA evaluated to 32×10^{-6} M.

Binding of n-octanoyl CoA and 2-Br-octanoyl CoA to G-AD was studied by adding varying amounts of each ligand to G-AD in Centricon-30 devices. The filtrates were analyzed by HPLC to determine the amount of each ligand, either as saturated or enoyl acyl CoA, was recovered. Attempts to fit the binding data to Scatchard-type plots (v/[L] vs. v) resulted in biphasic plots (Fig. 2, LHS) which were similar to "humped" plots characteristic of cooperative binding phenomena (4). Accordingly, plots of ln [ligand] vs. ln (n/v-1) were constructed to obtain H (Hill constant) and K_d. As G-AD is a tetrameric enzyme, plots were made assuming n = 4. Fractional saturation for interacting sites can be described by Eqn. 1:

$$\bar{y} = (x^{\alpha H}/K^{\alpha H}) \;/\; (1+ (x^{\alpha H}/K^{\alpha H})) \qquad \text{(Eqn. 1)}$$

where $\bar{y}$ is the fraction bound and x = [L] (4).

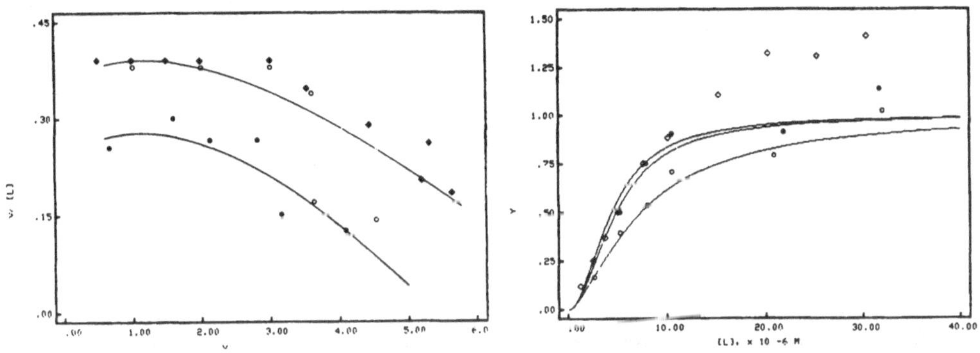

Figure 2.Binding of n-octanoyl- and 2-Br-octanoyl CoA to G-AD.

(LHS) Plot of v/[L] vs. v. n-Octanoyl CoA = (○,◆); 2-Br-octanoyl CoA = (●).

(RHS) Plot of y vs. [L]. Same data as in Fig. 2 (LHS). Curves were calculated according to Eqn. 1 using $\alpha H = 1.9$; $K = 4.8 \times 10^{-6}$ M (●) and $\alpha H = 1.92$; K = 4.33 (◇) for n-octanoyl CoA and $\alpha H = 1.5$; $K = 7.3 \times 10^{-6}$ M (○) for 2-Br-octanoyl CoA.

Values of K obtained for n-octanoyl CoA were 4.33 and 4.8 x 10^{-6} M; for 2-Br-octanoyl CoA, K was 7.3 x 10^{-6} M. This value is in contrast with K inferred from the spectral data. These indicate that the extent of spectral change seen upon addition of acyl CoA derivatives to G-AD is not necessarily related to binding _per se_, but rather to binding which can result in electronic interaction of the ligand with the flavin. The binding data suggest that there is cooperative interaction between substrate binding sites, and that some non-specific binding might also occur at higher (> than stoichiometric) levels of acyl CoA.

Acknowledgement

This research was supported by the Biomedical Research Support Grants Program.

References

1. Raaka, B. and J.M. Lowenstein. 1979. J. Biol. Chem. _254_, 6755.

2. Hall, C.L., J.D. Lambeth, and H. Kamin. 1979. J. Biol. Chem. _254_, 2023.

3. Benesi, H.A. and T.A. Hildebrand. 1949. J. Amer. Chem. Soc._71_, 2703.

4. Cantor, C.R. and P.R. Schimmel. 1980. In: Biophysical Chemistry Part III, Chapter 15, p. 849.

REACTION OF PIG KIDNEY MEDIUM CHAIN ACYL-CoA DEHYDROGENASE WITH β-OXIDATION INTERMEDIATES

P. J. Powell, S-M. Lau, and C. Thorpe

Department of Chemistry and Biochemistry, University of Delaware, Newark, Delaware 19716

Introduction

Mitochondrial β-oxidation involves a large number of acyl-CoA intermediates and a considerable number of distinct enzymes, and this poses problems of organization and coordination which are not well understood. In mammals, the enzymes representing the core activities of mitochondrial β-oxidation exist in at least two forms to complement the array of acyl chain lengths encountered. For example, there are three acyl-CoA dehydrogenases with overlapping substrate specificities. Recently, we have used alkyl-SCoA thioether analogs to probe the hydrocarbon binding site in the medium chain acyl-CoA dehydrogenase from pig kidney (1). This paper examines the response of the enzyme to various eight-carbon intermediates of β-oxidation (namely the trans-2-octenoyl-CoA, trans-3-octenoyl-CoA, L-3-hydroxyoctanoyl-CoA and 3-ketooctanoyl-CoA).

Results and Discussion

Trans-2-Octenoyl-CoA. Trans-2-Octenoyl-CoA binds tightly (K_d about 90 nM) to the medium chain acyl-CoA dehydrogenase with a red shift and increased resolution of the main absorbance band. No significant long wavelength band is observed on the addition of trans-2-octenoyl-CoA to the oxidized enzyme. In these experiments it is necessary to ensure that all contaminating enoyl-CoA hydratase activity is removed (2), because this enzyme can isomerize trans-2- to trans-3-enoyl-CoA derivatives (3).

Flavins and Flavoproteins 1987

Trans-3-Octenoyl-CoA. An intense long wavelength band (λ_{max} = 820 nm) is formed immediately upon addition of trans-3-octenoyl-CoA to the oxidized enzyme. This band may represent a charge transfer complex between oxidized flavin as the acceptor and an enolate formed upon α-proton abstraction as discussed elsewhere in this volume (4).

3-Ketooctanoyl-CoA. The spectral effects of the 3-keto- analog, acetoacetyl-CoA, were first described by Engel and Massey (5). 3-Ketooctanoyl-CoA exerts similar spectral changes with a K_d of approximately 55 nM (Table I). Titrations of the medium chain enzyme with 3-keto- derivatives give no evidence for half-site reactivity.

3-Hydroxyoctanoyl-CoA. This derivative partially reduces the enzyme FAD, as shown in Figure 1, with formation of a band at 570 nm. This reaction ($t_{1/2}$ = 45 sec) is inhibited by added 3-ketooctanoyl-CoA. Further, the partial reduction is driven to completion (Figure 1) by the addition of 3-hydroxyacyl-CoA dehydrogenase and excess NADH. Thus, the 3-ketooctanoyl-CoA, which accumulates during the oxidation of the L-3-hydroxy- analog, sequesters the remaining oxidized enzyme leading to

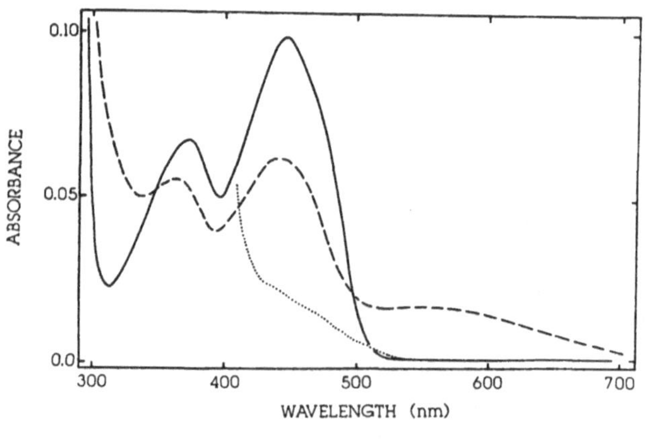

Figure 1. Reduction of enzyme by L-3-hydroxyoctanoyl-CoA. The oxidized enzyme (——) was mixed anaerobically with 1.3 equivalents of L-3-hydroxyoctanoyl-CoA and the spectrum recorded after fifteen minutes (----). The dotted curve is after eight hours in a reaction supplemented with NADH and 3-hydroxyacyl-CoA dehydrogenase.

the partial reduction seen in Figure 1. The reduction of the enzyme by L-3-hydroxyacyl derivatives was suggested earlier (5) and is consistent with the known stereochemistry of the acyl-CoA dehydrogenase (1).

Conclusion

Combining the results from the binding of thioether analogs with those of the various eight-carbon acyl-CoA thioesters suggests that all of the 28 intermediates in the catabolism of palmitoyl-CoA will bind to the medium chain length dehydrogenase from pig kidney. Table I summarizes the effects of a number of the thioesters studies on the activity of the dehydrogenase. All but the 3-hydroxy- derivative are effective inhibitors of the dehydrogenase in vitro. Should these metabolites accumulate in vivo they might represent feedback inhibitors of the β-oxidation pathway.

Table I. Effect of Thioesters on the Activity of Medium Chain Acyl-CoA Dehydrogenase. Activities determined using 30 μM octanoyl-CoA and 30 μM levels of potential inhibitors.

Potential Inhibitor	K_d	% Activity
None		100
Trans-2-	90 nM	38
Trans-3-	70 nM	33
L-3-OH-	not determined	93
3-keto-	55 nM	24
Acetoacetyl-	12 μM	99

Acknowledgment

This work was supported by NIH GM 26643.

References

1. Powell, P. J., Lau, S-M., Killian, D., Thorpe, C., Biochemistry (in press).

2. Lau, S-M., Powell, P. J., Buettner, H., Ghisla, S., Thorpe, C. 1986. Biochemistry 25, 4184-4189.

3. Stern, J. R. 1961. The Enzymes 5, 511-529.

4. Thorpe, C. This volume.

5. Engel, P. C., Massey, V. 1971. Biochem. J. 125, 889-902.

MEDIUM CHAIN ACYL-CoA DEHYDROGENASE FROM PIG KIDNEY: INACTIVATION OF THE REDUCED ENZYME BY 2-OCTYNOYL-CoA

Jing-Zhi Zhou and Colin Thorpe

Department of Chemistry and Biochemistry, University of Delaware, Newark, Delaware 19716

Introduction

The medium chain acyl-CoA dehydrogenase is a flavoprotein which catalyzes the first step of mitochondrial β-oxidation with the oxidation of acyl-CoA thioesters to their corresponding trans-2-enoyl-derivatives. We have found that the acetylenic thioester, 2-octynoyl-CoA:

$$CH_3-(CH_2)_4-C\equiv C-COSCoA$$

inactivates the enzyme by two distinct pathways depending on the redox state of the flavin prosthetic group. Inactivation of the oxidized dehydrogenase occurs with labelling of the protein and elimination of CoASH (1). This paper reports that incubation of the reduced dehydrogenase with 2-octynoyl-CoA yields an inactive reduced enzyme derivative, which probably represents a covalent adduct between acetylenic inhibitor and dihydroflavin.

Results and Discussion

The two-electron reduced enzyme was prepared by catalytic photoreduction (2; see inset to Figure 1) and mixed anaerobically with 3 equivalents of 2-octynoyl-CoA. After 2 hours, the enzyme was dialyzed versus 50 mM phosphate buffer, pH 7.6, without significant spectral change (Figure 1). The treated enzyme is stable in air, is only slightly reoxidized by

Flavins and Flavoproteins 1987

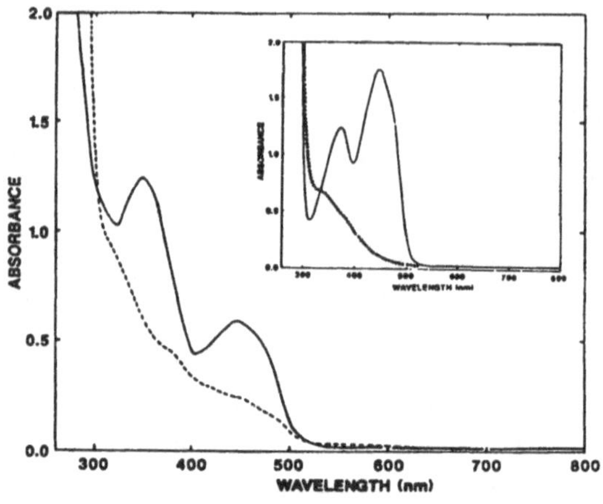

Figure 1. The inset shows oxidized (——) and photochemically reduced (...) dehydrogenase. The dashed line in the main figure is the reduced dehydrogenase treated for 2 h with 3 equivalents of 2-octynoyl-CoA, and dialyzed versus aerobic buffer, pH 7.6. The solid line is the spectrum of the supernatant after denaturation of the enzyme in 80% methanol.

electron transferring flavoprotein (20% in 20 hours), and resists displacement by trans-2-octenoyl-CoA. Inactivation of the reduced enzyme is rapid, with a half-time of less than 20 sec at pH 7.6, 25°C. No recovery of activity was observed upon dilution of the treated enzyme into a large excess of octanoyl-CoA during enzyme assay.

These observations are consistent with, but do not prove, the existence of a covalent dihydroflavin adduct in the treated enzyme. Denaturation of the reduced derivative in 80% methanol yields a mixture of products which was characterized spectrophotometrically, by reverse phase high pressure liquid chromatography and by radiolabelling experiments. Flavin was followed using FAD containing 2-^{14}C-riboflavin and 2-octynoyl-CoA was labelled either at C-1 of the acyl- moiety or using tritiated CoA. The spectrum of the supernatant, after removal of denatured protein and methanol, shows peaks at 450 and 345 nm (Figure 1). Approximately 20% of the flavin is recovered as oxidized FAD (largely accounting for the 450 nm absorbance of the supernatant in Figure 1), about 40% is retained covalently attached to the protein, while the remainder is distributed

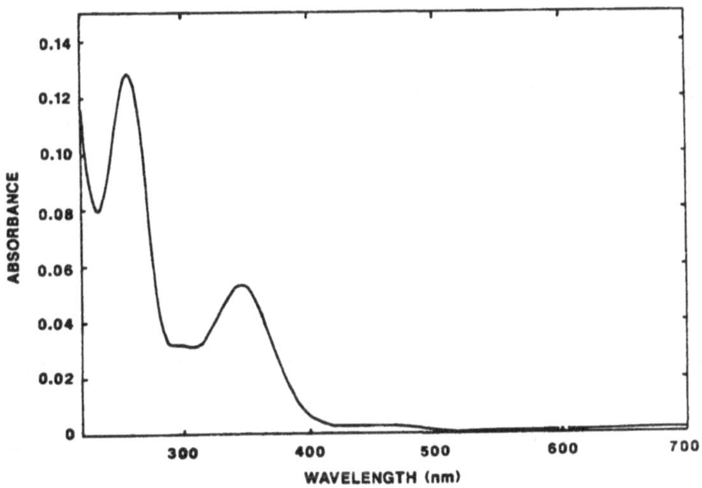

Figure 2. Spectrum of a species released on denaturation of the 2-octynoyl-CoA treated reduced enzyme (see text).

between at least 5 species eluting after FAD on HPLC. Only one of these species (10% of the total radioactivity) contained flavin, acyl-, and CoA moieties. The spectrum of this fraction (Figure 2) is clearly that of a reduced flavin adduct with maxima at 345 and 260 nm. It decays with reappearance of 450 nm absorbance at both acidic and basic pH values.

Thus we have described a second mode of inactivation of the medium chain acyl-CoA dehydrogenase by 2-octynoyl-CoA. Possibly the air stable product of 2-octynoyl-CoA treatment represents a single species. We have not proved that this putative primary enzyme-bound product represents a covalent flavin adduct, but it clearly rearranges to yield a variety of modified flavin species during denaturation of the enzyme. It should be noted that denaturation of the reduced flavin adduct formed upon treatment of the dehydrogenase with methylenecyclopropylacetyl-CoA also yields a number of covalently modified products (3). The adduct formed on the addition of 3,4-pentadienoyl-CoA to the native medium chain acyl-CoA dehydrogenase is too labile to be isolated (4).

Acknowledgments

This work was supported by NIH GM 26643. We would like to thank Dr. John Mizzer and Mr. Kurt Freund for performing some of the early experiments on this project.

References

1. Freund, K., Mizzer, J. P., Dick, W., Thorpe, C. 1985. Biochemistry 24, 5996-6002.

2. Thorpe, C., Matthews, R. G., Williams, C. H. 1979. Biochemistry 18, 331-337.

3. Wenz, A., Thorpe, C., and Ghisla, S. 1981. J. Biol. Chem. 256, 9809-9812.

4. Wenz, A., Ghisla, S., Thorpe, C. 1985. Eur. J. Biochemistry 147, 553-560.

OXIDATION-REDUCTION OF GENERAL ACYL-COA DEHYDROGENASE BY THE BUTYRYL-COA/CROTONYL-COA COUPLE. A NEW INVESTIGATION OF THE RAPID REACTION KINETICS

Lawrence M. Schopfer and Vincent Massey
Dept. of Biological Chemistry, University of Michigan Medical School
Ann Arbor, MI. 48109-0606 USA

Sandro Ghisla
Fakultät Biologie, Universität Konstanz, D7750 Konstanz, West Germany

Colin Thorpe
Dept. of Chemistry, University of Delaware, Newark, DE. 19716 USA

Introduction

General acyl-CoA dehydrogenase (GAD) (EC 1.3.99.3) catalyzes the insertion of a trans-double bond between C-2 and C-3 of acyl-CoA compounds. Previous mechanistic studies found that oxidized GAD reacts rapidly with butyryl-CoA to form an enzyme species which absorbs visible light (maximum at about 560 nm) and which slowly decays to a species resembling reduced enzyme (1). Rapid reaction studies (2-4) have shown that the rapid portion of this reaction is biphasic. We have undertaken a new rapid reaction investigation into the GAD/butyryl-CoA reaction mechanism. This work includes stopped-flow studies on the reactions of oxidized GAD with butyryl-CoA, reduced GAD with crotonyl-CoA, oxidized GAD with crotonyl-CoA, and reduced GAD with butyryl-CoA. We found the mechanism to be freely reversible and consistent with Scheme I.

Scheme I

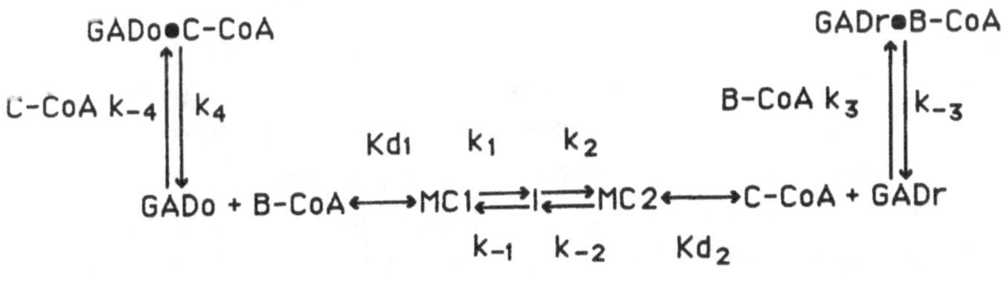

Flavins and Flavoproteins 1987

GADo is oxidized GAD, B-CoA is butyryl-CoA, C-CoA is crotonyl-CoA, GADr is reduced GAD, MC1 is a Michaelis-Menten complex between oxidized GAD and butyryl-CoA, I is a charge-transfer complex between reduced GAD and crotonyl-CoA, and MC2 is a Michaelis-Menten complex between reduced GAD and crotonyl-CoA.

Results and Discussion

Mixing butyryl-CoA with oxidized GAD (50 mM KPi pH 7.6, 4°C, anaerobically) gave a triphasic reaction, as has been seen before (1-4). The slowest phase (t1/2= 16 min) was about 250-fold slower than the preceeding phase. As such, it was considered to be a secondary process and was eliminated from consideration. The first, two, rapid phases were seen as a biphasic decrease in absorbance at 450 nm and an increase followed by a partial decrease at 560 nm, Figure 1. The faster phase resulted in formation of long wavelength absorbance. The apparent rate increased to saturation as the butyryl-CoA concentration was increased. These observations were sufficient to define steps Kd1 (37 µM), k1/k-1 (1.7 sec^{-1}/0.6 sec^{-1}) and species MC1 in Scheme I. The slower phase resulted in the partial decay of the long wavelength absorbance. The apparent rate decreased as the butyryl-CoA concentration increased. This qualitatively different type of substrate dependence indicated that butyryl-CoA entered the mechanism a second time to promote the decay of intermediate I, and led to the establishment of steps k2/k-2 (k2=0.07 sec^{-1}) and k3/k-3 (k-3=0.15 sec^{-1}).

Mixing reduced GAD with crotonyl-CoA (anaerobically) gave a biphasic increase in absorbance at 450 nm and an increase followed by a decrease at 560 nm. The fast phase resulted in formation of intermediate I. The apparent rate was directly dependent on the crotonyl-CoA concentration and led to a lower limit estimate for k-2 (220 sec^{-1}). The slower phase resulted in the formation of an oxidized enzyme spectrum, Figure 2. The apparent rate was independent of the crotonyl-CoA concentration and confirmed the value of step k-1 (0.5 sec^{-1}). The large difference in the apparent rates for formation and decay of intermediate I made determination of its spectrum possible, Figure 2. Intermediate I has very low absorbance in the 450 nm region and a pronounced long wavelength absorbance. This spectrum is consistent with the charge-transfer complex interpretation for intermediate I, where reduced GAD is the donor and crotonyl-CoA is the acceptor. Immediately after mixing, the spectrum of GAD is different from that of the starting photoreduced GAD, Figure 2.

The spectral change is consistent with the formation of a Michaelis-Menten complex and is taken as evidence for species MC2. From the overall equilibrium between species I and GADr and the values for k2/k-2, an upper limit estimate for Kd2 (250 μM) was obtained.

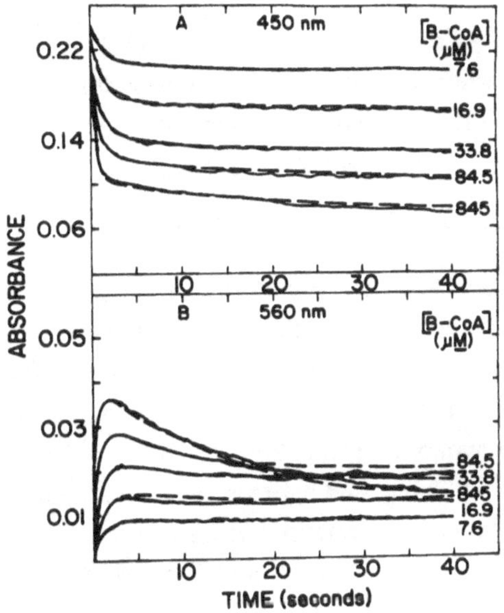

Figure 1

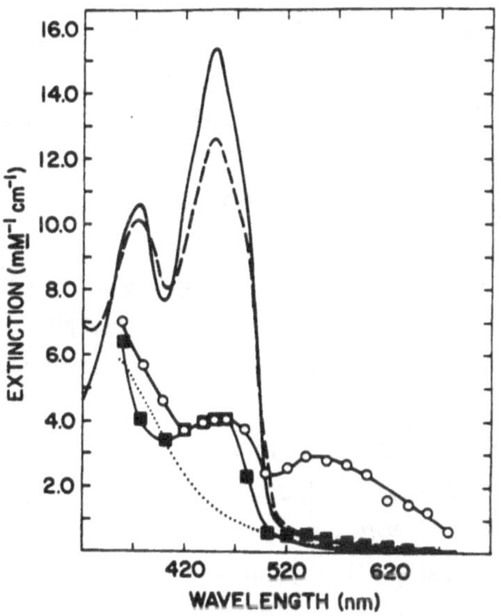

Figure 2

Figure 1 (on the left). The "rapid phases" in the reduction of oxidized GAD by butyryl-CoA. Conditions: oxidized GAD (16.4 μM) was anaerobically mixed with butyryl-CoA (15.2-1690 μM) in 50 mM KPi pH 7.6 at 4°C, using a stopped-flow spectrophotometer (2 cm optical pathlength). The indicated butyryl-CoA concentrations are after mixing. The solid lines are the measured time courses. The dashed lines are simulations based on Scheme I using the parameters described in the text.

Figure 2 (on the right). Spectra of the various species appearing during the oxidation of reduced GAD by crotonyl-CoA. Conditions: reduced GAD (16.2 μM) was anaerobically mixed with crotonyl-CoA (100 μM) in 50 mM KPi pH 7.6 at 4°C, using a stopped-flow spectrophotometer (2 cm optical pathlength). Spectra are: the initial oxidized GAD (solid line); photoreduced GAD (dotted line); the species immediately after mixing (open squares); the long wavelength intermediate (filled circles); and the final reaction mixture (dashed line).

Mixing reduced GAD with butyryl-CoA (anaerobically) gave a small, rapid rise in absorbance around the 450 nm region. The change in absorbance was similar to that seen upon the formation of MC2. This was taken to indicate formation of GADr•B-CoA, a complex originally suggested by the studies of Page and Beinert (1). The apparent rate was directly dependent on butyryl-CoA concentration and led to the description of k3 (3×10^6 M^{-1} sec^{-1}).

Mixing oxidized GAD with crotonyl-CoA, caused a marked red shift in the GAD spectrum, similar to that commonly encountered when ligands bind to oxidized flavoproteins. The apparent rate of formation for the complex was linearly dependent on crotonyl-CoA concentration and led to the characterization of k4/k-4 (20 sec^{-1}/9.8×10^5 M^{-1} sec^{-1}).

Thus we have obtained direct evidence for each enzyme species and kinetic parameter in Scheme I. The mechanism in Scheme I was then used to simulate the rate data. The measured kinetic constants presented above and a measured set of extinction coefficients served as initial estimates in making the simulations. These were adjusted by trial-and-error until a "best-fit" between the simulation and data was obtained, see Figure 1. Using this single set of kinetic and spectral parameters, all of tho measured rate traces were simulated. A close correspondence between the simulated and measured data traces resulted, which makes us confident that Scheme I accurately describes GAD.

Acknowledgement

Supported by U.S. Public Health Service grant GM-11106 (V.M.), GM-26643 (C.T.) and a grant from the Deutsche Forschungsgemeinschaft (Gh 2/44 to S.G.).

References

1. Beinert, H., E. Page. 1956. J. Biol. Chem. 225, 479-497.
2. Hall, C.L., J.D. Lambeth, H. Kamin. 1979. J. Biol. Chem. 254, 2023-2031.
3. Reinsch, J., A. Katz, J. Wean, G. Aprahamian, J.T. McFarland. 1980. J. Biol. Chem. 255, 9093-9097.
4. Pohl, B., T. Raichle, S. Ghisla. 1986. Eur. J. Biochem. 160, 109-115.

CATALYTIC VERSATILITY OF BUTYRYL-CoA DEHYDROGENASE FROM M.ELSDENII

Patricia A. Ellison and Paul C. Engel

Department of Biochemistry, University of Sheffield
Sheffield S10 2TN, England

Oxidase and dehydrogenase activity

Butyryl-CoA dehydrogenase (BCD) from the anaerobic bacterium *Megasphaera elsdenii* (1,2) is unusual amongst the acyl-CoA dehydrogenases in that it displays a considerable degree of oxygen reactivity (3). In our previous studies (4), kinetic parameters for this oxidase activity of BCD were measured using butyryl-CoA and two truncated analogues, S-butyryl-pantetheine and S-butyryl-N-propionyl cysteamine. All of these substrates gave significant oxidase turnover, the best oxidase substrate being the one with the least of the normal CoA structure i.e. S-butyryl-N-propionyl cysteamine.

The dehydrogenase activity of BCD is routinely measured under aerobic conditions with phenazine ethosulphate (PES) and dichlorophenol indophenol (DCPIP) as electron acceptors (5). However, given the oxygen reactivity of BCD, some proportion of the substrate-derived reducing equivalents might be diverted towards the reduction of O_2 rather than DCPIP. We have therefore compared the rates of dye reduction under aerobic and anaerobic conditions.

Aerobic and anaerobic measurements of dehydrogenase activity were made with butyryl-CoA, S-butyryl-pantetheine and S-butyryl-N-propionyl cysteamine. Table 1 summarises the steady-state kinetic parameters for the three substrates. The aerobic results agree with previous measurements (4): butyryl-CoA is the best dehydrogenase substrate with an apparent V of $350 min^{-1}$ and S-butyryl-N-propionyl cysteamine is the poorest substrate with an apparent V of $117\ min^{-1}$.

Flavins and Flavoproteins 1987

The anaerobic measurements of dehydrogenase activity gave more ambiguous results. For the two truncated analogue substrates, the anaerobic V values are, as predicted, significantly higher than those obtained under aerobic conditions. Furthermore, this effect is greatest with the best oxidase substrate; the anaerobic V for S-butyryl-N-propionyl cysteamine is twice the apparent value under aerobic conditions, whereas for S-butyryl-pantetheine,the increase in V on removal of O_2 is only about 30%. The results obtained with butyryl-CoA as substrate, however, are quite unexpected. The dehydrogenase V was not significantly increased under anaerobic conditions whilst the apparent Km is actually lower in the aerobic assays. Thus at sub-saturating substrate concentrations, oxygen increases the rate of dye reduction. We have no satisfactory explanation as yet for this surprising observation.

Table 1 Kinetic parameters for dehydrogenase activity of M.elsdenii BCD.

Substrate	Aerobic		Anaerobic	
	Km (μM)	V(min^{-1})	Km(μM)	V(min^{-1})
Butyryl CoA	17	350	31	350
S-butyryl-pantetheine	384	294	571	400
S-butyryl-N-propionyl cysteamine	2600	117	5200	250

Enoyl-CoA hydratase activity

We have also recently tackled the question of whether M.elsdenii BCD has intrinsic enoyl-CoA hydratase activity. Apparently pure preparations of BCD possess varying degrees of enoyl-CoA hydratase activity (3). As documented in more detail elsewhere (6), we have concluded that this is, at least in part, an intrinsic activity. Purified BCD, completely liganded with CoA persulphide, (7) was stable in 6M urea for several hours. During this incubation the enoyl-CoA hydratase activity of BCD declined to a constant level (fig. 1,) whereas a partially purified preparation of M.elsdenii enoyl-CoA hydratase was completely inactivated by similar urea treatment (data not shown).

These results suggest that the enoyl-CoA hydratase activity of typical preparations of BCD is due to a trace protein contaminant destroyed by urea treatment and an intrinsic activity resistant to urea. Intrinsic enoyl-CoA hydratase activity has now also been demonstrated for pig kidney medium chain acyl-CoA dehydrogenase (9).

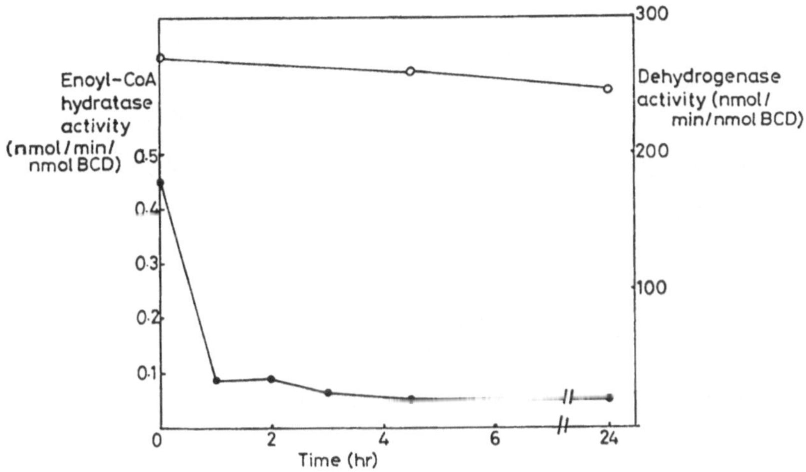

Figure 1. Time course of urea inactivation of BCD-associated enoyl-CoA hydratase activity (●) and dehydrogenase activity (0).

References

1. Baldwin, R.L. and Milligan, L.P. : Biochim.Biophys.Acta. 92, 431-432 (1964).

2. Engel, P.C. and Massey, V. : Biochem.J. 125, 879-887 (1971).

3. Engel, P.C. and Massey, V. : Biochem.J. 125, 889-902 (1971).

4. Ellison, P.A. ., Shaw, L., Williamson, G. and Engel, P.C. (1984) in "Flavins and Flavoproteins" (Bray, R.C., Engel, P.C. and Mayhew, S.G. eds.) Walter de Gruyter and Co. Berlin, New York, pp.413-416.

5. Williamson, G. and Engel P.C. : Biochem.J. 218, 521-529 (1984).

6. Ellison, P.A. and Engel, P.C. : Biochem.Soc.Trans. 14, 158 (1986).

7. Williamson, G. Engel, P.C., Mizzer, J.F., Thorpe, C. and Massey, V. : J.Biol.Chem. 257, 4314-4320 (1982).

8. Stern, J.R. : J.Biol.Chem. 221, 33-44 (1956).

9. Lau, S-M, Powell, P., Buettner, H., Ghisla, S. and Thorpe, C. : Biochem. 25, 4184-4189 (1986).

Acknowledgement

We are indebted to the Science and Engineering Research Council who supported this work through a training award to PAE.

OXIDATION-REDUCTION POTENTIAL OF GLUTARYL-CoA DEHYDROGENASE FROM PARACOCCUS DENITRIFICANS

Colleen M. Byron and Marian T. Stankovich
Department of Chemistry, University of Minnesota, Minneapolis, MN 55455

Mazhar Husain
Molecular Biology Division, Veterans Administration Medical Center, San Francisco, CA 94121

Introduction

Glutaryl-CoA dehydrogenase (GCD) from Paracoccus denitrificans catalyzes the oxidative decarboxylation of glutaryl CoA to crotonyl CoA. The paracoccus GCD shares several characteristics with mammalian acyl CoA dehydrogenases. It is a tetramer of four identical subunits with a subunit molecular weight of 43,000. It contains one FAD per subunit and forms spectrophotometrically detectable complexes with β-ketoacyl CoA and enoyl-CoA derivatives. Glutaryl-CoA dehydrogenase donates reducing equivalents to the electron transfer flavo-protein (ETF). Paracoccus GCD is also able to cross react enzymatically with mammalian ETF and ETF-Q-oxidoreductase (1). The present spectroelectrochemical study of the paracoccus GCD, it is hoped, will aid in the characterization of mammalian GCD which is present only in small amounts in mitochondria.

Results and Discussion

The spectoelectrochemical procedures employed here have been described elsewhere (2,3). Glutaryl-CoA dehydrogenase was purified according to the method of Husain and Steenkamp (1).

Flavins and Flavoproteins 1987

Coulometric reductions of free GCD were performed at pH values of 6.4 and 6.0 at 4°C. The coulometric titration at pH = 6.4 is shown in Figure 1. No blue neutral radical or red anionic radical was detected spectrally. This was also true at pH = 6.0.

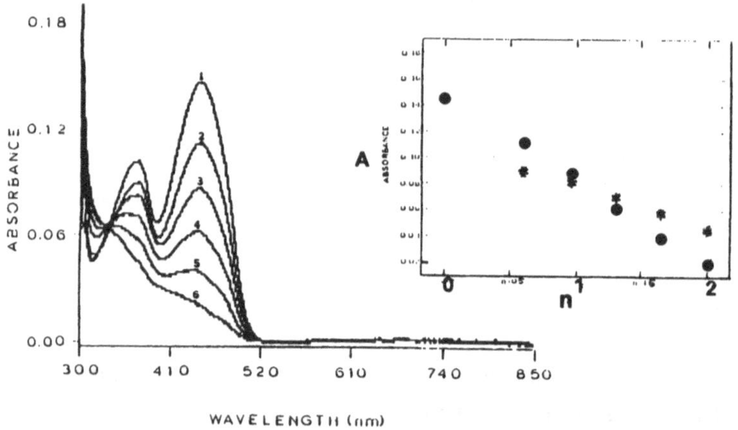

FIGURE 1. Coulometric Titration of Glutaryl-CoA Dehydrogenase. Enzyme concentration was 10.8 μM in in 0.05 M KPi, pH = 6.4, 4°C containing 10% ethylene glycol and 100 μM methyl viologen. Inset shows plot of absorbance vs. n, number of electrons transferred. ● 445 nm, ✱ 370 nm.

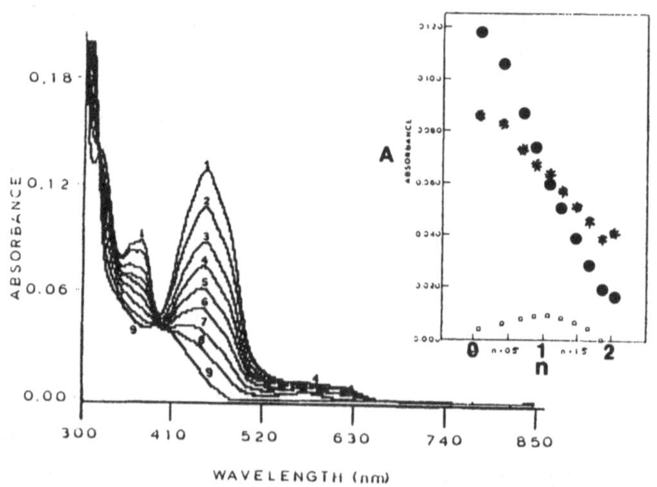

FIGURE 2. Coulometric Titration of Acetoacetyl-CoA Bound Glutaryl-CoA Dehydrogenase. Enzyme concentration was 9.5 μM in 0.05 M KPi, pH = 6.3, 4°C containing 10% ethylene glycol, 100 μM methyl viologen, and 35 μM acetoacetyl CoA. Inset shows plot of absorbance vs. n, number of electrons transferred. ● 445 nm, ✱ 370 nm, □ 590 nm.

The midpoint potential of free GCD at pH = 6.4, 4°C, was determined to be -0.085V vs. SHE using 1 μM 8-chororiboflavin and 1 μM lumiflavin-3-acetate as the redox indicators. A Nernstian slope of 0.027V vs. SHE was obtained. At pH = 6.0, 4°C, the midpoint potential of free GCD shifted positively to -0.064V vs. SHE in the presence of 1 μM 8-chlororiboflavin. Here, a Nernstian slope of 0.026V vs. SHE was obtained. Preliminary evaluation of this data suggests 0.060V/pH unit slope for the enzyme. Futhermore, the value -0.085 V. vs SHE determined here for paracoccus GCD is consistent with its role as the electron donor to ETF (-0.023 V. vs SHE) (4). Attempts to titrate GCD at pH values above 6.4 and below 6.0 were unsuccessful due to severe enzyme precipitation.

The spectra of the above titrations were corrected for turbidity. Scattering spectra were generated for each experiment by subtracting the initial oxidized spectrum from the reoxidized spectrum taken at the conclusion of the experiment.

The result of the addition of the inhibitor acetoacetyl-CoA to GCD in the coulometric titration at pH = 6.3, 4°C, is shown in Figure 2. As shown in the figure, acetoacetyl-CoA forms a charge transfer complex with the oxidized enzyme, and, upon reduction, the formation of blue neutral radical is observed spectrally. Both the charge transfer complex and the blue neutral radical appear to reduce completely by the end of the titration. A preliminary determination of the midpoint potential of acetoacetyl-CoA bound GCD was performed using the indicator dyes 8-chlororiboflavin, anthraquinone-1,5-disulfonate, and lumiflavin-3-acetate. Mid-tritration spectra were observed to overlap with anthraquinone-1,5-disulfonate, indicating a midpoint potential in the range of -0.110V to -0.120V vs SHE. Blue radical formation was detected spectrally, but only approximately 10% that formed in the coulometric titration.

Acknowledgments

This work was supported by grant GM29344 from the National Institutes of Health (GM 29344, HL-16251) and by the Veteran's Administration.

References

1. Husain, M. and Steenkamp, D.J. (1985) J. Bacteriol. 163, 709-715.

2. Stankovich, M.T. (1980) Anal. Biochem. 109, 295-308.

3. Stankovich, M.T. and Fox, B.G. (1983) Biochemistry 22, 4466-4472.

4. Husain, M., Stankovich, M.T. and Fox, B.G. (1984) Biochem. J. 219, 1043-1047

^{13}C-NMR STUDIES ON THE COMPLEXES AND REACTION INTERMEDIATES OF D-AMINO ACID OXIDASE

R. Miura, Y. Miyake

Department of Biochemistry, National Cardiovascular Center Research Institute, Suita, Osaka 565, Japan

Introduction

^{13}C-NMR spectroscopy with ^{13}C-labeled flavin has proved to be a powerful technique in the exploration of flavin-ligand and flavin-protein interactions and hence flavin functionality in flavoproteins and flavoenzymes. The potency of this technique has been demonstrated by accumulation of systematic data reported from several research groups including the present authors'.

D-Amino acid oxidase (DAO) from porcine kidney has been a target of investigation in biochemical, physicochemical and medical aspects, though its true physiological function is still veiled. DAO is known to form complexes with various ligands which induce characteristic absorption spectral changes. Among the known inhibitors, benzoate (BZ), and o-, m- and p-aminobenzoates (OAB, MAB and PAB, respectively) are the most extensively studied (1). Moreover, several reaction intermediates with peculiar spectral changes have been recognized and some of them can be stabilized for such a prolonged period suitable for detailed analysis. The present contribution concerns the flavin-environment as studied by ^{13}C-NMR spectroscopy applied to DAO which is reconstituted with ^{13}C-enriched flavin adenine dinucleotide.

Flavins and Flavoproteins 1987

Results and Discussion

DAO was reconstituted with [2-^{13}C]-, [4a-^{13}C]- and [4,10a-$^{13}C_2$]-labeled FAD according to the procedure described previously (2). The ^{13}C-NMR spectra of the reconstituted DAO in the oxidized state are shown in Fig. 1, which also includes the spectrum of the non-labeled DAO for comparison. The resonance signals corresponding to the enriched carbon atoms are marked with the arrow heads. Table 1 summarizes the chemical shift values of the carbon resonances in the oxidized DAO, DAO-inhibitor complexes and the reaction intermediates in which DAO is in the oxidized state. All the complexes (except DAO-BZ) and intermediates listed in Table 1 and Fig. 2 exhibit charge-transfer absorption bands in the long-wavelength region. Table 1 includes, for comparison, the corresponding values for FMN at infinite dilution (3) with correction for TSP reference (4).

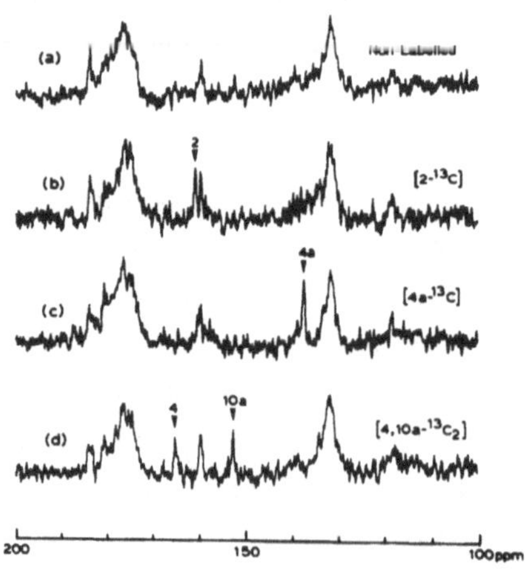

Fig. 1. ^{13}C-NMR spectra of the oxidized DAO reconstituted with [2-^{13}C]-, [4a-^{13}C]- and [4,10a-$^{13}C_2$]-FAD. Chemical shifts are relative to the methyl resonance of external 3-(trimethylsilyl)propionic acid-d_4 (TSP) at pH 8.0.

Table 1. ^{13}C-Chemical Shifts (ppm) of Oxidized DAO in the Ligand-Free Form, Complexes with Inhibitors and Reaction Intermediates

Species	Positions			
	2	4	4a	10a
FAD	161.0	164.8	137.3	153.2
DAO	160.7	165.0	137.3	152.5
DAO-OAB	162.0	163.5	137.4	152.9
DAO-MAB	161.7	162.9	136.9	152.1
DAO-PAB	161.4	163.9	136.4	152.3
DAO-BZ	161.5	163.3	136.9	152.2
DAO-D-BCNA a)	161.6	163.8	137.3	152.4
DAO-D-Pro b)	161.8	164.9	137.3	152.6
FMN c)	161.4	165.3	137.8	153.7

a) Intermediate, assigned as the complex with 2-amino-3-cyanoacrylate, in the reaction with 3-cyano-D-alanine (D-BCNA) (5).
b) Intermediate, assigned as the complex with Δ^1-pyrrolidine-2-carboxylate, in the reaction with D-proline (6).
c) Taken from Moonen et al. (3) with correction for TSP reference (4).

These chemical shift values are illustrated by a stick-bar diagram in Fig. 2. The DAO-aminobenzoate complexes make a good set of spectra for examining the charge-transfer interaction for the following reasons: (1) the inhibitors in common possess amino and carboxylate groups albeit with different orientation, (2) the complexes are characterized by the charge-transfer absorption bands, and (3) the oscillator strength and the rotaional strength decrease in the order of OAB, MAB and PAB (1,7). In the chemical shift variation for this set of complexes (Table 1 and Fig. 2), we found no systematic tendency which would characterize the charge-transfer interaction. However, we found subtle

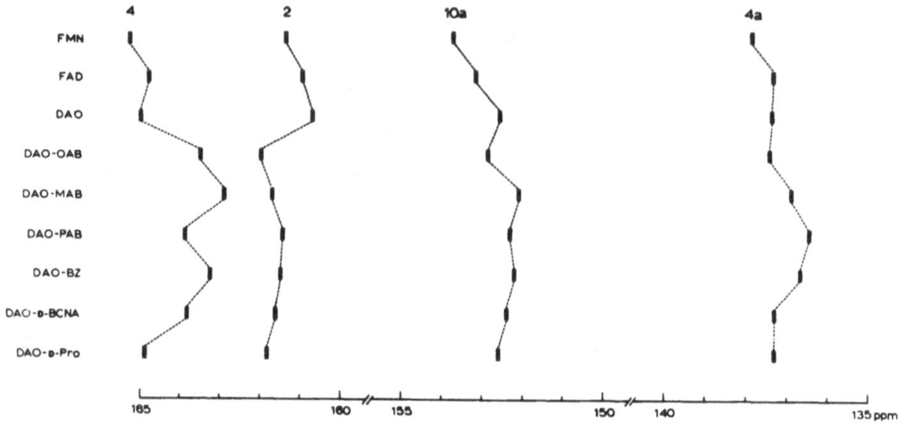

Fig. 2. Stick-bar diagram of the ^{13}C-resonances of the oxidized DAO in Table 1.

changes for the 4-^{13}C and, to a less extent, 2-^{13}C in the complexes and intermediates. We interpret these changes as the modulation of the hydrogen bondings between the protein moiety and the carbonyl oxygens at C(4)=O and C(2)=O of the flavin nucleus in the complexes and intermediates. Taking into account that hydrogen bondings at the isoalloxazine moiety are critical factors for changing flavin reactivity, we postulate that the hydrogen bondings at these positions are delicately tuned thereby regulating the flavin reactivity during the catalytic sequence. The 4a-^{13}C resonance is relatively insensitive to the charge-transfer interaction, as seen in Fig. 2. This casts a marked contrast to old yellow enzyme, in which 4a-^{13}C resonance undergoes remarkable changes upon formation of charge-transfer complexes with phenolic compounds (8,9). This difference may reflect the functional difference, hence the difference in flavoprotein category, between the two enzymes.

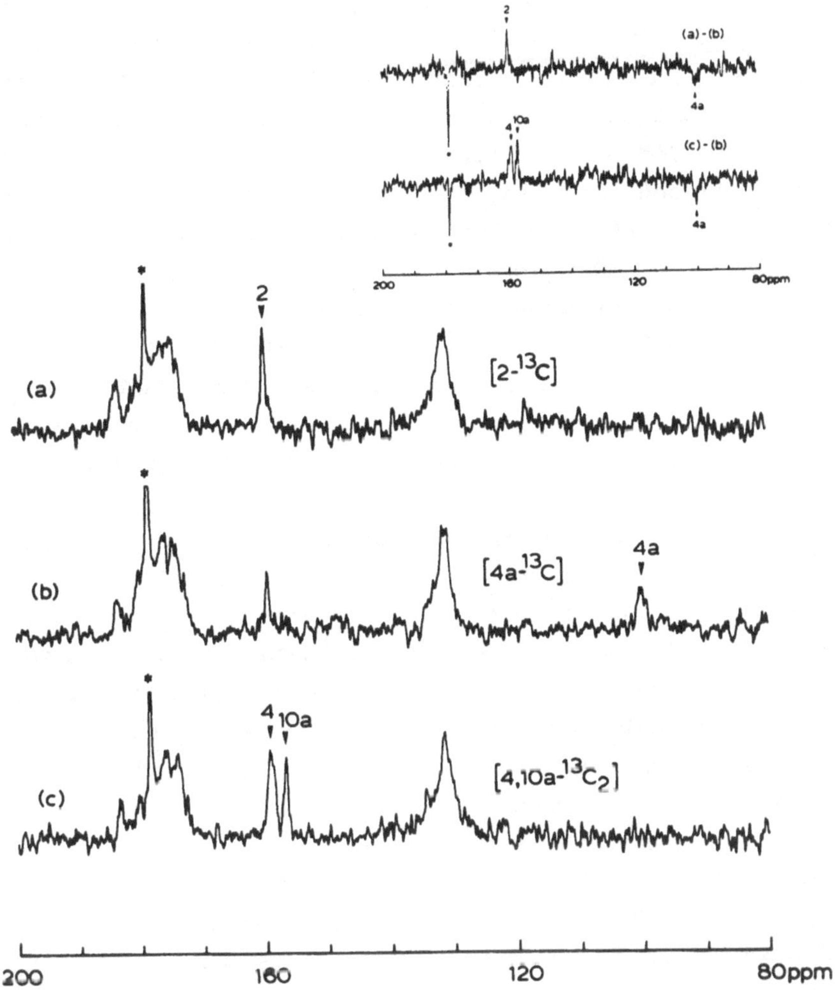

Fig. 3. ^{13}C-NMR spectra of reduced DAO reconstituted with [2-^{13}C]-, [4a-^{13}C]- and [4,10a-$^{13}C_2$]-FAD (a, b and c, respectively). Inset shows the difference spectra, a-b and c-b. The signal with an asterisk is derived from excess D-alanine used for reduction.

Figure 3 presents the ^{13}C-NMR spectra of DAO reconstituted with ^{13}C-FAD. The 2- and 4-^{13}C resonances were obscured by the natural abundance contribution from the protein moiety. But they became obvious in the difference spectra shwon in the inset. Comparison of these resonance positions (Table 2) with reduced FMN in the neutral ($FMNH_2$) and anionic ($FMNH^-$) states (10) indicates explicitly that FAD in the reduced DAO is in the anionic reduced form ($FADH^-$).

The ^{13}C-NMR spectra were measured with the purple intermediates with D-alanine and D-proline. The conditions for the purple intermediates were as described previously (2,12,13). The purple intermediate is a catalytically competent intermediate produced transiently during the enzyme reaction (11). It has been shown that this intermediate is a charge-transfer complex of the reduced FAD and an imino acid produced in the reductive half reaction (2,12,13). The chemical shift values for these intermediates are given in Table 2 and illustrated in the stick-bar diagram (Fig. 4). Comparison of the ^{13}C-resonances in the purple intermediates with those in $FMNH_2$ and $FMNH^-$ demonstrates unequivocally that FAD in the purple intermediates is in the reduced anionic form ($FADH^-$). The remarkable feature in the chemical shifts for the purple intermediates is the considerable upfield-shift of the 4a-^{13}C resonance relative to free reduced form. This reflects the elevated nucleophilicity at the 4a-carbon in the purple intermediate in comparison with the reduced state. This leads to a model in which molecular oxygen reacts at the C-4a in the reoxidation of the reduced flavin (oxidative half reaction). Our results, therefore, substantiate the reaction scheme on the molecular basis that the oxidation rate of the purple intermediate by O_2 is faster than that of free reduced state (14) and that the oxidative half reaction proceeds via the purple intermediate prior to product release (14,15).

Table 2. ^{13}C-Chemical Shifts of Reduced DAO and Purple Intermediates with D-Alanine and D-Proline

Species	Positions			
	2	4	4a	10a
DAO (reduced)	159.8	158.9	100.1	156.9
Purple (D-Ala)	160.0	160.6	97.7	155.9
Purple (D-Pro)	160.3	159.8	98.2	156.2
$FMNH^-$ a)	159.8	159.3	103.0	157.1
$FMNH_2$ a)	152.7	159.9	104.4	145.6

a) Taken from Vervoort et al. (10) with correction for TSP reference (4).

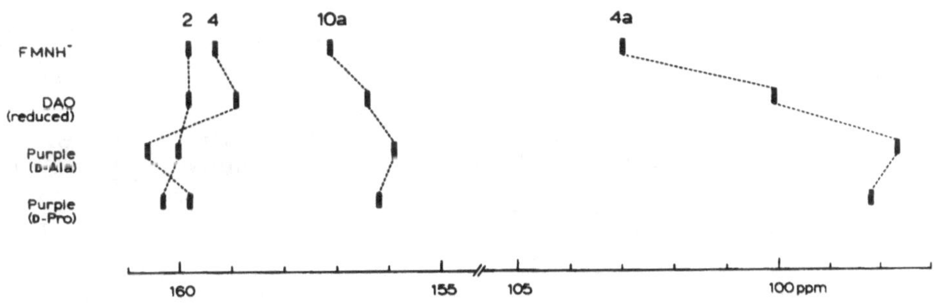

Fig. 4. Stick-bar diagram of the ^{13}C-resonances of DAO in the reduced form and purple intermediates with D-alanine and D-proline. Conditions for purple intermediates are as described previously (2,12,13).

References

1. Massey, V., H. Ganther. 1965. Biochemistry 4, 1161.

2. Miura, R., Y. Nishina, M. Ohta, H. Tojo, K. Shiga, H. Watari, T. Yamano, Y. Miyake. 1983. Biochem.Biophys.Res. Commun. 111, 588.

3. Moonen, C.T.W., J. Vervoort, F. Müller. 1984. Biochemsitry 23, 4859.

4. Miura, R., Y. Miyake. 1987. J.Biochem. 101, 581.

5. Miura, R., K. Shiga, Y. Miyake, H. Watari, T. Yamano. 1980. J.Biochem. 87, 1469.

6. Nishina, Y., R. Miura, H. Tojo, Y. Miyake, H. Watari, K. Shiga. 1986. J.Biochem. 99, 329.

7. Shiga, K., K. Horiike, A. Isomoto, T. Yamano. 1976. J.Biochem. 80, 1101.

8. Beinert, W.-D., H. Rüterjans, F. Müller, A. Bacher. 1985. Eur.J.Biochem. 152, 581.

9. Miura, R., T. Yamano, Y. Miyake. 1986. J.Biochem. 99, 907.

10. Vervoort, J., F. Müller, D.J. O'Kane, J. Lee, A. Bacher. 1986. Biochemistry 25, 8067.

11. Massey, V., Q.H. Gibson. 1964. Fed.Proc. 23, 18.

12. Nishina, Y., K. Shiga, H. Watari, R. Miura, Y. Miyake, H. Tojo, T. Yamano. 1982. Biochem.Biophys.Res.Commun. 106, 818.

13. Nishina, Y., K. Shiga, R. Miura, H. Tojo, M. Ohta, Y. Miyake, T. Yamano, H. Watari. 1983. J.Biochem. 94, 1979.

14. Porter, D.J.T., J.G. Voet, H.J. Bright. 1977. J.Biol.Chem. 252, 4464.

15. Koster, J.F., C. Veeger. 1968. Biochim.Biophys.Acta 151, 11.

^{13}C-, ^{15}N- AND ^{31}P-NMR STUDIES ON 6-HYDROXY-L-NICOTINE OXIDASE FROM ARTHROBACTER OXIDANS

Stefan Pust*, Jacques Vervoort°, Karl Decker*, Adelbert Bacher+ and Franz Müller°

*Biochemisches Institut der Albert-Ludwigs-Universität, Freiburg, FRG,
°Department of Biochemistry, Agricultural University, Wageningen, The Netherlands,
+Department of Organic Chemistry and Biochemistry, Technical University of Munich, Garching, FRG

Introduction

The interaction between the apoprotein of 6-hydroxy-L-nicotine oxidase (6-HLNO) from Arthrobacter oxidans and the prosthetic group, FAD, has been investigated by ^{13}C-, ^{15}N- and ^{31}P-NMR techniques. Selectively enriched ^{13}C- and ^{15}N-labeled riboflavin was added to the growth medium of a riboflavin-requiring mutant strain of A. oxidans. 6-HLNO was isolated essentially as described (1).

Results and Discussion

In the oxidized state the ^{13}C chemical shifts of the atoms in the xylene ring of the protein-bound flavin differ more from those of the free flavin in water (i.e. FMN) than the other atoms in the isoalloxazine ring (Table 1). Especially the resonances of the C(7) and of the C(8) atoms are shifted upfield by 3.7 and 3.4 ppm as compared to FMN. These chemical shifts indicate that (at least) the xylene ring is in a hydrophobic environment in the enzyme. The polarization of the protein-bound flavin can be ascribed to hydrogen bonds between the apoprotein and the two carbonyl groups. Although the hydrogen bond at the C(2) carbonyl group is comparable in strength to free flavin in water, this polarization does not lead to a downfield shift of the resonance at the C(8). This observation supports the idea of an apolar environment for the xylene ring (2). The downfield shift of the C(4) carbonyl group in the protein-bound flavin as compared to FMN indicates that

Flavins and Flavoproteins 1987

Table 1. ^{13}C and ^{15}N chemical shifts of various selectively enriched free flavins ((FMN and TARF) and FAD bound to 6-HLNO. ^{13}C chemical shifts are relative to TMS. ^{15}N chemical shifts are relative to liquid NH_3.

OXIDIZED STATE

Atom	6-HLNO	6-HLNO + inhibitor	FMN	TARF
C(2)	159.9	160.0	159.8	155.2
C(4)	164.9	164.8	163.7	159.8
C(4a)	136.9	138.3	136.2	135.6
C(5a)	136.9	136.3	136.4	134.6
C(6)	131.6	133.0	131.8	132.8
C(7)	137.7	137.4	140.4	136.6
C(7a)	20.7	20.8	19.9	19.4
C(8)	148.3	148.1	151.7	147.5
C(8a)	21.6	21.5	23.2	21.4
C(9)	120.3	120.0	118.3	115.5
C(9a)	135.5	135.2	133.5	131.2
C(10a)	151.5	151.7	152.1	149.1
N(1)	192.2	193.8	190.8	200.1
N(3)	160.6	160.1	160.5	159.6
N(5)	335.3	330.0	334.7	346.0
N(10)	162.2	161.7	163.5	151.9

REDUCED STATE

Atom	6-HLNO + dithionite	6-HLNO + substrate	FMNH-	$FMNH_2$	$TARFH_2$
C(2)	158.7	158.2	158.2	151.1	150.6
C(4)	160.0	157.1	157.7	158.3	157.0
C(4a)	98.1	99.3	101.4	102.8	105.2
C(5a)	135.2	134.8	134.2	134.4	136.0
C(6)	120.8	117.8	117.3	117.1	116.1
C(7)	134.5	131.8	133.0	134.3	133.6
C(7a)	19.9	20.5 (19.1 ?)	19.0	19.0	18.9
C(8)	129.7	128.5	130.3	130.4	129.0
C(8a)	21.4	21.4	19.4	19.2	18.9
C(9)	117.9	116.9	116.8	117.4	118.0
C(9a)	130.1	130.6	130.9	130.4	128.2
C(10a)	160.0	156.1	155.5	144.0	137.1
N(1)	192.7	192.5	181.3	128.0	121.1
N(3)	149.2	147.7	150.0	149.7	150.3
N(5)	52.7	60.5	58.4	58.0	59.8
N(10)	100.6	102.6	96.5	87.2	78.0

this carbonyl group is polarized to an even greater extent than in free flavin. The polarization of the two carbonyl groups suggests that the N(10) atom should be in the plane of the ring system and have a chemical shift comparable to FMN (2,3). This is indeed observed (Table 1). The chemical shift of the N(5) atom is comparable to the one found in FMN and is most probably due to hydrogen bond formation with bound water. The observation that the N(3) resonance is split up into a doublet indicates that the proton at the N(3) group is tightly bound and does not exchange with bulk water at the NMR timescale ($\nu << 100$ Hz.). The binding of 6-hydroxy-D-nicotine, a competitive inhibitor of the enzyme, strongly influences the resonances of the C(4a) and the N(5) atoms. The upfield shift of the N(5) resonance suggests the presence of a strong hydrogen bond, probably between the inhibitor and the N(5) atom.

On reduction of the protein almost every resonance is shifted upfield. In the presence of substrate C(4) and C(10a) are more strongly shifted upfield than with dithionite (Table 1). In the dithionite-reduced enzyme the chemical shift of N(5) indicates a considerable sp^3 character (Fig.1). On reduction with the substrate the resonance of the N(5) atom shifts strongly downfield, indicating a large increase in sp^2 character (2). In both the dithionite- and the substrate-reduced enzyme, the N(10) atom is sp^2 hybridized. Due to the bending of the N(5) atom in the dithionite-reduced enzyme and the sp^2 hybridized N(10) atom, electron density from the N(10) atom is reallocated at the C(4) carbonyl group giving rise to the large downfield shift for the C(4) atom, a large upfield shift for the C(4a) resonance and a large downfield shift for C(10a) as compared to free FMNH- (2). In the substrate-reduced enzyme part of the electron density from the N(10) atom is reallocated at C(7), C(9) and C(5a) (2,4). Due to the increased sp^2 character of the N(5) atom in the substrate-reduced enzyme, part of its electron density is reallocated at C(8) and C(6) (2,4).

From the chemical shift of the N(1) atom it can be unambigously concluded that the isoalloxazine ring is anionic in the reduced state. The strong downfield shift of the N(1) atom as compared to FMNH- indicates the presence of a negatively charged amino acid or helix dipole. Bulk water is excluded from the active center in both the dithionite- and

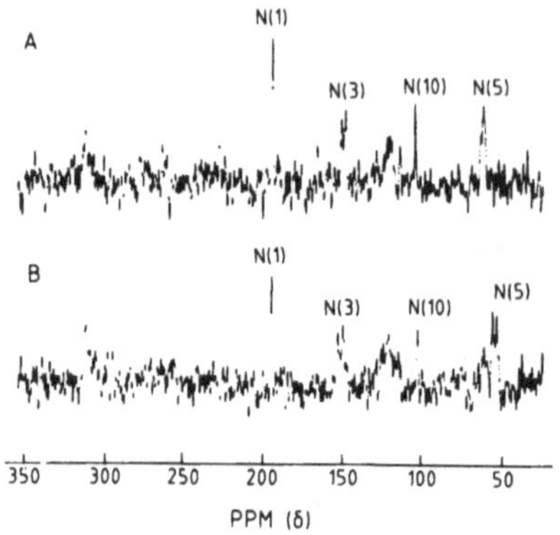

Fig. 1: ^{15}N-NMR spectra of 6-HLNO reduced with substrate (A) or dithionite (B)

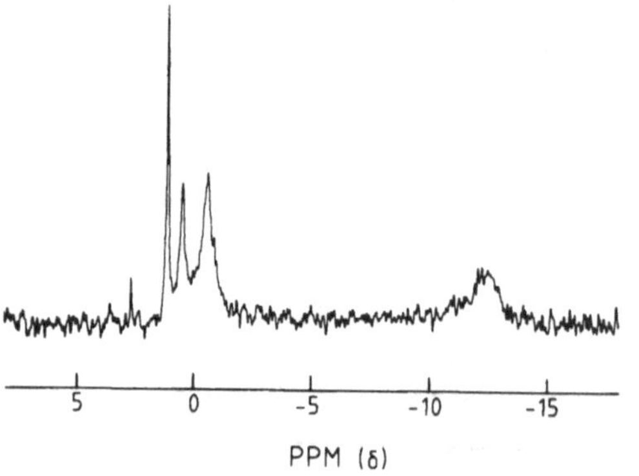

Fig. 2: ^{31}P-NMR spectrum of 6-HLNO in the oxidized state

substrate-reduced enzyme, as can be concluded from the doublet character of the N(5) and N(3) resonances. The resonances of the pyrophosphate of the bound FAD deviate slightly from free FAD. (Fig. 2). In contrast to other FAD-containing proteins the ^{31}P resonances of the pyrophosphate group are not resolved (5). The 4 resonances around 0 ppm belong to tightly (possibly covalently) bound phosphorus residues. The relative intensity ratio of these four resonances can differ, depending on the preparation, but the intensity ratio compared to the pyrophosphate resonances of the bound FAD show little difference.

References

1. Hinkkanen, A., Lilius, E.-M., Nowack, J., Maas, R., and Decker, K. (1983) Hoppe-Seyler's Z. Physiol. Chem. 364, 801-806.
2. Moonen, C.T.W., Vervoort, J., and Müller, F. (1984) Biochemistry 23, 4859-4868.
3. Vervoort, J., Müller, F., O'Kane, D.J., Lee, J., and Bacher, A. (1986) Biochemistry 25, 8067-8075.
4. Vervoort, J., Müller, F., Mayhew, S.G., Van den Bergh, W.A.M., Moonen, C.T.W., and Bacher, A. (1986) Biochemistry 25, 6789-6799.
5. Davis, M.D., Edmondson, D.E., and Müller, F. (1984) Eur. J. Biochem. 145, 237-243.

A STUDY ON THE INTERACTION BETWEEN REDUCED D-AMINO ACID OXIDASE AND LIGANDS

Yasuzo Nishina, Hidetaka Ushijima, Hiromasa Tojo,* and Kiyoshi Shiga

Department of Physiology, Kumamoto University Medical School, Honjo, Kumamoto, Kumamoto 860 and *Department of Biochemistry, Osaka University Medical School, Kita-ku, Osaka 530, Japan

Introduction

It has been suggested by the resonance Raman study that a catalytically important intermediate of D-amino acid oxidase (DAO), so called "purple intermediate", is a complex of reduced DAO with an imino acid derived from a substrate, D-amino acid (1). So, it is important to study the interaction between reduced DAO and ligands to elucidate the reaction mechanism of this enzyme. However, few such studies on reduced DAO have been reported, because ligands which are stable and have enough high affinity for reduced DAO are hardly known. Recently, we reported that pyridine carboxylates such as picolinate and nicotinate were the good ligands for reduced DAO (2). The purpose of this study is to report on the interaction between reduced DAO and these ligands by observing the pH dependencies of the equilibrium constants and rate constants (binding and dissociation rate constants).

The results strongly suggest that only the protonated cationic form at the nitrogen atom of picolinate or nicotinate can bind to reduced DAO, but the neutral form at the nitrogen atom can not bind.

Materials and Methods

Hog kidney D-amino acid oxidase (EC 1.4.3.3)(DAO) was prepared according to the method previously described (3). The dissociation constants between reduced DAO and picolinate or nicotinate were obtained at various pH's by the method of the spectroscopic titration (2). In the experiment, reduced DAO was prepared by reducing oxidized DAO in the presence of the substrate such as alanine or arginine. The values of the dissociation constants between reduced DAO and picolinate or nicotinate did not depend on the species of the reductant used. Also, the affinity of alanine or pyruvate for reduced DAO which was prepared by the photo-irradiation was very weak. So, the coexistence of the substrates (or products) did not hinder the interaction between reduced DAO and picolinate or nicotinate.

The binding and dissociation rate constants between reduced DAO and picolinate or nicotinate were obtained by a stopped-flow method at various pH's. Equal volumes of DAO reduced by a substrate and picolinate or nicotinate

Flavins and Flavoproteins 1987

solution which was maintained anaerobically by the glucose oxidase system (4) were mixed and the time course of the absorbance change was monitored by a stopped-flow spectrometer.

Results and Discussion

Accompaning the complex formation of reduced DAO with picolinate or nicotinate, the absorption spectral changes appear (2). Using these spectral changes, the titration curves of reduced DAO with picolinate or nicotinate were obtained at various pH's (pH 6.3-8.3) and a typical example was shown in Fig. 1. The analyses of these data by the Stockell (the typical result is shown in Fig. 2) and Scatchard plots (data are not shown) suggest that the binding of picolinate or nicotinate has a stoichiometry of 1:1 with respect to DAO.

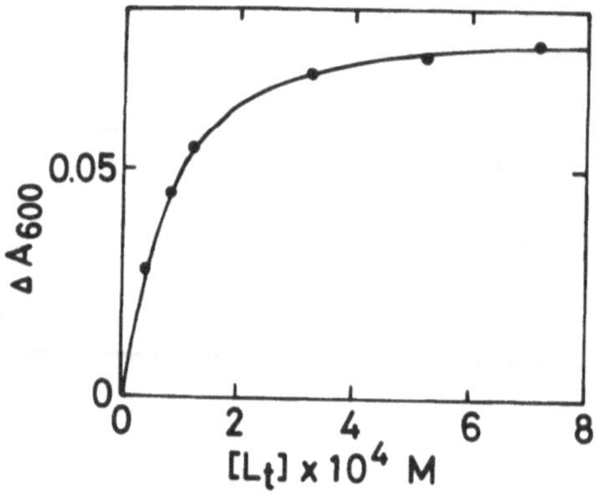

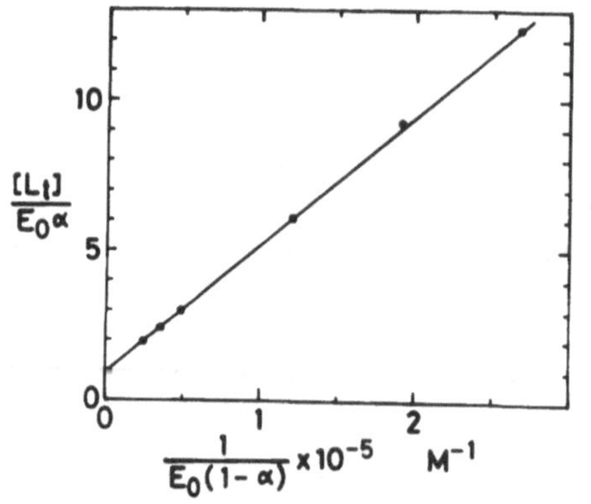

Fig. 1 (left) Titration curve of reduced D-amino acid oxidase with picolinate. The abscissa represents the total concentration of picolinate. The ordinate is the absorbance change at 600 nm. The concentration of reduced DAO is 61.5 μM and 16.4 mM D-arginine is used to reduce DAO in 0.1 M sodium phosphate buffer at pH 7.0, 25 ℃.

Fig. 2 (right) Stockell plot of data in Fig. 1
L_t represents the total concentration of picolinate and E_0 is the concentration of reduced DAO. α is the fractional saturation of reduced DAO with picolinate.

The values of the apparent dissociation constants Kd's were calculated from these plots at various pH's and are shown in Fig. 3 as the relationship between -log Kd and pH. As the slope of these plots is nearly minus one, only an ionizable group is associated with the reaction between reduced DAO and pyridine carboxylates in this pH region.
Previous Raman study suggested that the protonated cationic form at the

nitrogen atom of picolinate can bind to reduced DAO (2). So, the ionizable group mentioned above is mostlikely the proton at the nitrogen atom of picolinate or nicotinate and therefore, it is strongly suggested from Fig. 3 that the neutral form at the nitrogen atom of picolinate or nicotinate can not bind to reduced DAO, that is, only the protonated cationic form at the nitrogen atom of picolinate or nicotinate can bind. The time courses of the absorbance changes during the reaction of reduced DAO with picolinate or nicotinate were measured at various pH's (data are not shown).

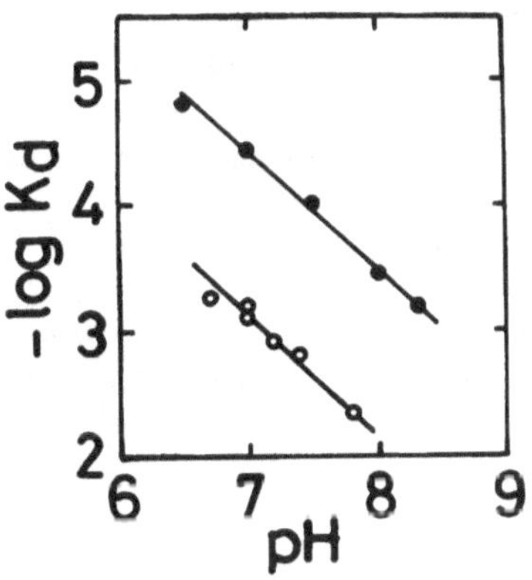

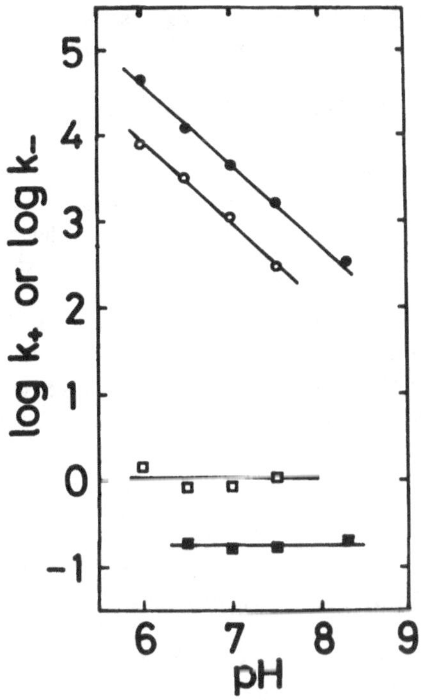

Fig. 3 (left) pH dependence of the apparent dissociation constant (Kd, M) at 25 ℃.
●, picolinate; ○, nicotinate

Fig. 4 (right) pH dependence of the apparent binding rate constant and the apparent dissociation rate constant at 25 ℃.
For the apparent binding rate constant (k_+, 1/(M sec)); ●, picolinate; ○, nicotinate
For the apparent dissociation rate constant (k_-, 1/sec); ■, picolinate; □, nicotinate

In the experiment, the concentration of picolinate or nicotinate used was always much higher than that of reduced DAO. So, we could determine the apparent binding rate constant (k_+) and apparent dissociation rate constant (k_-) at various pH's as the pseudo first order-first order reaction (in the case of picolinate, the values of k_+'s were obtained by the stopped flow experiment and those of k_-'s were calculated by Kd k_+). The results are shown in Fig.4 as the relationship between log k_+ or log k_- and pH. Again, the slope for the binding rate constant was nearly minus one. This was also consistent with that only the protonated cationic form at the nitrogen atom of picolinate or nicotinate can bind to reduced DAO, but the neutral form can not. On the other hand, the values of k_-'s did not depend on the value of pH, suggesting that the complex of reduced DAO and picolinate or nicotinate did not have any ionizable group in this pH region. This is consistent with that picolinate or nicotinate in the complex with reduced DAO exists only as the protonated cationic form.

As a whole, only the protonated cationic form at the nitrogen atom of picolinate or nicotinate can bind to reduced DAO. These results are different from the fact that even if picolinate is not protonated at the nitrogen atom, it can tightly bind to oxidized DAO (5). In the case of reduced DAO, the electrostatic interaction between the positively charged ligand at the nitrogen atom and the negatively charged flavin probably plays an important role.

Further, as picolinate or nicotinate can also bind to the oxidized (about picolinate, see the reference (5)) or semiquinoid form of DAO (data are not shown), the ligand may be a valuable tool to study the difference of the physicochemical properties among the oxidized, semiquinoid, and reduced forms of DAO.

References

1. Nishina, Y., K. Shiga, H. Watari, R. Miura, Y. Miyake, H. Tojo, T. Yamano. 1982. Biochem. Biophys. Res. Commun. 106, 818.

2. Nishina, Y., H. Tojo, K. Shiga. 1986. J. Biochem. 99, 673.

3. Shiga, K., Y. Nishina, K. Horiike, H. Tojo, T. Yamano, H. Watari. 1982. In: Flavins and Flavoproteins (V. Massey and C.H.,Jr. Williams, eds.) Elsevier North Holland, New York, p. 188.

4. Porter, D.J.T., J.G. Voet, H.J. Bright. 1973. J. Biol. Chem. 248, 4400.

5. Fitzpatrick, P.F., V. Massey. 1982. J. Biol. Chem. 257, 9958.

D-AMINO ACID OXIDASE FROM THE YEAST *Rhodotorula gracilis*

M. Pilone Simonetta, P. Casalin, L. Pollegioni, M.A. Vanoni and B. Curti.

Department of General Physiology and Biochemistry, University of Milano, 26 via Celcria, 20133 Milano, Italy.

Introduction

Kinetic and structural properties of D-amino acid oxidase (EC 1.4.3.3, DAAO) have been extensively studied using pig kidney as source of the enzyme (1). In the last years many efforts were devoted to the search of alternative sources of this enzyme, particularly among microorganisms. To date, the attempts to obtain DAAO from microorganisms in a homogeneous form with typical spectral properties of a flavoenzyme have been largely unsuccessful due to the low level of the enzyme present in the cells and to the instability of the protein. Our group has previously shown that synthesis of DAAO can be selectively induced in the cells of the oleaginous yeast *Rhodotorula gracilis* by the addition of D- or D,L-amino acids to the growth medium (2). In the present study we report the purification of DAAO from *Rhodotorula* cells and a preliminary characterization of this flavoprotein.

Results and discussion

D-amino acid oxidase from *Rhodotorula* has been purified ≃950-fold in comparison to the crude extract. The procedure gives a stable enzyme preparation with a specific activity of 58 U/mg prot at 37°C with D,L-alanine as substrate (Table 1) (3). DAAO from yeast is a FAD-containing oxidase with a well resolved visible spectrum and a E_{274}/E_{455} ratio of about 11 (Fig.1); a fluorescence emission maximum at 530 nm has also been detected.

Flavins and Flavoproteins 1987

Table 1. Purification of D-amino acid oxidase from Rhodotorula gracilis

Step	Total activity	Protein	Specific activity	Purification	Yield
	Units	mg	U/mg	Fold	%
1. Crude extract	1037	16929	0.061	1	100
2. 30-60% ammonium sulfate precipitate	1101	4042	0.272	4	106
3. DEAE-Sephacel chrom.	800	224	3.6	59	77
4. Ultrogel AcA34 chrom.	570	57	10.0	164	55
5. Phenyl-Sepharose CL-4B chromatography	303	6	50.5	828	29
6. Superose 12 chromatography	162	2.8	57.8	948	16

The activity was measured with D,L-alanine as substrate. Starting material: 320g of frozen cell paste.

Final DAAO preparations showed in disc-PAGE two major protein bands, one of which stains for activity. The same enzyme preparation migrated in SDS-PAGE as a major band which represented more than 90% of total protein; the molecular weight was determined to be 37±1.6 kDa. Molecular weight determination of purified enzyme on Superose 12 by FPLC gave a single symmetric peak of 79±3.5 kDa, thus suggesting that the

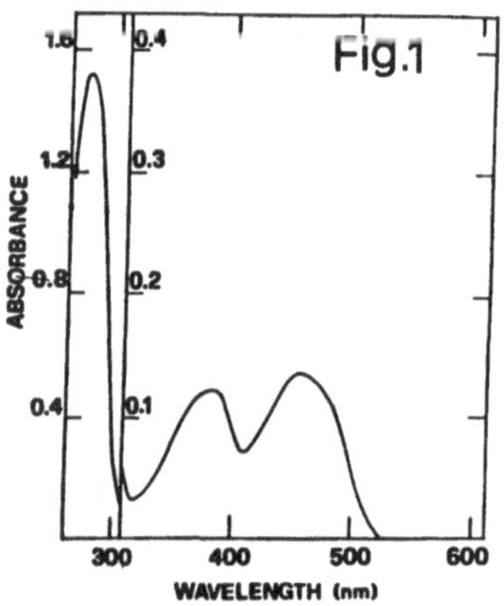

Fig. 1. Absorbance spectrum of a 0.94 mg/ml solution of FPLC Superose 12 chromatographed yeast DAAO in 20 mM K phosphate buffer, pH 7.4, containing 0.25 M NaCl, 10% glycerol and 0.3 mM EDTA.

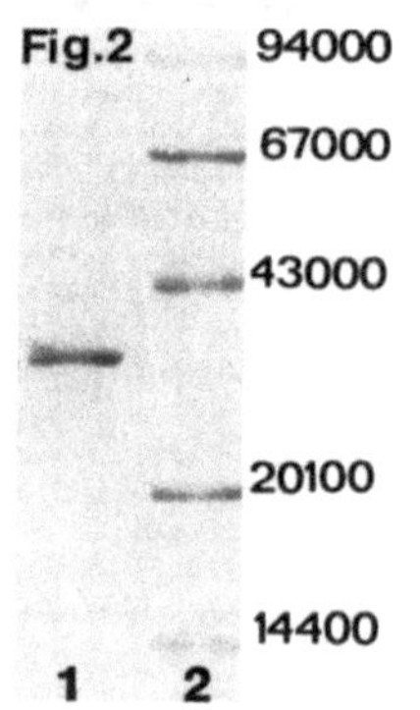

Fig.2. SDS-PAGE of DAAO after the last step of purification. Samples were incubated at 95°C for 12 min in 2% SDS and 2.5 mM 2-ME. 1)Superose 12 chromatographed enzyme 2)marker proteins (from the top): phosphorylase B, ovalbumin, soybean trypsin inhibitor, lactoalbumin (molecular weight in figure).

native enzyme is a dimer of identical subunits. On the basis of spectral data, subunit molecular weight and protein content determination a ratio of 0.5 mol FAD/mol monomer can be calculated. The present purification procedure showed some variability in the homogeneity of the final DAAO preparations, as reflected by the E_{274}/E_{455} ratio and the electrophoretic patterns. However, the second inactive protein band was always observed in disc-PAGE: this point is under current investigation. In all cases the kinetic properties and the reactivity of the enzyme were identical. DAAO from yeast used molecular O_2 as electron acceptor, producing H_2O_2, and it was strictly specific for D-amino acids. Highest activities were obtained with D-methionine, D-phenylalanine, D-alanine, D-tryptophane and D-cysteine; D,L-thiazolidine-2-carboxylic acid was instead a poor substrate. Benzoate inhibited only slightly the enzyme (K_i=1.04 mM). Addition of excess FAD to the assay mixture did not increase the activity, pointing to a tight binding of FAD to the protein. The enzyme-bound FAD was fully reduced in the presence of excess D-alanine under anaerobic conditions. Yeast DAAO showed the thermodynamic stabilization of the red anion radical, when photoreduced in the presence of 5-deazariboflavin (Fig.3). Moreover the enzyme exhibited a high reactivity with sulfite. On the basis of production of

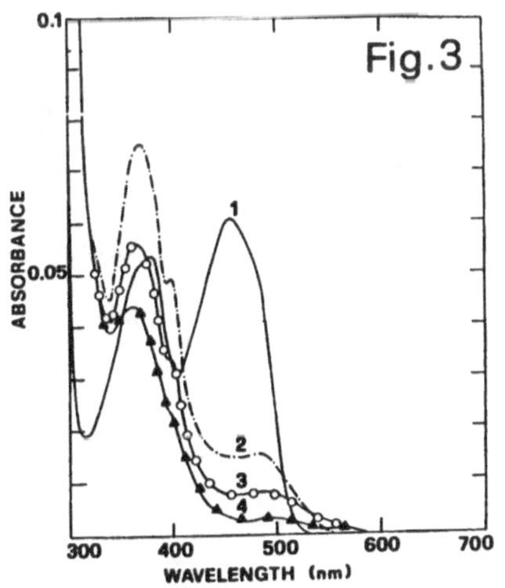

Fig.3. Photoreduction of DAAO in presence of EDTA and 5-deazariboflavin. Conditions: DAAO 11.92µM; 5-deazariboflavin 5.1µM; EDTA 6.7 mM; glycerol 10%; K phosphate 9.7 mM, pH 7.5, 14°C. Different periods of illumination were given using a 250 W lamp.
1) native enzyme;
2) after 7 min irradiation;
3) after 10 min irradiation;
4) after 20 min irradiation.

H_2O_2, the stabilization of the red radical and the formation of sulfite adduct, the yeast enzyme appears to be a true "flavo-oxidase" (4). More recently, a remarkable improvement in the isolation and purification of the enzyme has been obtained specially as concerns the reproducibility and the yield of the procedure. In particular, the use of a controlled growth medium containing D-alanine gave a 10-fold increase in the enzyme production by the Rhodotorula cells (5). The cross-reactivity between yeast DAAO and DAAO from pig kidney has been also examined using anti-yeast DAAO antibodies; both the results from inhibition experiments and from blot-transfer analyses point to an existing limited homology between the two purified enzymes.

The data insofar obtained concerning kinetic, structural and immunochemical properties of DAAO from Rhodotorula tend to indicate same interesting peculiarities of this enzyme in comparison to the DAAO from pig kidney.

Finally, the purification of DAAO from yeast will offer the possibility of novel studies on the characterization and regulation of this flavoprotein, besides the possible applications of this enzyme in analytical and industrial work.

Acknowledgements

This work was supported by a grant from Ministero della Pubblica Istruzione, Roma, Italy.

References

1. Bright, H.J. and Porter, D.J.T..1975. In: The Enzymes, vol.12 part B, Academic Press, New York. pp.445-456.

2. Pilone Simonetta, M., Vanoni, M.A. and Curti, B.. 1982. FEMS Microbiol. Letters 15, 27-31.

3. Pilone Simonetta, M., Vanoni, M.A. and Casalin, P.. Biochim. Biophys. Acta: in press.

4. Massey, V. and Hemmerich, P.. 1980. Biochem. Soc. Trans.. 8, 246-256.

5. Pilone Simonetta, M.: unpublished results.

PURIFICATION AND PROPERTIES OF D-ASPARTATE OXIDASE FROM BEEF KIDNEY

Armando Negri, Charles H. Williams, Jr. and Vincent Massey

Department of Biological Chemistry, University of Michigan and the Veterans Administration Medical Center, Ann Arbor, Michigan, 48109

D-aspartate oxidase (E.C. 1.4.3.1) catalyzes the oxidative deamination of dicarboxylic D-amino acids to the corresponding alpha-keto acids.
This reaction is formally equal to the reaction catalyzed by D-amino acid oxidase, which, however, has different substrate and inhibitor specificity (1, 2 and references therein).
D-aspartate oxidase was purified to homogeneity following a six step procedure including homogenization, heat treatment at pH 5.2, hydrophobic chromatography on Phenyl-Sepharose CL-4B, gel filtration chromatography on Sephacryl S 200 and ion-exchange chromatography on FPLC Mono Q and Mono S columns. This procedure led to the isolation of 7 mg of protein from 1 Kg of kidney cortex, with an overall yield of 25% and a purification factor of about 3200 fold (3). The turnover number, determined by monitoring the O_2 consumption in 50 mM KPi, 0.3 mM EDTA, pH 7.4, 25°C, was equal to 42 sec^{-1} at infinite concentrations of D-aspartic acid and O_2. The protein is a monomer of Mr = 39,000 as indicated by molecular weight determination in native and denaturing conditions. The enzyme as isolated is a mixture of a catalytically active form containing FAD and an inactive one containing 6-OH-FAD, accounting between 9 and 19% of total flavin in five different preparations. The visible absorption spectrum of the enzyme isolated from two different preparations is shown in Fig. 1. The ratio between the absorbance at 273 nm and at the maximum of the 450 nm peak varied between 8.5 and 8.7.

<u>Properties of 6-OH-FAD-D-aspartate oxidase</u>. The presence of 6-OH-FAD, suggested by the characteristic long-wavelength absorption band in the spectrum of the enzyme, was confirmed both by HPLC flavin analysis - following release of the

Flavins and Flavoproteins 1987

chromophores by heat denaturation of the protein - and by selective removal of FAD by dialysis against 3 M KBr in 0.1 M KPi, pH 7.0 (3). The latter procedure led to the material shown in Fig. 2, which also shows the spectrum of fully reconstituted 6-OH-FAD-D-aspartate oxidase obtained by incubating the previous material with excess 6-OH-FAD. The presence of variable amounts of this modified flavin in purified enzymes has been reported previously only for pig liver glycolate oxidase and Electron Transfer Flavoprotein from *Megasphaera elsdenii* (4).
6-OH-FAD-D-aspartate oxidase has no flavin fluorescence, is not reduced by

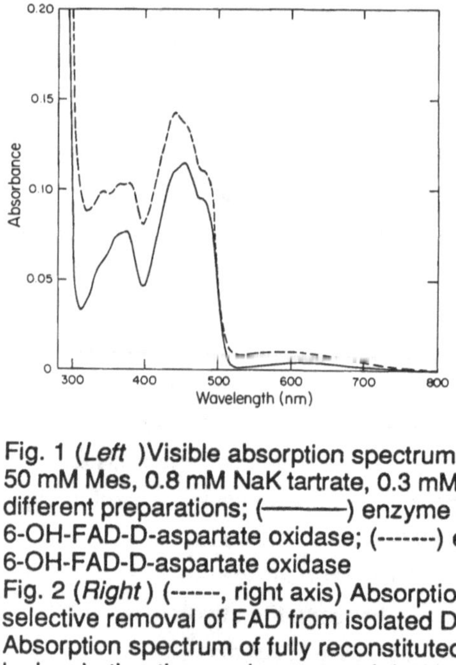

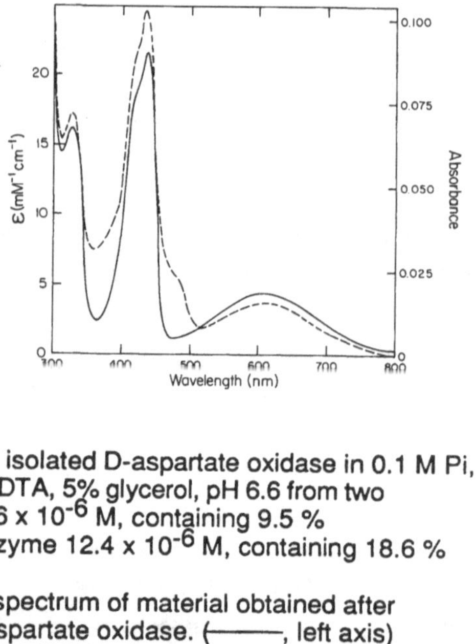

Fig. 1 (*Left*)Visible absorption spectrum of isolated D-aspartate oxidase in 0.1 M Pi, 50 mM Mes, 0.8 mM NaK tartrate, 0.3 mM EDTA, 5% glycerol, pH 6.6 from two different preparations; (———) enzyme 8.6 x 10^{-6} M, containing 9.5 % 6-OH-FAD-D-aspartate oxidase; (-------) enzyme 12.4 x 10^{-6} M, containing 18.6 % 6-OH-FAD-D-aspartate oxidase
Fig. 2 (*Right*) (------, right axis) Absorption spectrum of material obtained after selective removal of FAD from isolated D-aspartate oxidase. (———, left axis) Absorption spectrum of fully reconstituted 6-OH-FAD-D-aspartate oxidase obtained by incubating the previous material with 6-OH-FAD overnight and removing the excess flavin by gel filtration. Samples in 0.1 M Pi, 0.3 mM EDTA, pH 7.0, 4°C. The extinction coefficients for fully reconstituted 6-OH-FAD-D-aspartate oxidase have been calculated as described in Ref. 3

D-aspartic acid, does not stabilize any semiquinone upon photoreduction and is not affected by treatment with borohydride (see below); the enzyme stabilizes the green anionic benzoquinoid form of 6-OH-FAD, lowering its pKa below 4.5 (the precise value could not be determined due to denaturation of the enzyme at pH values lower than 4), which is more than two pH units below the value of 7.1 observed for the free flavin (4)

Properties of FAD-D-aspartate oxidase. FAD-D-aspartate oxidase was obtained by reconstitution of apoenzyme with FAD. Complete removal of flavin from the isolated enzyme was achieved by the method of Mayer and Thorpe (5). The dissociation constant for the binding of the coenzyme to the protein (Kd = 5×10^{-8} M in 0.1 M KPi, pH 7.4, 4°C) was determined by following the quenching of apoenzyme fluorescence or the perturbation of the flavin absorption spectrum upon titration of apo-D-aspartate oxidase with chromatographycally pure FAD.

FAD-D-aspartate oxidase is readily reduced by D-aspartic acid. It is fluorescent both in its oxidized (excitation maximum at 454 nm and emission maximum at 525 nm) and reduced forms (excitation maximum at 360 nm and emission maximum at 505 nm). Photoreduction in the presence of EDTA and 5-deazaflavin (6), in 50 mM NaPPi, pH 8.3, 4°C, resulted in the quantitative production of the red anionic semiquinone (absorption maxima at 373 and 488 nm); while the formation of the semiquinone was complete with 2.5 min irradiation, fully reduction was not achieved even after 110 min irradiation.

FAD-D-aspartate oxidase reacts with sulfite giving a flavin N(5)-sulfite adduct (Kd = 7.4×10^{-6} M at pH 7.0, 4°C) (7). The complex formation is reversed by binding of competitive inhibitors of the enzyme such as tartrate or oxalacetic acid.

Binding of tartrate to FAD-D-aspartate oxidase (Kd = 2.6×10^{-5} M at pH 7.0, 4°C for L(+) tartaric acid) causes perturbations of the visible spectrum of the enzyme similar to those induced by binding of benzoate to D-amino acid oxidase (8).

Reduction of FAD-D-aspartate oxidase with borohydride under aerobic conditions led to the formation of 3,4-dihydro-FAD-enzyme, with a characteristic fluorescence spectrum similar to that found with D-amino acid oxidase (9). This form is catalytically active, as was the case for the corresponding 3,4-dihydro-FAD forms of L- and D-amino acid oxidase (9).

All the studies described in the present report, both for the 6-OH-FAD and for the FAD forms of D-aspartate oxidase, are consistent with its classification in the dehydrogenase/oxidase class of flavoproteins and, particularly, with the presence of a positively charged amino acid residue near the N(1)C(2)=O locus of the flavin (10).

Acknowledgement

This work was supported by grants from the U.S. Public Health Service, GM 11106 and GM 21444, and by the Medical Research Service, Veterans Administration

Rererences

1) Still, J. L., and E. Sperling. 1950. J. Biol. Chem. 182, 585-589.
2) Hamilton, G. A. 1985. Adv. Enzymol. 57, 85-178.
3) Negri, A., V. Massey, and C. H. Williams, Jr. : J. Biol. Chem. (in press)
4) Mayhew, S. G., C. D. Whitfield, S. Ghisla, and M. Schuman-Jorns. 1974. Eur. J. Biochem. 44, 579-591.
5) Mayer, E. J., and C. Thorpe. 1981. Anal. Biochem. 116, 227-229.
6) Massey, V., and P. Hemmerich. 1978. Biochemistry. 17, 9-17.
7) Massey,V., F. Müller, R. Feldberg, M. Schuman, P. A. Sullivan, L. G. Howell, S. G. Mayhew, R. G. Matthews, and G. Foust. 1969. J. Biol. Chem. 244, 3999-4006.
8) Massey, V., and H. Ganther. 1965. Biochemistry 17, 9-17.
9) Massey, V., B. Curti, F. Müller, and S. G. Mayhew. 1968. J. Biol. Chem. 243, 1329-1332.
10) Massey, V., and P. Hemmerich. 1980. Biochem. Soc. Trans. 8, 246-257.

SPECTRAL AND KINETIC STUDIES ON *PSEUDOMONAS* L-PHENYLALANINE OXIDASE (DEAMINATING AND DECARBOXYLATING)

H. Koyama
Noda Institute for Scientific Research, Noda, Chiba 278, Japan

H. Suzuki, Y. Kawamura-Konishi
Department of Biophysical Chemistry, Kitasato University
School of Medicine, Sagamihara, Kanagawa 228, Japan

Introduction

Pseudomonas L-phenylalanine oxidase (deaminating and decarboxylating) contained 2 mol of FAD per mol of enzyme. The enzyme acted predominantly on L-phenylalanine (Phe) and stoichiometrical studies showed that both an oxidase reaction (Eq.1) and an oxygenase reaction (Eq.2) with Phe were catalyzed simultaneously by the enzyme under aerobic conditions as follows (1).

$$C_6H_5\text{-}CH_2\text{-}\overset{NH_2}{\overset{|}{C}H}\text{-}COOH + O_2 + H_2O \longrightarrow C_6H_5\text{-}CH_2\text{-}CO\text{-}COOH + NH_3 + H_2O_2 \quad (1)$$

$$C_6H_5\text{-}CH_2\text{-}\overset{NH_2}{\overset{|}{C}H}\text{-}COOH + O_2^{*} \longrightarrow C_6H_5\text{-}CH_2\text{-}\overset{NH_2}{CO^{*}} + CO_2 + H_2O^{*} \quad (2)$$

In addition to Phe, β-2-thienylalanine (βTA) and L-tyrosine (Tyr) served as substrates for both oxidation and oxygenation. However, L-methionine (Met) was exclusively oxidized by the enzyme (2). When the enzyme was mixed anaerobically with Phe, βTA, Tyr or Met, a spectral species (purple intermediate) with a broad absorption band in the long wavelength was observed with each substrate (3). To see if this spectral species (purple) is a catalytic intermediate, the Phe oxidase reaction has been analyzed by measuring the rate of the overall reaction and by the "stopped flow" methods. Furthermore, circular dichroism (CD) studies are described.

Flavins and Flavoproteins 1987

Results

The rate (v) of overall reaction per enzyme unit (e) was determined by measuring the oxygen uptake in the presence of various concentrations of substrates (Phe, βTA, Tyr and Met) and oxygen. The plots between e/v vs 1/[S] (or $1/[O_2]$) gave a series of straight parallel lines in the presence of various concentrations of oxygen (or substrates). From the plots, the values of the maximum velocity per enzyme unit, Vmax and Km were estimated and summarized in Table I.

Table I.
Values of Vm/e, Km, k_{red}^{max} and kox Estimated from the Overall Reaction Kinetics and the "Stopped Flow" Method.

Substrate	Vm/e(s^{-1})	Km(mM)	k_{red}^{max}(s^{-1})	kox($10^5 M^{-1} \cdot s^{-1}$)
L-Methionine	286	2.2	243	1.9
β-2-Thienylalanine	800	0.071	782	2.5
L-Phenylalanine	1,850	0.1	-	-
L-Tyrosine	3,570	4.0	-	-

A solution of enzyme and βTA (mostly an oxygenase substrate) or Met (an oxidase substrate) was mixed with oxygen by means of the "stopped flow" method and optical density was recorded against time. Typical recorder traces are given in Fig.1, showing that the enzyme was rapidly oxidized with oxygen and then reduced to the intermediary form after exhaustion of the added oxygen. The maximal absorbance changes (Pmax in Fig. 1) obtained with both substrates were plotted against various wavelengths. These spectra of βTA and Met were similar to those of 10 and 30 % oxidized form minus intermediary form, respectively. These results can be accounted for by assuming the cyclic change in the enzyme during the reaction such as,

$$X + O_2 \xrightarrow{kox} Eox, \qquad Eox \xrightarrow{kred} X$$

where, Eox and X represent the oxidized and intermediary forms of enzyme, respectively. The reaction constants, kox and kred

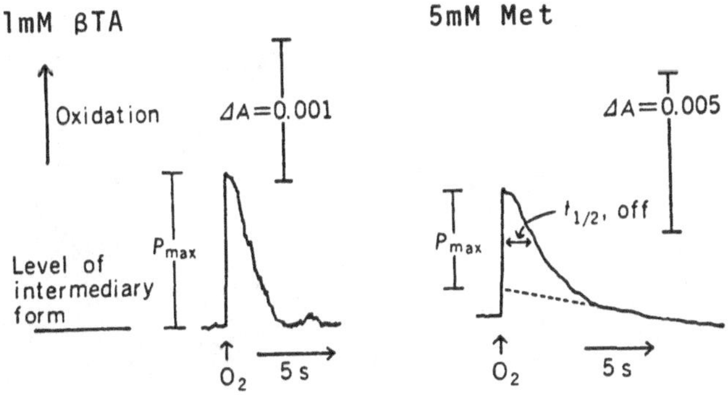

Fig. 1. The absorbance changes at 465 nm of Phe oxidase measured by the "stopped flow" method. A mixture of the enzyme and a substrate was mixed with 480 μM oxygen.

were determined from the absorbance change at 465 nm (Fig. 1, ref. (4)). The maximum rate constant of reduction, k_{red}^{max} was determined from the plots between 1/kred and 1/[S] and the values of kox and k_{red}^{max} were summarized in Table I. Since k_{red}^{max} values for βTA and Met were in good agreement with those of Vm/e obtained from the studies on the overall reaction kinetics, the scheme for the mechanism of both oxidation and oxygenation of Phe oxidase was expressed as follows.

$$\text{Eox} + \text{S} \rightleftharpoons \text{EoxS} \longrightarrow \text{X} \xrightarrow{\text{very slow}} \text{fully reduced state}$$

$$\text{X} + O_2 \xrightarrow{\text{kox}} \text{Eox} + \text{P}$$

where EoxS represents the enzyme unit (oxidized form)-substrate complex, X the purple intermediate, S the substrate and P the product. Kinetic studies indicate that oxidase and oxygenase reactions were explained by the same scheme and the same intermediate (X). Since CD is more sensitive than the absorbance to the change in the environment of the FAD moiety of enzyme, we measured the CD spectra of the enzyme after addition of various substrates. The spectrum was measured at 9°C to lower the rate of change of intermediate to the fully reduced form. The oxidized form was rapidly changed to the intermediate form by the addition of substrates. The intermediate spectra obtained by the addition of Phe, βTA and Met were very similar to each

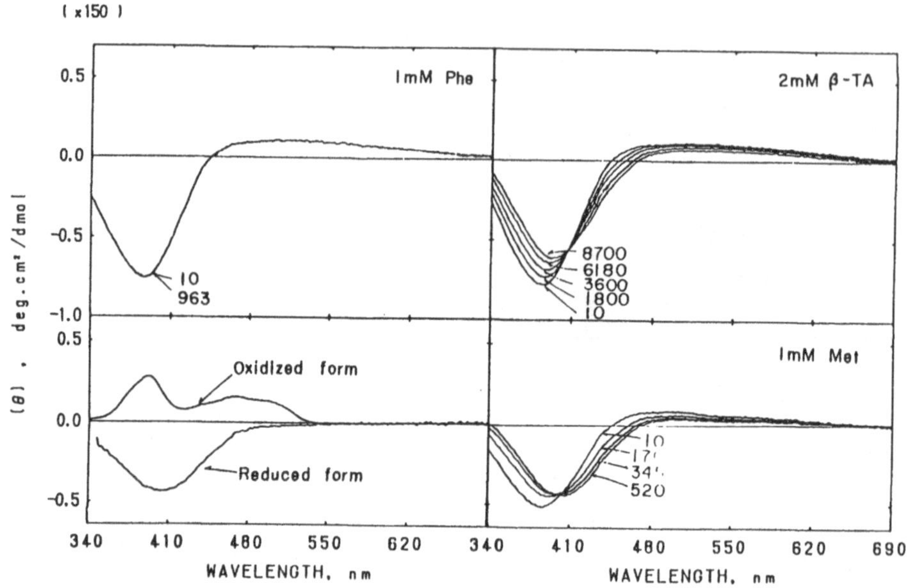

Fig. 2. CD spectra of Phe oxidase (1.13 x 10^{-4}M) during anaerobic reduction with substrates at 9°C. Numbers indicate the time (sec) of starting scan after addition of substrate. Lower left : CD spectra of the oxidized and reduced (dithionite) forms of the enzyme.

other (Fig. 2). The stability of the intermediate spectra was Phe>βTA>Met, and this order of stability agreed with that obtained from the absorbance change (3). These data confirm our above proposal that the enzyme catalyzes the reactions of both the oxidase and oxygenase substrates through the same intermediate.

References

1. Koyama, H. 1983. J.Biochem. 93, 1313-1319.

2. Koyama, H. 1984. J.Biochem. 96, 421-427.

3. Koyama, H. and H. Suzuki. 1986. J.Biochem. 100, 859-866.

4. Chance, B. 1957. Arch.Biochem. Biophys. 71, 130-136.

TIME-RESOLVED FLAVIN FLUORESCENCE OF PYRIDOXINE-5-P OXIDASE FROM SHEEP BRAIN

Antonie J.W. Visser

Department of Biochemistry, Agricultural University, Wageningen, The Netherlands

Francis Kwok, Jorge E. Churchich

Department of Biochemistry, University of Tennessee, Knoxville, TN 37996-0840 U.S.A.

Introduction

The oxidation of pyridoxine-5-P and pyridoxamine-5-P to pyridoxal-5-P is catalyzed by an enzyme that requires FMN as specific cofactor (1). Isolation of the pure enzyme from rabbit liver was first accomplished by Kazarinoff and McCormick (2). Subsequently, the enzyme was purified from pig brain (3,4) and rat brain (5).

Although it is well established that pyridoxine-5-P oxidase is a dimeric protein containing only one molecule of FMN, there is not information available on the mode of binding of FMN to the protein. We report here the results of time dependent anisotropy measurements designed to explore the rotational dynamics of the prosthetic group.

Results

When purified pyridoxine-5-P oxidase was analyzed using the technique of polyacrylamide 10-20% gradient gel electrophoresis, it was observed that the main protein band migrates as a dimer of 58 KD, but a minor component characterized by a

Flavins and Flavoproteins 1987

molecular weight of around 120 KD was also detected after staining for protein and enzymatic activity (Fig. 1). To further assess the macromolecular size of the oxidase in solution, HPLC analysis of purified samples of enzyme were conducted at several pH values.

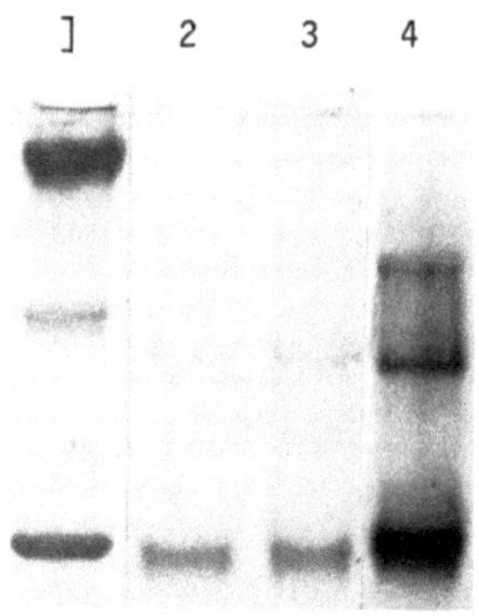

Figure 1. 10-20% gradient polyacrylamide gel electrophoresis of pyridoxine-P oxidase. Lane represents molecular weight markers; lanes 2 and 3 pyridoxine-P oxidase stained with commassie blue dye; and lane 4, pyridoxine-P oxidase stained for enzymatic activity.

At pH 5.5, only one protein peak of 58 KD was detected, whereas at pH 7.4 and pH 8.4, two protein peaks were discerned in the elution profiles. The major peak corresponded to the elution profile of a protein of 58KD, whereas the minor peak corresponded to the elution profile of a tetramer of 120 KD.

The shape of the elution profiles, observed at pH 7.4 and 8.4, indicate that the dimeric and tetrameric forms of the oxidase are in slow exchange under the chromatographic conditions employed.

In order to determine the dissociation constant of the following association process:

$$2 \text{ Dimer} \longleftrightarrow \text{Tetramer}$$

Samples of oxidase at various concentrations were injected to the molecular exclusion chromatography column (TSK 3000 SW-LKB) equilibrated with phosphate buffer (pH 7.4) and the concentrations of dimeric and tetrameric forms determined from the elution profiles. Using this experimental approach, it was possible to calculate an equilibrium dissociation constant $K_D = 4 \times 10^{-5}M$ for the equilibrium dimer/tetramer (Fig. 2).

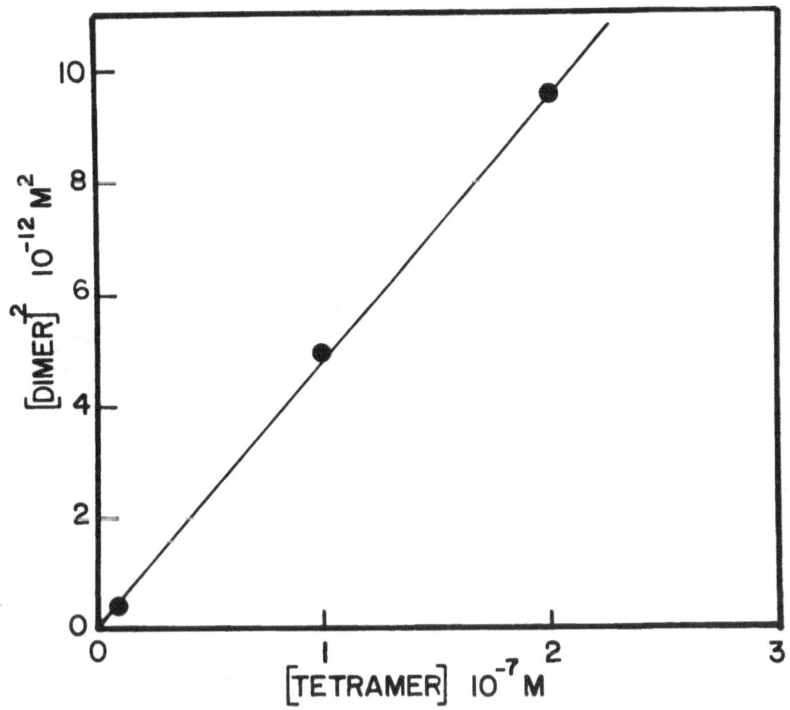

Figure 2. HPLC analysis of purified pyridoxine-5-P oxidase in 0.1 M phosphate buffer (pH 7.4).

Rotational correlation time measurements

The fluorescence decay of FMN bound to pyridoxine-5-P oxidase was analyzed using the experimental set up previously described (6).

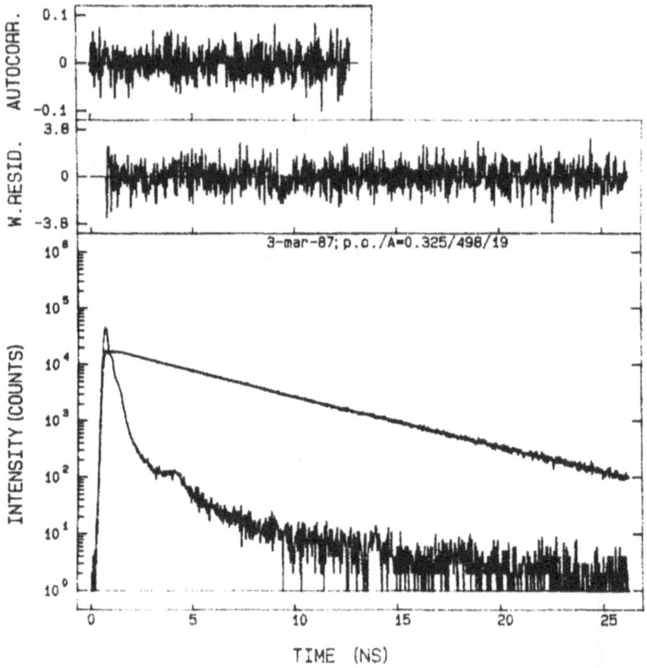

Figure 3. Emission decay of FMN bound to pyridoxine-5-P oxidase excitation 476.5 nm, emission 498 nm.

As shown in Figure 3, the emission decay of bound FMN exhibits 3 components that fit the equation:

$$F(t) = \sum_{i} \alpha_i \, e^{-t/\tau_i}$$

at excitation wavelength of 476.5 nm and detection wavelength of 498 and 544 nm (Table I).

The emission anisotropy decay of pyridoxine-5-P oxidase was investigated at two different protein concentrations in 0.01M sodium phosphate (pH 7.0). At a protein concentration of 2.8μM and excitation wavelength of 476.5 nm, the emission anisotropy decay function is clearly biexponential.

Table I. Fluorescence Decay of Bound FMN

Emission Wavelengths (nm)	α_1	α_2	α_3	τ_1 (ns)	τ_2 (ns)	τ_3 (ns)	$\langle\tau\rangle$ (ns)
498	0.078	0.009	0.914	4.70	1.47	0.025	4.35
544	0.294	0.043	0.662	4.83	1.61	0.035	4.60

Two rotational correlation times ϕ_1 = 0.15 ns and ϕ_2 = 0.49 ns with prexponential factors β_1 = 0.42 and β_2 = 0.05 fit the anisotropy decay function (Table 2).

Table II. Emission Anisotropy Decay of Bound FMN

Prot Concent (μM)	Emission Wavelengths (nm)	β_1	β_2	ϕ_1	ϕ_2 (ns)	Temp (ns)
2.8	498	0.42	0.05	0.15	0.49	19°C
2.8	544	0.36	0	0.19	0	19°C
28	544	0.31	0.03	0.21	4.79	19°C
28	544	0.32	0.02	0.36	5.70	4°C
28	498	0.36	0.04	0.21	1.68	19°C

The average rotational correlation time $\langle\phi\rangle$ = 0.24 ns is significantly shorter than the expected correlation time of 25 ns for the rotation of a macromolecule of 58,000 molecular weight. Thus, FMN bound to dimeric pyridoxine-5-P oxidase exhibits a high degree of rotational mobility in the subnanosecond time range.

As indicated in Table II, a ten fold increase in enzyme concentration influences the magnitude of the second correlation time component, particularly at detection wavelengths of 544nm. The presence of a longer component of 5 nanoseconds could be attributed to aggregation of the enzyme at high protein concentrations. In fact, it can be estimated that at a protein concentration of 28μM, the concentration of tetrameric species is approximately 25%.

Figure 4 is a typical biexponential decay of pyridoxine-5-P oxidase when the sample was excited at 476.5 nm.

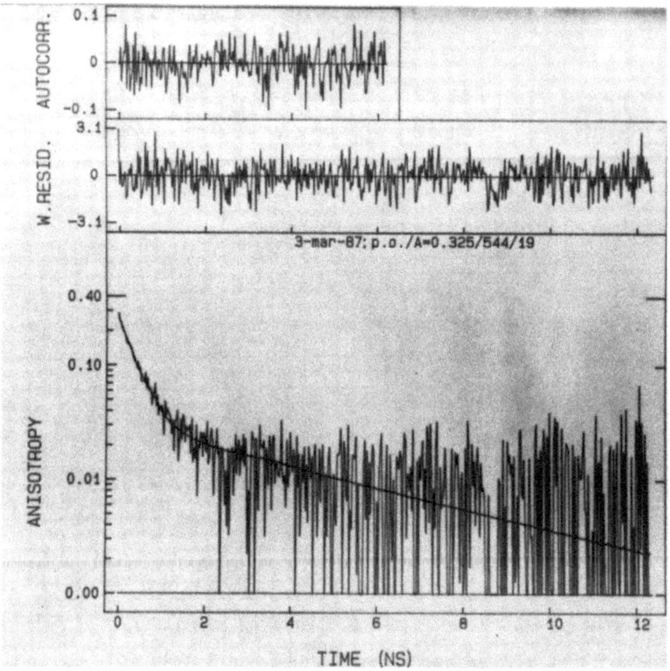

Figure 4. Emission anisotropy decay of FMN bound to pyridoxine-5-P oxidase. Excitation 476.5 nm, emission 544 nm protein concentration 28μM.

Discussion

The rotational dynamics of the prosthetic group was studied by time-dependent emission anisotropy measurements using time-correlated single photon counting (7). Analysis of the anisotropy decay function revealed the presence of two rotational correlation time components. The predominant contribution of a very short component ($\phi_1 \simeq 0.20$ns) was observed regardless of protein concentration.

In view of these results, it seems reasonable to conclude that sub-nanosecond motions of the protein influence the orientation of the prosthetic group at the catalytic binding site.

One interesting aspect of the catalytic site of pyridoxine-5-P oxidase is its ability to accomodate bulky substrates such as secondary amines carrying aromatic carboxylic acids covalently bound to P-pyridoxyl (4). It is conceivable that subnanosecond fluctuations at the level of the catalytic site facilitates penetration of bulky substrates as well as the release of the products.

Acknowledgement

This research was supported by NIH grant no. GM27639-04.

References

1. Wada, H., and Swell, E.E. (1961) J. Biol. Chem. 236, 2089-2095.

2. Kazarinoff, M.N., and McCormick, D.V. (1975) J. Biol. Chem. 250, 3436-3442.

3. Kwok, F., and Churchich, J.E. (1980) J. Biol. Chem. 255, 822-887.

4. Churchich, J.E. (1984) Eur. J. Biochem. 138, 327-332.

5. Cash, C.D., Maitre, M., Rumigny, J.F., and Mandel, P. (1980) Biochem. Biophys. Res. Commun. 96, 1755-1760.

6. Van Hoek, A., Vervoort, T., and Visser, A.J.W.G. (1983), J. Biochem. Biophys. Methods 1, 243-254.

7. Visser, A.J.W.G., Ykema, T., Van Hoek, A., O'Kane, D.J., Lee, J. (1985) Biochemistry 24, 1489-1496.

AN ACTIVE SITE THIOL IN LACTATE OXIDASE

David A. Giegel, Charles H. Williams, Jr. and Vincent Massey
Veterans Administration Medical Center and the Department of Biological Chemistry, University of Michigan, Ann Arbor, Michigan 48109-0606

Introduction

Previous modifications of lactate oxidase from either *Mycobacterium smegmatis* or *M. phlei* have shown evidence for the presence of an arginine (1), one to two histidines (2,3,4) and a two cysteines (5) in the active site. Inactivation of the enzyme by fluoro-2,4-dinitrobenzene (FDNB) was found to have modified one histidine and one equivalent of cysteine in the active site of the enzyme (3,4,5). The modified cysteine was found to be unequally distributed between two cysteines, both of which resided in the active site since they could be fully protected by competitive inhibitors. The histidine, however, was not protected but competitive inhibitors appeared to change the preference of the reagent for the two nitrogens on the imidazole ring of the histidine (5).

In order to determine the specific role of the cysteines in lactate oxidase, we have inactivated the enzyme with 5,5'-dithiobis(2-nitrobenzoic acid) (DTNB) and have examined the characteristics of the modified enzyme.

Results and Discussion

Lactate oxidase in 10 mM imidazole chloride pH 7 was reacted anaerobically at 25°C with DTNB. Inactivation of the enzyme proceeded monophasically above 2 mM reagent and was biphasic at lower concentrations (Fig. 1). The kinetics of inactivation indicate binding of DTNB prior to chemical reaction with the enzyme. After correcting for the slower phase of inactivation, a double-reciprocal plot (Fig. 2) of the observed rate of

Flavins and Flavoproteins 1987

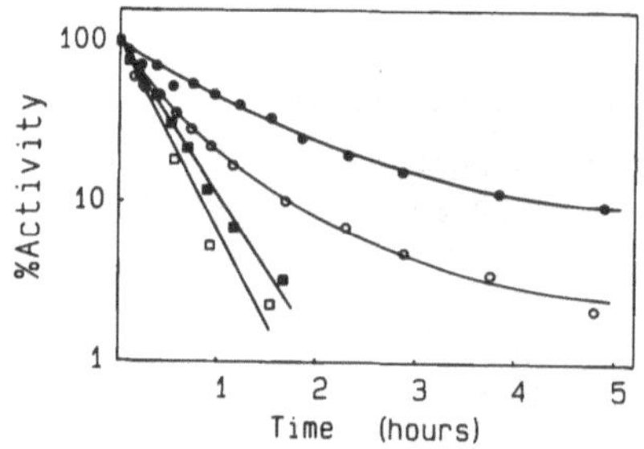

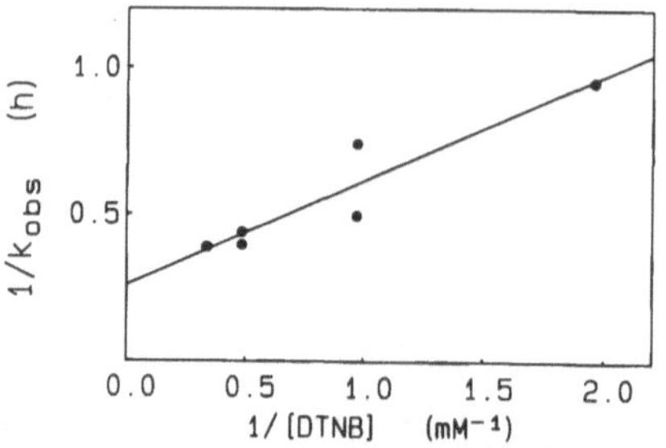

Figure 1, left. Inactivation of lactate oxidase with DTNB. Lactate oxidase (30 μM) in 10 mM imidazole chloride pH 7 at 25°C was reacted anaerobically with various concentrations of DTNB. Activity was assayed with the oxygen electrode (5) by removing an aliquot through the titrating port of the anaerobic cell. The initial activity of the sample was determined after anaerobiosis was completed. The final DTNB concentrations were: ● -0.51 mM; ○ -1.03 mM; ■ -2.06 mM; and □ -2.96 mM.

Figure 2, right. Double reciprocal plot for the inactivation of lactate oxidase by DTNB. The fast rates from Fig. 1 were corrected for the slow phase (where appropriate) then the reciprocal plotted against 1/[DTNB]. The line is a linear regression fit through the points and yields a K_d of 1.4 mM and a limiting rate for inactivation of 3.9 h^{-1}.

inactivation *versus* [DTNB] yields a dissociation constant for DTNB of 1.4 mM and a limiting rate of inactivation of 3.9 h^{-1} (0.065 min^{-1}). At 4°C, the inactivation of lactate oxidase by DTNB is very slow. The reagent does, however, perturb the spectrum of the enzyme-bound flavin in a manner similar to other competitve inhibitors (6). Plotting the change in absorbance at 502 nm as a function of [DTNB], a K_d of 0.53 mM for DTNB can be obtained, which is consistent with the K_d determined from the inactivation kinetics at 25°C.

The stoichiometry of incorporation of the thionitrobenzoyl (TNB) moiety into lactate oxidase was obtained by monitoring the release of TNB^- by the increase in A_{412} during the course of the inactivation. At 2.0 mM DTNB, full inactivation was obtained with the incorporation of 1.0 mol TNB/mole of flavin. The presence of oxalate, a potent inhibitor (7), fully protected the enzyme from inactivation over a period of two hours. The inactive enzyme (TNB-lactate oxidase) no longer showed any spectral perturbation upon addition of oxalate. Additionally, upon irradiation with light, the stable N(5)-carbonate adduct was not formed, as it is with native enzyme. Instead, under anaerobic conditions, irradiation of the

TNB-lactate oxidase/oxalate complex formed readily reoxidizable reduced enzyme.

When native lactate oxidase is reduced by L-lactate, a reduced flavin-pyruvate charge transfer complex is formed that has long wavelength absorption. This complex decays slowly to a fully reduced flavin spectrum (6). At 50 mM L-lactate, the rate of formation of the reduced enzyme-pyruvate complex is approximately 7000 min^{-1} (6). TNB-lactate oxidase can still be reduced by L-lactate, but at a rate that is much slower than native enzyme. At 50 mM L-lactate, the rate of reduction was 0.22 min^{-1} or approximately 30,000 times slower then native enzyme. No long wavelength species was observed at any time during the reduction of TNB-lactate oxidase by 1 mM L-lactate.

Native lactate oxidase stabilizes nearly 100% of the red flavin semiquinone upon photoreduction with deazaflavin and glycine (8). TNB-lactate oxidase exhibits only small amounts of semiquinone upon irradiation and after long exposure times only about 50% of the initial absorbance at 450 nm is bleached to give reduced enzyme. The enzyme rapidly reoxidizes to give back the original TNB-lactate oxidase spectrum.

DTNB has apparently modified one active site thiol, or as with FDNB, two thiols in a mutually exclusive fashion, and partially blocked access of the flavin. Small molecules like oxalate and L-lactate are still able to get into the active site and interact with the flavin, but the larger 1-deazaflavin molecule does not have ready access to the protein-bound flavin. Since modification is prevented by a competitive inhibitor, the thiol that has been modified is probably one of the same ones that were modified by FDNB (5).

A tentative structure for the active site of lactate oxidase can be proposed based upon the available protein and flavin modification data that (2-5,9) as well as on the stereochemical studies of Manstein, et al., (9). L-Lactate binds to the enzyme on the si side of the flavin with the carboxyl moiety probably binding to an arginine. The substrate is positioned such that the carboxyl group is oriented toward the C(6) and

C(7) positions of the flavin and extending toward position N(1). The histidine that is modified by FDNB (4-6) may reside near the flavin-N(1) and function as the positively-charged residue stabilizing the anionic flavin semiquinone as well as the negative charge that develops on the flavin during the reductive half-reaction. Both of the cysteines that are in the active site probably are situated such that L-lactate lies between the thiols and the flavin.

Acknowledgments

This work was supported by United States Public Health Service Grants GM 11106 (to V.M.) and GM 21444 (to C.H.W.).

References

1. Peters, R.G., W.C. Jones, T.H. Cromartie. 1981 Biochemistry 20, 2564-2571.

2. Choong, Y.S., M.G. Shepherd, P.A. Sullivan. 1977 Biochem. J. 165, 385-393.

3. Choong, Y.S., M.G. Shepherd, P.A. Sullivan. 1978 Biochem. J. 173, 255-262.

4. Murphy, C.J., M.G. Shepherd, P.A. Sullivan. 1983 Biochemistry 22, 1665-1669.

5. Giegel, D.A., V. Massey, C.H. Williams, Jr. 1987 J. Biol. Chem. 262, 5705-5710.

6. Lockridge, O., V. Massey, P.A. Sullivan. 1972 J. Biol. Chem. 247, 8097-8106.

7. Ghisla, S., V. Massey. 1977 J. Biol. Chem. 252, 6729-6735.

8. Massey, V., P. Hemmerich. 1978 Biochemistry 17, 9-17.

9. Ghisla, S., V. Massey, Y.S. Choong. 1979 J. Biol. Chem. 254, 10662-10669.

10. Manstein, D.J., E.F. Pai, V. Massey, S. Ghisla. (this volume)

11. Massey, V., S. Ghisla. 1983. In: Biological Oxidations (H. Sund and V. Ulrich, eds.) Springer-Verlag, pp. 114-139.

CAN METHYLENETETRAHYDROFOLATE REDUCTASE BE USED TO EVADE THE METHYL TRAP?

Jacalyn M. Green, David P. Ballou, and Rowena G. Matthews

Department of Biological Chemistry and Biophysics Research Division
The University of Michigan, Ann Arbor, Michigan, 48109

Introduction

Methylenetetrahydrofolate reductase from pig liver catalyzes the following two reactions (where CH_2-H_4folate and CH_3-H_4folate refer to methylenetetrahydrofolate and methyltetrahydrofolate, and $PteGlu_n$ refers to pteroylglutamate, folate with n glutamyl residues):

$$NADPH + CH_2\text{-}H_4folate \longrightarrow NADP^+ + CH_3\text{-}H_4folate \qquad [1]$$

$$CH_3\text{-}H_4PteGlu_n + CH_2\text{-}H_4PteGlu_m \longleftrightarrow CH_2\text{-}H_4PteGlu_n + CH_3\text{-}H_4PteGlu_m \qquad [2]$$

The first reaction commits these one-carbon units to the production of methionine and adenosylmethionine (AdoMet); the latter serves as the methyl donor in many biosynthetic reactions. The second reaction is a reversible exchange of reducing equivalents. Reactions [1] and [2] are mediated by the enzyme-bound FAD which in catalysis is reduced by NADPH or CH_3-H_4folate and reoxidized by CH_2-H_4folate (1).

When mammals are exposed to large amounts of the anesthetic, nitrous oxide, methionine synthase (the only other mammalian enzyme which reacts with CH_3-H_4folate) is inactivated, causing folate to accumulate as CH_3-H_4folate in the cell. This results in folate deprivation which can cause megaloblastic anemia in humans. Folate exists intracellularly in mammals with a tail of glutamyl residues in γ-linkage, but it is transported in the blood and enters the cell as CH_3-$H_4PteGlu_1$. Glutamyl residues are added to the folate by folyl-γ-

Flavins and Flavoproteins 1987

glutamate synthetase; $H_4PteGlu_1$ is a good substrate for this enzyme, while CH_3-$H_4PteGlu_1$ is a very poor substrate (2). Pig liver methionine synthase exhibits a marked preference for CH_3-H_4folate with a hexaglutamate tail (3). The conversion of the monoglutamate CH_3-H_4folate to a polyglutamylated form would therefore provide a much better substrate for the residual methionine synthase, and could provide a net conversion of CH_3-H_4folate to more useful forms of folate. It has been postulated that the exchange reaction catalyzed by methylenetetrahydrofolate reductase, in combination with the action of residual methionine synthase, could be used to relieve this trapping of folate as CH_3-H_4folate. This would be possible given reaction [2] above, where n=1 and m=6. These studies were undertaken to test this postulate.

Results

Does AdoMet inhibit the half-reactions necessary to support the exchange reaction, as shown in equation 2? Yes.

Stopped-flow techniques were used to measure the effect of AdoMet on the oxidation and reduction of the enzyme-bound flavin of methylenetetrahydrofolate reductase, which is monitored at 450 nm. All experiments were performed in a stopped-flow apparatus which was maintained under anaerobic conditions at 25° C. Reduced enzyme was generated by anaerobic titration with stoichiometric amounts of NADPH. Oxidized and reduced enzyme were mixed with 100 µM CH_3-$H_4PteGlu_1$ and 100 µM CH_2-$H_4PteGlu_1$, respectively. Results are summarized in Table 1. AdoMet is a potent inhibitor of the transfer of reducing equivalents between the enzyme-bound flavin of methylenetetrahydrofolate reductase and folate substrates.

Table 1

	no AdoMet	+110 µM AdoMet
E_{red} + CH_2-$H_4PteGlu_1$	47 s^{-1} (67%)* 3.0 s^{-1} (33%)	0.0037 s^{-1}
E_{ox} + CH_3-$H_4PteGlu_1$	6.1 s^{-1}	0.002 s^{-1}

*The reaction of E_{red} with CH_2–$H_4PteGlu_1$ was biphasic; the percentage values indicate what percent of the reaction was attributed to that phase.

Does exchange occur? Yes. In the presence of physiological concentrations of NADPH? No.

We used tritiated CH_3-H_4PteGlu$_1$ to measure the exchange of reducing equivalents between CH_3-H_4PteGlu$_1$ and CH_2-H_4PteGlu$_6$. Assays contained 50 mM phosphate buffer, pH 7.2, 0.3 mM EDTA, 10% glycerol, 20 mM 2-mercaptoethanol, 3 mM formaldehyde, 50 μM tritiated CH_3-H_4PteGlu$_1$ (68,000 dpm/nmole), 25 μM H_4PteGlu$_6$, and varying amounts of NADPH. CH_2-H_4PteGlu$_6$ was generated by nonenzymatic condensation of formaldehyde and H_4PteGlu$_6$. Assays were initiated by the addition of enzyme, and were heat-quenched in dimedone. Samples were cooled on ice, the dimedone-formaldehyde complex was extracted with toluene, and an aliquot was counted. The results are listed in Table 2. Each activity was determined from the slope of a line in which the amount of product was plotted versus time. One hundred percent activity represents the amount of activity observed with no added NADPH. NADPH is a potent inhibitor of the exchange reaction. In experiments not shown here, we have also observed potent inhibition of the exchange reaction by AdoMet and by a combination of AdoMet and NADPH.

Table 2.

[NADPH](μM)	% activity
0	100
15	74
30	35
50	14
100	0
200	0

Summary

1. Adenosylmethionine, the feedback inhibitor of methylenetetrahydrofolate reductase, is a potent inhibitor of the individual half reactions comprising the exchange reaction and of the exchange reaction itself. Since the concentration of AdoMet may be as high as 70 μM in the cell, the possibility of an exchange of reducing equivalents serving to evade the methyl trap seems unlikely.

2. NADPH is a potent inhibitor of the exchange of reducing equivalents between CH_3-H_4PteGlu$_1$ and CH_2-H_4PteGlu$_6$. Physiological concentrations of NADPH (200 μM) completely abolish this activity.

Conclusion

The exchange of reducing equivalents between folate substrates of methylenetetrahydrofolate reductase may be observed under *in vitro* conditions, but will not occur under physiological conditions. Therefore, methylenetetrahydrofolate reductase cannot be used to counteract the trapping of CH_3-H_4folate induced by a decrease in methionine synthase activity.

References

1. Vanoni, M.A., Ballou, D.P., and Matthews, R.G. Methylenetetrahydrofolate reductase: steady state and rapid reaction studies on the NADPH-methylenetetrahydrofolate, NADPH-menadione, and methyltetrahydrofolate-menadione oxidoreductase activities of the enzyme. 1983. J. Biol. Chem. 256, 11510-11514.

2. Cichowicz, D.J., and Shane, B. Mammalian folylpoly-γ-glutamate synthetase. 2. Substrate specificity and kinetic properties. 1987. Biochemistry. 26, 513-521.

3. Matthews, R.G., Ghose, C., Green, J.M., Matthews, K.D., and Dunlap, R.B. Folylpolyglutamates as substrates and inhibitors of folate-dependent enzymes. 1987. In: Advances in Enzyme Regulation 26, (G. Weber, ed.) Pergamon Press, New York. pp. 157-171.

THE CATALYTIC MECHANISM OF DNA PHOTOLYASE

Marilyn Schuman Jorns

Hahnemann University School of Medicine
Philadelphia, Pennsylvania 19102, U.S.A.

Introduction

The principal damage resulting from exposure of DNA to ultraviolet light is the formation of cyclobutane dimers between adjacent pyrimidine residues. Photoreactivating enzymes (DNA photolyases) repair UV-damaged DNA by splitting dimers in a rather unusual reaction that requires visible light. DNA photolyase from E. coli recognizes the dimer itself, rather than the deformation in DNA structure produced by the dimer (1). Binding of the enzyme is limited to a small region of DNA immediately surrounding the dimer and involves electrostatic contact of the protein with only 1 or 2 phosphate groups in the DNA backbone (1,2). In contrast to the sliding mechanism proposed for other proteins that bind to specific sequences in DNA, photolyase appears to locate dimers in a simple diffusion-controlled reaction (2). E. coli photolyase contains FAD plus a partially characterized second chromophore (3). This chapter focuses on the role of the chromophores in catalysis. A relevant model reaction is described as well as studies which relate to the chromophore composition of yeast photolyase.

Flavins and Flavoproteins 1987

Results and Discussion

Spectral properties of the chromophores in photolyase. The intense blue color of the isolated E. coli enzyme is due to a remarkably air-stable blue neutral FAD radical (3). The radical is solely responsible for the absorption of the enzyme at wavelengths longer than 450 nm since the second chromophore does not absorb in this region. The second chromophore, when bound to the enzyme, exhibits a maximum at 390 nm ($\mathcal{E} = 12.7 \times 10^3\ M^{-1}\ cm^{-1}$) and contributes to the intense absorption band of the isolated enzyme at 384 nm ($\mathcal{E} = 18.1 \times 10^3\ M^{-1}\ cm^{-1}$) (4,5). The protein-free second chromophore exhibits a

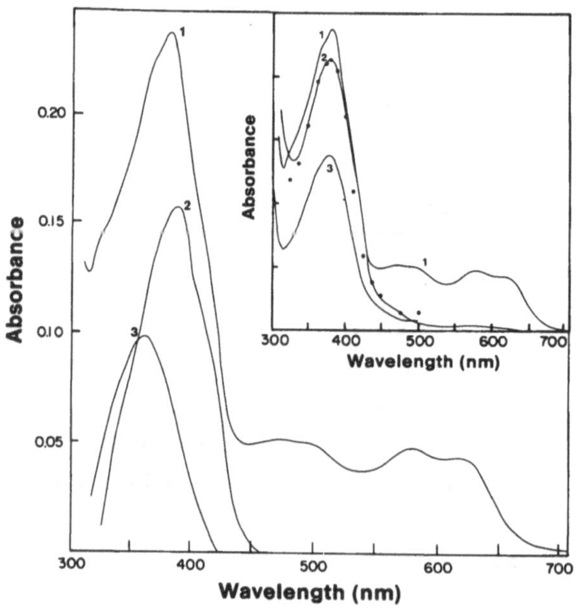

Figure 1. Spectral properties of E. coli photolyase and its second chromophore; comparison with yeast photolyase. The spectrum of the radical-containing form of E. coli photolyase is shown in curve 1. Spectra calculated (3,4) for the second chromophore, bound to the enzyme and free in solution, are shown in curves 2 and 3, respectively. INSET: Spectra recorded before (curve 1) and after (curve 2) anaerobic, dithionite reduction of the radical in E. coli photolyase are compared with the spectrum (23) of yeast photolyase (curve 3). The dots represent the spectrum of the yeast enzyme normalized to the same absorption at 385 as shown for the fully reduced E. coli enzyme.

maximum at 360 nm (Figure 1). The compound is not stable in aqueous solution except at acid pH (3). The isolated blue enzyme is fluorescent (Figure 2) with emission and excitation maxima at 480 nm and 390 nm, respectively. The fluorescence is entirely due to the second chromophore (3-5). The maximum fluorescence intensity observed for the second chromophore in blue enzyme (emission, 480 nm; excitation, 390 nm) is 20% of the maximum fluorescence observed with free FAD_{ox} (emission, 520 nm; excitation, 450 nm) (5). Conversion of the radical to fully reduced FAD ($FADH_2$) by reduction with dithionite results in a loss of the enzyme's blue color. The fully reduced enzyme exhibits a single absorption band in the visible region at 384 nm ($\mathcal{E}$ = 16.4 x 10^3 M^{-1}

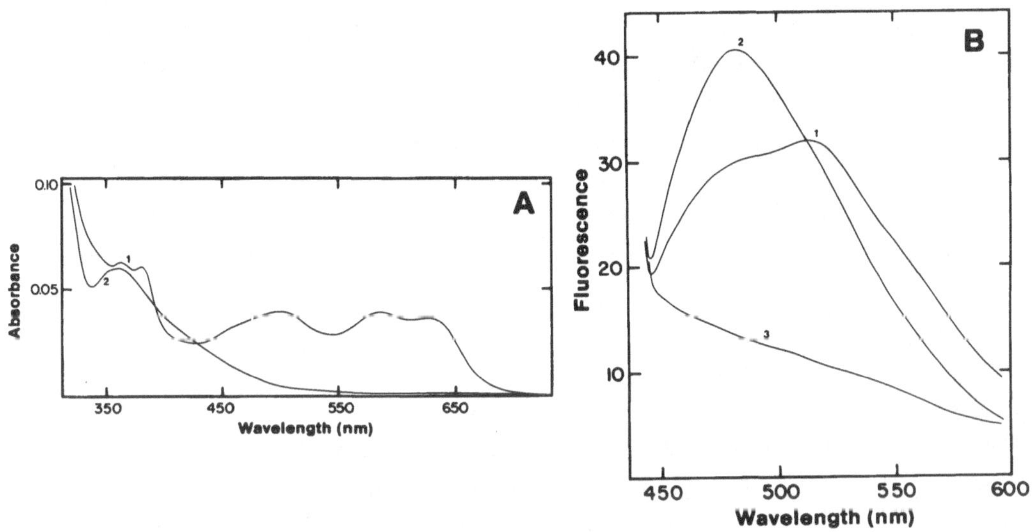

Figure 2. Spectral properties of the flavin chromophore in E. coli photolyase. PANEL A: The absorption spectrum of enzyme-bound radical (curve 1) was obtained after treating the isolated, blue enzyme with sodium borohydride to eliminate the absorption due to the enzyme's second chromophore (5). The spectrum of enzyme-bound $FADH_2$ (curve 2) was obtained after anaerobic, dithionite reduction of the radical in borohydride-treated enzyme. PANEL B: The fluorescence emission spectrum of enzyme-bound $FADH_2$ (curve 1) was obtained by treating the isolated blue enzyme (curve 2) with sodium borohydride to eliminate the fluorescence due to the second chromophore (curve 3), followed by reduction of the radical in borohydride-treated enzyme with dithionite.

cm^{-1}) (Figure 1) (4). Reduction of the radical is also accompanied by a 6-fold increase in fluorescence intensity (5). There is no change in the fluorescence excitation maximum but reduction does cause a small (5 nm), bathochromic shift in the emission maximum and the appearance of a shoulder at 515 nm in the fluorescence emission spectrum (5) (data not shown). The results indicate that the flavin radical in blue enzyme acts as a potent quencher of the fluorescence of the second chromophore. The fluorescence emission spectrum of the fully reduced enzyme is dominated by the enhanced fluorescence of the second chromophore but the shoulder at 515 nm is probably due to enzyme-bound $FADH_2$. The spectral properties of enzyme-bound $FADH_2$ and the absorption spectrum of enzyme-bound radical at wavelengths shorter than 450 nm have been determined using photolyase preparations where the spectral contribution due to the second chromophore is eliminated by prior reaction of the enzyme with sodium borohydride (vide infra) (5). Elimination of the absorption due to second chromophore in blue enzyme reveals underlying bands due to the flavin radical at 364 nm ($\varepsilon = 5.6 \times 10^3\ M^{-1}\ cm^{-1}$) and 382 nm ($\varepsilon = 5.3 \times 10^3\ M^{-1}\ cm^{-1}$) (Figure 2). A similar pair of absorption maxima are observed in this region for neutral flavin radicals bound to other flavoproteins (6). Enzyme-bound $FADH_2$ exhibits an absorption maximum at 360 nm ($\varepsilon = 5.5 \times 10^3\ M^{-1}\ cm^{-1}$) and is fluorescent (Figure 2). The intensity of the fluorescence emission band at 515 nm (excitation, 390 nm) is 14% of that observed for free FAD_{ox} (emission, 520 nm; excitation, 450 nm). The spectral properties of $FADH_2$ bound to photolyase are similar to those reported for 1,5-dihydroflavin derivatives bound to other flavoproteins (7). The results suggest that $FADH_2$ in fully reduced photolyase is 1,5-dihydroFAD.

Effect of the flavin redox state on catalytic activity. Insight regarding the effect of radical oxidation was obtained by studying the properties of green and yellow photolyase preparations (4). Green and yellow enzyme emerged from large scale enzyme purifications which required several additional days to complete as compared to smaller scale purifications which yield blue enzyme. Yellow enzyme was also obtained

TABLE I

Properties of Blue, Green and Yellow Photolyase Preparations[1]

Preparation	Total Flavin	Flavin Radical Control	Flavin Radical After Dithionite	Specific Activity Control	Specific Activity After Dithionite
	%	%	%	%	%
Blue enzyme	100	100	100	100	100
Green enzyme	109	55	107	51	89
Yellow enzyme	91	30	53	27	46

[1]Total flavin and radical content are estimated based on the ratios A_{280}/A_{450} and A_{280}/A_{580}, respectively. A_{450} is measured after denaturation. A_{580} is measured with intact enzyme. Dithionite treatment consists of reduction followed by reoxidation.For further details, see reference 4.

by storing an originally blue preparation for 6 months at -20^{0} C. The total flavin content of green enzyme is similar to blue enzyme but the radical content is only half of that observed for blue enzyme (Table I). The rest of the flavin in green enzyme is present as FAD_{ox}. The green color of the preparation is simply due to the presence of a mixture of blue radical plus yellow FAD_{ox}. The latter accounts for the prominent absorption band at 450 nm and the fluorescence emission band at 520 nm that are observed with green enzyme but not with blue enzyme. The specific activity of green enzyme, based on total protein, is approximately half of that observed for blue enzyme (Table I). Reaction of green enzyme with excess dithionite results in the immediate reduction of the flavin radical, followed by a slower, 2-electron reduction of FAD_{ox} that is complete in 3 h. Partial reoxidation of the fully reduced enzyme occurred after the excess dithionite was decomposed by air oxidation. The radical content and the specific activity observed for dithionite-treated green enzyme are nearly twice that observed with untreated green enzyme but very similar to that observed for blue enzyme (Table I). The results indicate that the binding sites occupied by oxidized flavin in untreated green enzyme retain the ability to stabilize the radical against air oxidation and that the loss of activity seen with untreated green enzyme is due to a reversible oxidation of the radical. Similar results were obtained in studies with yellow enzyme, where most of the flavin was present as yellow FAD_{ox}, except that about half the flavin sites in yellow enzyme had apparently lost the ability to stabilize the radical. As a consequence, dithionite treatment did not yield quantitative recovery of radical and activity (Table I). That the inactivation observed upon radical oxidation cannot be attributed to an inability to bind substrate is evidenced by the similar substrate binding affinity observed with yellow and blue enzyme (2).

To evaluate the effect of radical reduction on catalysis, the radical in blue enzyme was converted to $FADH_2$ with dithionite and the activity of the fully reduced enzyme was measured under anaerobic conditions (4). The activity of blue enzyme is unaffected by anaerobic conditions alone whereas reduction of the radical to $FADH_2$ with dithionite is accompanied by a 280% increase in photolyase activity. A similar activation (291%) is observed with dithiothreitol (DTT). DTT does not reduce the radical in the dark but a rapid photoreduction of the radical does occur when photolyase is illuminated in the presence of thiols (8). The effects of dithionite and DTT on activity are not additive. The results indicate that the activation observed with DTT is due to photoreduction of the enzyme when the assay mixture is exposed to photoreactivating light. In addition to DTT, other standard components of the assay mixture, such as Tris buffer and EDTA, will also cause photoreduction of the radical although the reactions are slower as compared to that observed with DTT (8). These observations led us to suspect that radical-containing enzyme might actually be catalytically inactive. The activity observed in the absence of DTT could be due simply to a less efficient photoreduction of the enzyme by other components in the assay mixture.

<u>Quantum yield and relative action spectra</u>. Further evidence regarding the catalytically and physiologically significant redox state of the flavin in photolyase was sought by determining the quantum yield and the

relative action spectrum for dimer repair with enzyme containing flavin radical or $FADH_2$ (9). Assuming that the quantum yield is wavelength-independent, the relative action spectrum should superimpose with the absorption spectrum of the sensitizer and the studies should also provide information regarding the role of the enzyme's chromophores in catalysis.

Studies by Harm have shown that dimer photorepair is a highly efficient process in E. coli, exhibiting a quantum yield of about 1 at 384 nm (10,11). In contrast, the isolated blue enzyme is inefficient and, even under optimal conditions, exhibits a quantum yield (ϕ_B) at 384 nm that is only 6% of that observed in in vivo studies. On the other hand, the fully reduced enzyme exhibits a quantum yield (ϕ_R) of 1.0 at 384 nm, similar to that obtained for in vivo dimer photorepair. The results indicate that the fully reduced enzyme is the physiologically significant form of photolyase in E. coli. That photolyase-overproducing E. coli cells are not blue, even though photolyase constitutes 15% of the total protein, can be explained by the fact that the enzyme present in these cells contains $FADH_2$ instead of the blue flavin radical. Photoreduction of the radical in the presence of DTT proceeds with a quantum yield (ϕ_P) of 0.1 (8). If dimer repair occurs only with enzyme containing fully reduced flavin, photoreduction of the radical must occur as a first step in reactions initiated with radical-containing enzyme. This is expected to cause a 95% reduction in the observed quantum yield for dimer repair as compared with reactions initiated with fully reduced enzyme. The predicted value for the quantum yield, 0.05 ($\phi = 0.5\ \phi_P\ \phi_R$), is in good agreement with the value of 0.06 observed with the radical form of the enzyme. The relative action spectrum and the absorption spectrum observed for radical-containing enzyme are similar in the 385 nm region but deviate sharply in the long wavelength region where a prominent band is seen only in the absorption spectrum (Figure 3). Since absorption in this region is solely due to the flavin radical, the results indicate that the flavin radical cannot act as a sensitizer. The data are consistent with the proposal that radical-containing enzyme is inactive until the radical is converted to fully reduced flavin. In contrast, the relative action spectrum observed for enzyme containing

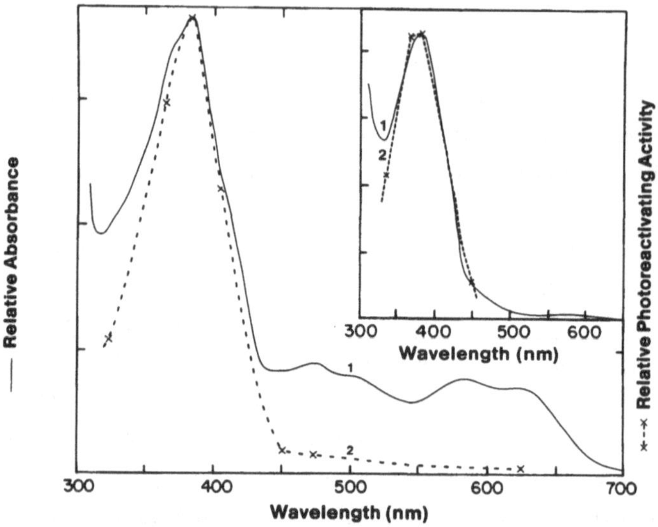

Figure 3. Relative action spectra for dimer repair with two redox forms of E. coli photolyase. Absorption and relative action spectra obtained for the blue, radical-containing enzyme are shown in curves 1 and 2, respectively. The INSET shows the corresponding data for the fully reduced, $FADH_2$-containing, enzyme.

$FADH_2$ superimposes with the absorption spectrum of the fully reduced enzyme (Figure 3). This result, in combination with the observed quantum yield of 1.0, means that every quantum of light absorbed by the fully reduced enzyme is used to split dimers and that either of the two chromophores in the fully reduced enzyme can act as the sensitizer.

Dimer repair via a second chromophore-independent pathway. Green and yellow enzyme contain decreased amounts of the second chromophore (17% and 7.7%, respectively) as compared with the level found in blue enzyme (4). The decrease in second chromophore content exceeds the decrease in activity observed with these preparations (see Table I). As previously discussed, the activity decrease is mainly due to radical oxidation and can be reversed by dithionite treatment. That dithionite-treated green enzyme exhibits a specific activity similar to blue enzyme, but has only 17% of the second chromophore content, suggested that the enzyme could repair dimers via a second chromophore-independent pathway. Further evidence for this hypothesis has been obtained by studying the reaction of blue enzyme with sodium borohydride (12). This reaction results in a loss of the absorption (Figure 4) and fluorescence (Figure 2) due to the second chromophore whereas the flavin radical is unaffected. Borohydride-treated enzyme does not contain any second chromophore, as judged by the SDS method (4), but does exhibit a specific activity that is 80% of that observed for untreated enzyme. The results clearly indicate that dimer repair can occur via a second chromophore-independent pathway. Since the second chromophore is not stable when free in solution at neutral pH, the spectral changes observed with borohydride might be due to a selective release of the second chromophore into

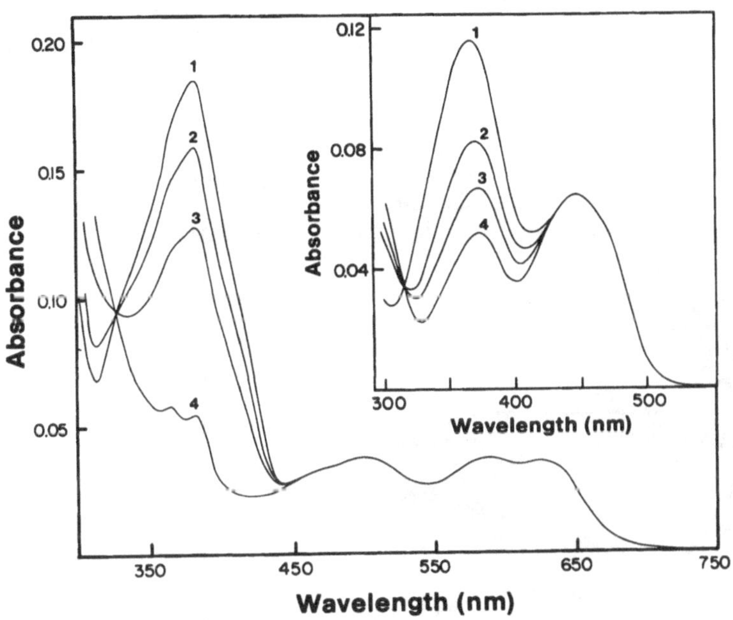

Figure 4. Reaction of E. coli photolyase and a protein-free extract with sodium (cyano)borohydride. Curve 1 is the spectrum of the isolated blue enzyme at pH 7.4 and 5.5° C. Curves 2-4 were recorded 30 min after adding sodium borohydride to a concentration of 4.2×10^{-4} M, 2.5×10^{-3} M and 1.6×10^{-2} M, respectively. INSET: Curve 1 is the spectrum of a protein-free extract, containing the second chromophore plus oxidized flavin, at pH 2.1 and 25° C. Curves 2-4 were recorded 5, 11 and 60 min, respectively, after adding 3.2×10^{-4} M sodium cyanoborohydride.

solution. However, a very similar reaction is observed when a protein-free extract, containing intact second chromophore plus FAD_{ox}, is reacted with sodium cyanoborohydride at acid pH (Figure 4). The reaction with cyanoborohydride results in a complete loss of the absorption due to the protein-free second chromophore whereas FAD_{ox} is unaffected. Since the second chromophore is stable in solution at acid pH, it is concluded that the observed spectral changes are due to reduction of the protein-free second chromophore by cyanoborohydride and that a similar reaction occurs when the intact enzyme is treated with borohydride at neutral pH.

Mechanistic considerations. Model studies show that the photosensitized cleavage of thymine dimers can involve either electron donation from or electron abstraction by the excited sensitizer (13-18). Since $FADH_2$

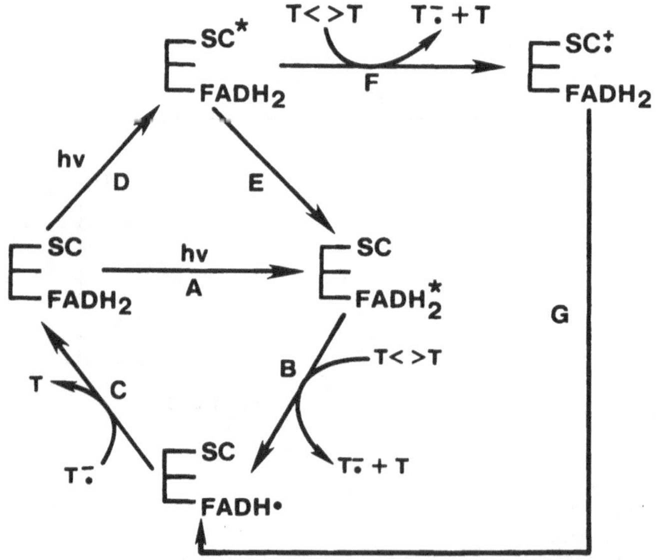

Scheme I. Dimer (T<>T) repair via the second chromophore (SC)-independent pathway is shown by the reaction sequence A,B,C. Dimer repair with the second chromophore as primary sensitizer, followed by energy transfer to $FADH_2$, is shown by the reaction sequence D,E,B,C. An alternate pathway with the second chromophore as sensitizer is shown by the reaction sequence D,F,G,C.

is unlikely to act as an electron acceptor, we propose that dimer cleavage by photolyase in the second chromophore-independent pathway proceeds via a mechanism involving electron donation from excited $FADH_2$ (Scheme I). This generates flavin radical plus pyridime dimer radical anion. The latter is unstable and spontaneously monomerizes to yield pyrimidine plus pyrimidine radical anion. The pyrimidine radical anion is oxidized by the flavin radical, regenerating $FADH_2$. Since every quantum of light absorbed by fully reduced enzyme is used to split dimers, there must exist a second pathway for dimer repair involving the second chromophore as sensitizer. The relative importance of the two pathways will clearly be wavelength-dependent since, for example, the second chromophore is the major chromophore at 384 nm whereas only $FADH_2$ absorbs at 450 nm. Since dimer repair is not observed with enzyme containing intact second chromophore plus FAD_{ox} (4), it is likely that the second pathway requires both the second chromophore and $FADH_2$ (Scheme I). The excited second chromophore may directly donate an electron to the pyrmidine dimer, generating second chromophore radical cation plus pyrimidine dimer radical anion. Electron transfer from $FADH_2$ regenerates the second chromophore in a reaction that may serve to prevent electron return from the dimer radical before monomerization can occur. The catalytic cycle is completed by monomerization of the pyrimidine dimer radical and subsequent regeneration of $FADH_2$. In this mechanism $FADH_2$ acts only as a redox catalyst, in contrast to the second chromophore-independent pathway, where its photochemical reactivity is also important. As an alternative second pathway, it is relevant to consider the possibility of energy transfer from the excited second chromophore to $FADH_2$ (Scheme I). In this case, dimer cleavage would proceed via the same mechanism proposed for the second-chromophore independent pathway except that excited $FADH_2$ is not generated by direct absorption of a photon.

Model studies. That photolyase can repair dimers via a pathway involving only $FADH_2$ prompted studies to determine whether an analogous reaction would occur in model studies with thymine dimers and reduced flavin derivatives (19). Among the four reduced flavins tested in an initial survey, the fastest rate of dimer cleavage was observed with reduced 1-deazariboflavin. Cleavage was also observed with reduced N(3)-methyllumiflavin whereas no cleavage was detected with the reduced form of 7,8-dimethyl-1,10-ethylene-isoalloxazinium perchlorate or with reduced 5-deazariboflavin. The following observations indicate that the dimer cleavage observed with reduced 1-deazariboflavin provides a good model for DNA repair as catalyzed by E. coli DNA photolyase: 1) Reduced 1-deazariboflavin is not consumed during dimer cleavage and hence acts as a catalyst. 2) The model reaction occurs at neutral pH. 3) Unlike results obtained with other oxidized flavin derivatives (18,19), dimer cleavage is not observed with oxidized 1-deazariboflavin. E. coli photolyase containing FAD_{ox} is also inactive. 4) Both the model and the photolyase reactions exhibit saturation kinetics. 5) Both reactions require light. In the initial stages of the reaction, the extent of dimer cleavage observed in the model and in the enzyme reactions is a linear function of the light dose. 6) Since oxidized 1-deazariboflavin is inactive, it is unlikely that the dimer cleavage observed with reduced 1-deazariboflavin is due to an impurity. The results indicate that reduced 1-deazariboflavin acts as a photosensitizer, similar to that proposed for $FADH_2$ in E. coli photolyase. Based on quantum yield measurements, the model reaction with reduced 1-deazariboflavin ($\emptyset = 9.2 \times 10^{-3}$) is about two orders of magnitude less efficient as compared with the reaction observed with fully reduced E. coli photolyase ($\emptyset = 1.0$). This suggests that the protein moiety may have a role in dimer cleavage.

Comparison of photoreactivating enzymes from different sources. E. coli photolyase is significantly different as compared with photolyase from Streptomyces griseus which contains an oxidized 5-deazaflavin derivative that apparently acts as the sensitizer in catalysis (20). Photolyase from yeast and E. coli exhibit considerable amino acid sequence homology (21,22). The pale yellow, yeast enzyme exhibits an absorption maximum at 385 nm (Figure 1) but does not exhibit significant absorption in the 500-700 nm region (23,24), unlike the isolated blue, E. coli enzyme. The yeast enzyme contains a reduced flavin derivative that is released as

FAD_{ox} upon aerobic denaturation (23,24). Iwatsuki, et al. (23) proposed that the yeast enzyme contained a 4a,5-reduced flavin derivative since other known types of reduced flavin would not be expected to exhibit an absorption maximum near 385 nm. However, it is pertinent that: 1) The yeast enzyme exhibits fluorescence properties similar to the isolated blue E. coli enzyme whose fluorescence is entirely due to the second chromophore (3,4); 2) The absorption spectrum of the yeast enzyme nearly superimposes with the spectrum obtained after reduction of the radical in the isolated E. coli enzyme to 1,5 dihydroFAD (Figure 1) (4). That the chromophore composition of yeast photolyase is similar to that observed for the fully reduced E. coli enzyme has recently been confirmed in studies which provide direct evidence for a chromophore in yeast photolyase with spectral and other properties similar to the second chromophore in E. coli photolyase (26). As compared with E. coli photolyase, $FADH_2$ in yeast photolyase must be in a somewhat different environment that makes it more stable towards air oxidation. The observed correspondence between the yeast enzyme and the fully reduced E. coli enzyme is particularly noteworthy since the latter is also the physiologically significant form of photolyase in E. coli.

References

1. Jorns, M.S., Sancar, G.B., Sancar, A.: Biochemistry 24, 1856-1861 (1985).
2. Sancar, G.B., Smith, F.W., Reid, P., Payne, G., Levy, M., Sancar, A.: J. Biol. Chem. 262, 478-485 (1987).
3. Jorns, M.S., Sancar, G.B., Sancar, A.: Biochemistry 23, 2673-2679 (1984).
4. Jorns, M.S., Baldwin, E.T., Sancar, G.B., Sancar, A.: J. Biol. Chem. 262, 486-491 (1987).
5. Jorns, M.S., Wang, B., Jordan, S.: submitted.
6. Müller, F., Brüstlein, M., Hemmerich, P., Massey, V., Walker, W.H.: Eur. J. Biochem. 25, 573-580 (1972).
7. Ghisla, S., Massey, V., Lhoste, J., Mayhew, S.G.: Biochemistry 13, 589-597 (1974).

8. Heelis, P.F., Sancar, A.: Biochemistry 25, 8163-8166 (1986).

9. Sancar, G.B., Jorns, M.S., Payne, G., Fluke, D.J., Rupert, C.S., Sancar, A.: J. Biol. Chem. 286, 492-498 (1987).

10. Harm, W.: Mutat. Res. 10, 277-290 (1970).

11. Harm, H., Rupert, C.S.: Mutat. Res. 10, 307-318 (1979).

12. Jorns, M.S., Wang, B., Jordan, S.: submitted.

13. Hélène, C., Charlier, M.: Biochem. Biophys. Res. Commun. 43, 252-256 (1971).

14. Charlier, M., Hélène, C.: Photochem. Photobiol. 21, 31-37 (1975).

15. Ben-Hur, E., Rosenthal, I.: Photochem. Photobiol. 11, 163-168 (1970).

16. Lamola, A.A.: Mol. Photochem. 4, 107-133 (1972).

17. Roth, H.D., Lamola, A.A.: J. Am. Chem. Soc. 94, 1013-1014 (1972).

18. Rokita, S.E., Walsh, C.T.: J. Am. Chem. Soc. 106, 4589-4595 (1984).

19. Jorns, M.S.: J. Am. Chem. Soc., in press.

20. Eker, A.B.M., Dekker, R.H., Berends, W.: Photochem. Photobiol. 33, 65-72 (1981).

21. Sancar, G.B.: Nucleic Acids Res. 13, 8231-8246 (1985).

22. Yasui, A., Langeveld, A.: Gene (Amst.) 36, 349-355 (1985).

23. Iwatsuki, N., Joe, C.O., Werbin, H.: Biochemistry 19, 1172-1176 (1980).

24. Eker, A.P.M., Yasui, A.: in Abstracts from the 13th International Congress of Biochemistry, p. 502, Elsevier Scientific Publishing Co., Amsterdam (1985).

25. Sancar, G.B., Smith, F.W.: submitted.

Flavoenzyme Structure and Electron Transfer Mechanisms

SEQUENCE AND STRUCTURE OF *Anacystis nidulans* FLAVODOXIN: COMPARISONS WITH FLAVODOXINS FROM OTHER SPECIES

David E. Laudenbach, Neil A. Straus
Department of Botany, University of Toronto, Toronto, Ontario M5S 1A1 Canada

Katherine A. Pattridge, Martha L. Ludwig
Biophysics Research Division, University of Michigan, Ann Arbor 48109 MI

Introduction

Flavodoxins have provided convenient models for more complex flavoprotein electron transferases, and they are the only flavoproteins for which several three-dimensional structures are available to correlate sequence variations with thermodynamic and kinetic properties. However, comparisons among flavodoxins, which differ significantly in the stereochemistry of their FMN binding sites, have been hampered by gaps in the chemical sequence of *A. nidulans* flavodoxin. DNA sequencing has now provided the missing data, enabling identification of critical residues in the FMN binding site. In this paper we review structure-function relationships in the flavodoxins with the perspective afforded by the new sequence information. Some aspects of models for electron transfer complexes have already been presented (1); here we focus on oxidation-reduction potentials.

Sequence Determination and Structure Refinement

Laudenbach, Reith, and Straus (2) have cloned the structural gene for flavodoxin on a Hind III fragment from a genomic library of *Anacystis nidulans* R2, a readily transformable cyanobacterial strain. Within this 1.2 Kb fragment is an open reading frame corresponding to the known partial protein sequences (3,4) of flavodoxin from *Anacystis nidulans* (*Synechococcus* PCC 6301). The DNA sequence predicts a polypeptide 169 residues in length, with the amino acid sequence shown in Figure 1. The deduced sequence has been used to add previously unidentified side chains to the three-dimensional model of flavodoxin from strain 6301.

Descriptions of the structure (Fig. 2) are based on coordinates resulting

Flavins and Flavoproteins 1987

from partial refinement with the programs of Hendrickson and Konnert (5). The first series of refinements included atomic coordinates obtained partly from earlier refinement (3), and partly by rebuilding sections of the segments 120-145 and 67-82, but excluded sidechains uniquely determined by the DNA sequence. For the 10997 reflections between 2.0 and 5.0 Å, the R-factor decreased from 0.45 to 0.33 after 21 cycles. When the DNA sequences became available, side chains were built into combined or $2|F_o|-|F_c|$ maps and major modifications were made at positions 26-30 and 75-82. In a second round of refinement, the initial R of 0.35 was reduced to 0.27 in 50 cycles computed on the Cray at the San Diego Supercomputer Center. No discrepancies between the protein and DNA sequences have been detected so far (Cf. Fig. 1), but because other proteins from strains 6301 and R2 are not identical (6), correspondences of the structure with the boldface sequences of Fig. 1 will need to be examined as refinement continues.

```
                        1       10        20        30        40        50        60
Anacystis nidulans R2   AKIGLFYGTQTGVTQTIAESIQQEFGGESIVDLNDIANADASDLNAYDYLIIGCPTWNVG
                        * ** * *  ** *   *  *      *              *   *  *  * **   *
Azotobacter vinelandii  AKIGLFFGSNTGKTRKVAKSIKKRFDDETMSDALNVNRVSAEDFAQYQFLILGTPTLGEG
                        * ** * *  ** *   *  *      *              *   *  *  * **   *
Klebsiella pneumoniae   ANIGIFFGTDTGKTRKIAKMIHKQLG-ELADAPVNINRTTLDDFMAYPVLLLGTPTLGDG

                                        70        80        90       100       110
Anacystis nidulans R2   EL--------QSDWEGIYDDLDSVNFQGKKVAYFGAGDQVGYSDNFQDAMGILEEKISSL
                         *           *             ** ** ** *** **  *       *
Azotobacter vinelandii  ELPGLSSDCENESWEEFLPKIEGLDFSGKTVALFGLGDQVGYPENYLDALGELYSFFKDR
                         *           *             ** ** ** *** **  *       *
Klebsiella pneumoniae   QLPGLEAGCESESWSEFISGLDDASLKGKTVALFGLGDQRGYPDNFVSGMRPLFDALSAR

                             120       130       140       150       160
Anacystis nidulans R2   GSQTVGYWPIEGYDFNESKAVRNNQFVGLAIDEDNQPDLTKNRIKTWVSQLKSEFGL
                        *    * *   ** *  * *     ****  * ***   *  *   *
Azotobacter vinelandii  GAKIVGSWSTDGYEFESSEAVVDGKFVGLALDLDNQSGKTDERVAAWLAQIAPEFGLSL
                        *    * *   ** *  * *     ****  * ***   *  *   *
Klebsiella pneumoniae   GAQMIGSWPNEGYEFSASSALEGDRFVGLVLDQDNQFDQTEARLASWLEEIKRTVL
```

Figure 1. Sequences of three 'long-chain' flavodoxins. Numbering corresponds to the sequence for A. nidulans R2 (2). Alignments with A. vinelandii and K. pneumoniae are those given by Drummond (7); * denotes positions that are identical in all three sequences. The residues in boldface are those determined solely from the DNA sequence of the flavodoxin gene from A. nidulans R2. Comparison with published amino acid sequences (3,4), which account for 131 residues, shows discrepancies at positions 54 and 78-80. However the latter peptide sequence has been revised to F-Q-G (G. Tarr, unpublished). Cysteine rather than serine at 54 is consistent with difference maps and with amino acid analyses (8).

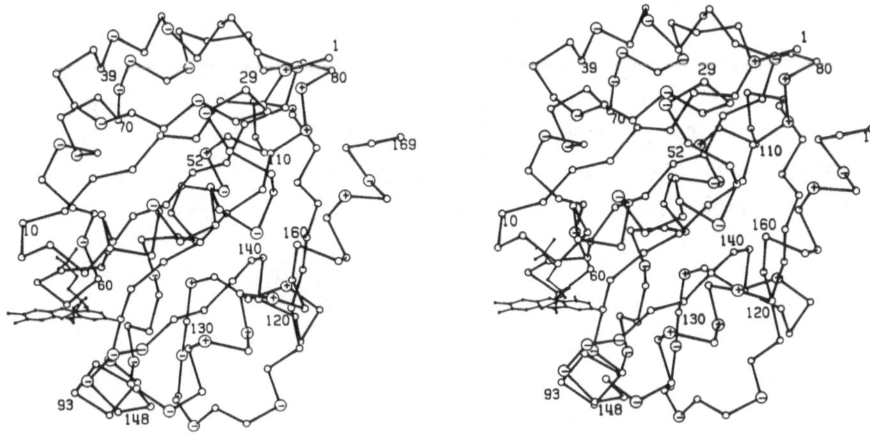

Figure 2. A stereoview of A. nidulans flavodoxin, based on the current model with the sequence of Fig. 1. Positions of charged residues are indicated by + or - at the C_a atoms. The preponderance of acidic residues is striking in this flavodoxin, which in contrast to A. vinelandii or K. pneumoniae (7,9) has no basic residues along the first helix. Positions of residues 27-29 and 76-80 are not well-determined; these parts of the molecule seem to be relatively mobile.

The Chain Fold

A. nidulans flavodoxin is a typical α/β protein with a parallel sheet at its core (Fig. 2). As additional flavodoxin sequences and structures are determined, the positions available for insertions or deletions that alter chain lengths can be defined more precisely. Location of insertions has been facilitated by superposition of the structures of Clostridium MP (10), Desulfovibrio vulgaris (11) and A. nidulans flavodoxins, whose chain lengths are 138, 148 and 169 residues (Fig. 3). The principal feature which distinguishes A. nidulans flavodoxin from the shorter chains found in D. vulgaris or clostridial species is an insertion of 20 residues in the fifth strand of the parallel sheet; conservation of this segment (120-139) in the three sequences of Fig. 1 suggests that it is a characteristic of long chain flavodoxins. A. vinelandii and Klebsiella flavodoxins contain additional inserts that must lie somewhere between positions 62 and 80 of the A. nidulans sequence, and would result in an excursion of the protein chain at the upper left of Fig. 2. Flavodoxins also vary in length near position 90. A. nidulans incorporates five extra residues in this region, compared to

Clostridium MP flavodoxin, whereas D. vulgaris flavodoxin resembles longer chain species at positions 90-100. Finally, the linker between the fifth strand of β-sheet and the C-terminal helix differs in length and orientation in Clostridium MP and A. nidulans flavodoxins; an acidic sequence which is highly conserved in the three flavodoxins of Fig. 1 occurs in this region, which has been implicated in interactions with electron donor/acceptors (12).

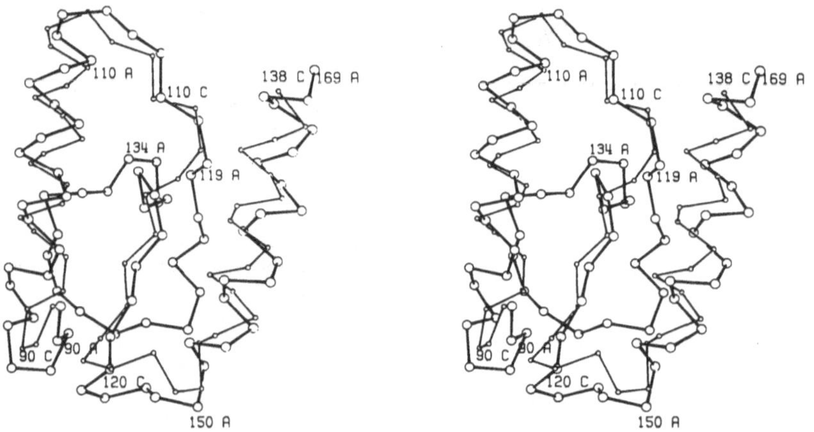

Figure 3. A drawing comparing the C-terminal halves of Clostridium MP (thick bonds) and A. nidulans flavodoxins. Major insertions occur in this half of the molecule, so that alignment of the C-termini of long- and short-chain sequences relies on the structures: A. nidulans flavodoxin residues 120-139 form an insert in the fifth strand of parallel sheet; the larger flavodoxin has 5 additional residues beyond 90 and two insertions in the connection leading to the last helix. The acidic sequence 144-148 of A. nidulans has no obvious homolog in short-chain flavodoxins. However, acidic residues or their amide analogs occupy very similar positions in the two molecules, even though the main chains take opposite directions at 145.

The FMN Binding Site in A. nidulans Flavodoxin

The environment and conformation of FMN in A. nidulans flavodoxin can now be described more fully. The flavin ring is enclosed by two aromatic residues (Fig. 4). The DNA sequence confirms the x-ray assignment of Trp at position 57, in contact with one side of the flavin ring, but indicates that the assignment of residue 94 as Trp was incorrect. We have reexamined the electron density in this region for independent verification that 94 is Tyr. When the side chain atoms are omitted from calculated structure factors, the shape of the density indeed matches Tyr better than Trp (Fig. 4).

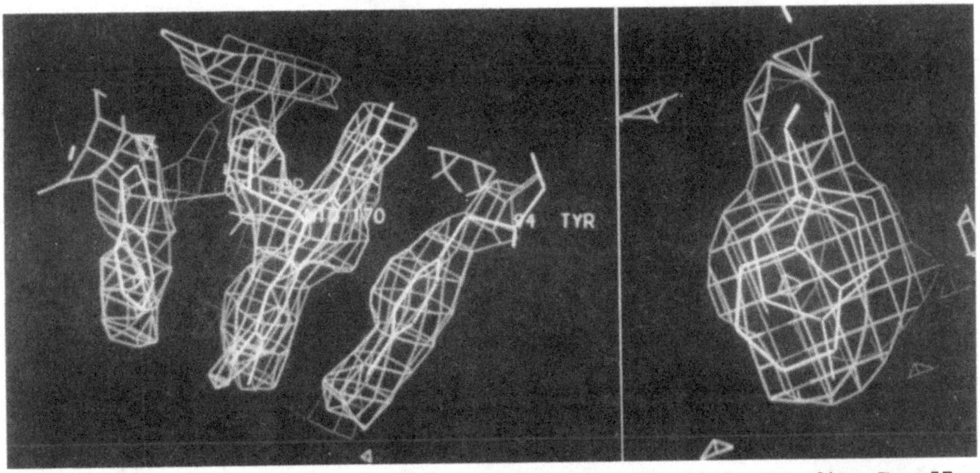

Figure 4. a) A view of the isoalloxazine ring and the surrounding Trp 57 (left) and Tyr 94 in a 2.5 Å electron density map, phased by isomorphous replacement (4). The atom positions have been adjusted by refinement.

b) Tyrosine is a better fit to the density at position 94 than is tryptophan. The electron density displayed here was calculated at 2 Å with model phases but excluding contributions from residue 94. The Trp ring extends beyond the contours, whereas the Tyr ring is well-accommodated in the density. Comparisons of amino acid sequences show that Tyr is the residue found most frequently at this position.

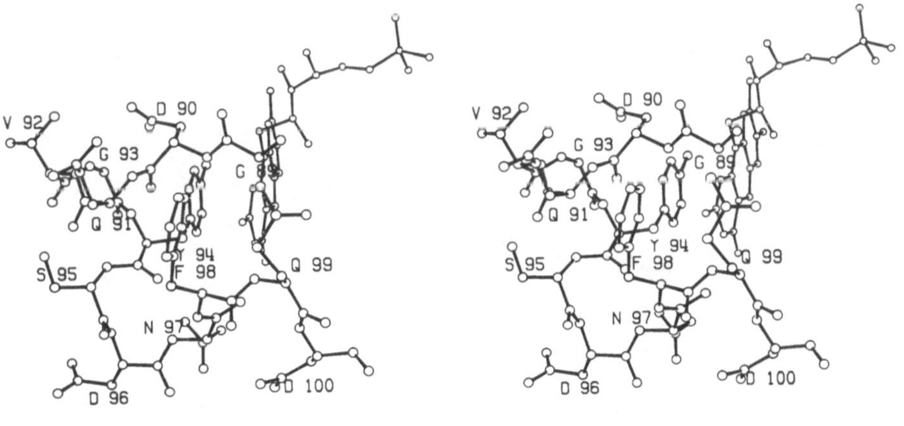

Figure 5. A stereoview showing the conformation of residues 89-100 of *A. nidulans* flavodoxin and interactions of these residues with FMN. Neighbors of N(1) are: O 56, 3.5 Å; N 90, 3.3 Å; of O(2): N 90, 3.2 Å; N 99, 3.3 Å; CD2 98, 3.5 Å; O 97, 3.5 Å; of N(3), O 97, 2.9 Å.

The isoalloxazine ring of oxidized A. nidulans flavodoxin is slightly bent. Restraints employed in refinement attempt to approximate a butterfly model in which atoms of each wing are coplanar; the resulting bending angle for the isoalloxazine ring shown in Figs. 5 and 6 is 8.8°.

Fig. 5 is a view of residues 89-100, identified by DNA sequencing, and their interactions with the isoalloxazine ring. The chain which wraps around the pyrimidine subnucleus contributes both hydrophobic and hydrogen-bonding contacts. It includes several acidic and amide side chains, some of them at the surface in positions where they may interact with other proteins. Residues Tyr 94 and Phe 98 provide hydrophobic contacts. Each of the pyrimidine N or O atoms is close to H-bond donor/acceptors: distances given in the figures indicate that the shortest hydrogen bonds are NH(3)-97 and O(4)-NH 60, whereas N(1) and O(2) are farther from potential donors. Asp at 90 occurs in all the sequences of Fig. 1, as well as in D. vulgaris flavodoxin. Asp may be conserved at this site not because of any interaction with FMN but because it stabilizes the backbone conformation by forming an Asx hydrogen bond (13) with N 93. Interactions of the flavin ring with residues 56-60 are shown in Fig. 6.

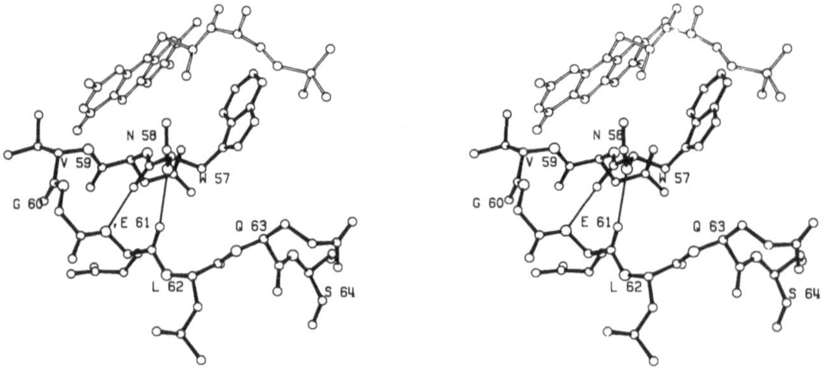

Figure 6. A stereoview of FMN and residues 57-65 in A. nidulans flavodoxin, oriented to show the three-residue bend at 58-60 and antiparallel hydrogen bonds between residues 57 and 61. The antiparallel interactions do not continue farther, since residues 62-65 veer away from 56-53, which become part of the central parallel sheet. Torsion angles ϕ, ψ at N 58, V 59 and G 60 are: (40,50); (50,80); and (90,10). The contact between O 56 and the ring can be seen in this view. Neighbors of O(4) are: N 59, 3.1 Å; N 60, 2.7 Å.

Oxidation-Reduction Potentials

Variations in the active site are an unusual feature of flavodoxins (14). Although the phosphate binding subsites near the N-terminus are very similar, sequences involved in isoalloxazine-protein interactions, residues 56-60 and 89-98 in A. nidulans flavodoxin, are not as well conserved. Figure 7 compares the lengths and compositions of these regions. The challenge is to relate the considerable differences in active center structures to species-dependent properties such as the oxidation-reduction potentials.

The species variations of potentials for flavodoxins (14-17) are substantial, ranging up to 150 mV. For the known examples, one can calculate (18,19) that the measured potentials require the FMN-protein association constant to be larger for FMN semiquinone than for oxidized FMN. On the other hand, comparison of the two-electron midpoint potentials and the sq/red potentials for the proteins with those for free flavin (20,21) shows the affinity of reduced FMN for the proteins to be less than that of either oxidized or semiquinone forms of FMN. In the case of A. nidulans flavodoxin at pH 7, with $E_{ox/sq} = -221$ mV and $E_{sq/red} = -447$ mV, K_{ox}/K_{red} is ~90 and K_{sq}/K_{red} is approximately 10^4 using the FMN potentials of Draper and Ingraham (20), or 10^5 using the values determined by Anderson (21).

The ox/sq potentials of Clostridium MP and A. nidulans flavodoxins are influenced by a conformation change which brings a backbone carbonyl oxygen close to the flavin NH(5) in the semiquinone structure (3,10). Species variations in potential may result from differences in the relative positions of NH(5), O(4), and the backbone carbonyl O in the semiquinones of different flavodoxins, or from varying contributions of the conformation changes. The conformational energies in particular are expected to depend on the amino acid sequence comprising the chain reversal. We have tried to assess the effect of the conformation change by treating the residues involved as the central pair of a 3_{10}, i.e. two residue, bend. The torsion angles deviate from those for ideal 3_{10} bends, but one can approximate the oxidized Clostridium MP or A. nidulans flavodoxins as Type III' and the semiquinone forms as Type II' (3). Because II' bends involve a conformation with ϕ=60, ψ=-120 at the first residue (Asn 58 in A. nidulans flavodoxin),

A. nidulans	^{57}W-N-V-G-E-L
A. vinelandii	^{57}L-G-E-G-E-L
K. pneumoniae	^{56}L-G-D-G-Q-L
Clostridium MP	^{56}M-G-D-E-V-L
D. vulgaris	^{60}W-G- -Q-L
	D-D-S-I

A nidulans	^{88}A-G-D-Q-V-G-Y-S-D-N-F-Q-D
A. vinelandii	^{96}L-G-D-Q-V-G-Y-P-E-D-Y-L-D
D. vulgaris	^{93}C-G-D- S-S -Y- E-Y -F-C-G
Clostridium MP	^{88}Y-G- -W-

Figure 7. Alignments of sequences which contact the isoalloxazine ring. For the three known structures, A. nidulans, Clostridium MP, and D. vulgaris flavodoxins, alignments are determined by spatial superpositions; the remaining alignments are derived from amino acid sequences. Glycine occurs frequently at positions corresponding to the first and third residues of three residue bends (upper sequences). The chain reversal in D. vulgaris flavodoxin contains a two-residue insert (3,11).

Gly is more acceptable than any other amino acid at this position. The analysis has suggested that Asn at 58, where Gly occurs in other structures (Fig. 7), contributes to the low ox/sq potential of A. nidulans flavodoxin; the magnitude of this effect can be tested by an N →G mutation.

A reappraisal of conformational energy contributions to the redox potentials is suggested by recent treatments of hairpin loops (22,23), which classify them according to the disposition of the final pair of antiparallel hydrogen bonds. Milner-White has identified the 57-59 turn in Clostridium MP flavodoxin as a three-residue loop (21); the refined coordinates of oxidized A. nidulans flavodoxin indicate that residues 58-60 of this flavodoxin also constitute a three-residue turn (Fig. 6). Preliminary surveys of known structures (23,24) reveal a tendency for the final residue in 3- or 4-residue loops to adopt angles near ϕ=90, ψ=0, in a region of conformational space preferred by Gly, and for the preceding residues to resemble Type I β-turns. Glycine is found in the third position in A. nidulans and Glu in Clostridium MP flavodoxin; both residues adopt the G1 bulge conformation with ϕ, ψ near 90,0. However, the preceding residues, as already noted, resemble III' bends in the oxidized state and II' bends in the semiquinone structures. A broader survey should be undertaken to see if flavodoxin loops are truly unusual, suggesting a perturbation accompanying FMN binding. The earlier conclusion that Gly at the first position will favor reduction to the semiquinone still seems valid; while Gly in the third position

(Cf. Fig. 7) could have a role in stabilizing three residue turns in flavodoxins, it is not clear that this Gly would affect the relative energies of the oxidized and semiquinone structures.

The very low potentials for the semiquinone/reduced equilibrium (-375 to -520 mV) are important for electron transfer reactions of flavodoxins. In efforts to rationalize these low potentials, the structure of reduced Clostridium MP flavodoxin (19,25) has been refined (26) to an R of 0.175 for data between 1.8 and 5.0 Å. The NH 89-N(1) distance of 3.0 Å is in agreement with the conclusions of Müller et al. (27) that the reduced species is the flavin anion: a hydrogen on N(1) would collide with the peptide H. The flavin ring bending angle is estimated at 4.2 to 8.6°, varying somewhat with the method of refinement (19). The atom positions are almost identical in this structure (Fig. 8) and in the semiquinone, yet the binding constant is decreased by at least three orders of magnitude. Why?

The planar flavin ring in reduced flavodoxin was initially thought to represent a high energy conformation (19,25,28) because several reduced model flavins were strongly bent about N(5)-N(10). NMR measurements by Müller and co-workers have subsequently questioned the presumption that free reduced flavins are non-planar. A reexamination of the "butterfly" motion (29) has concluded that ring inversion is actually rather facile (30), and hybridizations at C and N atoms of the reduced ring, deduced from chemical shifts and coupling constants, indicate that both model and protein-bound reduced flavin anions display a large degree of sp^2 character at N(10) and N(5) (27,31,32). The deviations from sp^2 hybridization, which differ somewhat in free flavins and flavodoxins (27), have been ascribed to local deformations rather than to overall ring bending. As an alternative to ring strain, electrostatic repulsion has been suggested (27,30) as a primary contributor to the decreased affinity of flavodoxins for reduced FMN, since the protein milieu offers no formal countercharge to the negatively charged N(1)..O(2) of the bound flavin anion.

The N(1) environment in flavodoxins can be compared with other sites that bind isolated anions: the phosphate site in flavodoxin, and the more dramatic sulfate site in sulfate binding protein (33). In these examples, the phosphate and sulfate dianions interact with backbone NH dipoles of α-helices. At the N(1) locus of flavodoxins, partial charge compensation can be provided by backbone amides, which are at 89 and 91 in Clostridium MP flavodoxin (Fig. 8), but the apolar contacts, the proximity of a carbonyl oxygen at 55, and the absence of a helix all combine to make this site appear less satisfactory for anion binding than the phosphate or sulfate sites. Species differences support the idea that the environment of N(1) may affect the sq/red potentials; e.g. in Clostridium MP flavodoxin a channel from solvent terminates near O(2) at a tightly bound water, whereas O(2) in A. nidulans flavodoxin, which has a lower potential (by 50 mV), is surrounded by protein atoms (Cf. Figs. 5 and 8).

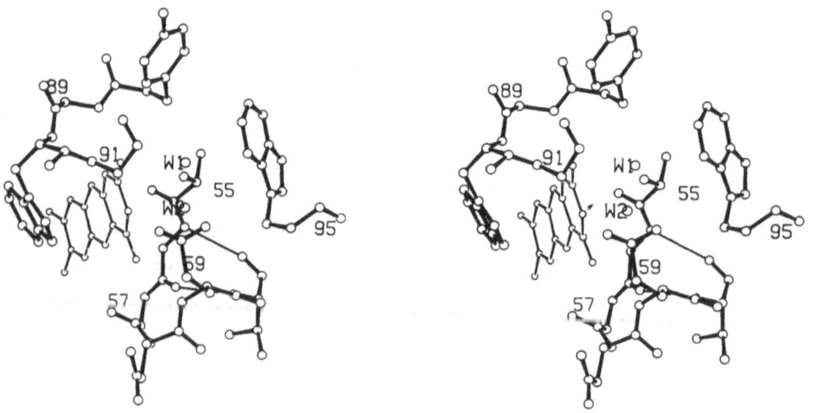

Figure 8. A stereoview of reduced flavodoxin from Clostridium MP flavodoxin, showing the surroundings of the N(1)..O(2) center, which carries a negative charge in the reduced flavin anion. The NH dipoles at 89 and 91 point toward these flavin atoms, with an N(1)-NH 89 and O(2)-NH 91 distances of 3.0 Å and 2.8 Å. The carbonyl oxygen of Ala 55 is 4.2 Å from N(1) and atom CD2 of Tyr 88 is in contact with O(2). A hydrogen bond connects N(5) and O 57, which are separated by 3.1 Å. Two bound water molecules, labelled W1 and W2, occupy sites near the carboxyl group of Glu 59; W1 is 3.1 Å from O(2). Ribityl and Met 56 side chains are omitted to simplify the drawing.

The pKs of modified flavins and of reduced FMN bound to flavodoxins should indicate electrostatic effects. Modified flavin probes such as 6-OH and 6-SH FMN display increased pKs in M. elsdenii flavodoxin (34), consistent with the interpretation that a negative charge at N(1)..O(2) is unfavorable. However, the pK for reduced FMN-flavodoxins, determined from redox potential measurements, is almost the same for Clostridium MP flavodoxin as for free flavin, and is lower than the free flavin value in M. elsdenii flavodoxin (14). Given the reasonable presumption that these pKs refer to protonation at N(1), the potential measurements are incompatible with large unfavorable effects of charging N(1), which ought to increase pKs significantly. The explanation of the low sq/red potentials may lie in the combination of many contributions, including changes of hybridization or ring mobility accompanying protein binding, differences in hydrogen bond strengths between reduced and semiquinone forms, and electrostatic effects.

Acknowledgments

This research has been supported by grants from the National Institutes of Health (GM 16429) and National Science Foundation (PCM 8401634) to M.L. and a grant from NSERC of Canada to N. S.

References

1. Cheddar, G., Meyer, T.E., Cusanovich, M.A., Stout, C.D., Tollin, G. 1986. Biochemistry 25, 6502.

2. Laudenbach, D.E., Reith, M.E., and Straus, N.A. 1987. J. Bacteriology (submitted).

3. Ludwig, M.L., Pattridge, K.A., Tarr, G. 1984. In: Flavins and Flavoproteins (R.C. Bray, P.C. Engel, S.G. Mayhew, eds.). deGruyter, p. 253.

4. Smith, W.W., Pattridge, K.A., Ludwig, M.L., Petsko, G.A., Tsernoglou, D., Tanaka, N., Yasunobu, K.T. 1983. J. Mol. Biol. 165, 737.

5. Hendrickson, W.A. and Konnert, J.H. 1980. In: Computing in Crystallography (R. Diamond, S. Rameseshan, and K. Venkatesan, eds.) Indian Acad. Sci., Bangalore, p. 13.01.

6. Lau, R.H., Alvarado-Urbina, G., Lau, P.C.K. 1987. Gene 52, 21.

7. Drummond, M.H. 1985. Biochem. J. 232, 891.

8. Smillie, R.M. and Entsch, B. 1971. Meth. Enz. 23, 504.

9. Drummond, M.H. 1986. Eur. J. Biochem. 159, 549.

10. Smith, W.W., Burnett, R.M., Darling, G.D., Ludwig, M.L. 1977. J. Mol. Biol. 117, 195.

11. Watenpaugh, K.D., Sieker, L.C., and Jensen, L.H. 1973. Proc. Nat. Acad. Sci. USA 70, 3857.

12. Weber, P.C. and Tollin, G. 1985. J. Biol. Chem. 260, 5568.

13. Baker, E.N. and Hubbard, R.E. 1984. Prog. Biophys. Mol. Biol. 44, 97.

14. Mayhew, S.G. and Ludwig, M.L. 1975. In: The Enzymes (P.D. Boyer, ed.) Academic Press, Vol. 12, p. 57.

15. Deistung, J. and Thorneley, R.N.F. 1986. Biochem. J. 239, 69.

16. Sykes, G.A. and Rogers, L.J. 1984. Biochem. J. 217, 845.

17. Klugkist, J., Voorberg, J., Haaker, H., Veeger, C. 1986. Eur. J. Biochem. 155, 33.

18. Mayhew, S.G. 1971. Biochim. Biophys. Acta 235, 289.

19. Smith, W.W., Ludwig, M.L., Pattridge, K.A., Tsernoglou, D., Petsko, G.A. 1978. In: Frontiers of Biological Energetics (P.L. Dutton, J.S. Leigh, A. Scarpa, eds.) Academic Press, Vol. 2, p 957.

20. Draper, R.D. and Ingraham, L.L. 1968. Arch. Biochem. Biophys. 125, 802.

21. Anderson, R.F. 1983. Biochim. Biophys. Acta 722, 158.

22. Milner-White, E.J. and Poet, R. 1986. Biochem. J. 240, 289.

23. Sibanda, B.L. and Thornton, J.M. 1985. Nature 316, 170.

24. Milner-White, E.J. 1987. Biochim. Biophys. Acta 911, 261.

25. Ludwig, M.L., Burnett, R.M., Darling, G.D., Jordan, S.J., Kendall, D.S., Smith, W.W. 1975. In: Flavins and Flavoproteins (T.P. Singer, ed.). Elsevier, p. 393.

26. Hoard, L.H. and Nordman, C.E. 1979. Acta Cryst. A35, 1010.

27. Vervoort, J., Müller, F., Mayhew, S.G., van den Berg, W.A.M., Moonen, C.T.W., and Bacher, A. 1986. Biochemistry 25, 6789.

28. Simondsen, R.P. and Tollin, G. 1980. Mol. Cell. Biochem. 33, 13.

29. Tauscher L., Ghisla, S., Hemmerich, P. 1973. Helv. Chim. Acta 56, 630.

30. Moonen, C.T., Vervoort, J., Müller, F. 1984. Biochemistry 23, 4868.

31. Moonen, C.T., Vervoort, J., Müller, F. 1984. Biochemistry 23, 4859.

32. Franken, H.-D., Rüterjans, H., Müller, F. (1984). Eur. J. Biochem. 138, 481.

33. Pflugrath, J.W. and Quiocho, F.A. (1985). Nature 314, 257.

34. Ghisla, S. and Massey, V. (1986) Biochem. J. 239, 1.

C-13, N-15 AND TWO-DIMENSIONAL NMR TECHNIQUES IN FLAVOPROTEIN RESEARCH

Franz Müller, Jacques Vervoort, Carlo P.M. van Mierlo, Steve G. Mayhew, Willem J.H. van Berkel and Adelbert Bacher

Department of Biochemistry, Agricultural University, De Dreijen 11, 6703 BC Wageningen, The Netherlands

Department of Biochemistry, University College, Belfield, Dublin 4, Ireland

Department of Organic Chemistry and Biochemistry, Technical University of Munich, D-8046 Garching, Federal Republic of Germany

Introduction

Our knowledge about the structure-function relationship of flavoproteins has increased considerably in the last two decades. These insights are mostly derived from kinetic and biochemical studies. Yet, we still do not understand well the principles responsible for the tuning of a particular flavoprotein for its specific biological action. Although several hypotheses have been put forward (1,2) only little experimental evidence is available in support of these hypotheses. It is obvious that direct evidence for the possible specific interactions between the prosthetic group and a particular apoflavoprotein can only be obtained by physical techniques yielding detailed information, preferably on an atomic level. There are only a few techniques fulfilling this requirement, i.e. X-ray studies if a high resolution can be achieved and NMR techniques. Other techniques such as light absorption, fluorescence, EPR and Raman spectroscopy yield less detailed information either because of lack of high resolution with respect to structural information or because of great complexity of the spectral content. In the following we will describe the use of various NMR techniques in flavoprotein research and discuss the structural information, which can be obtained by these techniques. A great disadvantage of NMR techniques is the fact that usually high concentrations of biological

Flavins and Flavoproteins 1987

materials are needed. This problem can partially be overcome by isotopic enrichment of e.g. the prosthetic group. Still concentrations in the range of 0.1-5 mM are required to obtain high quality C-13, N-15 and P-31 spectra. This in turn means that suitable methods for the preparation of apoflavoproteins for C-13 and N-15 NMR must be available and that the proteins must be highly soluble. The concentrations needed for NMR measurements seem high but they represent for many flavoproteins the natural occurrence in the cell. Therefore the information content of the NMR spectra is biologically relevant.

Results and Discussion

It is appropriate to discuss first briefly the basic principles of the interpretation of C-13 and N-15 NMR spectra. This shall be demonstrated with the aid of the free flavin molecule. The C-13 chemical shifts of a molecule are generally related to the π electron density of the corresponding atoms (3). It has been shown that this is also the case for oxidized and reduced flavin (4). Therefore the C-13 chemical shifts observed in protein-bound flavin can reliably be interpreted in terms of π electron changes with respect to free flavin. The nitrogen atoms in heterocyclic aromatic compounds are generally categorized as pyridine-like or β-type and pyrrole-like or α-type atoms (5). The N-15 chemical shifts of the former are very sensitive to hydrogen bonding interactions (upfield shift) whereas those of the latter are sensitive and show a small downfield shift on hydrogen bonding. In oxidized flavin the N(1) and N(5) atoms are of the β-type and the N(3) and N(10) atoms are of the α-type. In the reduced state all four nitrogen atoms of flavin are of the α-type.

It is an well established fact that oxidized flavin shows solvent polarity-dependent light absorption spectra, i.e. the spectrum is unresolved in H_2O, partly resolved in methanol and well resolved in $CHCl_3$. The C-13 and N-15 NMR spectra are also solvent dependent. Especially the chemical shift of the N(10) atom shows a remarkable dependence on the solvent. On going from $CHCl_3$ to H_2O the resonance of N(10) shifts downfield by about 12 ppm. Addition of methanol to the $CHCl_3$ solution causes a 2 ppm downfield shift. Since the N(10) atom cannot form a hydrogen bond with protic solvents, the large downfield shift can only be explained by a change of hybridization of the N(10) atom (6). This means that the N(10) atom is less sp^2 hybridized in $CHCl_3$ than in H_2O. Consequently the N(10) atom is somewhat out of the molecular plane in $CHCl_3$. Therefore the N(10) atom plays a central role in accomodating π electron density from the three-membered ring system or in releasing π electron density into the ring. The latter only occurs in solvents of high polarity, the charge being delocalized onto the C(4) carbonyl group. This interpretation is not solely based on the N(10) chemical shift but is also supported by the C-13 chemical shifts (6).

Reduced flavin has been considered to possess a bent structure. This idea was first based on the fact that reduced flavin contains formally eight π electrons in the central ring. Such a state is called anti-aromatic and represents an energetically unfavorable situation. The bent conformation of reduced flavin was later proven by X-ray studies (7). In these studies, however, heavily substituted flavins were used. However, C-13 and N-15 NMR data on neutral reduced flavin clearly show that the molecule possesses a nearly planar structure (6). The reason that reduced flavin can attain a planar structure is the fact that the lone pair of the N(10) atom is delocalized onto the C(4a) and C(5a) atoms in apolar solvents and onto the C(4) carbonyl group in polar solvents. This mesomerism allows the reduced flavin molecule to escape the anti-aromatic character and to attain a planar structure. However, if the molecule is substituted at N(1) and or N(5) then the structure becomes more bent. Furthermore NMR studies on the dynamics of the reduced flavin molecule indicate that nitrogen inversion does not take place, but ring inversion can be observed. In addition ring inversion is strongly dependent on the solvent polarity and hybridization of the N(5) atom (8). In this context we would like to emphasize that the molar extinction coefficient at 450 nm of reduced flavin is not correlated with the planarity of the molecule as previously suggested (9), but the molar extinction coefficient correlates with the degree of sp^2 hybridization of the N(5) atom. Ionization of reduced flavin leads to pH-dependent chemical shifts. The largest shifts are observed for C(10a) and C(2) (~ 10 ppm downfield) and for N(1) (~ 60 ppm downfield). The pK_a values calculated from C-13 NMR spectra (10) and N-15 NMR spectra (11,12) are 6.7.

These downfield shifts are caused by the negative charge on N(1) (electric field effect) and are valuable diagnostic tools for the evaluation of the molecular state of protein-bound reduced flavin.

Various flavodoxins form a red, protein-bound chromophore by a photochemical reaction in the presence of certain alcohols or acids (13). The structure of the chromophore has now been elucidated by C-13 and N-15 NMR. The relevant C-13 and N-15 chemical shifts of the red, the oxidized and the reduced forms of flavodoxin are compared in Table 1. Although several resonances are split in two peaks only the two peaks of the C(8) resonance are given. From Table 1 the following can be concluded:

1) the resonances due to C(4a), C(6), C(10a), N(1), N(3) and N(10) resemble those of reduced flavodoxin;
2) the resonances due to C(7) and N(5) resemble those of oxidized flavodoxin;
3) the C(8) resonance indicates that the atom is sp^3 hybridized. Thus the chromophore of the red flavodoxin possesses the structure of ionized 1,8-dihydroflavin (Fig. 1). In the photochemical reaction the generated alkyl residue from the substrate is added at C(8).

At high field in the C-13 NMR spectrum many resonance lines are observed. These lines are due to aliphatic carbon atoms and are absent in the spectrum of the native protein. Obviously many butyryl residues are also incorporated into the protein. In this context it would be tempting to speculate about the mechanism of the alkylation of the prosthetic group and the protein.

From the above it follows that the structure of a modified prosthetic group in a flavoprotein can relatively easily be elucidated by C-13 and N-15 NMR techniques, if a large variety of enriched flavin derivatives are available. The use of only one derivative, labeled at one particular position, is in general a rather unsafe approach, as has been shown recently by the determination of the structure of the dihydroperoxyflavin intermediate in bacterial luciferase (14,15). These studies also demonstrate that it is of utmost importance to run the proper blank NMR spectra when organic cosolvents are used. Whenever possible it is advisable to use the usual buffer systems only in order to avoid the measurement of resonances,

Table 1. Some C-13 and the N-15 chemicals shifts (in ppm) of flavodoxin from <u>M.elsdenii</u> in the oxidized, reduced and red state.

Atom	State		
	Oxidized	Reduced	Red
C(4a)	135.6	103.5	97.8
C(6)	133.0	112.4	117.2
C(7)	141.4	131.1	146.8
C(8)	153.0	125.9	48.5; 50.9
C(8α)	22.1	19.2	29.4
C(10a)	151.5	154.5	155.8
N(1)	185.0	183.4	184.1
N(3)	160.9	149.7	150.6
N(5)	349.3	61.3	310.7
N(10)	165.6	98.3	112.6

which represent artifacts (15).

In the last few years we have investigated the interaction of the prosthetic group with various apoflavoproteins. Flavodoxins from different sources were studied in detail (10,11,16,17). These studies show that information with regard to specific hydrogen bonding between the prosthetic group and the apoprotein are obtained. Moreover the perturbation of the electronic structure of protein-bound flavin as compared to that of free flavin, can be elucidated. The data show that there exists subtle differences in the π electron distribution of flavin in the different flavodoxins. These studies yield also great detail about the conformation of the oxidized and reduced protein-bound flavin. Surprisingly in several cases it has been observed that the N(10) atom in oxidized and the N(5) atom in the reduced state are usually sp^2 hybridized, but a considerable degree of sp^3 hybridization can in a few cases be observed (16). This change in hybridization is directly related to the above mentioned π electron distribution in the flavin molecule and depends strongly on the polarity of the environment of the prosthetic group and the possibility to form hydrogen bonds with C(2)O or C(4)O (6). Remarkable also is the fact

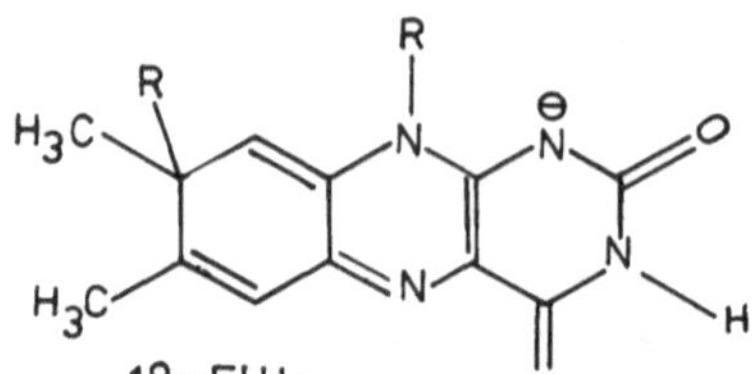

Fig. 1 The structure of the red chromophore in flavodoxin from M.elsdenii as deduced from NMR data.

that the reduced prosthetic group is ionized. This observation seems to be a rule rather than exception, since this fact has also been observed in flavoenzymes (21,22). The ionization of N(1) of flavin in flavoproteins led us to propose that the negative charge may play an important role in the regulation of the redox potential in flavoproteins (23).

More recently we have investigated lipoamide dehydrogenase, glutathione reductase and mercuric reductase. These enzymes have a disulfide group in the active center and form a red colored intermediate in the two-electron reduced state. The structural and electronic similarities of the three above mentioned enzymes in the three oxidation states are illustrated by the C-13 chemical shifts (Table 2).
The results indicate that the prosthetic group in all four proteins is polarized comparable to FMN in H_2O. In the EH_2 state the resonance of the C(4a) atom appears at lower field than that in the oxidized state. The same trend is observed for the green form of butyryl CoA dehydrogenase. The downfield shift of the resonance due to C(4a) is surprising. We expected an upfield shift as in phenol-complexed Old Yellow enzyme (18,19) due to charge delocalization from the ligand to the prosthetic group. Therefore the downfield shift is a consequence of the negatively charged persulfide group localized in the neighbourhood of the flavin (20).

Finally we wish to report briefly on results obtained by 2D NMR techniques (see paper by Van Mierlo et al., this volume). Fig. 2 shows on top a part of a stacked plot of Megasphaera elsdenii flavodoxin, which belongs to the class of proteins generally called "non-crystallizable". With 2D NMR techniques it is possible to unravel the three-dimensional structure of such

Table 2. The chemicals shifts of the carbon atoms C(2), C(4), C(4a) and C(10a) of lipoamide dehydrogenase, glutathione reductase, mercuric reductase and butyryl CoA dehydrogenase from M.elsdenii in the different redox states.

Atom	Mer.red.	Glut.red.	Lip.DH	Butyryl CoA DH (Yellow Form)
	Oxidized State			
C(2)	159.6	160.4	158.3	158.2
C(4)	163.6	163.3	163.2	-
C(4a)	137.5	135.7	137.6	138.7
C(10a)	152.5	152.3	152.2	-
	EH_2-State			Green Form
C(2)	160.1	161.6	158.6	158.0
C(4)	164.5	163.9	163.8	-
C(4a)	140.7	139.9	140.3	140.0
C(10a)	153.8	153.8	153.0	-
	EH_4-State			Reduced State (Yellow Form)
C(2)	160.1	161.0	159.4	158.3
C(4)	161.8	162.0	159.7	-
C(4a)	98.9	99.7	98.7	104.4
C(10a)	158.6	159.1	157.2	-

proteins. Presently the size of flavodoxin represents about the limit such proteins can be studied successfully. In the figure the aromatic part of the spectrum of reduced flavodoxin is shown. The diagonal respresents the conventional spectrum. It is obvious that from such a spectrum it would be rather tedious, if possible at all, to assign the peaks to individual atoms of the protein. However, using two frequencies, in our case the frequencies are identical, a two-dimensional spectrum can be created. The off diagonal peaks were obtained using the "spin-lock" COSY technique. The small peaks observed by this technique represent resonances which are connected to each other through one to several bonds. The same results are shown on bottom of Fig. 2 as a contour plot. As an example the resonance assignment of a tryptophan residue of the benzene subnucleus of tryptophan

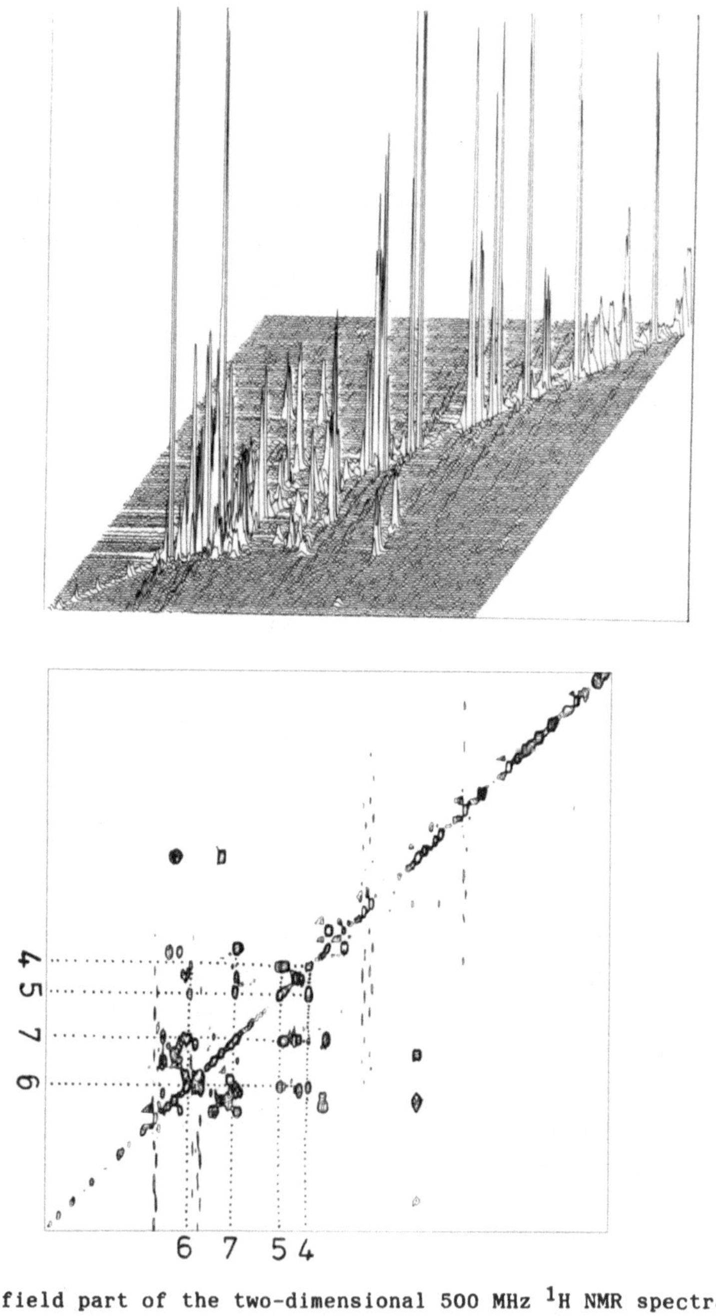

Fig. 2 Low field part of the two-dimensional 500 MHz ^{1}H NMR spectra of reduced M.elsdenii flavodoxin in deuterium oxide. Top: stacked plot, bottom: contour plot.

is indicated in Fig. 2. The usual COSY spectrum is not shown, but such a spectrum yields the resonances of directly coupled protons, up to three bonds. In this way the resonance assignments as shown in Fig. 2 have been obtained i.e. H(4), H(5), H(6) and H(7). As a matter of fact there was only one unique assignment possible for the pattern shown in Fig. 2. After the assignment of particular amino acid residues is ascertained this set can be further used to create NOESY spectra, which give information on the spatial arrangements of the residue with respect to other residues, i.e. three-dimensional structure. However, it should be mentioned that an usual COSY spectrum of a complex protein does not yield an unambiguous assignment. The assignment must be further supported by additional experiments, preferably by relay experiments in which magnetization transfer over at least three to six bonds can be observed. Such experiments yield the final assignments.

Acknowledgements

This study was done under the auspices of The Netherlands Foundation for for Chemical Research (SON) with financial aid from The Netherlands Organization for the Advancement of Pure Research (ZWO).

References

1 Müller, F.: Z.Naturforsch. 27b, 1023-1026 (1972).
2 Massey, V., Hemmerich, P.: Biochem.Soc.Trans. 8, 246-257 (1980).
3 Stothers, J.B.: Carbon-13 NMR Spectroscopy, pp. 313-385, Academic Press, New York (1972).
4 Van Schagen, C.G., Müller, F.: Helv.Chim.Acta 63, 2187-2201 (1980).
5 Witanowski, M., Stefaniak, L, Webb, G.A.: in Annual Report Nuclear Magnetic Resonance Spectroscopy 11B (G.A. Webb, ed.) pp. 1-439, Academic Press, London (1981).
6 Moonen, C.T.W., Vervoort, J., Müller, F.: Biochemistry 23, 4859-4867 (1984).
7 Kierkegaard, P., Norrestam, R., Werner, P.-E., Csöregh, I., Von Glehn, M., Karlsson, R., Leijonmark, M., Rönnquist, O, Stensland, B., Tillberg, O., Torbjörnssen, C.: in Flavins and Flavoproteins (H.

Kamin, ed.) pp. 1-22, University Park Press, Baltimore (1971).

8 Moonen, C.T.W., Vervoort, J., Müller, F.: Biochemistry 23, 4868-4872 (1984).

9 Dudley, K.H., Ehrenberg, A., Hemmerich, P., Müller, F.: Helv.Chim.Acta 47, 1354-1383 (1964).

10 Van Schagen, C.G., Müller, F.: Eur.J.Biochem. 120, 33-39 (1981).

11 Franken, H.-D., Rüterjans, H., Müller, F.: Eur.J.Biochem. 138, 481-489 (1984).

12 Moonen, C.T.W., Van den Berg, W.A.M., Boerjan, M., Müller, F.: Biochemistry 23, 4873-3878 (1984).

13 Mayhew, S.G., Massey, V.: in Flavins and Flavoproteins (R.C. Bray, P.C. Engel and S.G. Mayhew, eds.) pp. 261-267, W. de Gruyter, Berlin (1984).

14 Vervoort, J., Müller, F., Lee, J., Van den Berg, W.A.M., Moonen, C.T.W.: Biochemistry 25, 8062-8067 (1986).

15 Ghisla, S., Hastings, J.W., Favoudon, V., Lhoste, J-H.: Proc.Natl.Acad. Sci. USA 75, 5860-5863 (1978).

16 Vervoort, J., Müller, F., Mayhew, S.G., Van den Berg, W.A.M., Moonen, C.T.W., Bacher, A.: Biochemistry 25, 6789-6799 (1986).

17 Vervoort, J., Van Berkel, W.J.H., Mayhew, S.G., Müller, F., Bacher, A., Nielsen, P., LeGall, J.: Eur.J.Biochemi. 161, 749-756 (1986).

18 Beinert, W-D., Rüterjans, H., Müller, F.: Eur.J.Biochem. 152, 573-579 (1985).

19 Beinert, W-D., Rüterjans, H., Müller, F., Bacher, A.: Eur.J.Biochem. 152, 581-587 (1985).

20 Williamson, G., Engel, P., Mizzer, J.P., Thorpe, C. and Massey, V.: J.Biol.Chem. 257, 4314-4320 (1982).

21 Vervoort, J., Müller, F., O'Kane, D.J., Lee, J., Bacher, A.: Biochemistry 25, 8067-8075 (1985).

22 Van den Berg, W.A.M., Vervoort, J., Moonen, C.T.W., Müller, F., Carlberg, I., Mannervik, B.: in Flavin and Flavoproteins (R.C. Bray, P.C. Engel, and S.G. Mayhew, eds.) pp. 143-146, W. de Gruyter, Berlin (1984).

23 Moonen, C.T.W., Vervoort, J., Müller, F.: in Flavins and Flavoproteins (R.C. Bray, P.C. Engel and S.G. Mayhew, eds.) pp. 493-496, W. de Gruyter, Berlin (1984).

TWO DIMENSIONAL NMR STUDIES ON MEGASPHAERA ELSDENII FLAVODOXIN

Carlo van Mierlo, Jacques Vervoort and Franz Müller
Department of Biochemistry, Agricultural University, De Dreyen 11, 6703 BC Wageningen, The Netherlands.

Introduction

Megasphaera elsdenii flavodoxin is a "non-crystallizable" protein consisting of 137 amino acid residues. The protein in the oxidized and reduced state is currently under active study in our laboratory using phase-sensitive two-dimensional NMR techniques (1). It is one of the largest proteins tackled by these techniques. Recently developed two-dimensional NMR techniques (phase-sensitive double-quantum filtered Cosy and phase-sensitive "spinlock"-spectra (2)) are crucial in the study of proteins of this size. The spectra obtained from Megasphaera elsdenii flavodoxin are of a good quality and therefore promising.

Results and Discussion

The strategy used to study the flavodoxin by 2D-NMR is as follows:

a. Determination of individual amino acid systems (spinsystems) with the use of

   a.1 Phase-sensitive double-quantum-filtered Cosy-spectra: These spectra contain information about directly coupled protons within an amino acid. The cross-peak multiplet-structures give thereby also insight in the "passive" couplings between protons within an amino acid (3).
   In crowded spectra, as in our case and in case of small coupling constants, it is often difficult to unravel the "passive"-coupling-puzzle partly because of peak-cancellation.

Flavins and Flavoproteins 1987

a.2 Phase-sensitive relay-Cosy spectra (spinlock-spectra) (2): Protons which are not "directly" coupled within an particular spin system can be connected with each other via this type of spectra. These spectra do not suffer from peak-cancellation because there is a net magnetization transfer between coupled protons. By using variabel "locktimes", different relay connectivities can be observed.

Fig. 1 shows a part of a spinlock spectrum of the flavodoxin in an aqueous solution. Relay connectivities from the amide protons to Cα and Cβ protons can be observed. There is a good peak-resolution as compared with the double-quantum-filtered Cosy-spectra.

b. The assignment of positions to the individual amino acids in the primary structure of the flavodoxin with the use of Noesy-spectra: The spectra contain information about the distances between protons via their dipolar interactions and yield therefore information about the sequence of the protein.

c. The spatial distances between amino acids can be determined by Noesy-spectra. These make it possible, using the distance constrains, to generate the three-dimensional structure of the native flavodoxin.

At this moment 81 of the 137 amino acid residues of the reduced flavodoxin have been assigned. Among them are several amino acids, which are in the immediate neighbourhood of the flavin. They could be observed using the paramagnetic properties of the flavin in the semiquinone state. In this state the protons of amino acids in the neighbourhood of the flavin disappear by a broadening and a leakage mechanism (1). These mechanisms function effectivily during the mixing time in a Noesy and during the locktime in a spinlock experiment. Due to a fast electron-exchange between flavodoxin molecules, every molecule is affected by these mechanisms.
Previously we reported on the use of 2D difference spectra from flavodoxin in the reduced and partly reduced state in order to simplify the complex spectra and to assign the amino acids in the active center. In such spectra the resonances appear as positive cross-peaks. However,

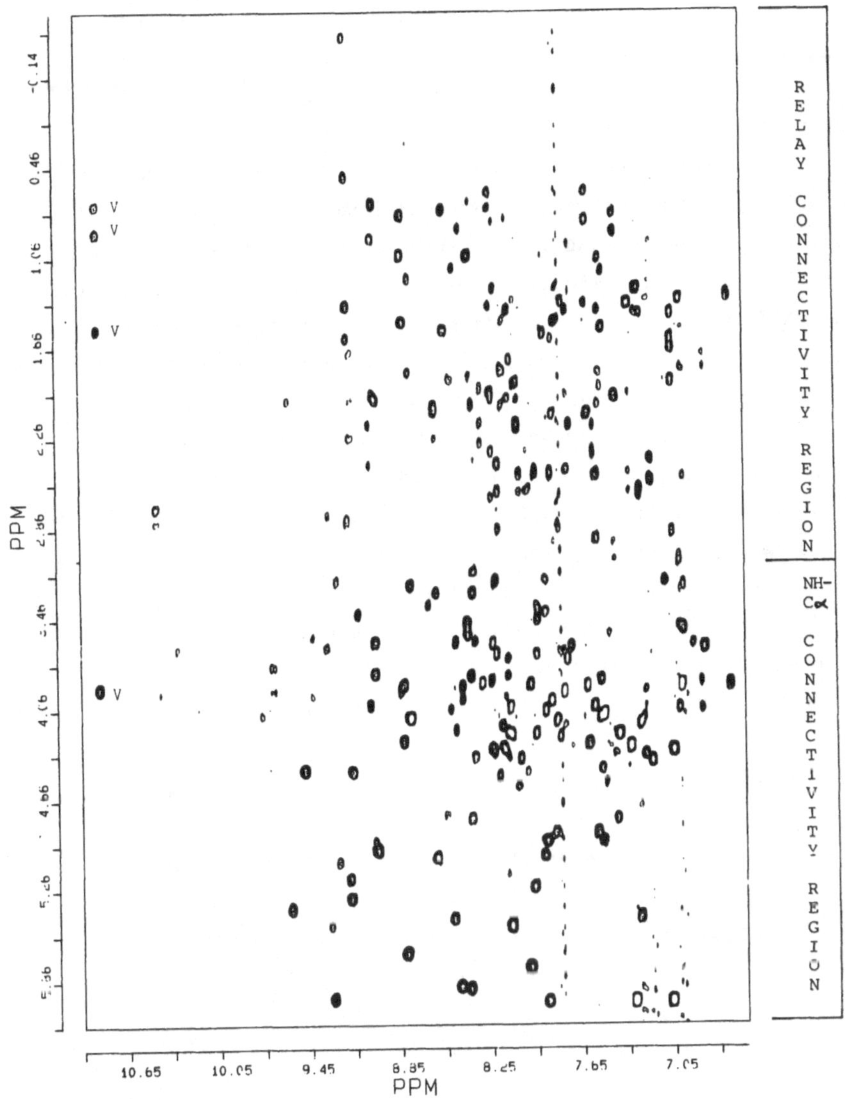

Fig. 1 Part of a 500 MHz phase-sensitive "spinlock"-spectrum of a 9 mM flavodoxin solution in 0.1 M potassium phosphate buffer (pH = 8.3). The solvent consists of 90% H_2O and 10% ^2H_2O. A Valine spinsystem is shown in the figure.

recently, we noticed also the appearance of negative crosspeaks, combined with positive crosspeaks in the difference spectra. Both peaks are the result of peakshifts of protons and are caused by a pH-change of the solution, which induces a slight conformational change of the protein. For example the NH-Cα-region in 2D NMR-spectra is highly characteristic for a particular conformational form of a protein ("finger-print" region) and is sensitive to minor changes in conformation (4). The pH-sensitivity of the spectra makes it also possible to find "hidden" amino acids in 2D spectra. We have not observed peakshifts for amino acids localized in the immediate neighbourhood of the flavin. It can therefore be concluded that the conformation around the flavin stays unaffected on minor pH changes. On the other hand proton resonance positions of amino acids in the neighbourhood of the flavin show great differences between the oxidized and the reduced state indicating a conformational change around the flavin on going from the reduced to the oxidized state.
Our study is at the threshold of making sequential resonance assignments. From these and other experiments the 3D structure of the oxidized and reduced flavodoxin will be obtained in the near future.

Acknowledgement

This study was done under the auspices of the Netherlands Foundation for Chemical Research (SON) with financial aid from the Netherlands Organization for the Advancement of Pure Research (ZWO).

References

1. Moonen, C.T.W., Scheek, R.M., Boelens R., Muller, F.: Eur.J.Biochem. 141, 323-330 (1984).
2. Bax, A., Davis, D.G.: J.Magn.Reson. 65, 355-360, (1985).
3. Neuhaus, D., Wagner, G., Vasák, M., Kägi, J.H.R., Wütrich, K.: Eur.J.Biochem., 51, 257-273, (1985).
4. Holak, T.A., Frederick, A.F., Prestegard, J.H.: J.Biol.Chem. 262, 3685-3689, (1987).

FLAVODOXIN FROM *ANABAENA* 7120: UNIFORM N-15 ENRICHMENT AND NITROGEN-15 AND PHOSPHORUS-31 NMR INVESTIGATIONS OF THE FMN BINDING SITE IN THE REDUCED AND OXIDIZED STATES

Brian J. Stockman and John L. Markley

Department of Biochemistry, 420 Henry Mall, University of Wisconsin-Madison, Madison, WI 53706

Introduction

The flavodoxin isolated from *Anabaena* 7120 has been uniformly enriched to 95% ^{15}N by growing the cyanobacteria on [^{15}N] KNO_3 as the sole nitrogen source. ^{15}N and ^{31}P NMR has been used to characterize the FMN binding site in the reduced and oxidized states.

Results

The ^{31}P NMR spectrum in both redox states shows a single resonance, corresponding to the phosphate of FMN. The chemical shift is 5.4 ppm (relative to phosphoric acid) in the oxidized protein. In the reduced protein the resonance is shifted downfield to 5.9 ppm. The chemical shifts indicate that FMN binds as the dianionic phosphate monoester in both redox states (1). A downfield shift upon reduction is characteristic of flavodoxins (2). The magnitude of the shift in this flavodoxin is greater by 0.4 ppm than that seen for any other flavodoxin yet reported (2). The size of the shift can be explained by a slight reduction in the smallest O-P-O bond angle (3), and thus suggests that a conformational change occurs at the phosphate binding site upon reduction.

Figure 1 shows direct-observation ^{15}N NMR spectra for flavodoxin in the oxidized and reduced states. Peak

Flavins and Flavoproteins 1987

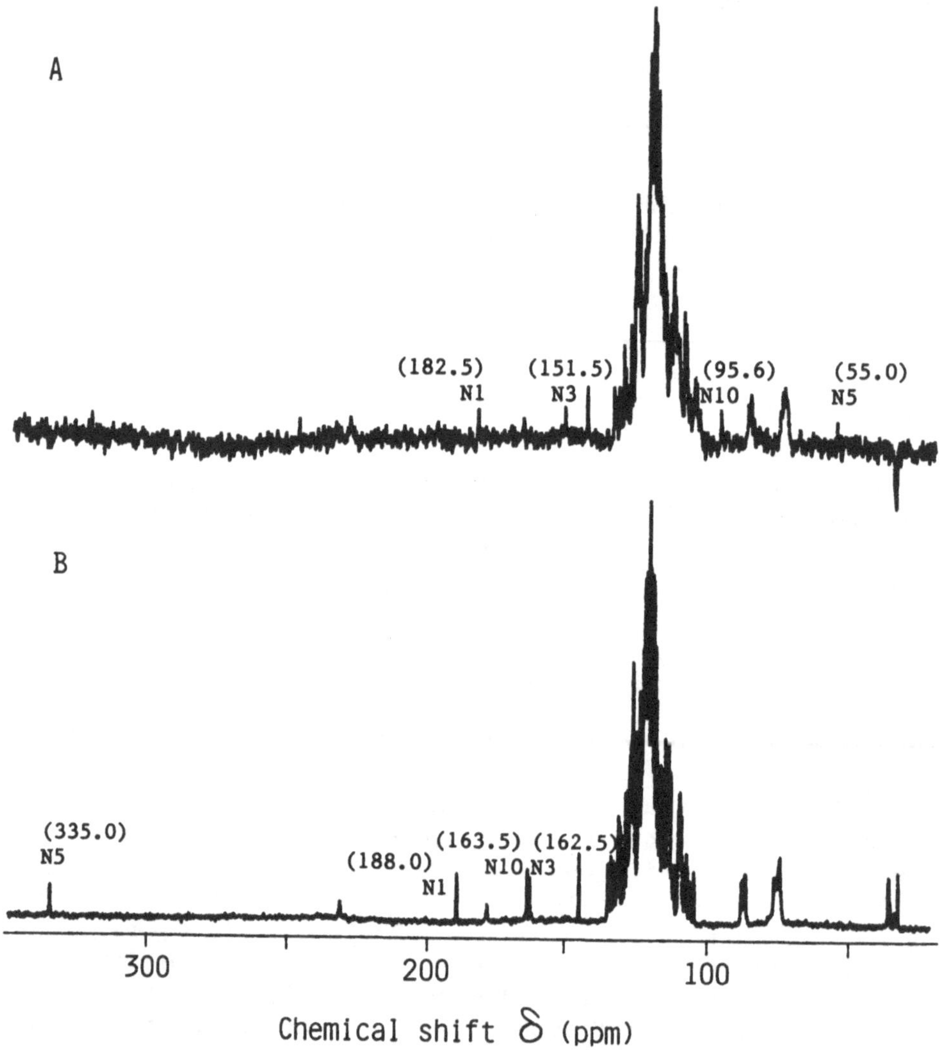

FIGURE 1. ^{15}N NMR (50.68 MHz) direct-observation spectra of [^{15}N]flavodoxin in the (A) reduced (6 000 transients) and (B) oxidized (10 500 transients) states. The reduced sample contained 2.1 mM flavodoxin in 100 mM phosphate at pH 7.0. The oxidized sample contained 4.4 mM flavodoxin in 100 mM phosphate at pH 7.5. The solvent for each was 90% H_2O/ 10% D_2O. Chemical shifts are referenced to liquid ammonia.

assignments to the isoalloxazine ring nitrogens are indicated in the figure. In the oxidized state, N3 and N10 were distinguished via an INEPT experiment (only the protonated nitrogen N3 was seen).

Discussion

The resonance positions of the isoalloxazine ring nitrogens are affected by hydrogen-bonding interactions with apoflavodoxin. Pyrrole-type nitrogens, such as N3 and N10 in oxidized and all nitrogens in reduced FMN, are shifted downfield 1-3 ppm by hydrogen bonding. Pyridine-type nitrogens, such as N1 and N5 of oxidized flavodoxin, are shifted upfield 5-15 ppm by hydrogen bonding (4). The hydrogen bonding in this flavodoxin is different from that seen in the other flavodoxins studied to date (2). N1 appears to accept a hydrogen bond in the oxidized state, although it is as weak or weaker than those in the other flavodoxins. In the reduced state, the chemical shift of N1 is dominated by the negative charge present. N3 of Anabaena 7120 flavodoxin in both redox states appears to form a stronger hydrogen bond than in the other flavodoxins studied (2). N5, however, shows the greatest chemical shift difference from the other flavodoxins: 6 ppm upfield of the resonances in other flavodoxins in both redox states. This is indicative of a very strong hydrogen bond in both redox states.

N3, whose protonation state is identical in both redox states, probably hydrogen bonds to a functional group in the apoprotein. N5, however, exists in different protonation states depending on the redox state. It thus donates a hydrogen bond in the reduced state and accepts one in the oxidized state. Since only small conformational changes occur upon reduction in other flavodoxins (5), it is unlikely that N5 hydrogen bonds to different apoprotein functional groups in each redox state. The dimethyl benzene edge of FMN is solvent accessible in other flavodoxins (6), thus hydrogen bonding to N5 may be from water molecules in one or both redox states.

The chemical shifts of the isoalloxazine ring carbon atoms will provide further information about the electronic environment of the FMN binding site in both redox states. These resonances will be assigned via a ^{13}C-^{15}N correlation experiment on flavodoxin doubly enriched with 20% ^{13}C and 95% ^{15}N.

Acknowledgments

We thank Dr. David W. Krogmann for supplying us with a culture of Anabaena 7120 and for supplying the flavodoxin isolation procedure prior to publication. Supported by USDA Competitive Research Grant 85-CRCR-1589. This study made use of the National Magnetic Resonance Facility at Madison, which is supported by NIH grants RR023021 and RR02781, NSF grant RR023021, the University of Wisconsin, and the U.S. Department of Agriculture.

References

1. Jones, R.A.Y. and Katritzky, A.R. 1962. Angew. Chem. Int. Ed. 1, 32-41.

2. Vervoort, J., Müller, F., Mayhew, S.G., van den Berg, W.A.M., Moonen, C.T.W., Bacher, A. 1986. Biochemistry. 25, 6789-6799.

3. Gorenstein, D. 1975. J. Am. Chem. Soc. 97, 898-900.

4. Witanowski, M., Stefaniak, L., Webb, G.A. 1981. Ann. Rep. NMR Spect. 11B, 1-502.

5. Ludwig, M.L., Burnett, R.M., Darling, G.D., Jordan, S.R., Kendall, D.S., Smith, W.W. 1975. In: Structure and Conformation of Nucleic Acids and Protein-Nucleic Acid Interactions (M. Sundaralingam and S.T. Rao, eds.). University Park Press, Baltimore. pp. 407-429.

6. Smith, W.W., Pattridge, K.A., Ludwig, M.L., Petsko, G.A., Tsernoglou, D., Tanaka, M., Yasunobu, K.T. 1983. J. Mol. Biol. 165, 737-755.

PULSE RADIOLYSIS STUDIES ON FLAVODOXIN

Robert F. Anderson

Gray Laboratory of the Cancer Research Campaign,
Mount Vernon Hospital, Northwood, Middlesex, HA6 2RN, U.K.

Vincent Massey and Lawrence M. Schopfer

Department of Biological Chemistry,
University of Michigan, Ann Arbor, Michigan, 48109, U.S.A.

Introduction

Both fast ($t_{1/2} < 7 \mu s$)(1,2) and slow ($t_{1/2} \sim 100 \mu s$ and > 3 ms)(1) spectral changes following one-electron reduction of flavodoxin have been ascribed to conformational changes in the protein. We have reinvestigated the radical chemistry of flavodoxin in both deaerated and aerated solutions and report the rapid formation of a species with features of a red flavosemiquinone ($EFl\cdot^-$) which converts quickly to the known (3) blue flavosemiquinone ($EFlH\cdot$). No subsequent spectral changes were observed up to 5 ms ruling out conformational changes and reaction between superoxide ($O_2\cdot^-$) and $EFlH\cdot$ in this time period.

Results

Aqueous solutions of flavodoxin (50 $\mu mol\ dm^{-3}$), purified by published methods (3) from Megasphaera elsdenii, were subjected to pulse radiolysis delivered by a 4 MV Van de Graaff accelerator. The methods of dose monitoring (calibrated against the aerated KSCN dosimeter (4)) and recording spectral changes have been described (5,6). Absorption spectra are presented as the product of the radiation chemical yield ($mol\ J^{-1}$) and the extinction coefficient ($dm^3\ mol^{-1}\ cm^{-1}$). All experiments were carried out in the presence of sodium formate (0.1 $mol\ dm^{-3}$) to convert the oxidizing $\cdot OH$ radical, and $H\cdot$ species to the reducing $CO_2\cdot^-$ radical.

Flavins and Flavoproteins 1987

$$H_2O \rightsquigarrow e_{aq}^-, \cdot OH, H\cdot, H_2O_2, H_3O^+, OH^- \quad (1)$$

$$\cdot OH(H\cdot) + HCOO^- \longrightarrow CO_2\cdot^- + H_2O(H_2) \quad (2)$$

The reaction of the e_{aq}^- with flavodoxin was studied in N_2-saturated solution at pH 6.1 where fast scavenging of e_{aq}^- ($k = 2.4 \times 10^{10}$ dm^3 mol^{-1} s^{-1}) was followed by the fast appearance ($t_{1/2} < 5$ µs) of EFlH·. The reaction of $CO_2\cdot^-$ radical with flavodoxin was studied at pH 9.15 in N_2O- saturated solution (to convert the e_{aq}^- to ·OH radicals which then react via reaction (2)) where the formation of EFlH· with a $t_{1/2}$ of 50 µs most likely arises from electron transfer from $CO_2\cdot^-$ ($k = 3 \times 10^8$ dm^3 mol^{-1} s^{-1}) followed by fast protonation. Both the e_{aq}^- and $CO_2\cdot^-$ radicals gave rise to a yield of EFlH· ca. 40% of the maximum possible yields of 0.3 µmol J^{-1} and 0.66 µmol J^{-1} for the two respective systems. As a low radiation dose was employed (7 Gy in 0.1 µs) losses through possible radical-radical reactions can be ignored and it is concluded that both e_{aq}^- and $CO_2\cdot^-$ radicals reduce other sites on the protein as well as FMN. This conclusion is supported by the observation of both rapid and slow spectral changes which are unrelated to the formation of EFlH·.

Experiments carried out on aerated solutions gave clear spectral results, where most of the e_{aq}^- and $CO_2\cdot^-$ radicals are scavenged by oxygen to form $O_2\cdot^-$ and any reduced sites on the protein of low redox potential will also be rapidly oxidized by oxygen. Hence within 2 µs of the pulse (70 Gy in 0.5 µs) only those e_{aq}^- which had reacted with the relatively high redox potential FMN moiety (ca. 10%) give rise to the initial transient spectrum. At both pH 6.1 and 9.15 this initial spectrum exhibits higher absorbance than oxidized flavoprotein in the 370 nm region, figure 1, indicating the formation of an anionic flavosemiquinone, EFl·⁻. This initial transient decayed rapidly ($k = 2.6 \pm 0.1 \times 10^5$ s^{-1} at pH 6.1; $1.1 \pm 0.2 \times 10^5$ s^{-1} at pH 9.15), insert A figure 1, independent of radiation dose (i.e. $O_2\cdot^-$ concentration) to yield the protonated EFlH·. This species is relatively stable over a 5 ms timeperiod, insert B figure 1, but decayed over a long time period back to the oxidized protein, figure 2 trace A. This slow oxidation was due to $O_2\cdot^-$ since it was completely inhibited by the addition of superoxide dismutase (0.5 mg ml^{-1}), figure 2 trace B.

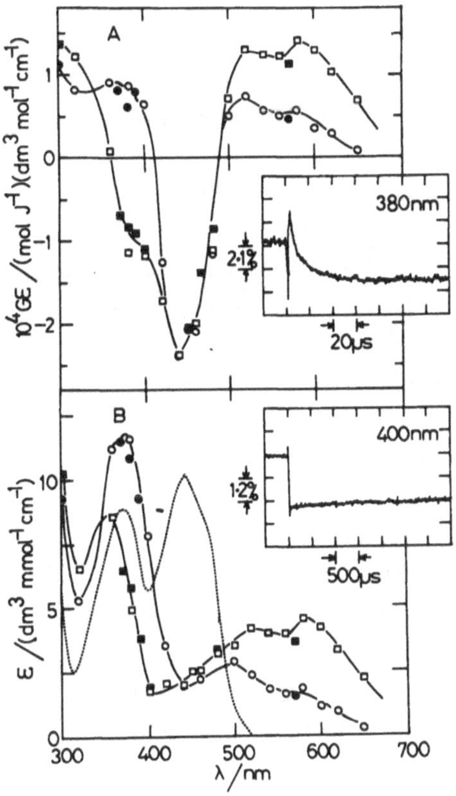

Figure 1. A: Absorption changes following pulse radiolysis (70 Gy in 0.5 μs) of aerated solutions of flavodoxin (50 μmol dm^{-3}) containing sodium formate (0.1 mol dm^{-3}). Open symbols are measurements at pH 9.15, O 2 μs, □ 50μs after the pulse. Closed symbols are measurements at pH 6.1.
B: Spectra obtained in A corrected for the loss in absorption of oxidized flavodoxin. Radical yields based on ε (580 nm) = 4500 dm^3 mol^{-1} cm^{-1} for EFlH· (ref. 3).
Inserts: Oscilloscope traces of change in percentage absorption with time following pulses of A, 68 Gy and B, 52 Gy.

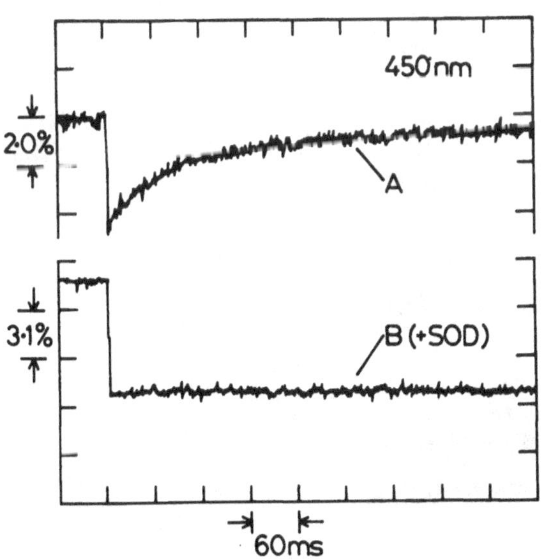

Figure 2. Oscilloscope traces of change in percentage absorption against time following pulse radiolysis of aerated solutions of flavodoxin (50 μmol dm^{-3}) containing sodium formate (0.1 mol dm^{-3}) at pH 6.1. Trace A is in the absence and trace B in the presence of superoxide dismutase (0.5 mg ml^{-1}).

Discussion

Our results show that the initially formed anionic flavosemiquinone of flavodoxin, $EFl\cdot^-$, undergoes rapid protonation to yield blue flavosemiquinone, $EFlH\cdot$. This reaction probably accounts for the previously postulated fast conformational change (1,2). No evidence was found to support the postulate that the $EFlH\cdot$ species undergoes a conformational change on the millisecond timescale (1). Some slow spectral changes were observed under anaerobic conditions which are probably related to electron attachment to sites on the protein other than the FMN moiety. Also it has been pointed out (2) that the use of tert-butanol as an $\cdot OH$ radical scavenger in the study where slow spectral changes were seen, leads to the formation of a radical which can undergo slow oxidation reactions.

The $EFlH\cdot$ did not react rapidly with the concomitantly formed excess concentration of $O_2\cdot^-$ to form a flavin C(4a)-hydroperoxide (EFlHOOH) as has been observed previously with glucose oxidase (7). These observations can be taken as support for the postulate (8) that whereas the flavin C(4a)-position is exposed to solvent in certain flavoprotein oxidases this is not the case for electron transferases such as flavodoxin.

References

1. Faraggi, M., Klapper, M.H.: J. Biol. Chem. 254, 8139-8142 (1979).
2. van Leeuwen, J.W., van Dijk, C., Veeger, C.: Eur. J. Biochem. 135, 601-607 (1983).
3. Mayhew, S.G., Massey, V.: J. Biol. Chem. 244, 794-802 (1969).
4. Schuler, R.H., Patterson, L.K., Janata, E.: J. Phys. Chem. 84, 2088-2089 (1980).
5. Vojnovic, B.: Ph.D. thesis, The University of London (1983).
6. Anderson, R.F., Hille, R., Massey, V.: J. Biol. Chem. 261, 15870-15876 (1986).
7. Anderson, R.F., Massey, V., Schopfer, L.M.: In Flavins and Flavoproteins, Bray, R.C., Engel, P.C., Mayhew, S.G. (eds.) pp 573-576, Walter de Gruyter, Berlin 1984.
8. Ghisla, S., Massey, V.: Biochem. J. 239, 1-12 (1986).

RESONANCE RAMAN INVESTIGATIONS OF FLAVINS AND FLAVOPROTEINS

Chris R. Lively, William G. Gustafson, and James T. McFarland

Department of Chemistry, University of Wisconsin-Milwaukee, Milwaukee, Wisconsin, 53201

Introduction

Resonance Raman (RR) spectroscopy has proven useful for the investigation of hydrogen bonding between flavin and protein and the determination of the structure of flavin charge-transfer complexes (1). One major challenge in the application of RR spectroscopy to flavin chemistry has been the difficulty of assignment of the vibrational spectrum because of the size and assymmetry of flavin. There are currently two different assignments of the vibrational bands of oxidized flavin (2,3). Structure-function studies on flavoproteins by RR spectroscopy require the investigation of reduced as well as oxidized flavin species. We report below a new normal coordinate assignment on lumiflavin and the RR spectra of the anion and neutral semiquinones of glucose oxidase.

Results

Table I shows a comparison of normal mode assignments for four of the major vibrational bands of flavin from two earlier investigations and the QCFF-π technique used in our laboratory. This table also compares the calculated isotope effects and predicted intensities with observations. Several important differences and similarities can be seen from the assignments of normal modes of RR bands II and X which show frequency shifts upon modification of hydrogen bonding between flavin and solvent. For Band II, the QCFF-π and Bowman and Spiro calculations show the stretching mode involving N_5 to predominate over that involving N_1 while the Kyogoku and Abe calculation indicates the opposite; this might result in greater sensitivity of Band II to hydrogen bonding at either of these two positions depending on which view proves correct.

Flavins and Flavoproteins 1987

Table 1 - Normal Coordinate Assignments of Major Flavin Bands

Band		$\nu(cm^{-1})$	Isotope effect[a]	I[b]	Assignment[c]
II	Observed	1582	(0,-3,-8,-13)	0.5	
	QCFF-π	1582	(-2, 0,-22,-36)	1.0	$\nu C_{4a}-N_5$, νC_4-C_{4a}, $\nu C_{10a}-N_1$, $\nu C_{10a}-C_{4a}$
	Bowman/Spiro	1510	(,-1,-11,-26)	-	$\nu N_{10}-C_{10a}$, $\nu C_{4a}-N_5$, $\nu C_{4a}-C_{10a}$, $\nu C_{10a}-N_1$
	Kyogoku/Abe	1586	(0, , ,)	-	νN_1-C_{10a}, $\nu N_{10}-C_{10a}$
III	Observed	1549	(3,-6,-5,-21)	.14	
	QCFF-π	1531	(-1, 0, 0,-2)	0.6	$\nu C_{5a}-C_6$, νC_8-C_9, νN_5-C_{5a}, νC_8-C_m
	Bowman/Spiro	1451	(,-6,-11,-16)	-	$\nu C_{4a}-N_5$, $\nu C_{10a}-N_1$
	Kyogoku/Abe	1548	(0, , ,)	-	νN_1-C_{10a}, $\nu C_{4a}-N_5$
VI	Observed	1408	(1,-3,-5,-12)	0.6	
	QCFF-π	1448	(0, 0,-5,-28)	0.0	$\nu C_{10a}-N_1$, $\nu N_{10}-C_{10a}$, $\nu C_{4a}-N_5$, νC_6-C_7
	Bowman/Spiro	1394	(,-4,-3,-7)	-	νC_8-C_9, $\nu C_{5a}-C_{9a}$, $\nu C_{5a}-C_6$, νN_1-C_2
	Kyogoku/Abe	----	(-, -, -, -)	-	
X	Observed	1256	(18,-14, 0, 17)	0.5	
	QCFF-π	1264	(3,-10,-1,-22)	.02	νN_3-C_4, νC_2-N_3, νC_4-C_{4a}, νN_1-C_2, $\delta C_4{=}O$
	Bowman/Spiro	1266	(-, -, -, -)	-	$\delta C_2{=}O$, $\delta C_4{=}O$, $\delta C_{4a}-H$, νN_1-C_{10a}, δN_3-H
	Kyogoku/Abe	1241	(14, , ,)	-	νC_4-N_3, νC_4-C_{4a}

[a]Isotope effects are given as differences in frequency in the following order: N_3-D, $^{13}C_2$, $^{13}C_{4a}$, $^{13}C_{2,4,4a,10a}$. [b]Intensities are given relative to the largest band. [c]Assignments are in decreasing order of contribution to the normal mode.

Similarly, for Band X, the QCFF-π claculation shows involvement of N_3 and $C_4{=}O$ but not $C_2{=}O$, while the Bowman and Spiro calculation indicates involvement of all three positions and the Kyogoku and Abe calculation shows involvement of N_3. Again interpretation of shifts in Band X with changes in hydrogen bonding will depend upon which of the assignments is finally judged to best fit the observed data.

Figure 1 shows the RR spectra of glucose oxidase anionic and neutral semiquinones. The RR spectrum of the neutral semiquinone is similar to that of P-450 Reductase (4). We observe bands at 1602 (II), 1511 (III), 1449, 1393 (VI), and

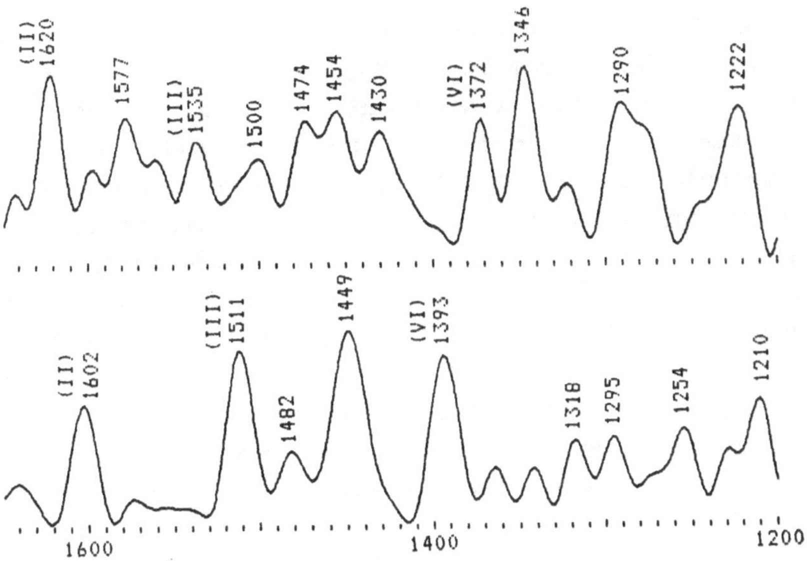

Figure 1-Resonace Raman Spectra of Anionic (A) and Neutral (B) Semiquinones of Glucose Oxidase

1295 cm^{-1}; we have listed, in parentheses, tentative assignments of the vibrational bands based on previous suggestions (4). The RR spectrum of the anionic semiquinone is quite similar to spectrum of the blue semiquinone with bands at 1620 (II), 1535 (III), 1474, 1454, 1430, 1372 (VI), 1346, 1290 and 1222 cm^{-1}. In interpreting our RR spectra, we have selected three RR bands for special attention, Bands II, III, and VI. These were selected because they involve vibrational modes at N_1 and N_5 (the two positions whose hydrogen bonding with protein is suggested to control the ionic form of the semiquinone) (5). As seen in Table I, QCFF-π and the Bowman and Spiro assignments indicate that Band II involves predominantly νC_{4a}-N_5 and to a lesser extent νC_{10a}-N_1; we suggest that the data are consistent with strong hydrogen bonding at N_1 in the anionic semiquinone and at N_5 in the neutral semiquinone. Increased hydrogen bonding is known to decrease the frequency of Band II, and since νC_{4a}-N_5 predominates in this band, we would expect that Band II would be at lower frequency in the blue neutral semiquinone as is observed. Futhermore, the relative positions of Bands III and VI in the two semiquinones also support the same conclusion. Band III consists of stretching modes involving νN_5-C_{5a} in the QCCFF-π calculation and νC_{4a}-N_5 in the Bowman and Spiro calculation; in either case Band III would move to lower frequency upon more extensive hydrogen bonding at N_5 in the neutral semiquinone. Again the prediction fits the observation. Finally Band VI involves predominantly νC_{10a}-N_1 and a lesser amount of νC_{4a}-N_5 in the QCFF-π calculation and νN_1-C_2 in the Bowman and Spiro calculation. In accord with observation, this band should occur at lower frequency in the red anionic semiquinone with its strong hydrogen bonding involving N_1. Thus, we tentatively conclude that N_1 hydrogen bonding is partcularly strong in the anionic semiquinone and N_5 hydrogen bonding in the neutral semiquinone. This conclusion must be tested with isotope effect experiments and normal coordinate calculations on the semiquinone species.

Acknowledgement

This research was supported by NIH grant GM-32504

References

1. McFarland, J.T. 1987. In: Biological Applications of Raman Spectrscopy 2 (Spiro, T.G.,Ed.). J. Wiley and Sons, p.211.

2. Bowman, W.D., Spiro, T.G. 1981. Biochemistry 20, p.3313.

3. Abe, M., Kyogoku, Y. 1987. Spectrochim. Acta. (in press)

4. Sugiyama, T., Nishimoto, Y., Mason, H.S., Loehr, T.M. 1985. Biochemistry 24, p.3012.

5. Massey, V., Ghisla, S., Moore, E.G. 1979. J. Biol. Chem. 254, p.9640.

SPECTRAL STUDIES ON HYDROGEN BONDED ALLOXAZINES

J. Kozioł, M.M. Szafran
Institute of Commodity Sciences, Academy of Economics
Marchlewskiego 146, 60-967 Poznań, Poland

P.F. Heelis
North East Wales Institute, Connah's Quay, Deeside
Clwyd, England

Introduction

Solvent polarity dependence of electronic spectra of alloxazines was interpreted in terms of non specific solute-solvent interactions (1), shifts of closely located $n-\pi^*$ and $\pi-\pi^*$ transitions (2) and specific interactions with molecules inducing phototautomerism (3). Similar solvent polarity dependence of isoalloxazines was recently ascribed by Yagi et al. (4) to hydrogen bond formation. We report here spectral properties of 1,3-dimethyllumichrome (1,3 MLc) in the presence of hydrogen donors.

Results

Solutions of 1,3 MLc (16.6 µM) in 1,2-dichloroethane (DCE) and for comparison riboflavintetrabutyrate (RTB, 16.6 µM) in CCl_4 were studied. Series of carboxylic acids (trifluoroacetic, TFA, trichloroacetic, TCA, dichloroacetic, DCA, monochloroacetic, MCA, and acetic, AA) and hexafluoroisopropanol (HFP) in concentrations up to 0.8 M were chosen as active hydrogen donors. Absorption spectra were taken with Cary 118 C spectrophotometer, fluorescence spectra with Perkin-Elmer MPF-44 A/E spectrofluorometer, all at 20° C.

Flavins and Flavoproteins 1987

The near UV part of absorption spectrum of 1,3 MLc in DCE shows maxima at 386 nm (max.I) and 336 nm (max.II). Upon addition of TFA (up to 0.064 M) max.I shifts gradually to 390 nm with slight hypochromicity, max.II shifts to 353 nm with strong hyperchromicity, the resulting spectrum resembles that observed in aqueous solutions and in the set of spectra clear cut isosbestic points are formed. Further increase of TFA concentration (up to 0.35 M) causes appearance of monocation (maxima at 400 and 360 nm) with increasing admixture of dication and (up to 0.8 M) of dication (broad band at about 470 and maximum at 405 nm), set of spectra shows new system of isosbestic points.

Emission spectrum of 1,3 MLc in DCE exhibits maximum at 435 nm. Addition of TFA causes sharp decrease of fluorescence intensity (by 86 % for 0.064 M) and no shift of maximum position. Instead a new band of low intensity appears with maximum at about 570 nm (for 0.8 M). This new emission has to be ascribed to dicationic species and can be detected by excitation at 440 nm starting from lowest TFA concentrations (0.002 M). Excitation spectra recorded by monitoring fluorescence at 435 nm match absorption spectra without TFA representing the species non hydrogen bonded to TFA, spectra monitored at 540 nm show the presence of dication. The sum of these excitation spectra does not match absorption spectra because of formation of non fluorescent monocation. The above data indicate that formation of hydrogen bonds at N-5 and N-10 by the very acidic TFA (pK_a 0.23) is hindered sterically and/or because TFA is present in form of unreactive dimers. Once hydrogen is formed it disproportionates to ion pairs which in turn split causing protonation of 1,3 MLc to mono- and dication already in the ground state.

Almost identical changes in 1,3 MLc spectra are observed upon addition of TCA (pK_a 0.70) and similar in character but much less pronounced for DCA (pK_a 1.48) which at 0.8 M concentration causes shift of max.II to 355 nm (absorption spectrum as in water) and in emission decrease of intensity at 435 nm by 94 % and appearance of new band of dication.

The least acidic carboxylic acids MCA (pK_a 2.83) and AA (pK_a 4.75) cause only small changes in absorption spectra (for 0.8 M concentration max.II shifts to 345 and 340 nm, respectively) and decrease in fluorescence intensity by 57 and 17 % but along with shift of the maximum to 446 and 445 nm, respectively. As in absorption and excitation spectra there are no traces of cationic forms the drop in fluorescence intensity can be ascribed to partial protonation of 1,3 MLc in the excited state. The observed shifts of emission maxima confirm the presence of hydrogen bonded (unprotonated) species.

Increasing concentration of HFP (up to 0.8 M) known as active hydrogen bonding reagent (pK_a 9.30) added to 1,3 MLc solution in DCE cause shift of max.I to 391 nm with slight hypochromicity, max.II to 352 nm with strong hyperchromicity (spectrum resembles that observed in aqueous solution), the set of spectra forms clear isosbestic points. In emission spectra strong shift of maximum to 462 nm is observed without any decrease in intensity. Excitation spectra match well absorption spectra. The extent of spectral changes caused by HFP indicate predominance of the hydrogen bonded forms, lack of steric hindrance for hydroxyl group approaching hydrogen acceptor sites in 1,3 MLc or low selfassociation tendency of HFP.

The spectral properties of 1,3 MLc in the presence of carboxylic acids and HFP justify some preliminary conclusions:

efficiency of H-bonds formation is related to donor acidity which may lead to protonation in ground and excited states,

carboxylic acids are less effective in H-bonding than HFP for steric reasons or their presence in form of inactive dimers,

under applied conditions system of multiequilibria should be considered: species non H-bonded $\rightleftharpoons$ H-bonded to solvent (DCE) $\rightleftharpoons$ H-bonded to added donor $\rightleftharpoons$ protonated,

decrease in fluorescence intensity in the presence of carboxylic acids is caused by protonation but not by hydrogen bonding itself,

hydrogen bonded species are more sensitive to micro- and macroscopic solvent polarity.

The order of hydrogen bonds formation at different sites in 1,3 MLc molecule is not known but it can be similar to that proposed for isoalloxazines (4). Preliminary results obtained with alloxazines unsubstituted at N-1 reveal importance of this site for hydrogen bonding.
Results obtained for RTB in CCl_4 in the presence of TFA and HFP confirm interpretations proposed above. The influence of added TFA was identical to reported by Yagi et al. (4) but at higher concentrations (0.8 M) excitation spectra do not match absorption indicating formation of nonfluorescent monocation responsible for drop in fluorescence intensity. In the presence of HFP (up to 0.1 M) absorption spectra change characteristically for hydrogen bond formation, in emission spectra no decrease in intensity and red shift of maximum was observed. Excitation spectra revealed changing participation of hydrogen bonded species.

Acknowledgement

This research was supported in part by Project CPBP.01.12. PAN and Project RP.II.13. MNSzW.

References

1. Kozioł, J. 1971. In: Methods in Enzymology (S.P. Colowick and N.O. Kaplan, eds.). vol. 18 B, Academic Press, New York, p. 253.

2. Kozioł, J. 1971. In: Flavins and Flavoproteins (H. Kamin, ed.). University Park Press, Baltimore, p. 54.

3. Koziołowa, A. 1979. Photochem. Photobiol. 29, 459.

4. Yagi, K., N. Ohishi, K. Nishimoto, J.D. Choi, P.S. Song. 1980. Biochemistry 19, 1553.

USE OF FLAVIN AND FLAVODOXIN SEMIQUINONES AS KINETIC PROBES TO ELUCIDATE THE FACTORS WHICH DETERMINE REACTION RATE CONSTANTS AND BIOLOGICAL SPECIFICITY FOR ELECTRON TRANSFER PROTEINS

Gordon Tollin, Glen Cheddar, James T. Hazzard, Terrance E. Meyer and Michael A. Cusanovich, Department of Biochemistry, University of Arizona, Tucson, AZ, U.S.A.

INTRODUCTION

We have used free flavin and flavodoxin semiquinones as kinetic probes of the electron transfer sites of redox proteins. These investigations have focussed on three types of systems: a) "simple" (i.e. single redox center) proteins such as cytochromes c and c', blue copper proteins (e.g. azurin, plastocyanin) and high potential ferredoxins (HiPIP's) (1-5), b) 1:1 complexes between two single-center redox proteins which are stabilized by electrostatic interactions at low ionic strength (e.g. flavodoxin-cytochrome c; ferredoxin-ferredoxin reductase) (6, 7), and c) more complex, multicenter, redox proteins such as xanthine oxidase and p-cresol methylhydroxylase (8, 9). Whenever possible, we have used structurally well-characterized proteins and, where available, structurally homologous series. The experimental techniques we have employed are stopped-flow spectrophotometry, in which flavodoxin semiquinone is anaerobically mixed with an oxidized redox protein, and laser flash photolysis. The latter method has been used with free flavins which, in the presence of a reductant such as EDTA, can be photoreduced in situ to the semiquinone by a laser flash and which subsequently can transfer an electron to an oxidized redox protein. These processes can be monitored spectrophotometrically, with a time resolution at least three orders of magnitude greater than for stopped-flow methodology. In all cases, the oxidant is present in large excess over the reductant, and thus pseudo-first order conditions apply. Values of the observed first order rate constant can be obtained over a range of oxidant concentration in order to characterize the kinetic properties of the system. The reactions can be studied as a function of wavelength, so as to follow multistep reaction sequences by means of time-resolved spectrophotometry.

Flavins and Flavoproteins 1987

In all of this work, we have focussed on elucidating the factors which control the rates and specificity of electron transfer between and within redox proteins. As will be illustrated below, in favorable cases we can quantitate the effects of redox potential differences, electrostatic potential surfaces and surface topography on reaction rates. Furthermore, less quantitatively, we have demonstrated that redox center distance and orientation, and the nature of the intervening medium between redox centers, also can play major roles in determining electron transfer rate constants. In what follows, we present an overview of these studies, which includes both published and as yet unpublished work. We will restrict ourselves to single redox center proteins and to 1:1 complexes between such proteins. Another contribution to this Conference (by Edmondson, Hazzard and Tollin) will present some new experiments on a multi-center protein (xanthine oxidase).

REACTIONS OF SINGLE-CENTER REDOX PROTEINS WITH FREE FLAVIN AND FLAVODOXIN SEMIQUINONES

Marcus electron transfer theory (10, 11), along with more general considerations, predicts that rate constants for reaction between two redox moieties are strongly influenced by several parameters: a) the redox potential difference between reactants (thermodynamic driving force), b) the reorganizational energy (which reflects the structural and solvation changes which must occur prior to electron transfer in order to reach the transition state), c) redox center accessibility or steric effects (which encompasses extent of orbital overlap and average distance and orientation between redox centers) and d) when the reactants are electrically charged, attractive or repulsive electrostatic forces which are involved in bringing them together within an intermediate complex. On this basis, one would expect that electron transfer between the electrically-neutral lumiflavin semiquinone (LfH·) and a structurally-homologous family of redox proteins would be dominated by thermodynamic driving force, at least to the extent to which steric and reorganizational factors remain constant within the group. We have indeed obtained this anticipated result for several such protein families (1-5, 12), as is shown in Fig. 1, left. Some additional kinds of conclusions can also be drawn from these results, e.g. relative

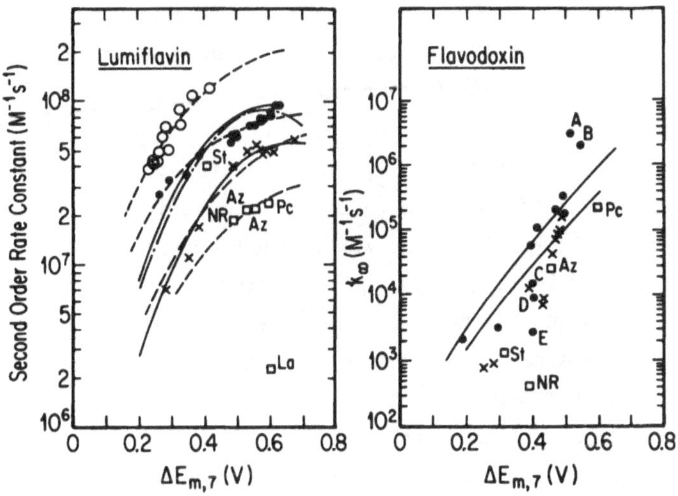

Fig. 1: Left
Semilog plots of second order rate constants for electron transfer from lumiflavin semiquinone to oxidized redox proteins vs. difference in redox potential between reactants. Each point corresponds to an individual protein (for identification of specific proteins, cf. 2,4,5,15).
○ cytochromes *c'*; ● cytochromes *c*; X high potential ferredoxins; □ copper proteins (Pc=plastocyanin; AZ=azurin; NR=nitrite reductase; La=laccase; St=stellacyanin)
Curves are theoretical plots using various equations (cf. 12).

Right
Semilog plot of electrostatically-corrected second order rate constants for electron transfer from *C. pasteurianum* flavodoxin semiquinone to oxidized redox proteins vs. difference in redox potential between reactants. Symbols as in upper box.
A: *Euglena* *c*-552; B: *Rhodospirillum* *c*-553; C: *Candida* *c*; D: tuna *c*; E: horse *c*.
Curves are theoretical plots; steeper curves, which would fit more of the points, can only be obtained using physically unreasonable parameters.

reactivities. Thus, as a group, the cytochromes *c'* are uniformly more reactive towards LfH· than are the cytochromes *c*. This result correlates with the greater degree of heme exposure in the former (13,14). In the case of the blue copper proteins, although our data are more limited, it is evident that plastocyanin, the two azurins, and nitrite reductase belong to a single reactivity class, in agreement with crystallographic and spectroscopic results, whereas the less reactive laccase and the more reactive stellacyanin (both structurally uncharacterized) belong to rather different classes (15). It is clear from these latter results that kinetic measurements allow taxonomic conclusions to be reached.

Another way in which the steric properties of the electron transfer site can be probed is to change the structure of the free flavin semiquinone. Riboflavin semiquinone (RfH•) is quite useful for this purpose, inasmuch as it retains the electronic characteristics of LfH• while at the same time having a bulkier side chain at the 10-position. Thus, for the cytochromes c, rate constants for reduction by RfH• are uniformly 30% smaller than for reduction by LfH•, whereas the decrease is 50% for the cytochromes c' (2,4). It is interesting that the effect is larger for the latter cytochromes, inasmuch as the comparative results with LfH• and the crystallography indicate that the c'- heme is less sterically occluded. However, as we shall see below, this can be rationalized in terms of the overall topography of the electron transfer site.

The electrostatic properties of protein redox centers can be readily probed by using the negatively-charged FMN semiquinone (FMNH•). In this situation, the ionic strength dependence of the reaction rate constant must be determined in order to assess the sign and magnitude of the active site charge. Some typical examples are shown in Fig. 2, left. A decreasing rate constant with increasing ionic strength implies a positively-charged electron transfer site (i.e. an attractive electrostatic interaction with FMNH•). The steepness of the curve is determined by the size of the active site charge. It is interesting that the apparent sign of the charge does not always correlate with the net charge on the protein. For example, with P. denitrificans cytochrome c_2 (cf. Fig. 2), the protein net charge is -7 but the active site charge is +3.6 (calculated from a theoretical fit to the data (16)). This is a consequence of a highly asymmetric charge distribution in which positive charges are localized in the vicinity of the solvent exposed heme edge. In general, local charge is more important than overall charge, although when the local charge at the electron transfer site is small, the apparent charge can be strongly influenced by the protein net charge (12,

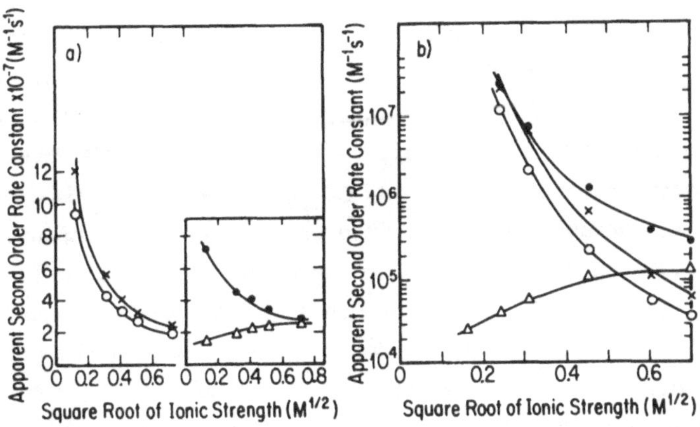

Fig. 2: (a) Ionic strength dependence of rate constants for electron transfer from FMNH• to c-type cytochromes. Solid lines are theoretical curves (cf. 16).
X Candida cyt. c (net charge 5+); O tuna cyt. c (net charge 7+); ● P. denitrificans cyt c_2 (net charge 7−); Δ Ps. aeruginosa cyt. c-551 (net charge 2−)
(b) Semilog plots of ionic strength dependence of rate constants for electron transfer from C. pasteurianum flavodoxin semiquinone to c-type cytochromes. Solid lines are theoretical curves. Symbols as in (a).

16). It is noteworthy that these kinetic ionic strength effects correlate well with theoretical calculations of the electrostatic potential at the protein surface (17, 18). As we shall see below, electrostatic attractive forces can lead to either favorable or unfavorable orientations for electron transfer, depending upon the system. Furthermore, in weakly-interacting systems, steric effects can strongly modulate the effects of electrostatic forces.

When flavodoxin semiquinone is used, the situation, in some but not all respects, becomes more complex. In Figs. 1 and 2 (bottom) we have plotted the kinetic results obtained with flavodoxin semiquinone for the same group of cytochromes c. The rate constant values in the bottom portion of Fig. 1 (k_∞) were obtained by extrapolating theoretical fits to ionic strength data (cf. Fig. 2) to infinite ionic strength, and thus correspond to electrostatically-corrected values. By comparing the right of Fig. 1 with the left, it is immediately apparent that the close correlations between rate constant and $\Delta E_{m,7}$ are no longer obtained (this is not due to the errors inherent in the extrapolation to infinite ionic strength, although this may contribute somewhat to the scatter). We attribute this to the much larger influence of the steric properties of the electron transfer site on the protein-protein than on the protein-lumiflavin interaction. In other words, small differences in surface topography, which were relatively inconsequential to lumiflavin, have a large effect on the flavodoxin reactions. It is also possible that reorganization energy differences are playing a role, but we have no way of testing for this at present. Further examples are to be found among the cytochromes *c'* and the copper proteins. In Table 1 are shown k_∞ values for electron transfer from FMNH· and flavodoxin semiquinone to a

Table 1: STERIC EFFECTS ON ELECTRON TRANSFER FROM FMN AND FLAVODOXIN SEMIQUINONES

OXIDANT/REDUCTANT	$k_\infty (M^{-1}s^{-1})$	Ref.
FMNH·		
CHLOROBIUM *C*-555	$1.3X10^7$	2
ALCALIGENES CYT. *C'*	$2.7X10^7$	4
FLAVODOXIN SEMIQUINONE		
CHLOROBIUM *C*-555	$3X10^3$	3
ALCALIGENES CYT. *C'*	4	4
FMNH·		
PLASTOCYANIN	$1.6X10^7$	15
STELLACYANIN	$2.2X10^7$	15
FLAVODOXIN SEMIQUINONE		
PLASTOCYANIN	$1.3X10^5$	15
STELLACYANIN	$1.3X10^3$	15

cytochrome *c* (*Chlorobium* *c*-555) and a cytochrome *c'* (*Alcaligenes* *c'*), which have approximately the same redox potential, and to plastocyanin and stellacyanin. As we have noted, with small flavins the cytochromes *c'* are more reactive than the *c*-types. However, this is dramatically reversed with flavodoxin. We attribute this (4) to the fact that, at least for the crystallographically-characterized cytochrome c' (13), the relatively exposed heme lies at the bottom of a cleft which would prevent flavodoxin, but not the smaller free flavins, from approaching closer than about 10Å without substantial rearrangement of protein side chains. Because of the expected exponential dependence of rate constant on distance (11), this can easily account for the otherwise anomalously slow reaction. The existence of this cleft can also explain the larger steric effects with RfH· noted above. A similar situation exists with stellacyanin and plastocyanin. Whereas the former protein is more reactive towards FMNH· than is plastocyanin, the order is reversed with flavodoxin semiquinone. Although we do not have a structure for stellacyanin, the kinetics predict that the copper site is sterically occluded for large substrates, i.e. that it lies within some type of surface depression.

It is also important to note that the overall influence of thermodynamic driving force is much greater for the flavodoxin reactions. As an example of this, the ratio of k_{∞} values for cytochromes c over an approximately 200 mV redox potential difference is about 4 for FMNH· and 1000 for flavodoxin semiquinone (2, 3). This is a consequence of a much larger reorganizational energy for the protein system, as well as a greater intrinsic reactivity for flavodoxin-bound FMN (3).

The ionic strength results (Fig. 2, right) demonstrate that the apparent sign of the charge on the electron transfer site is the same with flavodoxin (which is negatively-charged) as with FMN, even in the case of *P. denitrificans* c_2. However, the magnitude of the ionic strength effects are much larger with flavodoxin. This is to be expected for two reasons. First, the size of the charge on flavodoxin is greater than for FMN, which increases the strength of the electrostatic interaction. Furthermore, the larger flavodoxin experiences a greater fraction of the reactant surface. Electrostatic effects can also reverse relative reactivities when the ionic strength conditions are changed. Thus, at an ionic strength of 360 mM, *Pseudomonas* c-551 reacts approximately 4 times faster with flavodoxin semiquinone than does horse cytochrome c. On the other hand, at 50 mM ionic strength, the horse cytochrome reacts 1000 times more rapidly than does c-551. Furthermore, when the interaction is weak, electrostatics can be dramatically modified by steric effects. Thus, if one compares (19) the reactivities of C. pasteurianum, Anacystis and Azotobacter flavodoxins towards tuna cytochrome c, one finds that the electrostatics are quite comparable in all cases (i.e. strong minus-plus interactions), but the reactivity of the C. pasteurianum flavodoxin is greater than that of the other two flavodoxins, even though thermodynamics would predict the reverse. We attribute this to steric effects at the flavodoxin surface, which is in agreement with the crystal structure of Anacystis flavodoxin (20). With Pseudomonas c-551, for which the electrostatic interactions are quite weak, C. pasteurianum flavodoxin experiences a weak negative electrostatic potential during electron transfer, whereas both Anacystis and Azotobacter flavodoxins experience a weak positive potential. We interpret this to mean that, in the latter two cases, surface topography forces the interaction between the flavodoxin and the cytochrome to occur in a different (and oppositely charged) region of the cytochrome, even though this is less favorable to electron transfer, as shown by the fact that the rate constants are appreciably smaller than with C. pasteurianum flavodoxin.

REACTIONS OF ELECTROSTATICALLY-STABILIZED 1:1 REDOX PROTEIN COMPLEXES WITH FREE FLAVIN SEMIQUINONES

Laser flash photolysis allows several aspects of the structure and properties of preformed redox protein complexes to be examined. Thus, by comparing the reactivity of a protein redox center towards a free flavin semiquinone in both the complexed and uncomplexed state, any changes in the steric and electrostatic environments of the electron transfer site due to complex formation can be probed. In addition, if one can selectively reduce the lower potential member of a pair of proteins by a laser flash, it is possible to measure the rate constant for intracomplex electron transfer.

Several examples of redox center accessibilities to reduction by LfH$^{\cdot}$ within complexes are given in Table 2. All of these experiments were carried out at low ionic strength, under conditions in which the proteins were essentially 100% complexed. It is evident that quite variable results are obtained. For example, within the flavodoxin-cytochrome c complex, the flavin prosthetic group is effectively unobstructed, whereas the heme is significantly occluded. This result agrees qualitatively with a computer-generated model for this complex (6), although quantitatively one would have anticipated a greater degree of restriction for both of the redox centers. This raises the question of

Table 2: REDOX CENTER ACCESSIBILITY WITHIN 1:1 PROTEIN COMPLEXES AS DETERMINED BY LUMIFLAVIN SEMIQUINONE REDUCTION KINETICS

COMPLEX	COMPONENT REDUCED	k_{BOUND}/k_{FREE}	Ref.
FLAVODOXIN:CYT. C	FLAVODOXIN	1.0	6
FLAVODOXIN:CYT. C	CYT. C	0.4	6
FNR:FERREDOXIN	FNR	0.89	7
FNR:FERREDOXIN	FERREDOXIN	0.04	7
FNR:RUBREDOXIN	RUBREDOXIN	1.0	7
CCP(III):TUNA CYT. C	CYT. C	0.26	21
CCP(III):YEAST ISO-1 CYT. C	ISO-1	<0.10	22
CCP(III):YEAST ISO-1(ILE-18)	ISO-1	0.34	22
CCP(IV,$R^{+\cdot}$):YEAST ISO-1	ISO-1	0.68	22
CCP(IV,$R^{+\cdot}$):YEAST ISO-1(ILE-18)	ISO-1	1.00	22

CCP(IV,$R^{+\cdot}$) is the doubly-oxidized species formed by H_2O_2 treatment.

dynamics, i.e. to what extent do the motional properties of the molecules within the complex influence the ability of external reagents to gain access to the interface region? In the case of the flavodoxin-cytochrome c complex, we have evidence that such motions play an important role in controlling the access of external FMNH· to the cytochrome heme (6). A similar result has also been obtained for the cytochrome c-cytochrome c peroxidase (CcP) complex (21). This property of dynamic structural mobility may be physiologically significant in terms of the mechanism of electron flow in multi-step processes.

The results obtained with the ferredoxin-$NADP^+$ reductase (FNR) complexes (7) and the CcP complexes (21) raise an interesting point concerning biological specificity. In the physiologically-relevant complexes (i.e. FNR-ferredoxin and CcP-yeast iso-1 cytochrome c), the occlusion of the iron-sulfur and heme centers is much greater than it is with similar, but physiologically non-significant, redox proteins (compare with rubredoxin and tuna cytochrome c). Furthermore, in the CcP system, we have examined site-specific mutants of yeast cytochrome c (22). As an example of this, the replacement of an arginine residue at position 18 by isoleucine leads to a marked increase in heme accessibility, even though the free cytochromes behave identically. This particular residue has been suggested, based on the computer model, to play an important role in complex formation (23). Thus, relatively small changes at the interface between two proteins within a complex can produce marked structural effects which influence steric properties. It is worth noting that since the structure of FNR is not yet completely determined, the significance of the observation that the accessibility of the FAD moiety is almost unaffected by complex formation cannot be fully assessed. However, it does place some constraints on the nature of possible complexes, which may be of value in the future.

Another result from Table 2 which has potentially important physiological consequences is that changing the oxidation state of CcP by H_2O_2 addition markedly changes the cytochrome heme accessibility. This must be a consequence of a modification of the structural and/or dynamic properties of the complex interface. Such effects could be a significant component of the mechanism of reactions which, as is the case with CcP, have more than one electron transfer step occurring sequentially.

Table 3 presents some results on intracomplex electron transfer kinetics. The distances between redox centers given in the Table are estimates based either upon computer modelling or upon whatever structural information is available for a given system. Several points can be made based on these data. It is quite clear, for example, that no

Table 3: INTRACOMPLEX ELECTRON TRANSFER RATE CONSTANTS

REACTANTS	$E_{m,7}$ (mV)	EDGE-TO-EDGE DISTANCE (A)	$k(s^{-1})$	Ref.
TUNA CYT. C(II) → CCP(IV),$R^{+}\cdot$)	~0.75	~18	950	21
ISO-1 CYT. C(II) → CCP(IV,$R^{+}\cdot$)	~0.75	~18	250	22
ISO-1 CYT. C(ILE 18)(II) → CCP(IV,$R^{+}\cdot$)	~0.75	~18	1000	22
FNR$^{\cdot}$ → *PARACOCCUS* CYT. C_2	0.57	<15	700	24
FNR$^{\cdot}$ → *ECTO*. HIPIP	0.44	<10	23	24
FLAVODOXIN$^{\cdot}$ → HORSE CYT. C	0.40	3.5	85	25
FNR$^{\cdot}$ → RUBREDOXIN	6	<10	$2X10^3$	7
FERREDOXIN(II) → FNR	0.05	<10	$>4X10^3$	7
HORSE CYT. C(II) → HEME A	-0.04	?	>700	26

correlation exists between rate constant and thermodynamic driving force. Furthermore, although we cannot completely exclude distance as a factor, it would appear that redox center orientation and the nature of the intervening medium may play important roles. Finally, small structural changes at the complex interface (i.e. arg to ile at position 18 of yeast cytochrome c) can lead to rather large rate constant changes, again suggesting that orientation can be an important parameter. Another contribution to this Conference (by De Francesco, Tollin, Brown and Edmondson) will describe the effects of placing an imidazole ring in the interface between the flavin and heme centers of the flavodoxin-cytochrome c complex.

Among the FNR complexes, the physiologically-relevant one (i.e. ferredoxin) has the largest electron transfer rate constant, even though the driving force is quite small. In marked contrast, among the CcP complexes the physiological one has the __smallest__ rate constant. This raises an interesting question. The CcP results were obtained at low ionic strength within a preformed electrostatically-stabilized complex. Is such a complex in fact a good model for the reaction as it occurs

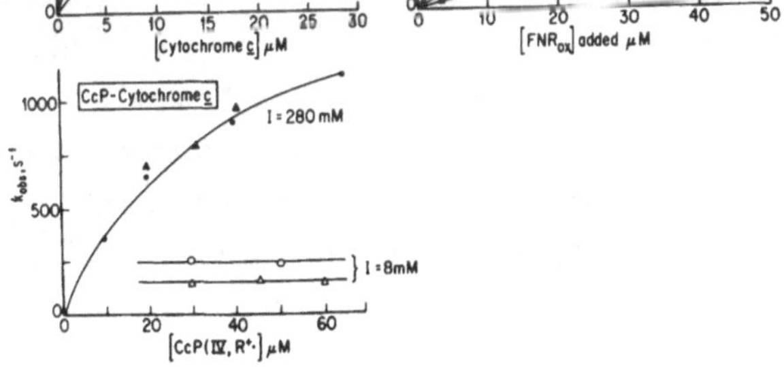

Fig. 3: Plots of k_{obs} vs. protein concentration at various ionic strengths.

flavodoxin semiquinone/cyt. c (data from ref. 25)

reduced ferredoxin/FNR (data from ref. 24)

cyt. c (II)/CcP(IV,R^{+}·): ○, ● yeast iso-I; △, ▲ yeast iso-2 (data from ref. 22).

under physiological conditions? The results shown for CcP-yeast cyt. c in Fig. 3 suggest a negative answer to this question. Thus, at high ionic strength, the limiting rate constant for electron transfer within a transient complex is much larger than the first order rate constant for intracomplex electron transfer within the preformed complex at low ionic strength. This strongly suggests that, in this case, electrostatic forces lead to the formation of a non-optimal complex. Such a conclusion is supported by the ile-18 substitution which, by eliminating an electrostatic interaction, greatly increases the rate constant for intracomplex electron transfer. On the other hand, the results with the flavodoxin and FNR complexes demonstrate that this is not always the case. Thus, for these latter two complexes, decreasing the ionic strength (and thus increasing the magnitude of the electrostatic effects) increases the limiting rate constants for electron transfer within the transient intermediate complexes. Presumably, electrostatic forces act to orient the molecules favorably for these systems.

CONCLUSIONS

The studies summarized here provide quantitative information concerning the contributions of thermodynamic driving force, protein surface topography and the electrostatic potential at the reaction site to rate constants for electron transfer reactions. By comparing the properties of related proteins, we have been able to draw taxonomic conclusions and to make structural predictions for proteins which are not well characterized structurally. These results have allowed us to obtain a clearer understanding of the factors involved in controlling the biological specificity of electron transfer processes. Thus, modulation (via evolutionary mechanisms) of redox potential, active site charge and surface topography together permit second order rate constant adjustments of as much as 10^9 for the optimization of reaction kinetics between two proteins. We have also obtained preliminary indications that protein dynamics, redox center orientation and the nature of the intervening medium may play major roles in determining limiting first order rate constants. Much work, however, will be required before a quantitative understanding of these latter parameters can be achieved.

Acknowledgement

The work described herein was supported in part by grants from the National Institutes of Health, AM15057 (to G.T.) and GM21277 (to M.A.C.).

REFERENCES

1. Meyer, T.E., Przysiecki, C.T., Watkins, J.A., Bhattacharyya, A., Simondsen, R.P., Cusanovich, M.A. and Tollin, G. (1983) Proc. Natl. Acad. Sci. USA 80, 6740-6744.

2. Meyer, T.E., Watkins, J.A., Przysiecki, C.T., Tollin, G. and Cusanovich, M.A. (1984) Biochemistry 23, 4761-4767.

3. Tollin, G., Cheddar, G., Watkins, J.A., Meyer, T.E. and Cusanovich, M.A. (1984) Biochemistry 23, 6345-6349.

4. Meyer, T.E., Cheddar, G., Bartsch, R.G., Getzoff, E.D., Cusanovich, M.A. and Tollin, G. (1986) Biochemistry 25, 1383-1390.

5. Przysiecki, C.T., Cheddar, G., Meyer, T.E., Tollin, G. and Cusanovich, M.A. (1985) Biochemistry 24, 5647-5652.

6. Hazzard, J.T., Cusanovich, M.A., Tainer, J.A., Getzoff, E.D. and Tollin, G. (1986) Biochemistry 25, 3318-3328.

7. Bhattacharyya, A., Meyer, T.E. and Tollin, G. (1986) Biochemistry 25, 4655-4661.

8. Bhattacharyya, A., Tollin, G., Davis, M. and Edmondson, D.E. (1983) Biochemistry 22, 5270-5279.

9. Bhattacharyya, A., Tollin, G., McIntire, W. and Singer, T. (1985) Biochem. J. 228, 337-345.

10. Marcus, R.A. (1964) Annu. Rev. Phys. Chem. 15, 155-196.

11. Marcus, R.A. and Sutin, N. (1985) Biochim. Biophys. Acta 811, 265-322.

12. Tollin, G., Meyer, T.E. and Cusanovich, M.A. (1986) Biochim. Biophys. Acta 853, 29-41.

13. Weber, P.C., Howard, A., Xuong, N.H. and Salemme, F.R. (1981) J. Mol. Biol. 153, 399-424.

14. Tollin, G., Hanson, L.K., Caffrey, M., Meyer, T.E. and Cusanovich, M.A. (1986) Proc. Natl. Acad. Sci U.S.A. 83, 3693-3697.

15. Tollin, G., Meyer, T.E., Cheddar, G., Getzoff, E.D. and Cusanovich, M.A. (1986) Biochemistry 25, 3363-3370.

16. Cusanovich, M.A., Meyer, T.E. and Tollin, G. (1987) Adv. in Inorg. Chem., in press.

17. Matthew, J.B., Weber, P.C., Salemme, F.R. and Richards, F.M. (1983) Nature 301, 169-171.

18. Weber, P.C. and Tollin, G. (1985) J. Biol. Chem. 260, 5568-5573.

19. Cheddar, G., Meyer, T.E., Cusanovich, M.A., Stout, C.D. and Tollin, G. (1986) Biochemistry 25, 6502-6507.

20. Ludwig, M.L., Pattridge, K.A., Smith, W.W., Jensen, L.H. and Watenpaugh, K.D. (1982) in "Flavins and Flavoproteins" (Massey, V. and Williams, C.D., eds.) pp. 19-27, Elsevier/North Holland, N.Y.

21. Hazzard, J.T., Poulos, T.L. and Tollin, G. (1987) Biochemistry, in press.

22. Hazzard, J.T., McLendon, G., Cusanovich, M.A. and Tollin, G., in preparation.

23. Poulos, T.L. and Finzel, B.C. (1984) Pept. Protein Rev. 4, 115-171.

24. Bhattacharyya, A.K., Meyer, T.E., Cusanovich, M.A. and Tollin, G. (1987) Biochemistry 26, 758-764.

25. Simondsen, R.P., Weber, P.C., Salemme, F.R. and Tollin, G. (1982) Biochemistry 21, 6366-6375.

26. Ahmad, I., Cusanovich, M.A. and Tollin, G. (1982) Biochemistry 21, 3122-3128.

FACTORS INFLUENCING THE RATE OF ELECTRON TRANSFER FROM FLAVODOXIN TO CYTOCHROME c

Raffaele De Francesco*, Gordon Tollin++, Kathleen Brown*, and Dale E. Edmondson*

Departments of Biochemistry, *Emory University, Atlanta, GA. 30322 and ++University of Arizona, Tucson, AZ. 85721

Introduction

The flavodoxin-cytochrome c system has proven to be a useful model for investigating protein-protein electron transfer reactions. Previous studies have shown the semiquinone form of A. vinelandii flavodoxin (A.V. Fld) to be considerably less reactive than C. pasteurianum flavodoxin (C.P. Fld) in electron transfer to cytochrome c (Cyt c) (1).

Both flavodoxins have been shown to form complexes with Cyt c which are detectable kinetically (1) as well as by spectroscopic (2) and crosslinking (3) studies. The reaction can be described as follows (4):

$$\text{FldH}\cdot + \text{Cyt } c_{ox} \underset{}{\overset{K_a}{\rightleftharpoons}} [\text{complex}] \xrightarrow{k_2} \text{Fld}_{ox} + \text{Cyt } c_{red}$$

Since the OX/SQ redox potential of A.V. Fld is approximately 30 mV less than that of C.P. Fld (1), the origin for the difference in reactivity must be in differences in the geometries of the respective complexes. This paper presents evidence for the alteration of heme enviroment of Cyt c on complexation with C.P. Fld which does not occur on complexation with the A.V. protein. Substitution of the FMN moiety of C.P. Fld with 8α-imidazolylFMN does not appreciably alter the K_a value for complex formation but does disrupt those interactions that lead to perturbation of the heme

Flavins and Flavoproteins 1987

enviroment and results in substantial reduction in the rate of electron transfer. A more detailed account of this work will appear elsewhere.

Results and Discussion

Fig. 1 shows the effect of cytochrome c concentration on the observed pseudo first order rate constants of Fld semiquinone oxidation for both C.P. and A.V. Fld. In both cases non-linear plots are observed which are due to intermediate complex formation. Analysis of the data show k_2 to be 236 s^{-1} for C.P. Fld and 62 s^{-1} for A.V. Fld with respective K_a values of 2.3 x 10^4 M^{-1} and 1 x 10^4 M^{-1}. Thus, the K_a values for complex formation are of similar magnitude whereas the rate of electron transfer is much faster for C.P. than for A.V. Fld.

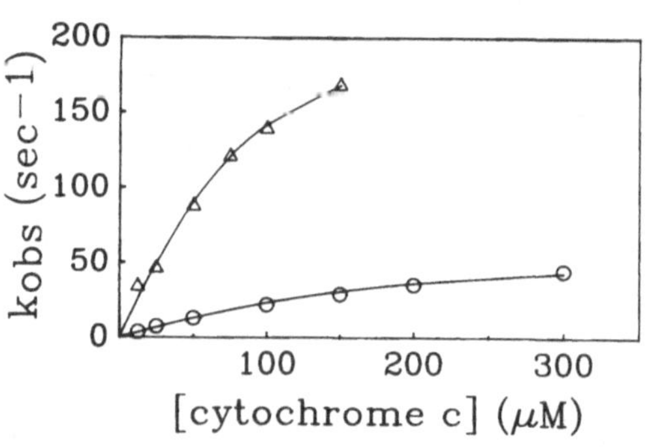

Fig.1 - Plots of observed pseudo-first order rate constants for Fld semiquinone oxidation vs. [Cyt c] at pH 8.25 in 5mM HEPES, 45mM KCl. The monitoring wavelength was 580 nm.

O - A.vinelandii

Δ - C.pasteurianum

Previous ^1H NMR studies of the C.P. Fld-Cyt c complex (2) showed extensive line broadening of the contact-shifted heme resonance to a value larger than that expected on simply increasing τ_c. Similar studies using A.V. Fld-Cyt c mixtures at 360 MHz (data not shown) gave the extent of line broadening expected for an increase in τ_c and not the extensive line broadening observed with C.P. Fld. These data suggested a

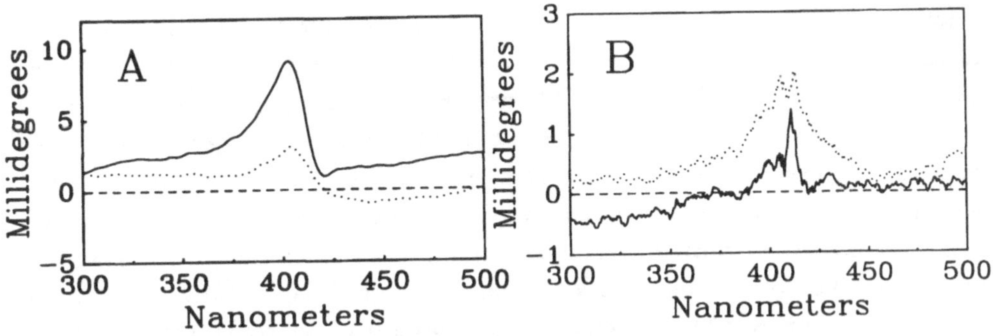

Fig. 2 - Difference CD spectra (mixture minus sum) of Cyt c/Fld mixtures. Buffer was 5mM phosphate, 0.1mM EDTA pH 7.0. Protein concentrations were 33μM each; cell pathlength .45 cm.

A - Solid line: native C.pasteurianum Fld
Dotted line: native A.vinelandii Fld
B - Solid line: 8α-imidazolylFMN substituted C.pasteurianum Fld
Dotted line: 8α-imidazolylFMN substituted A.vinelandii Fld

perturbation of the heme of the Cyt c on C.P. Fld binding but not on A.Z. Fld binding. The extensive line broadening observed with C.P. Fld binding could arise from Curie spin relaxation (5) although this remains to be experimentally verified. Circular dichroism (CD) spectral studies (Fig. 2) show that the amplitude of the ellipticity of the Soret band of the heme is increased on complexation of Cyt c with C.P. Fld to a much larger degree than on complexation with A.V. Fld. The difference amplitude is maximal at 1:1 stoichiometry and is dependent on the ionic strength of the medium. Similar small difference spectral amplitudes are observed for A.V. Fld-Cyt c mixtures over a 5-fold range in concentration, and are thus not due to binding constants differences. These data are in agreement with ^1H NMR data and show that the heme enviroment is altered on complexation of Cyt c with C.P. Fld, but that little effect is observed with A.V. Fld. Thus, the faster rate of electron transfer in the case of C.P. Fld is correlated with alteration of the heme environment on complexation. To test the above idea further, the FMN moiety in each flavodoxin was replaced with 8α-imidazolylFMN in which the 5' phosphate isomer was separated from the other isomers by HPLC. Neutral semiquinones of both reconstituted flavodoxin complexes are readily produced by the light-EDTA-deazaflavin method. Stopped-flow studies showed both reconstituted flavodoxins to react with Cyt c with k_2 values of 7 s^{-1}, which are ~ 10 to 40 times slower than the native forms. K_a values determined from the kinetic data were quite similar to those for the native flavodoxins. Preliminary redox potential data for the reconstituted A.V. complex showed the OX/SQ couple to be ~ 40 mV more positive than that of the native form. Difference CD spectral data (Fig. 3) showed that the amplitude of the Soret band perturbation was much less than that observed with native C.P. Fld, and of similar magnitude to that observed with native A.V. Fld. Thus, placing a bulky imidazole ring on the exposed 8α- position of the flavin alters the geometry of the complex (but not K_a) and results in alteration of the rate of electron transfer. Studies on the pH dependence of the electron transfer kinetics show that the native proteins exhibit no pH-rate dependence in the range of 5.5 - 8.7. Since the redox potential of Cyt c is invariant in this range (6) and the OX/SQ potential of flavodoxin does change, this suggests little dependence of rate on the difference in potential in the range tested. This indicates that proton release is much faster than electron transfer. Both 8α-imidazolyFMN flavodoxins exhibited a marked pH-rate dependence which followed a single proton ionization curve and was assigned to the protonation of the 8α-imidazole ring (Fig. 4). The pK_a values are determined to be 7.0 for 8α-imidazolylFMN semiquinone on either A.V. or on C.P. Fld. These values are 0.2 - 0.3 pH units higher than that observed for the oxidized form (7). The decrease in rate constant upon protonation of the imidazole ring probably reflects the difficulty of accomodating a positively-charged moiety within the non-polar interface region (4).

Conclusions

1. Spectral and kinetic data are consistent with the idea that rates of electron transfer from flavodoxin to cytochrome c can be correlated with alteration of heme environment on complex formation.
2. Placement of a bulky imidazole ring at the 8α-position of the flavin decreases both the rate of electron transfer and the observed perturbation of the heme enviroment.
3. The rate of electron transfer is pH-independent for the native flavodoxins, whereas the protonation state of the 8α-imidazole ring markedly influences the electron transfer kinetics.

Acknowledgement

Supported by NIH grants GM 29433 (D.E.E.) and AM 15057 (G.T.)

References

1. Cheddar, G., T.E. Meyer, M.A. Cusanovich, C.D. Stout, and G. Tollin. 1986. Biochemistry 25, 6502.

2. Hazzard, J.T. and G.Tollin. 1985. BBRC 130, 1281.

3. Dickerson, J.L., J.J. Kornue, and D.C. Reas. 1985. J. Biol. Chem. 260, 5175.

4. Simondsen, R.P., P.C. Weber, F.R. Salemme, and G. Tollin. 1982. Biochemistry 21, 6366.

5. Gueron, M. 1975. J. Mag. Res. 19, 58.

6. Moore, G.R., D.E. Harris, F.A. Leicht, and G.W.Pettigraro. 1984. Biochim. Biophys. Acta 764, 331.

7. Edmondson, D.E., and R. De Francesco: this volume.

A COMPUTER GRAPHICS MODEL OF THE COMPLEX FORMED BETWEEN FLAVODOXIN AND THE TETRAHEME CYTOCHROME C_3

D. E. Stewart, J. LeGall, H. D. Peck, Jr., J. E. Wampler
Department of Biochemistry, University of Georgia, Athens, Georgia, 30602

I. Moura, J.J.G. Moura, A.V. Xavier
Centro de Quimica Estrutural, Universidade Nova de Lisboa, Lisbon, Portugal

P.K. Weiner
Department of Chemistry, Rutgers University, New Brunswick, New Jersey, 08903

Introduction

A computer graphics model of the complex formed between the flavodoxin and the tetraheme cytochrome c_3 of Desulfovibrio vulgaris has been proposed (ref. 1 and Figure 1). This model is based on preliminary NMR data, electrostatic field calculations, and charge complementarity. It is free of steric interferences and represents a unique pairing of flavodoxin to one of the four hemes of the cytochrome.

Results and Discussion

The x-ray structures for D. vulgaris Miyazaki cytochrome c_3 at 1.8 Å resolution (2) and D. vulgaris (Hildenborough) flavodoxin at 2.0 Å resolution (3) were used to build the model complex (Figure 1). This model was built using an Evans

Flavins and Flavoproteins 1987

and Sutherland Corporation PS340 graphics workstation with the molecular graphics software MOGLI.

The complex is stabilized by five uniformly distributed hydrogen bonds formed between acidic residues on flavodoxin and lysines surrounding the heme of cytochrome c_3 (Table 1). These hydrogen bonds are well within hydrogen bonding limits, with none larger than 3.0 A. No steric interferences are present in the complex.

In the proposed complex, the flavin mononucleotide (FMN) group of flavodoxin and Heme 1 of cytochrome c_3 (numbering according to ref. 2) are nearly coplanar and parallel and are within van der Waals contact between atom C8 of FMN and atom C2B of the heme (4.18 A). The closest approach of the two pi systems is 5.5 A. Kinetic studies performed on Clostridium MP

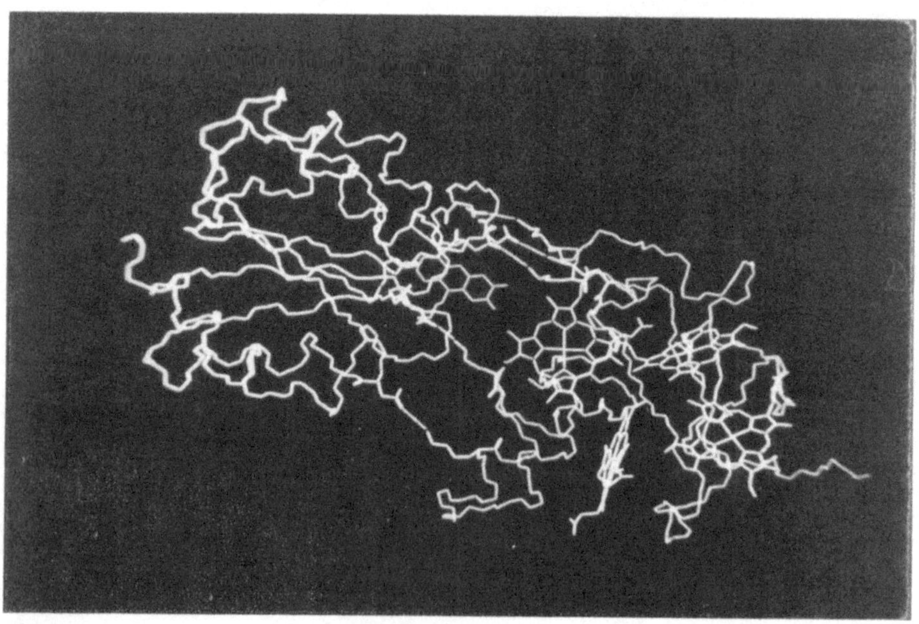

Figure 1. The proposed complex showing the main chain atoms, the prosthetic groups and the hydrogen bonding pairs.

TABLE 1. Cytochrome c_3 - Flavodoxin hydrogen bonding distances.

Flavodoxin	Cytochrome c_3	Distance (Å)
Asp - 69	Lys - 60	2.90
Glu - 66	Lys - 58	2.91
Asp - 62	Lys - 15	2.93
Asp - 95	Lys - 101	2.91
Asp - 129	Lys - 95	3.00

flavodoxin and tuna cytochrome c suggest that electron transfer occurs via an outer-sphere mechanism (4,5). Since these two hypothetical models are quite similar geometrically, a similar electron transfer mechanism can also be postulated for the flavodoxin - tetraheme cytochrome c_3 complex.

Electrostatic field calculations (details presented elsewhere, ref. 1) were performed on the regions surrounding the exposed portions of each prosthetic group using the molecular mechanics program AMBER (6) and related programs (P. K. Weiner, unpublished). Electrostatic potential calculations on flavodoxin revealed a very asymmetric surface charge distribution -- the region surrounding the FMN-binding site being completely negative and the other end of the molecule having regions of positive charge. This was also found for *Clostridium* MP flavodoxin (7). Electrostatic potential field calculations for the regions surrounding each of the cytochrome hemes reveal that three of the four hemes are surrounded by electrostatic fields consisting of positive and negative charges interspersed, while the other heme, heme 1, was surrounded by a region of completely positive charge. Thus, the strongest attraction, and least charge repulsion, would occur between the FMN binding region of flavodoxin and heme 1 of cytochrome c_3.

The results presented above show that 1) as flavodoxin and cytochrome c_3 approach each other, electrostatic forces would orient the two molecules properly for the complex described; 2) when the complex is formed, the two prosthetic groups are parallel and coplanar and are within van der Waals contact; and 3) the complex is consistent with electron transfer via an outer-sphere mechanism.

Acknowledgements

This work was supported in part by the Advanced Computational Methods Center, ETA Systems Inc. (1450 Energy Park Dr., St. Paul, Minnesota), the University of Georgia Research Foundation and by a grant (DMB-8602-789) from the National Science Foundation (to J. L.).

References

1. Stewart, D.E., J. LeGall, I. Moura, J.J.G. Moura, H.D. Peck, Jr., A.V. Xavier, J.E. Wampler, and P.K. Weiner. 1987. Manuscript in preparation.

2. Higuchi, Y., M. Kusunoki, Y. Matsuura, N. Yasoka, and M. Kakudo. 1984. J. Mol. Biol. 172, 109.

3. Watenpaugh, K.D., L.C. Sieker, and L.H. Jensen. 1976. In: Flavins and Flavoproteins (T.P. Singer, ed.). Elsevier New York, p. 405.

4. Simondsen, R.P., P.C. Weber, F.R. Salemme, and G. Tollin. 1982. Biochem. 21, 6366.

5. Jung, J. and G. Tollin. 1981. Biochem. 20, 5124.

6. Weiner, P.K. and P.A. Kollman. 1981. J. Comp. Chem. 2, 287.

7. Matthew, J.B., P.C. Weber, F.R. Salemme, and F.M. Richards. 1983. Nature 301, 169.

CLONING AND NUCLEOTIDE SEQUENCE OF A BACTERIAL FLAVODOXIN GENE: TOWARDS THE SITE-DIRECTED MUTAGENESIS OF THE FLAVIN BINDING SITE.

Grigorios D. Krey, Elio F. Vanin, and Richard P. Swenson
Department of Biochemistry, Ohio State University, Columbus, OH 43210

Flavodoxins are small bacterial flavoproteins that function as low potential electron transfer proteins. Several unique features have made this class flavoprotein the subject of intense research over the years. Firstly, the redox potential of the semiquinone/hydroquinone couple is the among lowest of any known flavoprotein (ca. -450 mV *vs* -175 mV for free FMN, pH 7), while that of the oxidized/semiquinone couple is generally more positive than that for free FMN (1). Secondly, the semiquinone state is unusually stable and is stabilized as the blue neutral flavin radical species. Finally, the three-dimensional structure of flavodoxins from three sources (*Clostridium MP* (1), *Desulfovibrio vulgaris* (2), and *Anacystis nidulans* (3)) have been determined, in some cases for all of the redox states. As a result, we have a clear picture of the specific flavin-apoprotein interactions that exist in this class of flavoprotein. The apparent conservation of many of the general types of interactions suggests that some or all must be intimately involved in the pronounced perturbation in the redox properties of the bound flavin cofactor.

With the advent of genetic engineering technology, specifically, *in vitro* site-specific mutagenesis, it should now be possible to systematically alter amino acid residues in or near the flavin binding site of this class of flavoprotein in order to gain a better understanding of how the redox properties of bound flavin cofactors are regulated when bound to proteins. Towards this end, we report here the identification, cloning, and nucleotide sequence of the structural gene

Flavins and Flavoproteins 1987

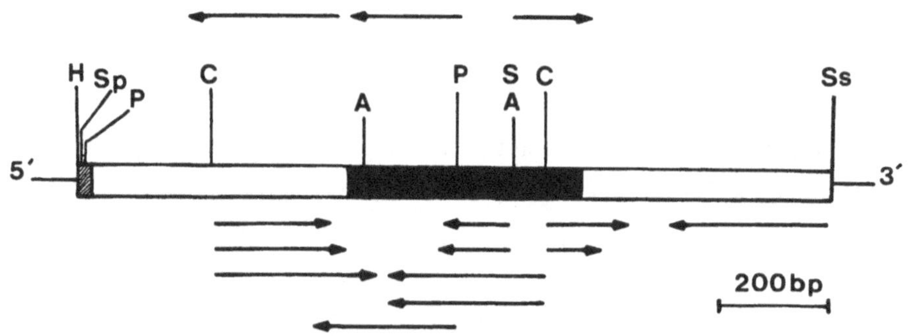

Figure 1. Partial restriction map and sequencing scheme for the 1.4 kb Xho I-Sac I restriction fragment. The darkened area represents the coding region for the flavodoxin polypeptide. The arrows represent the regions sequenced on each DNA strand. Restriction enzymes used: **A,** Acc I; **C,** Cla I; **H,** Hind III; **P,** Pst I; **S,** Sal I; **Sp,** Sph I; **Ss,** Sac I.

for the flavodoxin from the sulfate-reducing bacterium Desulfovibrio vulgaris.

DNA restriction fragments containing the flavodoxin gene were identified by direct hybridization of Southern blots of Sal I-digested genomic DNA with a synthetic heptadecanucleotide probe. The nucleotide sequences of this mixed probe were derived from the amino acid sequence between residues 98-102 of the reported protein primary structure, and was synthesized accordingly to contain the necessary redundancies in the genetic code. Three bands within the genomic digest, of sizes 4.8, 4.0, and 1.6 kb, consistently hybridized to the radiolabeled probe. Size-selected, Sal I restriction fragments corresponding to each of these hybridizing regions were cloned into the pUC19 vector and a representative sample of the transformants for each sized insert were rescreened with the mixed probe. A strongly hybridizing clone containing a DNA fragment from the 1.6 kb region was found to contain a nucleotide sequence which codes for the first 104 residues of the amino terminal portion of the flavodoxin protein sequence. A Pst I restriction fragment from this clone was used as a probe for the isolation of the entire gene from a partial Sau3A I library in

```
                                                     5' GATATCGATACGGAAGCGATCTCGGCCATCTGTGACCATAC

CTACCTCAAAGCATCCCGCAAATGATATTGCGCCCCCGCGTGATGTGCATTTTTATCTTGAAAACAAAAATCAATTTCA
                            -35'                    -10'    -35

ATATACACGAGTCACGCCCTTTCCACACCGGGCACAGCAAGGACAGGATGACGATCGAGAAGAACGACCGAACCACCCC
  -10

ATGCCCCACCCTCTGCCGGTGTACACCGCCGTCGCCCCGTGACGGGACGGCGGCATGAAAACACCCAGGAGTGCAGACC

Met Pro Lys Ala Leu Ile Val Tyr Gly Ser Thr Thr Gly Asn Thr Glu Tyr Thr Ala Glu
ATG CCC AAA GCC CTC ATC GTC TAC GGT TCC ACC ACA GGC AAC ACG GAA TAC ACC GCC GAA

Thr Ile Ala Arg Glu Leu Ala Asp Ala Gly Tyr Glu Val Asp Ser Arg Asp Ala Ala Ser
ACC ATC GCA CGT GAA CTT GCC GAT GCA GGG TAC GAA GTC GAT AGC CGG GAC GCG GCC TCT

Val Glu Ala Gly Gly Leu Phe Glu Gly Phe Asp Leu Val Leu Leu Gly Cys Ser Thr Trp
GTC GAG GCT GGC GGT CTC TTC GAA GGC TTC GAC CTC GTC CTT CTC GGA TGC TCG ACG TGG

Gly Asp Asp Ser Ile Glu Leu Gln Asp Asp Phe Ile Pro Leu Phe Asp Ser Leu Glu Glu
GGT GAC GAC TCC ATC GAA CTG CAG GAC GAC TTC ATT CCC CTT TTC GAC TCC CTC GAA GAG

Thr Gly Ala Gln Gly Arg Lys Val Ala Cys Phe Gly Cys Gly Asp Ser Ser Tyr Glu Tyr
ACG GGG GCG CAG GGC CGC AAG GTG GCC TGC TTC GGC TGC GGC GAC AGT TCC TAC GAG TAC

Phe Cys Gly Ala Val Asp Ala Ile Glu Glu Lys Leu Lys Asn Leu Gly Ala Glu Ile Val
TTC TGC GGG GCT GTC GAC GCA ATC GAA GAG AAG CTC AAG AAC CTC GGT GCG GAA ATC GTT

Gln Asp Gly Leu Arg Ile Asp Gly Asp Pro Arg Ala Ala Arg Asp Asp Ile Val Gly Trp
CAG GAC GGT CTT CGC ATC GAT GGC GAC CCC CGC GCC GCC CGG GAC GAC ATC GTC GGC TGG

Ala His Asp Val Arg Gly Ala Ile  *
GCG CAT GAC GTG AGG GGC GCC ATC TAG TCACACGCGGGCG 3'
```

Figure 2. Nucleotide sequence of the flavodoxin gene from Desulfovibrio vulgaris. The sequence shown is the nontranscribed stand. The translated amino acid sequence for the flavodoxin protein is indicated above the open coding region of the gene. The termination codon is indicated by *, the potential Shine-Dalgarno sequence is boxed, and two potential promoter sites are denoted by the solid and dashed underlines.

Charon 35. A single, positively hybridizing clone containing a 16 kb insert was isolated and characterized. The entire flavodoxin gene, including all probable initiation and termination sequences, was localized within a 1.4 kb Xho I-Sac I fragment of this clone. A partial restriction map of this clone is shown in Figure 1. It should be noted that the Pst I fragment was also used to reprobe the original Southern blot of Sal I-digested genomic DNA. Only the 1.6 kb fragment was found to hybridize suggesting that the 4.8 and 4.0 kb bands were nonspecifically binding to the small oligonucleotide probe and apparently are not structurally related to the flavodoxin gene. The sequencing strategy used to establish the complete nucleotide sequence of the flavodoxin gene is also indicated in Figure 1. The nucleotide sequence, including the protein sequence derived from the open reading frame found within it and the 5'- and partial 3'-flanking sequences, is shown in Figure 2. The derived protein sequence corresponds completely to the published protein sequence and allows for the assignment of several acid/amide ambiguities found there (4). Two potential promoter sites as well as a Shine-Dalgarno sequence were noted in the 5'-upstream flanking nucleotide sequence.

Acknowledgement

This research was supported in part by a grant from the Office of Research and Graduate Studies, The Ohio State University.

References

1. Mayhew, S.G. and M.L. Ludwig. 1975. In: The Enzymes, Vol. 12B, 3rd edition (P.D. Boyer, ed). Academic Press, New York, pp. 57-118.

2. Watenpaugh, K.D., L.C. Sieker, and L.H. Jensen. 1973. Proc. Natl. Acad. Sci. USA 70, 3857.

3. Smith, W.W., K.A. Pattridge, M.L. Ludwig, G.A. Petsko, D. Tsernoglou, N. Tanaka, K.T. Yasunobu. 1983. J. Mol. Biol. 165, 737.

4. Dubourdieu, M., J. LeGall, and J.L. Fox. 1973. Biochem. Biophys. Research Comm. 52, 1418.

FLAVIN BINDING SITE DIFFERENCES BETWEEN PHOTOBACTERIUM PHOSPHOREUM LUMAZINE PROTEIN AND MEGASPHAERA ELSDENII FLAVODOXIN. AN INVESTIGATION USING CIRCULAR DICHROISM AND FLUORESCENCE.

Antonie J.W.G. Visser, Arie van Hoek and Jacques Vervoort
Department of Biochemistry, Agricultural University
6703 BC Wageningen, The Netherlands

Dennis J. O'Kane and John Lee
Department of Biochemistry, University of Georgia
Athens, Georgia 30602

Introduction

Spectroscopic techniques have been widely employed to characterize the flavin binding site in various flavoproteins. We have applied these techniques in a native flavoprotein, Megasphaera elsdenii flavodoxin, and an artificial one, namely lumazine apoprotein from Photobacterium phosphoreum reconstituted with riboflavin. The chemical, physical and spectral properties of lumazine protein from Photobacteria have been described in a series of recent papers (1-6). It is demonstrated that both circular dichroism (CD) and fluorescence reveal characteristic structural and dynamic differences in flavin microenvironment, commensurate with the completely different functions of both proteins. A similar hybridization experiment, in which apoflavodoxin is recombined with lumazine phosphate (7), is reported here.

Materials and Methods

Lumazine protein was purified as described in (1). Flavodoxin was prepared according to (8). Lumazine phosphate was a generous gift of Dr. A. Bacher, Technical University of Munich. Recombination experiments were carried out as in (6,7). Spectroscopic measurements were performed as described in (6,9).

Flavins and Flavoproteins 1987

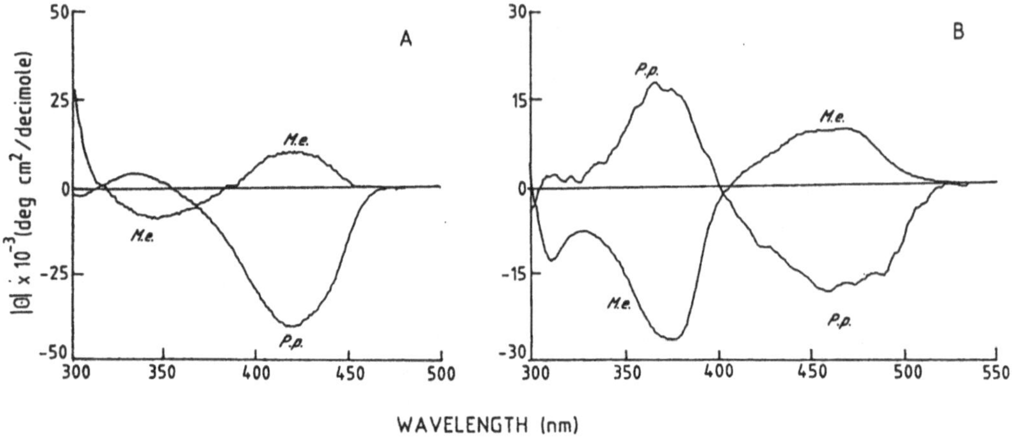

Figure 1. Circular dichroic spectra (20°C) of P.phosphoreum lumazine apoprotein and M.elsdenii apoflavodoxin with lumazine (A) or flavin (B) as bound ligands.

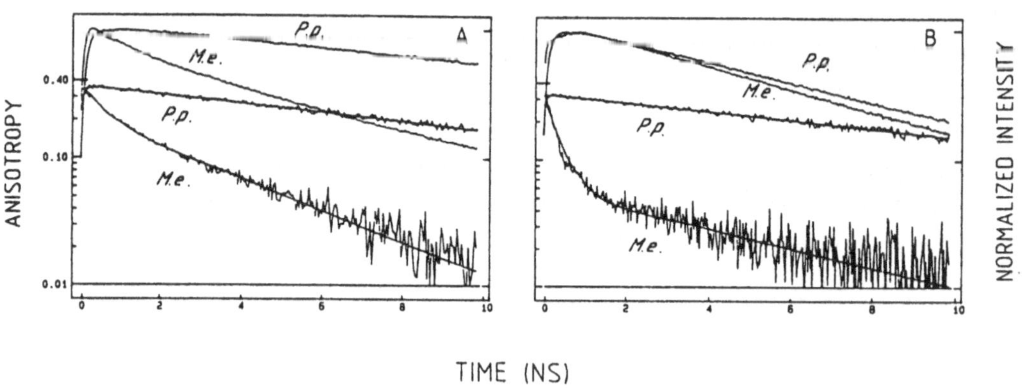

Figure 2. Fluorescence anisotropy decay (18°C) and fits of P.phosphoreum lumazine apoprotein and M.elsdenii apoflavodoxin with lumazine (A) or flavin (B) as bound ligands. Anisotropy parameters are listed in Table 2. Experimental fluorescence decay profiles, normalized to 1, are also presented to illustrate fluorescence quenching in the flavodoxin preparations. (Note the faster rise in flavodoxin).

Results and Discussion

Circular dichroism

In Fig.1A CD spectra are shown for flavodoxin recombined with lumazine phosphate and for native lumazine protein. Spectra of native flavodoxin and riboflavin-reconstituted lumazine protein are presented in Fig.1B. The spectra of the latter protein and of the native proteins are identical to published spectra (4,6,10). Irrespective of the bound ligand lumazine protein shows an intrinsic negative Cotton effect in the first absorption band, whereas flavodoxin exhibits positive dichroism. The second absorption band of the flavin shows opposite Cotton effects. A major conclusion is that the conformation of the ligand binding region is preserved to accomodate structurally related prosthetic groups. In order to demonstrate that the secondary structure is not changed relative to the parent protein, secondary structure parameters are listed in Table 1. The far UV CD spectrum of flavodoxin was identical to the previously reported one (11). It is clear that flavodoxin is more structured than lumazine protein.

Time-resolved fluorescence

In Figs.2A and B and in Table 2 the time-resolved and static fluorescence results are summarized. As with CD the fluorescence of both proteins shows opposite tendencies. The fluorescence properties of lumazine protein are similar as already reported (4-6). The 'artificial' flavoprotein gives off the largest amount of flavin fluorescence hitherto observed. Conversely, the fluorescence of native flavodoxin is heavily quenched and the fluorescence decay is heterogeneous with an important contribution of a very short fluorescence lifetime component (30-50 ps). The short lifetime must be ascribed to FMN interacting with electron-rich amino acids, which results in efficient radiationless deactivation of the singlet excited state. The short lifetime seems to be a general property of flavodoxins (12). The fluorescence of lumazine-bound flavodoxin is also quenched, but to a lesser extent. The fluorescence anisotropy decay of flavodoxin is biexponential. The short correlation time reflects internal motion, the longer correlation time is typical for overall flavodoxin rotation. However, the latter contribution decreases with lower flavodoxin concentration. A protein-protein association equilibrium must be involved to

Table 1. Secondary structural parameters of <u>P.phosphoreum</u> lumazine protein (LumP) and <u>M.elsdenii</u> flavodoxin (Fld) preparations at 20°C.

	% α-helix	% β-sheet	% other
LumP	17	35	48
Rfl-LumP	18	30	52
Fld	33	51	16
Lum-Fld	30	48	22

Table 2. Fluorescence properties of <u>P.phosphoreum</u> lumazine protein (LumP) and <u>M.elsdenii</u> flavodoxin (Fld) preparations.

	Q_f	λ_{ex} (nm)	λ_{em} (nm)	<τ> (ns)	β (-)	φ (ns)	Remarks
LumP	0.59	458	475	14.4	0.37	12.4	20 μM, 18°C
		410	480	14.6	0.35	20.8	Synchrotron, 4°C
Rfl-LumP	0.30	458	544	5.6	0.33	12.9	4 μM, 18°C
Fld	<0.01	476	544	1.2	0.22	0.2	0.23 mM, 18°C
					0.11	6.0	
		476	544	2.0	0.29	0.2	21 μM, 18°C
					0.06	5.7	
		476	544	3.6	0.34	0.2	2.8 μM, 18°C
					0.02	3.5	
Lum-Fld	0.03	458	475	2.4	0.17	0.4	25 μM, 18°C
					0.21	3.3	
		410	489	2.7	0.27	0.4	Synchrotron, 4°C
					0.07	4.1	8 μM
		410	489	4.7	0.27	0.4	Synchrotron, 4°C
					0.04	2.7	1.6 μM

explain the results. Our results at high flavodoxin concentration are in agreement with immobilized flavin as inferred from NMR experiments, which are carried out with even more concentrated samples (13). Also, the permanent presence of a mixture of protein conformations with either immobilized or flexibly bound flavin cannot be ruled out. With the sensitive fluorescence technique the more fluorescent protein population would be sampled.

Acknowledgement

Synchrotron experiments were made possible owing to a grant of the Netherlands Organization for the Advancement of Pure Research.
This work is also supported by NIH GM28139.

References

1. O'Kane, D.J., V.A. Karle, J. Lee. 1985. Biochemistry 24, 1467-1475.
2. O'Kane, D.J., J. Lee. 1985. Biochemistry 24, 1467-1475.
3. O'Kane, D.J., J. Lee. 1985. Biochemistry 24, 1484-1488.
4. Lee, J., D.J. O'Kane, A.J.W.G. Visser. 1985. Biochemistry 24, 1476-1483.
5. Visser, A.J.W.G., T. Ykema, A van Hoek, D.J. O'Kane, J. Lee. 1985. Biochemistry 24, 1489-1496.
6. Kulinski, T., A.J.W.G. Visser, D.J. O'Kane, J. Lee. 1987. Biochemistry 26, 540-549.
7. Harzer, G., S. Ghisla. 1979. In: Chemistry and Biology of Pteridines (R.L. Kisliuk and G.M. Brown, eds.). Elsevier North Holland, pp. 37-42.
8. Mayhew, S.G., V. Massey. 1969. J.Biol.Chem. 244, 794-802.
9. Vos, K., A. van Hoek, A.J.W.G. Visser. 1987. Eur.J.Biochem. 165, 55-63.
10. Edmondson, D.E., G. Tollin. 1971. Biochemistry 10, 113-124.
11. D'Anna, J.A., G. Tollin. 1972. Biochemistry 11, 1073-1080.
12. Karen, A., M. Takagi Sawada, F. Tanaka, N. Mataga. 1987. Photochem.Photobiol. 45, 49-53.
13. Vervoort, J., F. Müller, S.G. Mayhew, W.A.M. van den Berg, C.T.W. Moonen, A. Bacher. 1986. Biochemistry 25, 6789-6799.

FERREDOXIN AND FERREDOXIN-NADP$^+$ OXIDOREDUCTASE INTERACTIONS: MAPPING OF THEIR BINDING SITES

G. Zanetti, A. Aliverti, B. Curti

Dipartimento di Fisiologia e Biochimica Generali, Università di Milano, Italy

S. Ronchi

Istituto di Fisiologia Veterinaria e Biochimica, Università di Milano, Italy

Introduction

Recently, much progress has been made in the elucidation of protein--protein interactions, which are fundamental to molecular recognition in several biosystems. To gain insight in enzyme-substrate protein complexes, crystallographic studies are not always feasible, thus one must resort to differential chemical modification of protein groups or to cross-link formation between the proteins by means of bifunctional reagents. Carbodiimides, which are zero-length cross-linkers, have been shown to promote amide bond formation between carboxyl and amino groups that partecipate in protein-protein electrostatic interactions. In photosynthesis, the terminal part of the chloroplast electron transport requires the interaction between ferredoxin (Fd) and the reductase (FNR) for NADP$^+$ photoreduction. The formation of the covalent complex between the two purified proteins has been achieved by using a soluble carbodiimide (1). The cross-linked complex thus obtained, has been shown to possess many properties of the native complex: the 1/1 stoichiometry, the spectral perturbations, the NADPH-cyt c reductase activity, the isoelectric point, the CD and fluorescence spectra (1,2). Furthermore, the covalent complex had no reductase

Flavins and Flavoproteins 1987

activity towards exogenous Fd (1), suggesting that the Fd binding site on the flavoprotein is completely masked; thus, the covalent linkage between the two proteins should be considered highly specific. Here we report a structural characterization of the cross-linked complex between Fd and FNR which allowed to identify the peptide regions of the two proteins interfaced in the complex.

Results and Discussion

The best strategy to characterize the cross-linked regions appeared to be the protein fragmentation by cyanogen bromide: the cleavage at methionine residues would have yielded intact ferredoxin (which does not contain methionine in the primary structure (3)) with cross-linked peptide(s) from FNR, easily separable from other, smaller CNBr fragments of the flavoprotein by gel-filtration (4). The cross-linked complex, after carboxymethylation with radioactive iodoacetate to trace the -SH containing peptides (both proteins contain five cysteines), was cleaved by cyanogen bromide. As shown in Fig. 1A, the pattern of peptides as resolved in Nadodecilsulfate electrophoresis was difficult to interpret. Unfortunately, the Fd preparation contained a substantial amount (25%) of an isoform (Fd II) (which has two methionines in the sequence, Met 33 and Met 69 (5)) as also revealed by immunoblotting of the CNBr treated Fd (Fig. 1B). The major band at 31.6 KDa was highly radioactive and should arise from the cross-linking between Fd I (apparent M_r 22,000 in gel-electrophoresis (1)) and the peptide 61-155 of FNR (calculated M_r 10,130) which contains three cysteines. It should be recalled that all the other CNBr peptides from FNR (4) have $M_r \leq 3{,}700$. The hypothesis made was substantiated by the recognition of the band by an antibody anti-Fd in an immunoblotting experiment (Fig. 1B); the bands with M_r higher than 20,000 were positive to the immunoreaction, suggesting both a heterogeneity of cross-linking and possibly an incomplete

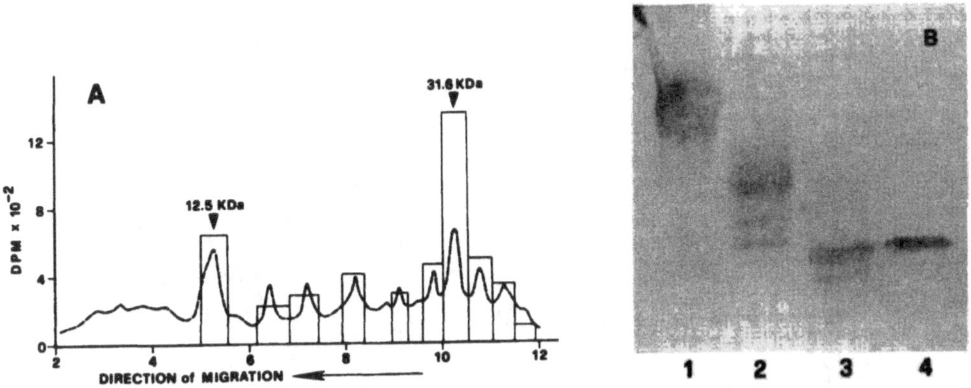

Fig. 1. SDS-PAG electrophoresis pattern of CNBr peptides obtained from the cross-linked complex and immunoblotting with anti-ferredoxin antibody. A, Each gel band (---) was cutted and the radioactivity due to carboxymethylcysteine was measured by scintillation counting (histogram). B, Western blot transfer and immunodecoration with anti-Fd antibody. Lanes: 1, cross-linked complex; 2, CNBr peptides of the cross-linked complex; 3, CNBr peptides of Fd; 4, Fd.

CNBr digestion. With respect to the 12.5 KDa band, it could contain either the unmodified FNR peptide 61-155 and/or a cross-linked peptide from Fd II.

A completely different approach to the problem was then attempted by submitting the cross-linked complex to tryptic digestion. The apoprotein, after carboxymethylation with radioactive iodoacetate, was hydrolyzed by trypsin and the resulting peptides separated by reverse phase HPLC (Fig. 2). A radioactive region of the chromatogram around 50 min substantially differed from the tryptic maps of the free proteins. Peak 57 and 58 contained mainly corresponding peptides of FNR; peak 63, the residues 5-40 of Fd. Peak 59 and 62 were further purified by reverse phase HPLC at pH 6: the main component of peak 59 was a cross-linked peptide, while peak 62 contained both the unmodified peptide 53-97 of Fd I and a different cross-linked peptide. Evidence for the presence of such cross-linked peptides was based

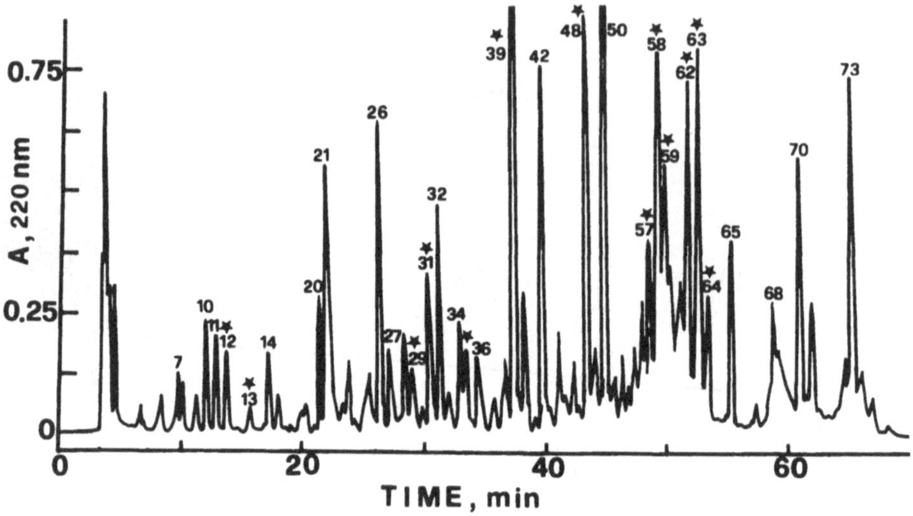

Fig. 2. HPLC separation of tryptic peptides from the cross-linked complex. An asterisk marks radioactive peptides containing carboxymethylcysteine.

Table 1: Analyses of the Tryptic Peptides n° 59 and n° 62 from the Cross--linked Complex.

Peak	Residues	
	Sequenced	Amino acid composition
59	Fd I 53-65 FNR 72-84	Fd I 53-97 FNR 72-91
62	Fd I 5-18 FNR -	Fd I 5-40 FNR 2-15

on amino acid composition and sequence analysis, as shown in Table 1. The interpretation given in Table 1 was substantiated by secondary digestions of the tryptic peptides 59 and 62 with chymotrypsin which should cleave only the Fd component of the cross-linked peptides. As expected, the protease hydrolyzed at (Phe 63), Trp 73 and Leu 75 of peptide 53-97 (peak 59), yielding a fragment which contained Fd residues 75-97 together with

Table 2: Secondary Digestion of the Purified Cross-linked Peptides with Chymotrypsin.

Chymotrypsin Treatment	
Before	After
Fd I 53-97	Fd I 76-97
FNR 72-91	FNR 72-91
Fd I 5-40	Fd I 5-37
FNR 2-15	FNR 2-15

FNR residues 72-91. Chymotrypsin cleaved the unmodified peptide 5-40 of Fd both at Tyr 23 and Tyr 37, while only the latter was susceptible to digestion in the cross-linked peptide of peak 62 (Table 2). Thus, two different sites of cross-linking between the two proteins were found; in quantitative terms, about 70% of the linkage was between the C-terminal moiety of Fd and residues 72-91 of FNR. The sequences of such peptides as deduced from the published primary structures of FNR and Fd (4,3) are the following:

```
                                   *         *
      72                           85        88       91
FNR   E-G-Q-S-V-G-V-I-P-D-G-E-D-K-N-G-K-P-H-K

                     *           *           * * *
      76             84          88            93        97
Fd I  T-C-A-A-Y-P-V-S-D-V-T-I-E-T-H-K-E-E-E-L-T-A
```

An involvement in the cross-linking of the FNR acidic residues (Glu 72, Asp 81, Glu 83 and Asp 84) could be ruled out because these residues were sequenced. Thus, FNR partecipated in the cross-link as the amino-donor: of the three lysines present, Lys 85 should be the most likely candidate to the role, because its bond was not cleaved by trypsin as it was the case for Lys 91. No data on the involvement of Lys 88 in the linkage could be obtained: the bond Lys 88-Pro 89 is trypsin resistent and the sequence Lys 85-Asn 86-Gly 87 is difficult to sequence as also reported by Karplus

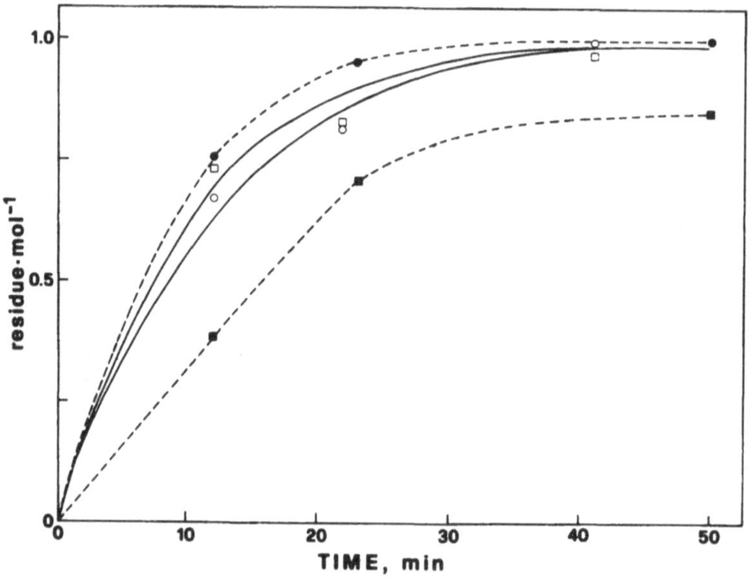

Fig. 3. Kinetics of carboxypeptidase Y release of C-terminal residues from an equimolar mixture of FNR+Fd and from the cross-linked complex. Free proteins:○—○, Tyr;□—□,Ala. Cross-linked complex:●- -●, Tyr;■- -■, Ala.

et al. (4). In the COOH-terminal region of Fd there are several side-chain carboxyl groups as well as the α-COOH of the C-terminal Ala. Digestion experiments with carboxypeptidase Y (Fig. 3), carried out both on an equimolar mixture of the free proteins and on the cross-linked complex, showed that the C-terminal Ala was actually released from the cross-linked complex nearly completely but at a quite different rate. Thus, the α-COOH of Fd was not directly involved in the cross-linking; the decreased rate of alanine removal from the cross-linked complex by carboxypeptidase could be due to steric hindrance by the cross-linked peptide suggesting that the site of binding could be situated nearby the C-terminal amino acid where three glutamic residues (Glu 92, 93, 94) are present. It is worthwhile to mention that Glu 92 is conserved in all the plant-type Fd sequenced (6); the Halobacterium-type protein which lacks such acidic residues in this region seems to interact much less effectively with spinach FNR (7).

The minor site of cross-linking was formed between the following peptides of FNR and Fd I:

```
*
2                          15
I-A-S-D-V-E-A-P-P-P-A-P-A-K         (FNR)

                    *       * *         *       * * *     *
5                   15      20          26        30      34    37
V-T-L-V-T-P-T-G-N-V-E-F-Q-C-P-D-D-V-Y-I-L-D-A-A-E-E-E-G-I-D-L-P-Y
```

The peptide 5-37 of Fd does not contain lysine, thus it can only act as COOH-donor in the linkage. Among the acidic residues, Glu 15 was sequenced; Glu 31 and Asp 34 could be also excluded as partecipants in the cross-link, on the basis of the cross-linked peptide derived from Fd II which contained the residues 5-31 (Fd II has a Lys at position 31). The lack of chymotryptic digestion at Tyr 23 would indicate a steric hindrance to the protease action due to modification of either Asp 20, 21 or 26. The last residue is again a conserved amino acid in the plant-type Fd (6). The choice of the amino-donor is limited to the α-NH_2 of FNR, because the only lysine present in the peptide is the C-terminal residue. This is in keeping with the fact that the peptide was not sequenced (Table 1). By amino acid determinations, isoleucine was found as the N-terminal residue in the cross--linked FNR peptide. This amino acid is the second residue in the published sequence (4): occurence of isoforms in FNR is a well known phenomenon and several N-terminal residues were found in different preparations (4).

The primary site of cross-linking is localized in a FNR region which contains a cluster of positive charges (Lys 85, 88, His 90, Lys 91, Arg 93) in the N-terminal domain of the flavoprotein which has been proposed as the FAD binding domain (8). Secondary structure prediction (9) made on FNR indicates that this segment is not structured. A hydropathy profile of the flavoprotein calculated according to Kyte and Doolittle (10) is shown in Fig. 4: the site of cross-linking is localized in a region highly hydrophylic and thus, most likely, exposed to solvent on the molecule surface.

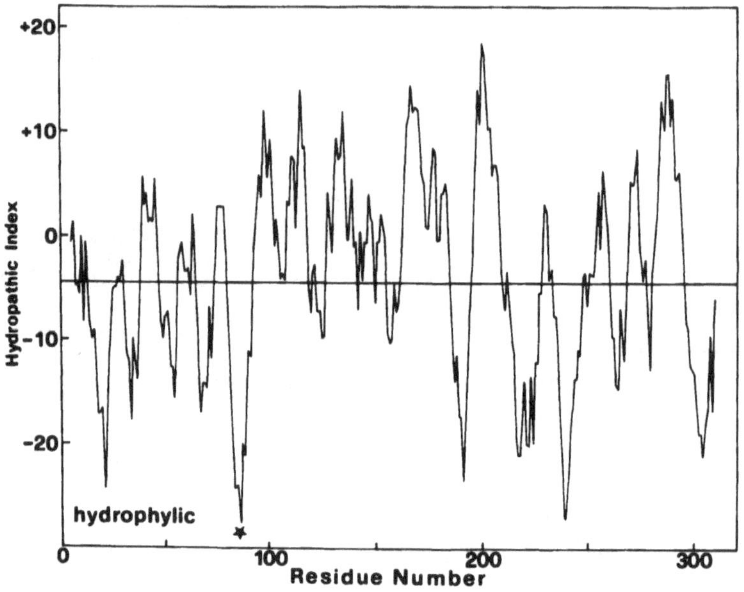

Fig. 4. Hydropathy plot of FNR. The hydropathy plot was calculated according to Kyte and Doolittle (10) using a span setting of 9 residues.

Plant-type ferredoxins are a highly homologous class of proteins: a X-ray structure of Spirulina platensis at 2.5 Å resolution (6) is available. By means of computer graphics (11), it has been produced a three-dimensional model of spinach Fd I by fitting the spinach sequence to the S. platensis X-ray structure (Fig. 5). The general outline of the molecule of the two proteins is very similar, because they have 69% identical residues. The iron-sulfur center and the lateral chains of Asp 20, 21, 26 and Glu 92, 93, 94 are shown. The residues most probably involved in the interaction with FNR are located in these acidic clusters. In this context, very recently Vieria et al. (12) have proposed the regions 26-30, 65-70 and 92-94 of Fd as important for binding to the reductase. The calculated distance between the conserved Asp 26 β-COOH and Glu 92 γ-COOH is 17 Å. The iron-sulfur cluster is nearer the Glu 92 -COOH (9.9 Å) than the Asp 26 -COOH (12 Å). In conclusion, it has been demonstrated on a structural basis, that carboxyl

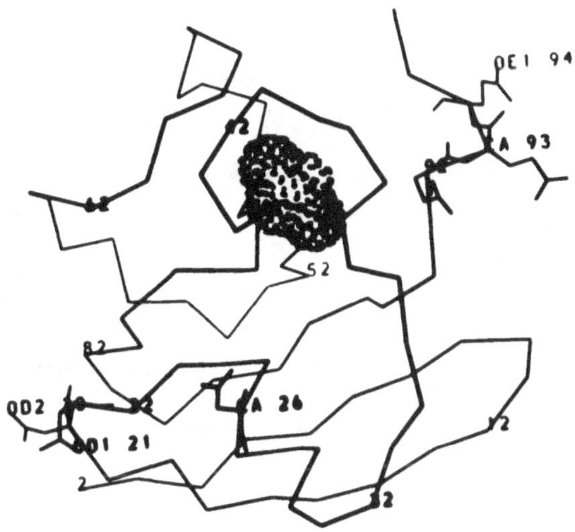

Fig. 5. Computer graphics model of spinach ferredoxin I.

groups of Fd interact with amino groups of the reductase; 70% of the cross--link is found between a FNR peptide region which is on the molecule surface in the flavin binding domain and a Fd peptide region near the iron--sulfur cluster, thus bringing in close proximity the two prosthetic groups for electron transfer to occur.

Acknowledgements

We would like to thank Dr Martino Bolognesi, University of Pavia, for performing the computer graphics of ferredoxin. This work was supported by grants from Ministero della Pubblica Istruzione of Italy.

References

1. Zanetti, G., A. Aliverti, B. Curti. 1984. J.Biol.Chem. 259, 6153.

2. Zanetti, G., B. Curti. 1984. In: Flavins and Flavoproteins (R.C. Bray, P.C. Engel and S.G. Mayhew, eds.). W. de Gruyter & Co., Berlin, p. 179.

3. Matsubara, H., R.M. Sasaki. 1968. J.Biol.Chem. 243, 1732.

4. Karplus, P.A., K.A. Walsh, J.R. Herriot. 1984. Biochem. 23, 6576.

5. Takahashi, Y., T. Hase, K. Wada, H. Matsubara. 1983. Plant Cell Physiol. 24, 189.

6. Tsukihara, T., M. Kobayashi, M. Nakamura, Y. Katsube, K. Fukayama, T. Hase, K. Wada, H. Matsubara. 1982. Biosystems 15, 243.

7. Werber, M.M., Y. Shahak, M. Avron. 1980. FEBS Lett. 113, 111.

8. Sheriff, S., J.R. Herriot. 1981. J.Mol.Biol. 145, 441.

9. Garnier, J., D.J. Osguthorpe, B. Robson. 1978. J.Mol.Biol. 120, 97.

10. Kyte, J., R.F. Doolittle. 1982. J.Mol.Biol. 157, 105.

11. Jones, T.A. 1978. J.Appl.Crystallogr. 11, 268.

12. Vieria, B.J., K.K. Colvert, D.J. Davis. 1986. Biochim.Biophys.Acta 851, 109.

COMPLEX FORMATION BETWEEN FERREDOXIN-NADP$^+$-OXIDOREDUCTASE AND FLAVODOXIN

C. Gomez-Moreno, J. Sancho, M. Fillat, J.J. Pueyo, D.E. Edmondson*

Departamento de Bioquimica y Biologia Molecular y Celular, Facultad de Ciencias, Universidad de Zaragoza, Spain and *Department of Biochemistry, Emory University, Atlanta, GA. 30322

Introduction

Although the interaction between ferredoxin-NADP$^+$-oxidoreductase (FNR) and ferredoxin has been studied extensively (1,2), little information is available on the interaction of FNR with flavodoxin since the initial demonstration (3) that the two proteins can form a tight complex via electrostatic interactions. Further study of the FNR-flavodoxin complex is of interest in view of sequence studies of mammalian Cytochrome P_{450} reductase which shows the C-terminal portion to exhibit homology with FNR and the N-terminal to exhibit sequence homology with flavodoxin (4). Detailed studies of the interaction of these two flavoproteins may provide insights into the general question of flavin-flavin electron transfer reactions. We report here studies of the interaction of Azotobacter flavodoxin with Anabaena variabilis FNR and the properties of the covalent, crosslinked complex of these two proteins. Both organisms share a common function in that they are both capable of nitrogen fixation. In addition, ^{31}P NMR studies are reported on the interaction of spinach FNR with Azotobacter flavodoxin.

Flavins and Flavoproteins 1987

Results and Discussion

The binding of <u>Azotobacter</u> flavodoxin to <u>Anabaena</u> FNR can be readily monitored by absorption spectral perturbations in the visible spectral region as a result of alterations in the vibronic bands of the FMN and FAD coenzymes on complex formation. Increasing the ionic strength of the medium abolishes the observed perturbations which suggests the complex to involve predominantely electrostatic interactions. A plot of spectral titration data is shown in Fig.1. Analysis of these data show a 1:1 stoichiometry and a K_d value of 7μM. Thus, flavodoxin binding to FNR is only 2-fold weaker than the binding constant determined for ferredoxin (5).

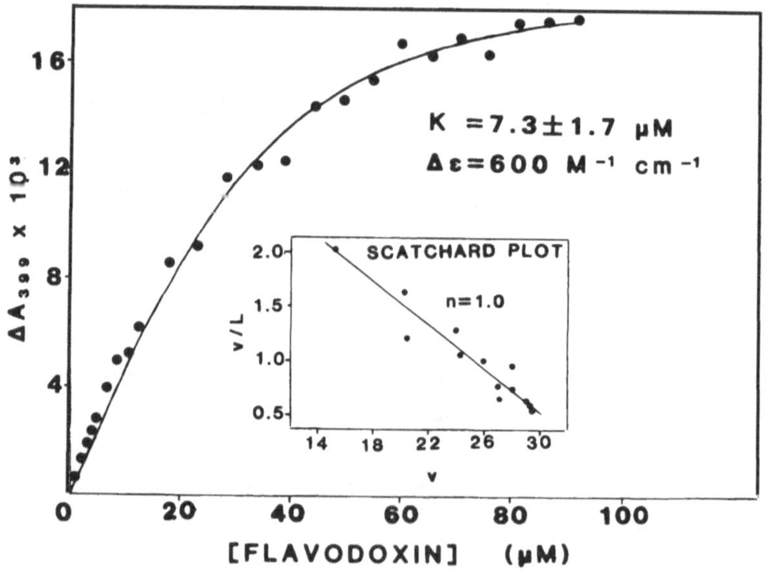

Fig.1. Spectral titration of the binding of <u>Azotobacter</u> flavodoxin to <u>Anabaena</u> FNR. Aliquots of flavodoxin were added to a solution of FNR in 50 mM phosphate, pH 7.0 and difference absorption spectra were measured after each addition. The inset is a Scatchard Plot of the data.

Treatment of the FNR-flavodoxin complex with N-ethyl-3-(3-dimethylaminopropyl)carbodiimide resulted in the formation of a covalently crosslinked complex of the two proteins as judged by SDS-gel electrophoresis. The crosslinked complex could be separated from unreacted proteins by gel filtration chromatography. Flavin analysis of the purified complex showed 1.23 moles FAD and 0.83 moles FMN per mole of complex (M.W.=59,000). The data in Table 1 shows a comparison of catalytic properties of flavodoxin crosslinked FNR with native FNR.

Table 1. Comparison of Catalytic Activities of FNR with Flavodoxin-Crosslinked FNR.

Activity	FNR	FNR-Flavodoxin Complex
A) Diaphorase		
K_m $Fe(CN)_6^{-3}$	230 μM	450 μM
K_m NADPH	66 μM	73 μM
Turnover Number	31,000 min^{-1}	16,000 min^{-1}
B) NADPH-Cytochrome c Reductase Activity		
Turnover Number (- Ferredoxin)	109 min^{-1}	170 min^{-1}
Turnover Number (+ Ferredoxin)	14,000 min^{-1}	170 min^{-1}

The results show the K_m for NADPH is unaffected on complex formation with flavodoxin while the K_m value for ferricyanide is increased by a factor of two. No stimulation of cytochrome c reductase activity by ferredoxin is observed for the complex and suggests that the bound flavodoxin occupies the same site as ferredoxin.

To further probe the interaction of FNR with <u>Azotobacter</u> flavodoxin, ^{31}P NMR studies were performed on the non-covalent complex. Spinach FNR was used in these studies since it was possible to obtain sufficient quantities of enzyme required for such experiments. The pyrophosphate resonances of the FAD moiety of FNR are observed in the -7 to -12 ppm region (Fig.2). The two phosphorus resonaces due to the FMN and covalent phosphate of flavodoxin in the complex are identical in chemical shift with those of the free form (6.3 and 1.3 ppm, respectively) (6). No significant alterations in chemical shift of any of the phosphorus resonances are evident on complex formation. Addition of Mn^{+2} to the solution results in line broadening of only the covalent phosphate resonance of the bound flavodoxin and has no affect on the linewidths of the flavin phosphate resonances. These data show the coenzyme phosphorus groups to be inaccessible to solvent while the covalent disubstituted phosphorus group of the flavodoxin is still exposed to solvent in the complex as is the situation for the free flavodoxin (6). Reduction of the FMN of the flavodoxin to its semiquinone form results in paramagnetic line broadening of only the FMN phosphate resonance but not

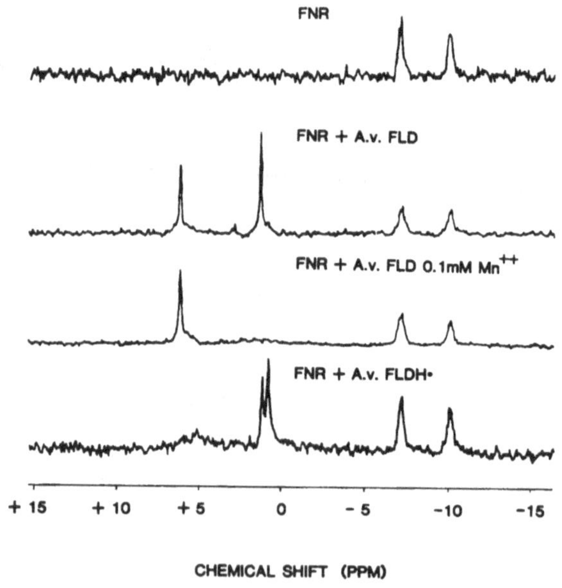

Fig.2. 81 MHz ^{31}P NMR specta of spinach FNR (0.8 mM) alone and in the presence of an equal molar quantity of Azotobacter flavodoxin. All samples were in 50 mM Tris-acetate, pH 8. All spectra were accumulated using ^{1}H broadband decoupling and 4K Hz spectral width. Flavodoxin semiquinone was generated by the addition of 0.5 moles dithionite/mole flavodoxin under an argon atmosphere.

of the FAD pyrophosphate resonances of FNR (Fig.2). These data show the FMN ring to be greater than 10 A from the side chain of the FAD moiety of FNR in the complex. The resonance due to the covalent phosphorus group of flavodoxin is split on FMN semiquinone formation. This is observed both in the free and complexed forms and suggests the covalent phosphorus group is sensing two different environments which do not rapidly interconvert on the NMR time scale.

Acknowledgements

This work was supported by a grant from the US/Spain Joint Committee for Scientific and Technical Cooperation, by a grant from the National Science Foundation (D.E.E.) and by grants from the Diputacion General de Aragon and Comision Asesora de Investigacion Cientifica y Tecnica (to C. G-M).

References

1. Nelson, N., J. Newman. 1969. J. Biol. Chem. 244, 1926.
2. Davis, D.J., A. San Pietro. 1977. Arch. Biochem. Biophys. 182, 266.
3. Foust, G.P., S.G. Mayhew, V. Massey. 1969. J. Biol. Chem. 244, 964.
4. Porter, T.D., C.B. Kasper. 1986. Biochemistry 25, 1682.
5. Sancho, J., M.L. Peleato, C. Gomez-Moreno, D.E. Edmondson. 1987. Arch. Biochem. Biophys. submitted for publication.
6. Edmondson, D.E., T.L. James. 1982. In: Flavins and Flavoproteins (V. Massey and C.H. Williams, eds.) Elsevier North Holland, p.111.

MECHANISM OF FERREDOXIN-NADP$^+$ REDUCTASE REACTION

NADP radical, semiquinone and fully reduced enzyme-NADP$^+$ charge transfer complex as the reaction intermediates

Shinya Yoshikawa, Satoshi Nitta, Kyoko Nishiyama, Norihiro Ohnishi, Katsuhito Takeshima, Yasuhiko Ishii, Makoto Nakamura* Hiroshi Matsuara** and Keiun Kodo***

Department of Biology, Konan University, Higashinada-Ku, Kobe 658, *Department of Biophysical Engineering and **Department of Biology, Osaka University, Toyonaka, Osaka 560, ***Spirulina Institute, Osaka Japan

Introduction

The terminal enzyme for the photosynthetic NADP$^+$ reduction is an FAD-containing protein, ferredoxin-NADP$^+$ reductase (FNR), which catalyzes the electron transfer reaction from reduced ferredoxin (Fd) to NADP$^+$. Extensive investigations for the transient kinetics of the enzyme system as well as the steady state kinetics (1) are indispensable for elucidation of the reaction mechanism. However, the electron transfer between the stoichiometric amounts of the enzyme and the substrates under optimum conditions is too fast to follow even with a stopped flow apparatus. Thus, recently, Batie and Kamin reported a stopped flow study on a multiple turnover system including excess amounts of reduced Fd and NADP$^+$, in which they claimed the enzymic reaction rate was lowered by the inhibitory effect of oxidized Fd (2). On the other hand, we examined the reaction under conditions slightly apart from the optimum, to control the reaction rate. Preliminary report has been given in the previous Symposium.

Flavins and Flavoproteins 1987

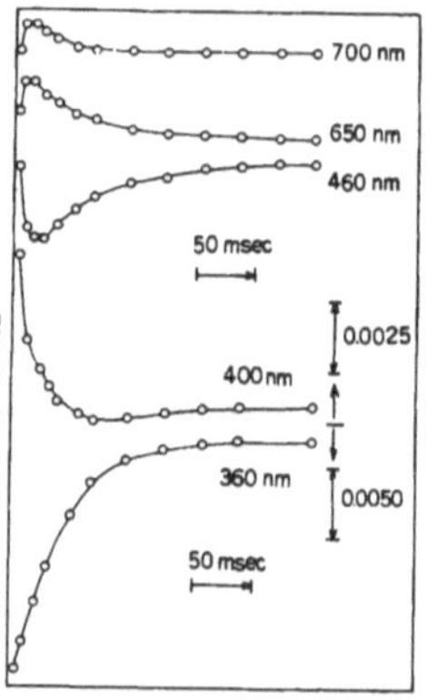

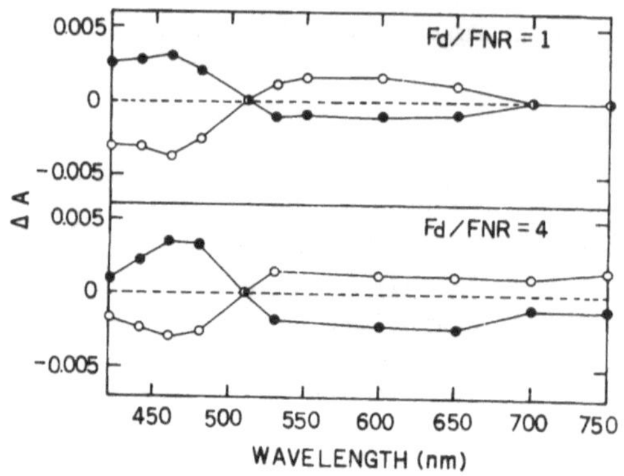

Fig. 1. (left) Absorbance changes for the complete reaction system. Absorbance change due to Fd oxidation was subtracted as described in the text.
Fig. 2. (right) Spectral change of FNR during the enzymic turnover. The absorbance changes in the initial rapid phases at various wavelengths (-o-) and the subsequent slower phases (-o-) are plotted against the wavelength.

Results and Discussion

1) Complete reaction system: Fd-FNR-$NADP^+$. Various concentrations of dithionite reduced Fd (8-46 μM) were mixed with oxidized FNR (8 μM) and $NADP^+$ (90 μM). Ferredoxin and FNR were isolated from a blue green alga, _Spirulina platensis_ (3). The reaction medium contained 50 mM Tris-HCl buffer, pH 7.4, and 40 mM $MgCl_2$ at 6°C. Figure 1 shows the absorbance changes of FNR at 16 μM Fd after mixing. The absorbance change due to oxidation of reduced Fd at each wavelength was estimated from the absorbance change where the absorbance change of FNR is negligible, and subtracted from the experimental curve. Three absorbances at longer wavelengths (700 nm, 650 nm and 460 nm) varied biphasically with time; rapid absorbance increases or decreases in the initial several msec after the dead time were followed by slower phases directed toward the original levels, respectively. However, at 400 nm, the initial rapid decrease was clearly more intense than the following slower increase. A fairly large monotonous increase was observed at 360 nm. The spectral changes in the two phases above 420 nm are given in Fig. 2.

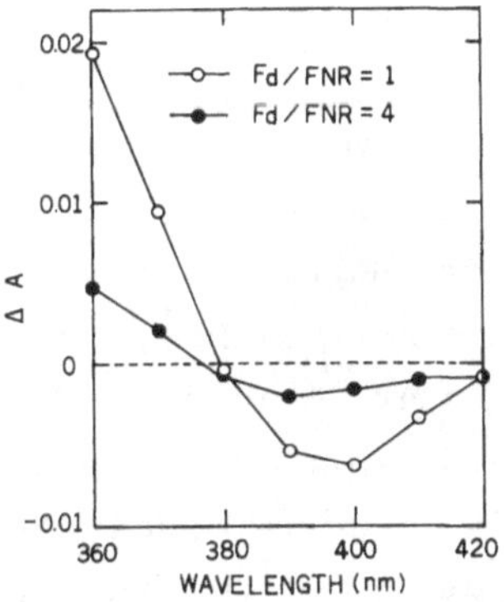

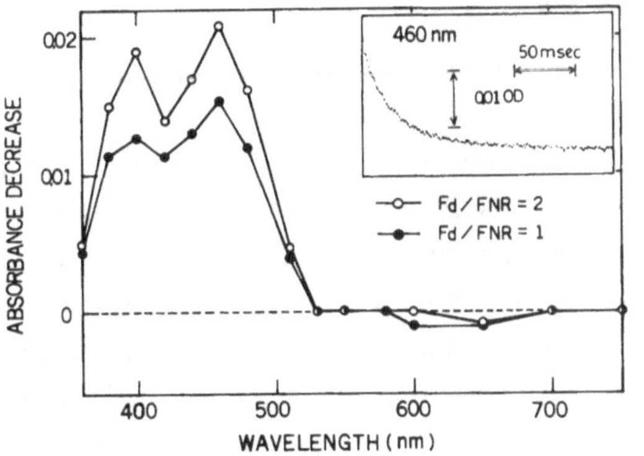

Fig. 3.(left) Overall absorbamce difference between the final and the initial levels. Any absorbance above 420 nm returns to the initial level within 250 msec, as shown in Fig. 1.

Fig. 4.(right) Spectral change of FNR in the absence of $NADP^+$. The inset shows an experimental curve from which the absorbance change of the Fd oxydation was subtracted. The absorbance decrease are plotted against the wavelength. The enzyme (8.6 μM) was mixed with reduced Fd of 15.8 μM(-o-) or 8.3 μM(-o-), respectively.

At Fd/FNR = 1, the spectral change in the first phase was consistent with that observed for the difference spectrum between semiquinoide and oxidized forms of FNR. The spectral increase between 530 nm and 650 nm and almost no change above 700 nm were diagnostic. The second slower phase was the reverse change. A clear absorbance change above 700 nm, in the presence of an excess amount of Fd, suggests a transient formation of the charge transfer complex between reduced FNR and $NADP^+$, as shown in Figs. 1 and 2. This change increased with the increase of the reduced Fd concentration. However, the spectral change between 420 nm and 650 nm was essentially independent of the reduced Fd concentration, as shown in Fig. 2. The absorbsnce change at 400 nm given in Fig. 1. involved an initial rapid decrease in addition to the biphasic absorbance change similar to that at 460 nm, but to a lesser extent. We consider that the initial absorbance level due to FNR should coincide with the final level after the completion of the reaction even below 420 nm. Then, the difference between the initial and the final absorbance values should be due to those other than FNR. These changes not related to FNR are dependent of reduced Fd concentration, as shown in Fig. 3. This spectral decrease around 400 nm and the increase below 380 nm indicate disappearance of NADP radical and appearance of NADPH, respectively(4).

2) Partial reaction system: When reduced Fd(4-8 μM) was reacted with the oxidized FNR(4 μM) in the absence of $NADP^+$, the absorbance of FNR decreased monotonously, as represented in Fig. 4, inset. The $t_{1/2}$ was independent of wavelength, but dependent of reduced Fd concentration. Essentially no absorbance change was observed above 530 nm. Spectral changes, as given in Fig. 4, indicate direct formation of reduced Fd from oxidized FNR without any detectable intermediate. The reduced FNR-$NADP^+$ charge transfer complex was formed from oxidized FNR and $NADP^+$ within the dead time.

3) Reaction mechanism: We propose the following mechanism based on the above results.

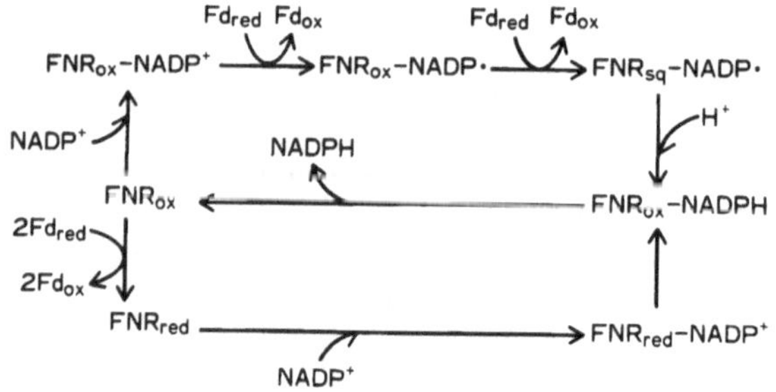

This mechanism includes one and two electron transfer cycles. Both cycles seems to be active under physiological conditions, because the reaction rate of reduced Fd with free enzyme is comparable to that of $NADP^+$, even in the absence of $MgCl_2$.

References

1. Masaki, R., Yoshikawa, S. and Mtsubara, H. 1982. Biochim. Biophys. Acta 700, 101-109.
2. Batie, C. J. and Kamin, H. 1984. J. Biol. Chem. 259, 11976-11986.
3. Wada, K., Tamura, T., Matsubara, H. and Kodo, K. 1983. J. Biochem. 94, 387-393.
4. Bielski, B. H. J. and Chan, P. C. 1980. J. Am. Chem. Soc. 102, 1713-1716.

IN VIVO DEACTIVATION/ACTIVATION OF FERREDOXIN-NADP$^+$-OXIDOREDUCTASE FROM THE CYANOBACTERIA *Anabaena variabilis*

M.F. Fillat, C. Gomez-Moreno

Departamento de Bioquimica y Biologia Molecular y Celular, Facultad de Ciencias, Universidad de Zaragoza, Spain

Introduction

Ferredoxin-NADP$^+$-oxidoreductase (FNR) plays a key role in the transport of reducing equivalents from photosystem I-reduced ferredoxin to NADP$^+$ (1,2). Its function of mediating the reducing power from photosynthesis to cellular metabolism is operative when illumination occurs. During dark cycles, the cell must rely on reducing power accumulated from this process. In spite of its central role, there is little information available on the regulation of FNR activity in algae and in higher plants (3). Vallejos *et al.* (4) have recently reported that light-energization of thylakoids produces a change in conformation that results in loss of activity of the membrane-bound FNR. Regulation of FNR activity could be required to prevent reverse electron transfer which would deplete the available pool of NADPH and thus be detrimental to cellular metabolic processes. We present here studies that demonstrate FNR activity in the nitrogen-fixing cyanobacterium *Anabaena variabilis* undergoes an *in vivo* deactivation/activation process that appears to be dependent on the light intensity incident on the cell.

Results and Discussion

FNR activity in crude extracts of cells harvested in the stationary phase of growth was found to be 10-20 % that found

Flavins and Flavoproteins 1987

in cell extracts from cells harvested in the logarithmic phase. Comparable levels of enzyme were found from both cell preparations using the purification procedure established for the active form of the enzyme (2). These results are shown in Table 1 and demonstrate that the activity of the enzyme isolated from cells grown to stationary phase is diminished independent of the chemical nature of the electron acceptor.

Table 1. FNR Activity In Different Phases Of *Anabaena variabilis* Growth

Growth Phase	μg FNR/g cells	FNR Activity			
		DCPIP		$Fe(CN)_6^{-3}$	Cyt c
		Units/g cells	TN*	TN*	TN*
Exponential	18	3.3	6,600	36,000	4,400
Stationary	14	0.5	1,300	12,500	1,800

* Turnover numbers are in min^{-1} and are based on enzyme-bound flavin of purified enzyme. [NADPH]=0.2 mM, Electron acceptor concentrations are at saturating levels.

These data show the low activity form of FNR is not a result of enzyme degradation in the stationary phase but to some modification of the enzyme which alters its activity.
One possible reason for FNR deactivation in cell cultures in stationary phase is that due to the high cell density, the effective illumination intensity on each cell would be reduced, and thus variations in activity levels could be a response to light limitations. This possibility is shown to be the case as demonstrated in Fig.1. Cell cultures exposed to white or to red light showed no decrease in FNR activity while cells deprived of light showed substantial losses of FNR activity. In separate experiments, the level of FNR activity was shown to increase (from 0.5 to 2.8 units/μg chlorophyll) in cells grown with increasing light intensities (from 1000 to

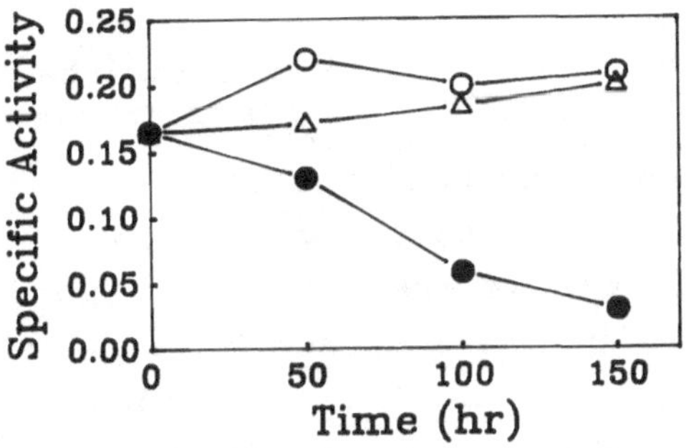

Fig.1. A culture of A. variabilis was grown to the exponential phase (A_{700}=1.0) at moderate light intensity (4000 lux). The culture was then divided into three portions: (o--o) control culture at 4000 lux, the optical density at 150 hrs. reached 2.5; (-Δ---Δ-) culture maintained under red light, optical density of 2.4 at 150 hrs.; and (-●---●-) culture maintained under dark conditions, final optical density equals 1.6. At the times indicated, aliquots were removed from each culture, cell-free extracts prepared, and FNR activity measured using DCPIP as electron acceptor.

5000 lux). FNR activity can be restored by exposing cells of low activity to saturating light levels simply by dilution into fresh media (Fig.2).

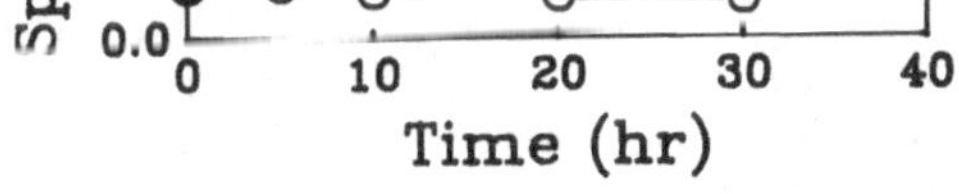

Fig.2. Reactivation of FNR activity. A cell culture grown to stationary phase was diluted to an optical density of 0.9 with fresh medium. FNR activity was monitored as described in Fig.1. The control culture was undiluted cells.

This process is more rapid than the rate of deactivation and shows the *in vivo* reversibility of the processes. This reactivation is not due to synthesis of FNR since similar results are observed in the presence or absence of chloramphenicol, an inhibitor of protein synthesis in procaryotic cells.

It remains for future work to determine the molecular basis of this deactivation/activation process. No alterations in FNR properties such as molecular weight, absorption spectrum, isoelectric point, or reactivity to antibodies prepared against the active enzyme are observed. To date, the major difference observed is an increase in K_m for NADPH for the deactivated form (135 μM) which is approximately twice that of the active form (62μM). No evidence for other possible alterations such as phosphorylation or thiol/disulfide interchange was found. Preincubation of the deactive form with ferricyanide did not alter the activity as described for other enzymes from higher plants and from algae (5).

Acknowledgements

This work was supported in part by grants from Diputacion General de Aragon and Comision Asesora de Investigacion Cientifica Tecnica, Spain. We wish to thank Dr. Dale E. Edmondson for useful discussions and for his help in writing the manuscript.

References

1. Forti, G. 1979. In: The Encyclopedia of Plant Physiology (M. Gibbs and E. Latzko, eds.). v.6, pp.222-226.

2. Sancho, J., M.L. Peleato, C. Gomez-Moreno, D.E. Edmondson. 1987. Arch. Biochem Biophys. , submitted.

3. Hug, D.H., P.S. O'Donnel, J.K. Hunter. 1980. Photochem. Photobiol. 32, 841.

4. Vallejos, R.H., E. Ceccarelli, R. Chan. 1984. J. Biol. Chem. 259, 8048.

5. Moreno, C.G., P.J. Aparicio, E. Palacian, M. Losada. 1972 FEBS Lett. 26, 11.

ISOLATION FROM DESULFOVIBRIO DESULFURICANS OF A FLAVOPROTEIN WITH NADH-H^+ OXIDOREDUCTASE ACTIVITY TOWARDS A PROTEIN CONTAINING A RUBREDOXIN-LIKE CENTER.

P. Saint-Martin, N. Galliano, G. Fauque

Section Enzymologie et Biochimie Bactérienne
A.R.B.S., C.E.N. Cadarache
13108 Saint-Paul-Lez-Durance Cedex, FRANCE

I. Moura, J.J.G. Moura

Centro de Quimica Estrutural. U.N.L. Lisbon, PORTUGAL

J. LeGall

Department of Biochemistry. University of Georgia.
Athens, GA, 30602 U.S.A.

Introduction

Rubredoxins (Rb) from anaerobic bacteria are small proteins with a molecular weight of 6,000 Da, containing a non-heme iron atom bound in a tetrahedral arrangement to the sulfur of four cysteinyl residues of the polypeptide chain. Their physiological role in anaerobic bacteria remains unknown. However, a NADH-H^+-rubredoxin oxidoreductase (NADH-H^+ Rb OR) has been purified in the sulfate-reducer Desulfovibrio (D.) gigas (1), as well as in some strains of Clostridia (2, 3) which catalysed the following reaction :

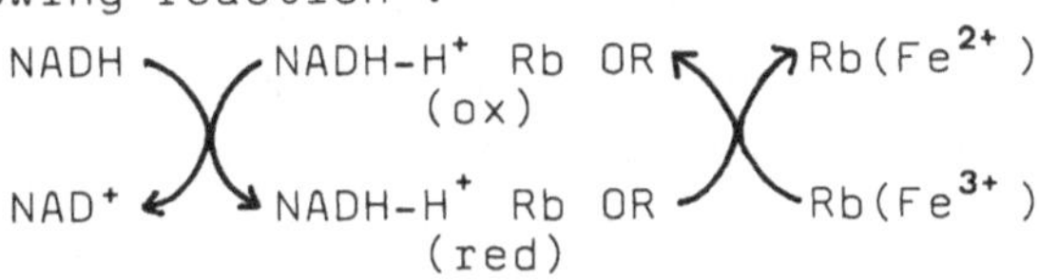

In Clostridium acetobutylicum, this protein had a molecular weight of 41,000 Da, and contained FAD as prosthetic group (2). We report here the purification from D. desulfuricans ATCC

Flavins and Flavoproteins 1987

27 774 of a flavoprotein with NADH-H^+ OR activity towards a protein containing a rubredoxin-like center.

Results

The purification was carried out from 1400 g of D. desulfuricans ATCC 27 774 cells grown on a lactate-sulfate medium at 37°C.
All the reductions were performed as followed : NADH and the electron acceptor were mixed with Tris-HCl buffer,0.1M, pH=7.6, and evolution of absorbance was followed during 10 min. Then 2 µg of oxidoreductase were added. Variations in optical density were recorded at the suitable wavelength.
The molecular weight was determined by polyacrylamide gel electrophoresis in the presence of SDS. For flavin extraction and analysis, the apoenzyme was precipitated by TCA 5% or by heating and removed by centrifugation. Supernatant was analysed by paper chromatography, using 5% Na_2HPO_4 as solvent.
In this purification, the small molecular weight Rb usually found could not be detected. Instead, a larger protein with a similar optical spectrum (absorption maxima at 490,375 and 280 nm) has been purified. It was found to be a dimer of about 24,000 Da per subunit. At a late step of this purification,the rubredoxin-like protein was separated from a yellow one, which showed a diaphorase activity towards O_2 in the presence of NADH. As this reductase seemed to be associated to the rubredoxin-like protein, the reduction of this latter one was tried with NADH-H^+ as electron donor and gave a positive result (fig. 1). However, no reduction was found with the small 6,000 Da rubredoxin from the same organism.
Other electron acceptors such as cytochrome c, myoglobin, potassium ferricyanide, are reduced by the enzyme in the presence of NADH-H^+ , but no reduction occurred with 2,6-DCPIP.
The optical spectrum of the oxidoreductase given in fig. 2 is typical of the presence of a flavin with a peak at 453 nm and

a molar extinction coefficient of 14,640 $M^{-1}.cm^{-1}$.

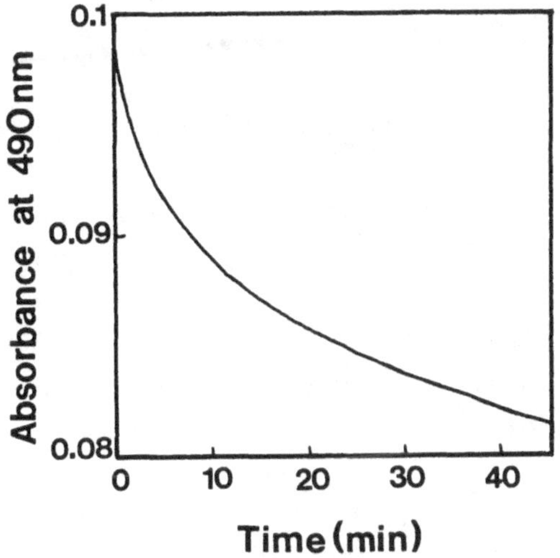

Fig. 1 : Reduction of the rubredoxin-like protein by the oxidoreductase from D. desulfuricans ATCC 27 774. The concentration of the rubredoxin-like protein was 20 µM. At t=0,2 µg of reductase were added.

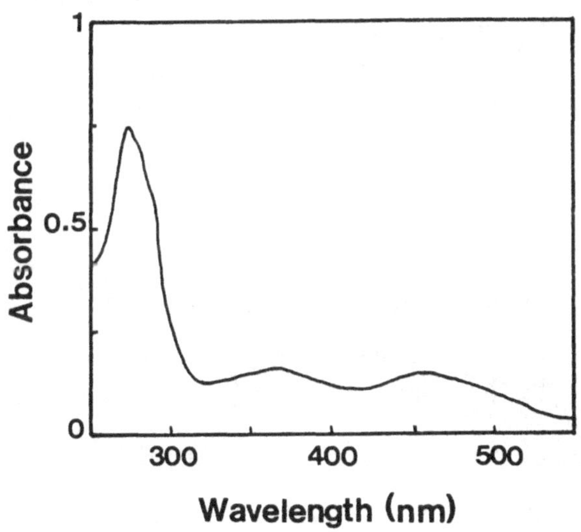

Fig. 2 : Absorption spectrum of D. desulfuricans ATCC 27 774 oxidoreductase. The protein concentration was 9.1 µM.

The molecular weight of the oxidoreductase was estimated to be 41,000 Da and the flavin prosthetic group identified as FAD. This protein contains 387 residues with a predominance of acidic amino-acids.
The presence of two different proteins with rubredoxin-like centers has already been reported in D. vulgaris Hildenborough (4). It has recently been found that the largest protein, similar to the protein from D. desulfuricans, contains, in addition to the rubredoxin-like center, another one bearing strong similarities with the iron center of hemerythrin (4). The presence of this extra redox center may explain the specificity of the reductase for this protein and the fact that the small rubredoxin is not reduced, assuming that the hemerythrin-like center is first reduced by the oxidoreductase. The physiological significance of these two proteins is still unknown.

Acknowledgements

This work was made possible by a special agreement between the University of Georgia and ARBS,and by a National Science Foundation grant DMB 8602789 to JLG. We thank the staff of the fermentation plant of UGA for growing bacteria used in this study.

References

1. LeGall, J. 1968. Ann. Inst. Pasteur, 114, 109-115.

2. Petitdemange, H., Marczak, R., Blusson, H. and Gay, R. 1979 Biochem. Biophys. Res. Comm., 91, 1258-1265.

3. Ballongue, J., Amine, J., Masion, E., Petitdemange, H. and Gay, R. 1986. Biochimie, 68, 575-580.

4. Huynh, B-H., Prickril, B., Moura, I., Moura, J.J.G. and LeGall, J. 1987 Recueil (J. Royal Netherlands chemical Society) in press (A non-heme iron protein that contains Rb-like centers and a hemerythrin-like spin-coupled binuclear iron cluster).

MICROSOMAL ELECTRON TRANSPORT SYSTEM FOCUSED ON NADPH-CYTOCHROME P-450 REDUCTASE.

Y. Nisimoto

Department of Biochemistry, Aichi Medical University, Nagakute Aichi 480-11, Japan.

Introduction

Hepatic Microsomal NADPH-cytochrome P-450 reductase has two different kinds of flavin, FAD and FMN, by which one electron each is transfered from NADPH to cytochrome P-450(P-450) in two separate steps. It is well known that low redox potential FAD accepts two electrons from NADPH, and only a fully reduced form of high redox potential FMN can reduce both P-450·substrate complex in step (I) and P-450·substrate·O_2 complex in step (II). Thus, on the basis of the fact that FMN is essential for reducing cytochrome c, b_5 or P-450, we assumed that the inter- and intramolecular flavin-flavin interaction between FAD and FMN takes place during P-450-linked mixed function oxidase system. From a series of our studies on chemical modification, resonance Raman spectroscopy and photo-CIDNP experiments using holoenzyme and FMN-depleted enzyme, this flavoprotein was observed to have structurally and functionally important different domains in the protein molecule.

Since the water-soluble carbodiimide, 1-ethyl-3-(3-dimethylaminopropyl) carbodiimide (EDC) has proven useful for the covalent cross-linking of charge paired functional protein-protein complexes such as adrenodoxin-adrenodoxin reductase and adrenodoxin-$P\text{-}450_{scc}$ (1), in the present report the author has focused on the cytochrome c-reductase complex in order to help define the location of the hemeprotein-binding domain on the flavoprotein molecule.

Flavins and Flavoproteins 1987

Results

A covalent 1:1 complex between the purified reductase and cytochrome c was formed in reconstituted phospholipid vesicles through cross-linking studies with EDC. A series of these studies suggested that the covalently cross-linked derivative is a valid model of the noncovalent electron transfer complex. Thus, the reductase-cytochrome c covalently cross-linked complex may define the location of the cytochrome c-binding domain on reductase.
The purified covalent complex was cleaved at Met residues with CNBr, and the CNBr fragments which contain heme c peptides were extensively digested with trypsin. This tryptic peptides were roughly separated by HPLC on a μBondapak C-18 column, and then a major peptide (Tp) which had covalently bound heme c was finally purified by reverse-phase HPLC on a SynChropak RPC-8 column. Approximately 30 residues were identified, of which only four were basic amino acid residues, and this Tp contains much hydrophobic and acidic amino acid residues. The tryptic peptide, Tp

Heme c

1 10 * 20 30
G D V E K G K K [I F V Q K C A Q C H T V E K] G G K H K T G P

161 171 ° 180 190
V D L T G V K F A V F G L G N K T Y E H F N A M G K Y V D Q R L E Q L G

200 * * *210 * * * 220 230
A Q R [I F E L G L G D D D G N L E E D F I T W R] E Q F W P A V C E F

Fig. 1. Amino acid Sequence of Cytochrome c, 1-30 (top) and NADPH-Cytochrome P-450 Reductase Fragment, 161-230 (bottom). Boxed sequence were recovered in the form of the covalently cross-linked peptide containing heme c. The possible location of the amino acid residues participating cross-link is indicated by an asterisk. The second FMN-binding domain of the protein is indicated on the reductase sequence by underlining (2).

having two different C-terminal residues and one N-terminal residue was sequenced to locate the cross-linked sites. When the amino acid sequence of peptide Tp was compared with the complete sequence of cytochrome c and reductase, it was assumed that cross-linking had occurred between the sequence 9-22 of cytochrome c and the sequence 200-220 of the reductase (Fig. 1). Therefore, the acidic region of 207-215 must be the most important for the complementary charge-pairing with the lysyl residues around the exposed heme edge at the surface of the cytochrome c. Besides the Tp fragment, a few cross-linked peptides which have heme c were detected in HPLC, however, they could not be isolated in the present study because their amounts were not sufficient for amino acid sequence analysis.

Thus, as shown in Fig. 2, we have suggested that one of the cytochrome c-interacting domain (207-215 acidic cluster involving Asp and Glu residues) is likely to be adjacent to the second FMN-

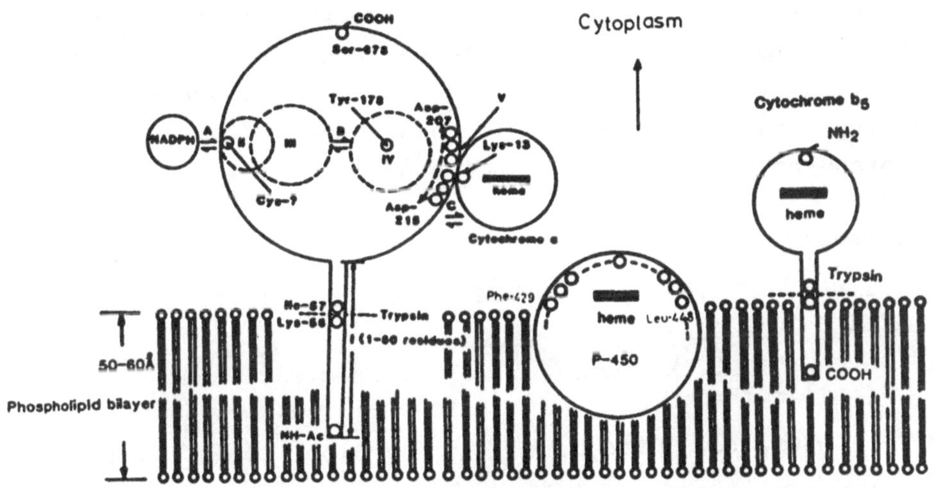

Fig. 2. Proposed Model of the Five Domains of Membrane-Binding NADPH-Cytochrome P-450 Reductase. Domain I contains both membrane binding and P-450 (or cytochrome b_5) interacting sites. A Cys residue is essential for the NADPH-binding domain II, which interacts with FAD-binding domain III. Tyrosyl residue is involved in the stacking with FMN-isoalloxazine ring in FMN-binding domain IV. Six clustered carboxyl groups (Asp, 207-209 plus Glu and Asp, 213-215) are possible site for the interaction with cytochrome c (cytochrome binding domain V).

binding region (residues 165-202) (2). This FMN-prosthetic group binding domain has been reported to be stabilized with tyrosyl or tryptophanyl residues through stacking interactions between the flavin isoalloxazine ring and its aromatic side chain (3,4). Both cytochrome c and P-450-binding domains on the reductase molecule could be expected to be in close proximity and they may have elements in common. Whereas, the FAD-binding domain would be involved in the reaction with NADPH-binding site in which a specific Cys residue has an essential role for catalytic function (3,5). In the light of these characteristic difference between the two flavin binding domains in reductase molecule, it could be assumed that the flavin-flavin interaction is crutially important step in the process of the P-450-linked mixed function oxidase system.

In addition, a high degree of hydrophobic regions are also required for membrane binding and interaction with P-450 and cytochrome b_5. On the basis of a hydropathy profile calculated according to the methods of Kyte and Doolittle (6), apparently, two groups (residues 27-43 and 523-544) of reductase show strong hydrophobicity, suggesting possible candidates for the membrane-bound cytochrome binding domains (7).

References

1. Lambeth, J.D., Seybert, D.W., Kamin, H. 1979. J.Biol.Chem. 254, 7255-7264.
2. Porter, T.D., Kasper, C.B. 1985. Proc.Natl.Acad.Sci. U.S.A. 82, 973-977.
3. Nisimoto, Y., Shibata, Y. 1982. J.Biol.Chem. 257, 12532-12539.
4. Nisimoto, Y., Hayashi, F., Akutsu, H., Kyogoku, Y., Shibata, Y. 1984. J.Biol.Chem. 259, 2480-2483.
5. Haniu, M., Iyanagi, T., Legesse, K., Shively, J.E. 1984. J.Biol.Chem. 259, 13703-13711.
6. Kyte, J., Doolittle, R.F., 1982. J.Mol.Biol. 157, 105-132.
7. Haniu, M., Iyanagi, T., Miller, P., Lee, T.D., Shively, J.E. 1986. Biochemistry 25, 7906-7911.

TIME-RESOLVED FLUORESCENCE SPECTROSCOPY ON NADPH-CYTOCHROME P-450 REDUCTASE

Peter Bonants, Phil Bastiaens, Ton Visser and Franz Müller
Dept. of Biochemistry, Agricultural University , De Dreyen 11,
6703 BC Wageningen, The Netherlands

Introduction

Fluorescence spectroscopy is among the available techniques one of the most sensitive for studying structure and dynamics of macromolecules. Since the development of the subnanosecond pulse techniques the fluorescence spectroscopy is no longer restricted to steady state fluorescence measurements. We measured the decay of fluorescence and the decay of fluorescence anisotropy of NADPH-Cytochrome P-450 reductase . This flavoprotein contains FAD and FMN and plays a role in the detoxification of xenobiotics(1). Both flavins are fluorescent although they are heavily quenched in the protein. Experiments are performed on different states of the enzyme : holo, FMN-depleted, FMN-recombined and one-electron reduced reductase. A considerable degree of energy-transfer between the two flavins is observed. The results allow the estimation of the distance between FAD and FMN.

Results and Discussion

Fig.1 and 2 show the fluorescence- and anisotropy decay of the holo-reductase, respectively. The parameters used to fit the experimental curves are summarized in Table 1 for the different samples. The fluorescence decay was fitted with a four-exponential decay law for all samples, except for FMN-recombined reductase.
Depending on the sample the fluorescence lifetime varies considerably. The average lifetime $\bar{\tau}$ of the FMN-depleted enzyme is longer than that of the holoenzyme and in fact represents the fluorescence life-time of the

Flavins and Flavoproteins 1987

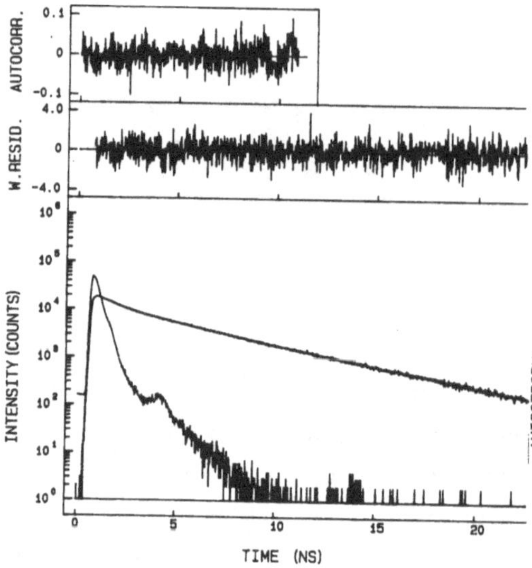

Figure 1

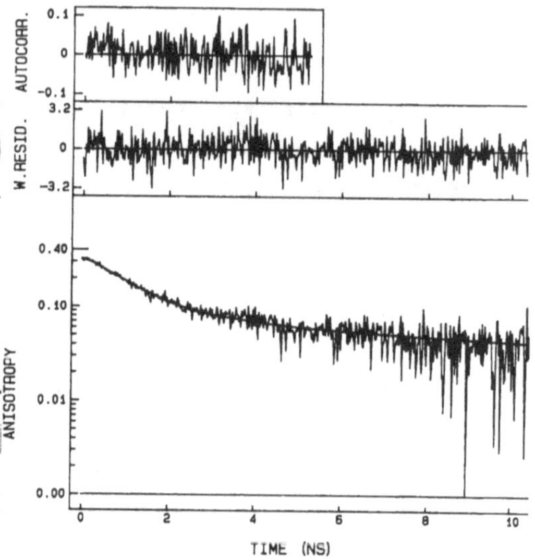

Figure 2

Experimental and calculated fluorescence (Fig. 1) and anisotropy (Fig. 2) decay curves of the holoreductase The weighted residuals and the autocorrelation function of the residuals indicate the quality of the fit.

TABLE 1:
Fluorescence- and anisotropy decay parameters of different reductase samples.

Sample	λexc (nm)	τ (ns)	β1	φ1 (ns)	β2	φ2 (ns)	β3	φ3 (ns)	R(tot)
Holo (17 μM)	458	1.14	0.23	0.81	0.08	2.76	0.04	65.2	0.13
Holo (17 μM)	514	0.97	0.13	2.76	0.23	65.2	----	----	0.29
FMN-depleted (11 μM)	458	1.62	0.14	0.67	0.09	5.93	0.08	65.2	0.16
SQ* anaer. (17 μM)	458	0.75	0.16	2.61	0.18	93.6	----	----	0.26
SQ* aer. (17 μM)	458	0.76	0.22	1.71	0.12	57.5	----	----	0.20
REC (20 μM)	458	3.49	0.34	0.43	0.03	5.04	----	----	0.05
free FMN (20 μM)	458	4.88	0.35	0.46	----	----	----	----	0.03

* : semiquinone

protein-bound FAD. On the other hand the lifetime of FAD in the "air-stable" semiquinone state is shorter than that in the FMN-depleted form. These results suggest that the conformation in the FAD binding site alters upon removal of FMN. This interpretation is supported by 31-P NMR measurements. The average fluorescence lifetime of FMN-recombined reductase is much longer than that of the holoenzyme indicating that FMN is bound differently in the two samples. This conclusion is deduced from the fact that the lifetime differs from that of free FMN and therefore cannot be assigned to the presence of free FMN. We have also observed that free FMN itself exhibits a slightly heterogeneous fluorescence decay, in contradiction to published results. More interesting are the time-dependent anisotropy data (Table 1). The experimental data of the holoenzyme can be fitted with three exponentials,i.e. one of about 65 ns, representing the rotational correlation time of the enzyme, and two rather short values. However on excitation at 514 nm instead of 458 nm the decay curve can be fitted with two exponentials. The total anisotropy (Rtot) of the holoenzyme excited at 458 nm is smaller than that excited at 514 nm. We ascribe the difference to energy transfer which is present when excited at 458 nm and absent when excited at the red edge of the absorption spectrum (514 nm). The relatively low value of R(tot) (Table 1) for the apo-sample is probably due to the less firmly bound FAD, which causes more depolarization than in the holoenzyme. The anisotropy parameters of the semiquinone sample show that the longest correlation time $\phi 2$ is about the same as the rotation correlation time of the protein; different values are found for the holo-and apo-samples ($\phi 2$=25.3 and 20.5 ns, respectively). The total anisotropy of the semiquinone sample decreases from 0.26 to 0.20 if air is admitted to the sample. The correlation times become shorter and the amplitude of the shortest component is increased. This indicates that FMN becomes partially oxidized and energy transfer occurs.

The values for the recombined reductase show clearly that FMN is not bound in the same manner as in the holo-enzyme. A comparison with the values for free FMN suggests that FMN in the recombined sample possesses a high degree of mobility. Activity measurements show that activity returned for nearly 100% after recombination suggesting that the strength of the FMN binding is not important for activity to be observed. The fluorescence depolarization of the holo (exc. 458 nm) can be described by three processes: overall rotation of the protein (ϕ3), wobbling of the flavins (ϕ2) (2) and energy transfer (ϕ1). At 514 nm excitation only the first two processes are observed (no energy transfer). The anisotropy decay for the 458 nm excitation becomes then :

R(t)=R(0) x R(wobbling) x R(protein) x R(energy-transfer)

With the help of literature-equations (2-5) and using the anisotropy-parameters of the two decays we can calculate (4) the boundaries for the orientation factor k**2 (0.14 and 2.9). The distance between the two flavins is calculated to be 0.8-1.5 nm.
Our results are partly in agreement with the data of Blumberg et al (6). More details will be published elsewhere.

Acknowledgement

We thank A. van Hoek for the assistance in the measurements. This study was done under the auspices of the Netherlands Foundation for Chemical Research (SON) with partial financial support from the Netherlands Organization for the Advancement of Pure Research (ZWO).

References

1. Yasukochi,Y., Masters,B.S.S. : J.Biol.Chem. 251, 5337-5344, (1976)
2. Lipari,G., Szabo,A. : Biophys.J. 30, 367-546, (1980)
3. Tanaka,F., Mataga,N. : Photochem. Photobiol. 29, 1091-1097, (1979)
4. Dale,R.E., Eisinger,J. : In Biochemical Fluorescence: Concepts, Eds. Chen, R.F. and Edelhoch, H. Vol. 1, pp 115-284, Dekker Inc. New York (1975)
5. Visser,A.J.W.G., Santema,J.S., van Hoek A. : Photobiochem. Photobiophys. 6, 47-55, (1983)
6. Blumberg,W.E., Nisimoto,Y., Mason,H.S.: In Oxygenases and Oxygen metabolism, pp 333-343, Academic Press New York, (1982)

STUDIES ON THE REACTIONS OF FLAVIN- AND NICOTINAMIDE NUCLEOTIDE ANALOGUES WITH A BACTERIAL ELECTRON-TRANSFERRING FLAVOPROTEIN

Eoin M. Ó Nualláin and Stephen G. Mayhew
Department of Biochemistry, University College Dublin, Belfield, Dublin 4, Ireland

Introduction

The electron-transferring flavoprotein (ETF) from *Megasphaera elsdenii* is a heterodimer that functions in the production of short-chain fatty acids (1). It differs from ETF purified from other sources (e.g. 2,3) in containing two molecules of FAD, rather than a single molecule, and in catalysing the oxidation of NADH. Massey observed that 8-halogenated flavins react covalently with this ETF (4,5). These flavins have been used to further study the flavin-binding sites of the protein, while the site of NADH binding is being sought with the photo-affinity analogue arylazido-β-alanyl NAD^+.

Results and Discussion

When the apo-protein of ETF, prepared by gel filtration of native ETF in guanidine-HCl (1), was mixed with 8-Cl-FAD or 8-F-FAD that was equimolar with the flavin-binding sites, the absorption spectrum of the flavin changed during several hours. The covalent reaction was followed by measuring the increase in $A_{480\ nm}$. The kinetic traces could be described by two first order reactions; an initial reaction ($k \simeq 0.02\ min^{-1}$), followed by a much slower reaction during about 12 hours. Comparison of the final spectra (Fig. 1) with those obtained by reaction of the halogenated flavins with model compounds (5,6) suggested

Flavins and Flavoproteins 1987

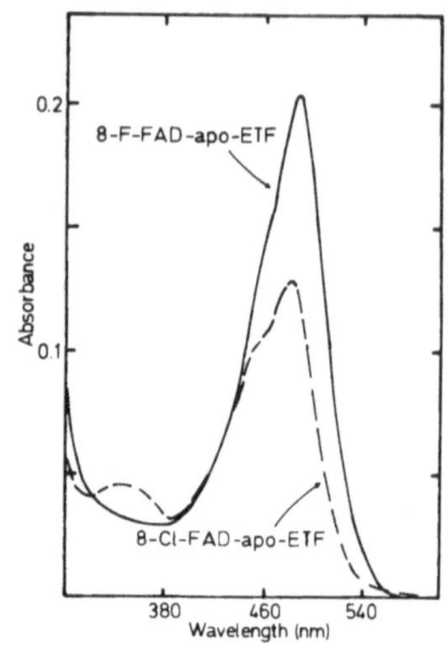

Fig. 1. Spectra of 8-Cl-FAD-apo-ETF (7.8 µM) and 8-F-FAD-apo-ETF (10 µM) in 60 mM K phosphate buffer pH 6.2

that while all of the 8-F-FAD had reacted covalently with the protein, only one of the two molecules of 8-Cl-FAD had become covalently bound. Treatment of the complexes with trichloroacetic acid showed that none of the flavin was extracted from the complex with 8-F-FAD, but that 48-55% was extracted from the 8-Cl-FAD complex. The failure of one of the bound 8-Cl-FAD molecules to react covalently suggests that only one of the flavin sites contains a sulphydryl group suitably positioned for reaction with the flavin.

Electrophoresis (SDS-PAGE) of the complexes separated the two protein subunits; in both cases only the small subunit showed yellow fluorescence, indicating that the covalent reactions had all occurred with this subunit, and suggesting that the two flavin sites in this ETF are on the same subunit. An earlier report that both subunits contained covalently-bound flavin after reaction with 8-F-FAD (7) was based on the observation of two fluorescent bands, and has not been confirmed; one of the bands may have arisen by proteolysis during the lengthy incubation of apo-ETF and the halogenated flavin.

The complexes formed with the halogenated flavins were both

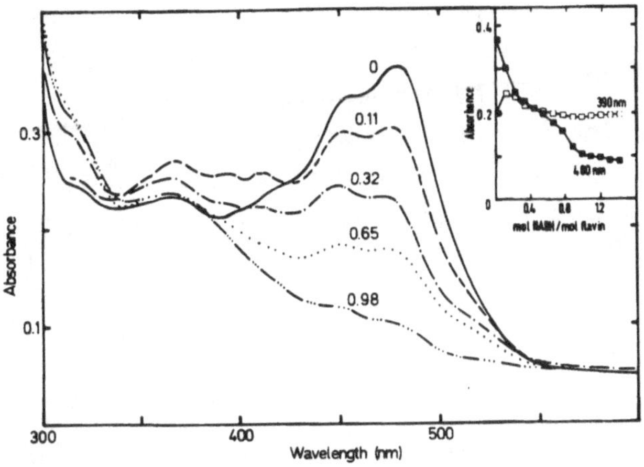

Fig. 2. Anaerobic titration of 8-Cl-FAD-apo-ETF with NADH. Molar ratios of NADH to enzyme-bound flavin are shown

reduced by NADH (Fig. 2), suggesting that the flavin-protein interactions had occurred specifically at the native flavin-binding sites. The complexes also catalysed NADH oxidation with DCPIP as electron acceptor. However, they failed to couple NADH oxidation to the reduction of crotonyl-CoA in the presence of butyryl-CoA dehydrogenase. Removal of the non-covalently bound 8-Cl-FAD with guanidine-HCl caused a loss of diaphorase activity; the activity was partly restored by adding back either 8-Cl-FAD or unmodified FAD. Interestingly, the mixed complex of covalently-reacted 8-Cl-FAD and unmodified FAD showed activity with butyryl-CoA dehydrogenase.

When thiophenol was added to the complex with 8-Cl-FAD, $A_{482\ nm}$ increased rapidly and $A_{350\ nm}$ decreased, indicating that 8-phenyl-S-FAD had formed. These observations show that the 8 position of the non-covalently bound 8-Cl-FAD in M.elsdenii ETF is accessible to solvent. A similar conclusion has been made for the single flavin in pig kidney ETF (8).

The photo-affinity analogue arylazido-β-alanyl NADH has been found to be a substrate for ETF. The enzyme is inactivated by photo-irradiation in the presence of the analogue, and the inactivation is prevented by NADH. As judged by the colour of the protein precipitated by subsequent addition of trichloroacetic acid, the inactivation is accompanied by covalent reaction of the analogue with the protein. The analogue labelled with tritium is being used to determine the site(s) of its reaction.

Acknowledgement

We thank Dr. K. Matsui and Dr. V. Massey for providing gifts of 8-F-riboflavin.

References

1. Whitfield, C.D., S.G. Mayhew. 1974. J.Biol.Chem. 249, 2801-2801.

2. Gorelick, R.J., J. Mizzer, C. Thorpe. 1982. Biochemistry, 21, 6937-6942.

3. Husain, M., D.J. Steenkamp. 1985. J.Bacteriol. 163, 709-715.

4. Massey, V., P. Hemmerich. 1980. Biochem.Soc.Trans. 8, 246-257.

5. Kasai, S., K. Sugimoto, R. Miura, T. Yamano, K. Matsui. 1983. J.Biochem. 93, 397-402.

6. Moore, E.G., E. Cardemil, V. Massey. 1978. J.Biol.Chem. 253, 6413-6422.

7. Ó Nualláin, E.M., S.G. Mayhew. 1986. Biochem. Soc. Trans. 14, 440-441.

8. Gorelick, R.J., Thorpe, C. 1986. Biochemistry 25, 7092-7098.

KINETICS OF LIGAND BINDING, AMINO ACID SEQUENCE AND CRYSTALLIZATION OF CHLOROBIUM AND CHROMATIUM FLAVOCYTOCHROME C

T.E. Meyer, H.M. Holden, I. Rayment, R.G. Bartsch and M.A. Cusanovich

Department of Biochemistry, University of Arizona, Tucson, Arizona 85721

J. Van Beeumen

Laboratory of Microbiology and Microbial Genetics, State University of Ghent, Ghent, Belgium

Introduction

Flavocytochromes *c* have been found in several species of phototrophic purple and green sulfur bacteria where they appear to function as sulfide dehydrogenases (1). Nucleophilic ligands such as sulfite, thiosulfate, cyanide, and mercaptans form adducts with the flavin of these proteins and the adducts in turn form charge transfer complexes with the protein (2). We previously measured the kinetics of adduct formation in *Chromatium vinosum* flavocytochrome *c* and showed that the reaction was complex (3). With sulfite, there is a rapid second-order bleach of flavin absorbance (with the associated appearance of a charge transfer band at 660 nm) followed at high pH by a slower recoloring reaction. Reaction with both sulfite and thiosulfate is apparently facilitated by protonation of a postulated histidine residue.

We have extended detailed characterization of the reaction of sulfite with flavocytochrome *c* to examples from *Chlorobium thiosulfatophilum* and *Chromatium gracile* to determine the extent of similarity and difference to *Chromatium vinosum*.

Flavins and Flavoproteins 1987

Results and Discussion

For all three species of phototrophic bacterial flavocytochrome c, we observed a rapid bleach of flavin absorbance (adduct formation) by sulfite, which at low pH results in a relatively stable species, but at high pH, the protein recolors in a first-order process. The rapid bleach was second order at low sulfite concentrations, but became first order at high. C. gracile is intermediate between Chlorobium and C. vinosum in terms of the magnitude of both second- and first-order reactions with sulfite. Both reactions were pH dependent. This new finding of a change in the rate limiting step in the rapid bleach is a consequence of the higher sulfite concentrations used in the present study. Typically K_d ranges from 6-24 μM, complex formation from $9x10^3$ to $6x10^4$ $M^{-1}s^{-1}$ and the first-order isomerization from 100 to 200 s^{-1} for the three flavocytochromes c studied.

The recoloring reaction which was observed with sulfite at high pH in C. vinosum flavocytochrome also occurred with the Chlorobium and C. gracile proteins. The same pH effect was observed, but the C. gracile reaction was 3 times more rapid than C. vinosum and Chlorobium was a factor of 2 slower. This first-order reaction reflects a conformational change which places the flavin in an environment, which results in decomposition of the adduct perhaps through loss of the charge transfer interaction. This is a consequence of the reaction of sulfite at another site in the protein, such as at a disulfide. The pK of 8.4 previously observed in the C. vinosum reaction suggested the involvement of a cysteine residue (3). As a consequence of examining even lower sulfite concentrations than in the previous study, we find that with C. vinosum at pH 9, the recoloring reaction begins to show concentration dependence. This suggests that the postulated reaction with a disulfide is becoming rate limiting.

The recoloring reaction is over in less than two minutes at pH 8 and above, but at the highest sulfite concentrations this reaction is followed by a very slow rebleach of flavin absorbance complete in about 30 min or longer. This reaction occurs with all three species of protein; it is pH dependent, and also concentration dependent. This reaction is not likely to have physiological relevance, but it illustrates yet another conformation which the flavin can assume.

Available data suggests that the flavocytochromes can exist in several forms. These are summarized in equation 1 where for the present FC_1 and FC_2 as well as $adduct_1$ and $adduct_2$ represent species with different reactivities, and complex represents a collision complex prior to an isomerization leading to adduct formation.

$$FC_1 \overset{SO_3^{=}}{\rightleftarrows} \text{complex} \rightleftarrows adduct_1 \overset{SO_3^{=}}{\rightleftarrows} FC_2 - \overset{SO_3^{=}}{\rightleftarrows} adduct_2 \qquad (1)$$

A correlation between redox potentials of model flavins and reactivity with sulfite (4), and with flavoproteins (5) has been shown, which suggests that the flavin of flavocytochrome c has a different redox potential in each of its proposed conformations. Thus it can be argued that the flavin moiety of FC_1 and FC_2 have different redox potentials due to the presence of a sulfite modified disulfide near the flavin. This would explain the different kinetics of reaction with sulfite of FC_1 and FC_2.

We determined the amino acid sequence of the Chlorobium flavocytochrome heme subunit. It consists of 89 residues and is thus about the same size as that of Pseudomonas putida flavocytochrome c (78 residues), (6). From the location of the heme binding cysteine, histidine, and methionine residues, it is apparent that these proteins are homologous, although there is very little similarity otherwise (about 21%). In fact, the similarity is so low as to suggest convergence. In spite of the fact that flavocytochromes are quite rare in nature, our results suggest that the chance association of cytochrome and flavoprotein to form a stable dimer may have occurred more than once over the course of evolution.

In a further attempt to compare these proteins and to determine the basis for the differences in their properties, we crystallized all three phototrophic flavocytochromes. *C. vinosum* flavocytochrome crystallization was reported by Salemme (7). Although that report was optimistic, the crystals were not suitable for high resolution X-ray analysis. Large, single crystals of the flavocytochrome isolated from *Chromatium gracile* were grown at room temperature. While the crystals were physically robust rods, they diffract only to very low resolution. *Chlorobium thiosulfatophilum* flavocytochrome was crystallized with large rod-shaped crystals appearing overnight. Again these crystals show diffraction maxima to only very low resolution. The poor quality of the flavocytochrome crystals, in terms of their diffraction patterns, suggests possible disorder in the crystalline lattices. Such disorder can often arise from microheterogeneity in the protein sample itself. Conformational heterogeneity is another possible source of disorder particularly if the protein under investigation is quite flexible. With this in mind, we are currently exploring approaches which might force the protein to adopt a single conformation.

Acknowledgments

This work was supported by a grant from the National Science Foundation DMB-84-16849 (M.A.C.) and National Institutes of Health GM-35186 (I.R.).

References

1. Yamanaka, T. and Y. Kusai. 1986. In: Flavins and Flavoproteins (T.P. Singer, ed.). Elsevier, Amsterdam, pp. 292-302.

2. Meyer, T.E. and R.G. Bartsch, 1976. In: Flavins and Flavoproteins (T.P. Singer, ed.). Elsevier, Amsterdam, pp. 312-317.

3. Cusanovich, M.A. and T.E. Meyer. 1982. In: Flavins and Flavoproteins (V. Massey, C.H. Williams, Jr., eds.). Elsevier, Amsterdam, pp. 844-848.

4. Muller, F. and V. Massey. 1969. J.Biol.Chem. 244, 4007-4016.

5. Schuman-Jorns, M. and L.B. Hersh. 1976. J.Biol.Chem. 251, 4872-4881.

6. McIntire, W., T.P. Singer, A.J. Smith and F.S. Mathews. 1986. Biochem. 25, 5975-5981.

7. Salemme, F.R.. 1982. J.Mol.Biol. 159, 551-552.

REDOX PROPERTIES OF FLAVOCYTOCHROME C

M.A. Cusanovich, R.G. Bartsch, T.E. Meyer
Department of Biochemistry, University of Arizona, Tucson, Arizona 85721

Introduction

Flavocytochrome c from the photrophic bacteria is an unusual flavoprotein in that it contains both covalently bound heme and flavin. We have recently shown that electron transfer from added electron sources proceeds to the flavin moiety, typically to make a flavin semiquinone and then via rapid intramolecular electron transfer ($>10^4$ s^{-1}) to the heme iron (2). Interestingly, the flavin moiety of flavocytochrome c will react with a variety of ligands, for example, thiosulfate, cyanide and sulfite, forming a stable adduct (at pH 6.0) which has its principal absorption band at greater than 600 nm (1). At high pH (7 and greater), some of the adducts (e.g., sulfite) are not stable and the flavin absorbance returns (recolors) with a corresponding loss of the change transfer band (3). The recolored protein has spectral properties identical to the native proteins; however, its reactivity with ligands is substantially altered, thus is a distinct chemical species. During the course of studies on adduct formation, it became clear that flavocytochrome c was reactive with a variety of mercaptans. For example, glutathione forms a stable adduct (2). However, in some cases (mercaptoethanol, cysteine) heme reduction also occurred. Since the presumed function of flavocytochrome c is that of a sulfide dehydrogenase (4), it is of substantial interest to determine the details of the reaction mechanism with reduced sulfur compounds. The kinetics

Flavins and Flavoproteins 1987

of interaction of mercaptoethanol with three different examples of flavocytochrome c have been studied and are reported here.

Results and Discussion

At 480 nm the reaction of *C. vinosum* flavocytochrome c with mercaptoethanol is biphasic with the fast phase representing approximately one half of the observed absorbance change. At 552 nm, the reaction is also biphasic but with the fast phase representing only 10-15% of the total absorbance change. The fast phase has an apparent second-order rate constant of $8.8 \times 10^4\ M^{-1}s^{-1}$ at pH 7.0 irrespective of the wavelength monitored. Importantly, if the kinetics of reaction are followed at 650 nm an increase in absorbance is observed with a rate constant of $8.8 \times 10^4\ M^{-1}s^{-1}$. Thus, the fast bleach at 480 nm and the small increases at 552 and 650 nm are consistent with the formation of the change transfer band typical of adduct formation by flavocytochrome c. The fast phase is followed by a slower increase in absorbance at 552 nm (a further decrease at 480 nm) consistent with the formation of reduced heme. The absorbance change at 480 nm due to heme reduction is expected to be 25-30% of the change at 552 nm based on the spectral properties of c-type cytochromes. Thus, the observed second-order reduction ($k \sim 25\ M^{-1}s^{-1}$) occurs at both 480 and 552 nm. It is important to note that there is a lag phase which precedes heme reduction and the length of this lag phase is concentration dependent, increasing with decreasing concentration. Taken together the results are consistent with three consecutive second-order reactions as shown in equation 1.

$$FC \xrightarrow{k_1[ME]} adduct \xrightarrow{k_2[ME]} intermediate \xrightarrow{k_3[ME]} reduced\ FC \qquad (1)$$

where FC and ME represent flavocytochrome c and mercaptoethanol, the adduct is a species with bleached absorbance and a corresponding change transfer band and the intermediate is a form of the protein whose formation is accompanied by no or a very small absorbance change which accounts for the lag phase.

The pretreatment of flavocytochrome with sulfite at pH 7.0, substantially alters the observed kinetics of mercaptoethanol reduction. Sulfite treatment results in adduct formation followed by a recoloring reaction (that is, recovery of the flavin absorbance and loss of the change transfer band) to form a species which is spectrally identical with untreated flavocytochrome c but unreactive with additional sulfite at this pH and concentration. When mercaptoethanol is added to sulfite treated flavocytochrome, the fast phase is absent and the rate constant for second-order reduction has increased ($\sim$100 $M^{-1}s^{-1}$). Interestingly, the lag phase observed in the absence of sulfite is still present but is now independent of mercaptoethanol concentration.

To obtain further insight into the mechanism, the reaction of mercaptoethanol with flavocytochrome c from *Chromatium gracile* and *Chlorobium thiosulfatophilum* has been investigated. The kinetics of reaction of *C. gracile* flavocytochrome c with mercaptoethanol are essentially identical to those of the *C. vinosum* proteins except that the rate constants for the fast and slow phases are somewhat larger (3.6 and 1.3 fold respectively). In contrast, *C. thiosulfatophilum* flavocytochrome c demonstrates substantial kinetic differences. Although a fast and slow phase are present, no detectable lag phase is observed. Moreover, both phases although yielding rate constants which are dependent on the mercaptoethanol concentration are not truly second-order. That is, a change in rate limiting step appears to occur at high reductant concentration. Estimated values for the fast and slow phase rate constants are $6x10^3$ and 40 $M^{-1}s^{-1}$ respectively. Thus, the C. thiosulfatophilum flavocytochrome c kinetics although in some sense similar to the Chromatium proteins are quite distinct and when fully resolved may provide useful information on the reaction mechanism.

Conclusions

The flavocytochromes c show complex kinetics in their reaction with mercaptoethanol. Available data are consistent with a minimum of three second-order processes: adduct formation, interconversion to a new form of protein and heme reduction by mercaptoethanol. Presumably heme reduction takes place thru rapid intramolecular electron transfer from the reduced flavin. The effect of sulfite is consistent with the reaction scheme proposed as it removes the requirement for adduct formation. However, it is not clear presently how sulfite alters the lag phase. Moreover, the C. thiosulfatophilum reaction requires further analysis to establish whether it follows the same mechanism although available data do not require a new mechanism to be proposed. To the best of our knowledge, the flavin chemistry shown here is unique and suggests that a new family of reactions result from the binding of flavin to the flavocytochromes. Although more studies are required to fully understand the reaction mechanism, new insights into flavoprotein chemistry are expected to result.

Acknowledgments

This research was supported by NSF Grant DMB-84-16849.

References

1. Cusanovich, M.A. and T.E. Meyer. 1982. In: Flavins and Flavoproteins (V. Massey, C.H. Williams, Jr., eds.). Elsevier, Amsterdam, pp. 844-848.

2. Cusanovich, M.A., T.E. Meyer and G. Tollin. 1985. Biochemistry 24, 1281-1287.

3. Meyer, T.E. and R.G. Bartsch. 1976. In: Flavins and Flavoproteins (T.P. Singer, ed.). Elsevier, Amsterdam, pp. 312-317.

4. Yamanaka, T. and Y. Kusai. 1976. In: Flavins and Flavoproteins (T.P. Singer, ed.). Elsevier, Amsterdam, pp. 292-302.

Electron Transfer Kinetics of Phthalate Oxygenase Reductase, an Iron Sulfur Center Containing Flavoprotein

Christopher J. Batie and David P. Ballou, Dept. of Biological Chemistry, University of Michigan, Ann Arbor, MI 48109

Introduction

Phthalate oxygenase reductase (POR) is the electron-transferring component of a bacterial dioxygenase system which catalyzes the dihydroxylation of o-phthalate (1,2-dicarboxybenzene) to yield a 4,5-cis-dihydrodiol. POR is a 34 kDa protein containing both FMN and a spinach Fd-type [2Fe-2S] center (1). The reductase passes electrons from NADH to the oxygenase, which contains both a Fe^{2+} center and a Rieske-type [2Fe-2S] center. We presume that the path of electron flow is NADH --> FMN --> [2Fe-2S] (POR) --> Rieske-type [2Fe-2S] (PO) --> Fe^{2+}-O_2. Thus the flavin and [2Fe-2S] moieties of POR effect a 2 to 1 electron step-down analogous to the role of the P 450 reductase systems (2). This work examines the kinetics of the reaction of NADH with POR.

Methods and Materials

All experiments were conducted at 4 °C. Except where noted otherwise, all experiments utilized 0.1 M HEPES buffer at pH 8.0. POR was isolated by a method similar to that reported in (1). NAD and NADH were purchased from Sigma. Isotope effect experiments utilized NADH and (4R-^{2}H)-NADH (NADD) which were prepared in parallel by a method very similar to that of (3). NADH was purified by ion exchange and gel filtration chromatography prior to use.

Flavins and Flavoproteins 1987

Results and Discussion

Figure 1 shows that soon after mixing POR with a 5-fold molar excess of NADH a spectrum is observed that is the same (at wavelengths > 400 nm) as that of two-electron reduced POR observed during titrations. The character of this spectrum indicates that both FMN and [2Fe-2S] are each reduced by one electron equivalent. Reductive titrations with dithionite or NADH at neutral pH show that the redox potential of the FMN_{ox}/FMN_{sq} couple (E_1) is nearly equal to that of the $[2Fe-2S]_{ox}/[2Fe-2S]_{red}$ couple ($E_{m-Fe/S}$). In the presence of excess NADH, two-electron reduced POR is very slowly reduced by one additional electron to yield the 20 min. spectrum.

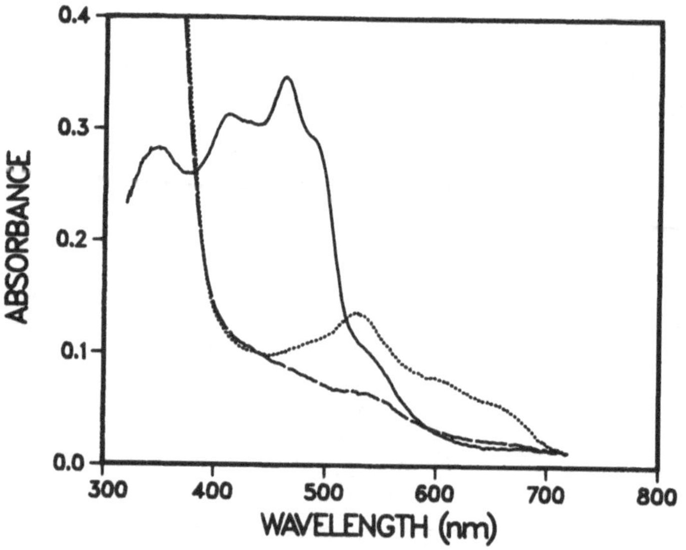

Figure 1 REACTION OF OXIDIZED POR WITH NADH -- POR (21 μM) was mixed in a stopped flow apparatus with 100 μM NADH (or buffer) and spectra were recorded. The spectra presented include: POR + buffer (———); POR + NADH, 3 sec. after mixing (·········), POR + NADH, 20 min. after mixing (------). T = 4 °C, pH = 8.0.

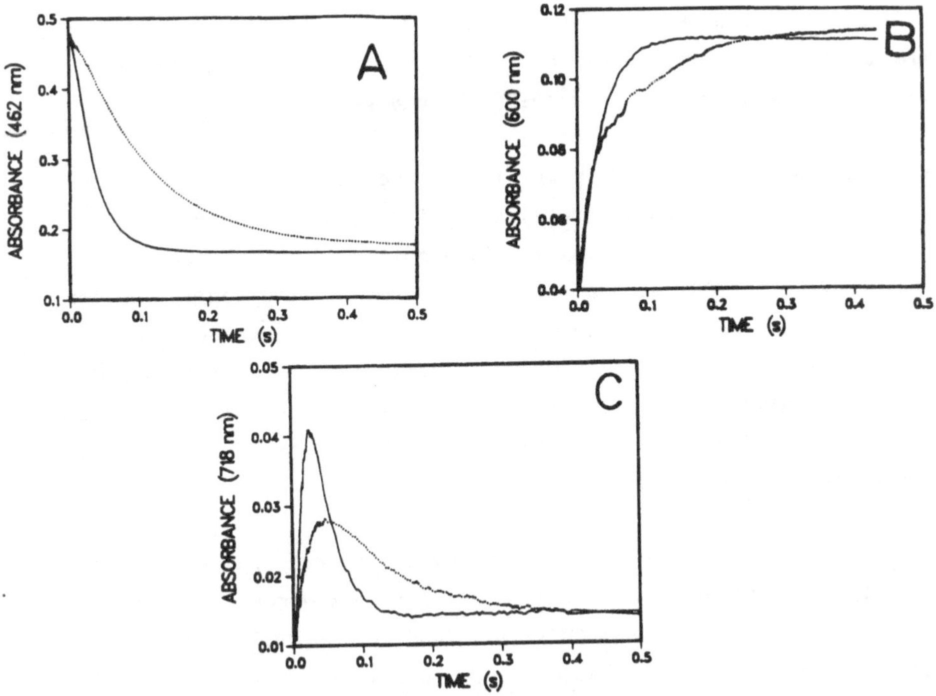

Figure 2 KINETICS OF REACTION OF POR AND NADH. NADH (——) or 4R-NADD (.......) were mixed with POR as in Fig. 1. (A) 462 nm, (B) 600 nm, (C) 718 nm.

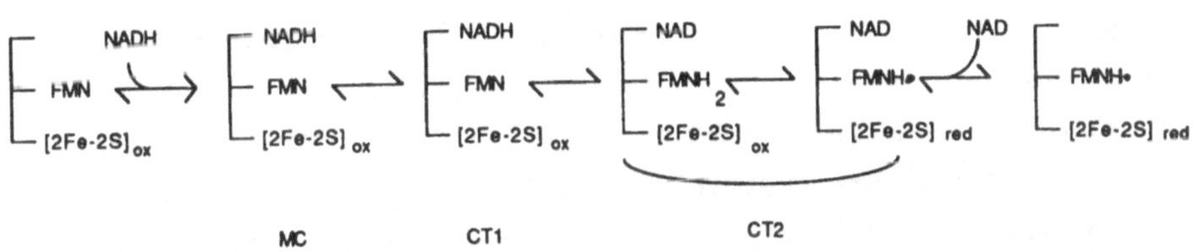

The kinetics of the reduction of POR_{ox} by NADH were determined (Fig. 2). The scheme above summarizes our understanding of the reductive half-reaction of POR. We found that electron transfer from NADH to POR (340 nm) and intramolecular electron transfer from FMN to [2Fe-2S] (600 nm) occurred concurrently. Thus the intramolecular electron transfer must be fast compared to the hydride transfer from NADH. A transient increase in absorbance at long wavelength (718 nm) preceded the appearance of two-electron reduced POR; this long-wavelength band is indicative of an NADH-POR_{ox} complex with charge-transfer character. At very high NADH (or NADD) concentrations (>500 μM) the rate of increase in 718 nm absorbance was independent of NADH concentration; thus another NADH-POR complex must precede the charge-transfer complex.

NADH specifically deuterated in the 4-pro-R position (NADD) was found to inhibit the NADH-cytochrome c reductase activity of POR with a V_{max} effect of about 3 (data not shown). Examination of the NAD product by NMR has shown that POR removed the pro-R hydrogen, thus confirming the stereochemistry. NADD also reduced the rate of POR reduction (Fig. 2). Kinetic simulations indicate a 10-fold kinetic isotope effect on hydride transfer, but since other steps in the overall reaction occur at rates similar to that of electron transfer from NADH (see below) only a small steady state kinetic isotope effect is observed.

Comparison of the reactions of NADH and NADD with POR indicates that two different charge-transfer complexes occur during the reductive half-reaction of POR: NADH-POR_{ox} (CT1) and NAD-POR_{2e-} (CT2). With NADH the transient increase in 718 nm absorbance results from both CT1 and CT2, but with NADD CT1 is the predominant species. The kinetic behavior of these phases indicates that NADH and POR must form an initial complex prior to the appearance of CT1. Numerical simulations which utilize the scheme shown above indicate that several steps in the mechanism have similar microscopic rate constants (50-90 s^{-1}); these include formation of CT1, electron transfer from NADH to FMN (^{1}H-NADH), and dissociation of NAD. Preliminary data indicate that CT2 is a complex involving NAD and the FMN semiquinone. Thus POR presents an unusual example of charge donation from a flavin semiquinone to a pyridine nucleotide.

Bibliography

1. Batie, C.J., LaHaie, E., and Ballou, D.P. (1987) J. Biol. Chem. 262, 1510-1518..

2. Lambeth, J.D., Seybert, D.W., Lancaster, J.R., Salerno, J.C., and Kamin, H. (1982) Mol. Cell. Biochem. 45, 13-31.

3. Arnold, L.J. Jr., You, K-S. (1978) Methods Enzymol. 54, 223-232.

CD AND POTENTIOMETRY OF THE FAD OF NITRATE REDUCTASE

M.J. Barber, C.J. Kay and L.P. Solomonson

Department of Biochemistry, University of South Florida, College of Medicine, Tampa, FL 33612.

Introduction

NADH:nitrate reductase (NR) catalyzes the reduction of nitrate to nitrite, the initial and rate-limiting step in the process of nitrate assimilation. The enzyme purified from the green alga *C. vulgaris*, is a homotetramer of subunits (100 kDa), each of which contains FAD, b_5-type cytochrome and Mo-pterin cofactors in a 1:1:1 stoichiometry (1). Limited proteolysis of NR using corn inactivator protease has enabled separation into two fragments, a small monomeric FAD-containing fragment (28 kDa) which exhibits NADH:ferricyanide reductase activity and a large tetrameric (283 kDa) fragment that contains heme, the Mo-pterin cofactor and retains reduced methyl viologen:nitrate reductase activity (2).

Results and Discussion

The contribution of the heme to the visible spectrum of NR (2), both in the oxidized and reduced states, effectively masks contributions due to FAD and Mo-pterin. However, the absence of heme in the small fragment has enabled changes in the visible spectrum of FAD, typically a bleaching of the visible absorption, to be detected during oxidation and reduction In contrast, all cofactors exhibited optical activity using CD spectroscopy in the range 300-650 nm (Fig. 1). Utilizing the small and large fragments in combination with NR samples poised at

Flavins and Flavoproteins 1987

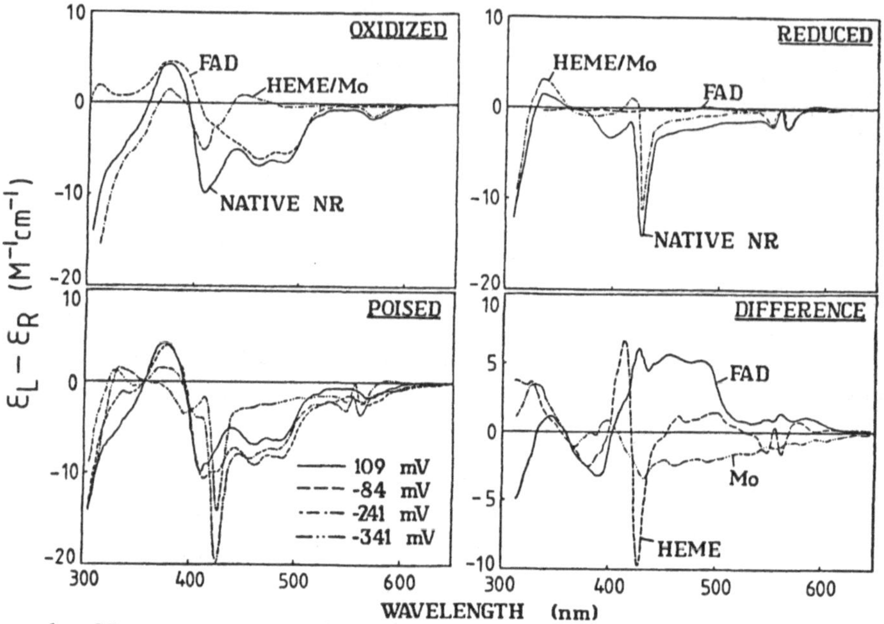

Fig. 1. CD spectra of nitrate reductase. Spectra for oxidized and reduced NR, small and large fragment are shown together with spectra obtained for the enzyme poised at controlled potentials and the difference spectrum for each of the cofactors.

controlled potentials in the presence of optically-inactive dye mediators, CD bands have been assigned to each chromophore and suitable wavelengths have been determined to monitor cofactor redox changes. Enzyme-bound FAD was optically active in the oxidized state giving a major CD band at 375 nm and unresolved bands at 460 and 489 nm. Reduction of FAD resulted in loss of optical activity. Heme was optically active in both oxidized and reduced states with major bands at 415 nm and 570 nm, respectively, corresponding to the Soret and α peaks. Mo-pterin contributed the weakest component to the CD spectrum, most easily discerned as a shoulder at 333 nm. The sum of the CD spectra of the large and small domains closely resembled the spectrum of NR suggesting proteolysis had little effect on chromophore structure. In addition, CD difference spectra for the FAD, heme and Mo-pterin centers, were obtained by subtracting the oxidized and reduced species obtained from NR samples

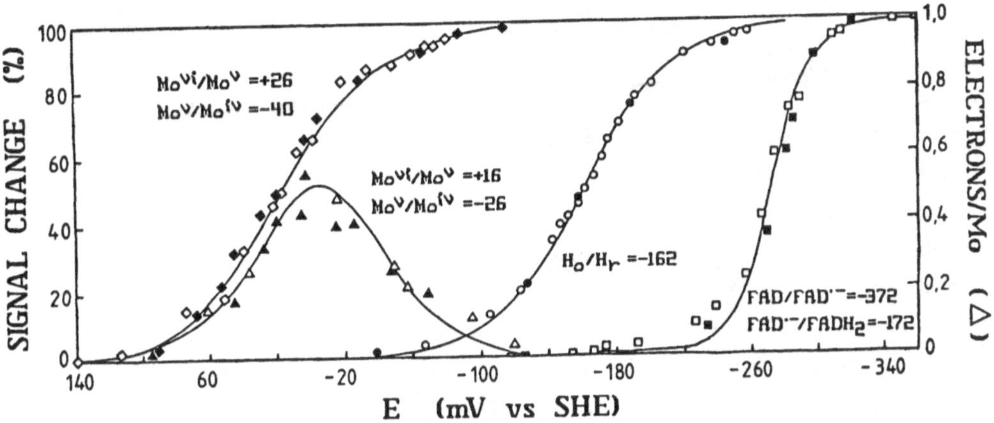

Fig. 2. Potentiometric titrations of nitrate reductase. Potentiometric titrations of native enzyme (10-40 µM heme) in the presence of dye mediators (10-30 µM) were performed at 25°C in 50 mM MOPS, 0.1 mM EDTA buffer, pH 7.0. CD titrations of Mo (◇ 333 nm), heme (○ 423-416 nm) and FAD (□ 460 nm). Room temperature epr titration of Mo^V (△).

poised at controlled potentials.

To avoid problems due to temperature-dependent electron redistributions (3), visible, CD and room temperature epr spectroscopy in the presence of dye-mediators, were used to determine the redox potentials (pH 7) for the FAD, heme and Mo centers present in NR. The results of typical titrations are shown in Fig. 2. Visible potentiometric titrations of the flavin domain and CD titrations of the FAD in NR both yielded n=2 redox processes with midpoint potentials of -283 mV and -272 mV, respectively. Both redox titrations suggested concerted two electron reduction indicating a wide separation between the quinone/semiquinone and semiquinone/hydroquinone redox couples with the formation of very little flavin semiquinone. In addition, agreement between both determinations indicated limited proteolysis had minimal effect on the flavin potentials. Quantitation, using room temperature epr, of the amount of $FAD^{\cdot -}$ (low temperature epr has shown the flavin radical has a 15 gauss linewidth (4)) formed at the midpoint of the titration indicated that less than 1% of the FAD was converted to the semiquinone form, indicating midpoint potentials for the $FAD/FAD^{\cdot -}$ and $FAD^{\cdot -}/FADH_2$ couples were approximately -372 mV and -172 mV, respectively. Visible and CD potentiometric titrations of the heme center both yielded n=1 redox processes with midpoint potentials of -162 mV. Cleavage and separation of the large and small fragments had no effect on the heme potential. CD titrations of the Mo center yielded a biphasic curve. Approximately 80% of the overall absorbance change corresponded to reduction of Mo^{vi} to Mo^{v} and 20% corresponded to reduction of Mo^{v} to Mo^{iv} with midpoint potentials for the two processes of +26 mV and -40 mV, Potentiometric room temperature epr measurements indicated maximum formation of Mo^{v} detected at -5 mV, corresponded to 55% conversion of the Mo to the pentavalent form yielding values of +16 mV and -26 mV for the Mo^{vi}/Mo^{v} and Mo^{v}/Mo^{iv} couples, in agreement with the values obtained by CD. Thus the room temperature measurements are 14-60 mV higher than the values of -34 mV and -54 mV obtained from the low temperature epr measurements (4) suggesting different entropies of reduction [$dE_m/dT=\Delta S/(nF)$] for the Mo couples.

Acknowledgments

Supported by grants from N.I.H. and the U.S. Dept. Agriculture.

References

1. Howard, W.D. and Solomonson, L.P. 1982. J. Biol. Chem. 257, 10243-10250.

2. Solomonson, L.P., Barber, M.J., Robbins, A.P. and Oaks, A. 1986. J. Biol. Chem. 261, 11290-11294.

3. Palmer, G. and Olson, J. 1980. In: Molybdenum and Molybdenum-Containing Enzymes (M.P. Coughlan, ed). Pergamon, pp189-220.

4. Solomonson, L.P., Barber, M.J., Howard, W.D., Johnson, J.L. and Rajagopalan, K.V. 1984. J. Biol. Chem. 259, 849-853.

KINETIC STUDIES ON GLUTAMATE SYNTHASE FROM *Azospirillum brasilense*

S. Ratti, M.A. Vanoni, R. Grandori, G. Zanetti, B.Curti
Dipartimento di Fisiologia e Biochimica Generali, Università di Milano, Via Celoria 26, 20133 Milano, Italia

Glutamate synthase together with Glutamine synthetase plays a major role in ammonia assimilation in nitrogen fixing bacteria. The enzyme catalyzes the NADPH-dependent reductive transfer of the amino group of L-glutamine to the α-carbon of 2-oxoglutarate to yield 2 molecules of L-glutamate. Alternatively ammonia can serve as the amino donor. The enzyme also exhibits glutaminase activity and can catalyze the L-glutamate-dependent reduction of $NADP^+$ to yield 2-oxoglutarate and L-glutamine or 2-oxoglutarate and ammonia.

reaction 1: $NADPH + H^+ + 2\text{-OG} + \text{L-GLN} \longrightarrow NADP^+ + 2\ \text{L-GLU}$

reaction 2: $NADPH + H^+ + 2\text{-OG} + NH_3 \longrightarrow NADP^+ + \text{L-GLU} + H_2O$

reaction 3: $\text{L-GLN} + H_2O \longrightarrow \text{L-GLU} + NH_3$

reaction 4: $NADP^+ + 2\ \text{L-GLU} \longrightarrow NADPH + H^+ + 2\text{-OG} + \text{L-GLN}$

or: $NADP^+ + \text{L-GLU} \longrightarrow NADPH + H^+ + 2\text{-OG} + NH_3$

Glutamate synthase has been recently purified to homogeneity from *Azospirillum brasilense* Sp6 grown under limiting concentration of ammonia (1). The purified enzyme is a tetramer of 740 kDa. The protomer contains two dissimilar subunits of 135 and 50 kDa, respectively and 1 FAD, 1 FMN, 8 non-heme iron and 8 sulfur atoms per 185 kDa.

Flavins and Flavoproteins 1987

Results and Discussion

During a preliminary characterization of the enzyme, the pH dependence of its activities were determined at constant concentration of the substrates. The L-GLN-dependent reaction was carried out in Hepes/Pipes buffer and showed a pH optimum of about 7.5. The reverse reaction proceeded optimally at pH 9.8. The pH dependence of reaction 2, in the presence of a constant concentration of NH_4Cl in borate buffer reflected the increase of free ammonia concentration at increasing pH values, indicating that the enzyme binds this form of the substrate; if the experiment was repeated in the presence of a constant concentration of NH_3, an apparent pH optimum of 9.0 was obtained. The relative rates for the L-GLN-dependent, NH_3-dependent, and reverse reaction, each one calculated at its pH optimum, were 100, 7.7, and 2.7%, respectively. The steady-state kinetic parameters of the L-GLN- and NH_3-dependent reactions were determined at their pH optimum and at 25°C. Double reciprocal plots of the initial velocities as a function of the varied substrate were in all cases linear. Parallel lines were obtained when NADPH was the varied substrate and 2-OG or L-GLN were held constant at several concentrations, while lines were intersecting when 2-OG was the varied substrate in the presence of different L-GLN concentrations, and constant NADPH. The same pattern was obtained when ammonia substituted for L-GLN. The overall reaction can be described in all cases by a ping-pong Uni Uni Bi Bi kinetic mechanism, according to the following scheme:

NADPH ↓ NADP$^+$ ↑ 2OG, L-GLN(or NH_3) ↓ L-GLU, L-GLU(or H_2O) ↑

E_{ox} —— E_{red} —— E_{ox}

The kinetic parameters determined for Glutamate synthase reaction are summarized in Table I.

TABLE I. Steady-state kinetic parameters of Glutamate synthase reactions.

Substrate concentration								
NADPH (μM)	2-OG (μM)	L-GLN (μM)	NH_3 (μM)	TN (min^{-1})	K_{NADPH} (μM)	$K_{2\text{-}OG}$ (μM)	$K_{L\text{-}GLU}$ (μM)	K_{NH_3} (μM)
∞	∞	∞		1,600	6.3	8.6	184	
∞	100		∞	255	1.9			8.6
20	∞		∞	290		65.0		3.5

Several glutamine analogs were tested as inhibitors of both L-GLN and NH_3-dependent activities. L-methionine, L-methionine sulphoximine and L-methionine sulphone are competitive inhibitors of L-GLN and the observed K_I were 1.05, 0.058 and 0.0046 mM, respectively. However, when the NH_3-dependent reaction was examined at pH 9.0, no inhibitory effect of the same compounds was observed.

It has been reported that although L-GLN-dependent activity of Glutamate synthase requires the flavin coenzymes, the NH_3-dependent activity does not. Furthermore, an acidic treatment of E.Coli enzyme was found to increase its NH_3-dependent activity (2,3). To determine whether flavins are required to carry out such reaction in A.brasilense enzyme, the preparation of reconstitutable deflavo-Glutamate synthase was attempted through several methods. Protein precipitation at pH 4.6 (3) or with ammonium sulphate, pH 1.6 (4) led in all cases to greater than 98% L-GLN and NH_3-dependent activity loss. Extensive dialysis against buffer containing 2M KBr and activated charcoal, at pH

7.5, led to 70% loss of all activities with ⩽ 7% reactivation in the presence of added flavins. It thus seems that flavins are required for both L-GLN and NH_3-dependent activities of Glutamate synthase even when the integrity of the iron sulphur centers seems preserved. The involvement of Glutamate synthase redox centers in NH_3-dependent reaction was also confirmed by the experiment shown in fig. 1. Glutamate synthase was made

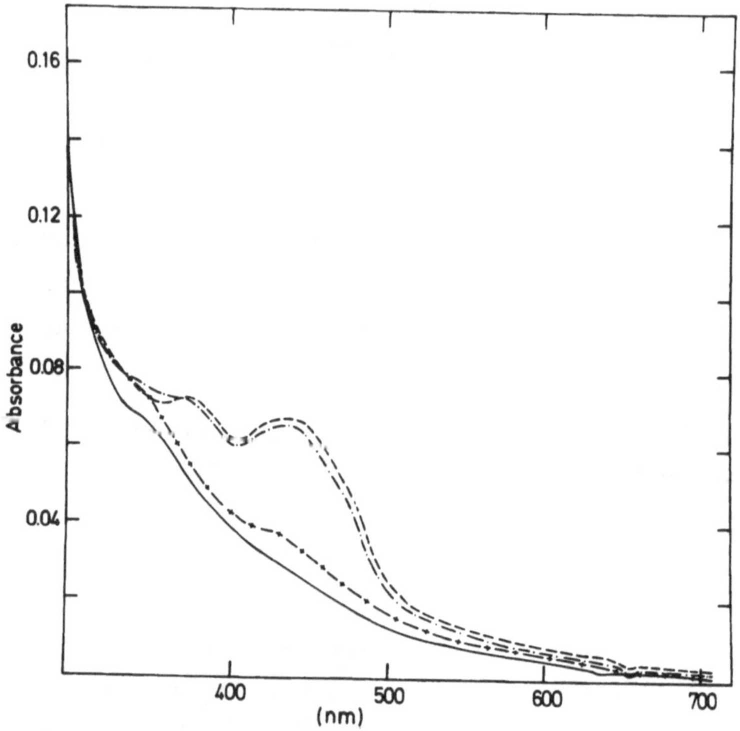

Fig. 1 - Photochemical reduction of Glutamate synthase. Conditions: 1.63 nM Glutamate synthase in 25 mM Hepes, pH 8.0, containing 10% glycerol, 0.3 mM EDTA, 1 mM DTT and 0.5 nM 5-deazaflavin.
(– – – – –), oxidized enzyme;
(————), fully reduced enzyme;
(–+–+–), after addition of 100-fold excess 2-oxoglutarate;
(–·–·–·), 10 min after addition of 100-fold excess NH_3.

anaerobic and photoreduced by the EDTA/deazaflavin system. The addition of 100-fold excess 2-OG did not produce significant spectral changes, while the enzyme was fully reoxidized when 100-fold excess NH_3 wal also added. A similar result was obtained when L-GLN substituted for NH_3. These results are also consistent with the kinetic mechanism proposed above for Glutamate synthase reactions.

Acknowledgement

This research was supported by a grant from Ministero della Pubblica Istruzione, Roma, Italia.

References

1. Ratti, S., Curti, B., Zanetti, G., Galli, E. 1985. J.Bacteriol. 163, 724.

2. Mantsala, P., Zalkin, H. 1976. J.Biol.Chem. 251, 3300.

3. Matsuoka, K., Kimura, K. 1986. J.Biochem. 99, 1087.

4. Husain, M., Massey, V. 1978. Methods Enzymol. 53, 429.

REDOX INTERACTIONS OF THE FAD IN XANTHINE OXIDASE WITH OTHER CENTERS OF THE PROTEIN

Russ Hille

Department of Physiological Chemistry, Ohio State University, Columbus, OH 43210

Introduction

Xanthine oxidase contains in each of its two independent subunits a molybdenum center, two iron-sulfur centers of the 2Fe/2S variety, and flavin adenine dinucleotide. The reductive half-reaction of the catalytic cycle, where xanthine is hydroxylated to form uric acid, takes place at the molybdenum center, and the oxidative half-reaction, where the pair of reducing equivalents thus obtained by the enzyme are passed to molecular oxygen to form either peroxide or superoxide, at the FAD. The transfer of electrons from molybdenum to flavin within xanthine oxidase is thus an integral aspect of the catalytic cycle and the enzyme provides a unique opportunity to study the interaction of FAD with other-active centers in a well-defined environment. In the present work, we examine these interactions and how they influence the thermodynamic and kinetic redox behavior of xanthine oxidase.

Results

It is now generally accepted that reducing equivalents distribute themselves among the four redox-active centers of partially reduced xanthine oxidase (i.e., enzyme containing fewer than six reducing equivalents per active site)

Flavins and Flavoproteins 1987

according to the relative reduction potentials of the centers (1). Thus at any level of enzyme reduction the electron distribution may be calculated from the potentials. Our own efforts aimed at demonstrating the validity of this model have included a study of enzyme species in which the native FAD cofactor has been replaced in the enzyme by chemically modified FAD derivatives of varying oxidation-reduction potential (2). In these enzyme species, the expected thermodynamic behavior is invariably observed. Thus, when enzyme containing a flavin derivative of high potential (e.g., 8-Cl or 2-thio FAD) is titrated with sodium dithionite, flavin reduction precedes that of the iron-sulfer centers. Conversely, when enzyme containing a low-potential flavin (e.g., 6-OH, 8-OH or 1-deaza FAD) is used, iron-sulfur reduction precedes that of the flavin. These results demonstrate that the flavin substitution technique is a valid approach to the study of xanthine oxidase.

It was originally hypothesized that the equilibration of electrons among the several centers in the enzyme was rapid relative to turnover (1). That this is the case is suggested by the kinetic behavior of a series of enzyme species containing a range of flavin potentials in the reaction with 100 µM xanthine under anaerobic conditions. Figure 1 gives plots of the difference between the normalized time courses observed at two wavelengths, one where the reduction of th iron-sulfur centers predominates the observed spectral change and the other where the reduction of flavin predominates. In these plots a positive deflection indicates reduction of flavin prior to that of the iron-sulfur centers, and a negative deflection the reverse. It can be seen that the observed kinetic behavior is strongly dependent on the flavin potential. When a high-potential flavin is present in the enzyme (2-thio and 8-Cl), flavin reduction precedes that of the iron-sulfur centers, whereas when a flavin of low potential is present (6-OH or

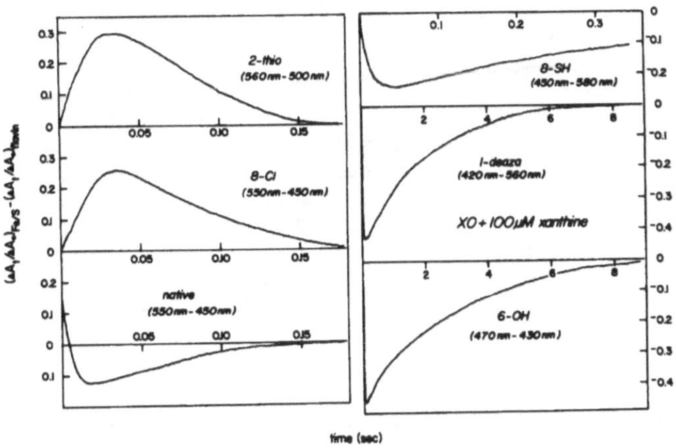

Figure 1. The reaction of xanthine oxidase containing a series of FAD analogs of varying oxidation-reduction potential with 100 μM xanthine. The specific wavelengths used in the subtraction procedure (see text) for each enzyme species are indicated.

1-deaza) iron-sulfur reduction occurs first. These results are readily explained by assuming that reducing equivalents partition rapidly (on a time scale of 50-100 msec) between the iron-sulfur centers and the flavin according to the relative potentials of the centers once having been introduced into the enzyme via the molybdenum center. Evidence that the equilibration of reducing equivalents within xanthine oxidase is rapid is also provided by steady-state studies with this series of flavin-substituted enzyme species. Figure 2 shows the dependence of the observed V_{max} on the flavin midpoint potential (relative to that of the molybdenum center). The sigmoidal relationship is striking and interpretable in terms of a rapid equilibration of reducing equivalents between the flavin and molybdenum centers (the iron-sulfur centers need not be explicitly taken into account). In the steady state, partially reduced enzyme will have its reducing equivalents partition between the flavin and molybdenum according to the relative midpoint

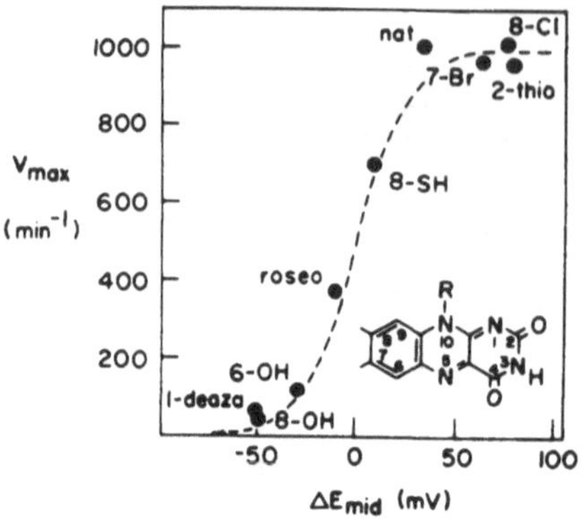

Figure 2. The dependence of the observed V_{max} on flavin midpoint potential for xanthine oxidase containing a series of FAD analogs. The flavin midpoint potential is expressed relative to that of the enzyme in the presence of substrate.

potentials of the two centers. To the extent that molybdenum reduction predominates over flavin, the enzyme must be catalytically incompetent as neither center is in the appropriate oxidation state for participation in its respective half-reaction. The dashed line in Figure 2 shows the response of V_{max} to the flavin potential expected on the basis of this argument, and it can be seen that the fit to the data is extremely good. In calculating this curve, a 50 mV increase in the midpoint potential of the molybdenum center on binding xanthine has been taken into account. That the enzyme species containing flavins of low midpoint potential are less active for this reason and not because of intrinsically low chemical reactivity of the flavin is reflected in the kinetics observed in the reductive and oxidative half-reactions. The kinetics of 6-OH xanthine oxidase are representative. It can be seen that the reaction with 50 µM xanthine is multiphasic, with approximately half of the spectral change observed at 430 nm taking place in the first 100 ms (Figure 3A). The remainder of the spectral change takes place on a timescale of 15-20 sec. The wavelength dependence of the slow phase indicates that it is primarily due to the reduction of the 6-OH FAD. It is to be noted that the dependence of the fast phase of the reductive half-reaction indicates a K_D of 10 µM and a limiting rate at high xanthine concentration of approximately 30 s^{-1} (data not shown), in good agreement with the values expected for an individual reaction in the three-step reduction of xanthine oxidase by xanthine (10 μM and 30 s^{-1}, respectively; 1). This indicates that flavin substitution has a negligible effect on the basic kinetics of the reductive half-reaction at the molybdenum center. The reaction of reduced 6-OH xanthine oxidase with 200 μM oxygen is also multiphasic (Figure 3B). The fast phase exhibits a rate of 17 s^{-1} and its wavelength dependence indicates that it is due to reoxidation of the 6-OH FAD. Thus 6-OH $FADH_2$ exhibits high reactivity toward molecular oxygen when bound to xanthine oxidase. The wavelength dependence of the slow phase indicates that it is primarily due to the reoxidation of the iron-sulfur centers.

The rapid kinetic results for both the reductive and oxidative half-reactions are entirely consistent with the conclusion from the steady-state experiments that the enzyme species containing flavin analogs of low midpoint potential exhibit low activity because of an unfavorable electron distribution within partially reduced enzyme encountered in the steady-state. This phenomenon has been termed thermodynamic control of enzyme activity (3).

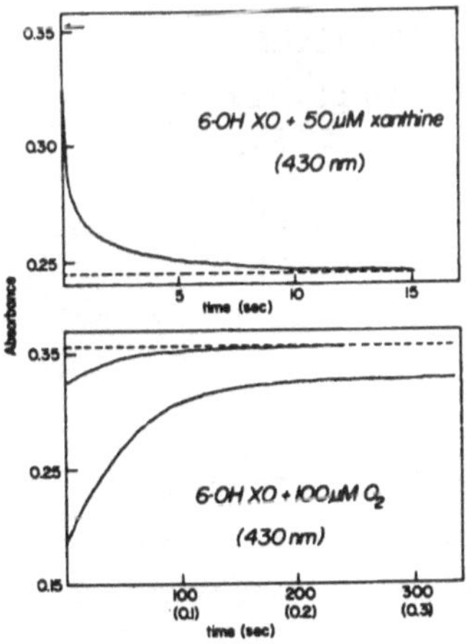

Figure 3. The reaction of 6-OH xanthine oxidase with 50 μM xanthine (Upper Panel) and 100 μM oxygen (Lower Panel). The kinetics were monitored at 430 nm.

The above observations provide indirect evidence in support of fast equivalents within xanthine oxidase, and are in disagreement with flash photolysis studies in which it was concluded that electron equilibration within xanthine oxidase was for the most part a rather slow process compared to turnover (5). We therefore began an independent investigation of the rate(s) of internal electron transfer within xanthine oxidase. At the Eighth Flavin Symposium we presented the results of pH jump experiments in which advantage was made of the known pH dependence of the reduction potentials of xanthine oxidase to monitor equilibration of reducing equivalents within the enzyme (6). In brief, we found that by rapidly jumping the pH of a solution of partially reduced enzyme it was possible to perturb the distribution of reducing equivalents within the enzyme and monitor the approach to the new equilibrium position spectrophotometrically. The observed process was found to reflect the equilibration of a reducing equivalent between the flavin and the iron-sulfur center designated Fe/S I. The k_{obs} was found to be 160 s^{-1} at pH 7 and 330 s^{-1} at pH 8.5, approximately tenfold greater than k_{cat} in each case. Because k_{obs} is the sum of rate constants for the forward and reverse processes, and the ratio of these rate constants known from the known reduction potentials for the iron-sulfur center and the FAD/FADH·couple, it is possible to calculate the rate constants at a given pH (Table 1). It can be seen that pH has relatively little effect on the rate of electron transfer from the iron-sulfur center to the flavin, but that the reverse reaction decreases by more than an order of magnitude as the pH is decreased from 8.5 to 7.0. This has the result of removing any obvious dependence of the rate constant on the thermodynamic driving force of the reaction. Electron transfer from the flavin semiquinone to Fe/S I decreases by a factor of approximately 15 concommitant with a decrease of 75 mV in going from pH 8.5 to 7.0, but the reverse reaction

Table 1. Calculated rate constants for the forward and reverse reactions of electron equilibration between the flavin and Fe/S I of xanthine oxidase.

	potentials (mV vs. SHE) Fe/S $I_{ox/red}$	FAD/FADH·	K_{eq}	k_{obs}	$k_{Fe/S\rightarrow FAD}$	$k_{FADH^{\cdot}\rightarrow Fe/S}$
pH 8.5	-350	-355	1	330 s^{-1}	165 s^{-1}	165 s^{-1}
pH 7.0	-290	-220	15	160 s^{-1}	150 s^{-1}	10 s^{-1}

exhibits no comparable increase, and in fact decreases slightly. For this reason, it appears likely that under the experimental conditions the prototropic equilibria giving rise to the pH dependence of the enzyme reduction potentials contribute in some way to limiting the rate of electron transfer.

We have also undertaken a pulse radiolysis study of electron transfer within oxidase (9). In this technique, anaerobic solutions of enzyme are exposed to microsecond pulses of MeV-energy electrons to generate strongly reducing chemical species which may then react with the enzyme. The strategy is much the same as in the flash photolysis experiment, namely to rapidly introduce reducing equivalents into the enzyme and monitor any subsequent internal equilibration spectrophotometrically. At high radiolytic doses, the expected spectral changes are observed when $CO_2^{\cdot-}$ (formed by the reaction of HO· with formate) is used as reductant. The time course for this reaction at 550 nm (Figure 4A) exhibits a very rapid increase in absorbance, followed by an exponential decay to an endpoint that is below the initial absorbance of the fully oxidized enzyme. The rate constant for this process is 290 s^{-1}. At 375 nm the time course is biphasic with much of the absorbance change taking place with a rate constant of 65 s^{-1} (Figure 4B). The wavelength dependence of the spectral change is shown in Figure 5. Above 420 nm the results are consistent with the formation of the neutral flavin semiquinone within 30 µs, in the

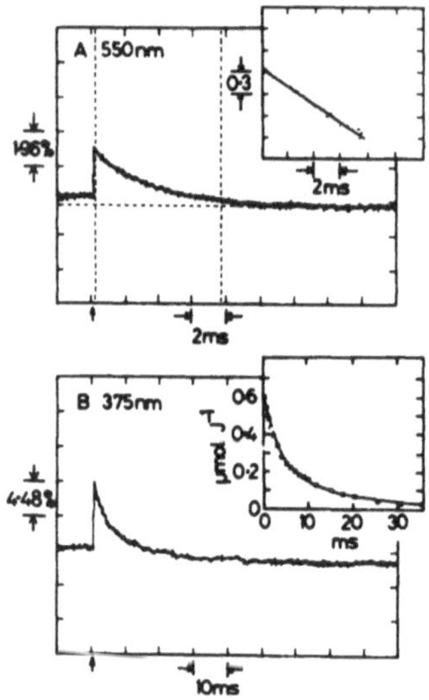

Figure 4. The time course of the reaction of 30 µM xanthine oxidase with approximately 10 µM CO_2^- at 550 nm (A) and 375 nm (B). The inset to A shows a semi-logarithmic plot of the absorbance change indicating that it is due to a single exponential process (k_{obs}=290 s^{-1}). The inset to B shows the fit to the data using the sum of two exponentials (k_1=290 s^{-1}, k_2=60 s^{-1}). The experimental conditions were 0.02 M pyrophosphate buffer, pH 8.5, 20°C, 0.1 M formate, an under 1 atm of N_2O.

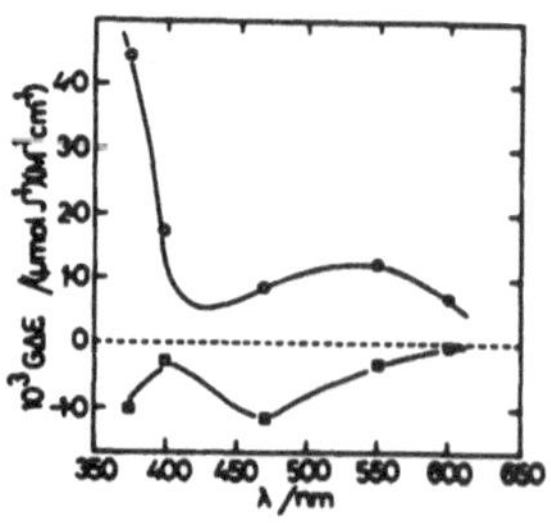

Figure 5. The wavelength dependence of the reaction shown in Figure 4. Circles, the absorbance at 30 µs after pulse, squares, absorbance change 35 ms after pulse.

initial reduction of enzyme by $CO_2^{\cdot-}$, followed by the transfer of this electron to one of the iron-sulfur centers of the enzyme. This is precisely the pair of sites involved in the process observed by the pH jump technique, and the observed rate constant of 290 s^{-1} is in excellent agreement with the pH jump results when the ambient temperature of the pulse radiolysis experiment (20°C) is taken into account. Below 400 nm an absorbance increase in the 30 µs spectral change is observed which cannot be attributed to FADH·. $CO_2^{\cdot-}$ is known to reduce protein disulfides to form an $RSSR^{\cdot-}$ species that absorbs strongly below 400 nm (7). It is known that xanthine oxidase has a particularly redox-active disufide bond, and that this functional group can be blocked (without effect on enzyme activity) by reaction with iodoacetate in the reduced dithiol state (8). That reduction of this disulfide is associated with the absorbance below 400 nm (and therefore implicated in the slow kinetic process observed in this spectral region) is indicated in the titrations shown in Figure 6. With native xanthine oxidase, a pronounced lag is observed in the reduction of enzyme with γ-irradiation from a calibrated ^{60}Co source, and a total of eight reducing equivalents are required to completely bleach the enzyme. When disulfide-blocked enzyme is used, there is no lag in the titration and only six equivalents are required to effect complete reduction. When disulfide-blocked enzyme is used in the pulse experiments, no anomalous absorbance below 400 nm is observed immediately after the pulse, and the slow kinetic phase is not seen (data not shown).

The results shown in Figure 6 are consistent with the transient accumulation of a one-electron reduced disulfide species in the course of the radiolytic titrations (and by implication the pulse experiments as well). To examine this possibility, the EPR properties of radiolytically pulsed xanthine oxidase (sufficient to generate one reducing equivalent per enzyme active site) have been investigated. The EPR signal seen with enzyme treated in this way is shown

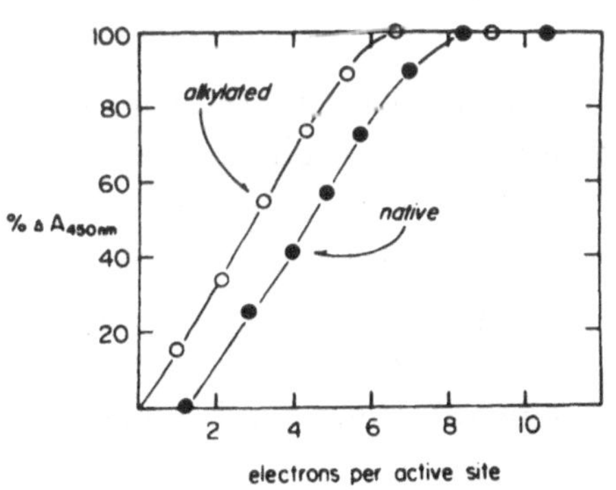

Figure 6. Titration of native and disulfide-blocked xanthine oxidase with radiolytically generated reducing equivalents. Upper Panel, the spectra observed after the radiation exposure shown (given in seconds) for native enzyme. Lower Panel, plots of the fractional absorbance change at 450 nm for native (● circles) and disulfide-blocked (○ circles) enzyme versus the cumulative reducing equivalents generated.

in Figure 7. The signal is axially symmetric, exhibits extremely narrow linewidths, and is extremely sensitive to power saturation at liquid nitrogen temperatures. The signal-giving species is indefinitely stable at room temperature under anaerobic conditions, but disappears with 60 sec on exposure to air. It remains for future work to unambiguously demonstrate that this signal is due to the reduction of a disulfide bond in xanthine oxidase, but all the data to date are consistent with this interpretation.

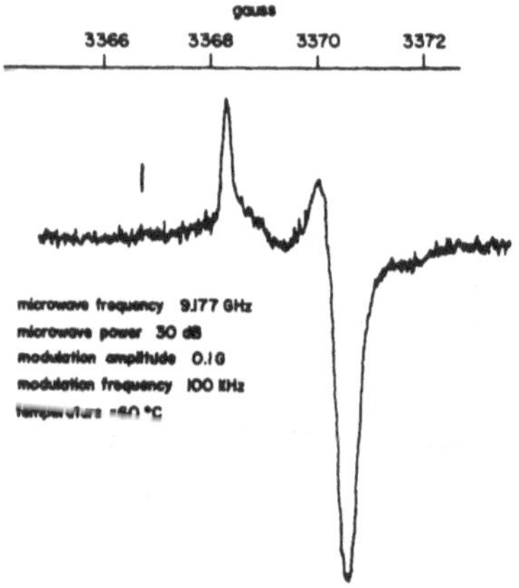

Figure 7. EPR signal of radiolyzed xanthine oxidase. Experimental conditions were as shown.

Conclusions

In conclusion, our pH jump and pulse radiolysis experiments provide strong support for the hypothesis that reducing equivalents equilibrate rapidly among the several redox-active centers of xanthine oxidase. Under certain conditions the distribution of reducing equivalents may be such that the rapid and steady-state kinetics of the enzyme are slowed. It must be emphasized, however, that all the evidence points to this being a thermodynamic rather than kinetic effect.

Acknowledgement

This research was supported by grants from the NSF (DMB 8521645) and the American Heart Association (861290).

References

1. Olson, J.S., D.P. Ballou, G. Palmer, V. Massey. 1974. J. Biol. Chem. 249 4363-4382.

2. Hille, R., J.A. Fee, V. Massey. 1981. J. Biol. Chem. 256, 8933-8940.

3. Palmer, G., and J.S. Olson. 1980. In: Molybdenum and Molybdenum-Containing Enzymes (M. Coughlan, ed.). Pergamon Press, New York. pp. 187-220.

4. Hille, R., R.C. Stewart. 1984. J. Biol. Chem. 259, 1570-1576.

5. Bhattacharyya, A., G. Tollin, M. Davis, D.E. Edmondson. 1983. Biochem. 22, 5270-5279.

6. Hille, R. 1985. In: Flavins and Flavoproteins (R.C. Bray, P.C. Engel, S.G. Mayhew, eds.). Walter de Gruyter, New York. pp. 683-686.

7. Bisby, R.H., R.B. Cundall, J.L. Redpath, G.E. Adams. 1976. J. Chem. Soc. Faraday Trans. 72, 51-63.

8. Hille, R., V. Massey. 1982. J. Biol. Chem. 257, 8898-8901.

9. Anderson, R.F., R. Hille, V. Massey. 1986. J. Biol. Chem. 261, 15870-15876

LASER FLASH PHOTOLYSIS STUDIES OF INTRAMOLECULAR ELECTRON TRANSFER BETWEEN THE FAD AND Fe/S II CENTERS IN XANTHINE OXIDASE

Dale E. Edmondson*, James T. Hazzard, and Gordon Tollin
Departments of Biochemistry, *Emory University, Atlanta, GA. and University of Arizona, Tucson, AZ.

Introduction

The direct determination of the rates of electron transfer among the redox centers (Mo, FAD, Fe/S I, and Fe/S II) of xanthine oxidase (XO) has been the subject of a number of investigations following the suggestion (1) that the distribution of reducing equivalents among the centers is very rapid and governed by their respective redox potentials. A previous flash photolysis study (2) observed a biphasic rate of Fe/S reduction (77 s^{-1} and 12 s^{-1}) on reaction with 5-deazaflavin semiquinone (5-Dzfl·). Only a small amount of FADH· formation was observed and this species apparently did not decay rapidly. The reaction of 5-Dzfl· with the enzyme was shown to be second order ($k=6.6 \times 10^7\ M^{-1}s^{-1}$) and appeared to react mainly with the functional Mo center. From these and other data, the intrinsic rate constants were calculated to range from 10 s^{-1} to 38 s^{-1} (2). Another approach to the problem used " pH-jump" stopped-flow kinetic experiments in which the rate of re-equilibration of reducing equivalents were determined as a consequence of differentially altering the redox potentials of the centers by rapidly changing the pH (3). Spectral changes were observed to occur at rates of 155 s^{-1} at pH 6 and 330 s^{-1} at pH 8.5 (3). These rate constants are approximately 10-fold faster than k_{cat} under the same conditions. Recent pulse radiolysis studies (4) showed the

Flavins and Flavoproteins 1987

FAD center in XO to react with $CO_2^{-}\cdot$ to form the semiquinone and to subsequently react with an Fe/S center with an observed rate constant of 290 s^{-1} at pH 8.5. Evidence was found for the reduction of a disulfide bond by radiolytically-generated low potential viologen radicals (4). These studies resulted in the suggestion (4) that the discrepancy between the laser photolysis data (2) and other studies (3,4) could arise from the reduction of a disulfide bond by the photo-generated 5-DzFl·. This could then explain the slow rates of Fe/S reduction observed in the laser photolysis experiments. This paper presents results of a reinvestigation of the kinetics of XO reduction by laser-generated 5-DzFl· using native XO and an XO derivative in which the suspect disulfide bond was reduced and carboxymethylated (5). The results show that the disulfide does not influence the observed rates of Fe/S reduction on reaction of XO with 5-DzFl·. Further data show that FADH· reduction of Fe/S II does in fact occur, with a rate constant somewhat larger than the 77 s^{-1} estimated previously (2). The 12 s^{-1} reduction of Fe/S was again observed using both 5-DzFl· and methyl viologen radical as reductants.

Results and Discussion

Fig. 1 shows the kinetics of enzyme reduction at 483 nm for native XO (50% functional) and for CMXO (also 50% functional). Both samples show identical rates of bleaching on reaction with 5-Dzfl·. In agreement with previous data (2), the reduction shows a biphasic behavior with the rapid phase (k~ 100 s $^{-1}$) showing a larger amplitude than the slow phase (k= 10-12 s^{-1}). These data demonstrate that reduction and carboxymethylation of the reducible disulfide bond (5) does not influence the reduction rate on reaction with 5-DzFl· as suggested previously (4).

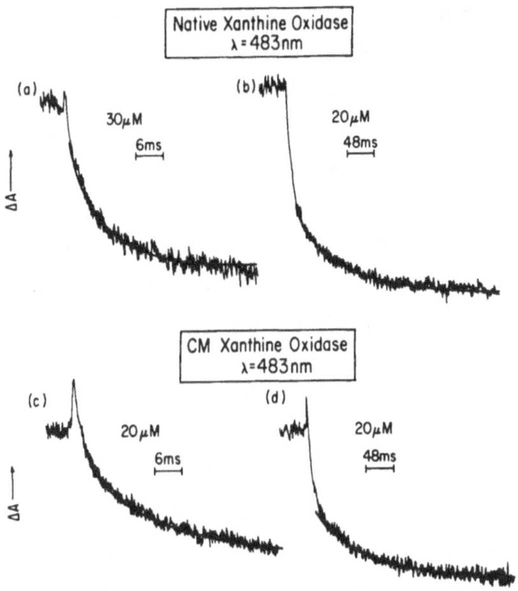

Fig.1 - Comparison of reduction kinetics of native and carboxymethylated XO upon laser flash photolysis of 5-DzFl (70μM) in 20mM P_i - 20mM EDTA, pH 7.2.

In an effort to investigate further the reason for the observed differences in rates of reduction obtained with the different experimental approaches, laser flash experiments were performed in the presence of 2mM methyl viologen (MV). Fig. 2 demonstrates that 5-DzFl· reacts very rapidly with MV to form $MVH^{+}\cdot$ (less than 10 μs). No decay of the MVH+· is observed in the absence of XO. However, a rapid decay is observed in the presence of XO (Fig. 2). Since $MVH^{+}\cdot$ has an E_m of -440mV at pH 7.2 and the lowest potential center on the enzyme is -300 mV (6), the observed rate is largely due to the forward electron transfer reaction. A small equilibrium level of $MVH^{+}\cdot$ is apparent on completion of the reaction since the absorbance at

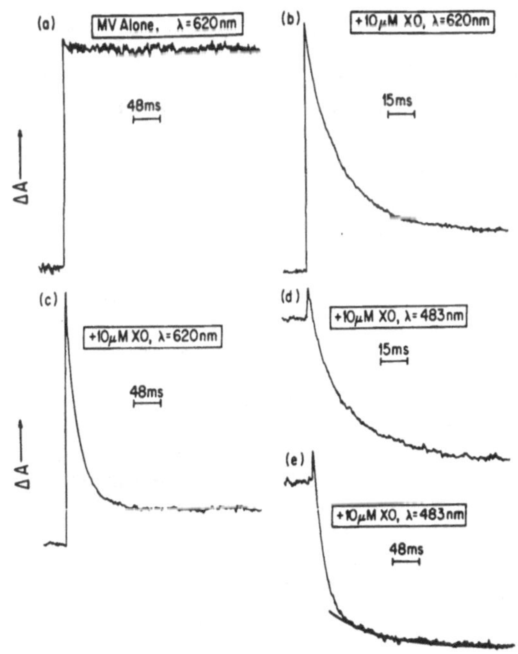

Fig.2 - Transients observed after the laser flash in the presence of 2mM methyl viologen. The other experimental conditions are as in Fig.1.

620 nm does not return to the baseline value (Fig. 2c). In addition to absorbance changes at 620 nm, Fe/S center reduction is evident from the observed bleaching at 483 nm (Fig. 2d,e). The rapid phase exhibits the same rate of decay as $MVH^{+\cdot}$ oxidation and both rates are dependent on XO concentration (Fig.3) (k= $4x10^6$ $M^{-1}s^{-1}$). The slow phase of Fe/S reduction is independent of XO concentration, the rate constant being 10-12 s^{-1}, the same as in experiments where MV is absent (Fig.1). When the FAD moiety was rendered redox inactive by alkylation with iodoacetamide (7), $MVH^{+\cdot}$ reacted with the enzyme more slowly and in a biphasic manner in which both phases were observed to be second order (k=$3x10^5$ and k=$5x10^4$ $M^{-1}s^{-1}$) (Fig.4). These data suggest that $MVH^{+\cdot}$ reacts rapidly with the FAD and more slowly with the Fe/S centers in the enzyme, although it is also possible that FAD alkylation causes appreciable structural changes which modify the properties of the other centers. At this point, we have no direct evidence concerning the reactivity of the Mo center with $MVH^{+\cdot}$. The fact that Fe/S reduction occurs at a rate coincident with $MVH^{+\cdot}$ oxidation and that the FAD is the site of electron entry into the enzyme suggests that electron transfer from FADH· to Fe/S is faster than the observed rates of $MVH^{+\cdot}$ oxidation (Fig.3). Since this rate is faster than that previously estimated (2), the reaction of 5-DzFL· with

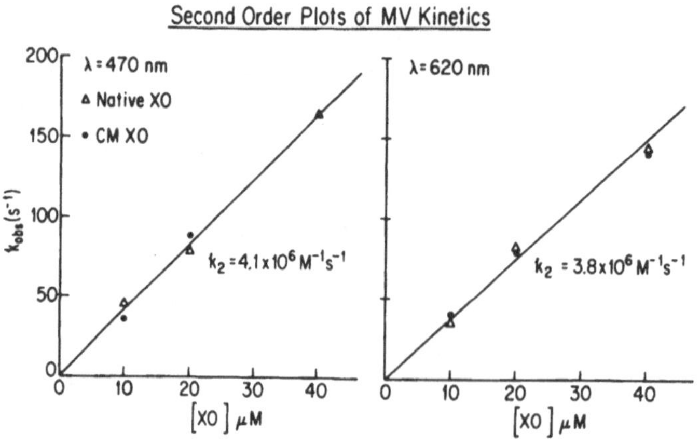

Fig.3 - Second order plots of the reaction of native and carboxymethylated XO with $MVH^{+\cdot}$ generated in the flash apparatus.

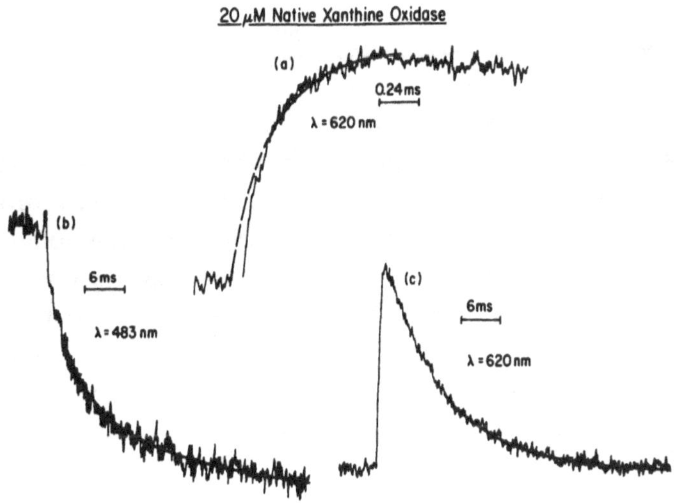

Fig.4 - FADH· formation and decay and Fe/S center reduction on reaction of native XO with 5-DzFl·. Buffer conditions were as in Fig.1 but the 5-deazaflavin concentration was 130μM.

the FAD center in the enzyme was reinvestigated. Using an improved optical design of the apparatus and higher 5-DzFl· concentrations(130μM), we were able to directly determine the rate of formation and decay of FADH (Fig.4). The second-order reaction of 5-DzFl· with the FAD was found to be very fast ($k=2x10^8\ M^{-1}s^{-1}$) at pH 7.2. This value is slightly larger than that previously determined by monitoring 5-DzFl· absorption ($k=7x10^7\ M^{-1}s^{-1}$) (2). The FADH· generated then decayed at a rate independent of XO concentration ($k=118\ s^{-1}$). The kinetics of the absorbance changes at 483 nm due to Fe/S reduction (Fig.5b) were identical to those of FADH· oxidation thus suggesting that the Fe/S center is the direct oxidant of FADH·. Fe/S II is the most obvious candidate since it is the only center with a redox potential higher than the OX/SQ FAD couple (6). Similar experiments at pH 8.5 showed an observed rate of FADH· oxidation-Fe/S reduction of 100 s^{-1}. With the values of the redox potentials of the two interacting centers, it is possible to calculate the intrinsic rate constants for electron transfer:

$$\text{pH 7.2} \quad \text{FADH}\cdot + \text{Fe/S II}_{ox} \underset{21s^{-1}}{\overset{97s^{-1}}{\rightleftharpoons}} \text{FAD} + \text{Fe/S II}_{red}$$

$$\text{pH 8.5} \quad FADH^{\cdot} + Fe/S\ II_{ox} \underset{2s^{-1}}{\overset{98s^{-1}}{\rightleftharpoons}} FAD + Fe/S\ II_{red}$$

It should be noted that a slower phase of Fe/S reduction ($k=10-12s^{-1}$) is observed in all cases. The slow reduction of an Fe/S center (presumably the low potential Fe/S I cluster) may be a consequence of equilibration with a reduced center of higher potential upon one-electron reduction of the enzyme. If we revise our previous assignment that the more rapidly reduced Fe/S center is Fe/S II rather than Fe/S I, then the rate constant of Mo V reduction of Fe/S II would be approximately 100 s^{-1}. Further work is underway to test this.

Acknowledgement

This work was supported by NIH grants GM 29433 (D.E.E.) and AM 15057 (G.T.).

References

1. Olson, J.S., D.P. Ballou, G. Palmer, and V. Massey. 1974. J. Biol. Chem. 249 , 4363.

2. Bhattacharyya, A., G. Tollin, M.Davis, and D.E. Edmondson. 1983. Biochemistry 22, 5270.

3. Hille, R. and V. Massey. 1986. J. Biol. Chem. 261, 1241.

4. Anderson, R.F., R. Hille, and V. Massey. 1986. J. Biol. Chem. 261, 15870.

5. Hille, R. and V. Massey. 1982. J. Biol. Chem. 257, 8898.

6. Porras, A.G. and G. Palmer. 1982. J. Biol. Chem. 257, 11617.

7. Komai, H. and V. Massey. 1971. In: Flavins and Flavoproteins (H. Kamin, ed.). University Park Press, 399.

MECHANISM OF ACTION OF CHICKEN LIVER XANTHINE DEHYDROGENASE

Takeshi Nishino and Keizo Tsushima
Dept. of Biochemistry, Yokohama City University School of Medicine
Yokohama 236, Japan

Lawrence M. Schopfer and Vincent Massey
Dept. of Biological Chemistry, University of Michigan Medical School
Ann Arbor, MI. 48109-0606 USA

Introduction

Chicken liver xanthine dehydrogenase (XDH)(EC 1.2.1.37) is an enzyme, very similar to milk xanthine oxidase (XO) (EC 1.2.3.2). Both are dimers of molecular weight about 300,000; have wide substrate specificities; and contain one FAD, one molybdenum and two Fe_2S_2 centers per monomer (1). They differ in their reactivity toward molecular oxygen and pyridine nucleotides. XDH reacts poorly with oxygen, while XO reacts readily. On the other hand, XDH is readily oxidized by NAD, while XO is not. In order to better understand the basis of the XDH reactivity, we have examined both its response to stepwise, anaerobic reduction by sodium dithionite and its rapid reaction kinetics, at lower temperature (4°C) than previous studies with turkey liver enzyme (2).

Results and Discussion

The anaerobic titration of XDH with sodium dithionite is summarized in Panel A of Figure 1. Reduction can be divided into four phases. Introduction of the first electron resulted in the loss of about 50% of the 520 nm absorbance, a marked loss of absorbance at 450 nm and a small loss at 620 nm. This is indicative of tho reduction of one Fe_2S_2 center. The next two electrons caused a pronounced rise in absorbance at 620 nm, continued loss at 450 nm, and minimal change at 520 nm. This suggests that formation of flavin semiquinone (FADH$^{\bullet}$) is occuring. Since FADH$^{\bullet}$ requires only one electron per mole of XDH (active site), the consumption of the second electron in this phase argues that molybdenum is also being reduced. The next two electrons caused a precipitous drop in absorbance at 450, 520 and 620 nm. This is indicative of the simultaneous reduction of FADH$^{\bullet}$ and the second Fe_2S_2 center. Finally, there are minor losses in absorbance at all wavelengths for the last electron, suggesting that

Flavins and Flavoproteins 1987

the second reduction of molybdenum occurs in this phase. These results argue strongly for the following sequence of oxidation-reduction potentials for the active site groups of XDH:

$$Fe_2S_2 > FAD/FADH^{\bullet} \approx Mo^{vi}/Mo^{v} > FADH^{\bullet}/FADH_2 \approx Fe_2S_2 \approx Mo^{v}/Mo^{iv}$$

This pattern would be consistent with the values of Barber et.al. (3) except for the second Fe_2S_2 center being moved to a much lower potential. Addition of aminopyridine-adenine dinucleotide (AAD) to XDH prior to reduction resulted in formation of $FADH^{\bullet}$ earlier in the titration (Figure 1, Panel B), suggesting that the $FAD/FADH^{\bullet}$ potential is shifted away from the Mo^{vi}/Mo^{v} couple toward the first Fe_2S_2 value. In both of these examples, the sequence in which the active site groups are reduced is strikingly different from that seen with xanthine oxidase (4). In addition, $FADH^{\bullet}$ is much more stable in XDH than it is in XO.

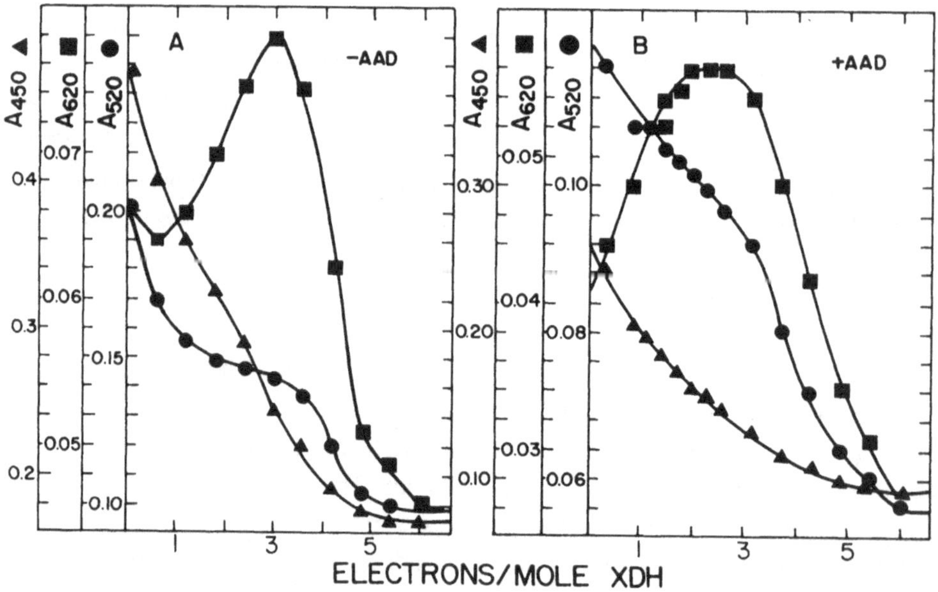

Figure 1. Plots of absorbance versus electron equivalents added for dithionite titrations of oxidized XDH, plus-and-minus amino-pyridine-adenine dinucleotide (AAD). Conditions: Both titrations were done in 50 mM KPi buffer pH 7.8 containing 0.2 mM EDTA, at 25°C. Panel A, XDH (11.8 μM) was made anaerobic and titrated with 3.9 mM sodium dithionite. Panel B, AAD was added to XDH (7.1 μM) to a final concentration of 0.3 mM. The mixture was made anaerobic and titrated with 2.8 mM sodium dithionite. Dithionite was standardized by titrating a sample of lumiflavin-3-acetate. The points were measured from spectra taken after each addition of dithionite: filled circles, 520 nm; filled squares, 620 nm; and filled triangles, 450 nm.

Formation of FADH˙ in the early stages of XDH reduction is also seen during NADH dependent reduction. Mixing excess NADH with oxidized XDH anaerobically in the stopped-flow spectrophotometer, gave rise to a biphasic reaction which led to partial reduction. The apparent rate of the fast phase displayed a reversible, second-order dependence on NADH concentration (kon=1.1×10^6 M^{-1} s^{-1}, koff=59 s^{-1}). This phase showed a rise in absorbance at 620 nm together with a loss in absorbance at both 450 and 520 nm, suggesting both reduction of an Fe_2S_2 center and formation of FADH˙. The apparent rate for the second phase was essentially independent of the NADH concentration (k=5.3-6.6 s^{-1}). This phase resulted in loss of absorbance throughout the spectrum, indicating reduction of FADH˙ as well as Fe_2S_2. The end-point of this reaction, at all NADH concentrations, was approximately that expected for 4-electron reduction of XDH.

Oxidation of reduced XDH by NAD was monophasic and stopped at a level of partial oxidation which was essentially identical to that reached upon reduction of oxidized XDH by NADH. The apparent rate of oxdation exhibited a saturating dependence on the NAD concentration, with a limiting rate of 27 s^{-1}.

That oxidation of reduced XDH by NAD and reduction of XDH by NADH both ended at essentially the same level of XDH reduction, suggests that the end-point for the reaction of XDH with pyridine nucleotides is controlled by the oxidation-reduction potentials of the reactants. This suggestion was confirmed by the observation that 3-acetyl-pyridine NAD (AcPyNAD) caused nearly complete oxidation of reduced XDH. The potential of the AcPyNAD/AcPyNADH couple is 73 mV more positive than the NAD/NADH couple, making AcPyNAD more readily reducible.

Though reduction of XDH by NADH was only partial, reduction of XDH by its physiological substrate, xanthine, appears to be complete. Mixing xanthine with XDH (anaerobically) gave a tetraphasic reaction. The fastest phase (k1=36 s^{-1}) gave rise to spectral changes (increase around 360 nm) consistent with reduction of the molybdenum center. The second phase (k2=1.8 s^{-1}) involved formation of FADH˙ and reduction of an Fe_2S_2 center. The spectral changes in the third phase were consistent with formation of 6-electron reduced XDH. This was the only phase in which the apparent rate showed a dependence on xanthine concentration. The saturating dependence gave a limiting rate, k3=0.89 s^{-1}. The fourth phase was much slower than the first three and was generally consistent with the transfer of electrons from active, reduced XDH to inactive, oxidized XDH. Full, 6-electron reduction of XDH by xanthine would be required for xanthine/NAD turnover to occur, since NAD appears to be capable of oxidizing 6-electron reduced enzyme only.

During the turnover between 4- and 6-electron reduced XDH, the steady state level of XDH would be expected to be the 4-electron reduced state because oxidation of 6-electron reduced XDH by NAD (limiting rate of 27 s^{-1}) is much faster than reduction of 4-electron reduced XDH by xanthine (limiting rate of 0.89 s^{-1}). In order to confirm this, enzyme (3.9 μM) was mixed with an equal volume of the same buffer containing 500 μM xanthine and 1090 μM NAD under anaerobic conditions, at 4°C. The reaction reached steady state in about 10 seconds and continued to produce NADH for 80 seconds. The spectrum of the enzyme after 10 seconds was effectively the same as that at the end of the reaction. Since NAD was in excess, the enzyme would have come to a final level of reduction determined by the NAD. From the earlier discussion, NAD will oxidize XDH to a final, 4-electron reduced level. Thus the steady state level of XDH was also at the 4-electron reduced level.

Acknowledgement

Supported by grant from the National Science Foundation (DMB-8500291) (V.M.) and by a Grant-in-Aid for scientific research from the Japanese Ministry of Education, Science and Culture.

References

1. Rajagopalan, K.V., and P. Handler. 1967 J. Biol. Chem. 242, 4097-4107.
2. Fhaoláin,I.N., M.J. Hynes, and M.P. Coughlan. 1978. Biochem. J. 171, 83-88.
3. Barber, M.J., M.P. Coughlan, M. Kanda, and K.V. Rajagopalan. 1980 Arch. Biochem. Biophys. 201, 468-475.
4. Olson, J.S., D.P. Ballou, G. Palmer, and V. Massey. 1974. J. Biol. Chem. 249, 4363-4382.

INTERCONVERSION BETWEEN O_2-DEPENDENT AND NAD-DEPENDENT TYPES OF RAT LIVER XANTHINE DEHYDROGENASE

T. Saito, T. Nishino, K. Tsushima
Department of Biochemistry, Yokohama City University School of Medicine, 3-9 Fukuura, Kanazawa-ku, Yokohama 236, Japan

Introduction

Rat liver xanthine dehydrogenase exists as NAD-dependent type(D-type) in the freshly prepared sample, but through the procedures of extraction and purification most of the enzyme molecules are converted to O_2-dependent type(O-type). This conversion occurs either through proteolysis of the enzyme molecule or through oxidation of sulfhydryl groups of cysteine residues (1,4). In the latter case this conversion is reversible and we can obtain mostly D-type enzyme(D/O ratio is around 10) by incubating the freshly purified enzyme with dithiothreitol(DTT) (2). In order to know the difference in characters of NAD-dependent and O_2-dependent types, we made experiments of dithionite titration of both types and replacement of Enzyme-FAD with artificial flavins.

Results and Discussion

We carried out the anaerobic titration of both D-type and O-type enzyme with sodium dithionite. In the case of O-type enzyme, FAD was preferentially reduced in the early stage of titration, then two Fe_2S_2 centers were reduced in the later stage. FAD semiquinone was not formed during titration of O-type enzyme. But in the case of D-

Flavins and Flavoproteins 1987

type enzyme, one of the two Fe_2S_2 centers was reduced in the first stage, then FAD was reduced to semiquinone, and finally semiquinone and the other Fe_2S_2 center were reduced (data not shown here). These results show that the relative order of redox potential of FAD and two Fe_2S_2 centers is changed through the D-O conversion, and that in the D-type enzyme semiquinone can be stabilized well. To know how the redox potential of flavin influences the NAD-dependent and O_2-dependent activities we prepared deflavo-enzyme of rat liver xanthine dehydrogenase and reconstituted the enzymes with FAD and artificial FAD analogues having various redox potentials. The deflavo-enzyme was prepared according to the method of Komai (3), with minor modification. After purified rat liver xanthine dehydrogenase was incubated with 2M $CaCl_2$(Suprapure, obtained from Merck) at pH 8.0 at 20 °C for 90 min, the incubation mixture was applied on Sephadex G-25 column to remove $CaCl_2$ and released flavin. About 1.5% of original xanthine(X)-NAD+O_2 activity was retained in the deflavo-enzyme and this retained activity was close to that estimated from the content of FAD in the sample. When the deflavo-enzyme was incubated with an excess concentration of FAD at 25 °C for 15 min, more than 90% of X-NAD+O_2 activity was recovered. For the reconstitution of the enzymes the following FAD analogues were used; 8-OH FAD(-340mV), 8-NH_2 FAD(-310mV), 1-deaza FAD(-280mV), 6-OH FAD(-260mV), 8-Cl FAD(-160mV), and 2-thio FAD(-130mV) (values in the parentheses are their redox potentials). To freshly prepared deflavo-enzyme 3 to 4 molar excess of FAD or each FAD analogue was added and the mixture was incubated at 0 °C overnight at pH 8.0. Then an aliquot of the incubation mixture was subjected to the determination of X-NAD+O_2 and X-O_2 activities. The results are shown in Fig.1. The NAD-dependent activity was given as the difference between X-NAD+O_2 activity and X-O_2 activity. FAD enzyme, of which sulfhydryl groups remained to be oxidized, showed much lower NAD-dependent activity than O_2-dependent activity. Even though the enzymes were reconstituted with flavins with lower redox potential(6-OH,1-deaza,8-NH_2 and 8-OH FAD), they showed very low NAD-dependent activity.

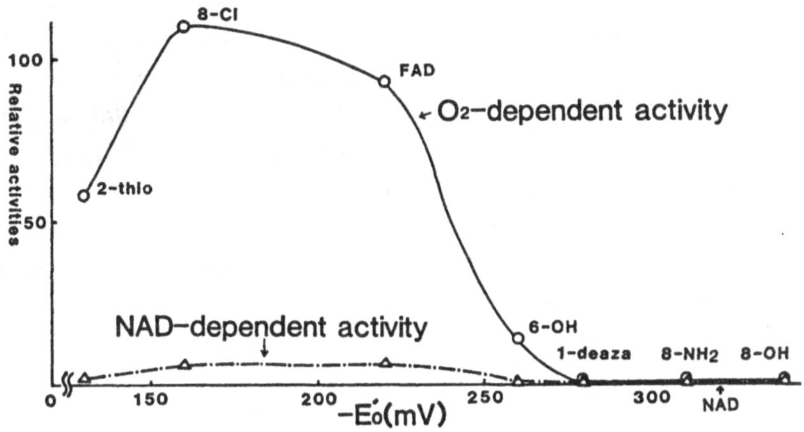

Fig.1 Activities of O-type enzymes reconstituted with FAD or various FAD analogues

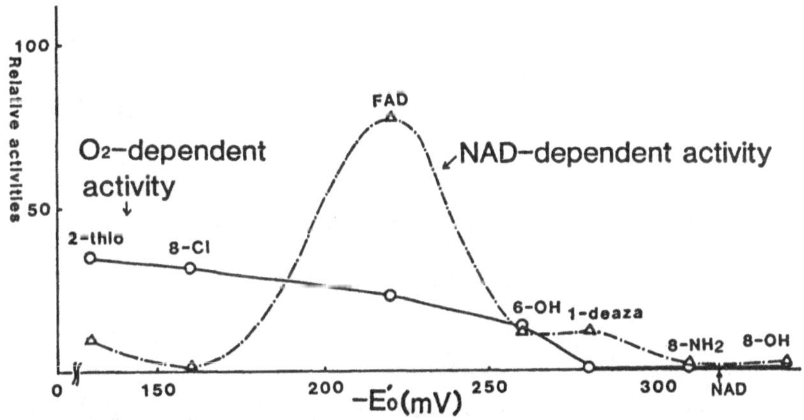

Fig.2 Activities of D-type enzymes reconstitued with FAD or various FAD analogues (after reconstitution each enzyme was incubated with DTT)

However, when each reconstituted enzyme was incubated with 5mM DTT for 2.5 hours at 25 °C, the reconstituted enzyme was converted to D-type. As shown in Fig.2, NAD-dependent activity became prominent with FAD enzyme and, though to a lesser degree, also with 6-OH and 1-deaza FAD enzymes.

These data show that the sulfhydryl groups are important for NAD-dependent activity independently of the relative order of redox potential of chromophores. It is also unlikely that sulfhydryl groups play a role in NAD-binding site, because modification of NAD-binding site with fluorosulfonylbenzoyl adenosine or 2,3-butandione did not result in conversion of D-type enzyme to O-type one. These modification resulted in inactivation of NAD-dependent activity without increase of O_2-dependent activity, while oxidation of sulfhydryl groups of 2,3-butandione-modified enzyme caused the increase of O_2-dependent activity. It is possible that reduced sulfhydryl groups are important for NAD-dependent activity as stabilizer of semiquinone effecting on FAD chromophore directly or indirectly.

Acknowledgment

We thank to Professor Vincent Massey, the University of Michigan, for generous gift of artificial FAD analogues. This work was supported by a Grant-in-Aid for Scientific Research from the Japanese Ministry of Education, Culture and Science and by a Research Grant for the Intractable Diseases from the Japanese Ministry of Health and Welfare.

References

1. Della Corte, E., and Stirpe, F. (1972) Biochem. J. 126, 739-745
2. Ikegami, T., Nishino, T. (1986) Arch. Biochem. Biophys. 247, 254-260
3. Komai, H., Massey, V., Palmer, G. (1969) J. Biol. Chem. 244, 1692-1700
4. Waud, W.R., and Rajagopalan, K.V. (1976) Arch. Biochem. Biophys. 172, 365-379

REACTIVITY OF CHICKEN LIVER XANTHINE DEHYDROGENASE CONTAINING ARTIFICIAL FLAVINS

Tomoko Nishino, Takeshi Nishino and Keizo Tsushima
Dept. of Biochemistry, Yokohama City University School of Medicine
Yokohama 236, Japan

Lawrence M. Schopfer and Vincent Massey
Dept of Biological Chemistry, University of Michigan Medical School
Ann Arbor, Michigan 48109-0606 USA

Introduction

Chicken liver xanthine dehydrogenase (XDH) has high reactivity with NAD, and very low reactivity with oxygen. The low oxygen activity of chicken XDH is not changed by chemical modification or proteolytic digestion. In order to gain insight into the XDH mechanism, we examined the oxygen and NAD reactivity of XDH in which the FAD was replaced by FAD analogues of different oxidation-reduction potentials.

Materials and Methods

XDH was purified from the liver of chickens fed a high protein diet, by the procedure of Nishino (1). The enzyme which we obtained had an AFR_{25} value (1) between 300 and 350.

Preparation of XDH apoprotein: All procedures were performed at 4°C unless otherwise indicated. Under anaerobic conditions, enzyme in 0.1 M Tris-Cl buffer pH 8.0 containing 10 mM salicylate and 0.2 mM EDTA was reduced with 2 mM NADH. Then 5 M $CaCl_2$ (Merck, suprapure) was added to give a final concentration of 2 M. The enzyme was kept in ice for 45 minutes. To remove the small molecular weight compounds ($CaCl_2$, NADH, NAD, flavin, etc), the sample was passed through a Sephadex

Flavins and Flavoproteins 1987

G-25 column equilibrated with 0.1 M Tris-Cl pH 8, containing 0.4 mM EDTA. The pooled apoenzyme was concentrated with a Centricon 30 microconcentrator and incubated with 10 mM cysteine overnight in ice, then passed through a second Sephadex G-25 column, equilibrated with 50 mM KPi buffer pH 7.8 containing 0.2 mM EDTA.

The artificial flavins were prepared as previously described., 6-OH-FAD (2), 8-Cl-FAD (3), 2-thio-FAD (4) and 4-thio-FAD (5). To prepare the reconstituted FAD or artificial FAD enzymes, the apoenzyme was incubated with a 2-5 fold molar excess of flavin for 100 minutes. The preparation was then passed through a G-25 column, equilibrated with 50 mM KPi buffer pH 7.8 containing 0.2 mM EDTA, in order to remove excess free flavin.

Results and Discussion

When apoprotein, prepared from XDH (AFR_{25} = 350) was reconstituted with FAD, 89% of the original activity was regained (AFR_{25} = 310). Table 1 shows the activity of XDH reconstituted with various flavins.

Activity of Flavin Reconstituted XDH

Flavin Potential	Relative Activity[a]		Redox
	Xan-NAD	Xan-O_2	mV
6-OH-FAD	55	43	-305[b]
FAD	100	100	-219
8-Cl-FAD	14	96	-152
2-thio-FAD	36	480	-126[c]
4-thio-FAD	0	87	-55

a) Relative activities were determined aerobically at pH 7.8 in 50 mM KPi buffer containing 0.15 mM xanthine and with or without 0.5 mM NAD, at 25°C.
b) Measured at pH 8.5 to reflect the 6-O$^-$-FAD form.
c) Measured at the riboflavin level.

The activity of XDH is clearly perturbed by changing the potential of the flavin, however no consistent pattern appears.

Titration of either 2-thio- or 4-thio-FAD reconstituted XDH with dithionite occured in two phases. The first two electron equivalents were taken up primarily by the flavin, causing first a rise and then a fall in the absorbance at wavelengths greater than 600 nm (for both cases). Thus the flavin semiquinone was stabilized similarly to native XDH. Stabilization of the semiquinone is a property of XDH which is not shared by xanthine oxidase (XO). The remaining four electron equivalents caused a steady loss of absorbance at all wavelengths, as the Fe_2S_2 and molybdenum centers were reduced. This behavior marks a distinct departure from the case of native XDH, where an Fe_2S_2 center is the first to be reduced and molybdenum comes between flavin semiquinone and fully reduced flavin (6). The observed differences are consistent with the more positive oxidation-reduction potentials of 2-thio- and 4-thio-FAD, compared to native FAD (see Table 1), and are similar to results with XO reconstituted with modified flavins (7).

A reorganization of the oxidation-reduction potentials for the cofactors in XDH provides an opportunity to address some issues in the mechanism of XDH. For example, the interaction of NAD/NADH with XDH appears to be controlled by the relative potentials of the reactants. Raising the potential of the flavin in XDH would be expected to shift the equilibrium of the pyridine nucleotide reaction toward reduced XDH. In fact, mixing 6-electron reduced 4-thio-FAD XDH with NAD (anaerobically) caused negligible oxidation of the enzyme. The same reaction with reduced 2-thio-FAD XDH resulted in minor, but detectable oxidation. The relative extent of oxidation in each case is qualitatively consistent with that expected, based on the order of the flavin potentials. Both of these observations support the concept that the XDH/NAD reaction is controlled by the potentials of the reactants.

Anaerobic reduction of 2-thio-FAD XDH by NADH adds further strength to this hypothesis. The reaction, as expected, proceeds to nearly full reduction. In addition, this reaction provides yet more evidence for the concept of rapid electron transfer between groups within the XDH active site. The reaction is triphasic with the fastest phase resulting in the formation of semiquinone. Maximum measured rates for this phase were 400 to 500 s^{-1}. Since NADH transfers two electrons at a time to the FAD of XDH, the transfer of one electron from the flavin to other sites within XDH, leaving the semiquinone, must occur at a rate greater than the observed 500 s^{-1}.

Another feature of XDH is its poor reactivity toward oxidation by molecular oxygen. This is of special interest in view of the fact that XO, an enzyme similar to XDH in many respects, is very reactive toward oxygen. By using modified flavins in XDH, we can address the possibility that oxygen reactivity is controlled simply by the potential of the flavin. Oxidation of 6-electron reduced 2-thio-FAD XDH by molecular oxygen is biphasic. The apparent rates of both phases show a second-order dependence on oxygen concentration (k_{fast} =5100 M^{-1} s^{-1}, k_{slow} = 710 M^{-1} s^{-1}). Both of these rates are faster than the corresponding rate in the oxidation of native enzyme (k_{fast} = 1900 M^{-1} s^{-1}, k_{slow} = 111 M^{-1} s^{-1}). (It should be kept in mind that oxidation of native XDH by oxygen exhibits a third, slow phase.) Oxidation of 6-electron reduced 4-thio-FAD XDH by molecular oxygen is slower. The reaction is biphasic. Again the rates of both phases were directly dependent on oxygen concentration (k_{fast}=170 $M^{-1}s^{-1}$, k_{slow}=17.6 $M^{-1}s^{-1}$). With both the 6-electron reduced 2-thio-FAD XDH and 4-thio-FAD XDH, the biphasic reoxidations with oxygen involve the formation of considerable amounts of semiquinone in the fast phase and its subsequent disappearance in the slow phase. No simple correlation exists between the potential of the flavin (see Table 1) and the reactivity of the reduced enzyme toward oxygen.

Acknowledgement

This work was supported by a grant from the National Science Foundation DMB-8500291 (V.M.) and by a Grant-in-Aid for scientific research from the Japanese Ministry of Education, Science and Culture.

References

1. Nishino, T. 1974. Biochim. Biophys. Acta 341, 93-98.
2. Krauth-Siegel, R.L., R.H. Schirmer, and S. Ghisla. 1985. Eur. J. Biochem. 148, 335-344.
3. Moore, E.G., E. Cardemil, and V. Massey. 1978. J. Biol. Chem. 253, 6413-6422.
4. Claiborne, A., V. Massey, P.F. Fitzpatrick, and L.M. Schopfer. 1982. J. Biol. Chem. 257, 174-182.
5. Massey, V., A. Claiborne, M. Biemann, and S. Ghisla. 1984. J. Biol. Chem. 259, 9667-9678.
6. Nishino, T., K. Tsushima, L.M. Schopfer and V. Massey. in this edition.
7. Hille, R., J.A. Fee, and V. Massey. 1981. J. Biol. Chem. 256, 8933-8940.

THE MOLYBDENUM COFACTOR OF MILK XANTHINE OXIDASE

Kenneth M. Fish and Vincent Massey
Dept. of Biological Chemistry, University of Michigan,
Ann Arbor, Michigan 48109

Introduction

Xanthine Oxidase is a dimeric enzyme containing an FAD molecule, 2 Fe_2S_2 centers and a molybdenum center per monomeric unit. The environment of the molybdenum atom is not well understood; however, an organic cofactor has been proposed to be ligated to it. On denaturing a variety of molybdenum-containing enzymes including milk xanthine oxidase, pterin derivatives have been isolated and chemically characterized.[1] The structures below have been proposed.

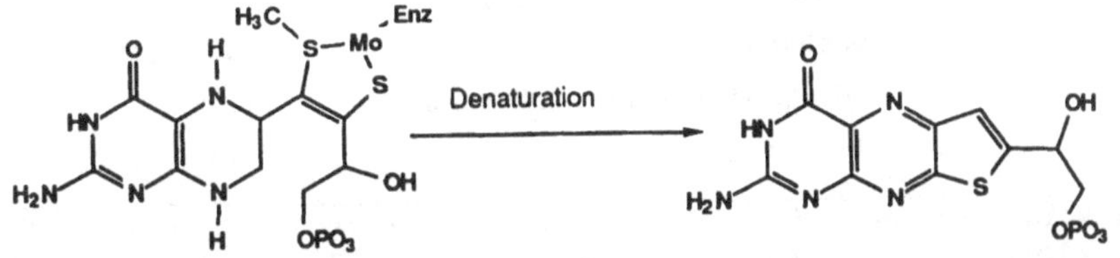

Recently, our work has been directed toward two important problems associated with the molybdenum cofactor. First, the isolation of cofactor derivatives from molybdenum-containing enzymes resulted in less than stoichiometric amounts of products.[1] Therefore, the amount of cofactor per enzyme subunit was uncertain. Since the molecule was proposed to contain a phosphate monoester, it should be possible to quantitate the cofactor by comparison of the phosphorus content relative to that of FAD. Second, attempts at the structural elucidation of the cofactor with pterin characteristics involved indirect chemical modification.[1] We considered it desirable to isolate the cofactor itself, or a nearly identical derivative. In the proposed structure above, the mercaptan functional group is probably responsible for the oxygen sensitivity of the cofactor. Thus with only slight modification, the R-SH group could be protected and more direct structural information about the cofactor obtained.

Flavins and Flavoproteins 1987

Results and Discussion

Quantitation and structural elucidation of the molybdenum cofactor required that it be released in high yield by enzyme denaturation. Therefore a study of was made of the results of using different denaturing techniques. The relative fluorescence intensities (λ_{ex} 365 nm, λ_{em} 455 nm) recorded from various denaturation methods are given in Table 1.

Table 1

Method	Rel. Yield[a]
HCl[b]	0.51
Heat[c]	1.00
Methanol	0.02
6M Guanidine hydrochloride	1.10
2% SDS	1.11

[a]All solutions were adjusted to pH 8.5 before measuring fluorescence. [b]HCl added to reach pH 2.0. [c]Enzyme heated in a boiling water bath for 5 min.

Because of the high yield of fluorescent product, and removal of the denatured protein as a precipitate, heat treatment proved to be most convenient for quantitation experiments. Dissolving the precipitate in 5% SDS revealed that only 4-6% of the total FAD was adsorbed on or trapped in the solid. The molybdenum cofactor could not be detected in the precipitate.

Xanthine oxidase (2 μMol) was denaturated in 0.1 M Tris HCl pH 8.0. The supernatant was collected and a ^{31}P NMR spectrum recorded (Figure 1). Resonances at 3.09 and -10.5 ppm integrate to a 1.2:2 ratio respectively. Comparison with standard spectra under identical conditions shows that the doublet at high field is due to the phosphodiester linkage of FAD. Chromatographic behavior of the denatured enzyme supernatant, coupled with ^{31}P NMR analysis is consistent with these resonances arising from molybdenum cofactor and FAD only. These results clearly establish that the phosphorus-containing molybdenum cofactor is released from xanthine oxidase in amounts which are stoichiometric with FAD.

The structure of the molybdenum cofactor was probed by protecting the reactive mercaptan substituent. Xanthine oxidase (2 μMol) was exposed to iodoacetamide and then reduced with sodium dithionite prior to anaerobic denaturation with 2% SDS in 0.1M sodium pyrophosphate pH 8.5. The solution was nonfluorescent until it was made aerobic. The cofactor derivative was purified by ultrafiltration, gel filtration, reverse phase chromatography, and finally anion exchange chromatography. The ^{31}P NMR spectrum of the purified material is shown in Figure 1.

The product of these manipulations resulted in isolation of a material exhibiting absorption maxima at 290 and 367 nm and with a strong fluorescence; λ_{ex} 367 nm, λ_{em} 455 nm. The ^{1}H NMR spectrum is shown in Figure 2. The visible, fluorescence and NMR spectra are consistent with the structure initially proposed.[1] The aromatic resonance at 8.79 ppm is characteristic of a proton at the 7 position of a 6-substituted pterin. Due to non-first order coupling in the NMR spectrum, further analysis is required to definitively assign -C-H protons of the side chain (integrated to 5.16 protons). However, the 3-4 ppm region is consistent with a ***saturated*** four carbon chain in the proposed cofactor. Resonances at 3.12 ppm and 1.86 ppm are consistent with $R-S-CH_2C(O)NH_2$ and $R-SCH_3$ substituents respectively. Therefore we propose the following structure for the molybdenum cofactor and product of denaturation (under the above conditions).

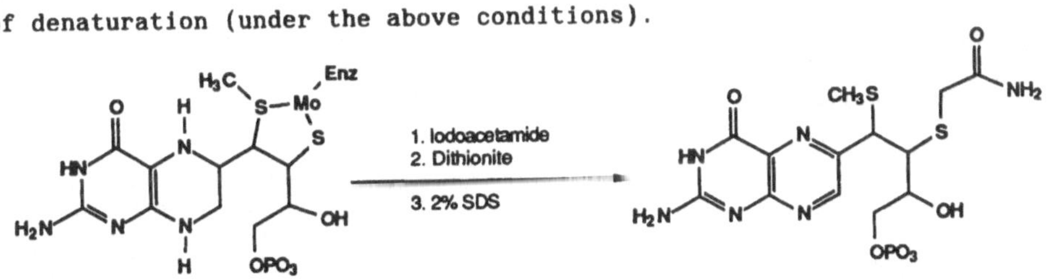

Acknowledgements

This research was supported by a grant from the National Science Foundation (DMB-8500291). The GE 500 MHz spectrometer was partially funded by NIH grant RR02415.

Reference

1. Hageman, R. Y.and Rajagopalan, K. V. 1986. Methods of Enzymology, 122, 399-412, and references therein.

FIG 1. QUANTITATION OF THE MOLYBDENUM COFACTOR OF MILK XANTHINE OXIDASE. ALL SAMPLES WERE DISSOLVED IN 0.1 M TRIS CHLORIDE PH 8.0 AND DATA RECORDED AT 25 DEG. C. AN EXPONENTIAL BROADENING OF 10 HZ WAS APPLIED PRIOR TO FOURIER TRANSFORMATIONS. A. PURIFIED MOLYBDENUM COFACTOR—ACETAMIDE ADDUCT; 10240 ACQUISITIONS. B. SUPERNATANT FROM CENTRIFUGED, HEAT DENATURED (5 MIN. BOILING WATER) XANTHINE OXIDASE; 5120 ACQUISITIONS. C. STANDARD FAD (0.5 MM); 2048 ACQUISITIONS.

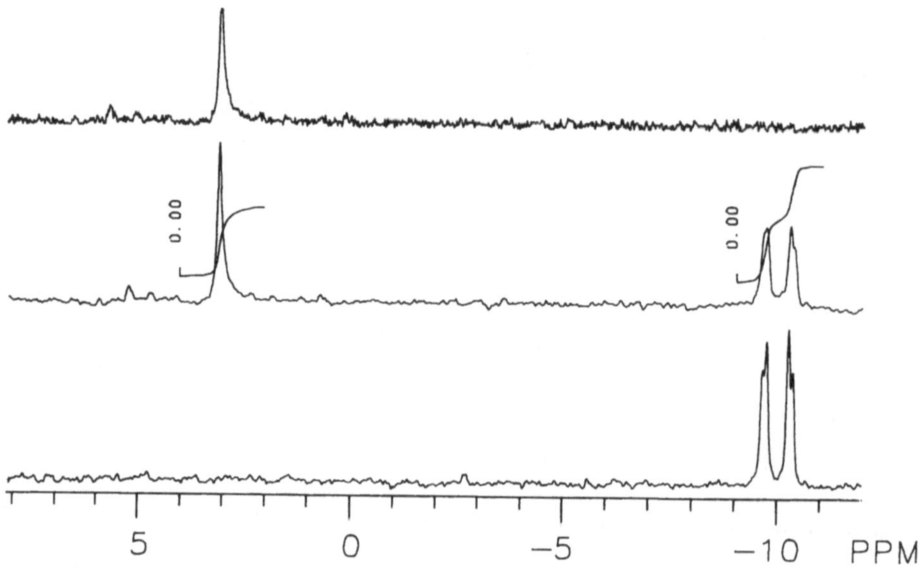

FIG 2. PROTON NMR OF THE MOLYBDENUM COFACTOR—ACETAMIDE ADDUCT. PURIFIED MATERIAL WAS DISSOLVED IN 0.1 M PHOSPHATE PH 7.6. AN EXPONENTIAL BROADENING OF 0.05 HZ WAS USED PRIOR TO FOURIER TRANSFORMATION (2500 ACQUISITIONS).

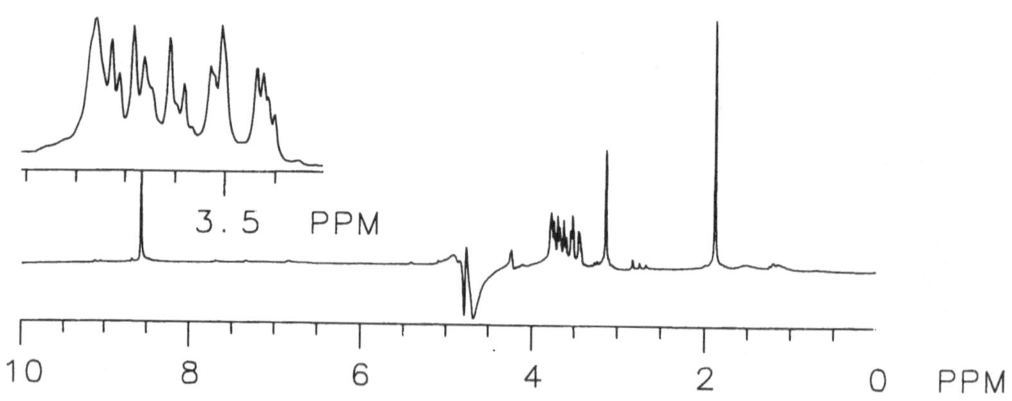

ISOLATION OF cDNA SEQUENCE CODING FOR A PART OF XANTHINE DEHYDROGENASE FROM RAT LIVER

K. Yamazaki, T. Nishino and K. Tsushima

Department of Biochemistry, Yokohama City University School of Medicine, Fukuura 3-9, Kanazawa-ku, Yokohama 236, Japan

Introduction

Xanthine dehydrogenase (EC 1.2.1.37) is composed of two apparently identical subunits (Mr=150,000), each of which contains one atom of molybdenum, one molecule of flavin, and two separate Fe_2S_2 clusters as redox centers (1-6). To understand the molecular mechanisms of the electron transfer between these redox centers, it is essential to elucidate the primary structure of the constituent protein. Using recombinant DNA techniques, we have isolated the cDNA sequences coding for the peptides reactive with antiserum to purified rat liver xanthine dehydrogenase (XDH).

Experimental Procedures

An expression library of cDNA clones constructed with poly(A)-containing RNA from rat liver is commercially available (Clontech Laboratories, Inc.). The library was screened immunologically according to the previous paper(7). Determination of the size of mRNA for XDH was performed as described by Thomas (8).

Flavins and Flavoproteins 1987

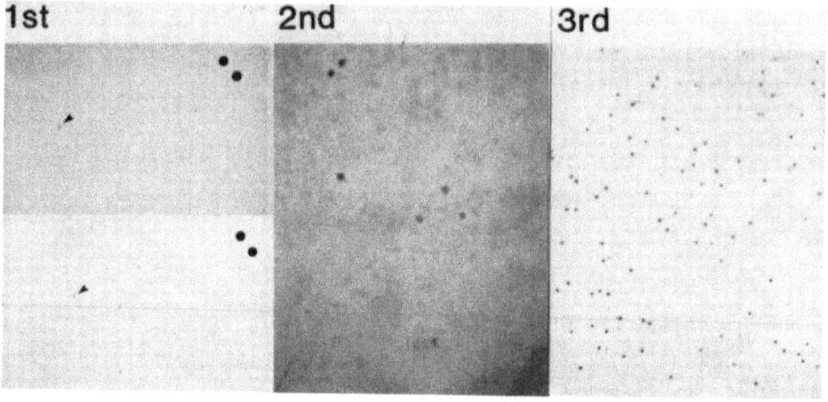

Fig. 1. Screening of λgtRXD4 by antibody to purified rat liver XDH. 1st: First screening, 5×10^4 pfu (of bacteriophages)/plate. 2nd: Second screening, 2×10^3 pfu/plate. 3rd: Third screening, 3×10^2 pfu/plate.

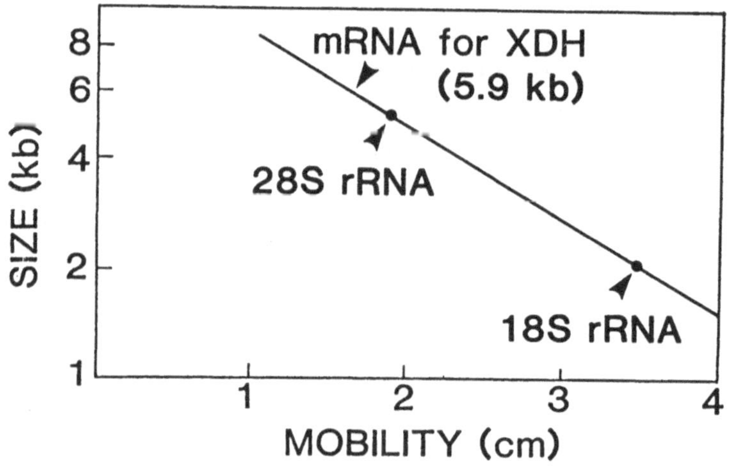

Fig. 2. Determination of the size of mRNA for XDH from rat liver. Total rat liver RNA (18 µg) was separated on a formamide agarose gel electrophoretically, transferred onto a nylon membrane, and hybridized with ^{35}S-labeled EcoRI DNA fragment (750 bp) of cDNA insert of clone λgtRXD4. The size markers used were rat liver rRNA (18S and 28S).

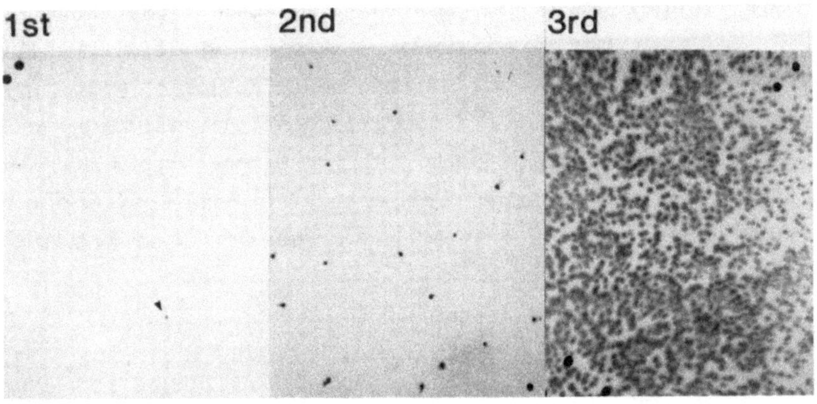

Fig. 3. Screening of λgtRXD13 by EcoRI DNA fragment (750 bp) of cDNA insert as described in Fig. 2. 1st: First screening, $1x10^5$ pfu/plate. 2nd: Second screening, $2x10^3$ pfu/plate. 3rd: Third screening, $2x10^3$ pfu/plate.

Results and Discussion

Screening of the cDNA library with rabbit antiserum to purified rat liver XDH yielded six immunologically positive clones from $2x10^5$ pfu of recombinant bacteriophages. Phage DNAs of two (λgtRXD4 and λgtRXD7) of the six clones were isolated (Fig. 1), and the cDNA insert (1 kbp) of each clone was subjected to nucleotide sequence analysis. Two clones produced their specific antigens as fusion polypeptides which were recognized by specific antibody probe (to rat XDH). The antigen produced by E. coli cells containing recombinant bacteriophages, λgtRXD4 or λgtRXD7, was not recognized by rabbit antiserum to purified chicken liver XDH (data not shown). When total RNA from rat liver was subjected to northern blot hybridization analysis with the cDNA insert of clone λgtRXD4 as probe, an RNA species of 5,900 nucleotides was hybridizable (Fig. 2).

To clone longer cDNA sequences, we used the same cDNA library as described previously. Screening of this library by hybridization with the EcoRI DNA fragment (750 bp) of clone λgtRXD4 yielded 4 hybridization-positive clones from $2x10^5$ transformants (Fig. 3). Clone λgtRXD13, which apparently harboured the largest cDNA insert among the 4 clones, contained 3.7 kbp of cDNA insert (data not shown).

Acknowledgement

This work was supported in part by a Grant-in-Aid for scientific research from the Japanese Ministry of Education, Science and Culture.

References

1. Bray, R.C.. 1975. In: The Enzymes (P.D. Boyer, eds.) 12, p. 299. Academic Press, New York.

2. Massey, V., P.E. Brumby, H. Komai, G. Palmer. 1969. J.Biol.Chem. 244, 1682.

3. Waud, W.R., F.O. Brady, R.D. Wiley, K.V. Rajagopalan. 1975. Arch.Biochem.Biophys. 169, 695.

4. Rajagopalan, K.V., P. Handler. 1967. J.Biol.Chem. 242, 4097.

5. Nishino, T. 1974. Biochim.Biophys.Acta. 341, 93.

6. Cleere, W.F., M.P. Coughlan. 1975. Comp.Biochem.Physiol. 50B, 311.

7. Young, A.R., W.R. Davis. 1983. Proc.Natl.Acad.Sci.USA. 80, 1194.

8. Thomas, P.S.. 1980. Proc.Natl.Acad.Sci.USA 77, 5201.

Flavin Biosynthesis, Metabolism, and Medical Aspects

BIOSYNTHESIS OF FLAVINS AND DEAZAFLAVINS

Adelbert Bacher[a], Rainer Volk[a], Paul J. Keller[b], Heinz G. Floss[b], Quang Le Van[a], Wolfgang Eisenreich[a], and Bruno Schwarzkopf[a]

[a]Department of Chemistry, Technical University of Munich, D-8046 Garching, Federal Republic of Germany, and [b]Department of Chemistry, The Ohio State University, 43210 Columbus, Ohio, USA

The biosynthesis of riboflavin (**7**) is summarized in Fig. 1 (for review see Refs. 1-3). Starting from GTP (**1**), the release of formate and pyrophosphate yields the pyrimidine intermediate **2**. Reduction of the ribose side chain and deamination of the pyrimidine ring yields **5** via **3** (in yeasts) or via **4** (in bacteria).

The conversion of **5** to 6,7-dimethyl-8-ribityllumazine (**6**) requires the addition of a 4C-moiety which has been the object of numerous investigations. A variety of precursors such as acetoin, diacetyl, and various carbohydrates had been advocated as the source of the 4C-moiety. We have studied this problem by *in vivo* isotope incorporation with *Ashbya gossypii* using a variety of singly and multiply ^{13}C-labeled precursors such as acetate, succinate, glucose, ribose, and glycerol (3-5). Rather surprisingly, these studies suggested the formation of the 4C-moiety from a pentose precursor by intramolecular elimination of C-4. In agreement with these data, Shavlovsky and his coworkers have reported on the formation of riboflavin in vitro from GTP and ribose phosphate using crude cell extract of the flavinogenic yeast, *Candida guilliermondii* (6-8). We could subsequently show that this *in vitro* system catalyzes the transfer of C1 - C3 and C-5, but not C-4 of the pentose phosphate into riboflavin, in agreement with our earlier *in*

Flavins and Flavoproteins 1987

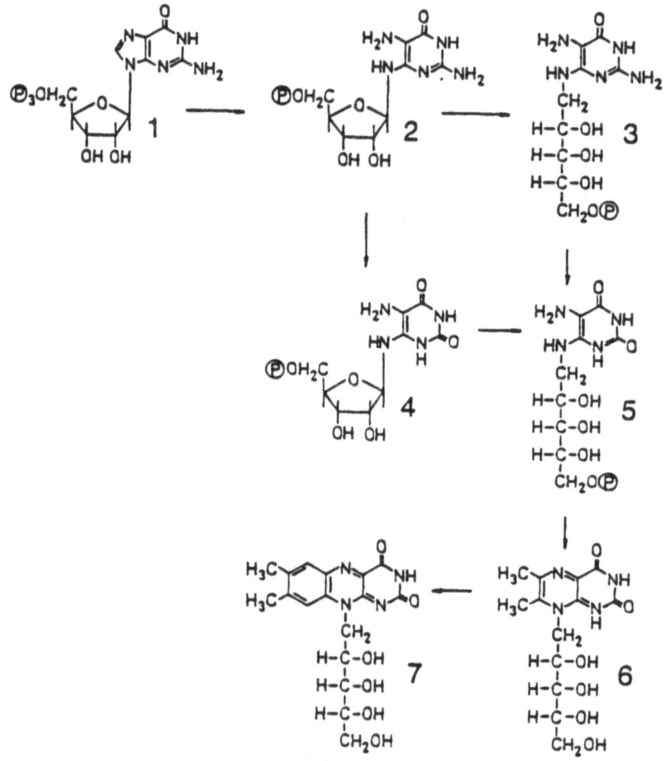

Fig. 1. Biosynthesis of riboflavin

vivo studies (9). We could also show the transfer of hydrogen atoms from positions 1 and 5 of the carbohydrate phosphate to the methyl groups of **6**.

On this basis we started the search for a biosynthetic intermediate between a pentose/pentulose phosphate and the lumazine **6**. We found that mutants of *Candida guilliermondii* deficient in the formation of the lumazine could be divided into two complementation groups by in vitro complementation experiments. Using these mutants, we were able to demonstrate the formation of a carbohydrate phosphate, compound X, which is the committed 4C-precursor for the formation of **6** (10). The

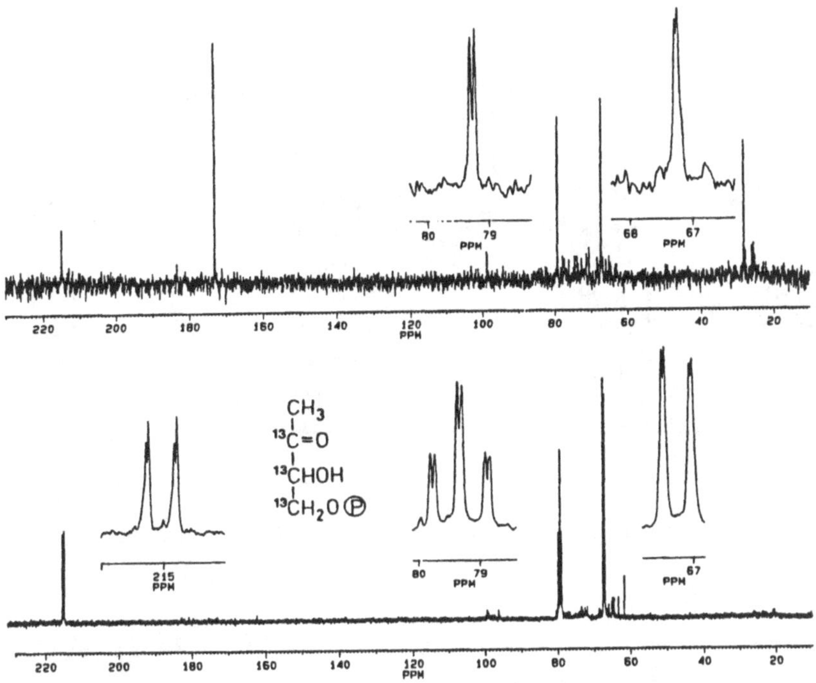

Fig. 2. ^{13}C-NMR spectra of compound X. Top, from natural abundance ribulose phosphate; bottom, from [2,3,5-$^{13}C_3$]ribose phosphate via [2,3,5-$^{13}C_3$]ribulose phosphate.

formation of compound X from the pentose phosphate required Mg^{++} but no other cofactors. The enzymatic condensation of compound X with the pyrimidine intermediate could proceed in the presence of EDTA and required no cofactors.

The enzyme catalyzing the formation of compound X was subsequently purified about 1000-fold by chromatographic procedures. Whereas the crude enzyme extracts could utilize a variety of pentose and pentulose phosphates as substrates for the formation of compound X, the highly purified enzyme specifically required ribulose 5-phosphate as substrate.

Attempts to purify compound X were frustrated by its low stability. However, using the purified enzyme, we could follow

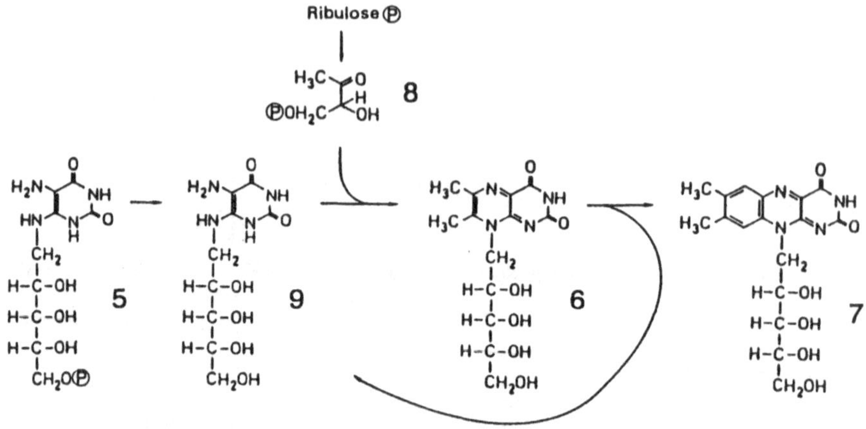

Fig. 3. Late reactions in the pathway of riboflavin biosynthesis

the formation of compound X by ^{13}C and ^{1}H NMR spectroscopy. Incubation of ribulose 5-phosphate with the purified compound X synthase afforded the ^{13}C-NMR spectrum shown in Fig. 2 (top). The signal at 173 ppm was easily identified as formate (^{13}C-NMR and ^{1}H-NMR using internal standardization with formate). The ^{13}C-NMR signals at 28.1, 67.2, 79.3 and 214.9 ppm were tentatively assigned the structure of a 1-deoxytetrulose 4-phosphate (**8**, Fig. 3) on the basis of proton multiplicities as deduced from a DEPT experiment and of ^{31}P coupling of the signals at 67.2 and 79.3 ppm.

Treatment of compound X with phosphatase shifted the ^{13}C-NMR signals to 28.4, 65.1, 80.4 and 215.1 ppm, respectively, with loss of the ^{31}P coupling. These signals could be conclusively assigned to 1-deoxytetrulose by internal standardisation with an authentic sample (^{13}C- and ^{1}H-NMR measurements).

Further confirmation of the proposed structure was obtained by enzymatic formation of compound X from [2,3,5-$^{13}C_3$]ribulose 5-phosphate (generated enzymatically from [2,3,5-$^{13}C_3$]-ribose phosphate by the action of pentose phosphate isomera-

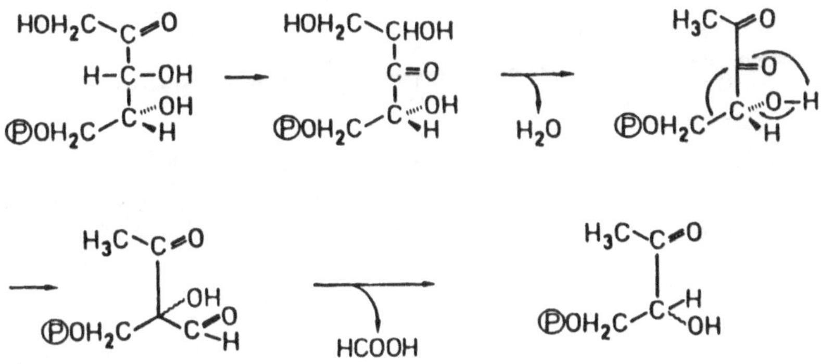

Fig. 4. Hypothetical mechanism for the formation of compound X

se). In this experiment, the signal of C-3 of compound X appears as a triplet originating from the coupling with two neighbouring ^{13}C atoms (C-2 and C-4), thus confirming the formation of compound X by elimination of the unlabeled C-4 from ribulose phosphate (Fig. 2, bottom). A hypothetical mechanism for this unusual reaction is suggested in Fig. 4.

The condensation of compound X with 5-amino-6-ribitylamino-2,4(1H,3H)-pyrimidinedione (**9**, Fig. 3) is catalyzed by the β subunits of the lumazine synthase/riboflavin synthase complex from *Bacillus subtilis* (11). The structure of this unusual enzyme is discussed in an accompanying paper (12). Briefly, 60 β subunits form a capsid with icosahedral 532 symmetry. The core space of this capsid is occupied by 3 α subunits (12, 13). The formation of **6** can be catalyzed by the native enzyme and by isolated β subunits. The ^{13}C-NMR spectrum of **6** generated enzymatically from [2,3,5-$^{13}C_3$]ribose phosphate (via ribulose phosphate and compound X) showed 3 contiguous ^{13}C atoms at positions 6, 7 and 7α in full agreement with our earlier in vivo studies.

5-Amino-6-ribitylamino-2,4(1H,3H)pyrimidinone 5'-phosphate (**5**) cannot serve directly as substrate for the condensation

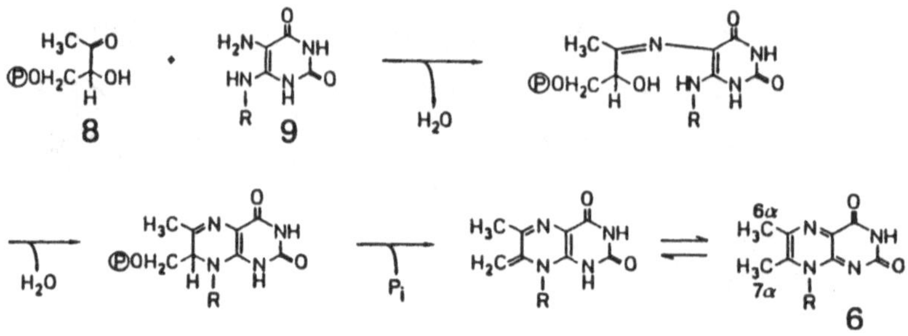

Fig. 5. Hypothetical mechanism for the formation of 6,7-dimethyl-8-ribityllumazine

with compound X (11). It follows that this biosynthetic intermediate must first be dephosphorylated. It is not known whether this reaction requires a specific enzyme or can be carried out by a general phosphatase.

A hypothetical mechanism for the formation of **6** is suggested in Fig. 5. The carbonyl group of the deoxytetrulose phosphate **8** may form a Schiff base with the more reactive 5 amino group of the pyrimidine substrate **9**. Ring closure can then occur by elimination of water. The elimination of phosphate yields the exomethylene structure **6a** which is a tautomeric form of **6**.

Methanogenic bacteria utilize FAD and also the more recently discovered deazaflavin coenzyme F_{420}. The chromophoric moiety of coenzyme F_{420} is 7,8-didemethyl-8-hydroxy-5-deazaflavin (**10**, also designated as factor F_0). We have studied the formation of both riboflavin and factor F_0 by *in vivo* isotope incorporation studies using cultures of *Methanobacterium thermoautotrophicum*. Such studies are hampered by the poor utilization of complex exogeneous metabolites in this microorganisms; however, acetate is utilized reasonably well at pH 6.

The results of 3 separate experiments with [1-^{13}C]acetate, [2-^{13}C]acetate and [1,2-$^{13}C_2$]acetate are summarized in Figs. 6 and 7. The interpretation of the labeling patterns of

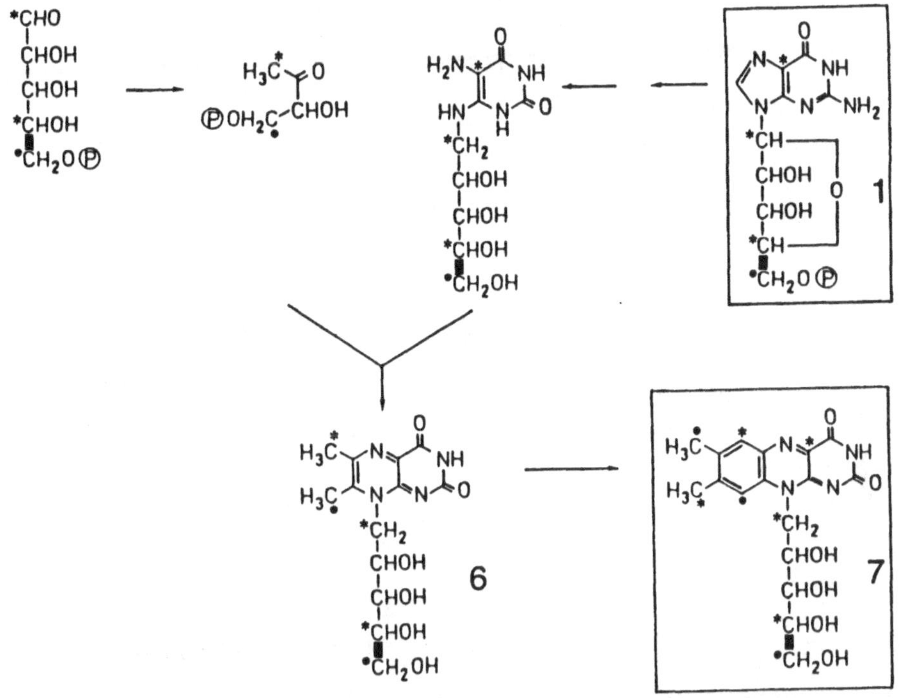

Fig. 6. Biosynthesis of riboflavin in *Methanobacterium thermoautotrophicum*. *, atoms labeled with ^{13}C from [1-^{13}C]acetate; •, atoms labeled with ^{13}C from [2-^{13}C]acetate; bars, transfer of contiguous pairs of ^{13}C atoms from [1,2-$^{13}C_2$]acetate. Compounds in boxes were isolated and measured. The other labeling patterns were infered indirectly.

riboflavin and factor F_0 is greatly facilitated by comparison with other metabolites obtained from the same fermentation. Thus, it is apparent that the ribityl groups of both riboflavin and factor F_0 show the same labeling pattern as the ribose moiety of guanosine. Another salient feature is the correspondence in the labeling pattern of the pyrimidine ring in all 3 metabolites. It follows that the pyrimidine ring and the ribityl side chain have the same origin in riboflavin and

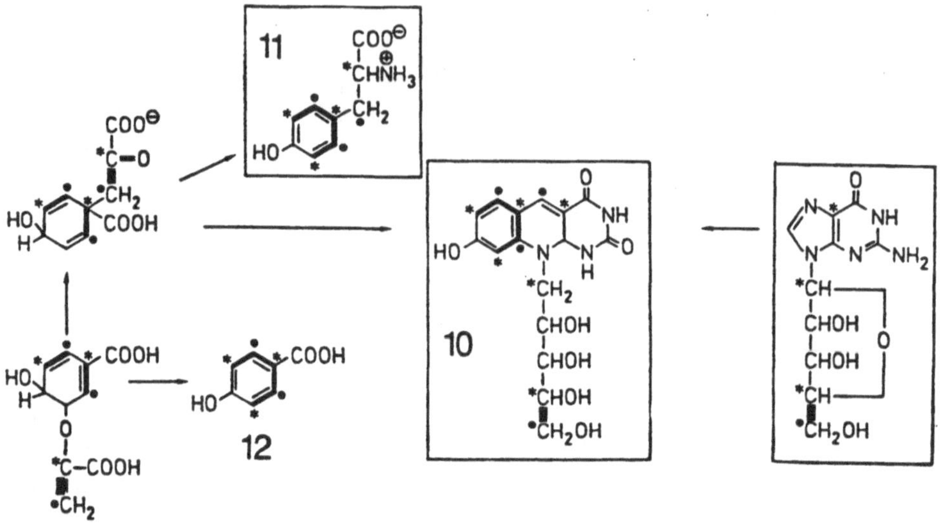

Fig. 7. Biosynthesis of factor F_0 (**10**) in *Methanobacterium thermoautotrophicum*. For details see Fig. 6.

ribityl side chain have the same origin in riboflavin and factor F_0. The labeling pattern of the xylene ring of riboflavin is fully in line with the pathway described above, i.e. loss of C-4 from ribulose phosphate.

The labeling pattern of the phenolic ring of factor F_0 (**10**) is entirely different from the benzenoid ring of riboflavin. Most conspicuous is the observation of ^{13}C-^{13}C-coupling involving all benzene ring atoms except C-8. This situation reflects accurately the labeling pattern of the aromatic ring in tyrosine (**11**) and suggests the formation of the phenolic ring from a shikimate derivative. 4-Hydroxybenzoic acid (**12**) can be ruled out as the specific precursor since C-5 of F_0 should be unlabeled in this case. On the other hand, tyrosine has been ruled out as precursor by Schönheit et al. (15). This suggests that the phenolic ring is derived from an shikimate pathway intermediate prior to tyrosine.

Acknowledgement

This work was supported by grants from the Deutsche Forschungsgemeinschaft and the Fonds der Chemischen Industrie to A. B. and by NSF to H. G. F. and P. J. K.

References

1. Plaut, G.W.E. 1971. In: Comprehensive Biochemistry Vol. 21 (M. Florkin, and E.H. Stotz, eds.). Elsevier Scientific Publishing Co., Amsterdam. pp. 11-45.
2. Brown, G.M., M. Williamson. 1982. Adv. Enzymol. 53, 345-381.
3. Bacher, A., Q. Le Van, P.J. Keller, H.G. Floss. 1983. J. Biol. Chem. 258, 13431-13437.
4. Bacher, A., Q. Le Van, P.J. Keller, H.G. Floss. 1985. Amer. Chem. Soc. 107, 6380.
5. Le Van, Q. P.J. Keller, D.H. Bown, H.G. Floss, A. Bacher. 1985. J. Bacteriol., 1280-1284.
6. Logvinenko, E.M., G.M. Shavlovsky, A.E. Zakal'sky, I.V. Zakhodylo. 1982. Biokhimiya 47, 931-936.
7. Logvinenko, E.M., G.M. Shavlovsky, N.Y. Tsarenko. 1984. Biokhimia 49, 43-50.
8. Logvinenko, E.M., G.M. Shavlovsky, N.Y. Tsarenko. 1985. Biokhimia 50, 744-748.
9. Nielsen, P., G. Neuberger, I. Fujii, D.H. Bown, P.J. Keller, H.G. Floss, A. Bacher. 1986. J. Biol. Chem. 261, 3661-3669.
10. Neuberger, G., A. Bacher. 1985. Biochem. Biophys. Res. Comm. 127, 175-181.
11. Neuberger, G., A. Bacher. 1986. Biochem. Biophys. Res. Comm. 139, 1111-1116.
12. Ladenstein, R., K. Schott, M. Schneider, A. König, A. Bacher, this volume

13. Bacher, A., H. C. Ludwig, H. Schnepple, Y. Ben-Shaul. 1986. J. Mol. Biol. 187, 75-86
14. Ladenstein, R., B. Meyer, R. Huber, H. Labischinski, K. Bartels, H.-D. Bartunik, L. Bachmann, H. C. Ludwig, A. Bacher. 1986. J. Mol. Biol. 187, 87-100
15. Jähnchen, R., P. Schönheit, R.K. Thauer. 1985. Arch. Microbiol. 137, 362

THE ICOSAHEDRAL β_{60} CAPSID OF HEAVY RIBOFLAVIN SYNTHASE (6,7-DIMETHYL-8-RIBITYLLUMAZINE SYNTHASE/RIBOFLAVIN SYNTHASE)

Rudolf Ladenstein[a], Robert Huber[a], Karin Schott[b], Monika Schneider[a], Astrid König[b], and Adelbert Bacher[b]

[a]Max-Planck-Institut für Biochemie, D-8033 Martinsried, Federal Republic of Germany and [b]Department of Chemistry, Technical University of Munich, D-8046 Garching, Federal Republic of Germany

Heavy riboflavin synthase of *Bacillus subtilis* is a bifunctional enzyme complex catalyzing the formation of riboflavin from 5-amino-6-ribitylamino-2,4(1H,3H)-pyrimidinedione and a deoxytetrulosephosphate via 6,7-dimethyl-8-ribityllumazine (1, 2). The 1.000.000 Dalton protein consists of 60 ß subunits which form an icosahedral capsid containing 3 α subunits in the central core (3-5).

Crystals of space group $P6_322$ grown from 1.3 M phosphate pH 8.7 in the presence of the substrate analogue, 5-nitroso-6-ribitylamino-2,4(1H,3H)-pyrimidinedione yield X-ray reflections extending to 3.2 Å resolution (6). Intensity data of native crystals and 3 isomorphous heavy atom derivatives were collected using synchrotron radiation (EMBL outstation Hamburg). Single isomorphous replacement phases were calculated from the heavy atom positions of the $K[Au(CN)_2]$ derivative. This phase set was used to calculate an electron density map at 5 Å resolution which was subsequently improved by 10-fold

Flavins and Flavoproteins 1987

icosahedral averaging. The phases were extended by cyclic averaging from 5 to 3.3 Å resolution.

The electron density map was interpreted using the chemically deduced sequence (154 residues) of the ß subunit (7). The main peptide chain of the ß subunit forms into a 4-stranded parallel ß sheet structure flanked on both sides by 2 α helical segments (Fig. 1A, B). This folding patterns resembles the folding of flavodoxins, although no sequence homology was observed. The overall folding pattern shows no similarity with

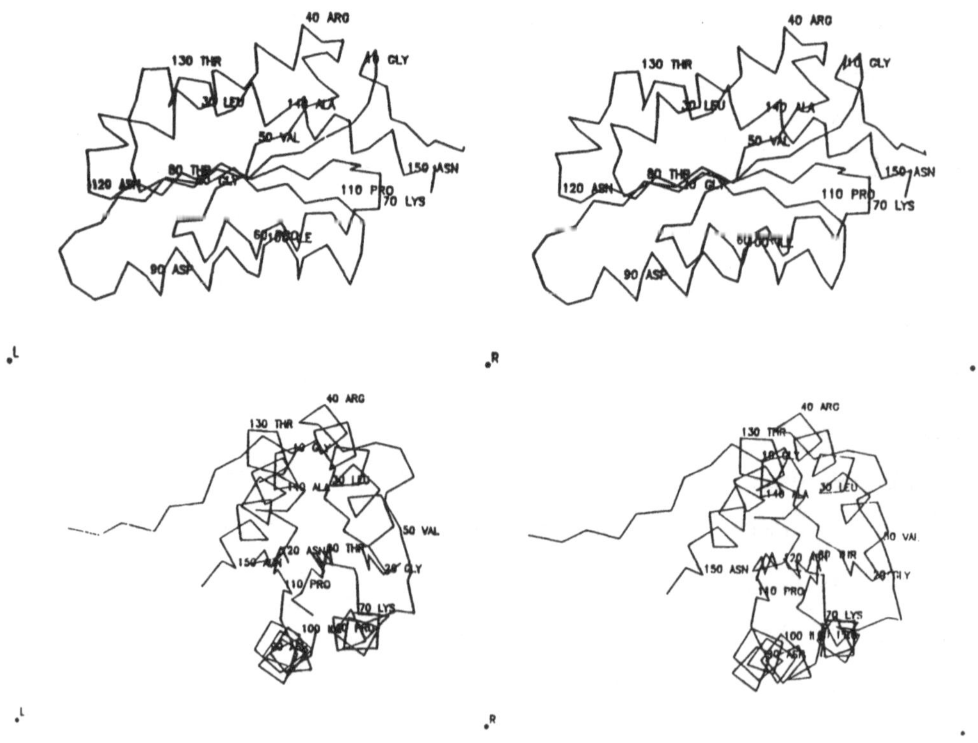

Fig. 1: C_α model of heavy riboflavin synthase of ß subunit. Top, projection tangential to capsid surface; bottom, projection perpendicular to capsid surface as viewed from outside.

the recently described structures of several icosahedral plant and animal viruses (8).

In the ß subunit pentamer, the N-terminal segment of each monomer is attached to the neighbouring monomer and comprises the fifth strand of a parallel sheet (Fig. 2). Along the fivefold axis of each pentamer a channel is formed by the helix α_3 and its 4 symmetry-related neighbours. These helices are twisted into a large superhelical motive, a socalled α-helical coiled coil structure, thus forming a pore structure with a diameter of 9 Å and an approximate length of 30 Å. The inner wall of the channel is coated primarily by polar amino acid side chains providing for a hydrophilic surface. These pores may play a role in substrate and product exchange.

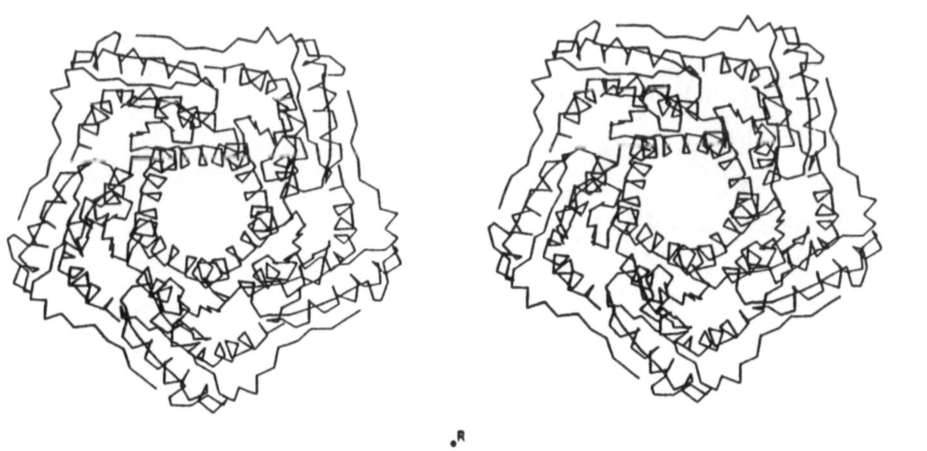

Fig. 2: C_α model of ß subunit pentamer viewed along an icosahedral fivefold axis.

The present electron density contains no information on the α subunits. However, it has been shown earlier that the α subunits are located in the center of icosahedral capsid (4).

Acknowledgement

This work was supported by the Deutsche Forschungsgemeinschaft and the Fonds der Chemischen Industrie.

References

1. Neuberger, G., and A. Bacher. 1986. Biochem. Biophys. Res. Commun. 139, 1111-1116

2. Schott, K., F. Lottspeich, and A. Bacher, this volume

3. Bacher, A., R. Baur, U. Eggers, H. Harders, M.K. Otto, and H. Schnepple. 1980. J. Biol. Chem. 255, 632-637

4. Bacher, A., H.C. Ludwig, H. Schnepple, and Y. Ben-Shaul. 1986. J. Mol. Biol. 187, 75-86

5. Ladenstein, R., B. Meyer, R. Huber, H. Labischinski, K. Bartels, H.-D. Bartunik, L. Bachmann, H.C. Ludwig, and A. Bacher. 1986. J. Mol. Biol. 187, 87-100

6. Ladenstein, R., H.C. Ludwig, and A. Bacher. 1983. J. Biol. Chem. 258, 11981-11983

7. Ludwig, H.C., F. Lottspeich, A. Henschen, R. Ladenstein, and A. Bacher. 1987. J. Biol. Chem. 262, 1016-1021

8. Liljas, L. 1986. Progr. Biophys. Mol. Biol. 48, 1-36

CHARACTERIZATION OF HUMAN RIBOFLAVIN-BINDING IMMUNOGLOBULINS

A. H. Merrill, W. S. A. Innis-Whitehouse and D. B. McCormick

Department of Biochemistry, Emory University School of Medicine, Atlanta, Georgia 30322 (USA)

Introduction

Immunoglobulins were first identified as a new class of riboflavin-binding proteins by Farhangi and Osserman (1) in studies of the yellow hair and skin of an elderly patient with multiple myeloma. Substantial amounts of monoclonal IgG were isolated from this patient's serum and the binding site for riboflavin was shown to reside in the Fab fragment.

Riboflavin-binding immunoglobulins were next detected in plasma from healthy individuals by affinity chromatography (2). Assessment of the relative contribution of serum proteins to riboflavin binding in circulation (3) established that immunoglobulins were the major species responsible for individual variations, and may contribute to the abnormal handling of this micronutrient by cancer patients (4).

These analyses suggested that at least a subfraction of the human riboflavin-binding immunoglobulins were IgG and IgA that bind riboflavin with moderate affinity. To further understand these components of normal human circulation, they have been purified using affinity chromatography (3).

Isolation of riboflavin-binding immunoglobulins

Human plasma from multiple donors was obtained from the Cell Separator Department of Emory University. The plasma was subjected to affinity chromatography on N3-, 8α-, and N10-flavinyl agaroses essentially as described previously (3,5). Each isolated fraction migrated on PAGE-electrophoresis as a broad band coincident with immunoglobulins; no other proteins were visible on native or SDS-PAGE gels.

Flavins and Flavoproteins 1987

Riboflavin binding was measured by equilibrium dialysis. The protein(s) isolated using the N3-agarose bound 185 pmol [^{14}C]riboflavin/mg protein, compared to 51 & 32 for the proteins isolated using 8α- and N10-agaroses, respectively.

Immunochemical characterization

The proteins isolated by each type of matrix were identified by immunoelectrophoresis and quantitated by nephelometry. Each contained multiple classes and both kappa and lambda chains (see McCormick et al., this volume). The immunoglobulins from the N3-flavinyl agarose were mainly IgG (78%), whereas the 8α-column yielded mostly IgM (74%), and a mixture of all three classes (IgG, 53%; IgA, 11%; & IgM, 36%) were in the N10-isolate. Subclass analyses (6) revealed that all three isolates had a similar distribution (Table 1).

Table 1. Distribution of immunoglobulins among subclasses

Isolate	IgG_1	IgG_2	IgG_3	IgG_4
		% of total IgG		
N3	21	77	1	0.1
8α	21	77	2	0.5
N10	29	66	4	1

Localization of the binding site

For localization of the flavin binding site, the purified riboflavin-binding IgG from the N3-flavinyl agarose were separated from other immunoglobulin subtypes by ion-exchange chromatography. The IgG were treated with papain (2 mg/100 mg of IgG) for 8h, then the fragments separated by ion-exchange chromatography (7). The three eluates were found by Ouchterlony immuno-double diffusion to be undigested IgG, the Fc fragment plus some undigested IgG, and the Fab fragment. The majority of the binding was recovered in the Fab eluate; and the small amount in the other fractions (ca. one third) was probably due to undigested IgG.

Flavin binding

Double reciprocal plots of the [^{14}C]riboflavin bound by the isolate from the N3-agarose beads revealed a single line (r=0.99) corresponding to an apparent Kd of 0.23 uM. The other isolates yielded more complex curves, with evidence for some of the proteins having Kd less than 1 uM.

Table 2. Competition for [^{14}C]riboflavin binding*

Isolate:	N3	8α	N10
Flavin:	Inhibition of [^{14}C]riboflavin binding (%)		
3-methyl riboflavin	10	14	0
Lumichrome	44	37	22
Lumiflavin	34	30	43
FMN	39	49	64
FAD	16	39	38
8α-Cysteinyl riboflavin	11	40	9
8α-Histidyl riboflavin	6	23	30
8-hydroxy riboflavin	37	100	77

*Measured with [^{14}C]riboflavin and flavin analogs at 0.2 uM.

The flavin binding specificity differed among the three isolates; however, the derivatives and analogs generally were not able to compete effectively with riboflavin at equal concentrations, with the exception of 8- hydroxyriboflavin, and in some instances FMN, FAD, lumiflavin & lumichrome (Table 2).

Summary

These findings confirm that serum from normal individuals contains immunoglobulins that bind riboflavin. The immunoglobulins encompass multiple classes and exhibit heterogeneous binding, as often exhibited by the antibodies produced against an antigenic substance. Since binding could be located in the Fab fragment; it is possible that these riboflavin-binding immunoglobulins represent an (apparently mild) autoimmune response to riboflavin or a flavin derivative.

Acknowledgements

We thank Drs. C. Papadea and I. Check for analyzing the immunoglobulin subclasses. Supported by NIH grant AM26746.

REFERENCES

1. Farhangi, M., and Osserman, E. F. 1976. New Engl. J. Med. 294, 177.

2. Merrill, A. H., Jr., Froehlich, J. A., and McCormick, D. B. 1981. Biochem. Med. 25, 198.

3. Innis, W. S. A., McCormick, D. B., and Merrill, A. H., Jr. 1985. Biochem. Med. 34, 151.

4. Innis, W. S. A., Nixon, D. W., Murray, D. R., McCormick, D. B., and Merrill, A. H., Jr. 1986. Proc. Soc. Exp. Biol. Med. 181, 237.

5. Merrill, A. H., and McCormick, D. B. 1978. Anal. Biochem. 89, 87.

6. Papadea, C., Check, I. J., and Reimer, C. B. 1985. Clin. Chem. 31, 1940.

7. Porter, R. R. 1959. Biochem. J. 73, 119.

RIBOFLAVIN UPTAKE BY ISOLATED RAT KIDNEY CELLS

Delores M. Bowers-Komro and Donald B. McCormick
Department of Biochemistry, Emory University, Atlanta, GA 30322

INTRODUCTION

In several mammalian species there is evidence which suggests that riboflavin is transported bidirectionally through the renal tubules (1). At low plasma riboflavin concentrations, reabsorption must occur to conserve the vitamin. At high plasma riboflavin concentrations, clearance of riboflavin is higher than glomerular filtration, indicating a secretory process. In the rat, Christensen (2) has shown that this secretory process is active at plasma riboflavin concentrations from 1 to 50 μM and is probenecid sensitive. Spector (3) has also shown that accumulation of riboflavin in rabbit kidney slices is an energy-dependent saturable process. We now report the characterization of the accumulation of riboflavin into isolated renal proximal tubular epithelial cells of rat.

METHODS AND MATERIALS

Renal proximal cells were isolated from male, Sprague-Dawley rats weighing 200-300 g by the recirculating collagenase perfusion technique described by Jones *et al.* (4). An aliquot of the cells was counted and the viability determined by exclusion of 0.2% trypan blue. Experiments were performed within two hours after isolation (viability ≥ 87%) using dim light conditions. To characterize cellular uptake of riboflavin, 1 ml of modified Krebs-Henseleit buffer containing 10^6 cells was incubated in open plastic vials secured in a shaking water bath at 37 °C. The cells were preincubated for 5 min before the reaction was initiated by the addition of a

Flavins and Flavoproteins 1987

predetermined amount of purified [2-^{14}C]riboflavin (30 µCi/ mol). The uptake was stopped by rapidly removing 0.8 ml of the mixture, filtering under mild suction through 8 µm membrane filters and washing twice with cold Krebs-Henseleit buffer. Filters and cells were transferred to scintillation vials and radioactivity was determined. All samples were corrected for nonspecific adsorption of [^{14}C]riboflavin and quenching. To determine the purity and the fate of riboflavin, chromatography was performed using published HPLC methods (5).

RESULTS AND DISCUSSION

The uptake of [^{14}C]riboflavin by renal proximal cells is biphasic in nature as shown in Fig. 1. The process proceeds

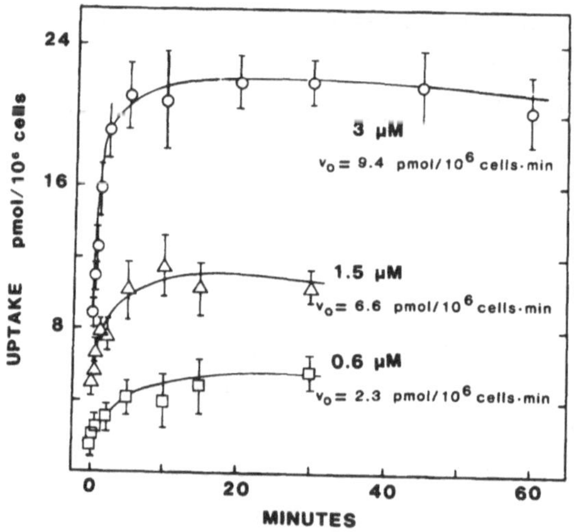

Fig. 1 Time course of riboflavin uptake by isolated renal proximal epithelial cells. Incubations in modified Krebs-Henseleit buffer with 12.5 mM Hepes (pH 7.4), 12 mM glucose and 7 mM glutamate (10^6 cells/ml) were conducted at 37°C at external riboflavin concentrations of 3.0, 1.5 and 0.6 µM. Values given are the mean ± SEM of 15, 3 and 4 cell preparations respectively. V_o values were calculated from least squares fit of data collected during the first half minute.

through a rapid initial phase lasting about 2 min with very little additional uptake after 5 min. The accumulation of riboflavin at equilibrium is about 2.5-fold greater than external concentration which is similar to that observed for the isolated rat liver cells but less than that reported for enterocytes of guinea pig (see Table III, Ref. 6). The 0.5-min rates of uptake measured at various concentrations, Fig. 2, indicate that above 15 μM diffusion predominates.

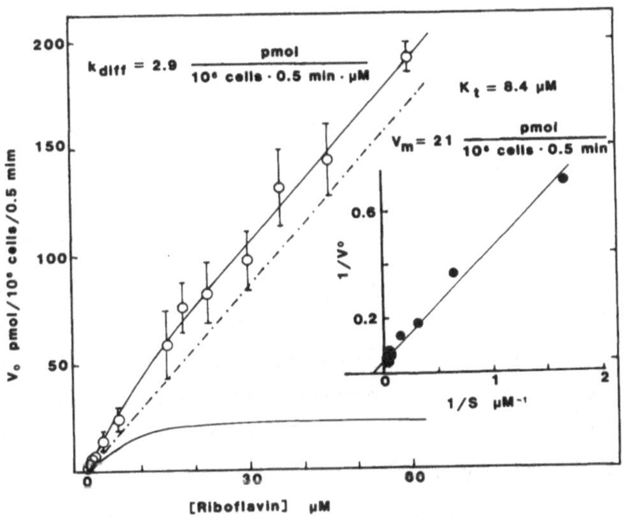

Fig. 2. Effect of concentration on the 0.5-min uptake of riboflavin measured from 0.3 to 60 μM at 37°C. Results are mean ± SEM of 6 separate cell preparations. At high riboflavin concentrations (> 15 μM) diffusion predominates (k_{diff}). Subtracting this linear component (dashed line) from the overall curve (O) gives a saturable component (solid line). Insert is a Lineweaver-Burk plot of corrected uptake data.

Subtracting this component from the overall curve reveals a saturable component with very high affinity. At low concentrations of riboflavin typical of normal physiological states, this saturable component may reflect the role of flavokinase in trapping the vitamin by phosphorylation, since the measured K_t is similar in magnitude to the K_m of flavokinase for riboflavin (7). HPLC separation and radioactivity determination on cellular extracts from cells incubated with 3 and 12 μM riboflavin for 2, 15 and 30 min indicate that riboflavin is internalized and metabolically converted first to FMN then to FAD. This is consistent with data presented by Spector (3) on the accumulation of riboflavin into rabbit kidney slices. Variations of incubation conditions (Table III, Ref. 6) indicate that the accumulation of riboflavin by the cells is sensitive to temperature and is sodium ion-dependent. Preincubation for 20 min with 1 mM ouabain did not lower initial uptake or ultimate accumulation, thus riboflavin uptake does not appear to be driven by the sodium-ion gradient associated with ATP-hydrolysis but may be cotransported with sodium ion in some manner. A 20-min preincubation with 10 mM ethionine showed some lowering of the initial rate but no noticeable effect on accumultion. Addition of 1 μM FCCP lowered both initial velocity and accumulation. These findings demonstrate that there is an ATP requirement but that it may not be the driving force behind transport, however, ATP is certainly involved in phosphorylation of the vitamin. Since some of the riboflavin is metabolized to FMN and FAD, inhibition by flavin analogs was tested (see Table I). 7,8-Dichlororiboflavin, a flavokinase substrate, shows the greatest reduction in riboflavin uptake. Acid anions such as p-aminohippurate (PAH) and penicillin G (Pen G) were also tested, since Sepctor (3) had reported that these cyclic organic acids inhibited riboflavin uptake into rabbit kidney slices. We do not see the influence of these acids on the isolated cells. Thus it appears that riboflavin is conserved by the kidney by a facilitated diffusion process and metabolically trapped by flavokinase within the cell.

TABLE I. Factors influencing the uptake of 3 µM [^{14}C]riboflavin into proximal cells of kidney.

Effector		Initial velocity at 0.3 min (%)	Accumulation at 5 min (%)
Unlabeled riboflavin			
15 µM		89.8 ± 12.7 (8)	72.0 ± 4.4 (8)
30 µM		82.9 ± 6.1 (10)	56.8 ± 4.7 (10)
60 µM		68.9 ± 7.4 (6)	52.6 ± 4.6 (7)
Analogs of riboflavin			
7,8-Dichloro	15 µM	69.5 ± 5.8 (9)	67.3 ± 10.5 (5)
$N(3)CH_3$	15 µM	84.4 ± 21.8 (3)	63.9 ± 8.8 (2)
$N(3)CH_2COOH$	15 µM	85.1 ± 20.8 (3)	76.1 ± 2.1 (2)
$N(3)CH_2COOC_2H_5$	15 µM	92.9 ± 8.1 (3)	81.4 ± 9.4 (2)
$N(10)(CH_2)_4COOH$	15 µM	74.9 ± 1.4 (3)	73.4 ± 5.9 (2)
$N(10)CH_2CH_2OH$	15 µM	100.1 ± 4.4 (3)	82.8 ± 6.8 (2)
Lumiflavin	15 µM	85.9 ± 7.9 (7)	75.1 ± 5.9 (3)
8α-S-CH_2CH_2COOH	15 µM	90.3 ± 8.0 (3)	89.9 ± 6.7 (2)
Acid anions			
PAH, 1 mM		89.4 ± 11.9 (4)	94.6 ± 0.4 (2)
Pen G, 1 mM		93.6 ± 3.2 (4)	91.0 ± 5.5 (2)

Error is expressed as SEM with the number of determinations in parentheses.

REFERENCES

1. Jusko, W.J., Levy, G.: in Riboflavin (Rivlin, R.S., ed.) pp 99-152, Plenum Press, New York (1975).

2. Christensen, S.: Acta Pharmacol. et Toxicol. 29, 428-440 (1971).

3. Spector, R.: J. Pharmacol. Exp. Ther, 221, 394-398 (1982).

4. Jones, D.P., Sundby, G.B., Ormstad, K., Orrenius, S.: Biochem. Pharmacol. 28, 929-935 (1979).

5. Oka, M., McCormick, D.B.: J. Nutr. 115, 496-499 (1985).

6. McCormick, D.B., Innis, W.S.A., Merrill Jr., A.H., Bowers-Komro, D.M., Oka, M., Chastain, J.L.: in Flavins and Flavoproteins (McCormick, D.B., Edmondson, D.E., eds.) this volume.

7. Merrill Jr., A.H., McCormick, D.B.: J. Biol. Chem. 255, 1335-1338 (1980).

PURIFICATION AND CHARACTERIZATION OF THE BIFUNCTIONAL ENZYME FAD SYNTHETASE FROM BREVIBACTERIUM AMMONIAGENES

Dietmar J. Manstein and Emil F. Pai

Department of Biophysics, Max-Planck-Institute for Medical Research, Jahnstrasse 29, D-6900 Heidelberg, FRG

Introduction

5'-phosphorylation of riboflavin followed by the adenylylation of FMN to FAD are the two final steps in the biosynthesis of the flavin coenzymes. Only partial purification of the enzymes catalyzing one or both of these transformations have been reported. Sources were bacteria (1,2), yeast (3), higher plants (4), and rat liver (5).

(I) riboflavin kinase: riboflavin + ATP → FMN + ADP

(II) FMN adenylyltransferase: FMN +ATP → FAD + PP_i

The catalytic activity from B. ammoniagenes thought to reside in a protein complex has recently been of high interest to researchers working with modified flavins. After it had been applied in the conversion of 5-deaza-riboflavin to 5-deaza-FAD (1) it became widely used in the preparation of the coenzyme forms of flavin analogues (6). Its ability to catalyze both reactions with a broad variety of riboflavin isosteres as substrates together with its intriguing stability made this enzyme a standard reagent (7,8). It was for this reason that we started to work with this enzyme. When we tried to improve the purification procedure for FAD synthetase and also to optimize the conversion of 8-demethyl-8-OH-5-deaza-riboflavin to the FAD, evidence

Flavins and Flavoproteins 1987

grew that both activities are located on a single polypeptide chain. Therefore we decided to try to isolate this bifunctional enzyme of only moderate molecular weight (9).

Results and Discussion

The course of a typical preparation of B. ammoniagenes FAD synthetase is shown in Table 1. The cells were broken by prolonged sonication. An ammonium sulfate cut followed by column chromatography on DEAE-Sepharose, Blue Sepharose, and ATP-Agarose resulted in an about 7000-fold purification of the enzyme. The yield was close to 50% and the procedure was very reproducible.

Table 1. Purification of FAD Synthetase from B. ammoniagenes

Step	Vol.	Protein	Tot. Act.	Spec. Act.	Recovery	Purification
	ml	mg	U	U/mg	%	-fold
Crude extract	750	10,500	200	0.02	(100)	(1)
50-70% $(NH_4)_2SO_4$	235	2,490	210	0.1	(105)	(5)
DEAE-Sepharose	25	163	175	1.1	88	54
Blue Sepharose	5.5	21	134	6.4	67	320
ATP-Agarose	2.4	0.7	96	137	48	6,850

One enzyme unit (U) catalyzes the formation of 1 nmol of FAD/min at 37°C. Numbers in parentheses could not be accurately determined due to the presence of phosphatases and phosphodiesterases.

Riboflavin kinase and FMN adenylyltransferase activities copurified together. Both, SDS-PAGE and nondenaturing gel electrophoresis indicated a molecular weight of M_r=38,000. In two different systems the isoelectric point was determined to 4.6.

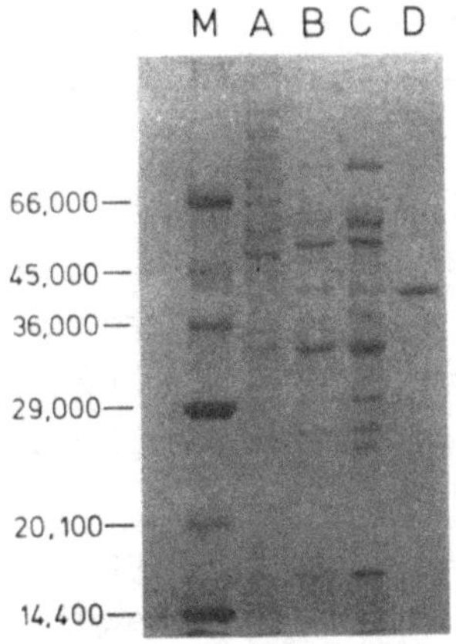

Fig. 1. SDS-polyacrylamide gel electrophoresis of FAD synthetase at various stages of purification. Lane A, 50-70% ammonium sulfate fraction (50 µg); lane B, DEAE-Sepharose fraction (40 µg); lane C, Blue Sepharose fraction (70 µg); lane D, ATP-Agarose fraction (7 µg); lane M, molecular weight markers.

Automated Edman degradation of intact protein gave a single N-terminal amino acid sequence (Scheme I). We take these results as evidence that FAD synthetase from B. ammoniagenes is a monomeric bifunctional enzyme (10).

Met-Asp-Ile-X-Tyr-Gly-Thr-Ala-Ala-Val-Pro-Lys-Asp-Leu-X-Asn- -

Scheme I: N-terminal Amino Acid Sequence of FAD Synthetase from Brevibacterium ammoniagenes.

FAD synthetase is specific for ATP as the donor substrate. With up to 3 mM ADP, GTP, CTP, ITP, or UTP no turnover could be observed. The Michaelis constants for Mg·ATP were approximately 5 µM for the first and 160 µM for the second reaction, respectively. While 2'-dATP as well as both diastereomers of ATPαS were good substrates in the kinase reaction, conversion of FMN to FAD could not be observed. The divalent cations of Zn, Cd, Co, and Mn could, in part, replace Mg^{2+} as cofactor of the FAD synthetase. However, as the two activities are influenced quite differently in this respect the overall behaviour is very complex. It is interesting to note that even in the presence of 5 mM EDTA there was still 10% of the 5'-phosphotransferase activity which could be measured when 200 µM $MgCl_2$ was present. The enzyme exhibits a broad specificity for the flavin substrates. It accepts oxidized and reduced isoalloxazines, but seems to have a preference for neutral ring systems, as was noticed with 8-OH-5-deaza-riboflavin. This analogue was converted with reasonable velocity only when the pH of the reaction mixture was at or below the pK of the 8-hydroxyl substituent. For use as a reagent in producing FAD from corresponding riboflavins in most cases the DEAE-fraction can be used which is stable when stored at -20°C. From our experience we recommend to use 1-2 mM Zn^{2+} as cofactor instead of Mg^{2+} together with a 2-fold excess of ATP. For initial tests we use pH 6.5 to 8.5, specific analogues might need lower pH-values.

References

1. Spencer, R., J. Fisher, C. Walsh. 1976. Biochem. 15, 1043.
2. Kearney, E.B., J. Goldenberg, J. Lipsick, M. Perl. 1979. J.Biol.Chem. 254, 9551.
3. Schrecker, A.W., A. Kornberg. 1950. J.Biol.Chem. 182, 795.
4. Giri, K.V., N. Appaji-Rao, H.R. Cama, S.A. Kumar. 1960. Biochem.J. 75, 381.
5. Gomes, B., D.B. McCormick. 1983. Proc.Soc.Exp.Biol.Med. 172, 250.
6. Walsh, C., J. Fisher, R. Spencer, R.W. Graham, W.T. Ashton, J.E. Brown, R.D. Brown, E.F. Rogers. 1978. Biochem 17, 1942.
7. Krauth-Siegel, R.L., R.H. Schirmer, S. Ghisla. 1985. Eur.J.Biochem. 148, 335.
8. Light, D. R., C. Walsh, M.A. Marletta. 1980. Anal.Biochem. 109, 87.
9. Manstein, D.J., E.F. Pai, L.M. Schopfer, V. Massey. 1986. Biochem. 25, 6807.
10. Manstein, D.J., E.F. Pai. 1986. J.Biol.Chem. 261, 16169.

AN UPDATE ON FLAVIN METABOLISM IN RATS AND HUMANS

D.B. McCormick, W.S.A. Innis, A.H. Merrill, Jr., D.M. Bowers-Komro, M. Oka, and J.L. Chastain
Dept. of Biochemistry, School of Medicine, Emory University,Atlanta, GA 30322, U.S.A.

General Introduction

Within the past few years, mainly since the last Symposium on Flavins and Flavoproteins (1), there has been considerable progress along several lines as relate to flavin metabolism in the broad sense of total disposition of natural flavins and derivatives. Much of the work has been done with mammalian systems and includes aspects of transport, cellular uptake, coenzyme biosynthesis, catabolism, excretion and secretion. Our contributions to current understanding, especially most recent findings, will be outlined in the following in context with main points about what is known and what remains to be determined.

Circulatory Transport

Transport of flavins by the plasma fraction of blood is known to involve both loose association with albumin, which by sheer amount contributes modestly to the total flavin carried, and tight associations with some globulins. Among the latter, we have identified immunoglobulins as the major binding proteins for riboflavin in serum from normal humans (2,3) and from patients with certain types of cancer (4). More recently we have used flavins attached via the N3-, N10-, or 8α-positions to aminoalkyl agarose beads as affinity materials to purify riboflavin-binding immunoglobulins. Though more details are given elsewhere in this volume (5), isolates obtained from normal human plasma were constituted by immunoglobulin subclasses IgG, IgM, and IgA, with both kappa and lambda light chains. Relative amounts of the subclasses depended on the type of flavin agarose, and structural

Flavins and Flavoproteins 1987

specificity for flavin binding differed among the three isolates. Though binding was heterogeneous, a substantial portion of the species in each isolate had a Kd of < 1 μM for riboflavin. Other flavins and derivatives were less effectively bound with the exception of the natural catabolite 8α-hydroxyriboflavin and in some instances the coenzymes FMN and FAD and the terminal photoproducts lumichrome and lumiflavin. Papain cleavage of the immunoglobulins yielded Fab fragments that still bound riboflavin. Hence, at least a portion of the antigenic binding site may be involved. This suggests that flavin-binding immunoglobulins may represent a mild immune response to the vitamin or certain metabolic derivatives. A summary of the principal characteristics and quantities of flavin-binding immunoglobulins in human serum is given in Table I.

TABLE I. Immunoglobulins isolated by different flavinyl agaroses[1]

Agarose	Immunoglobulins			Light Chains[2]	
	IgG	IgA (mg/dL)	IgM	Kappa	Lambda
N3	28 (78%)	2.7 (7%)	5.3 (15%)	+	+
8α	14 (22%)	2.7 (4%)	46.7 (74%)	+	+
N10	15 (53%)	3.0 (11%)	10.3 (36%)	+	+

[1]The original ammonium sulfate precipitate of plasma contained approximately 1.6 g of immunoglobulin/dL with 77% IgG, 12% IgA, and 12% IgM; [2]Results with each subclass.

Cellular Uptake

Uptake of flavins by mammalian cells has some characteristics in common, but clearly there are both qualitative and quantitative differences among the different cell types. Both isolated enterocytes (6) and hepatocytes (7) exhibit an initial rapid uptake followed by slower passive diffusion of riboflavin into cells where the vitamin is metabolically trapped by flavokinase-catalyzed phosphorylation. However, uptake in enterocytes is dependent upon Na^+ and inhibited by ouabain, reflecting an ATPase-involved active cotransport system, whereas intake by hepatocytes is relatively insensitive to both Na^+ and ouabain, reflecting a facilitated, probably carrier-mediated system. Our recent work with proximal tubular epithelial cells from rat kidney demonstrates that the faster, facilitated phase of flavin uptake exhibits Na^+-dependence (like small intestine but not liver) but is insensitive to ouabain (like liver but not small intestine). More details on the renal system are given elsewhere in this volume (8), but Table II summarizes the main features of uptake of flavin by renal cells in comparison to enterocytes and hepatocytes.

Coenzyme Biosynthesis

Enzymes responsible for the conversions of vitamins to coenzymes typically require thousands of fold purification, usually from cytosol, to achieve homogeneity. This is not surprising inasmuch as only small but essential amounts of coenzymes are needed as the prosthetic functions of catalysts. This was shown by us to be the case with the kinase and oxidase that catalyzed the two-step conversion of the natural vitaminic forms of B_6 to pyridoxal 5'-phosphate. It is likewise the case for the synthetase-requiring formation of FAD with the intermediacy of FMN formed by flavokinase. Though the formation of FMN and FAD by crude preparations from a range of pro- and eucaryotic organisms had been noted, it was only within the last decade that a homogeneous preparation of flavokinase was achieved, first from rat liver in our laboratory using an N^{10}- flavinyl agarose (9), and more recently from Brevibacterium ammoniagenes using an N^6-ATP-agarose step preferred by Manstein and Pai (10). In this latter case, the molecular weight of the bacterial kinase (ATP:riboflavin 5'-phosphotransferase) was found to be

Table II: Characteristics of flavin uptake by isolated mammalian cells

	Guinea Pig Enterocytes (6) jejunal	Guinea Pig Enterocytes (6) ileal	Rat Hepatocytes (7) control	Rat Hepatocytes (7) deficient	Rat Renal Proximal (8) Tubular Cells
Kinetic Parameters					
K_t	70 nM	87 nM	12 μM	12 μM	8.4 μM
V_{max}	1.6 $\frac{pmol}{mg\ protein \cdot min}$	1.8 $\frac{pmol}{mg\ protein \cdot min}$	82 $\frac{pmol}{10^6\ cells \cdot min}$	119 $\frac{pmol}{10^6\ cells \cdot min}$	21 $\frac{pmol}{10^6\ cells \cdot 0.5\ min}$
V_{dif}					2.9 $\frac{pmol}{10^6\ cells \cdot 0.5\ min \cdot \mu M}$
$\frac{Intracellular}{Extracellular}$	6.3-fold	7.2-fold	2.5-fold	5.2-fold	2.5-fold
	Relative % accumulation at equilibrium (or 10 min) using standard buffer at 37°C as reference				
Temperature (°C)					
27		36	41	68	
20	43	32			
0–4	9	8	2	4	37
Buffer (replacing Na^+)					
K^+	12	24			
$Choline^+$			100	100	77
Metabolic Inhibitors					
Ouabain 0.2 mM	27	27			
1.0 mM			100	100	95
Antimycin A 5 μg/ml	32	29			
FCCP 1 μM			45	60	86
Ethionine 10 mM			52	72	97
Competitor/Riboflavin Ratio					
FMN 187	9	2			
10			100	100	
FAD 187	18	7			
10			100	100	
Lumiflavin 5			66	—	
7,8-Dichloro-flavin 2			38	51	75
5					
2′-Hydroxyethyl-flavin 3			51	25	67
5					83

approximately 38,000 whereas we had found 28,000 for the monomeric rat enzyme. A preference for Zn^{2+} was shown in both cases. An interesting additional finding, however, was that the bacterial enzyme appeared bifunctional with synthetase (ATP:FMN adenylyltransferase) activity reflected when Mg^{2+} replaced Zn^{2+}. An apparently homogeneous preparation of the synthetase, recently purified from rat liver in our laboratory using an FMN-agarose step (11), does separate from kinase, as reflected by activity ratios given in Table III.

TABLE III: Ratios of flavokinase to FAD synthetase activities during purification of the latter from rat liver

Step	Kinase/synthetase[1]
$(NH_4)_2SO_4$ fraction (after dialysis)	(5.10)[2]
DEAE-Sephadex (A-50) concentrate	1.99
Chromatofocusing pool	1.07
FMN-agarose concentrate	0.044
Sephadex G-200 concentrate	--

[1]Activities expressed as Units/mg protein (9,11).
[2]Numbers in parentheses could not be accurately determined due to the presence of phosphatases and pyrophosphatases.

Moreover, the mammalian synthetase has a larger molecular weight as illustrated in Fig. 1 with comparison to marker proteins subjected to molecular sieving where approximately 100,000 is noted, or when such a dimer is dissociated by SDS to yield subunits of about 50,000 M.W.

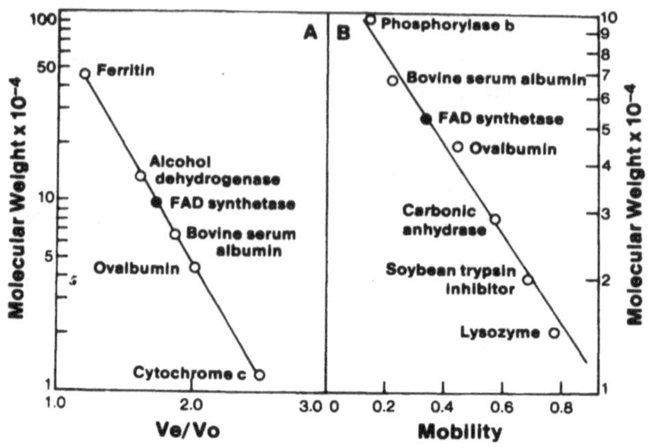

Fig. 1. Molecular weight determination of synthetase. Estimates of molecular weight were determined using gel filtration, panel A, and SDS-polyacrylamide gel electrophoresis, panel B. The following proteins served as molecular weight markers: cytochrome c (12,400), lysozyme (14,400), soybean trypsin inhibitor (20,100), carbonic anhydrase (29,000), ovalbumin (45,000), bovine serum albumin (66,000), phosphorylase b (97,400), alcohol dehydrogenase (150,000), and ferritin (440,000).

The liver synthetase also strongly prefers Mg^{2+}, instead of Zn^{2+} which optimizes kinase activity. The pH optimum of the synthetase with Mg^{2+} is near the physiological range. Such buffers as phosphate, which affects divalent metal ion and phosphorylated substrate binding, exhibit a strong concentration dependence, whereas Tris and Hepes do not, though all maximize activity near pH 7.1 as shown in Fig. 2.

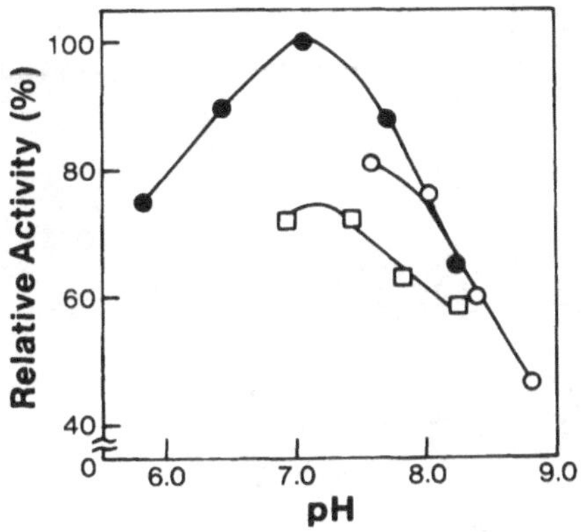

Fig. 2. Effect of pH on the synthetase activity. Buffers at 50 m<u>M</u> each were: ●, potassium phosphate; o, Tris-Cl; □, Hepes.

A comparison of primary reported properties of the kinase/synthetase from <u>B</u>. <u>ammoniagenes</u> versus those for the separately and differently isolated flavokinase and FAD synthetase from rat liver are given in Table IV.

Catabolism and Excretion

Our more recent studies of urinary flavins from rat (12) extend identification, clarify origin, and lead to fairly complete quantitation. As regards methyl oxidation attributable to tissue-level microsomal mixed-function oxidase, we have found both 7- and 8-hydroxymethylriboflavins that previously were reported in human urine by Yagi and associates (13), and also confirmed carboxylumichromes. The presence of the latter, which reflect side-chain cleavage as well as more extensive methyl oxidation, was decreased by inclusion of 1% succinyl sulfathiazole in the diet as well as by decrease in riboflavin ingested. Clearly microbial action in the gastrointestinal tract contributes to cleavage of the ribityl side chain as we had shown much earlier (14). Moreover, the 10-hydroxyethylflavin, first noted in ruminant urine by Owen

TABLE IV: Reported properties of bacterial and mammalian flavokinase and FAD synthetase

Source	Enzyme (activity)	Purification (-fold)	Molecular weight	Spec. act. (units/mg)	T.N. (min^{-1})	Preferred Me^{2+}	Km(μM) Flavin	Km(μM) ATP
Brevibacterium	Flavokinase/				36	Zn^{2+}	--	7.5[4]
ammoniagenes	FAD synthetase[1]	6,850	38,000	137	27	Mg^{2+}	--	160
Rat liver	Flavokinase[2]	7,090	28,000	142	5	Zn^{2+}	4.0	9.8
	FAD Synthetase[3]	9,873	100,000 (dimer)	133	13	Mg^{2+}	9.6	5.3

[1]Manstein and Pai (10); [2] Merrill and McCormick (9); [3]Oka and McCormick (11); [4]Estimated

and West (15), is also present in rat urine. Trace amounts of lumiflavin and lumichrome occur and can result from photochemical events at dermal level, but lumichrome per se does not arise by somatic enzyme action (16). A scheme updating these diverse events is given in Fig. 3

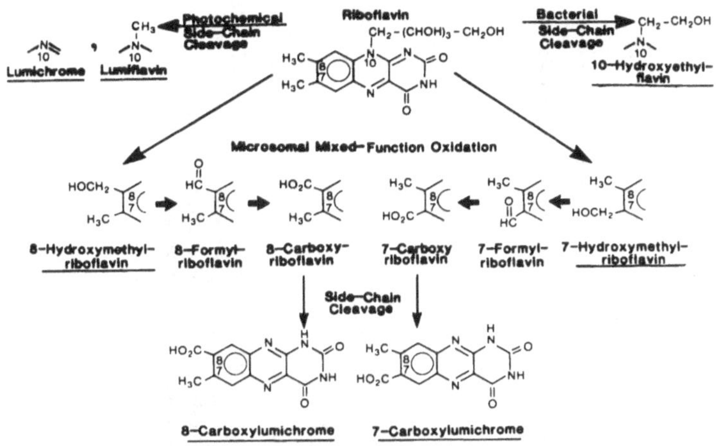

Fig. 3. Metabolism and photodegradation of riboflavin that leads to rat urinary catabolites (underlined).

In human urine, we have found small amounts of 10-hydroxyethylflavin as well as 7- and 8-hydroxymethylriboflavins reported earlier (17) and additionally found a riboflavin glycoside. A new catabolite not previously recognized from any source is an 8α-sulfonylriboflavin (18) that probably arises from the 8α-cysteinyl flavin of mitochondrial monoamine oxidase. This oxidation product of an 8α-thioether linked flavin we had earlier synthesized (19) may in part arise from ingestions of foods, notably meat and milk, since there are suggestive changes in urinary level relating to such exogenous supply, but some may reflect endogenous turnover of the covalently bound FAD. Updated information on the types and quantities of urinary flavins from adult humans is presented in Table V.

TABLE V. Total flavin concentration and urinary flavins expressed as percent.[1]

	Total flavin[2]	7,8-Dimethyl types[3]				7α,8- or 7,8α-Substituted, methyl types		
Subject	µg/g Creatinine	Ribityl	Glycoside	Hydroxyethyl	Lumiflavin	7α-Hydroxy, 8-methyl	7-Methyl 8α-hydroxy	7-Methyl 8α-sulfonyl
Females (n=2)	263 ± 85	64.5 ± 2.2	4.1 ± 0.9	2.7 ± 0.5	2.0 ± 1.2	12.2 ± 2.2	6.1 ± 1.1	8.4 ± 1.8
Males (n=12)	187 ± 36	66.8 ± 3.6	3.9 ± 1.3	1.9 ± 0.9	1.7 ± 0.8	11.4 ± 2.0	5.6 ± 1.6	8.7 ± 3.2
Male[4]	484 ± 88	82.2 ± 5.7	1.0 ± 0.7	0.5 ± 0.2	0.1 ± 0.1	10.0 ± 3.6	2.5 ± 0.5	3.7 ± 0.8

[1]Mean ± SD (4 urine samples per subject); [2]Based on fluorescence characteristics comparable to riboflavin; [3]Common names of compounds are as follows: ribityl = riboflavin; glycoside = riboflavin glycoside; hydroxyethyl = 10-hydroxyethylflavin; 7α-hydroxy, 8-methyl = 7α-hydroxyriboflavin; 7-methyl,8α-hydroxy = 8α-hydroxyriboflavin; 7-methyl,8α-sulfonyl = 8α-sulfonylriboflavin; [4]Subject received a daily riboflavin supplement of 1.7 mg/d.

Secretion in Milk

Earlier assessments of flavins in milk have usually relied on microbiological assays for "riboflavin", often not differentiating other biologically active flavins, or fluorometric (AOAC) assays that do not adequately distinguish the different kinds of flavin. Only recent application of HPLC technique have been applied to such foods and even then using treatments for extraction that may cause breakdown of more labile coenzymes and catabolites (20). As complete documentation of the total flavin economy in a lactating mammal must take into account the types and quantities of flavin secreted in milk, as well as the fact that accurate quantitation of riboflavin level in this important food should be known, we have begun the systematic identification and quantitation of flavins in milk from both cows and humans. Careful extractions using phenol for typical isoalloxazines followed by HPLC has led to such information as representatively illustrated in Fig. 4.

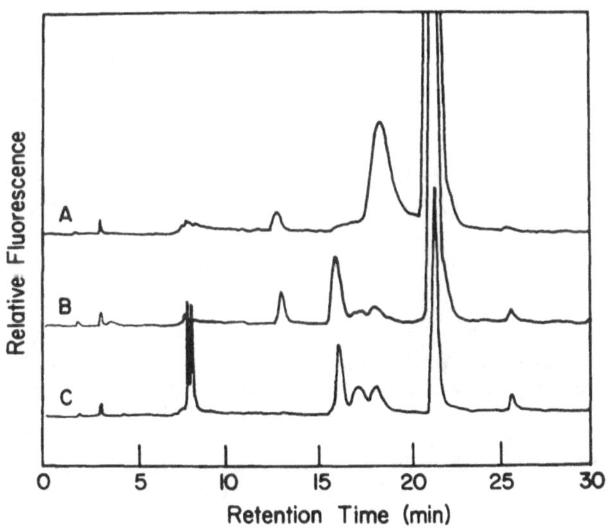

Fig. 4. HPLC profiles for phenol-extracted flavins from processed cow's milk (A), raw cow's milk (B) and fresh human milk (C). Use of standards and fluorescence/absorption properties indicate that peaks with retention times near 21,18,17,16, and 13 are riboflavin, FMN, FAD, perhaps 8α-sulfonylriboflavin, and 7α-hydroxyriboflavin, respectively.

In addition to riboflavin as a predominant flavin, lesser amounts of FMN, FAD, and a flavin with retention time suggestive of the 8α-sulfonylriboflavin are present in fresh milk from both human and cow. Some 7α-hydroxyriboflavin also appears in cow's milk. Processing of this latter to yield the typical commercial product results in breakdown of most of the FAD and putative sulfonyl flavin.

Gaps in Knowledge

Though a much clearer picture of overall flavin processing in the mammal is beginning to emerge, there are still several gaps where more work is needed. As indications where we and perhaps others may add to our understanding, some specific points can be made: It is still not clear whether specific, e.g. pregnancy-induced, carrier proteins are a main vehicle for transport and especially for fetal delivery. The facilitated, probably carrier-mediated importing of flavin by cells should be elucidated at molecular level with isolation and reconstitution ultimately involved. The phylogenic differences among flavokinases and FAD synthetases should be better understood as well as elucidation of their mechanism and control. Some further characterization of flavin metabolites is needed and their origins and contribution to secretion as well as excretion elaborated upon. Further challenges await.

References

1. McCormick, D.B., Innis, W.S.A., Merrill, Jr., A.H., Lee, S.-S.: Flavins and Flavoproteins (R.C. Bray, P.C. Engel and S.G. Mayhew, eds.) pp. 833-846, Walter de Gruyter, Berlin, 1984.

2. Merrill, Jr., A.H., Froehlich, J.A., McCormick, D.B.: Biochem. Med. 25, 198-206, 1981.

3. Innis, W.S.A., McCormick, D.B., Merrill, Jr., A.H.: Biochem. Med. 34, 151-165, 1985.

4. Innis, W.S.A., Nixon, D.W., Murray, D.R., McCormick, D.B., Merrill, Jr., A.H.: Proc. Soc. Exp. Biol. Med. 181, 237-241, 1986.

5. Merrill, Jr., A.H., Innis-Whitehouse, W.S.A., McCormick, D.B.: Flavins and Flavoproteins (D.B. McCormick and D.E. Edmondson, eds.) this volume.

6. Hegazy, E., Schwenk, M.: J. Nutr. 113, 1702-1707, 1983.

7. Aw, T.Y., Jones, D.P., McCormick, D.B.: J. Nutr. 113, 1249-1254, 1983.

8. Bowers-Komro, D.M., McCormick, D.B.: Flavins and Flavoproteins (D.B. McCormick and D.E. Edmondson, eds.) this volume.

9. Merrill, Jr., A.H., McCormick, D.B.: J. Biol. Chem. 255, 1335-1338, 1980.

10. Manstein, D.J., Pai, E.F.: J. Biol. Chem. 261, 16169-16173, 1986.

11. Oka, M., McCormick, D.B.: J. Biol. Chem. 262, 7418-7422, 1987.

12. Chastain, J.L., McCormick, D.B.: J. Nutr. 117, 468-475, 1987.

13. Yagi, K., Ohishi, N., Ohkawa, H.: Flavins and Flavoproteins (R.C. Bray, Engel and S.G. Mayhew, eds.) pp. 819-832, Walter de Gruyter, Berlin, 1984.

14. Yang, C.S., McCormick, D.B.: J. Nutr. 93, 445-453, 1967.

15. Owen, E.C., West, D.W.: Vitamins and Coenzymes, Methods Enzymol. (D.B. McCormick and L.D. Wright, eds.) Vol. 18, pp. 574-579, Academic Press, New York, 1971.

16. Oka, M., McCormick, D.B.: J. Nutr. 115, 496-499, 1985.

17. Ohkawa, H., Ohishi, H., Yagi, K.: J. Biol. Chem. 258, 5623-5628, 1983.

18. Chastain, J.L., McCormick, D.B.: Amer. J. Clin. Nutr., in press, 1987.

19. Falk, M.C., Johnson, P.G., McCormick, D.B.: Biochemistry 15, 639-645, 1976.

20. Johnson, H., Branzell, C.: Internat. J. Vit. Nutr. Res. 57, 53-58, 1987.

EFFECTS OF QUINACRINE, TETRACYCLINE, ADRIAMYCIN, AND CHLOROQUINE UPON FLAVIN METABOLISM IN RATS

P. Dutta, J.T. Pinto, G.B. Raiczyk, and R.S. Rivlin

Memorial Sloan-Kettering Cancer Center and Cornell University Medical College
New York, New York 10021, USA

Introduction

Dietary riboflavin deficiency has long been associated with resistance to malaria (1-3). In light of these findings, we proposed that, in addition to dietary riboflavin deficiency, hormonal disorders and drugs which inhibit riboflavin metabolism should provide protection against malaria (4). We have proposed further that pharmacological actions of antimalarial drugs structurally analogous to riboflavin may be related to their ability to inhibit various aspects of riboflavin metabolism.

In this investigation, we examined whether quinacrine, an antimalarial drug with a tricyclic ring, tetracycline, a broad-spectrum antibiotic with antimalarial properties, and chloroquine, a widely used antimalarial drug with a bicyclic ring, each inhibits biosynthesis of flavin adenine dinucleotide (FAD). We also investigated whether adriamycin, an antineoplastic agent with close structural similarities to tetracycline, inhibits FAD biosynthesis because of our preliminary observations that this drug inhibits multiplication of Plasmodium falciparum (FCR3) in cell culture. Since our previous studies have shown that cardiac and skeletal muscle from rats is particularly sensitive to drugs which affect flavin metabolism (5), we utilized these two tissues exclusively in this study.

Materials and Methods

Animals and Diet. All experiments were performed on adult male Holtzman rats (Holtzman Rat Co., Madison, WI, USA) weighing 200-300 g. All animals were maintained on tap water ad libitum, housed individually in stainless steel metabolic cages, and consumed rat chow ad libitum purchased from Ralston Purina Company, St. Louis, MO, USA.

Drugs, chemicals and isotopes. Tetracycline hydrochloride (Achromycin) was purchased from Lederle Laboratories, Pearl River, NY, USA; Quinacrine dihydrochloride, chloroquine diphosphate, nonradiolabeled riboflavin, flavin adenine dinucleotide (FAD), and ATP from Sigma Chemical Company, St. Louis, MO, USA; Adriamycin from Adria laboratories Inc., Columbus, OH, USA; and (^{14}C)Riboflavin (28 mCi/mmol) from the Amersham/Searle Corporation, Arlington Heights, IL, USA.

Flavins and Flavoproteins 1987

Drug treatments. Rats were distributed into five groups (6-9 animals per group): saline control, tetracycline-, quinacrine-, chloroquine-, and adriamycin-treated. Each group received i.p. injections of drug twice daily for 3 days (20 mg/kg body weight). Controls received an equivalent volume of isotonic saline.

Analysis of (^{14}C) FAD formation in tissues. Food was removed from all cages 18 h prior to sacrifice. One h prior to sacrifice, each animal received an s.c injection of (^{14}C)riboflavin, 25 uCi/kg body weight. After decapitation, heart and skeletal muscle surrounding the femur were excised and stored at -20°C until assayed for (^{14}C)FAD formation.

Tissues were analysed for (^{14}C)FAD formation using reverse isotope dilution and anion exchange column chromatography (DEAE-Sephadex A-25, Pharmacia Fine Chemicals, Inc., Piscataway, NJ, USA) as previously established in this laboratory (6). Aliquots of tissue (200 mg heart and 600 mg skeletal muscle) were homogenized in a solution containing 7 ml of absolute methanol and 1 ml each of freshly prepared solutions of nonradiolabeled riboflavin and FAD (100 ug/ml). (^{14}C)FAD synthesized (dpm/100 mg tissue) was determined and results expressed as percent of control.

Results

As shown in Table 1, conversion of (^{14}C)riboflavin into (^{14}C)FAD was significantly diminished ($p < 0.01$) in hearts of animals injected with either tetracycline or quinacrine. Furthermore, adriamycin decreased biosynthesis of (^{14}C)FAD in heart to an even greater degree. By contrast, the bicyclic drug, chloroquine, did not interfere significantly with (^{14}C)FAD formation in heart.

The formation of (^{14}C)FAD in skeletal muscle was significantly inhibited ($p < 0.05$) by tetracycline, quinacrine, and adriamycin as shown in Table 1. As in heart, chloroquine did not inhibit (^{14}C)FAD formation.

Table 1. Incorporation of (^{14}C)riboflavin into (^{14}C) FAD in heart and skeletal muscle of rats injected with various drugs.

	Heart		Skeletal Muscle	
Drug	FAD Synthesis (% control)	Signif. of Difference	FAD Synthesis (% control)	Signif. of Difference
Tetracycline	89.1 ± 2.4*	$p < 0.01$	90.7 ± 2.3	$p < 0.05$
Quinacrine	83.2 ± 2.7	$p < 0.01$	87.0 ± 3.1	$p < 0.05$
Adriamycin	76.7 ± 3.5	$p < 0.01$	73.0 ± 4.7	$p < 0.02$
Chloroquine	97.4 ± 7.0	N.S.	98.9 ± 5.2	N.S.

*All data were determined as mean ± SEM of dpm/100 mg tissue/h, with 6-9 animals per group, and expressed as % of values in saline-treated controls.

Discussion

The isoalloxazine ring of riboflavin shares a number of important structural features with various groups of drugs, e.g., the tricyclic phenothiazine derivative, chlorpromazine, and the tricyclic antidepressants, imipramine and amitriptyline. We showed previously that each of these drugs inhibits the conversion of riboflavin to FAD (6). The antiparasitic agent, quinacrine, the anthracycline antibiotic, tetracycline, as well as the anti-cancer drug, adriamycin, also have structural features in common with those of riboflavin. Each of these agents also has antimalarial properties.

We have postulated that drugs with diverse actions but which are structurally similar to riboflavin may derive their antimalarial efficacy by inhibiting FAD biosynthesis (4). This study demonstrates that quinacrine and tetracycline, as well as adriamycin, each inhibits riboflavin conversion to FAD in heart and in skeletal muscle of rats, findings which possibly may explain, at least in part, their antimalarial efficacy.

Chloroquine, a well known antimalarial drug, does not inhibit FAD biosynthesis in either cardiac tissue or skeletal muscle of rats. The inability of chloroquine to inhibit FAD formation suggests that, unlike quinacrine, tetracycline and adriamycin, this bicyclic drug probably does not compete effectively with the isoalloxazine ring of riboflavin for binding. The mechanism of action of chloroquine may reside at sites other than those affecting riboflavin metabolism.

To show conclusively that the antimalarial properties of quinacrine and tetracycline relate directly to their ability to interfere with FAD biosynthesis, subsequent experiments will examine whether supplementation of riboflavin antagonizes the antimalarial potency of these drugs. Similarly, other studies are presently in progress in our laboratory to explore whether cardiotoxicities associated with adriamycin treatment can be reversed by riboflavin supplementation.

Summary

The acridine derivative, quinacrine, the broad-spectrum antibiotic, tetracycline, and the 4-aminoquinoline, chloroquine, all have antimalarial properties. Based on our recent proposal that antimalarial effects of certain drugs may be related to their inhibition of riboflavin metabolism, this study determined whether quinacrine, tetracycline and chloroquine act as riboflavin antagonists. In addition, we investigated effects of adriamycin, which also possesses antimalarial properties, on FAD biosynthesis. Compared to results in controls, rats treated with either tetracycline, quinacrine, or adriamycin exhibit diminished formation of (^{14}C)FAD in heart and in skeletal muscle. By contrast, chloroquine, which was also studied in heart and skeletal muscle had no significant inhibitory effect upon (^{14}C)FAD formation in either organ. These data provide evidence that the tri-

and tetracyclic drugs, i.e., quinacrine, tetracycline, and adriamycin, act as riboflavin antagonists. Chloroquine does not appear to be a riboflavin antagonist. This study provides further evidence supporting a possible link between the actions of certain tricyclic and tetracyclic drugs upon decreasing the rate of synthesis of FAD and their effectiveness against malaria.

Acknowledgement

This research supported by NIH grants no. CA 29502, CA 08748, and Training Grant HL 07379, and by grants from the American Cancer Society, Parenteral Drug Association Foundation, Inc., Stella and Charles Guttman Foundation, and The General Foods Fund.

References

1. Seeler, A.O., Ott, W.H. 1944. J. Infect. Dis. 75: 175-178.

2. Thurnham, D.I., Oppenheimer, S.J., Bull, R. 1983. Trans. Royal Soc. Trop. Med. Hyg. 77: 423-424.

3. Kaikai, P., Thurnham, D.I. 1983. Trans. Royal Soc. Trop. Med. Hyg. 77: 680-686.

4. Dutta, P., Pinto, J., Rivlin, R.S. 1985. Lancet ii: 1040-1043.

5. Raiczyk, G.B., Dutta, P., Pinto, J. 1985. The Physiologist 28: 322.

6. Pinto, J., Huang, Y.P., Rivlin, R.S. 1981. J. Clin. Invest. 67: 1500-1506.

ROLE OF FLAVOPROTEINS IN THE NEUROTOXIC ACTION OF MPTP

Thomas P. Singer and Rona R. Ramsay
Molecular Biology Division, Veterans Administration Medical Center, San Francisco, California 94121 and Departments of Biochemistry/Biophysics and Pharmacy, University of California, San Francisco, California 94143

Introduction

MPTP (N-methyl-4-phenyl-1,2,3,6-tetrahydropyridine), an impurity in an illicit meperidine analog used by drug abusers, causes parkinsonian symptoms in humans, sub-human primates, and some other species. The connection between MPTP and the neurological disorders it causes was recognized (1) in 1982 when a number of relatively young heroin addicts were hospitalized in the San Francisco Bay Area with the sudden onset of parkinsonian symptoms closely resembling those which develop progressively in elderly patients with Parkinson's disease, all of whom showed transient relief on administration of L-dopa. The neurologist in charge eventually discovered that the cause was "new heroin," manufactured in a clandestine laboratory. This designer drug was intended to contain MPPP (1-methyl-4-propionoxypiperidine), an uncontrolled substance having the pharmacological effects of meperidine, a controlled substance. while meperidine is stable, MPPP breaks down on heating and at acid pH to MPTP (Fig. 1). The latter substance in faulty batches of "new heroin" was clearly the agent inducing Parkinsonism, since systemic administration to primates and mice resulted in dopamine depletion, neurological disorders, and degeneration of the nigrostriatal cells, as in Parkinson's disease (2-4). These reports generated great excitement since they suggested that a long sought animal model for the disease had been found. They also created great alarm when the Center for Disease Control found that over 300 people had been exposed to MPTP in Northern California alone.

Flavins and Flavoproteins 1987

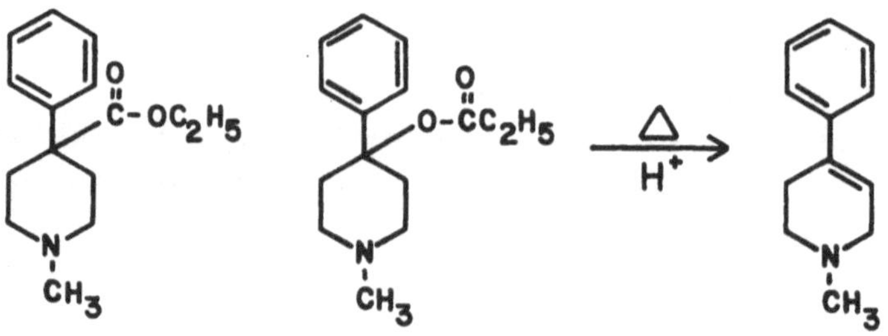

Fig. 1. Structures of meperidine, its designer drug analog, MPPP, and MPTP.

Results and Discussion

Bioactivation of MPTP to the neurotoxic form

Being neutral, MPTP crosses the blood-brain barrier. The absence of chemically reactive groups in the molecule suggests that its extreme toxicity is due to a metabolic transformation. Initial efforts to demonstrate bioactivation were based on the expectation that, like other cyclic tertiary amines, the action of microsomal P-450 would be involved, until Chiba *et al.* (5) discovered that brain mitochondria, rather than microsomes, were responsible for the formation of the neurotoxic species. Using brain mitochondria, they demonstrated the progressive oxidation of MPTP to the dihydropyridinium form and then to MPP^+ (1-methyl-4-phenylpyridinium), the species now widely believed to be the neurotoxic agent (Fig. 2). Being a powerful electrophile, its reactivity with proteins is to be expected. Further, MPP^+ is the product which accumulates in the nigrostriatum of MPTP-treated mice and monkeys. The enzyme responsible for the oxidation appeared to be MAO-B, since deprenyl, a selective inhibitor of the B type of MAO, but not clorgyline, a MAO-A selective inhibitor, blocked the oxidation (5). A collaboration between the laboratories of Castagnoli, Trevor and ours confirmed the basic observations but also brought to light some unexpected facts. Until this time, tertiary amines were thought to be slowly or not at all oxidized by MAO. Using highly purified MAO-A and MAO-B and initial rate measurements, MPTP was found to be rapidly oxidized by the B form and reasonably fast by the A form, with Km values comparable to those of the best known substrates of the enzyme (6). In a recent extension of these studies, some MPTP analogs synthesized by Youngster (7) were found to be excellent substrates for MAO-A or -B (Table I). Although the turnover numbers are lower, the V_{max}/K_m value for N-methyl-4-benzyltetrahydropyridine far exceeds that for benzylamine for MAO-B and that for the 2'-ethyl analog is almost as high as that for kynuramine for MAO-A. Gratifyingly, prevention of the *in vivo* toxicity of MPTP congeners which are primarily B substrates requires deprenyl alone, that of A-B substrates, like 2'-methyl MPTP, requires both deprenyl and clorgyline, and clorgyline alone blocks most of the neurotoxic action of 2'-ethyl MPTP, a predominantly A substrate (7). There is thus a gratifying agreement between *in vivo* findings and the enzyme studies. One area of seeming disagreement between enzymological and toxicological studies was the finding that inhibitors of MAO B alone blocked the neurotoxic action of MPTP, although the compound was also oxidized by the A enzyme. An examination of the kinetics of MPTP and $MPDP^+$ oxidation by the two enzymes revealed the resolution of the paradox (8,9). Both the 2-e and 4-e oxidation products of MPTP oxidation were excellent competitive inhibitors of the A enzyme but were relatively ineffective inhibitors of

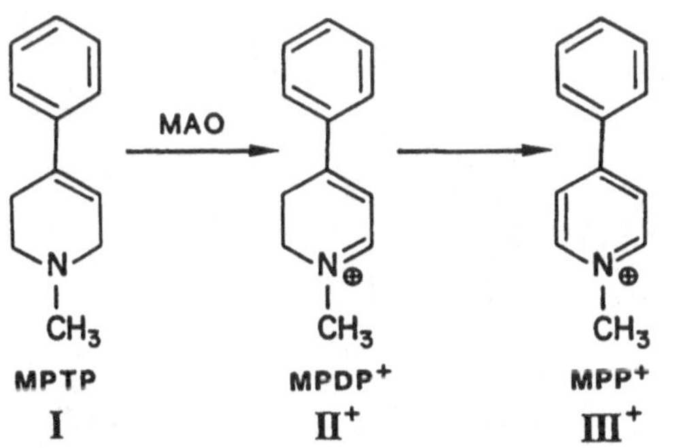

Fig. 2. Oxidation of MPTP to MPP^+.

TABLE I. REACTIVITIES OF MPTP ANALOGUES WITH MAO

COMPOUND	V_{max}/K_m MAO A	MAO B
KYNURAMINE	800	
BENZYLAMINE		1580
MPTP	100	657
2'-CH_3-MPTP	480	1490
2'-CH_2-H_5-MPTP	743	295
3'-CH_3-MPTP	76	650
N-CH_3-CYCLOHEXYLTETRAHYDROPYRIDINE	270	688
N-CH_3-4-BENZYLTETRAHYDROPYRIDINE	103	2675

TABLE II. INHIBITION OF NADH OXIDASE ACTIVITY IN INNER MEMBRANE PREPARATIONS

INHIBITOR	I_{50}
1-METHYL-4-PHENYLPYRIDINIUM (MPP^+)	7.2
1-METHYL-3-PHENYLPYRIDINIUM	20
1-METHYL-2-PHENYLPYRIDINIUM	11
1,2-DIMETHYL-4-PHENYLPYRIDINIUM	NOT REACHED
1-METHYL-4-T-BUTYLPYRIDINIUM	NOT REACHED
4-PHENYLPYRIDINE	0.30
$MPDP^+$	0.60
MPTP	0.92
4-PHENYL-1,2,3,6-TETRAHYDROPYRIDINE	1.1
2,3,4,9,9-PENTAHYDRO-2-METHYL-1-H-INDENO [2,1-C]PYRIDINE	0.75
4-PHENYLPIPERIDINE	1.3

the B form. Thus, oxidation of MPTP by MAO-A ceases soon after the first few catalytic cycles because of competitive inhibition by the products, while oxidation by MAO-B continues.

Mechanism based inactivation of MAO by MPTP and $MPDP^+$

Both types of MAO also undergo mechanism based inactivation by MPTP and $MPDP^+$, but not by MPP^+, of course (8,9). The characteristics of the process are those of a typical suicide inhibitor with a large partition coefficient, so that the overwhelming majority of the product dissociates, instead of forming a covalent adduct with the enzyme. This may be the reason why about 4 mols of radioactive product were found per mol of enzyme after denaturation, when [^{3}H]-MPTP was used for inactivation of MAO-B, some of which must represent recombination of the product with nucleophiles on the enzyme. Interestingly, dialysis of the inactive enzyme for several days results in partial recovery of the catalytic activity, suggesting the formation of a slowly dissociated covalent adduct at the active site.

Kinetic mechanism of MAO revealed by the use of MPTP

The high reactivity of MPTP with MAO and the relative chemical stability of $MPDP^+$, the primary oxidation product, has permitted re-examination of the kinetic mechanism of MAO-B. Prior work in our laboratory (10) and elsewhere (11) suggested that, depending on the substrate, either a binary complex or a ternary complex of reduced enzyme-product-O_2, but not both, is found, as with other flavoprotein oxidases (12). Using synthetic $MPDP^+$ to form the ternary complex, it has now become clear that the mechanism for a given substrate is not exclusive but is determined by competition between alternate pathways. Moreover, the substrate can also bind to the reduced flavoprotein and form a ternary complex with O_2 (13). A schematic representation of the three competing pathways is given in Fig. 3. In contrast, in stopped-flow studies with pure MAO-A, using MPTP or kynuramine as substrates, only a binary complex mechanism was evident.

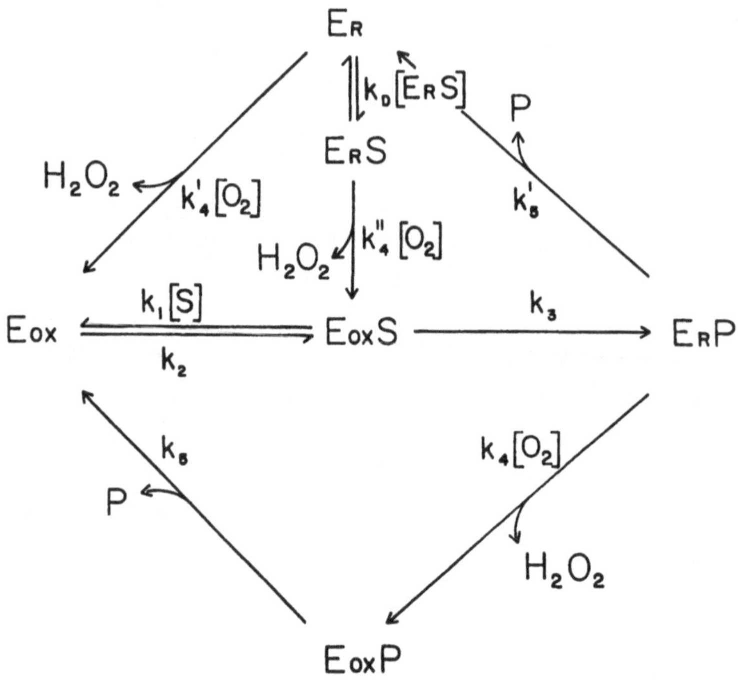

Fig. 3. Alternate kinetic mechanisms for the action of MAO-B. E_R = reduced enzyme, E_{ox} = oxidized enzyme, S = substrate, P = product, K_d is a dissociation constant.

Accumulation of MPP^+ in the dopaminergic neurons

Until recently, it was widely believed that most of the MAO-B in the brain is localized in the astrocytes. This posed the problem of how the positively charged and reactive MPP^+ (or its precursors) get from their site of formation to the dopaminergic cells in the nigrostriatum, where MPP^+ eventually accumulates (Fig. 4). Studies in Snyder's and Heikkila's laboratories (14,15) illuminated an unexpected reaction between MPP^+ and the dopamine reuptake system on the pre-synaptic side of dopaminergic nerve functions. They demonstrated that the reuptake system takes up MPP^+ avidly, competing with dopamine, and that the uptake of both compounds is inhibited to comparable extents by "reuptake inhibitors," such as mazindol, and both are Na-dependent. This would provide an efficient collecting mechanism for MPP^+ formed in the astrocytes. More recent studies point to the presence of appreciable MAO-B activity in the neurons of the substantia nigra, but it is not known whether this is sufficient to account for the formation of all the MPP^+ which gets eventually localized in the nigrostriatum. The fact that reuptake inhibitors protected certain species from the neurotoxic effect of MPTP seems to suggest that the reuptake system does play an important role in the progress of the neuronal loss initiated by the administration of MPTP.

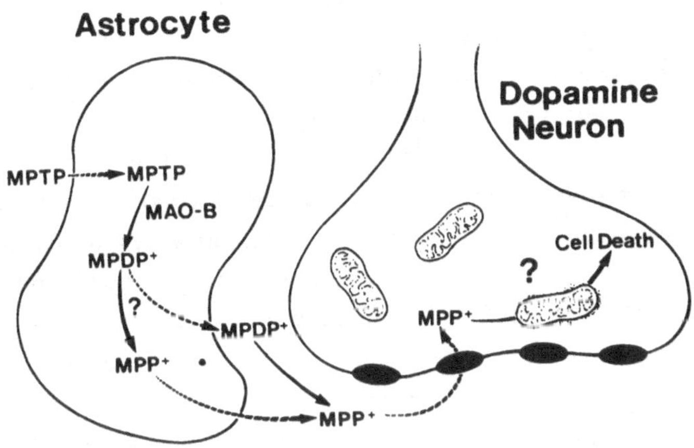

Fig. 4. Schematic representation of the events leading from the oxidation of MPTP in the astrocytes to nigrostriatal cell death.

Accumulation of MPP^+ in the mitochondria

From the standpoint of the biochemist, the key question remained: how and why does MPP^+ destroy the neurons of the substantia nigra. The first clue came from the work of Nicklas *et al.* (16) who reported that the incubation of rat liver or brain mitochondria with a moderately high concentration (0.2 to 0.5 mM) of MPP^+ resulted in progressive cessation of ADP-stimulated (State 3) respiration on NAD^+-linked substrates, with no effect on succinate oxidation. They concluded that MPP^+ somehow blocks NADH-dehydrogenase, halting oxidative phosphorylation. The hypothesis was appealing, because the resulting ATP depletion would lead to cell death. Brain cells are well known to depend uniquely on aerobic oxidations for ATP synthesis and, once damaged or destroyed, do not regenerate. The only problem apparent was that the concentration of MPP^+ required to inhibit the oxidation of glutamate + malate was considerably higher than the 30 µM concentration found by others in the nigrostriatum on autopsy of subhuman primates and mice exposed to MPTP. In the mitochondria inverted by brief sonication to expose the substrate binding site of the dehydrogenase to the outside of the inner membrane, preincubation with 0.5 mM MPP^+ resulted in very little inhibition of NADH oxidation. Further comminution of the mitochondria to isolated inner membrane (ETP) and then to Complex I similarly showed virtually complete loss of the effect seen in intact mitochondria, although at much higher concentrations of MPP^+ (> 10 mM) inhibition was extensive (17,18). Having ruled out the obvious possibilities that one of the substrate carriers or the pyridine nucleotide dehydrogenases themselves, rather than NADH dehydrogenase, were the targets of MPP^+, we were lead to postulate the existence of an energy-dependent concentration mechanism for MPP^+ in the intact inner membrane. By measuring $[^3H]$-MPP^+ uptake into the matrix, we found a highly efficient concentration mechanism for MPP^+ in liver and midbrain mitochondria (19,20). The system could provide an exceedingly high (20-30 mM) concentration of MPP^+ in the matrix, even with a 30 µM external steady-state concentration. It could be energized by succinate, NAD^+-linked substitutes, or ATP. Uncouplers blocked this action: if 2,4-DNP was added in flight while MPP^+ was being accumulated in the matrix, the MPP^+ inside flowed out with the concentration gradient (Fig. 5). Valinomycin plus K^+ inhibited the "carrier" completely while nigericin did not, showing that the electrical gradient of the membrane supports the process. Although in terms of all properties examined (Km, energy of activation, etc.) this pumping mechanism is typical of mitochondrial metabolite transporters, perhaps it is premature to call it a carrier, though the involvement of a protein with rigid structural requirements for binding and a concentration process dependent on the electrical potential are well established.

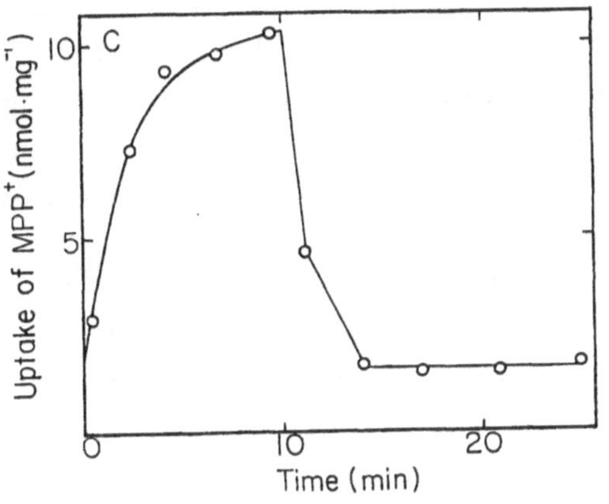

Fig. 5. Energy dependent uptake of MPP^+ into the matrix and its reversal by uncouplers. Uptake by rat liver mitochondria was followed for 10 min, at which time DNP was added, causing efflux with the concentration gradient.

Inhibition of mitochondrial NADH dehydrogenase by pyridine derivatives

The concentration of MPP^+ reached in the matrix by the action of the pumping system is more than sufficient to block NADH dehydrogenase extensively or completely. Table II presents I_{50} values for some substituted pyridines. It is seen that substitution on the pyridine ring or removal of aromatic ring at the 4-position decreases the reactivity with the enzyme. The same may not be true of substitition on the aromatic ring; however (21). As seen in Table II, MPTP, $MPDP^+$, 4-phenylpyridine, and 4-phenyl-1,2,3,6-tetrahydropyridine are all far more inhibitory to NADH oxidation than is MPP^+. MPTP is not pumped into the mitochondria by the carrier described, nor does it react with the dopamine reuptake system. This is also why compounds like 4-phenylpyridine are not <u>acutely</u> neurotoxic, since passive diffusion probably does not result in the intramitochondrial concentrations needed to bring about extensive blockade of oxidative phosphorylation. As we shall presently see, however, <u>chronic</u> exposure to such compounds would be quite another story (21). For the biochemist, the entire phenomenon has many interesting, even puzzling effects. The first question is the mechanism of the inhibition. The electrophilic nature of MPP^+ is not necessarily relevant, for reactive nucleophiles are present in many mitochondrial enzymes, including succinate dehydrogenase, none of which is inhibited. Moreover, the inhibition of NADH dehydrogenase is completely reversible on dilution (22), showing that covalent adduct formation is rather unlikely. A related question is where in the complex NADH dehydrogenase molecule, containing FMN and at least 4 Fe-S clusters MPP^+ blocks electron transport. In collaborative studies with Dr. M.K. Johnson's laboratory, (21,22), we approached the problem in two ways: by the use of enzyme assays involving electron acceptors which tap off electrons at known points in the electron transfer sequence and by the use of low temperature EPR, comparing the signal intensity and line shape of untreated preparations with those completely inhibited by MPP^+ or 4-phenylpyridine. Both approaches suggested that the inhibition occurs on the O_2 side of the highest potential Fe-S cluster; i.e., at the junction of the dehydrogenase and Q (Fig. 6). This is the site where rotenone, barbiturates, and piericidin A block electron transfer. For the chemists, justifiably discomfited by such vague descriptions of a non-covalent

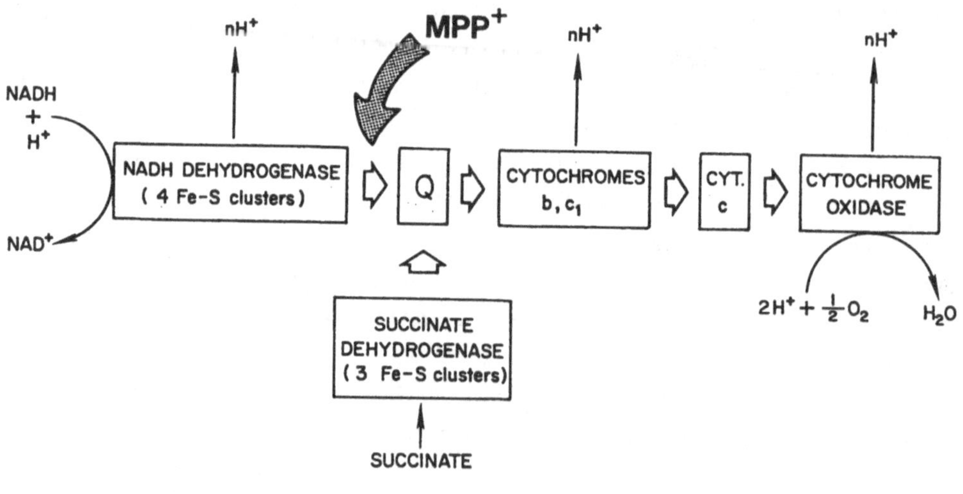

Fig. 6. Proposed inhibition site of MPP^+ in the respiratory chain.

chemical reaction, a reminder is the best we can offer: over 30 years after the localization of the reaction site of these classical inhibitors of NADH dehydrogenase, no viable theory has emerged as to why they are inhibitory.

Relation to idiopathic parkinsonism

The sequence of events depicted, involving the action of two flavoproteins and two biochemical pumping mechanisms, seems to provide a rational framework for explaining the neurotoxicity of MPTP. Several unexplained or poorly understood facts remain, however. One is the striking species specificity. Primates, certain strains of mice, and beagle dogs develop neurological symptoms on exposure to MPTP, which vary from permanent to long-lasting. In other species, the symptoms are transient or do not ever appear. A case in point is the laboratory rat, which is remarkably resistant to MPTP. One theory holds that sensitivity to MPTP is a function of the neuromelanin concentration in the nigrostriatum, prominent in susceptible species, undetectable in insensitive ones. Neuromelanin binds (23,24) MPP^+ and thus may serve as a repository until the pumping system concentrates it all in the matrix. Other theories invoke differences in the metabolic removal of MPP^+. Another uncertainty is the organ specificity: why does only a small part of the brain succumb to MPTP exposure, and not other tissues with high levels of MAO-B and of the MPP^+ carrier? The prevalent speculation, again, invokes detoxification mechanisms. In the long run, however, as biochemists, we must ask whether these stimulating findings have a bearing on the etiology and eventual elimination of Parkinson's disease. Volumes of speculations have been written on the causes of the human disease, one which was first described in the modern area. Perhaps for this reason, there has been a tendency to blame its appearance on the industrial revolution and the environmental pollution it caused. Opponents of this notion have been vocal in pointing out that regions of the world which are not industrialized contain pockets of rampant parkinsonism. An alternative was to look for genetic or infectious agents, but none has emerged. It has been emphasized that 4-phenylpyridine, for example, is present in both industrial emissions and in some foodstuffs (25). With such a possibility in mind, statistical correlations with a given factor may be exceedingly difficult to establish. Some of us in the field share the view that the tragic events which brought MPTP to our consciousness may be, in the long run, balanced by the potential benefit of arriving at a rational explanation of the etiology of the human disease. If we can identify the agents, we can at least compile a list of potentially hazardous substances. The analogy between the neurological, anatomical, and pharmacological findings in MPTP-poisoned primates and patients afflicted with Parkinson's disease is too close to be a coincidence. What the biochemist has to offer is a battery of simple tests readily carried out in any competent biochemical laboratory for detecting substances structurally related to pyridines, exposure to which may put a person at risk.

Acknowledgments

These studies were supported by the National Institutes of Health (HL 16251 and NS 23066), the National Science Foundation (DM 84967) and the Veterans Administration.

REFERENCES

1. Langston, J.W., P. Ballard, J.W. Tetrud, I. Irwin. 1983. Science 219, 979.

2. Burns, R.S., C.C. Chiueh, S.P. Markey, M.H. Ebert, D.M. Jacobowitz, I.J. Kopin. 1983. Proc. Nat. Acad. Sci. USA 80, 4546.

3. Heikkila, R.E., A. Hess, R.C. Duvoisin. 1984. Science 224, 1451.

4. Langston, J.W., L.S. Forno, C.J. Robert, I. Irwin. 1984. Brain Res. 292, 390.

5. Chiba, K., A. Trevor, N. Castagnoli, Jr. 1984. Biochem. Biophys. Res. Commun. 120, 574.

6. Salach, J.I., T.P. Singer, N. Castagnoli, Jr., A. Trevor. 1984. Biochem. Biophys. Res. Commun. 125, 831.

7. Heikkila, R.E., S.K. Youngster, B.A. Sieber, P.K. Sonsalla, K. McKeown, T.P. Singer. 1987. Abstracts, Society for Neuroscience.

8. Singer, T.P., J.I. Salach, D. Crabtree. 1985. Biochem. Biophys. Res. Commun. 127, 707.

9. Singer, T.P., J.I. Salach, N. Castagnoli, Jr., A. Trevor. 1986. Biochem. J. 235, 785.

10. Husain, M., D.E. Edmondson, T.P. Singer. 1982. Biochemistry 21, 595.

11. Pearce, L.B., J.A. Roth. 1985. Biochemistry 24, 1821.

12. Bright, H.J., D.J.T. Porter. 1975. Enzymes, 3rd ed. 12, 421.

13. Ramsay, R.R., S.C. Koerber, T.P. Singer. 1987. Biochemistry 26, 3045.

14. Javitch, J.A., R.J. D'Amato, S.M. Strittmatter, S.H. Snyder. 1985. Proc. Nat. Acad. Sci. USA 82, 2173.

15. Heikkila, R.E., S.K. Youngster, L. Manzino, F.S. Cabbat, and R.C. Duvoisin. 1985. J. Neurochem. 44, 310.

16. Nicklas, W.J., I. Vyas, R.E. Heikkila. 1985. Life Sci. 36, 2503.

17. Ramsay, R.R., J.I. Salach, T.P. Singer. 1986. Biochem. Biophys. Res. Commun. 134, 743.

18. Ramsay, R.R., J.I. Salach, J. Dadgar, T.P. Singer. 1986. Biochem. Biophys. Res. Commun. 135, 269.

19. Ramsay, R.R., T.P. Singer. 1986. J. Biol. Chem. 261, 7585.

20. Ramsay, R.R., J. Dadgar, A.J. Trevor, T.P. Singer. 1986. Life Sci. 39, 581.

21. Singer, T.P., N. Castagnoli, Jr., R.R. Ramsay, A.J. Trevor. 1987. J. Neurochem. 49, 1.

22. Singer, T.P., A.J. Trevor, N. Castagnoli, Jr. 1987. TIBS 12, 266.

23. D'Amato, R.J., Z.P. Lipman, S.H. Snyder. 1986. Science 231, 987.

24. Wu, E.Y., K. Chiba, A.J. Trevor, N. Castagnoli, Jr. 1986. Life Sci. 39, 1695.

25. Snyder, S.H., R.J. D'Amato. 1985. Nature 317, 198.

BIOSYNTHESIS OF FLAVOPROTEINS

Kunio Yagi, Taisuke Inagaki, Chikako Ichihara,
Takuya Koshizaka, Nobuko Ohishi

Institute of Applied Biochemistry, Yagi Memorial Park,
Mitake, Gifu 505-01, Japan

Sandro Ghisla

Fakultät für Biologie, Universität, D-7750 Konstanz, FRG

Norihiro Tsukagoshi, Shigezo Udaka

Department of Food Science and Technology, Faculty of
Agriculture, Nagoya University, Nagoya 464, Japan

Morimitsu Nishikimi, Takayuki Ozawa

Department of Biomedical Chemistry, Faculty of Medicine,
University of Nagoya, Nagoya 466, Japan

Introduction

Although many investigations have been made to clarify the mechanism of reaction of flavoproteins, their biosynthesis, especially the mechanism of the incorporation of flavin into covalently bound flavoproteins, has remained to be elucidated. We intended to address this problem by using techniques such as in vitro protein synthesis and molecular cloning. Flavoproteins studied here were mitochondrial general acyl CoA dehydrogenase (GAD) and electron transfer flavoprotein (ETF), peroxisomal D-amino acid oxidase (DAO), and microsomal L-gulono-γ-lactone oxidase (GLO). Although experiments with these enzymes are still in progress in our laboratories, some useful results for a better understanding of the problem have already been obtained. This paper reports our initial findings.

Flavins and Flavoproteins 1987

Results and Discussion

Specificity of rabbit antibodies against flavoproteins

In the present study, antibodies against flavoproteins were used for detection of synthesized proteins and screening of the cDNA's. Thus, we examined the specificity of the antibodies prior to those experiments. Immunoblotting was carried out using the crude extract prepared from the homogenate, as well as mitochondrial and microsomal fractions, of kidney or liver. In the experiment using antisera against GAD, ETF, DAO, and GLO, the main band which was visualized on the blot by the peroxidase method was shown to coincide exactly with that of the respective purified flavoprotein (1-5). This indicates that the antibodies used here are specific for each flavoprotein.

In vitro synthesis of flavoproteins

For the *in vitro* synthesis of flavoproteins, total RNA's of pig kidney and liver were prepared by SDS-phenol extraction (6) and those of rat liver were prepared by guanidinium/cesium chloride extraction (7). Poly(A)$^{+}$RNA was separated from total RNA's by oligo-dT cellulose column chromatography.

Cell-free translation in a wheat germ extract or rabbit reticulocyte lysate system was carried out at 30°C for 60 min. Immunoprecipitation of the translational products was performed using protein A-Sepharose (8). The solution containing the immunoprecipitated products was heated at 100°C for 5 min and subjected to 10% SDS-PAGE according to the method of Laemmli (9). After electrophoresis, the gel was stained with Coomassie Brilliant Blue R and then subjected to autoradiography or fluorography.

When translational products in the wheat germ extract system were immunoprecipitated with anti-GAD antiserum, and analyzed by SDS-PAGE/autoradiography, a main band with an approximate molecular weight of 43K was detectable, as shown in Fig. 1. This band was identified immunologically as GAD, since its appearance could be prevented by the addition of an excess amount of the purified enzyme during immunoprecipitation. This indicates that GAD in pig kidney is synthesized as a precursor having a larger molecular weight (10).

Figure 2 shows SDS-PAGE/fluorography of GLO synthesized in the rabbit reticulocyte lysate system (11). The apparent molecular weight was estimated to be 51K, indicating that GLO was synthesized with the same molecular weight as the mature enzyme (5).

We also attempted in vitro synthesis of DAO and ETF using the wheat germ extract system. However, no bands corresponding to these proteins were detectable. On the other hand,

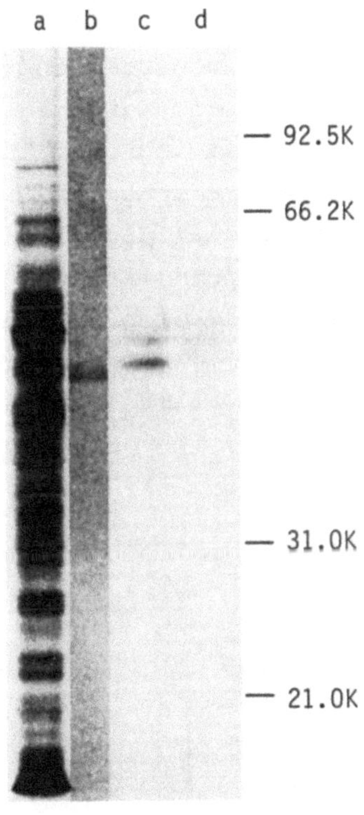

Fig. 1. In vitro synthesis of general acyl CoA dehydrogenase. Lane a, total translation products; b, the purified enzyme stained with Coomassie Brilliant Blue R; c, immunoprecipitates formed in the absence of the purified enzyme; d, immunoprecipitates formed in the presence of the purified enzyme.

when the rabbit reticulocyte lysate system was used for translation, a faint band with a molecular weight of 38K corresponding to DAO was detected (12). The appearance of this band was prevented by the addition of an excess amount of the purified enzyme, indicating that the synthesized protein is indeed DAO and undergoes little, if any, processing once initially synthesized. Thus, the information directing DAO to peroxisomes is likely to reside somewhere in its primary structure. Fukui et al. (13) also reported a similar result for the *in vitro* synthesis of pig kidney DAO.

The band corresponding to ETF was scarcely detectable.

Effect of methylmercury hydroxide on the biosynthesis of DAO and ETF

Payvar and Schimke (14) have reported that treatment of conalbumin mRNA with the nucleic acid denaturant methylmercury hydroxide prior to translation improved the translational efficiency. We, therefore, examined the effect of the reagent on the translational activity of pig kidney and liver poly(A)$^+$-RNA in order to enhance the translational yield. As shown in Fig. 3, translational activity of pig kidney poly(A)$^+$RNA was

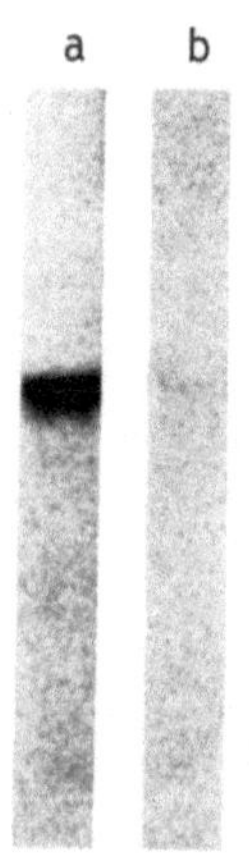

Fig. 2. *In vitro* synthesis of L-gulono-γ-lactone oxidase. Lane a, immunoprecipitates formed with anti-GLO antiserum; b, immunoprecipitates formed with normal rabbit serum.

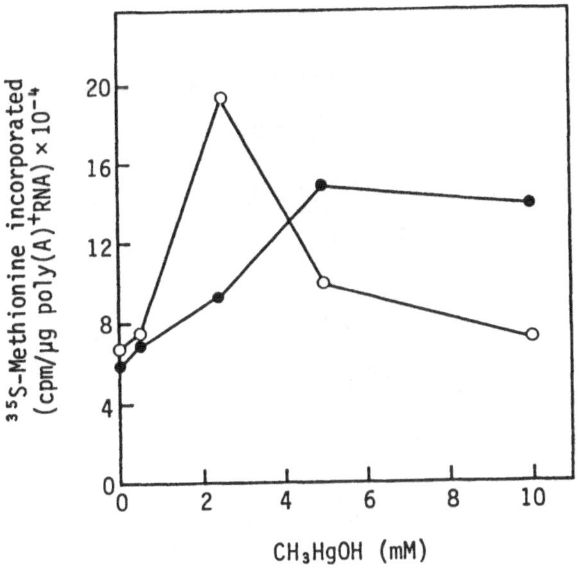

Fig. 3. Effect of methylmercury hydroxide on the translational activity of pig kidney and liver poly(A)$^+$RNA. The activity was assayed with the rabbit reticulocyte lysate. Open circles, pig kidney poly(A)$^+$RNA; closed circles, pig liver poly(A)$^+$RNA.

increased about 3-fold at 2.5 mM and the activity of pig liver poly(A)$^+$RNA was increased 2-fold at 5 mM.

After pig kidney poly(A)$^+$RNA was treated with 2.5 mM methylmercury hydroxide, it was incubated in the rabbit reticulocyte lysate system; and the translational products were immunoprecipitated with anti-DAO antiserum. The immunoprecipitates were analyzed by SDS-PAGE/fluorography. The band of DAO became much more distinct by this treatment (12).

Immunoprecipitates obtained with anti-ETF antiserum from the total translational products in the rabbit reticulocyte lysate system after the treatment of pig liver poly(A)$^+$RNA with methylmercury hydroxide were also analyzed by SDS-PAGE/fluorography. As shown in Fig. 4, lane b, three main bands were detected and their molecular weights were estimated to be 37K, 31K, and 29K, respectively. The bands became more distinct when the poly(A)$^+$RNA was denatured with higher concentrations

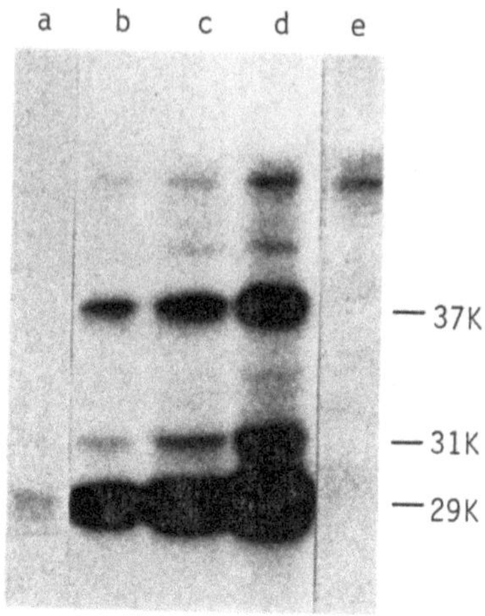

Fig. 4. Effect of methylmercury hydroxide on the production of electron transfer flavoprotein directed in rabbit reticulocyte lysate by poly(A)$^+$RNA from pig liver. Translation was performed in the presence of 0 (a), 0.5 (b), 2.5 (c), and 5 mM (d) methylmercury hydroxide. In lane e, the translation was performed in the presence of 2.5 mM methylmercury hydroxide and immunoprecipitation was carried out in the presence of the purified ETF.

of methylmercury hydroxide (lanes c and d). It was noted that the appearance of these three bands was prevented by the addition of an excess amount of the purified ETF, indicating that all these proteins synthesized were immunologically assigned to ETF. From these results we assume that the synthesized protein having a molecular weight of 37K would be a precursor of the α-subunit (mature protein: ∿34K) and that both or one of the two proteins synthesized having a molecular weight of 31K or 29K would be responsible to the β-subunit (mature protein: ∿27-31K). Ikeda et al. (15) reported on rat liver ETF that the α-subunit was synthesized as a precursor and the β-subunit as the mature protein.

The presently observed enhancement of the synthesis of ETF by treatment of poly(A)$^+$RNA with methylmercury hydroxide could be ascribed to the structure of the mRNA in its isolated form or to its aggregation during preparation, as in the case of

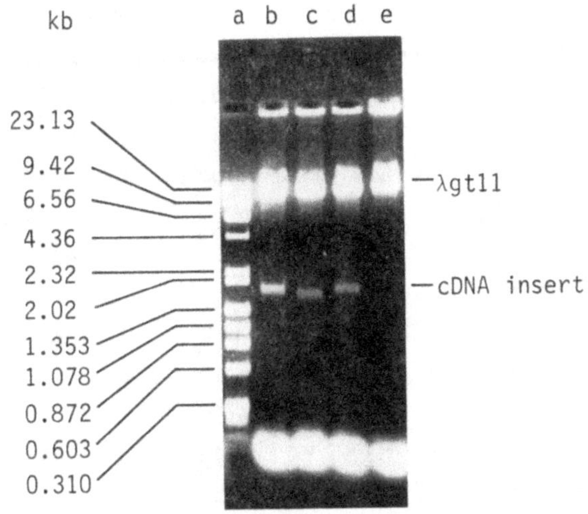

Fig. 5. Agarose gel electrophoresis of λ-GAD 1, 2, and 3 after digestion with EcoRI. Lane a, marker DNA; b, λ-GAD 1; c, λ-GAD 2; d, λ-GAD 3; e, λgt11 alone.

DAO (12,16). This seems not to be the case for GAD, since the synthesis of GAD was not enhanced by the treatment with this reagent at 2.5 mM.

Isolation of the cDNA clones for GAD and DAO

The cDNA clones for GAD and DAO were isolated by immunoscreening a rat liver cDNA library in λgt11, in which antisera against the respective flavoproteins were used according to the method of Young and Davis (17). Immunopositive clones were visualized by the horseradish peroxidase method using goat anti-rabbit IgG-horseradish peroxidase conjugates.

Three positive clones for GAD were obtained after the third immunoscreening. In order to determine the insert sizes of the cDNA clones, the phage DNA's were digested with EcoRI and subjected to 1% agarose gel electrophoresis. From this, the sizes of the inserts were estimated to be 1.5-1.6 kb (Fig. 5).

For the analysis of the fusion protein, E. coil Y1090 was infected with plaque-purified immunopositive phage and induced with 10 mM isopropylthiogalactoside for fusion protein production. Proteins extracted from bacteria were subjected to SDS-PAGE. After electrophoresis, the proteins on the gel was transferred to a nitrocellulose membrane and the blot was subjected to immunostaining. Fusion proteins of approximately 155-160K produced in the bacteria were detected on the immunoblots. As the fusion protein consists of GAD and β-galactosidase (114K), the net molecular weight of GAD was estimated to be 40-50K. This indicates that the cDNA clones have presumably a full length of the GAD structure gene, because the molecular weight of the precursor was shown to be approximately 43K (see Fig. 1).

By the same immunoscreening procedure, we isolated two positive clones for DAO. Immunoblotting analysis showed that the cDNA's having 1.65-1.7 kb have only a partial structure gene for DAO followed by a relatively long-tailed 3' non-coding region of 1.3-1.4 kb.

In Table 1, the results of the present study are summarized.

Table 1. *In vitro* synthesis of flavoproteins and molecular cloning of their cDNA's

	Molecular weight		Translation		Enhancement with methylmercury hydroxide	Size of cDNA (kb)
	Mature	Synthesized	Wheat germ	Rabbit reticulocyte		
GAD	42K	43K	+	+	No	1.6
ETF	34K	37K	-	+	Yes	ND
	27-31K	27,31K	-	+	Yes	
DAO	38K	38K	-	+	Yes	1.65-1.7 (partial) 3.0-3.5 (putatively native form)
GLO	51K	51K	ND	+	ND	ND

ND, not determined.

GAD was synthesized as a precursor protein having a slightly larger molecular weight than that of the mature form and the size of the cDNA was 1.6 kb. ETF was shown to be quite different in that its α-subunit was synthesized as a protein having a larger molecular weight, whereas its β-subunit was synthesized as protein(s) having the same molecular weight as the mature protein. Furthermore, the biosynthetic efficiency was affected markedly by denaturation of the mRNA. DAO was synthesized as a protein having the same molecular weight as that of the mature protein and its biosynthesis was also enhanced by denaturation of the mRNA, which was found to have a long-tailed 3' non-coding region. GLO was synthesized as a protein having the same molecular weight as that of the mature enzyme.

References

1. Thorpe, C., R. G. Matthews, C. H. Williams, Jr. 1979. Biochemistry 18, 331.
2. Husain, A., D. J. Steenkamp. 1983. Biochem. J. 209, 541.
3. Yagi, K., M. Naoi, M. Harada, K. Okamura, H. Hidaka, T. Ozawa, A. Kotaki. 1976. J. Biochem. 61, 580.
4. Jenkins, W. T., A. Behlen, P. J. Rogers. 1979. Anal. Biochem. 94, 105.
5. Nishikimi, M., B. M. Tolebert, S. Udenfriend. 1976. Arch. Biochem. Biophys. 175, 427.
6. Palmiter, R. D. 1974. Biochemistry 13, 3606.
7. Glišin, V., R. Crkvenjakov, C. Byus. 1974. Biochemistry 13, 2633.
8. Watanabe, A., C. A. Price. 1982. Proc. Natl. Acad. Sci. USA 79, 6304.
9. Laemmli, O. K. 1970. Nature 227, 680.
10. Inagaki, T., N. Tsukagoshi, C. Ichihara, N. Ohishi, S. Udaka, S. Ghisla, K. Yagi. 1986. Biochem. Biophys. Res.

Commun. 137, 1049.

11. Koshizaka, T., M. Nishikimi, T. Ozawa, K. Yagi. Biochem. Int. in press.

12. Inagaki, T., N. Ohishi, N. Tsukagoshi, C. Ichihara, S. Udaka, K. Yagi. 1986. Biochem. Int. 13, 1045.

13. Fukui, K., K. Momoi, F. Watanabe, Y. Miyake. 1986. Biochem. Biophys. Res. Commun. 141, 1222.

14. Payvar, F., R. T. Schimke. 1979. J. Biol. Chem. 254, 7636.

15. Ikeda, Y., S. M. Keese, K. Tanaka. 1986. J. Clin. Invest. 78, 997.

16. Inagaki, T., N. Ohishi, T. Koshizaka, C. Ichihara, K. Yagi, N. Tsukagoshi, S. Udaka: This volume.

17. Young, R. A., R. W. Davis. 1983. Science 222, 778.

Biological and Medical Aspects of D-Amino Acid Oxidase
— Biogenesis and *in vivo* reaction with D-propargylglycine —

Yoshihiro Miyake, Kiyoshi Fukui, Kyoko Momoi, Fusao Watanabe and Toshihiro Shibata.

Department of Biochemistry
National Cardiovascular Center Research Institute
Fujishiro-dai, Suita, Osaka 565, Japan

Introduction

D-amino acid oxidase (E.C. 1.4.3.3., DAO) is one of the representative flavoproteins with flavin adenine dinucleotide (FAD) as the prosthetic group, that catalyzes the oxidative deamination of D-amino acids. Many studies have been carried out to clarify the physicochemical properties and reaction mechanism of this enzyme. Histochemical studies also revealed the existence of DAO in various tissues, i.e. liver, kidney proximal tubules, certain parts of brain (1) and granules of neutrophilic leukocyte (2). Since D-amino acids are rare in mammalian proteins and they are not found as normal metabolic intermediates, the biological significance of DAO remains to be elucidated. In this respect, it is interesting to note that D-propargylglycine (D-PG, Figure 1.) binds covalently to DAO by dynamic affinity labelling. Thus, the reaction with D-PG results in inactivation of DAO and simultaneous formation of 2-amino-4-hydroxy-2,4-pentadienoate γ-lactone, a main unusual reaction product as DAO reaction (3). On the basis of this *in vitro* reaction, the effect of D-PG on rat has been investigated by intraperitoneal injection of D-PG.
In this communication, we describe our approach to study the biological and medical aspects of DAO with special emphasis on the reaction of the enzyme with D-PG.

Flavins and Flavoproteins 1987

Results and Discussion

<u>In vivo synthesis of DAO and its reaction with D-PG</u>

Intraperitoneal injection of D-PG to rat resulted in massive glucosuria, amino aciduria and polyuria. Almost all the amino acids increased in the urine, and Lys, Gly, and Ala were predominant (4). Urinary sodium and lysozyme excretions were also enhanced by D-PG injection but potassium level remained almost normal. The symptoms disappeared in a week without leaving any histological changes in the kidney tissue. These observations indicated the disorder in renal proximal tubles, since they contain high level of DAO and uptake glucose and amino acids in the nephron.

A functionally differentiated cell line LLC-PK_1 (5), derived from pig kidney proximal tubules, was selected to study this symptom further at cellular level. DAO activity decreased greatly by trypsinization of the cell, although significant amount of DAO was induced in the cell when the confluent monolayer was formed. Biosynthetic labelling of the cell and subsequent immunoprecipitation with anti-hog DAO antibody detected a single band on sodium dodecyl sulfate polyacrylamide gel electrophoresis (SDS-PAGE).

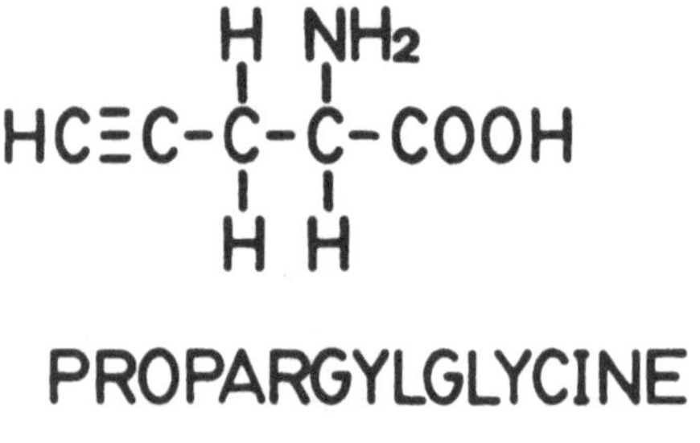

Figure 1. Propargylglycine

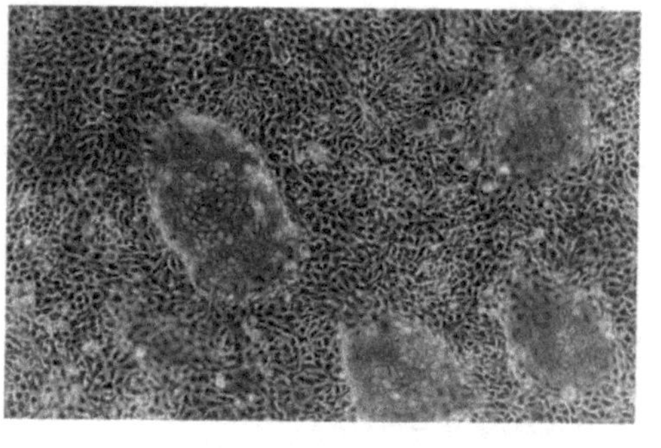

Figure 2. LLC-PK_1 cells 14 days after seeding (phase contrast microscopy).

To investigate the effect of exogenous D-PG on the *in vivo* synthesized DAO in cultured cell, these cells were incubated with D-PG. Intracellular DAO activity was effectively inhibited by D-PG, but no change was detected by L-PG. Both D-PG and L-PG had no effect on intracellular catalase activity. LLC-PK_1 cells exhibit unidirectional glucose transport, characteristic of epithelium of renal proximal tubules. This sodium dependent and phlorizin sensitive glucose transporting activity was evaluated by α-methyl-D-glucoside uptake, but no change was observed even 0.1 M D-PG nor L-PG was incubated with the cells. On the contrary, a chromatographically isolated oxidation product of D-PG after *in vitro* reaction with purified DAO inhibited the glucose uptake very efficiently. These results suggested the possibility that accumulation of D-PG oxidation product produced by DAO at proximal tubules caused inhibition of transepithelial transporting function of the kidney.

In vitro synthesis of DAO and molecular analysis of essential amino acid residues

We have already established that DAO was synthesized *in vitro* using hog kidney mRNA and free polysomes and also *in vivo* using ^{35}S-methionine and LLC-PK_1 cells (6). In this previous study, we did detect DAO protein, but could not observe the enzymatic activity. Therefore, we constructed *in vitro* DAO synthesizing system. We determined the complete nucleotide sequence of the complementary DNA (cDNA) for pig kidney DAO (7). The restriction enzyme map of the cDNA insert is shown in Figure 3, and the deduced amino acid sequence of the enzyme is shown in Figure 4. With this cloned cDNA as a molecular probe, we tried to synthesize functional enzyme *in vitro*. Restriction enzyme fragment of cDNA containing whole coding region of the protein was ligated with the pSP64 vector to generate SP6 promotor-directed RNA transcript. *In vitro* translation was carried out in rabbit reticulocyte lysate with the RNA transcript capped by guanylyltransferase.

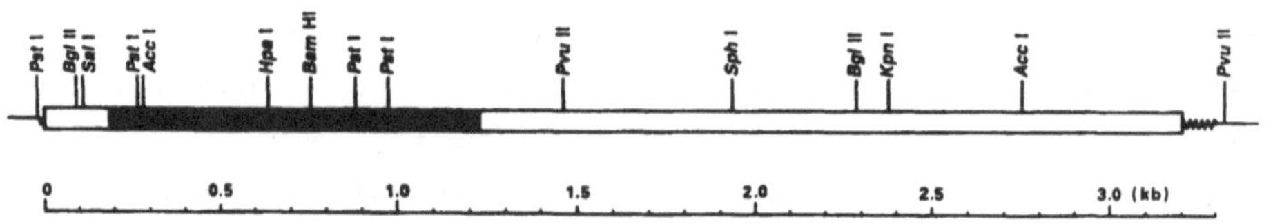

Figure 3. Restriction enzyme map of cDNA for pig kidney DAO. The map displays only the relevant restriction enzyme sites. A solid line represents the vector DNA; a zigzag line represents the oligo(dG)(dC) tail; a solid box represents the coding region for DAO; an open box represents the untranslated region.

The translation product labelled in vitro with ^{35}S-methionone was detected as a single band in SDS-PAGE. The catalytic activity of the in vitro synthesized DAO protein was assayed by coupling with LDH using D-alanine as substrate. In vitro translation product showed significant enzymatic activity that was inhibited specifically by sodium benzoate, a well-known potent inhibitor of DAO. Kinetic properties of this in vitro synthesized DAO were comparable to the purified enzyme from pig kidney.

In the next step to use the cDNA clone as a molecular probe, we searched for the active site of DAO by oligonucleotide-directed mutagenesis. Chemical modification of DAO has already been performed by the use of many reagents. In these studies much attention was given to the primary structure of the DAO protein and to the function of specific amino acid residues.(reviewed in 8,9) Those reagents modified DAO and inactivated DAO partially or completely. However, it is ambiguous whether the modifications occur at the active site or not. Chemical modification studies revealed two alkylation sites of DAO by D-PG, 307-histidine and 228-tyrosine. Site-directed mutagenesis was carried out for

1
Met-Arg-Val-Val-Val-Ile-Gly-Ala-Gly-Val-Ile-Gly-Leu-Ser-Thr-Ala-Leu-Cys-Ile-His-
Glu-Arg-Tyr-His-Ser-Val-Leu-Gln-Pro-Leu-Asp-Val-Lys-Val-Tyr-Ala-Asp-Arg-Phe-Thr-
Pro-Phe-Thr-Thr-Thr-Asp-Val-Ala-Ala-Gly-Leu-Trp-Gln-Pro-Tyr-Thr-Ser-Glu-Pro-Ser-
Asn-Pro-Gln-Glu-Ala-Asn-Trp-Asn-Gln-Gln-Thr-Phe-Asn-Tyr-Leu-Leu-Ser-His-Ile-Gly-
Ser-Pro-Asn-Ala-Ala-Asn-Met-Gly-Leu-Thr-Pro-Val-Ser-Gly-Tyr-Asn-Leu-Phe-Arg-Glu-
Ala-Val-Pro-Asp-Pro-Tyr-Trp-Lys-Asp-Met-Val-Leu-Gly-Phe-Arg-Lys-Leu-Thr-Pro-Arg-
Glu-Leu-Asp-Met-Phe-Pro-Asp-Tyr-Arg-Tyr-Gly-Trp-Phe-Asn-Thr-Ser-Leu-Ile-Leu-Glu-
Gly-Arg-Lys-Tyr-Leu-Gln-Trp-Leu-Thr-Glu-Arg-Leu-Thr-Glu-Arg-Gly-Val-Lys-Phe-Phe-
Leu-Arg-Lys-Val-Glu-Ser-Phe-Glu-Glu-Val-Ala-Arg-Gly-Gly-Ala-Asp-Val-Ile-Ile-Asn-
Cys-Thr-Gly-Val-Trp-Ala-Gly-Val-Leu-Gln-Pro-Asp-Pro-Leu-Leu-Gln-Pro-Gly-Arg-Gly-
Gln-Ile-Ile-Lys-Val-Asp-Ala-Pro-Trp-Leu-Lys-Asn-Phe-Ile-Ile-Thr-His-Asp-Leu-Glu-
228
Arg-Gly-Ile-Tyr-Asn-Ser-Pro-Tyr-Ile-Ile-Pro-Gly-Leu-Gln-Ala-Val-Thr-Leu-Gly-Gly-
Thr-Phe-Gln-Val-Gly-Asn-Trp-Asn-Glu-Ile-Asn-Asn-Ile-Gln-Asp-His-Asn-Thr-Ile-Trp-
Glu-Gly-Cys-Cys-Arg-Leu-Glu-Pro-Thr-Leu-Lys-Asp-Ala-Lys-Ile-Val-Gly-Glu-Tyr-Thr-
Gly-Phe-Arg-Pro-Val-Arg-Pro-Gln-Val-Arg-Leu-Glu-Arg-Glu-Gln-Leu-Arg-Phe-Gly-Ser-
307
Ser-Asn-Thr-Glu-Val-Ile-His-Asn-Tyr-Gly-His-Gly-Gly-Tyr-Gly-Leu-Thr-Ile-His-Trp-
Gly-Cys-Ala-Leu-Glu-Val-Ala-Lys-Leu-Phe-Gly-Lys-Val-Leu-Glu-Glu-Arg-Asn-Leu-Leu-
347
Thr-Met-Pro-Pro-Ser-His-Leu

□ Tyr-228,His-307 (D-PG)

Figure 4. Primary structure of porcine kidney DAO predicted from cDNA sequence (7).

both amino acid residues with the use of oligonucleotide and M13 vector. His-307 was converted to Leu and Tyr-228 was changed to Phe. These mutant DNAs were cloned into pSP64, transcribed and translated _in vitro_ as described above. Both mutant DAOs with one amino acid substitution showed no catalytic activity. Based on these results, two amino acid residues His-307 and Tyr-228 that are modified by D-PG seemed to be quite essential for the enzyme activity of DAO.

DAO gene expression in mice

In the course of study searching for the animal model of DAO deficiency from the medical point of view, we chose DDY strain mouse (10) for the molecular biological analysis of the DAO gene expression .

Homogenate of kidneys from DDY strain showed various DAO activity ranginig from 2.0 μmol D-Ala oxidized / min / g kidney to almost undetectable level. Western blot analysis

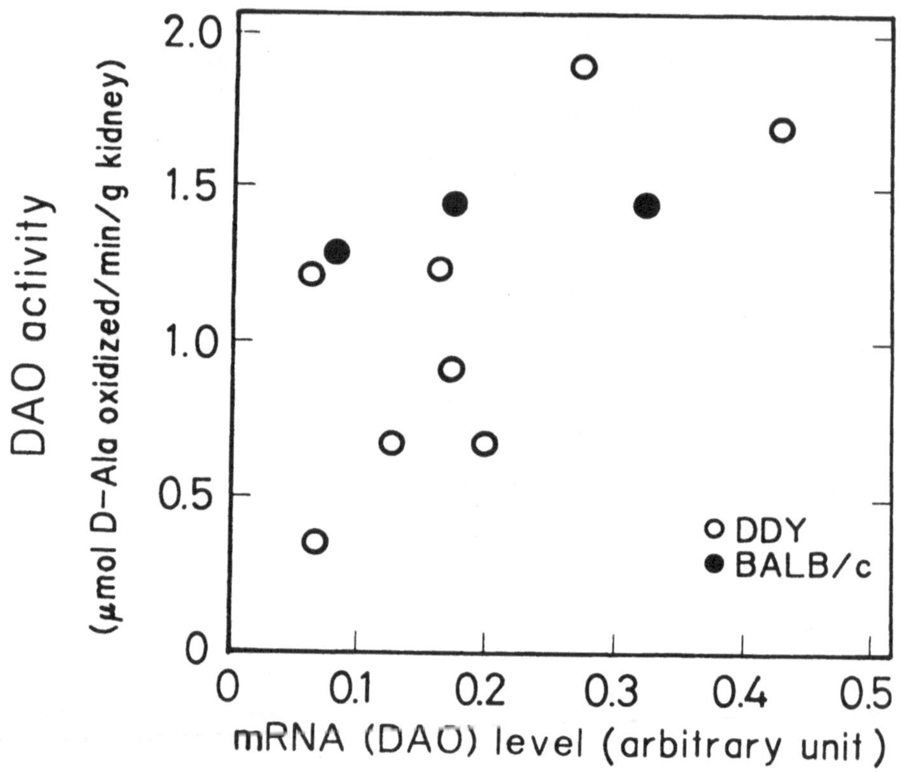

Figure 5. Correlation between DAO activity and its mRNA level in mouse kidney. Each circle represents enzyme activity and mRNA level of DAO in the kidney of individual mouse. mRNA level was estimated relatively by the densitometric comparison of the hybidizing band in Northern blot.

of the kidney homogenates of DDY as well as BALB/c with anti-hog DAO antibody detected a single band in SDS-PAGE. Genomic organization of the DAO gene was analyzed with the cDNA clone for pig kidney DAO. A single copy of mouse genomic DNA fragment from all DDY and BALB/c mice examined so far showed strong homology with the cDNA in Southern blot. No deletion was detected in this analysis, but some restriction fragment length polymorphism was observed among DDY strain mice. Northern blot showed a single band of RNA transcript in the mouse, although variations of the intensity of the hybridizing band were observed. When the catalytic activity of the kidney homogenate of the individual mouse was compared with the relative intensity of the hybridizing band, positive correlation was observed. From the data obtained thus far, variations of the DAO activity in DDY strain mouse were suspected to be due to the transcriptional control of the gene expression.

Acknowledgement

This study was supported in part by a Research Grant for Cardiovascular Disease (62A-1) from the Ministry of Health and Welfare of Japan, and Grants-in-Aid for Scientific Research (No. 61570151) and for Special Project Research (No. 61132005) from the Ministry of Education, Science and Culture of Japan.

References

1. Horiike, K., Arai, R., Tojo, H., Yamano, T., Nozaki, M., Maeda, T. 1985. Acta. Histochem. Cytochem. 18, 539-550.

2. Cline, M.J., Lehrer, R.I. 1969. Proc. Natl. Acad. Sci. U.S.A. 62, 756-763.

3. Marcotte, P., Walsh, C. 1978. Biochemistry 17, 5613-5619.

4. Nakajima, H., Ohta, M., Yamano, T., Miyake, Y. 1981. Biomed. Res. 2, 154-165.

5. Hull, R.N., Cherry, W.R., Weaver, G.W. 1976. In Vitro 12, 670-677.

6. Fukui, K., Momoi, K., Watanabe, F., Miyake, Y. 1986. Biochem. Biophys. Res. Commun. 141, 1222-1228.

7. Fukui, K., Watanabe, F., Shibata, T., Miyake, Y. 1987. Biochemistry (in press)

8. Ronchi, S., Minchiott, L., Galliano, M., Curti, B., Swenson, R., Williams, C.H.J., Massey, V. 1982. In: Flavins and Flavoproteins (V. Massey and C.H. Williams,

eds.). Elsevier North Holland, p. 66-72.

9. Williams, C.H.J., Arscott, L.D., Swenson, R.P., 1984. In: Flavins and Flavoproteins (R.C. Bray, P.C. Engle and S.G. Mayhew, eds.). Walter de Gruyter, Berlin·New York, p. 95-109.

10. Konno, R., Yasumura, Y., 1983. Genetics 103, 277-285.

MOLECULAR AND CELLULAR BIOLOGY OF D-AMINO ACID OXIDASE

Kiyoshi Fukui, Fusao Watanabe, Toshihiro Shibata, Kyoko Momoi and Yoshihiro Miyake

Department of Biochemistry
National Cardiovascular Center Research Institute
Fujishiro-dai, Suita, Osaka 565, Japan

Introduction

Systematic studies of D-amino acid oxidase (E.C. 1.4.3.3., DAO) in various tissues revealed its existence in liver, kidney, brain (1). Intracellular localization of this enzyme is reported to be in peroxisomes and DAO is one of the principal and characteristic enzymes of the peroxisomes of porcine kidney. In this paper, in order to facilitate the elucidation of the biological function of DAO we first describe the study of biosynthesis of DAO with the use of pig kidney cell line and in vitro translation system. Then, cloning of cDNA for DAO and the molecular biological analyses were also presented.

Results and Discussion

Initially, in vitro DAO synthesis was carried out using pig kidney mRNA as well as free and membrane-bound polysomes in cell-free translation system. mRNA and free polysomes, but not membrane-bound polysomes, directed the synthesis of immunoreactive DAO protein. In vivo DAO synthesis was also examined in a pig kidney cell line LLC-Pk_1, derived from proximal tubules (2). The same molecular weight of 38,000 was estimated for both the in vivo and in vitro synthesized DAO, as that of the purified enzyme. Based on these results,

Flavins and Flavoproteins 1987

DAO was considered to be synthesized on free ribosomes and transferred to the interior of peroxisomes without any proteolytic modification (3).

Complementary DNAs for DAO were isolated from porcine kidney cDNA library constructed by Okayama and Berg vector. The cDNA library was screened by the use of synthetic oligonucleotide probes corresponding to the partial amino acid sequences. Analysis of the nucleotide sequence of the cDNA insert revealed a complete 3211 nucleotide sequence with a 5'-terminal untranslated region of 198 nucleotides, 1041 nucleotides of an open reading frame that encoded 347 amino acids, and a 3'-untranslated region of 1972 nucleotides (4). The deduced amino acid sequence agreed with the reported sequence of the mature enzyme (5). One ambiguous residue Asx at position 192 in the reported sequence was concluded to be Asp based on the nucleotide sequence determination. These results demonstrate directly the absence of a cleavable presequence at the amino terminal of the primary translation product of DAO mRNA, indicating the post-translational transfer of DAO into peroxisomes without any proteolytic modification. It would seem, therefore, reasonable to speculate the presence of a recognizable signal for translocation into peroxisomes within the polypeptide. Analysis of the relative hydrophobicity at each amino acid position in DAO was carried out according to the Kite-Doolittle method (6), and revealed a highly hydrophobic region of the N-terminal 17 amino acids (Figure 1). This region showed significant homology with other electron-transporting enzymes and nucleotide-binding proteins as shown in Figure 2. This sequence might function as a signal for DAO to associate with peroxisomal membrane, or FAD molecule. Southern blot analysis showed that DAO was coded by a single gene in the porcine genome. On the other hand, the analysis of the expression of this gene in pig kidney by RNA blot hybridization revealed the presence of three mRNAs that differed in size. Comparison of the sequences of the cDNAs

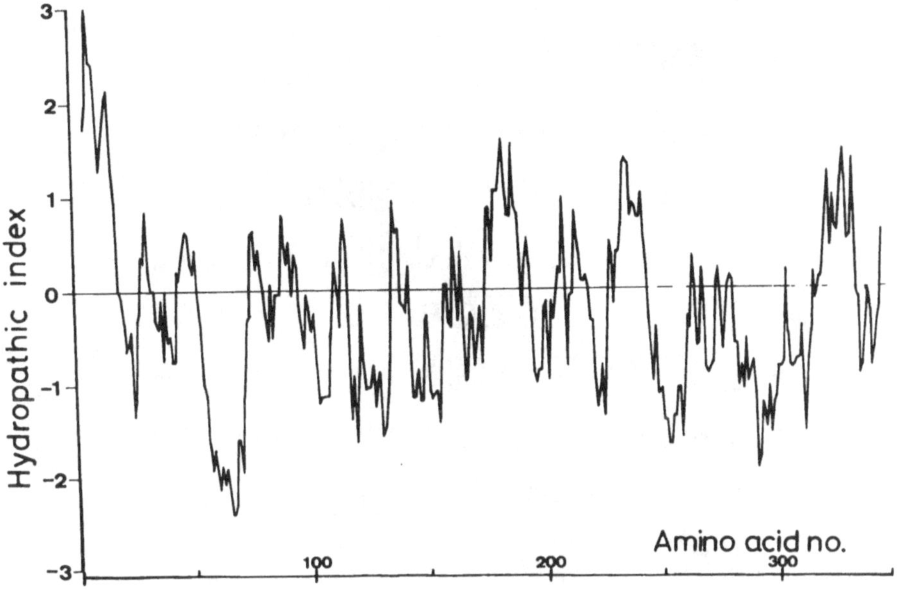

Figure 1. Hydropathy profile of porcine kidney DAO. The averaged hydropathic index of a nonadecapeptide composed of amino acid residues n-9 to n+9 has been plotted against n, where n represents amino acid number; the hydropathy indices of individual amino acids have been taken from ref.(6).

for DAO revealed that multiple polyadenylation signal sequences (ATTAAA and AACAAA) are present in the 3'-untranslated region, producing the different mRNA species (4). The efficiencies of 3' processing of the RNA were quite different between the two signals, ATTAAA and AACAAA.

The present study opens the way to a molecular analysis of the regulation of this enzyme and biogenesis of peroxisomes. Moreover, molecular studies of the enzyme structure and its function would, hopefully, lead us to the elucidation of the biological role of DAO.

References

1. Dunn, J.T., Perkoff, G.T. 1963. Biochim. Biophys. Acta. 73, 327-331.

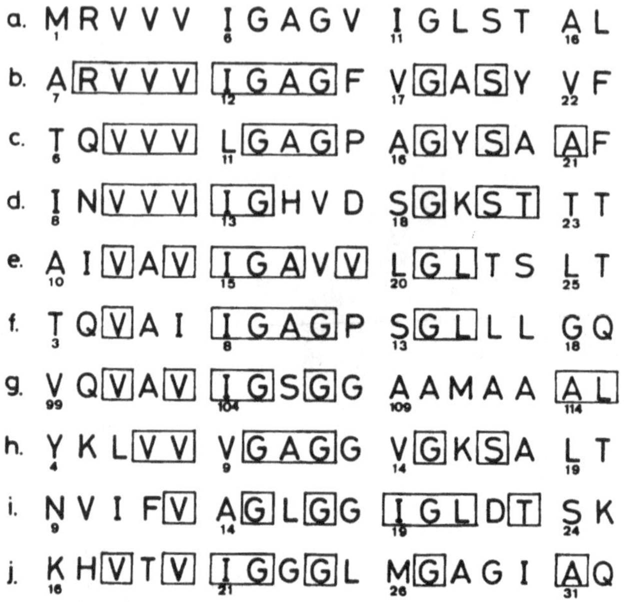

Figure 2. Homologies of the N-terminal amino acid sequence with other proteins. NBRF-PDB Protein Database (Release 8.0) was surveyed for the homology using the program developed by Software Development Co. Amino acid residues identical with DAO are boxed. a, DAO (Pig); b, Lactate dehydrogenase (Bacillus); c, Lipoamide dehydrogenase (E. coli); d, Elongation factor 1-α A (Bakers' yeast); e, Transforming protein ros (Avian sarcoma virus); f, p-Hydroxybenzoate hydroxylase (Pseudomonas); g, Mercuric reductase (Pseudomonas); h, Transforming protein H,K,N-ras (Human); i, Alcohol dehydrogenase (Drosophila); j, 3-Hydroxyacyl-CoA dehydrogenase (Pig).

2. Hull, R.N., Cherry, W.R., Weaver, G.W. 1976. In Vitro 12, 670-677.

3. Fukui, K., Momoi, K., Watanabe, F., Miyake, Y. 1986. Biochem. Biophys. Res. Commun. 141, 1222-1228.

4. Fukui, K., Watanabe, F., Shibata, T., Miyake, Y. 1987. Biochemistry (in press)

5. Ronchi, S., Minchiotti, L., Galliano, M., Curti, B., Swenson, R.P., Williams, C.H.J., Massey, V. 1982. J. Biol. Chem. 257, 8824-8834.

6. Kite, J., Doolittle, R.F. 1982 J. Mol. Biol. 157, 105-132.

LOCALIZATION OF D-AMINO ACID OXIDASE IN RAT CEREBELLUM

Kihachiro Horiike, Hiromasa Tojo*, Ryohachi Arai**, Toshio Yamano*, Mitsuhiro Nozaki, and Toshihiro Maeda**

Departments of Biochemistry and **Anatomy, Shiga University of Medical Science, Seta, Ohtsu, Shiga 520-21, Japan, and *Department of Biochemistry, Osaka University Medical School, Nakanoshima, Kita-ku, Osaka 530, Japan

Introduction

In the mammalian brain, D-amino acid oxidase is biochemically found to be almost completely confined to the hindbrain, particularly the cerebellum (1-7). However, many contradictory results as to the histochemical localization and the identification of oxidase-containing cells in the mammalian cerebellum have been presented (5-8). To solve these problems, and to facilitate elucidation of the physiological role of the oxidase, we carried out a histochemical study on the oxidase in rat cerebellum by a new coupled peroxidation method (9).

Experimental Procedures (9)

Under sodium pentobarbital anesthesia, adult male rats were perfused with 8% glutaraldehyde in 0.1 M sodium phosphate (pH 7.4). Vibratome sections (50 μm) of the cerebellum were floated in the incubation medium at 20°C for 12-16 hr: the incubation medium consisted of 0.1% horseradish peroxidase (Grade III; Toyobo, Osaka), 0.005% 3, 3'-diaminobenzidine tetrahydrochloride, 0.065% sodium azide, 0.6% nickel ammonium sulfate and 22 mM D-proline as a substrate in 0.05 M Tris-HCl, pH 7.6. For observation under an electron microscope, the

Flavins and Flavoproteins 1987

sections were postfixed in 1% osmium tetroxide.

Results and Discussion

The enzyme activity was histochemically observed in all gross regions of the cerebellum. Staining in the cortex was much more intense than that in the medulla (Fig. 1A). On both light and electron microscopic observation, the oxidase activity was found to be present only in astrocytes and Bergmann glial cells (Golgi epithelial cells), and not in any neuronal components, endothelial cells or ependymal cells.

In the granular layer, deposits of electron dense material can be exclusively seen in the glial elements (Fig. 1B). The astrocyte somata were stained weakly. Reaction products were present in the processes and end-feet of astrocytes, which filled the spaces between various neuronal elements. The cerebellar glomeruli were enveloped by the heavily stained processes of astrocytes. However, the glomeruli were not completely surrounded by the deposits, and the staining around them was not of equal intensity. Heavy deposits were frequently seen in the glial sheaths, through which Golgi cell axons and granule cell dendrites penetrated into the glomeruli.

In the Purkinje cell layer, Bergmann cell somata were generally stained weakly, and their processes were heavily stained. The oxidase-containing processes of Bergmann glial cells enveloped basket cell axons, their terminals, the terminals of the recurrent collaterals of Purkinje cell axons and Purkinje cell bodies which were invariably unstained. Deposits were also found around the synapses between climbing fibers and Purkinje cell dendrites.

In the molecular layer, the ascending processes of Bergmann glial cells were deeply stained (Fig. 1C). The clusters of synapses between the dendritic thorns of Purkinje cells and parallel fiber varicosities were found to be covered by heavily deposit-containing processes of Bergmann glial cells (Fig. 1D). In these complexes of synapses, the terminals of other neurons, such as basket cells and/or stellate cells, were frequently seen. In addition, deposits were present in the neuroglial spaces around the synapses on Golgi cell dendrites.

In the white matter, the reaction products were deposited in typical astrocytes, the end-feet of which were in contact with the blood vessels.

In the present study, we showed clearly, on both light and electron microscopy, that the oxidase is distributed throughout the entire cerebellum of rat, and that the oxidase activity is present only in glial elements, viz. astrocytes and Bergmann glial cells, and not in neuronal components, endothelial cells or ependymal cells. The greater proportion of the oxidase activity is present in the glial spaces around

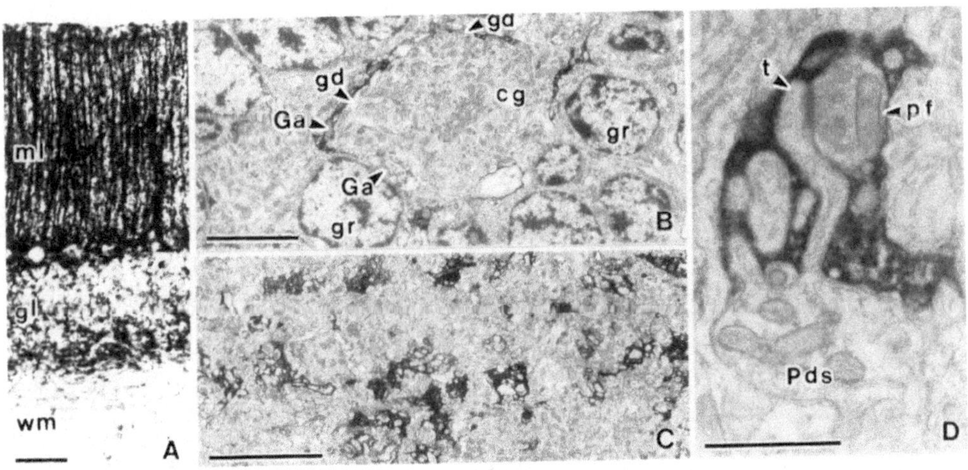

Fig. 1. Histochemical staining of D-amino acid oxidase activity in rat cerebellum. A, Photomicrograph of a sagittal section, scale bar = 50 μm. B-D, Electron micrographs of the granular layer (B, scale bar = 5 μm) and the molecular layer (C, a horizontal section, scale bar = 5 μm; D, a frontal section, scale bar = 1 μm). ml, molecular layer; gl, granular layer; wm, white matter; cg, cerebellar glomerulus; Ga, Golgi cell axon; gd, granule cell dendrite; gr, granule cell; t, dendritic thorn of Purkinje cell; Pds, spiny branchlet of Purkinje cell dendrite; pf, parallel fiber varicosity.

various kinds of synapses, but not only around a single kind of neuron or synapses at which a particular transmitter is released. Based on the above characteristic localization of the oxidase, the possibility that the oxidase distinguishes the various types of synapses in terms of their functions by surrounding the synapses is considered. And, the oxidase-containing glial cell may facilitate the smooth and successful transmission of neural information at the synapse by a neuron.

Acknowledgements

This research was supported by a Grant-in-Aid for Scientific Research (No. 60580150) to K.H. from the Ministry of Education, Science and Culture of Japan, and by a grant from the Takeda Science Foundation. One of us (K.H.) wishes to acknowledge the Yamada Science Foundation for a travel grant.

References

1. Neims, A.H., W.D. Zieverink, J.D. Smilack. 1966. J.Neurochem. 13, 163.
2. Goldstein, D.B. 1966. J.Neurochem. 13, 1011.
3. De Marchi, W.J., G.A.R. Johnston. 1969. J.Neurochem. 16, 355.
4. Yusko, S.C., A.H. Neims. 1973. J.Neurochem. 21, 1037.
5. Gaunt, G.L., C. de Duve. 1976. J.Neurochem. 26, 749.
6. Weimar, W.R., A.H. Neims. 1977. J.Neurochem. 28, 559.
7. Weimar, W.R., A.H. Neims. 1977. J.Neurochem. 29, 649.
8. Arnold, G., L. Liscum, E. Holtzman. 1979. J.Histochem.Cytochem. 27, 735.
9. Horiike, K., R. Arai, H. Tojo, T. Yamano, M. Nozaki, T. Maeda. 1985. Acta Histochem.Cytochem. 18, 539.

BIOSYNTHESIS OF D-AMINO ACID OXIDASE AND MOLECULAR CLONING OF THE cDNA

Taisuke Inagaki, Nobuko Ohishi, Takuya Koshizaka, Chikako Ichihara, Kunio Yagi

Institute of Applied Biochemistry, Yagi Memorial Park, Mitake, Gifu 505-01, Japan

Norihiro Tsukagoshi, Shigezo Udaka

Department of Food Science and Technology, Faculty of Agriculture, Nagoya University, Nagoya 464, Japan

Introduction

D-Amino acid oxidase (DAO), one of the peroxisomal flavin enzymes, has been extensively studied and characterized, especially in terms of the mechanism of its enzymatic activity. On the other hand, the mechanism of biosynthesis of the enzyme and the structure of the gene remained to be examined. We, therefore, attempted *in vitro* synthesis of the enzyme and isolation of the cDNA clone in order to address the problem. We describe herein that the enzyme was synthesized as a protein having the same molecular weight as that of the mature enzyme and that its mRNA has a relatively long-tailed 3' non-coding region.

Results and Discussion

For *in vitro* translation, poly(A)$^+$RNA was isolated from freshly excised pig kidney by SDS-phenol extraction (1) followed by oligo-dT cellulose column chromatography; and antibodies were raised against DAO purified according to the

Flavins and Flavoproteins 1987

procedure of Yagi et al. (2) and Jenkins et al. (3).

When translational products synthesized in a wheat germ lysate were subjected to immunoprecipitation using anti-DAO antiserum according to the method of Watanabe and Price (4), and the immunoprecipitates were analyzed by SDS-PAGE/fluorography, no band corresponding to DAO was detectable, even though the dried gel was exposed to X-ray film for 40 days at -80°C. On the other hand, when the rabbit reticulocyte lysate system was used for the translation, a band of protein with a molecular weight of 38K was detected by a 3-day exposure to the X-ray film as shown in Fig. 1, lane a. The appearance of the band was prevented by the addition of an excess of the purified DAO during the immunoprecipitation

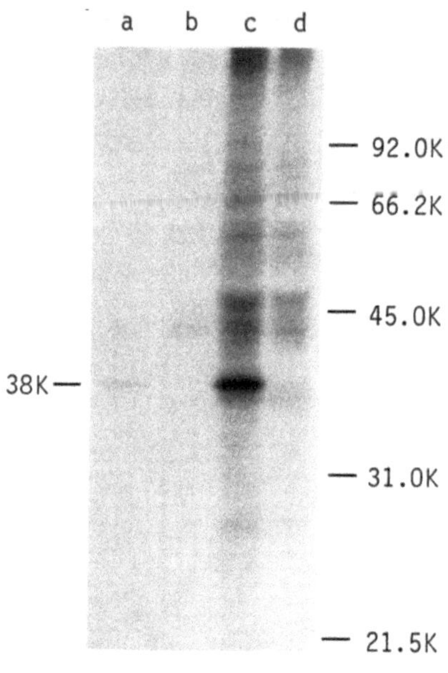

Fig. 1. In vitro synthesis of D-amino acid oxidase. The dried gel was exposed to X-ray film for 3 days at -80°C. Lanes a and c, immunoprecipitates formed in the absence of the purified enzyme; lanes b and d, those formed in the presence of the purified enzyme (20 µg). In the cases of lanes c and d, prior to translation, pig kidney poly(A)$^+$RNA was incubated with 2.5 mM methylmercury hydroxide for 3 min at 25°C.

Table 1. Sizes of the cDNA inserts and net molecular weights of expressed DAO

	Size of cDNA* (kb)	net M. W. of expressed DAO** (dalton)
λ-DAO 1	1.70	9,000 ± 2,000
λ-DAO 2	1.65	8,000 ± 2,000

* estimated by 0.8% agarose gel electrophoresis.
** estimated by 5% SDS-PAGE/Western blotting.

(Fig. 1, lane b), indicating that the synthesized protein with a molecular weight of 38K was indeed DAO (5). The present result revealed that DAO was synthesized as a protein having the same molecular weight as that of the mature enzyme. In fact, most of the peroxisomal proteins have been shown to be synthesized as the same-size precursor molecules (6). Similar results for in vitro synthesis of pig kidney DAO was reported by Fukui et al. (7)

When the poly(A)$^+$RNA was treated with 2.5 mM methylmercury hydroxide according to Payvar and Schimke (8), the band became much more distinct (Fig. 1, lane c); and its appearance was also prevented in the presence of the purified DAO (Fig. 1, lane d).

The enhancement of the synthesis of the enzyme by treatment of the mRNA with methylmercury hydroxide may be ascribed to a tight conformational structure of the mRNA or to propensity of the molecules to interact with each other or with other mRNA's. We can speculate that the native mRNA for DAO in vivo is much more efficient in synthesis than the isolated one in vitro, since a considerably large amount of the enzyme is normally synthesized in vivo.

We therefore intended to isolate the cDNA clone for DAO in order to elucidate the structure of the mRNA. The DNA clone encoding the enzyme gene was screened with anti-DAO antibody according to the method of Young and Davis (9). We screened

5×10^5 recombinants from a λgt11 rat liver expression library. Two positive clones were plaque-purified and designated λ-DAO 1 and 2. To determine the insert sizes of the cDNA clones, the cDNA's were digested with *EcoRI* and subjected to 1% agarose gel electrophoresis. The results indicate that the sizes of the inserts were 1.65 and 1.7 kb, respectively (Table 1). The net molecular weight of DAO in the fusion protein (DAO followed by β-galactosidase) expressed in *E. coli* was determined by SDS-PAGE/immunoblotting. As shown in Table 1, the net molecular weight was estimated to be 9-10K. This indicates that cDNA clones have a partial structure gene for DAO. The result revealed that the size of the structure gene for DAO was approximately 0.3 kb and that the remaining region of 1.4 kb was 3' non-coding. The relatively long-tailed 3' non-coding region might be characteristic of the mRNA for DAO.

References

1. Palmiter, R. D. 1974. Biochemistry *13*, 3606.
2. Yagi, K., M. Naoi, M. Harada, K. Okamura, H. Hidaka, T. Ozawa, A. Kotaki. 1976. J. Biochem. *61*, 580.
3. Jenkins, W. T., A. Behlen, P. J. Rogers. 1979. Anal. Biochem. *94*, 105.
4. Watanabe, A., C. A. Price. 1982. Proc. Natl. Acad. Sci. USA *79*, 6304.
5. Inagaki, T., N. Ohishi, N. Tsukagoshi, C. Ichihara, S. Udaka, K. Yagi. 1986. Biochem. Int. *13*, 1045.
6. Miura, S., M. Mori, M. Takiguchi, M. Tatibana, S. Furuta, S. Miyazawa, T. Hashimoto. 1984. J. Biol. Chem. *259*, 6397.
7. Fukui, K., K. Momoi, F. Watanabe, Y. Miyake. 1986. Biochem. Biophys. Res. Commun. *141*, 1222.
8. Payvar, F., R. T. Schimke. 1979. J. Biol. Chem. *254*, 7636.
9. Young, R. A., R. W. Davis. 1983. Science *222*, 778.

CHARACTERIZATION OF A PARTIALLY PURIFIED NITROARYL REDUCTASE

Christopher Bryant, Tracy Martin, Marlene DeLuca
Department of Chemistry, University of California, San Diego
LaJolla, Calfornia 92093

Introduction

Nitroaromatic compounds are widely used in a number of industrial processes including the production of munitions, plastics, pharmaceuticals, and agricultural chemicals (1). The toxic and genotoxic effects of nitroaromatic compounds on biological systems has long been recognized. Several nitroaromatic compounds have been identified as mutagenic using bacterial and mammalian assay systems, and have been found to be carcinogenic in experimental animals (2). The bacterial nitroreductases associated with the mammalian intestinal microflora have been implicated in the biotransformation of these compounds to their proximate mutagenic metabolites (3). Thus, these enzymes can be directly linked to the potential health hazards associated with nitroaromatic compounds.

Although several flavoprotein nitroreductases have been identified and characterized to a limited extent, a detailed description of their enzymatic mechanism and the key structural elements involved in the reduction of nitroaromatics is yet to be made. It is our feeling that such a description at the structure-function level will yield information of value in understanding how these

Flavins and Flavoproteins 1987

enzymes mediate the biological effects of nitroaromatic compounds. Towards this end we have recently developed a purification protocol for the isolation of a flavoprotien nitroreductase from the enteric bacteria Enterobacter cloacae (ATCC# 43560).

Results

Our purification protocol is outlined in Table 1. The final enzyme preparation was found to be greater than 85% pure as judged by silver stained SDS-PAGE. In addition constant specific activity was observed for the peak fractions collected during TSKG2000 size exclusion chromatogrophy. The enzyme could be inactivated by repeated extractions with ammonium sulfate and reconstituted with FMN but not FAD. The molecular weight and subunit structure were

TABLE 1. Purification of E.cloacae Nitro-Aryl Reductase.

SAMPLE	PROTEIN (mg)	ACTIVITY (IUnits)	Sp.ACTIVITY (U/mg)	Yield %
Crude Extract	1690	1217	0.7	---
Acetone Preicpitation	742	1258	1.7	103
High Speed Centrifugation (24 hrs. @ 135,000g)	382	1283	3.4	105
Q300 Anion Exchange (Tris Buffer)	38.1	921	24.2	76
Propyl HIC	1.26	550	437	45
MONO Q Anion Exchange (Phosphate Buffer)	0.67	440	660	36
TSKG200 SEC	0.39	335	850	28

Enzyme activity was determined using a standard spectrophotometric assay utilizing Trinitrotoluene (TNT) and NADH as co-substrates. Activity is expressed as µM NADH oxidized per minute as determined by absorbance at 340nm. Protein concentrations were determined by the BCA method(Pierce).

determined by size exclusion chromatography and SDS-PAGE analysis (figure 1). These results indicate that the nitro-aryl reductase is monomeric in structure with an apparent molecular weight of 26,000 daltons.

Preliminary steady state kinetic analysis of various substrates indicate that NADH is the prefered pyridine nucleotide electron donor and that tri- nitroaromatics are more readily reduced than similarly substituted di-nitroaromatics (figure 2).

Conclusions

We have developed a purification protocol for the isolation of a near homogeneous preparation of the Nitro-Aryl

Figure 1. Molecular weight determination by SDS-PAGE analysis. A 12% gel was loaded with a SDS7 molecular weight standards (Sigma Chem Co.) adjacent to a 10 µg sample of the purified Nitroaryl reductase. The molecular weight was determined from a standard curve based on the relative mobility of the markers.

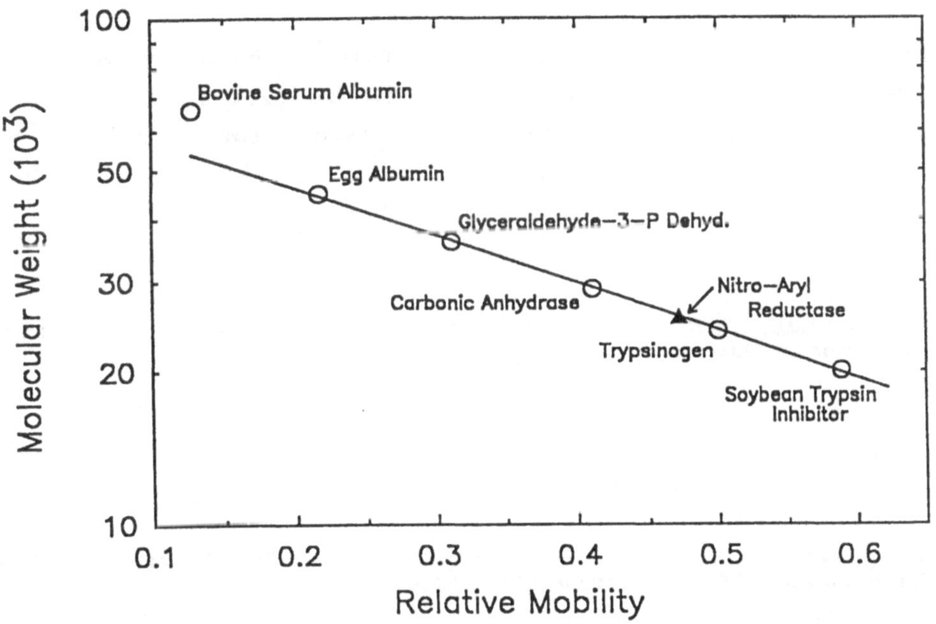

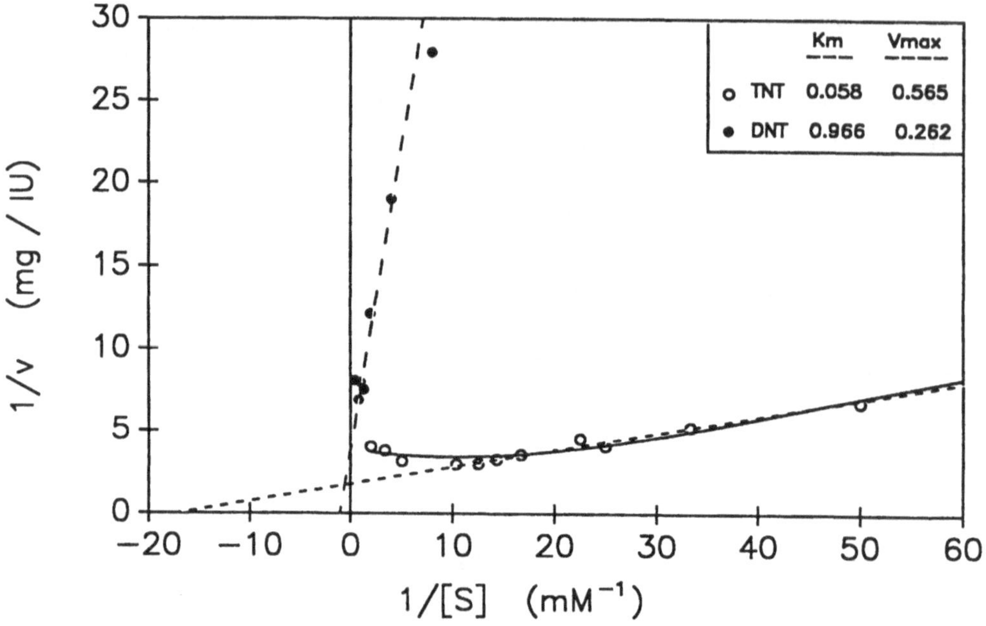

Figure 2. Steady state kinetic analysis of 2,4,6-TNT and 2,4-DNT as nitroaromatic substrates for E.cloacae Nitroaryl reductase. TNT seems to exhibit substrate inhibition at high concentrations

Reductase from E. cloacae. This method allows the preparation of sufficient quantites of protein for studies involving substrate levels of enzyme for physical and kinetic characterizations.

References

1. Hartter, D.R..1982. Toxicity of Nitroaromatic Compounds. Hemisphere Pub. Corp.

2. Rickert, D.E., B.E. Butterworth, J.A. Popp. 1983. CRC C R Toxic. 13(3),217.

3. Mirsalis,J.C., T.E. Hamm Jr., J.M Sherrill, B.E. Butterwoth. 1982. Nature 295,322.

Flavoenzyme Hydroxylases

CRYSTALLOGRAPHIC STUDIES ON THE MECHANISM OF P-HYDROXYBENZOATE HYDROXYLASE FROM *PSEUDOMONAS FLUORESCENS*

Herman A. Schreuder, Jan M. van der Laan, Marjolein M.G.M. Thunnissen, Kor H. Kalk, Myra B.A. Swarte, Wim G.J. Hol and Jan Drenth

Laboratory of Chemical Physics, University of Groningen, Groningen, The Netherlands

Introduction

The flavin monooxygenase p-hydroxybenzoate hydroxylase (PHBH) from *Pseudomonas fluorescens* catalyzes the incorporation of molecular oxygen into p-hydroxybenzoate (pOHB):

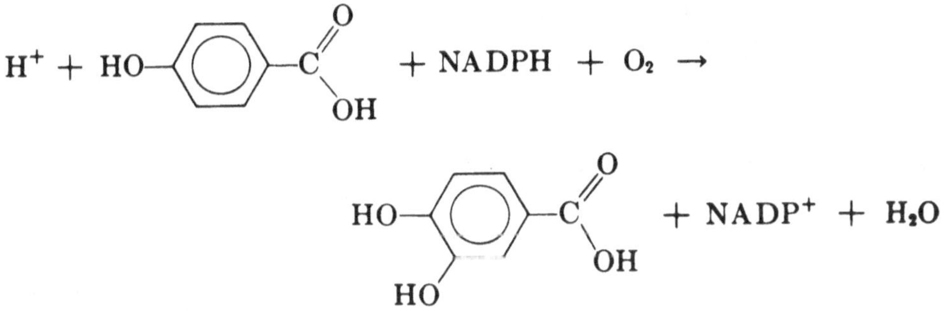

One oxygen atom is combined with the substrate whereas the other oxygen atom is reduced to water. PHBH belongs to the class of mixed function oxidases. The flavin containing mixed function oxidases are unique in that they do not require a metal ion as cofactor for catalysis (1).

The Pseudomonas species have developed a wide variety of aromatic hydroxylases, because the substituted aromatic compounds, which serve as substrates, are produced in bulk quantities during the biodegradation of lignin (2). Since PHBH has to deal with two substrates (pOHB and O_2) and two cofactors

Flavins and Flavoproteins 1987

(FAD and NADPH), its reaction mechanism is more complex than that of many other enzymes. A scheme with the various complexes occurring during catalysis by PHBH is given in figure 1.

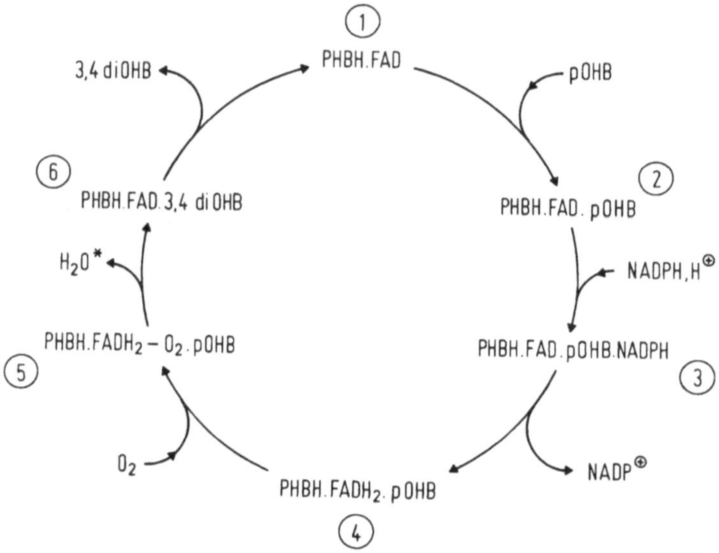

Figure 1 Intermediates of the catalytic cycle of PHBH

After substrate binding, the FAD of PHBH is reduced by NADPH. The reduced FAD ($FADH_2$) rapidly reacts with oxygen to form a covalent adduct, followed by oxygen transfer to the substrate and release of water. Using a fast stopped flow spectrophotometer, Massey and coworkers have proposed the following scheme for oxygen transfer (3):

$FADH_2$.pOHB + O_2 → I → II → III → FAD.3,4-diOHB

where I, II and III are the following molecules

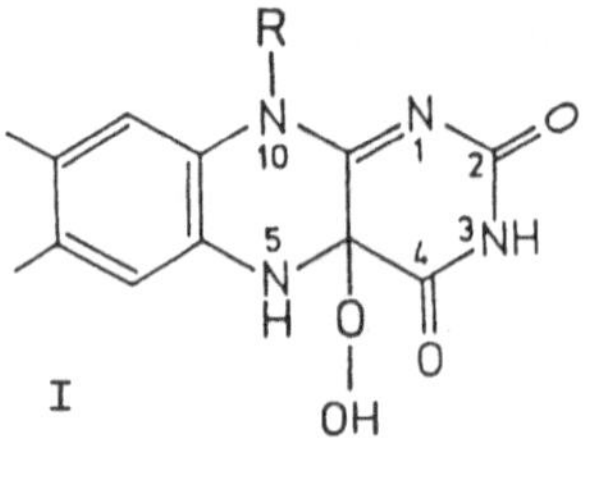

I

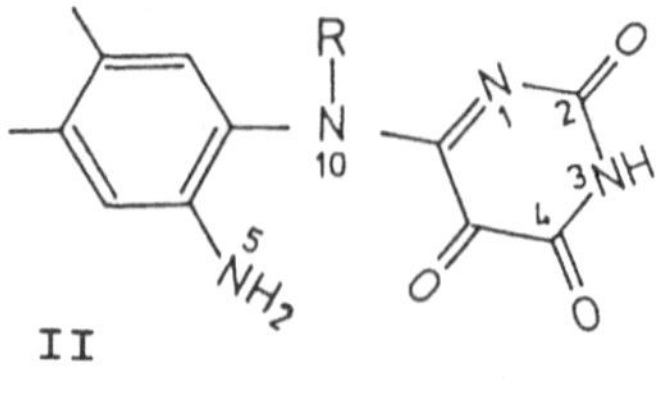

II

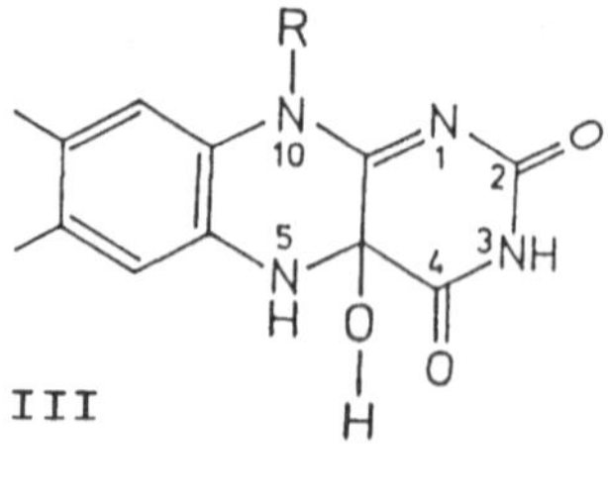

III

The intermediates I and III are well established now, the nature of intermediate II is still questionable (4). Oxygen transfer takes place via an electrophylic attack of the distal peroxide oxygen on the substrate. This implies that the oxygen needs to have some positive charge which can be achieved by the electron withdrawing properties of the flavin ring, modulated for this specific purpose by the surrounding protein.

The complexes 1, 2, 4 and 6 in figure 1 have now been obtained in crystalline form and will be described briefly in the following sections.

PHBH crystals

After the elucidation of the structure of the PHBH.FAD.pOHB complex by Wierenga et al. (5), the progress of the research on the X-ray structure of PHBH has been delayed considerably by the fact that no crystals were obtained anymore. With use of HPLC techniques, Van der Laan et al. (6) were able to solve this problem.

The original PHBH crystals (7) were obtained with protein purified according to Howell et al. (8). This purification procedure included many steps of ion exchange chromatography. Later on this purification scheme was replaced by a scheme of Müller et al. (9) introducing a cibacron blue affinity chromatography step. This resulted in a much higher yield. After detailed biochemical studies on the properties of the purified PHBH it appeared that only one very pure homogeneous fraction of dimers with reduced sulfhydryl groups can be used succesfully for crystallization experiments that reproduce the crystals published by Wierenga et al. With these newly grown crystals we could continue the X-ray work on PHBH. By soaking, cocrystallization or even catalytic action in the crystals several intermediates of the catalytic cycle of PHBH (see figure 1) were obtained in crystalline state. The crystallographic data are summarized in table 1.

Table 1 Data of several types of PHBH crystals with spacegroup $C222_1$

Intermediate	number of unique reflections	resolution (Å)	crystallographic R-factor (%)
PHBH.FAD	11,562	3.0	24.9
PHBH.FAD.Cl^-	5,372	3.3	20.8
PHBH.FAD.pOHB	14,019	2.4	21.2
PHBH.$FADH_2$.pOHB	14,880	2.3	20.8
PHBH.FAD.3,4-diOHB	15,546	2.3	19.3
PHBH.ADPR.pOHB	11,017	2.7	22.9

PHBH without substrate

Two types of substrate-free PHBH crystals have been obtained: PHBH.FAD and PHBH.FAD.Cl^- complexes. Here we will report only about the PHBH.FAD.Cl^- complex, obtained by soaking PHBH.FAD. pOHB crystals in mother liquor without pOHB but in the presence of 100 mM KCl.

Upon substrate binding, the rate of reduction of the FAD by NADPH is enhanced 4 to 5 orders of magnitude (8,10) and it is therefore very interesting to compare the structure of the PHBH.FAD.pOHB complex with the structure of the PHBH.FAD.Cl^- complex. This comparison shows that there are no significant changes in the protein part. The only significant changes concern the flavin ring (see fig. 2) which moves into the now empty substrate binding site: The N5 of the flavin ring moves 0.8 Å. This N5 is thought to receive a hydride ion from NADPH(11). Model building (12) showed that in the PHBH.FAD.pOHB complex

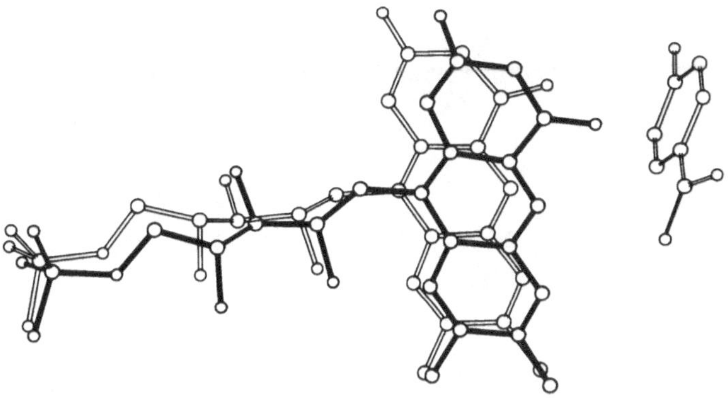

Figure 2 Superposition of the PHBH.FAD.pOHB and the PHBH.FAD.Cl^- structure. Open lines: PHBH.FAD.pOHB; solid lines: PHBH.FAD.Cl^-. Only the flavin ring and substrate are shown.

the nicotinamide ring can be properly oriented with respect to the N5, but in the PHBH.FAD.Cl^- complex Pro 293 makes this impossible. In the empty substrate binding pocket there is a water molecule which makes a hydrogen bond to the O4 of the flavin and to the side chain of Ser 212. Further experimental details, in particular that of the unknown PHBH.FAD.pOHB.NADPH complex are required to make a more definite statement.

At the current resolution of 3.3 Å we were not able to locate a bound chloride ion. The measurements are now extended towards higher resolution and if a chloride ion is indeed bound, we expect to locate it in a higher resolution map.

The PHBH.FAD.pOHB complex

The structure of the PHBH.FAD.pOHB has been described by Wierenga et al. (5,13). A schematic drawing of the active site is given in figure 3. The plane of substrate is almost perpendicular to the plane of the flavin ring. In the upper left corner of figure 3, a part of helix H10 is visible, with the positive end of its dipole pointing towards N1. This helix may serve a double function: stabilization of the negative charge

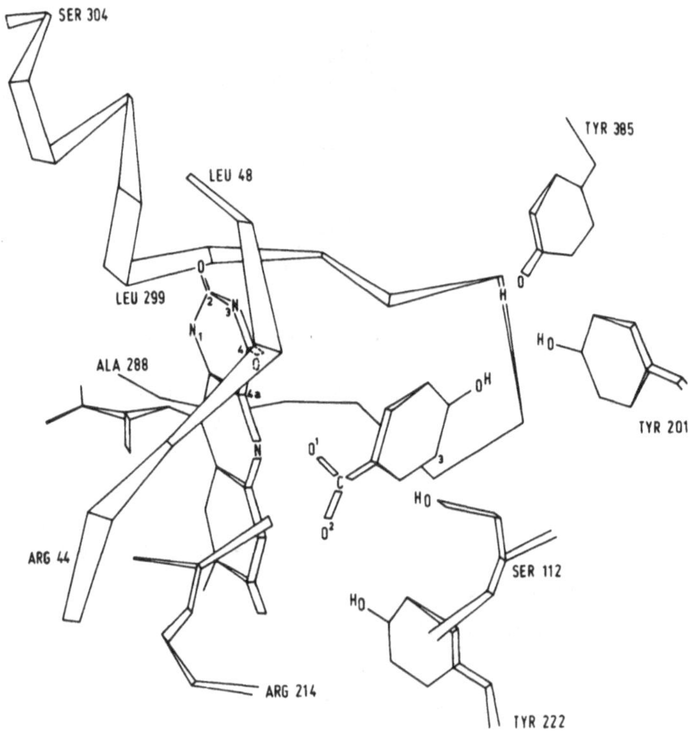

Figure 3 Active site in the PHBH.FAD.pOHB complex

which develops on the N1 upon reduction and removal of the positively charged $NADP^+$ from the active site after reduction. The loop from Ala 288 to Leu 299 contains a type I β-turn and residues from this loop probably interact with the peroxide oxygens of the flavin 4a-hydroperoxide.

The PHBH.$FADH_2$.pOHB complex

When yellow PHBH.FAD.pOHB crystals are soaked anaerobically in mother liquor containing NADPH, the crystals become colourless, indicating that the flavin in the crystals is reduced. This was confirmed by single crystal spectroscopy (14).

The reduced flavin ring in the PHBH.$FADH_2$.pOHB complex is not bent as in low molecular weight compounds (15) but planar. After superposition of the PHBH.FAD.pOHB and the PHBH.$FADH_2$.pOHB complexes, the RMS deviation between the two structures is 0.25 Å indicating little differences in conformation.

The PHBH.FAD.3,4-diOHB complex

The substrate bound to the PHBH.FAD.pOHB crystals was converted to the product 3,4-diOHB by allowing the reaction to proceed in the crystals. In addition, a high concentration of 3,4-diOHB was used in the soaking solution (16).

The electron density obtained from these crystals clearly indicated which of the two possible meta positions was hydroxylated (figure 4) and constrained the number of possible oxygenated flavin intermediates.

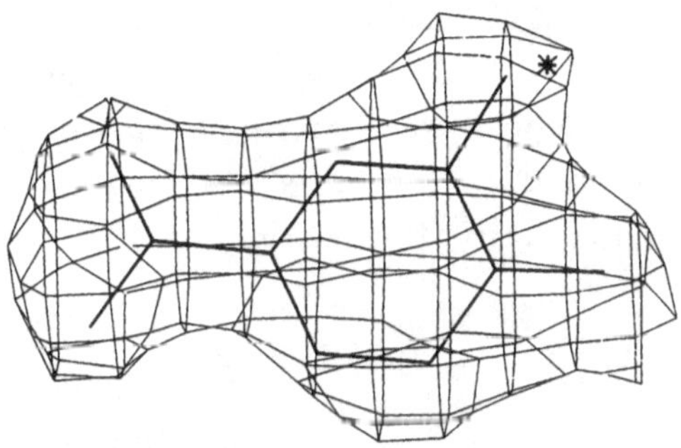

Figure 4 Electron density of the 3,4-diOHB in the PHBH.FAD. 3,4-diOHB complex. The hydroxyl group attached by the enzyme is indicated with an asterisk.

The way the product is bound is illustrated in figure 5.

The carbonyl group of Pro 293 is almost certain catalytically important. It makes a rather short hydrogen bond with the 3-OH group of the product. Because the 3-OH is the hydroxyl group

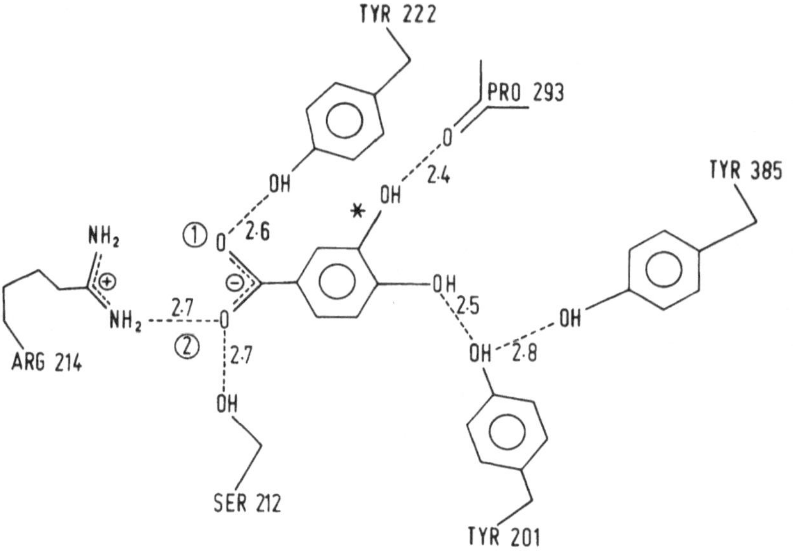

Figure 5 Schematic picture of the binding of the 3,4-diOHB molecule in the active site of PHBH.

inserted by the enzyme, the carbonyl of Pro 293 most likely also interacts with this oxygen prior to and during the oxygen transfer. The partial negative charge of the carbonyl could polarize the FlHO-OH bond to enhance the reactivity of the peroxide and hence the rate of catalysis.

The flavin 4a-hydroperoxide intermediate

Although the lifetimes of the oxygenated flavin intermediates in PHBH are much too short to be studied by conventional protein crystallography (16,17), molecular modelling yielded some interesting results. We used the structure of 4a,5-epoxyethano-3-methy-4a,5-dihydrolumiflavin as a model for the flavin 4a-hydroperoxide intermediate. This compound was chosen because its X-ray structure is known (18) and its optical spectra most closely resemble those of the oxygenated flavin intermediates observed in PHBH (19). Moreover, the ^{13}C NMR chemical shifts of this compound are almost identical to the shifts of the flavin 4a-hydroperoxide bound to luciferase (20,21).

The model compound and the part which was fitted to the active site is given in figure 6.

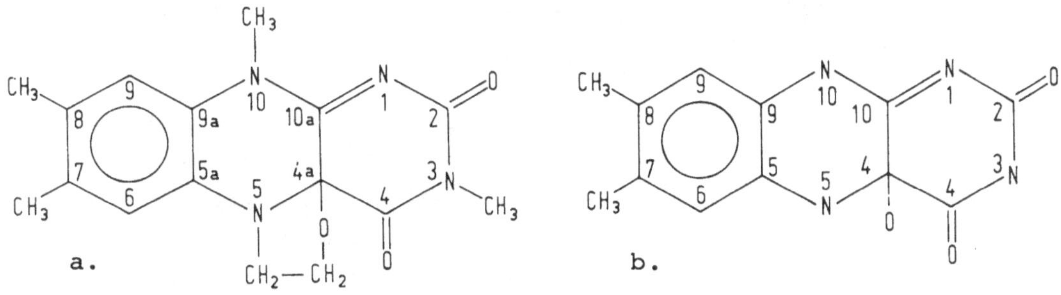

Figure 6 Compound used for modelling. a: 4a,5-epoxyethano-3-methyl-4a,5-dihydrolumiflavin. Its structure was determined by Bolognesi et al. (18). b: part of this model which was fitted in PHBH.

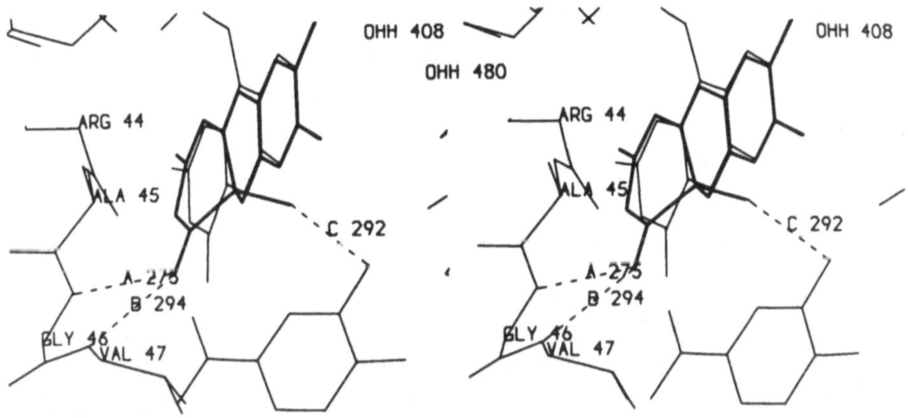

Figure 7 View of the flavin 4a-hydroperoxide model in the active site of PHBH. The dashed lines A and B indicate hydrogen bonds between the O4 of the model and main chain NH groups. Line C indicates the distance of the model O and the 3-OH group of the 3,4-diOHB molecule. The distances indicated are in pm.

The fit of the model compound is remarkably good (see figure 7). The C4A carbon has now 4 substituents and becomes somewhat tetrahedral. This results in a twist of the pyrimidine ring and a movement of 0.8 Å of the O4. At its new position O4 is perfectly accommodated by an "oxyanion hole" consisting of the

NH groups of Gly 46 and Ala 47. In this way not only the flavin peroxide intermediate is stabilized, but the electronegativity of O4 is enhanced as well, leading to a further polarization of the FlHO-OH bond.

The position of one oxygen of the peroxide is known from the model just described and the position of the other oxygen from the hydroxyl group which is attached to the substrate by the enzyme. It is significant that the distance between both oxygens is only 2.9 Å indicating that minimal shifts are required for the actual oxygen transfer step.

Discussion

Based on these structural results, it is possible to speculate about the mechanism of PHBH. If we assume that O2 enters the active site via the same channel as the nicotinamide moiety of NADPH is believed to enter (12), then the distal O is close to the C10A when the proximal O reacts with the C4A. The now negatively charged distal O could take a proton from water 408 which is located near the N10. Next the distal O rotates in the nicotinamide binding pocket towards the substrate and ends up in the "hydroxylation cavity". Now the dipole moment of helix H10 and hydrogen bonds of protein with N1, O2, N3 and O4 of the flavin ring, all work together to enhance the electronegativity of the flavin ring in this way polarizing the FlHO-OH bond. Further polarization of this bond is probably achieved by the dipole moment of the peptide of Pro 293. The proximal oxygen may now be sufficiently activated for the hydroxylation reaction to take place.

Although this proposal is highly speculative, it could nicely explain the longer lifetimes of the oxygenated flavin intermediates observed when the inhibitor azide is added (3). This compound could hinder the rotation of the distal oxygen by binding in the nicotinamide pocket.

This work has been carried out under the auspices of the Netherlands Foundation for Chemical Research (SON) with financial aid from the Netherlands Organization for the Advancement of Pure Research (ZWO).

We thank Dr. Franz Müller and his coworkers for continuous interest and support.

References

1. Bruice, T.C. 1984. Isr. J. Chem. 24, 54.

2. Dagley, S. 1982. In: Flavins and Flavoproteins (V. Massey and C.H. Williams, eds.). Elsevier Scientific Publishing Co.; Amsterdam. p. 311.

3. Entsch, B., D.P. Ballou, V. Massey. 1976. J. Biol. Chem. 251, 2550.

4. Wessiak, A., J.B. Noar, T.C. Bruice. 1984. Proc. Natl. Acad. Sci. USA 81, 332.

5. Wierenga, R.K., R.J. de Jong, K.H. Kalk, W.G.J. Hol, J. Drenth. 1979. J. Mol. Biol. 131, 55.

6. Laan, J.M. van der, M.B.A. Swarte, W.G.J. Hol, J. Drenth. (in preparation)

7. Drenth, J., W.G.J. Hol, R.K. Wierenga. 1975. J. Biol. Chem. 250, 5268.

8. Howell, L.G., T. Spector, V. Massey. 1972. J. Biol. Chem. 247, 4340.

9. Müller, F., G. Voordouw, W.J. van Berkel, P.J. Steennis, S. Visser, P.J. van Rooijen. 1979. Eur. J. Biochem. 101, 235.

10. Husain; M., V. Massey. 1979. J. Biol. Chem. 254, 6657.

11. Ballou, D.P. 1984. In: Flavins and Flavoproteins (R.C. Bray, P.C. Engel and V. Massey, eds.). Walter de Gruyter & Co., Berlin, New York. p. 605.

12. Berkel, W.J.H. van, E.J. Bakx, F. Müller, W.J. Weijer, P.A. Jekel, J.J. Beintema, H.A. Schreuder, J.M. van der Laan, R.K. Wierenga, J. Drenth, W.G.J. Hol. This volume.

13. Wierenga, R.K., K.H. Kalk, J.M. van der Laan, J. Drenth, J. Hofsteenge, W.J. Weijer, P.A. Jekel, J.J. Beintema, F. Müller, W.J.H. van Berkel. 1982. In: Flavins and Flavoproteins (V. Massey and C.H. Williams, eds.). Elsevier Scientific Publishing Co., Amsterdam. p. 11.

14. Laan, J.M. van der. 1986. Ph.D. Thesis, University of Groningen, The Netherlands.

15. Kierkegaard, P., R. Norrestam, P.-E. Werner, I. Tsöregh, M. von Glehn, R. Karlsson, M. Leyonmark, O., Rönnquist, B. Stensland, O. Tillberg, L. Torbjörnson. 1971. In: Flavins and Flavoproteins (M. Kamin, ed.). University Park Press, Baltimore. p. 1.

16. Schreuder, H.A., J.M. van der Laan, W.G.J. Hol, J. Drenth. submitted to J. Mol. Biol.

17. Hiromi, K. 1979. In: Kinetics of fast enzyme reactions. Halsted Press, New York. p. 287.

18. Bolognesi, M., S. Ghisla, L. Incoccia. 1978. Acta Cryst. B34, 821.

19. Ghisla, S., B. Entsch, V. Massey, M. Husain. 1977. Eur. J. Biochem. 76, 139.

20. Vervoort, J., F. Müller, J. Lee, W.A.M. van den Berg, C.T.W. Moonen. 1986. Biochemistry 25, 8062.

21. Lhoste, J.M., V. Favoudon, S. Ghisla, J.W. Hastings. 1980. In: Flavins and Flavoproteins (K. Yagi and T. Yamano, eds.). University Park Press, Baltimore, M.D. p. 131.

APPLICATIONS OF DEUTERIUM AND TRITIUM ISOTOPE EFFECTS TO THE ELUCIDATION OF KINETIC MECHANISMS OF FLAVOPROTEIN HYDROXYLASES

Shiao-Chun Tu, Lee-Ho Wang, and Yimin Yu

Department of Biochemical and Biophysical Sciences, University of Houston-University Park, Houston, Texas 77004

In recent years, kinetic isotope effects have been established to be a powerful approach for the delineation of enzyme kinetic and chemical mechanisms. For flavoenzymes, the fruitful outcome of kinetic isotope effect studies has also been demonstrated in several investigations [e.g. References 1–3]. However, this valuable approach has not yet gained any wide applications in the general area of flavoenzymes. During the past several years, we have been utilizing techniques of deuterium and tritium isotope effects to elucidate the kinetic mechanisms of several flavoprotein external monooxygenases. Some salient points of our studies on *Pseudomonas cepacia* salicylate hydroxylase (SAL HOase), *P. cepacia* 3-hydroxybenzoate 6-hydroxylase (MHB HOase) and *Vibrio harveyi* luciferase are described in this report to update our understanding of the kinetic mechanisms of these enzymes and to demonstrate the usefulness of the application of kinetic isotope effects to studies of flavoenzymes.

(1) Salicylate Hydroxylase

Salicylate hydroxylase was the first external flavoprotein monooxygenase ever discovered [4]. This enzyme utilizes FAD as a cofactor and catalyzes the decarboxylative hydroxylation of salicylate to form catechol as shown below. We have previously established that the site of decarboxylation and that for hydroxylation are at the same position in the ring [5]. This enzyme also exhibits an oxidase activity in the absence of salicylate for

Flavins and Flavoproteins 1987

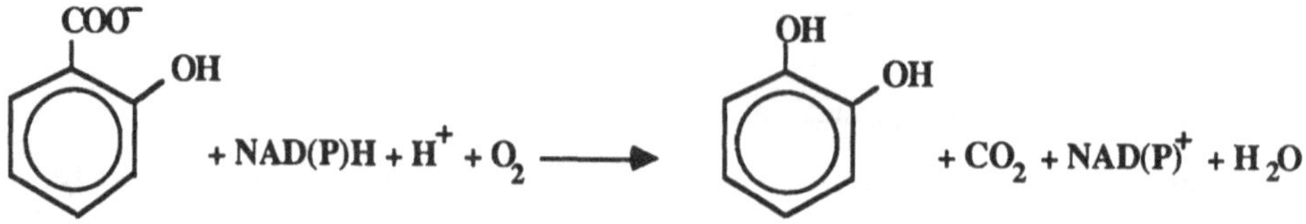

catalyzing, at a much slower rate, the oxidation of reduced pyridine nucleotide to form H_2O_2 [6]. Benzoate markedly enhances the oxidase activity and itself remains unaltered. With other salicylate derivatives, the oxidation of NADH leads to partial substrate hydroxylation and partial H_2O_2 formation [6].

Our work on SAL HOase was carried out using the enzyme isolated from *P. cepacia* [7, 8]. This hydroxylase utilizes either NADH or NADPH as the external reductant. The kinetic mechanism was first elucidated by a series of steady-state kinetic studies. The principle of such studies [9] is illustrated in Fig. 1. For double reciprocal plots of initial velocities versus concentrations of the variable substrate A (with a second substrate B kept constant), the "apparent first-order rate constant" of reaction at very low levels of A (defined as $V_{max}/K_{m,A}$ and abbreviated as V/K) is equal to the reciprocal of the slope. For sequential mechanisms (cases 1 and 2), increases in the concentration of the fixed substrate B lead to increases of V/K (or decreases in the slope). In case 1, a higher [B] facilitates the removal of EA, thus enhancing the net rate of combination of E and A at low concentrations of A. In case 2, a

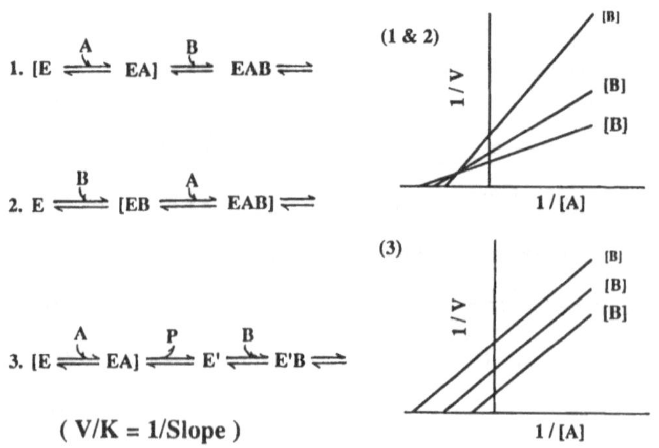

Fig. 1. Effect of [B] on the double reciprocal plot of initial rate versus [A].

higher [B] produces more EB, thus pushing for a faster net reaction rate of EB with A. Consequently, one observes a series of converging lines with each obtained at a different [B]. For ping-pong mechanism (case 3), changes in [B] affect neither the formation of E nor the removal of EA. Plots of 1/V versus 1/A at several fixed [B] thus show parallel lines.

For SAL HOase initial rate measurements [8], concentrations of salicylate were first varied at a constant level of oxygen and several fixed levels of NADH. Plots of 1/V versus 1/[salicylate] showed a family of converging lines, indicating a sequential binding of salicylate and NADH by the hydroxylase to form a ternary complex. On the other hand, parallel lines were observed for double reciprocal plots of initial velocity versus [NADH] at a fixed concentration of salicylate and several levels of oxygen. Parallel lines were also obtained for plots of 1/V versus 1/[salicylate] at a constant level of NADH and several fixed concentrations of oxygen. These results indicate a ping-pong type of reaction in which a product is first released from the E:Sal:NADH complex and then O_2 binds to an enzyme form which is distinct from the original oxidized holoenzyme.

Results described thus far, however, do not provide any diagnosis for the sequence of bindings of salicylate and NADH. Although other approaches could also be applied, we chose to determine the binding sequence by analyses of tritium isotope effects following the principle [10, 11] depicted in Fig. 2. When the isotope effect on V/K_B is plotted against the

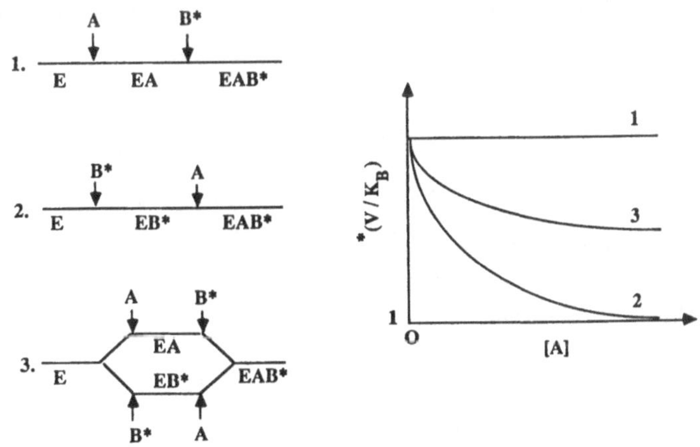

Fig. 2. Effect of [A] on $^*(V/K_B)$. A is non-labeled and B* is the isotopically labeld substrate.

concentration of the non-labeled A, the two cases of fixed-order sequential bindings (cases 1 and 2) and the random order binding (case 3) all exhibit distinct patterns, thus allowing an identification of the reaction mechanism. We have demonstrated that *P. cepacia* SAL HOase is stereo-specific for the abstraction of the hydrogen at the 4R-position of NADH and exhibits primary deuterium and tritium isotope effects using (4R)-[4-^{2}H]- and (4R)-[4-^{3}H]NADH, respectively, as a substrate. When (4R)-[4-^{3}H]NADH and benzoate were used as substrates, the $^{T}(V/K)$ effect decreased at increasing [benzoate] and reached a final level 3.16 at [benzoate]$\rightarrow \infty$ (Fig. 3). Similar results were also obtained with salicylate as the non-labeled substrate. These findings indicate that the enzyme binds the benzenoid substrate and NADH in a random order. Although the sensitivity of the above described isotope effect test may vary from one enzyme to another depending upon the relative stickiness of the labeled and non-labeled substrates, this approach is straightforward in principle and experimentally far less cumbersome than other approaches. On the basis of the above described results a minimal kinetic mechanism has been formulated as shown in Scheme 1. Symbols are: E and EH_2, for oxidized and fully reduced enzyme, respectively, S for substrate (or pseudosubstrate); SOH for the hydroxylated product; NH and N for NAD(P)H and NAD(P)$^{+}$, respectively.

Using salicylate and (4R)-[4-^{2}H]- or (4R)-[4-^{3}H]NADH as substrates, the following isotope effects have been determined [8]: $^{D}k_9 = 5.57$; $^{D}V = 1.69$; $^{D}(V/K) = 1.97$; and

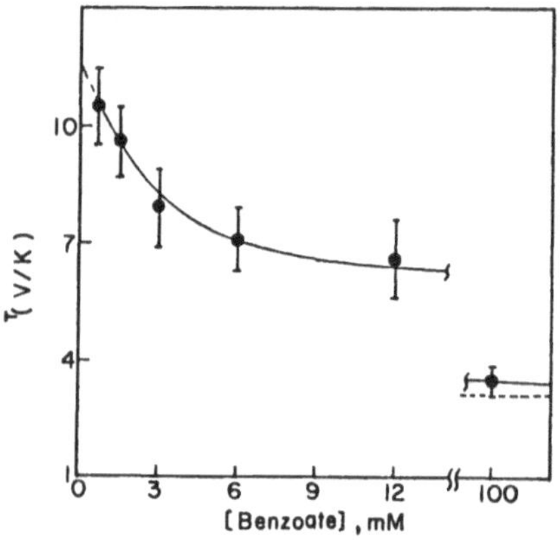

Fig. 3. Effect of [benzoate] on the $^{T}(V/K)$ of (4R)-[4-^{3}H]NADH in the SAL HOase-catalyzed reaction.

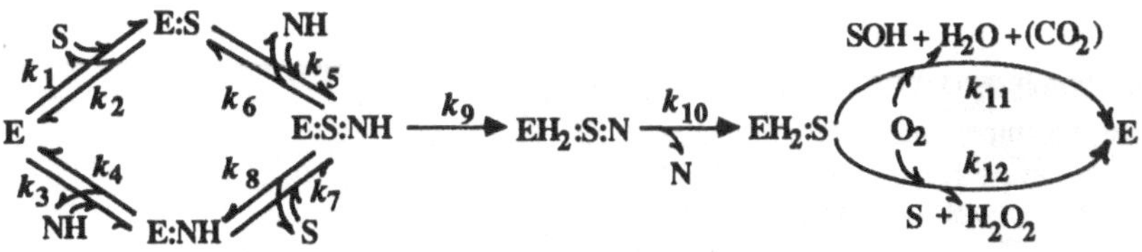

Scheme 1

$^T(V/K) = 3.36$. Further analyses of these isotope effects have revealed additional interesting kinetic information. (A) It has been pointed out [1] that if the first irreversible reaction is a step after the enzyme reduction by NAD(P)H, the observed $^T(V/K)$ will be significantly larger than the $^T(V/K)$ calculated from values of $^D(V/K)$ and $^Dk_9$ based on Eq. 1

$$\frac{^D(V/K) - 1}{^T(V/K) - 1} = \frac{^Dk_9 - 1}{(^Dk_9)^{1.44} - 1} \qquad (1)$$

Furthermore, we have previously shown that the SAL HOase-bound $FADH_2$ formed during catalysis is exposed to medium [7] and will most probably exchange hydrogen with water. If the reduction of oxidized SAL HOase is reversible, then the forward reaction will be associated with an isotope effect but the isotope effect for the back reaction will be abolished or greatly reduced because the heavy hydrogen on the reduced flavin can be washed out by the medium. Consequently the equilibrium isotope effect will be > 1 and Eq. 1 cannot be validated. In the present case, the calculated value of 3.30 for $^T(V/K)$ corresponded well with the value of 3.36 determined at 0.14 mM salicylate. When benzoate was used in place of salicylate, the value of 4.04 for $^T(V/K)$ calculated using $^Dk_9 = 5.96$ and $^D(V/K) = 2.25$ determined at 30 mM benzoate again agreed well with the $^T(V/K)$ of 4.36 determined at the same benzoate concentration [8]. These findings clearly indicate that the reduction of bound FAD by NADH is the first irreversible step in the reaction scheme. (B) Since the value of ^DV is substantially smaller than $^Dk_9$, there must be at least one slow reaction step after the enzyme reduction step. (C) A series of salicylate analogs were used as substrates for testing the fraction of NADH oxidation that was channelled into H_2O_2 formation and the ^DV effect

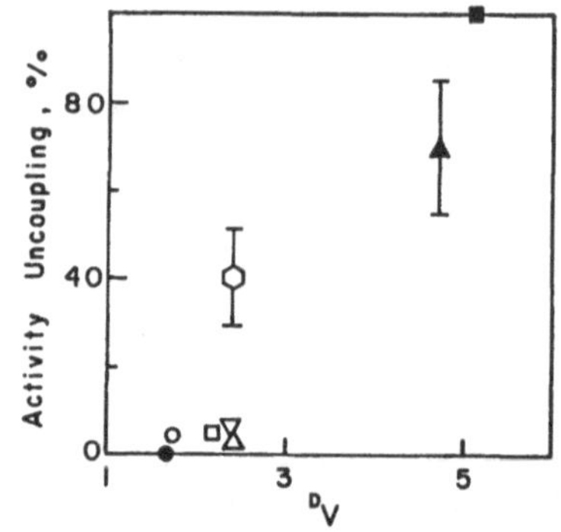

Fig. 4. Correlation of % activity uncoupling with the ^{D}V effect for various benzenoid substrates. (4R)-[4-^{2}H]NADH was used as a cosubstrate in all cases. The % activity uncoupling is defined as the % of NADH oxidation that is channelled into H_2O_2 formation. Symbols are: ●, salicylate; ○, 3-methylsalicylate; □, 2,6-dihydroxybenzoate; ▽, 2,3-dihydroxybenzoate; △, 2,4-dihydroxybenzoate; ⬡, *p*-aminosalicylate; ▲, *p*-hydroxybenzoate; and ■, benzoate.

of (4R)-[4-^{2}H]NADH. For a first approximation, the higher efficiency the H_2O_2 is formed the better the ^{D}V is expressed (Fig. 4). We concluded that, in comparison with the step of enzyme reduction, the substrate hydroxylation reaction was slow whereas the oxidation of reduced enzyme:substrate complex to form H_2O_2 was fast.

(2) 3-Hydroxybenzoate 6-Hydroxylase

Recently we have succeeded in inducing the synthesis of a new species of FAD-dependent hydroxylase, namely MHB HOase, in *P. cepacia* and isolated this enzyme to apparent homogeneity [12]. This flavo-hydroxylase is unique in that it utilizes NADH and NADPH with similar efficiencies (in terms of both V_{max} and K_m) and it inserts a new hydroxyl group to a benzenoid substrate at a position *para* instead of *ortho* to an existing one [12]. A minimal kinetic mechanism, essentially identical to Scheme 1, has been recently

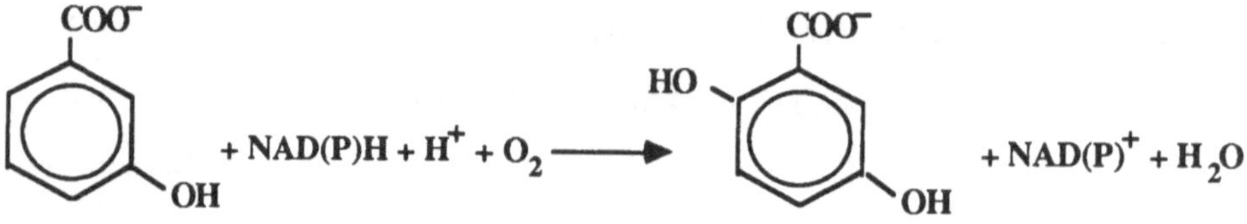

formulated for this MHB HOase [13]. Several important features of this scheme are based on analyses of kinetic isotope effects. *P. cepacia* MHB HOase also shows a stereo-specific hydrogen abstraction from the 4R-position of NADH. In an experiment similar to that

Table I: Deuterium and tritium isotope effects for MHB HOase

Isotope effect	$^{T}(V/K_{NH})_{MHB\to 0}$	$^{T}(V/K_{NH})_{MHB\to\infty}$	$^{D}(V/K_{NH})$	$^{D}(V/K_{MHB})$	^{D}V
Value	12.5	9.6	3.9 ± 0.5[a]	3.0 ± 0.6[b]	1.4 ± 0.1[a] 1.6 ± 0.3[b]

[a]Determined at 1.5 mM *m*-hydroxybenzoate and variable NADH or (4R)-[4-^{2}H]NADH.
[b]Determined at 0.6 mM NADH or 1.3 mM (4R)-[4-^{2}H]NADH and variable *m*-hydroxybenzoate.

described for Fig. 3, the $^{T}(V/K)$ effect of (4R)-[4-^{3}H]NADH was determined at different concentrations of *m*-hydroxybenzoate (MHB). Again, the $^{T}(V/K)$ effect decreased at increasing [MHB]. The limiting values of $^{T}(V/K)$ at [MHB] approaching zero and infinity are 12.5 and 9.6, respectively (Table 1). The enzyme has thus been shown to bind MHB and NADH in a random order. This conclusion is further confirmed by the comparison of $^{D}(V/K_{NH})$ with $^{D}(V/K_{MHB})$ (Table 1). Kiick et al. [14] have pointed out that a random binding case can be identified for a bi- or tersubstrate reaction if finite $^{D}(V/K)$ effects are observed for all substrates. For MHB HOase, values of 3.9 and 3.0 were obtained for $^{D}(V/K_{NH})$ and $^{D}(V/K_{MHB})$, respectively, thus providing another line of evidence in support of the random binding of MHB and NADH. The following kinetic expressions can be derived on the basis of Scheme 1:

$$^{D}(V/K_{NH}) = (^{D}k_9 + k_9/k_6)/(1 + k_9/k_6) \qquad (2)$$

$$^{D}(V/K_{MHB}) = (^{D}k_9 + k_9/k_8)/(1 + k_9/k_8) \qquad (3)$$

The $^{D}k_9$ for MHB HOase was not directly determined but the upper limit of $^{D}k_9$ was estimated to be 10.2 based on Eq. 1. This value together with those determined for $^{D}(V/K_{NH})$ and $^{D}(V/K_{MHB})$ can be used to obtain a ratio of $k_6/k_8 = 1.7$. This indicates that NADH dissociates slightly faster than MHB from the E:S:NH complex. This conclusion still holds even if $^{D}k_9$ is indeed smaller than 10.2; assuming a value of 6 for $^{D}k_9$, the corresponding value of k_6/k_8 would be 2.1. Furthermore, a comparison of the ^{D}V effect with the estimated ${}^{D}k_9$ effect indicates the existence of at least one slow step after the isotopically sensitive enzyme reduction step.

(3) Bacterial Luciferase

Bacterial luciferase catalyzes the following light-emitting monooxygenation reaction:

$FMNH_2 + CH_3\text{-}(CH_2)_n\text{-}CHO + O_2 \rightarrow FMN + H_2O + CH_3\text{-}(CH_2)_n\text{-}COOH + h\nu$

The enzyme activity was usually determined by the "standard assay" in which an $FMNH_2$ solution was injected into a reaction vial containing luciferase, aldehyde, and bovine serum albumin in air-saturated buffer. The bioluminescence intensity rapidly reaches a peak and then decays exponentially for at least 4 to 5 halflives. Using [1-^{2}H]-labeled decanal, dodecanal, and tetradecanal as substrates in the standard assay, the deuterium isotope effects on the decay rate (${}^{D}k$) have been previously reported to be 1.7, 1.2, and 1.0, respectively [15]. By measuring absorbance changes at 380 nm in stopped-flow experiments, a deuterium isotope effect of 5.4 for [1-^{2}H]decanal was also reported [16]. However, the former study was carried out under conditions of significant aldehyde substrate inhibition [17] and the latter investigation was compromised by the formation and decay of catalytically inactive semiquinone [18]. We have now determined isotope effects of [1-^{2}H]decanal on both the decay rate of the bioluminescence emission (${}^{D}k$) and the total quantum yield (${}^{D}Q$), all at 23°C, using a "dithionite assay" [17, 19] which does not result in any aldehyde

Table II: Deuterium isotope effects of [1-^{2}H]decanal on bioluminescence

Buffer (pH 7.0)	Reduced flavin	${}^{D}k$	${}^{D}Q$
0.02 M Pi	FMN	1.4	1.1
0.2 M Pi	FMN	1.4	1.3
0.05 M Tris	FMN	1.5	1.7
0.2 M Tris	FMN	1.6	2.6
0.02 M Pi	2,3-DiacetylFMN	1.0	1.4
0.02 M Pi	Riboflavin	1.0	1.4

inhibition. These isotope effects can be analyzed according to the following scheme, in which II is the 4a-hydroperoxyflavin, III is the II:aldehyde conjugate, * is the emitter, ϕ is the emission efficiency of *, q is quantum, and F_{ox} is the oxidized flavin product. The net rate constants for the light-emitting decay and the dark decay of III are represented by k_1 and

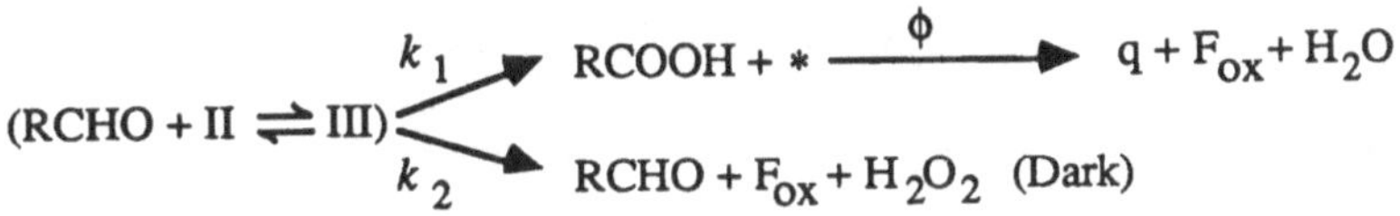

k_2, respectively. The bioluminescence intensity (I) at time t can be expressed as:

$$I = \frac{dq}{dt} = \frac{\phi d(*)}{dt} = \phi k_1 (III)_o e^{-(k_1 + k_2)t}$$

and the quantum yield (Q) can be expressed as $Q = \phi \cdot (k_1)/(k_1 + k_2)$. It then follows that

$$^{D}k = (k_{1H} + k_2)/(k_{1D} + k_2)$$

$$^{D}Q = (k_{1H})(k_{1D} + k_2)/(k_{1D})(k_{1H} + k_2)$$

If the light pathway dominates ($k_1 >> k_2$) then $^{D}k \cong k_{1H}/k_{1D}$ and $^{D}Q \cong 1.0$. When the dark pathway dominates, $^{D}Q \cong k_{1H}/k_{1D}$ and $^{D}k \cong 1.0$. If both light and dark pathways are significant, $k_{1H}/k_{1D} > {}^{D}k$; $^{D}Q > 1.0$. Results in Table 2 indicate that, in phosphate buffer with $FMNH_2$ as a substrate, light reaction was the major pathway but a minor dark (H_2O_2 forming) reaction also existed in the presence of saturating aldehyde, consistent with our previous product analysis results [20]. The partition between the two pathways was apparently sensitive to the type and concentration of the buffer in use. With reduced riboflavin and 2,3-diacetylFMN each as a substrate, the dark pathway apparently became the major reaction, again consistent with our previous product analysis [20]. The earlier product analysis atudy involved intermediate isolation and temperature variation. Experimentally it was far more labor intensive and required much more enzyme than the present isotope effect investigation.

Acknowledgement: This work was supported by grants GM25953 from the National Institute of General Medical Sciences and E-1030 from the Robert A. Welch Foundation.

References

1. Ryerson, C. C., D. P. Ballou, C. Walsh. 1982. Biochemistry 21, 1144-1151.

2. Ghisla, S. 1984. In: Flavins and Flavoproteins (R. C. Bray, P. C. Engel, and S. G. Mayhew, eds.) Walter de Gruyter, Berlin, pp. 385-401.

3. Vanoni, M. A., R. G. Matthews. 1984. Biochemistry 23, 5272-5279.

4. Katagiri, M., S. Yamamoto, O. Hayaishi. 1962. J. Biol. Chem. 237, 2413-2414.

5. Hamzah, R. Y., S.-C. Tu. 1981. J. Biol. Chem. 256, 6392-6394.

6. White-Stevens, R. H., H. Kamin. 1972. J. Biol. Chem. 247, 2358-2370.

7. Tu, S.-C., F. A. Romero, L.-H. Wang. 1981. Arch. Biochem. Biophys. 209, 423-432.

8. Wang, L.-H., S.-C. Tu. 1984. J. Biol. Chem. 259, 10682-10688.

9. Cleland, W. W. 1970. In: The Enzymes (P. D. Boyer, ed.) Academic Press, New York, 3rd Ed., Vol. 2, pp. 1-65.

10. Klinman, J. P., H. Humphries, J. G. Voet. 1980. J. Biol. Chem. 255, 11648-11651.

11. Cook, P. F., W. W. Cleland. 1981. Biochemistry 20, 1790-1796.

12. Wang, L.-H., R. Y. Hamzah, Y. Yu, S.-C. Tu. 1987. Biochemistry 26, 1099-1104.

13. Yu, Y., L.-H. Wang, S.-C. Tu. 1987. Biochemistry 26, 1105-1110.

14. Kiick, D. M., B. G. Harris, P. F. Cook. 1986. Biochemistry 25, 227-236.

15. Shannon, P., R. B. Presswood, R. Spencer, J. E. Becvar, J. W. Hastings, C. Walsh. 1978. In: Mechanisms of Oxidizing Enzymes (T. P. Singer and R. N. Ondarza, eds.). Elsevier North Holland, pp. 69-78.

16. Presswood, R. P., J. W. Hastings. 1979. Photochem. Photobiol. 30, 93-99.

17. Meighen, E. A., R. E. MacKenzie. 1973. Biochemistry 12, 1482-1491.

18. Kurfürst, M., S. Ghisla, R. Presswood, J. W. Hastings. 1982. Eur. J. Biochem. 123, 355-361.

19. Tu, S.-C., J. W. Hastings. 1975. Biochemistry 14, 4310-4316.

20. Tu, S.-C. 1982. J. Biol. Chem. 257, 3719-3725.

CHEMICAL MODIFICATION OF TYROSINE-38 IN P-HYDROXYBENZOATE HYDROXYLASE FROM PSEUDOMONAS FLUORESCENS BY 5'-P-FLUOROSULFONYLBENZOYLADENOSINE: A PROBE FOR THE ELUCIDATION OF THE NADPH BINDING SITE?

Willem J.H. van Berkel, Edwin J. Bakx, Franz Müller, Wicher J. Weyer, Peter A. Jekel, Jaap J. Beintema, Herman A. Schreuder, Jan M. van der Laan, Rik K. Wierenga, Jan Drenth and Wim G.J. Hol

Department of Biochemistry, Agricultural University, De Dreijen 11, 6703 BC Wageningen and Departments of Biochemistry and Chemical Physics, State University of Groningen, Nijenborgh 16, 9747 AG Groningen, The Netherlands

Introduction

p-Hydroxybenzoate hydroxylase from Pseudomonas fluorescens is an FAD containing monooxygenase which catalyzes the NADPH dependent conversion of p-hydroxybenzoate into 3,4-dihydroxybenzoate. A three dimensional model of the enzyme-substrate complex is available at a resolution of 0.25 nm (1). Chemical modification studies have provided much information about the role of specific amino acid residues involved in the binding of substrate and coenzyme (2).
In order to support the crystallographic attempts to resolve the NADPH binding site we have investigated the covalent modification of the enzyme by the adenine nucleotide analog 5'-p-fluorosulfonylbenzoyl adenosine (FSO_2BzAdo).

Results and Discussion

N-ethylmaleimide prelabeled enzyme (E-MalNEt) is inactivated at a very slow rate ($k_i \leqslant 2 \times 10^{-2}$ $min^{-1}mM^{-1}$) by FSO_2BzAdo. The reaction does not obey pseudo-first-order kinetics during its whole course owing to the instability of the reagent which can be mainly attributed to the hydrolysis of the fluorosulfonyl moiety (3).
The inactivation reaction is strongly inhibited by NADPH and 2',5'-ADP and in a much lesser degree by p-hydroxybenzoate or p-fluorobenzoate.

Flavins and Flavoproteins 1987

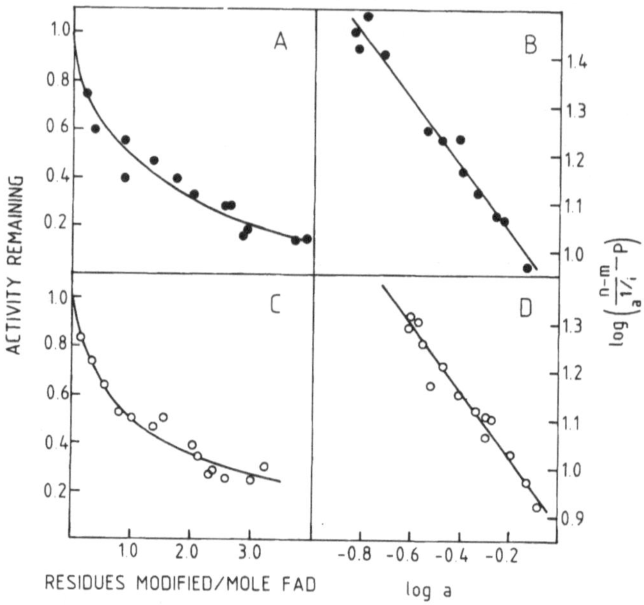

Fig. 1. Correlation between the residual activity of N-ethylmaleimide prelabeled p-hydroxybenzoate hydroxylase and the number of residues modified by p-fluorosulfonylbenzoyladenosine.
A,B : E-MalNEt   C,D : E-MalNEt.p-hydroxybenzoate complex.

Radioactive labeling using [^{14}C]-FSO_2BzAdo shows that in the presence of NADPH at least one residue is protected from modification. The correlation between the residual activity (a) and the number of residues (m) that are modified in free E-MalNEt (Fig. 1A) and E-MalNEt-p-hydroxybenzoate complex (Fig. 1C) can be interpreted according to the statistical method developed by Tsou (4):

$$\log [(n-m)/a^{1/i}]-p = \log (n-p) + [(\alpha-1)/i] \log a$$

where n = total number of residues modified, p=residues of which i are essential and react with a rate constant k, and n-p=residues reacting with a rate constant αk. In both cases (Fig. 1B, 1D) the best fit is obtained for n = 8 (α = 0.32) and p=i=1 indicating that only one residue is responsible for the loss of activity caused by modification by FSO_2BzAdo.

The amino acid residue has been identified by sequencing radioactive tryptic peptides purified by reverse phase HPLC.

O-(4-carboxybenzene sulfonyl) tyrosine-38 was found as the main product of modification in purified peptide CB3T4 spanning residues 34-42 (Gln-Thr-Pro-Asp-Tyr-Val-Leu-Gly-Arg).

The visible absorption spectrum of FSO_2BzAdo modified E-MalNEt is identical with that of native enzyme. Fluorescence titration experiments show that the dissociation constants of modified E-MalNEt-p-hydroxybenzoate and modified E-MalNEt-p-hydroxybenzoate-gold cyanide complex are the same as those of the native enzyme complexes. Gold cyanide is a very strong inhibitor of the enzyme with respect to NADPH (2). The binding site of gold cyanide is probably also the site where the nicotinamide part of NADPH binds (1). NADPH shows a decreased affinity for the modified enzyme but FAD can still be reduced under anaerobic conditions. Steady-state kinetics show that the activity of modified E-MalNEt can be increased by high concentrations of NAPDH. 2',5'-ADP can also be bound to modified E-MalNEt as found by affinity chromatography.

The results of modification and sequencing studies are used for model building of the structure of the modified enzyme using computer graphics. Starting point of model building was a bound NADPH molecule fitted into the enzyme-p-hydroxybenzoate structure (1). The nicotinamide moiety of NADPH is bound parallel to the flavin ring (Fig. 2), just as in glutathione reductase (5), between the flavin ring and a loop formed by the residues 293-297. The pyrophosphate forms salt links with Arg-269 and His-162. The 2'-phosphate of the adenosine makes a salt link with Arg-166. The adenosine is bound in a hydrophobic pocket made up by the carbon atoms of the side chains from Phe-271, Val-176, Ala-177 and Arg-166. The NH_2 group at the 6 position is near the side chain of Glu-178.

The adenosine part of FSO_2BzAdo was superimposed onto the adenosine part of the NADPH molecule. The sulfonyl benzoyl moiety of FSO_2BzAdo could be fitted between the 5'O of the adenosine and the OH of Tyr-38 (Fig. 2). The distance between the gold cyanide position and the OH of Tyr-38 is 1.4 nm. The results strongly indicate that FSO_2BzAdo in p-hydroxybenzoate hydroxylase is bound in an extended conformation far away from the active site and modifying a residue that is not directly involved in NADPH binding.

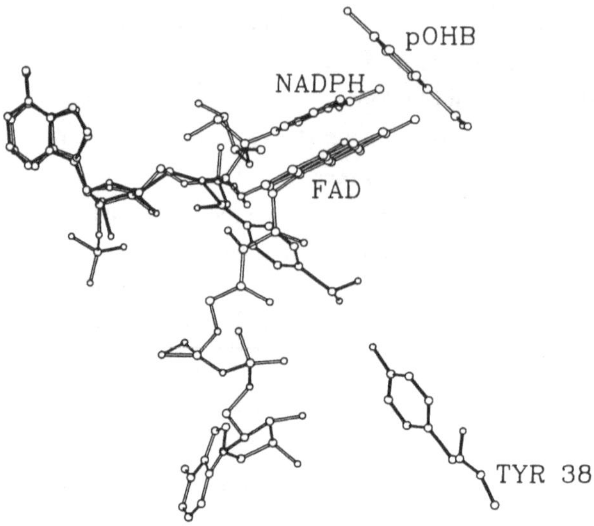

Fig. 2. Position of the FSO_2BzAdo and NADPH molecules with respect to the FAD and substrate (POHB) molecule.

Aoknowledgements

This study was carried out under the auspices of the Netherlands Foundation for Chemical Research (SON) with financial aid from the Netherlands Organization for the Advancement of Pure Research (ZWO).

References

1. Weijer, W.J., Hofsteenge, J., Beintema, J.J., Wierenga, R.K., Drenth, J.: Eur.J.Biochem. 133, 109-118 (1983).
2. Wijnands, R.A., Müller, F., Visser, A.J.W.G.: Eur.J.Biochem. 163, 535-544 (1987).
3. Colman, R.F.: Annu.Rev.Biochem. 52, 67-91 (1983) .
4. Tsou, C.L.: Sci.Sin. (Engl. Ed.) 11, 1535-1558 (1962).
5. Pai, E.F., Schultz, G.E.: J.Biol.Chem. 258, 1752-1757 (1983).

PHENOL COMPLEXES OF PHENOL HYDROXYLASE

Torsten Sejlitz and Halina Y. Neujahr

Department of Biochemistry and Biotechnology, The Royal Institute of Technology, S-100 44 Stockholm, Sweden

INTRODUCTION

Phenol hydroxylase (EC 1.14.13.7; Phenol, NADPH: oxygen oxidoreductase, 2-hydroxylating) is dimeric (M_r=2x76000) and contains 2 FAD (1). The phenolic substrates of phenol hydroxylase act as effectors, causing conformational changes, with a concomitant increase in the reactivity towards NADPH (2,3) and spectral perturbation (4). Substrate inhibition is observed with many phenolic substrates, including phenol itself (2). Steady-state kinetics are consistent with the Bi Uni Uni Bi ping-pong mechanism proposed for other aromatic hydroxylases. The mechanism is ordered with phenol binding before NADPH followed by reduction of the flavin and release of $NADP^+$. Molecular oxygen then reacts with the reduced enzyme-substrate complex (3). Partial uncoupling of oxygen reduction from hydroxylation with concomitant production of H_2O_2 (oxidase activity) occurs with all hydroxylatable substrates (2).

BINDING OF PHENOL MEASURED BY EQUILIBRIUM DIALYSIS

Equilibrium dialysis alone (as well as other equilibrium methods) cannot distinguish negative cooperativity from the occurrence of non-identical binding sites (5). Fitting a non-cooperative model to the dialysis data (Fig 1) yielded two "high-affinity" sites with K_d=1.3 and 19 µM, and a group of "low-affinity" sites. The data suggest that there are not less than two low-affinity sites and that their K_d is in the 100-500 µM range. If the two high-affinity sites are non-equivalent, this would imply two non-equivalent subunits. Any attempt to separate the subunits by chromatography or electrophoresis have been unsuccessful (unpublished). Single-turnover experiments and stopped-flow measurements of flavin intermediates suggest two functionally eqivalent catalytic sites (3,6). On the basis of all these observations we assumed that phenol hydroxylase has two functionally equivalent subunits. This assumption, together with our binding data (Fig 1),

Flavins and Flavoproteins 1987

implies negative cooperativity among the high-affinity sites. Assuming two equivalent and negatively cooperative high-affinity sites, the number of phenol molecules bound can be expressed as:

$$n = \frac{2\cdot\frac{[S]}{K_{dA}} + 2\cdot\frac{[S]^2}{K_{dA}K_{dB}}}{1 + 2\cdot\frac{[S]}{K_{dA}} + \frac{[S]^2}{K_{dA}K_{dB}}} + \frac{N_3\cdot\frac{[S]}{c_3}}{1 + \frac{[S]}{c_3}} \qquad (1)$$

where K_{dA} is the intrinsic dissociation constant of any of the high-affinity sites in the free enzyme and K_{dB} is the intrinsic dissociation constant of the remaining site when one site is bound. N_3 and c_3 are the number of sites and the apparent dissociation constant in the low-affinity group. The best fit was obtained with K_{dA}=2.6±0.4 μM, K_{dB}=11.6±2.0 μM (N_3 and c_3 as above).

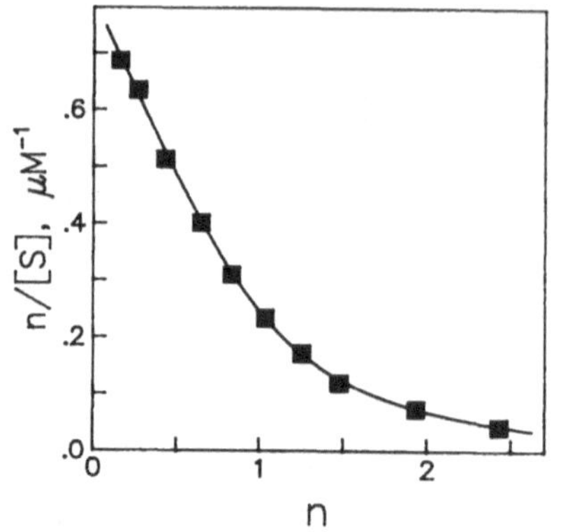

Fig 1. Scatchard plot of the binding of [^{14}C]-phenol to phenol hydroxylase.

RELATION BETWEEN BINDING OF PHENOL AND SPECTRAL PERTURBATION

The spectral perturbation was measured as δ, defined as normalized mean absorbance change over the range 410-440 nm, relative to the free enzyme. Titration data (Fig 2A) suggested that the binding of phenol to the low-affinity sites had no effect on the spectrum. High phenol levels partly reverse the spectral perturbation (4). To compensate for the reversal in the evaluation of the high-affinity sites, it was treated by analogy with noncompetitive inhibition. If each type of complex has a characteristic spectral perturbation, i.e. Δ_A (mono-phenol complex) and Δ_{AB} (di-phenol complex), the total spectral perturbation can be expressed as:

$$\delta = \frac{2\cdot\Delta_A\cdot\frac{[S]}{K_{dA}} + \Delta_{AB}\cdot\frac{[S]^2}{K_{dA}K_{dB}}}{1 + 2\cdot\frac{[S]}{K_{dA}} + \frac{[S]^2}{K_{dA}K_{dB}}} \quad \frac{1 + X_R\cdot\frac{[S]}{K_R}}{1 + \frac{[S]}{K_R}} \qquad (2)$$

where K_R is the apparent "reversal constant", and X_R is the fraction of the spectral perturbation retained at infinite phenol concentration. The best fit was obtained with K_{dA}=2.6 μM; Δ_A=0.080; K_{dB}=10.6 μM; Δ_{AB}=0.10. K_{dA} and K_{dB} agrees well with the binding constants from the equilibrium dialysis.

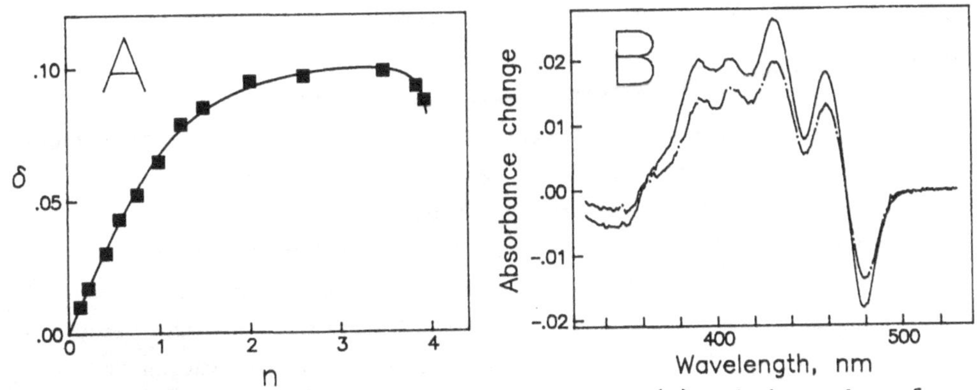

Fig 2. (A) Relation between spectral perturbation (δ) and the number of equivalents phenol bound (n). (B) Calculated spectra of the mono-phenol (—·—) and of the di-phenol complex (——).

Spectra of the different complexes were calculated using these dissociation constants (Fig 2B). Since Δ_A/Δ_{AB}=0.8, this shows that the binding of one phenol per enzyme dimer was sufficient to effect the main part of the maximum spectral perturbation.

RELATION BETWEEN THE LEVELS OF THE ENZYME-PHENOL COMPLEXES AND OVERALL REACTION RATES

The hydroxylase as well as the oxidase activity were studied as a function of phenol concentration. The data in Fig 3 show that the oxidase activity coincided with the mono-phenol complex and that the hydroxylase activity seems to be associated with both the mono-phenol and the di-phenol complexes. The substrate inhibition of the hydroxylase activity coincided with the binding to the low-affinity sites. Reaction rates of the enzyme-phenol complexes were calculated using the same method as was used for the spectral perturbation (Table 1).

DISCUSSION

From the data presented we propose a binding model involving two identical and negatively cooperative effector/substrate sites per enzyme dimer. A major part of the spectral perturbation as well as of the increase in the NADPH oxidation rate is achieved by one phenol bound (Fig 2, Table 1). This suggests that the binding of phenol to one subunit and the accompanying conformational change may also affect the conformation of the opposite

subunit and thus its reactivity towards NADPH.
The phenol dependence of the oxidase activity (Fig 3) suggests that the choice of reaction path reflects an association-dissociation equilibrium between enzyme and phenol. We suggest, that the high oxidase activity at low phenol levels arises from the formation of the mono-phenol complex. Due to the cooperativity, the binding of one phenol may accelerate the reduction of both FAD's. Consequently, the FAD of the free site in the mono-phenol complex will be readily reduced, and enter the oxidative half-reaction without bound substrate. This is consistent with steady-state reaction rates, which show that the oxidase activity coincides with the mono-phenol complex.

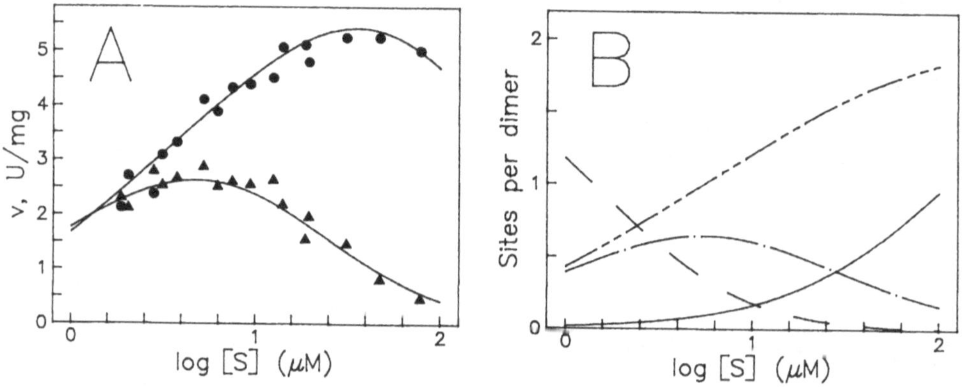

Fig 3. (A) Hydroxylase (•), and oxidase activity (▲) at 1 mM NADPH and varying phenol. (B) Calculated equilibrium levels of sites per dimer in different states: Free high-affinity sites in free enzyme (— —), free high-affinity sites in mono-phenol complex (—•—), total bound high-affinity sites (—••—), and total bound low-affinity sites (——).

Table 1. Overall reaction rates of the enzyme phenol complexes. Reaction rates were calculated using K_{dA}=3μM and K_{dB}=10μM. K_I for the substrate inhibition was 170 μM.

	Oxidase (U/mg)	Hydroxylase (U/mg)
E	0.3	0
ES	4.1	3.9
ES_2	<0.1	8.2

REFERENCES

1. Neujahr,H.Y. & Gaal,A. (1975) Eur.J.Biochem. 58, 351-357
2. Neujahr,H.Y. & Kjellén,K.G. (1978) J.Biol.Chem. 253, 8835-8841
3. Detmer,K. & Massey,V. (1984) J.Biol.Chem. 259, 11265-11272
4. Neujahr,H.Y. (1983) Biochemistry 22, 580-584
5. Dahlquist, F.W. (1978) Meth.Enzymol. 48, 270-299
6. Detmer,K. & Massey,V. (1985) J.Biol.Chem. 260, 5998-6005

THE INVOLVEMENT OF SH-GROUPS IN COOPERATIVITY OF PHENOL-BINDING SITES IN PHENOL HYDROXYLASE

Halina Y. Neujahr

Dept of Biochemistry, The Royal Institute of Technology,
S-100 44 Stockholm, Sweden

INTRODUCTION

Phenol hydroxylase (EC 1.14.13.7) from the soil yeast Trichosporon cutaneum (1,2), is a dimer containing 2 FAD per 2x76 Kdal. The two subunits do not separate upon molecular sieving, ion exchange HPLC or electrophoresis. As in the procaryotic aromatic hydroxylases, the effector function of the phenolic substrate in phenol hydroxylase is expressed in perturbation of the absorption spectrum (3), quenching of flavin fluorescence (3) and in drastically increased affinity towards the reduced pyridine nucleotide cosubstrate (4). Phenol hydroxylase undergoes, in addition, a number of effector-triggered changes, not reported for procaryotic aromatic hydroxylases. Thus, in the presence of one equivalent of phenol, a lysyl residue involved in the binding of NADPH becomes more reactive towards modifying reagents as well as towards NADPH (5), whereas the most reactive SH-groups of the enzyme become less accessible (2) and FAD becomes less prone to be displaced by chaotropic agents (3). Phenol hydroxylase contains 16 cysteinyl residues, but no disulfide bridge. Reduced -SH groups are essential for the retention of FAD (2), but have no obvious function in any redox reaction essential for the catalytic cycle (6). This paper provides indications of slow conformation changes during the overall reaction, which seem to be speeded up by added thiol. The results suggest that the reduced SH groups of phenol hydroxylase, though not participating in any demonstratable redox reactions, are important for the cooperativity of phenol binding sites, whereby the anionic S^- forms may be involved.

Flavins and Flavoproteins 1987

RESULTS

In both, the presence and the absence of added thiol-EDTA, equilibrating the enzyme with phenol prior to the addition of NADPH, results in higher initial rates (Fig 1) and longer range of linearity (not shown), than when NADPH is added first. The effect is more pronounced with no thiol added. In the absence of added thiol, the pH-optimum is affected by the order of

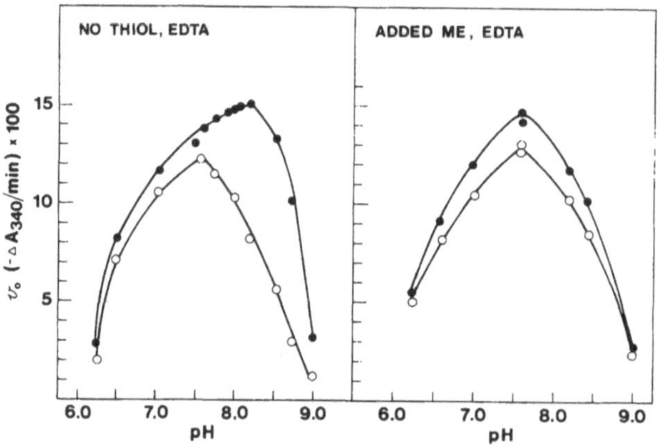

Fig 1. The order of addition of phenol and NADPH and pH-dependence of v_o. Phenol first (●); NADPH first (○)

addition of phenol and NADPH. The spectral perturbation by phenol increases with pH in the pH range 7.0-8.7 and it is not significantly affected by the presence of added mercaptoethanol-EDTA (Fig 2). Fig 2 also shows the

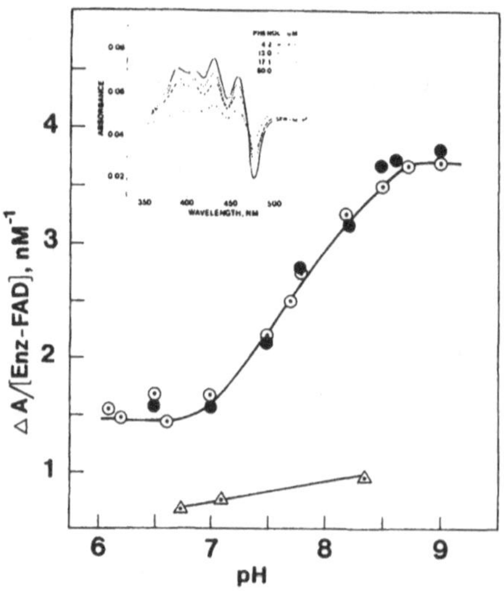

Fig 2. Main diagram: The pH dependence of the spectral perturbation ($A_{478-458}$) by saturating concentrations of phenol, without (⊙) or with (●) mercaptoethanol 2 mM and EDTA 0.2 mM. Perturbation by resorcinol (△, no thiol-EDTA) is shown for comparison. Enzyme FAD: 8-14 μM phenol: 400 μM below pH 7, 100 μM above pH 7. Inset: difference spectrum

spectral perturbation by resorcinol, a good substrate of phenol hydroxylase (1,4) but apparently, a rather poor effector.

Spectrophotometric titrations of phenol hydroxylase with phenol were carried out at several pH values in the range pH 6.5-9.0, in the absence and in the presence of added thiol-EDTA. The perturbations obtained were used to calculate the levels of enzyme-phenol complexes and, by substraction from the total concentration of phenol, those of free phenol. Fig 3 shows Scatchard plots constructed from such data. Numerical values of Hill coefficients deduced from a number of similar experiments with the native enzyme were consistently above 1.0 (1.2-1.6) when computed from experiments in the pH range 6.1-7.6 without added thiol- EDTA, but below 1.0 (0.6-0.8) when computed from those at pH 8.3-9.0. In the presence of added mercapto-ethanol-EDTA, these coefficients never exceeded the value 0.8 in the entire pH range that was studied. These results, as well as the curved profiles of Scatchard plots (Fig 3), suggest interacting sites of phenol binding, a property not observed with procaryotic hydroxylases. The cooperativity is positive below pH 8, but negative at higher pH values, with a transition

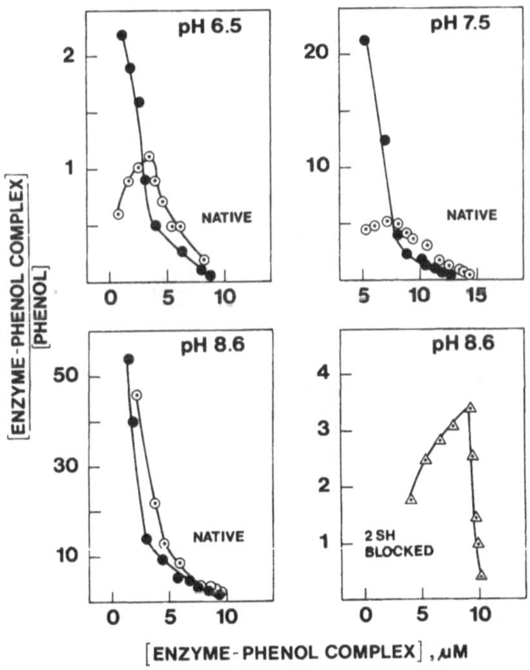

Fig 3. Profiles of saturation curves (Scatchard plots) of enzyme-phenol complexes at various pH, in the absence (empty symbols) and in the presence (filled symbols) of mercaptoethanol 2 mM and EDTA 0.2 mM

around pH 8.3. However, in the presence of added thiol-EDTA, the cooperativity is negative throughout the investigated pH range. Conversely, the cooperativity of phenol binding to the enzyme with 2 blocked SH-groups is always positive (Fig 3). The enzyme was subjected to turnover, freed from excess reagents and then tested for reactivity of its SH groups towards Ellman's reagent. The reactivity of the most reactive SH groups was significantly decreased after turnover, as compared to the native enzyme. Prolonged incubation of the enzyme solution at 4°C increased this reactivity slowly. Original reactivity could be restored more rapidly by incubating the enzyme with thiol-EDTA-FAD.

The cooperative properties of phenol hydroxylase discussed here may be inherent in the dimeric nature of the enzyme. They would thus distinguish this eucaryotic enzyme from the extensively studied procaryotic aromatic hydroxylases which are usually described as monomers containing 1 FAD per single polypeptide chain (6,7,8). However, a recent re-assessment of the p-hydroxybenzoate hydroxylase (EC 1.14.13.2) from Ps.fluorescens suggests that also this enzyme exists mainly as a dimer (9).

REFERENCES

1. Neujahr,H.Y. and Gaal,A. (1973) Eur.J.Biochem. 35, 386-400
2. Neujahr,H.Y. and Gaal,A. (1975) Eur.J.Biochem. 58, 351-357
3. Neujahr,H.Y. (1983) Biochemistry 22, 580-584
4. Neujahr,H.Y. and Kjellén,K.G. (1978) J.Biol.Chem. 253, 8835-8841
5. Neujahr,H.Y. and Kjellén,K.G. (1980) Biochemistry 19, 4967-4972
6. Detmer,K. and Massey,V. (1984) J.Biol.Chem. 250, 11265-11272
7. Higashi,N., Shoun,H., Yano,K., Arima,K. and Hiromi,K. (1972) Z.Naturforschung 27, 1172-1175
8. Husain,M. and Massey,V. (1979) J.Biol.Chem. 254, 6657-6666
9. Müller,F., Voordouw,G, van Berkel,W.J.H., Steennis,P.J., Visser,S. and van Rooijen,P.J. (1979) Eur.J.Biochem. 101, 235-244

6-THIOCYANATE-FAD AND 6-MERCAPTO-FAD AS ACTIVE SITE PROBES OF PHENOL HYDROXYLASE

M. G. Taylor and V. Massey

Department of Biological Chemistry, University of Michigan, Ann Arbor, MI 48109

Introduction

Phenol hydroxylase(1), isolated from the soil yeast, *Trichosporon cutaneum*, is an FAD-containing enzyme which catalyzes the production of numerous catechols from their corresponding phenols with the concomitant utilization of NADPH and molecular oxygen. Detmer and Massey(2) replaced the native flavin with a number of FAD analogs and established the large degree of accessiblity to the 7 and 8-positions of the flavin moiety. In addition, they showed the relative inaccessibility of the flavin 2 and 4-positions. In extension of previous studies(3), we have substituted 6-thiocyanate-FAD and 6-mercapto-FAD for the native flavin of phenol hydroxylase. By examining the reactivity, and hence the accessibility of the flavin 6-position, we have further defined the flavin binding pocket, as well as provided insights into the actual protein environment of the active site.

Materials and Methods

6-Amino-riboflavin was the generous gift of Sandro Ghisla. This was converted to the FAD form and subsequently to both 6-thiocyanate-FAD and 6-mercapto-FAD by the method of Ghisla et al.(4) FAD-synthetase was partially purified by following the method of Manstein and Pai(5). The apo-enzyme of phenol hydroxylase was generated by a slight modification of the procedure of Mayer and Thorpe(6). Reconstitution was initiated by the addition of a slight excess of the appropriate FAD analog. An incubation for 30 minutes at 4°C was followed by a second incubation at 20°C of varying time (0-20 minutes). The sample was then centrifuged to clarify the solution, concentrated by Centricon ultrafiltration, and loaded onto a G-25 column equilibrated with 50mM KP_i/3mM EDTA/pH7.0 to eliminate any residual free flavin. All reactions were performed at pH=7.0 and 20°C and were monitored at 444 nm by use of a Cary 219 Spectrophotometer, HP8452A Diode Array Spectrophotometer, or stopped-flow spectrophotometer interfaced to a Nova II minicomputer.

Flavins and Flavoproteins 1987

Results and Discussion

Upon addition to the apoprotein of phenol hydroxylase, the spectrum of 6-thiocyanate-FAD is slightly modified with a shift of λ_{max} from 446nm to 444nm and a modest increase of extinction coefficient, ε_{max}, from 12,400 $\underline{M}^{-1}cm^{-1}$ to 13,200 $\underline{M}^{-1}cm^{-1}$. The spectrum of the 6-thiocyanate-FAD enzyme spontaneously decays to that of authentic 6-mercapto-FAD phenol hydroxylase with a rate of 5 x 10^{-3} min^{-1} (Figure 1). This conversion can be rationalized by the presence of an enzyme nucleophile, probably a thiol, near the flavin 6-position. Such an active site thiol could conceivably participate in the elimination of cyanide by one of the two mechanisms shown in Scheme I. The resultant bound 6-mercapto-FAD spectrum showed a shift of λ_{max} from 445nm to 443nm, and a decrease in ε_{max} from 18,900 $\underline{M}^{-1}cm^{-1}$ to 18,100 $\underline{M}^{-1}cm^{-1}$ relative to the free flavin. In addition, the long wavelength band shifted from 610nm to 660nm with a decrease in ε_{max} from 2,900 $\underline{M}^{-1}cm^{-1}$ to 2,200 $\underline{M}^{-1}cm^{-1}$. These differences are in part due to a slightly increased pK_a when enzyme bound (data not shown here).

Scheme I

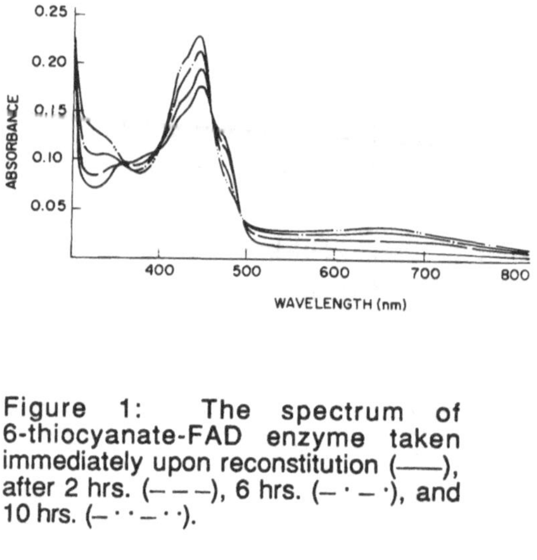

Figure 1: The spectrum of 6-thiocyanate-FAD enzyme taken immediately upon reconstitution (——), after 2 hrs. (– – –), 6 hrs. (– · – ·), and 10 hrs. (– · · – · ·).

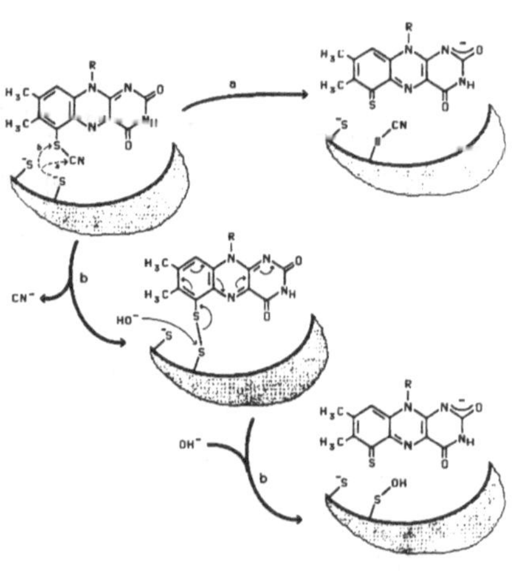

The spontaneously formed 6-mercapto-FAD enzyme was treated with MMTS, rapidly forming a methyl disulfide enzyme derivative (k=2.7 x 10^5 $\underline{M}^{-1}min^{-1}$) (Table 1) whose spectrum is similar to that of the thiocyanate form. This reaction, however, was slowly, but spontaneously, reversed, regenerating the original mercapto-form when only a slight molar excess of MMTS was used initially. A second nucleophile, again presumably a thiol, could account for such a reversal.

Table 1

REACTION	k, reaction rate (M^{-1} min^{-1}) (unless otherwise specified) free flavin	6-mercapto enzyme w/o phenol	6-mercapto enzyme w/ phenol
Flavin-S^- + MMTS ⟶ Flavin-$SSCH_3$	>>4×10^4	2.7×10^5	1.1×10^5
Flavin-$SSCH_3$ ⟶(spontaneous) Flavin-S^-	—	2.3×10^{-3} min^{-1}	—
Flavin-S^- + NEM ⟶ Flavin-5,6-adduct	≥4×10^4	1.9×10^4	5×10^5
Flavin-5,6-adduct + DTT ⟶ Flavin-S^-	—	1.8×10^3	2.4×10^3
Flavin-S^- + H_2O_2 ----> Flavin-SO_2^-(H^+)	dimerizes	53	8
Flavin-S^- + m-CPBA ----> Flavin-SO_2^-(H^+)	—	6×10^3	8×10^3
Flavin-S^- + ICH_2CONH_2 ⟶ Flavin-SCH_2CONH_2	150	41	18
Flavin-S^- + $ICH_2CO_2^-$ ⟶ Flavin-$SCH_2CO_2^-$	—	2.7	—

Treatment of the mercapto- FAD enzyme with NEM results in extensive bleaching of the spectrum, similar to that found with the free 6-mercaptoflavin, presumably to produce a flavin- 5,6-adduct(7) (k=1.9 x 10^4 $M^{-1}min^{-1}$). Addition of excess DTT readily reversed this reaction with a rate of 1.8 x 10^3 $M^{-1}min^{-1}$. A possible mechanism for these reactions is shown in Scheme II.

Scheme II

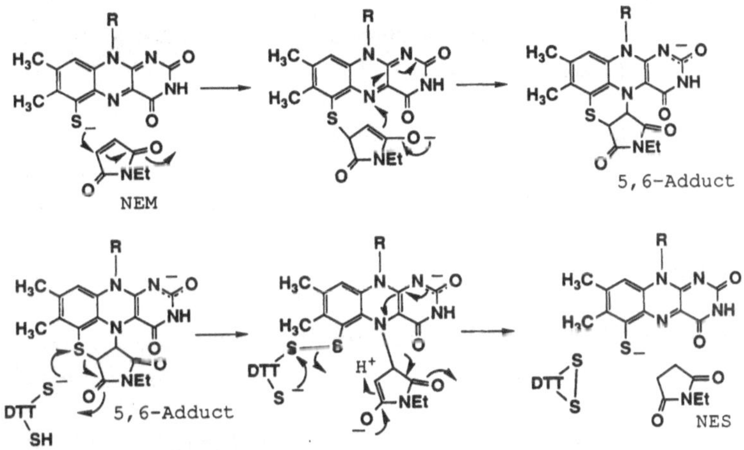

The standard thiol-specific reagents, iodoacetamide and iodoacetate, were used to alkylate the flavin 6-position of 6-mercapto-FAD phenol hydroxylase. As expected, iodoacetamide reacted somewhat faster than the anionic iodoacetate with a reactior rate of 41 $M^{-1}min^{-1}$ for the former, compared with 2.7 $M^{-1}min^{-1}$ for that of the latter

Oxidation of the 6-mercapto-FAD enzyme was initiated by the addition of H_2O_2 or m-chloroperbenzoic acid (m-CPBA). Both reagents reacted relatively rapidly (k=53

$\underline{M}^{-1}min^{-1}$ and k= 6 x 10^3 $\underline{M}^{-1}min^{-1}$, respectively) to produce a species whose spectrum includes peaks at 446nm (ε_{max} of 15,300 $\underline{M}^{-1}cm^{-1}$) and at 376nm (ε_{max} of 11,700 $\underline{M}^{-1}cm^{-1}$). The latter peak is indicative of an anionic sulfur species(3). This derivative is not, however, reconverted to the original mercapto form with addition of excess DTT, suggesting that a deprotonated sulfinic or sulfonic acid has been generated.

In the presence of 400μM phenol, several reaction rates were affected. The bleaching noted with NEM occurred 30-fold faster when substrate was bound. Reactions involving MMTS, iodoacetamide, and H_2O_2, however, showed small but significant decreases in rate (2.3, 2.3, and 6.6-fold, respectively) (Table 1).

It is apparent that 6-mercapto-FAD phenol hydroxylase reacts relatively rapidly when treated with a wide variety of thiol-specific reagents. Comparison of these rates with those found with free 6-mercapto-FAD, as well as those found with several other 6-mercaptoflavin substituted proteins(3), strongly suggests a high degree of accessibility to the flavin 6-position. The presence of two enzyme thiols in the environment of the flavin 6-position is consistent with the spontaneous conversion of 6-thiocyanate-FAD enzyme to the mercapto form and the spontaneous reversal of the reaction seen with MMTS.

Acknowledgements

This research was supported by U. S. Public Health Service grant no. GM-11106.

References

1. Neujahr, H. Y. and A. Gaal. 1973. Eur. J. Biochem. 35, 386-400 .
2. Detmer, K. and V. Massey. 1984. Flavins and Flavoproteins, Proc. Eighth Int. Symp. pp.765-768.
3. Massey, V., S. Ghisla, and K. Yagi. 1986. Biochemistry. 25, 8103-8112.
4. Ghisla, S., V. Massey, and K. Yagi. 1986. Biochemistry. 25, 3282-3289.
5. Manstein, D. J. and E. F. Pai. 1986. J. Biol. Chem. 261, 16129-16173.
6. Mayer, E. J. and C. Thorpe. 1981. Anal. Biochem. 116, 227-229.
7. Steenkamp, D. J., W. McIntire, and W. C. Kenney. 1978. J. Biol. Chem. 253, 2818-2824.

HYSTERETIC BEHAVIOR OF ANTHRANILATE HYDROXYLASE

Justin Powlowski, David Ballou, and Vincent Massey
Department of Biological Chemistry, University of Michigan, Ann Arbor, Michigan 48109-0606, U.S.A.

Introduction

Anthranilate hydroxylase catalyzes the conversion of anthranilate to 2,3-dihydroxybenzoate, with the insertion of an atom of oxygen from O_2 into the 3-position, and an atom of oxygen from H_2O into the 2-position (1), with expulsion of ammonia. Investigations of the enzyme kinetics indicate hysteretic behavior. We present evidence that after the first turnover with oxygen, the enzyme is in a different form which reacts much more rapidly with anthranilate, and thus NADPH, than the "resting" enzyme. This appears to be a result of conformational changes undergone upon enzyme reduction and ligand binding.

Results and Discussion

Lags in NADPH oxidation were observed in multiple turnover experiments when the enzyme was mixed with anthranilate (or a non-substrate effector such as salicylate), and NADPH (2). The lag corresponded to slow initial bleaching of the enzyme flavin and thus appeared to be related to the initial enzyme reduction. Since the difference in rates of NADPH reduction of the ligand-free enzyme and the enzyme-anthranilate complex was at least three orders of magnitude, the rate of substrate binding appeared to be of possible importance in the transient slow turnover.

Flavins and Flavoproteins 1987

Anthranilate binding was monitored by the spectral perturbation at 492 nm (1), while salicylate perturbed enzyme absorbance maximally at 393 nm. The rate of anthranilate binding to E_{ox} was far too slow to account for the observed rates in multiple turnovers; at 125 μM anthranilate, k_{obs}=0.03 sec^{-1}. Reaction of the reduced enzyme-anthranilate (125 μM anthranilate) complex with O_2 at pH 7.4 was biphasic, and an intermediate spectrum corresponding to the flavin 4a-hydroperoxide observed with other flavin monooxygenases was seen (3). The k_{obs} for the first phase was dependent on $[O_2]$ with k_f=3.4 x 10^4 $M^{-1}sec^{-1}$ and k_r=2.3 sec^{-1}, while the k_{obs} of the second phase was independent of $[O_2]$, and was 2.2 sec^{-1} at 3.5°. At the end of the reactions with O_2, the enzyme spectrum was that of the enzyme-substrate complex (1). This indicated that anthranilate bound rapidly either to one of the oxygenated flavin intermediates, or to a form of E^* generated after the initial turnover, since this binding event was much more rapid than that to E_{ox}.

Results obtained for the oxidative half-reactions of ligand-free enzyme suggest the latter possibility. Thus, when reduced enzyme was mixed with oxygen, a single intermediate was observed. This intermediate (k_f=3.7 x 10^4 $M^{-1}sec^{-1}$, k_r=2.2 sec^{-1}) had a highly perturbed E_{ox} spectrum which relaxed to the normal E_{ox} spectrum with a k_{obs}= 1.5 sec^{-1} at 3°. E_{ox}-anthranilate complex was observed at the end of the second phase when anthranilate was included in the oxygen-containing syringe (Fig. 1). Since very similar results were obtained when $K_3Fe(CN)_6$ was the oxidant, substrate binding to an oxygenated flavin intermediate was ruled out. Therefore, enzyme reduction appears to result in some E^* species after oxidation which is capable of binding the aromatic ligand much more rapidly than resting E_{ox}.

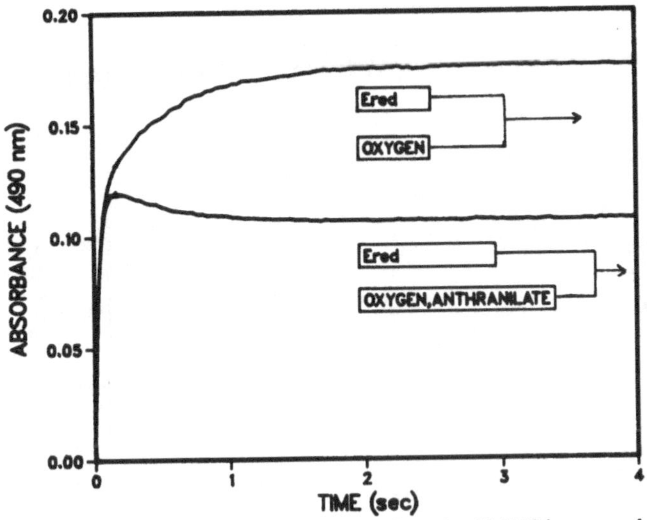

Figure 1. *Dithionite-reduced enzyme (10.3 μM FAD) was mixed with oxygenated buffer (620 μM O_2, pH 7.4) with and without anthranilate (200 μM) at 4°. At this [anthranilate], k_{obs}=0.046 sec^{-1} for binding to "resting" E_{ox}. The rate of the reaction of E_{red} with O_2 is 64 x k_{obs} for anthranilate binding to E_{red}. All concentrations are after mixing.*

Substrate also appears to stabilize a different enzyme conformation. Thus, when salicylate, a non-substrate effector, was mixed with enzyme-anthranilate complex, much more rapid binding of salicylate was observed than with uncomplexed enzyme (Fig. 2). An E_{ox}-anthranilate/E'_{ox}-anthranilate equilibrium must be postulated, with E' representing a form of enzyme which rapidly binds ligand. A more direct indication of conformational change upon substrate binding is suggested by the large increase in accessibility of the solvent (4) to the 8-position of the flavin in the presence of bound anthranilate. Thus, free 8-Cl FAD reacted with thiophenol at 3.5^o with a second order rate constant of 680 M^{-1} min^{-1}, while the reaction of 450 μM thiophenol with free 8-Cl FAD enzyme was less than 10% complete at the end of 24 hours. The E_{ox}-anthranilate complex, however, reacted with thiophenol with a second order rate constant of 360 M^{-1} min^{-1}, which is more similar to the free flavin reaction.

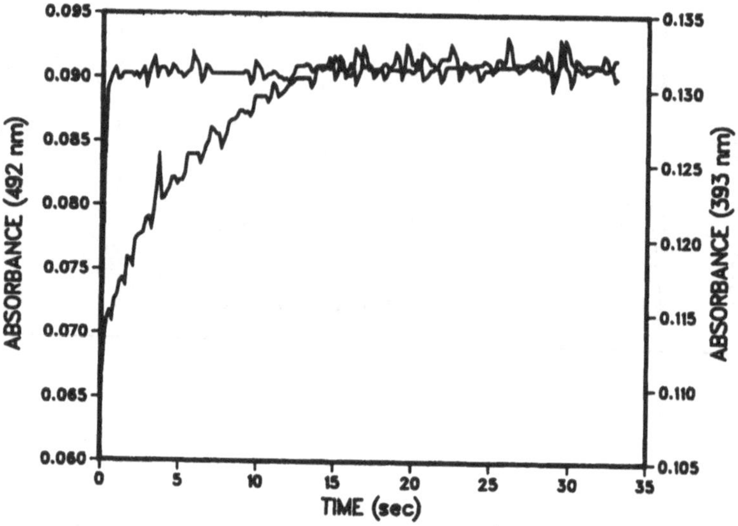

Figure 2. *Lower trace*: *E_{ox} (9.8 µM FAD) was mixed with 1 mM salicylate at 3°, and the reaction was monitored at 393 nm (k_{obs}=0.19 sec^{-1}). All concentrations are after mixing.*

Upper trace: *as above, except E_{ox} was pre-complexed with saturating anthranilate (400 µM), and the reaction was monitored at 492 nm where the perturbation by anthranilate is far greater than that caused by salicylate (k_{obs}=7 sec^{-1}).*

Acknowledgement

This work was supported by USPHS grants GM 20877 (to D.B.) and GM 11106 (to V.M.).

References

1. Powlowski, J., Dagley, S., Massey, V., and Ballou, D.P. 1987. J. Biol. Chem. 262, 69-74.

2. Powlowski, J., Ballou, D.P., and Massey, V. 1985. Biochemistry 24, 3379 (Abstr. 129).

3. Entsch, B., Ballou, D.P., and Massey, V. 1976. J. Biol. Chem. 251, 2550-2563.

4. Schopfer, L.M., Massey, V. and Claiborne, A. 1981. J. Biol. Chem. 256, 7329-7337.

REDOX POTENTIALS OF SALICYLATE HYDROXYLASE

G.H. Einarsdottir and M.T. Stankovich
Department of Chemistry, University of Minnesota, Minneapolis, MN 55455

L.-H. Wang and S.-C. Tu
Department of Biochemical and Biophysical Sciences, University of Houston, Houston, TX 77004

Introduction

Salicylate hydroxylase (E.C.1.14.13.1) catalyzes the decarboxylative hydroxylation of salicylate to form cathecol. In the absence of salicylate, the enzyme catalyzes the oxidation of NADH to form NAD^+, at a much slower rate. Benzoate acts as an effector to increase the rate of NADH oxidation and remains unchanged in the reaction. We report here the reduction potentials of the free, salicylate bound and benzoate bound salicylate hydroxylase from Pseudomonas cepacia (1).

Results

The spectroelectrochemical method used has been described (2). When the free salicylate hydroxylase is reduced in 20 mM phosphate buffer, pH 7.6 at 10°C with methyl viologen as a reducing mediator, a maximum of 35% of the anionic semiquinone is formed. However, in a dithionite reduction a maximum of 75% of the semiquinone is formed at pH 7.6 and a maximum of 80% is formed at pH 8.0. The molar absorptivity of the anion semiquinone estimated from the dithionite reduction is 16,800 M^{-1} cm^{-1} at 370 nm. In a potentiometric titration about 5% anion

Flavins and Flavoproteins 1987

semiquinone is stabilized at pH 7.6 and a midpoint potential, E_m = -0.094V was obtained, with a 0.034V Nernst slope. The presence of the semiquinone at thermodynamic equilibrium was verified with EPR (3). Based on the amount of semiquinone formed, the potential for the two individual electrons can be determined. The individual electron transfers are:

$EFl_{ox} + e^- \rightleftharpoons EFl^{\bar{\cdot}}$ $\quad E_1^{o\prime}$ = -0.150V vs. SHE

$EFl^{\bar{\cdot}} + e^- + H^+ \rightleftharpoons EFl_{red}H^-$ $\quad E_2^{o\prime}$ = -0.040V vs. SHE

The proton is transferred with the second electron in agreement with the anion semiquinone formed in the potentiometric titration. Potentiometric measurements were made over the pH range 6.2 - 8.0. Over this pH range the enzyme exhibits a potential pH dependence of 0.026V per pH unit which is consistent with a two-electron, one proton transfer (see figure 1).

When salicylate hydroxylase is reduced in the presence of 0.2 mM salicylate in 20 mM phosphate buffer, pH 7.6 at 10°C, a long wavelength absorbing species is formed. This long wavelength absorbance was proven to be due to the formation of the neutral semiquinone using EPR. The molar absorptivity for the neutral semiquinone calculated from the EPR experiment is 6,600 M^{-1} cm^{-1} at 570 nm. The neutral semiquinone was also observed in potentiometric titrations. At pH 7.6 the midpoint potential obtained was -0.114V with a 0.030V Nernst slope, and a maximum of 7% of the semiquinone was stabilized. Based on the amount of semiquinone formed, the potential for the two individual electrons can be determined. The individual electron transfers are:

$EFl_{ox}Sal + e^- + H^+ \rightleftharpoons EFlH^{\cdot}Sal$ $\quad E_1^{o\prime}$ = -0.160V vs. SHE

$EFlH^{\cdot}Sal + e^- \rightleftharpoons EFl_{red}HSal$ $\quad E_2^{o\prime}$ = -0.067V vs. SHE

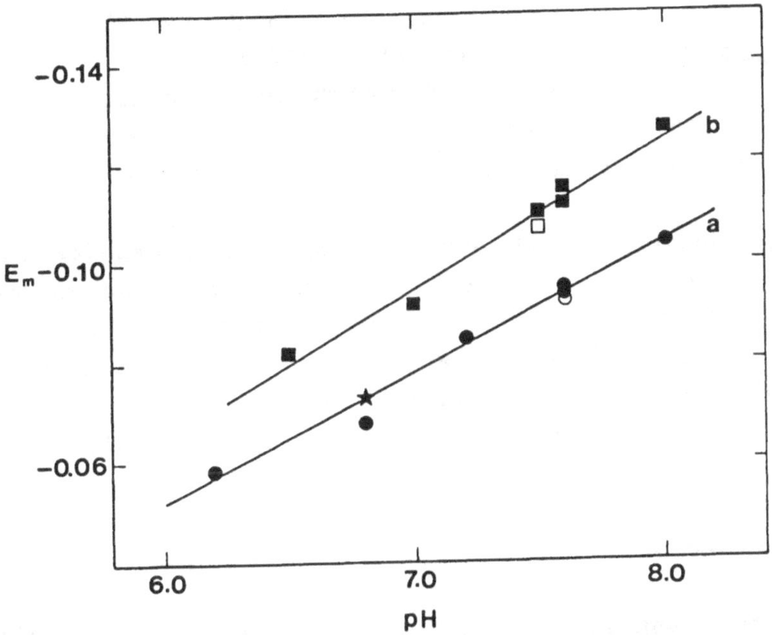

FIGURE 1. Potential vs. pH behavior of salicylate hydroxylase in 20 mM phosphate buffer at 10°C. Line A represents the free enzyme and Line B the enzyme in presence of 0.2 mM salicylate. Redox mediator dyes used were (●) indigodisulfonate (reductive direction), (○) indigodisulfonate (oxidative direction), (★) 2-hydroxy 1,4-naphthoquinone (reductive direction), (■) 9-azariboflavin (reductive direction) and (□) 9-azariboflavin (oxidative direction).

In the presence of salicylate the proton is transferred with the first electron. Over the pH range 6.5 - 8.0 (see figure 1) the substrate bound enzyme exhibits a potential pH dependence of 0.032V per pH unit which is consistent with a two-electron, one-proton transfer.

Substrate binding shifts the E_m for the salicylate bound enzyme 0.020V more negative than the E_m for the free enzyme. This potential shift indicates that the ratio of binding constants differs by a factor of 5 for reduced and oxidized salicylate hydroxylase, with the oxidized form of the enzyme being more tightly bound to salicylate. From binding constants measured for salicylate hydroxylase from P. sp. 29351 (4) and Pseudomonas putida (5), shifts of 0.036V and 0.020V would have been expected, respectively, which is in reasonably good agreement with the results reported here, especially since these three hydroxylases are not identical (1,4,5).

In the presence of 35 mM benzoate in 20 mM phosphate buffer, pH 7.6 at 10°C, the E_m for the benzoate bound enzyme is shifted 0.030V negative of that of the free enzyme, to -0. 126V. This indicates that the ratio of binding constants for reduced and oxidized enzyme differs by a factor of 14, with the oxidized form of the enzyme being more tightly bound to benzoate.

Acknowledgements

We wish to thank Dr. J.D. Lipscomb and Mr. A.M. Orville for help in running the EPR experiments. Support was provided by a grant from NIH (GM 29344) to G.H.E. and M.T.S. and grants from NIH (GM 25953) and the Robert Welch Foundation (E 1030) for L.-H.W. and S.-C.T.

References

1. Wang, L.-H., and Tu, S.-C. (1984) J. Biol Chem. 259, 10682-10688.

2. Stankovich, M. (1980) Anal. Biochem., New York, 109, 295-308.

3. Beinert, H., Orme-Johnson, W. and Palmer, G. (1978) Methods in Enzymol. 54, 111-132.

4. Presswood, R.P. and Kamin, H. (1976) in Flavins and Flavoproteins (Ed. Singer, T.P.), pp. 145-154, Elsevier, Amsterdam.

5. Takemori, S., Nakamure, M., Suzuki, K., Katagiri, M. and Nakamure, T. (1972) Biochim. Biophys. Acta 285, 382-392.

THE OXIDATIVE HALF-REACTION OF 2-METHYL-3-HYDROXYPYRIDINE-5-CARBOXYLIC ACID OXYGENASE FROM *PSEUDOMONAS sp. MA-1*

Pierre Brissette, David P. Ballou and Vincent Massey
Department of Biological Chemistry, University of Michigan
Ann Arbor, Michigan 48109

Introduction

The flavoprotein, MHPC oxygenase, catalyzes the oxygenation of 2-methyl-3-hydroxypyridine-5-carboxylate (MHPC) to yield α-(N-acetylaminomethylene) succinic acid (AAMS). This conversion also requires NADH and molecular oxygen as shown below (1).

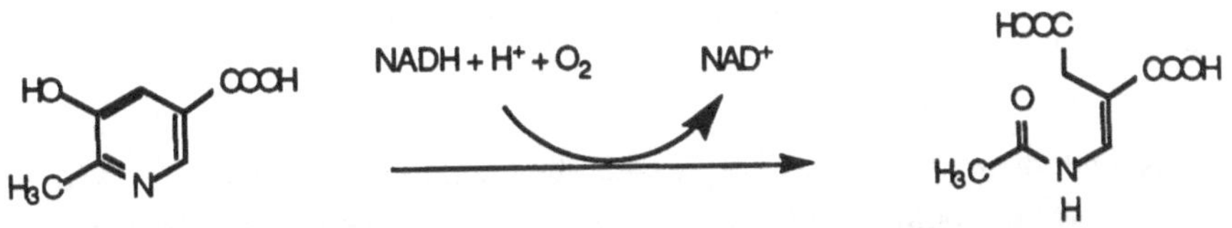

The overall reaction is the sum of two partial reactions (1): a) a MHPC-dependent reduction of enzyme-bound FAD by NADH and b) reoxidation of reduced FAD by molecular oxygen accompanied by oxygenation of MHPC to produce AAMS. In this paper rapid kinetic studies were carried out to elucidate the oxidative half-reaction and to characterize possible reaction intermediate(s).

Results and Discussion

Reoxidation of reduced enzyme in the absence of its substrate effector MHPC. When a solution of anaerobic reduced enzyme was reacted with solutions containing varying excess concentrations of oxygen, reoxidation of the flavin cofactor was monophasic at all wavelengths observed in the range of 320 nm to 520 nm. No flavin semiquinone could be detected, as there were no absorbance changes at wavelengths greater than 550 nm. Thus no flavin intermediates could be detected spectrally. In the absence of substrate the kinetics of this reaction were first order with respect to oxygen concentration. The second order rate constant for the reaction between reduced enzyme and oxygen determined from a linear plot of the apparent rate of

Flavins and Flavoproteins 1987

reoxidation versus oxygen concentration was 480 $M^{-1}s^{-1}$ ±4%. In the absence of substrate the oxygenase functioned poorly as a NADH oxidase where oxygen is reduced to hydrogen peroxide (2).

Reoxidation of reduced enzyme in the presence of its substrate effector MHPC . In order to minimize the kinetic complexity of the reaction, it was important to carry out these experiments under conditions of saturating substrate concentration. This ensured a homogeneous enzyme population. Addition of small aliquots of an anaerobic solution of MHPC to anaerobic reduced enzyme caused a spectral perturbation in the 365-420 nm region. From the decrease in absorbance at 380 nm associated with binding of the substrate to the reduced enzyme the dissociation constant was determined to be 18 μM ±6% with a stoichiometry of one mole of MHPC per mole of flavin.

Reduced enzyme in complex with a saturating concentration of MHPC was mixed with oxygenated buffer solution in the stopped flow spectrophotometer and the reoxidation process was monitored at several wavelengths. At wavelengths greater than 450 nm a monophasic increase in absorbance occurred at a rate of 0.95 s^{-1} ±19%. In the range of 440 nm to 360 nm the reoxidation is biphasic (upward) with fast and slow phases having apparent rates of 61 s^{-1} ±8% and 0.95 s^{-1} ±19% respectively. Between 350 nm and 320 nm the two phases are still represented, but the slow phase in this region is associated with a decrease in absorbance. Representative traces of the reoxidation at selected wavelengths are shown in Fig. 1.

Qualitative inspection of the results clearly indicated the transient formation of an intermediate flavin species during the course of reoxidation. Maximal formation of the intermediate occurred at 75 ms with its subsequent disappearance leading to oxidized enzyme. The apparent rate for the fast phase is dependent on oxygen concentration and is associated with the formation of the transient flavin species, whereas the slow phase is independent of oxygen concentration and is associated with its decay. A double reciprocal plot of the apparent rate of formation of the intermediate versus oxygen concentration is linear and the positive Y-intercept indicated the presence of a spectrally silent ternary complex between reduced enzyme, MHPC and oxygen prior to formation of the intermediate(3). The dissociation constant for oxygen to this complex was estimated to be 1.9 mM from the value of the slope to intercept ratio. The limiting first order rate constant for the formation of the intermediate was evaluated as 135 s^{-1}. The above rate and dissociation constants were obtained from a rapid kinetic experiment performed at 2°C.

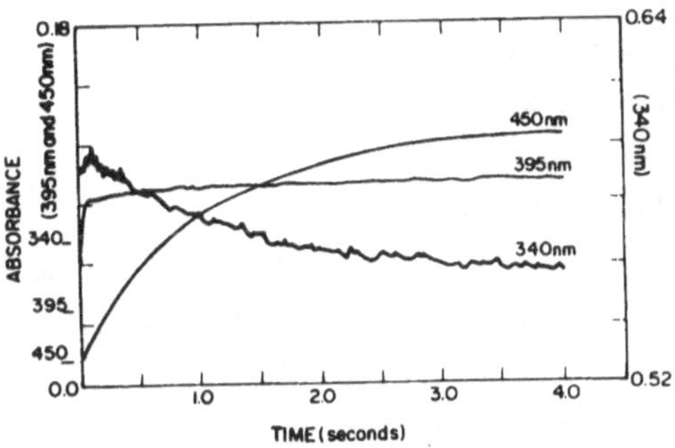

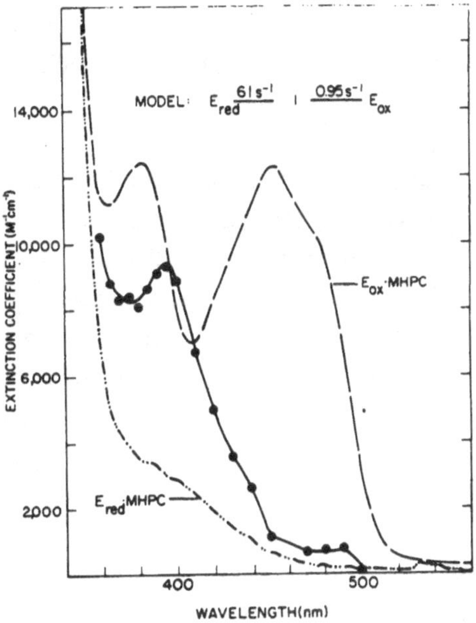

Fig. 1 (on the left). Reoxidation of reduced MHPC oxygenase in the presence of MHPC. MHPC oxygenase was purified by a modification of the procedure of Sparrow et al. (1). MHPC was chemically synthesized as described by Palm et al. (4). MHPC oxygenase (11 µM) was made anaerobic and reduced with dithionite. MHPC at a final concentration of 372 µM was added to the reduced enzyme (95% saturation before mixing). Enzyme solution was then reacted in an anaerobic stopped-flow apparatus at 5°C with an equal volume of buffer previously equilibrated with 1.9 mM oxygen. Absorbance changes upon mixing were recorded at wavelengths ranging from 320 to 550 nm. Buffer was 50 mM NaPi pH 6.8 and saturation of reduced enzyme after mixing was 91%. The short arrows point to the initial absorbance value observed in the stopped-flow apparatus.

Fig. 2 (on the right). Calculated spectrum of the transient flavin intermediate present during reoxidation of reduced enzyme▪MHPC complex. Conditions were the same as those in Figure 1. Using the apparent rates of formation and decay of the intermediate and assuming that reoxidation occurred as a 2-step irreversible sequential mechanism its absolute spectrum was calculated as described in (5,6). E_{red}▪MHPC complex (_.._.._..), E_{ox}▪MHPC (_ _ _ _), intermediate (I) (______).

Using the apparent rates observed during reoxidation (61 s^{-1} and 0.95 s^{-1}) and assuming a 2-step irreversible sequential mechanism for the reoxidation it is possible to calculate the absolute spectrum of the observed intermediate (5,6). This spectrum has an absorbance maximum at 395 nm with an extinction coefficient of 9300 $M^{-1}cm^{-1}$ (Fig. 2). The intermediate is spectrally similar to the C4a-hydroperoxide flavin adduct previously observed with all known flavin monooxygenases (7).

Oxygen transfer. Our rapid kinetic studies demonstrated that decay of the flavin hydroperoxide intermediate is concomitant with oxygen transfer to the substrate. The absorbance of MHPC in the near UV region allowed us to monitor its disappearance during single turnover experiments. Results indicated that the absorbance of MHPC at 340 nm disappeared at a rate equivalent to that at which the flavin hydroperoxide decayed. However, at this time the mechanism of oxygen transfer is unclear. Early tracer experiments with ^{18}O (1) were unable to discriminate between a mono- or dioxygenase type of mechanism. Generation of a flavin C4a-hydroperoxide is an intrinsic part of the oxygen transfer mechanism in all flavin-dependent monooxygenases studied to date (7). These hydroperoxyflavins decayed to the hydroxyflavins through a heterolytic cleavage in the presence of the appropriate oxygenatable substrate (7). Experiments are planned to determine if MHPC oxygenase behaves in a similar fashion.

Acknowledgments

This work was supported by grants from the U.S. Public Health Service, GM 11106 (VM) and GM 20877 (DPB).

References

1. Sparrow, L.G., P.P.K. Ho, T.K. Sundaram, D. Zach, E.J. Nyns, and E.E. Snell.1969. J. Biol. Chem. 244, 2590-2600.
2. Nelson, M.J.K., and E.E. Snell.1986. J. Biol. Chem. 261, 15115-15120.
3. Strickland, S., G. Palmer, and V. Massey.1975. J. Biol. Chem. 250, 4048-4052.
4. Palm, D., A.A. Smucker, and E.E. Snell.1967. J. Org. Chem. 32, 826-828.
5. Schopfer, L.M., and V. Massey.1980. J. Biol. Chem. 255, 5355-5363.
6. Entsch, B., D.P. Ballou, and V. Massey.1976. J. Biol. Chem. 251, 2550-2563.
7. Ballou, D.P..1984. In: Flavins and Flavoproteins (R.C. Bray, P.C. Engel, S.G. Mayhew, eds.). Walter de Gruyter, Berlin New York. pp. 605-618.

LIPID STABILIZATION OF THE FLAVINHYDROPEROXIDE IN THE MICROSOMAL FLAVIN-CONTAINING MONOOXYGENASE

L. L. Poulsen, T. Nagata

Clayton Foundation for Biochemical Research, University of Texas, Austin, Texas 78712

K. L. Taylor

Department of Medicine & Biochemistry, Emory University, Atlanta, Georgia 30322

Introduction

The Microsomal Flavin Containing Monooxygenase (MFO) catalyses the NADPH dependent oxidation of a great variety of nucleophilic nitrogen and sulfur containing compounds (1,2). The catalytic mechanism involves reduction of the enzyme flavin, oxidation of the flavin to form a stable flavinhydroperoxide (E(FAD-OOH)) and finally oxidation of the nucleophillic center by the flavinhydroperoxide (3). The rate limiting step is presumed to be regeneration of the oxidized flavin by dehydration of the flavin pseudobase (4). In the absence of an oxidizable substrate, the enzyme exists as a stable protein bound flavinhydroperoxide which breaks down to H_2O_2 and oxidized flavin. The half life of the E(FAD-OOH) with the purified enzyme is in excess of 1 min at 25°. The effect of phospholipids on stability of the E(FAD-OOH), flavin reduction and overall turnover are reported.

Results

During purification of the monooxygenase (Table 1), turnover per flavin decreases over 50 percent even though the specific activity increases. There is also a 10 fold increase in NADPH dependent oxygen uptake in the abscence of substrate and in

Flavins and Flavoproteins 1987

Table 1. Specific Activity and Turnover of the Flavin-Containing Monooxygenase During Purification

	Specific Activity	Turnover(min^{-1})
Microsomes	16	47
6-14% PEG ppt from Triton extract	29	53
1st DEAE column eluate	80	42
2nd DEAE column eluate	189	22

some preparations, this rate approaches that of the substrate dependent rate. Since nonsubstrate dependent oxygen uptake is a function of the flavinhydroperoxide stability, studies were carried out to determine the factors responsible for E(FAD-OOH) stability.

DEAE chromatography decreased turnover (Table 1) and removed significant amounts of phospholipid. Reconstitution of enzyme with phosphatidylserine (PS) increased the half life (Table 2) of the flavinhydroperoxide by almost 4 fold while phosphatidylethanolamine (PE) decreased it by 40 percent. Nonionic detergents and all of the other phospholipids tested, increased stability less than two fold.

Table 2. Effects of Phospholipid on the Stability of E(FAD-OOH)

Phospholipid	$t_{1/2}$(min)[a]
none	1.4
Phosphatidylserine	5.9
Phosphatidylcholine	1.8
Phosphatidylethanolamine	0.8
Phosphatidylinositol	2.3
Phosphatidylinositol-4- Phosphate	2.1
Phosphatidylinositol-4,5-diphosphate	2.1

a) Half lives were determined by following the rate of appearance of oxidized flavin at 450 nm at pH 7.5 at 25° after stoichiometric reduction of partially purified enzyme.

Table 3. Effect of Phospholipids on DCPIP Reductase Activity of the Microsomal Flavin-Containing Monooxygenase

Phospholipid	Turnover (min^{-1})[a]
none	70.1
Phosphatidylserine	66.7
Phosphatidylcholine	64.8
Phosphatidylethanalamine	67.0
Phosphatidylinositol	60.4
Phosphatidylinositol-4-phosphate	29.1
Phosphatidylinositol-4,5-diphosphate	21.3

a) NADPH and enzyme dependent reduction of DCPIP was followed anaerobically at 600 nm in phosphate buffer at pH 7.5 at 37°.

Although the phosphoinositides increased stability of the E(FAD-OOH), they severely inhibited flavin reduction (Table 3) when measured anaerobically with dichlorophenolindophenol (DCPIP) as an electron acceptor instead of oxygen. None of the other phospholipids tested, including phosphatidylserine, had any significant affect on flavin reduction.

In addition to stabilizing the flavinhydroperoxide, incubation with phophatidylserine restored flavin turnover. In the

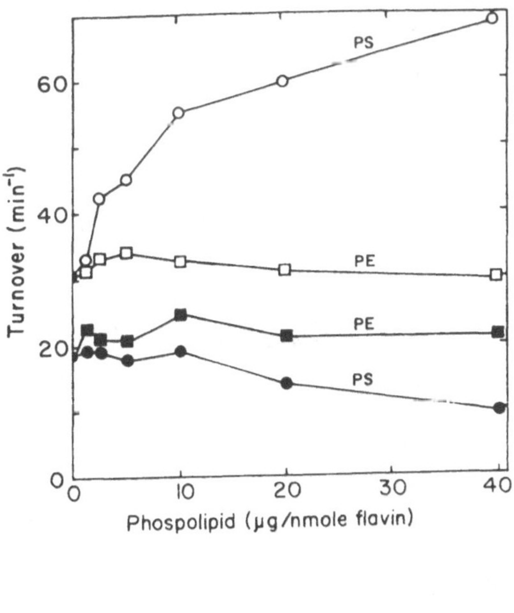

Figure 1. Concentration Dependence for Phospholipid Reactivation of the Flavin-Containing Monooxygenase. Enzyme was incubated for 30 min at 0° with the indicated amounts of phospholipid and the activity was measured in 0.1 M phosphate buffer at 37° in the presence, open symbols, and abscence, closed symbols, of 3.5 mM octylamine.

Table 4. Effect of Phosphatidylserine on Partially Purified Enzyme as a Function of pH and Octylamine. Enzyme was incubated for 30 min at 0° with and without 0.5 mg phosphatidylserine per mg protein. Activity was measured polarigraphically in 0.1 M phosphate (ph 7.5) or 0.025 M glycine - 0.04 M sodium pyrophosphate (pH 8.4) buffers at 37°.

	Turnover	
	Control	+3.5mM Octylamine
pH 7.5		
-PS	27	47
+PS	22	99
pH 8.4		
-PS	44	81
+PS	2	170

presence of octylamine (Figure 1), the turnover increases with PS concentration. In the absence of octylamine, PS is inhibitory. PE, as shown, and all other phospholipids tested had no effect on turnover. At pH 8.4, the pH optimum for the monooxygenase (4), the effect in the presence of octylamine was the same, however, much greater inhibition was found (Table 4).

Acknowledgement

The assistance of the personel at the J & B Food Co., Waelder, Texas in securing fresh hog liver and the continued support and interest of Dr. D. M. Ziegler in this project are gratefully acknowledged.

References

1. Ziegler, D.M. 1980. In: Enzymatic Basis of Detoxication (W.B. Jakoby ed.). Academic Press, N.Y., Vol. 1, Chapter 9, p. 201.

2. Ziegler, D.M. 1982. In: Metabolic Basis of Detoxication (W.B. Jakoby, J.R. Bend and J. Caldwell eds.) Academic Press , New York, Chapter 9, p. 171.

3. Poulsen, L.L. and D.M. Ziegler. 1979. J. Biol. Chem. 254, 6449.

4. Beaty, N.B. and D.P. Ballou. 1981. J. Biol. Chem. 256, 4619.

Flavin-Dependent Bioluminescence

ONE-ELECTRON TRANSFERS IN FLAVIN SYSTEMS: RELEVANCE TO THE POSTULATED CIEEL MECHANISM IN BACTERIAL BIOLUMINESCENCE*

Humphrey I. X. Mager[§] and Shiao-Chun Tu

Department of Biochemical and Biophysical Sciences, University of Houston-University Park, Houston, Texas 77004

1. INTRODUCTION

The chemiluminescence shown by some types of organic peroxides in the presence of a catalytic "chemiluminescent activator" has been formulated by Schuster [1] to follow a *Chemically Initiated Electron Exchange Luminescence* (CIEEL) mechanism. Essential steps in this mechanism are:

1) The activator reduces the peroxide **P** to a radical anion $\mathbf{P^{\dot{-}}}$. **2**) The radical anion $\mathbf{P^{\dot{-}}}$ undergoes an efficient rearrangement to produce a new radical anion $\mathbf{D^{\dot{-}}}$. **3**) $\mathbf{D^{\dot{-}}}$ is a powerful electron donor which pumps the activator from the radical cation form up to the excited state. **4**) The excited activator returns to its ground state with the emission of light.

$$P + Act \longrightarrow [P\text{---} Act] \longrightarrow P^{\dot{-}} + Act^{\dot{+}} \quad (1)$$

$$P^{\dot{-}} \longrightarrow D^{\dot{-}} + product(s) \quad (2)$$

$$D^{\dot{-}} + Act^{\dot{+}} \longrightarrow D + Act^* \quad (3)$$

$$Act^* \longrightarrow Act + light \quad (4)$$

With reference to the bacterial bioluminescence, Schuster speculated on the possibility that *"the dihydroflavin nucleus may serve as an intramolecular activator"* but no structural details were given. Starting from the dihydroflavin 4a-peroxyhemiacetal, Mager and Addink

---

*This work was supported by a grant from the Texas Advanced Technology Research Program to S.C.T. through the University of Houston, and by grants E-1030 from the Robert A. Welch Foundation and GM25953 from NIH.

§On sabbatical leave from September 1986 until September 1987. Permanent address: 34 de Ruyterstraat, 2266 KZ LEIDSCHENDAM, The Netherlands.

Flavins and Flavoproteins 1987

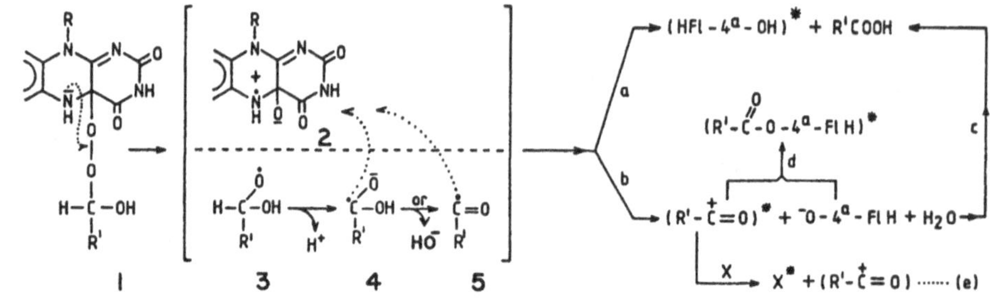

Scheme 1

[2] have formulated a mechanism for bacterial bioluminescence as given in Scheme **1**. **Step 1**: The N^5-flavin moiety donates an electron to the peroxide bond resulting in the formation of a radical pair **2, 3. Step 2:** The loss of a proton **(3→4)** produces the radical pair **2, 4,** or, eventually, the radical pair **2, 5. Step 3:** The back transfer of an electron from either **4** or **5** to the flavin radical **2** would give the excited pseudobase HFl-4a-OH* (pathway **a**) or, alternatively, an excited acylium cation $(R'\text{-}\overset{+}{C}{=}O)^*$ (pathway **b**). Subsequent *energy transfer* (pathway **c, d** or **e**) might result in either the excited pseudobase, or the excited dihydroflavin ester or an excited fluorophore **X**, respectively.

This scheme requires the occurrence of a hypothetical key intermediate **2** which, structurally, can be considered as a 4a-hydroxyflavin radical cation derived from the flavin pseudobase by the loss of one electron:

$$\text{HFl-4a-OH} \rightleftharpoons (\text{HFl-4a-OH})^{\dot{+}} + e \qquad (5)$$

Proving the existence of this novel type of flavin radicals was considered as the first step in studying the validity of the CIEEL theory in flavin mediated luminescence. Since it is very unlikely that **2** will have an appreciable life-time, it seems essential to look for probably better detectable derivatives in the 5- and the 1-blocked flavin model series (cf. **9, 13;** Scheme 2).

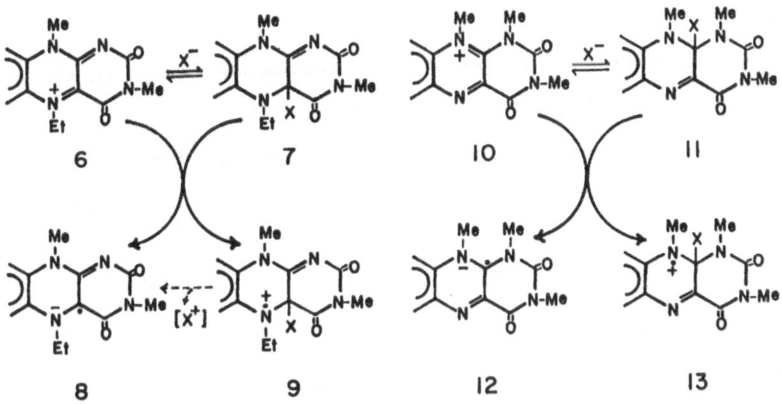

Scheme 2

2. THE FORMATION AND THE DETECTION OF 4ª- AND 10ª-SUBSTITUTED FLAVIN RADICAL CATIONS

As indicated in Scheme **2**, we attempted to generate the new radicals in *intermolecular* one-electron transfers by letting flavin adducts react as the one-electron donors and the flavinium cations as the one-electron acceptors.

The first spectacular results were obtained on investigating the preparation and the properties of flavinium salts [3,4] of which the anionic part was derived from organic acids ($X^- = A^-$ = the anion derived from e.g.: TFA, TCA, AcOH, α-ketoglutaric acid, salicylic acid, β-naphtalenesulfonic acid). On dissolving the flavinium salts in low polar solutions, *O_2 was observed to be consumed in spite of the fact that the salts were flavin derivatives in the fully oxidized state!* This result implied that some reduced flavin species had spontaneously been formed.

The intermediacy of covalent adducts (**7, 11**) and the spontaneous formation of a flavosemiquinone (**8**) were shown. This indeed indicated the occurrence of the equilibria and the one-electron transfers as depicted in Scheme 2.

Although much slower, covalent adducts like the 4ª-hydroxyflavin pseudobase or the

4ª-methoxyflavin adduct (**7**; X = OH, OCH_3) showed a similar conversion:

$$\text{5-EtFl-4}^{a}\text{-X} \rightleftharpoons \text{5-EtFl}^{+}_{ox} + X^{-} \qquad (6)$$

$$\text{5-EtFl}^{+}_{ox} + \text{5-EtFl-4}^{a}\text{-X} \longrightarrow \text{5-EtFl}\cdot + (\text{5-EtFl-4}^{a}\text{-X})^{\dot{+}} \qquad (7)$$

$$2\ \text{5-EtFl-4}^{a}\text{-X} \xrightarrow[X^{-}]{} \text{5-EtFl}\cdot + (\text{5-EtFl-4}^{a}\text{-X})^{\dot{+}} \qquad (8)$$

The use of an acid, in substoichiometric amounts, considerably accelerated the process resulting in a mixture of 5-EtFl· and two new radicals of type **9** (Eqs. 9-11). $(\text{5-EtFl-4}^{a}\text{-A})^{\dot{+}}$ was formed as the primary product which, in the presence of small amounts of H_2O was hydrolyzed to $(\text{5-EtFl-4}^{a}\text{-OH})^{\dot{+}}$. This gave a recovery of HA explaining the catalytic effect of substoichiometric amounts of acid [2].

$$\text{5-EtFl-4}^{a}\text{-X} + HA \longrightarrow \text{5-EtFl}^{+}_{ox}, A^{-} + HX \qquad (9)$$

$$2 \times [\text{5-EtFl}^{+}_{ox}, A^{-} \rightleftharpoons \text{5-EtFl-4}^{a}\text{-A}] \xrightarrow[A^{-}]{} \text{5-EtFl}\cdot + (\text{5-EtFl-4}^{a}\text{-A})^{\dot{+}} \qquad (10)$$

$$(\text{5-EtFl-4}^{a}\text{-A})^{\dot{+}} + H_2O \longrightarrow (\text{5-EtFl-4}^{a}\text{-OH})^{\dot{+}} + HA \qquad (11)$$

Starting from 5-EtFl-4ª-OH in benzene (~3 x 10^{-5} M, Fig. 1 curve **a**), the proceeding of one-electron transfer giving **8** and **9** (X = OH) was indicated by the appearance of spectra similar to curve **b**. The absorption between 500 to 700 nm was entirely due to the

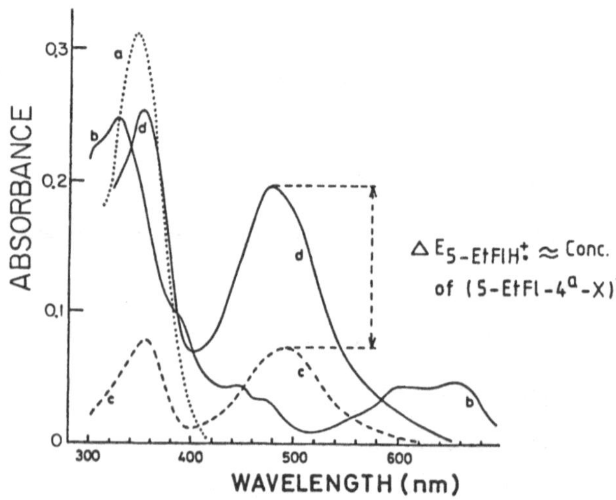

Fig. 1. (**a**) 5-EtFl-4ª-OH, freshly dissolved in benzene. (**b**) Final absorbance after the one-electron transfer (Eq. 8) and the partial removal of 5-EtFl·. (**c**) Calculated absorbance for $\text{5-EtFlH}^{\dot{+}}$ corresponding to the amount of 5-EtFl· determined by the 500-700 nm region of curve **b**. (**d**) The actual amount of $\text{5-EtFlH}^{\dot{+}}$ obtained on acidification of **b**.

accumulation of 5-EtFl•. The simultaneous accumulation of $(5\text{-EtFl-4a-X})^{\dot{+}}$ was detected on adding an excess of acid. The amount of 5-EtFl•, represented by curve **b**, should have given an amount of $5\text{-EtFlH}^{\dot{+}}$ (Eq. 12) corresponding to the calculated curve **c**. Instead, acidification gave considerably higher amounts of 5-EtFlH+ (curve **d**). The difference between the curves **d** and **c** ($\Delta E_{5\text{-EtFlH}^{\dot{+}}}$) revealed the accumulated amount of $(5\text{-EtFl-4a-X})^{\dot{+}}$ produced according to Eq. 13.

$$5\text{-EtFl}\bullet + H^+ \rightleftharpoons 5\text{-EtFlH}^{\dot{+}} \qquad (12)$$

$$(5\text{-EtFl-4a-OH})^{\dot{+}} + H^+ \longrightarrow 5\text{-EtFlH}^{\dot{+}} + H^+ + [O] \qquad (13)$$

A mixture of $(5\text{-EtFl-4a-OH})^{\dot{+}}$ and $(5\text{-EtFl-4a-A})^{\dot{+}}$ can be analyzed by the different rates of respective acidic decompositions. The acidic decomposition of $(5\text{-EtFl-4a-A})^{\dot{+}}$ appeared to be relatively slower since it first had to undergo a rate-limiting hydrolytic step (Eq. 11).

Note:curve **b** represents the typical result of experiments in which the final yields of 5-EtFl• (calculated yield 50%) were lowered to 30-35% by a subsequent oxidation with traces of O_2. 5-EtFl• could even be completely removed without influencing the 45-50% yields obtained for $(5\text{-EtFl-4a-X})^{\dot{+}}$ (calculated yield 50%).

In the 1-alkylflavin model series, the occurrence of a flavin radical cation $(1\text{-MeFl-10a-OH})^{\dot{+}}$ (cf. **13**, X = OH; Scheme 2) was detected by its reductive rearrangement to a blue-colored 10,10a-ring opened derivative [3, 4]. This only occurred when the 1-MeFl• and $(1\text{-MeFl-10a-OH})^{\dot{+}}$ arose in the presence of O_2!. We assume that the formation of $O_2^{\dot{-}}$ and/or 1-MeFl-10a-OO• was followed by an electron transfer to $(1\text{-MeFl-10a-OH})^{\dot{+}}$ resulting in a 10,10a-ring opened isomer of the flavin pseudobase (Eq. 14).

$$1\text{-MeFl}\bullet + O_2 \longrightarrow 1\text{-MeFl}^+_{ox} + O_2^{\dot{-}} \rightleftharpoons 1\text{-MeFl-10a-OO}\bullet$$

$$\xrightarrow{(1\text{-MeFl-10a-OH})^{\dot{+}}} O_2 + 1\text{-MeFl}^+_{ox} + \qquad (14)$$

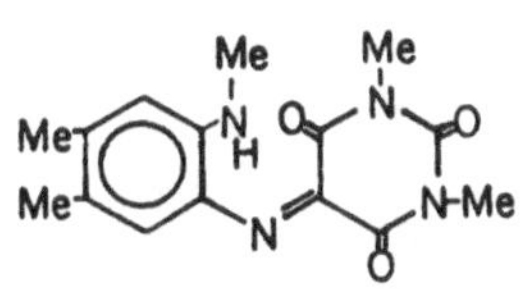

14

We have observed that the process giving the radicals **8** and **9** (Scheme 2) also proceeds in aqueous solutions [5]. As mentioned in the literature [6], a rapid equilibrium was found to exist in H_2O:

$$\text{5-EtFl}^+_{ox} + H_2O \rightleftharpoons \text{5-EtFl-4a-OH} + H^+ \quad (15)$$

Identical spectra were obtained by dissolving either the flavinium salt or the flavin pseudobase in aqueous solution at a given pH. The influence of pH is illustrated by the spectra shown in Fig. 2. At pH 7.00 (curve **a**), the equilibrium was shifted to the right giving practically quantitative amount of 5-EtFl-4a-OH. In contrast, a practically complete shift of the equilibrium to 5-EtFl$^+_{ox}$ was obtained in strongly acidic solution (curve **b**). In between pH 3.00 and 7.00, the occurrence of both species was indicated by the spectra composed of **a** and **b**. This is illustrated by curve **c** giving both species in a 1:1 ratio.

These aqueous solutions showed *spontaneous* productions of the flavosemiquinone at room temperature in the dark (Eq. 7, X = OH). In contrast with a significant accumulation in low polar solutions, (5-EtFl-4a-OH)$^{\dot{+}}$ rapidly decomposed in aqueous solutions to give an additional amount of 5-EtFl• (such as by Eq. 13) by which the overall yield of the latter could even be increased to 98% under strictly anaerobic conditions. Finally, spectra were

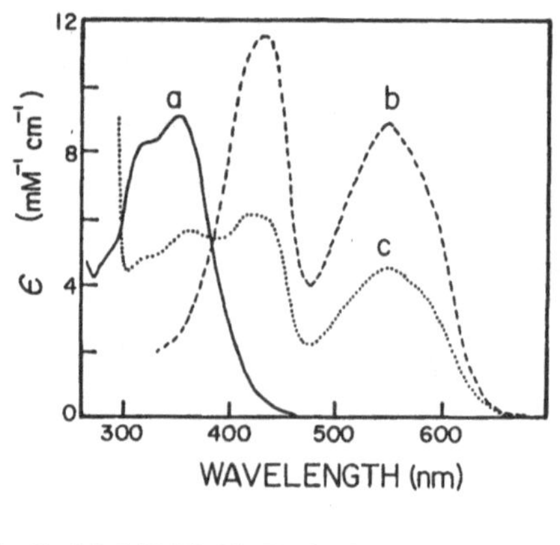

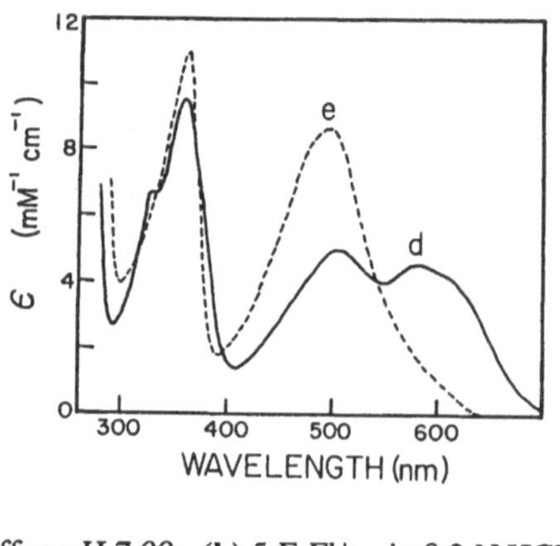

Fig. 2. (**a**) 5-EtFl-4a-OH in 0.1 M phosphate buffer, pH 7.00. (**b**) 5-EtFl$^+_{ox}$ in 0.2 N HCl. (**c**) [5-EtFl-4a-OH] / [5-EtFl$^+_{ox}$] (1:1) in 0.1 M citrate buffer, pH 4.15.
Fig. 3. (**d**) 5-EtFl• in 0.1 M phosphate buffer, pH 7.00. (**e**) 5-EtFlH$^{\dot{+}}$ in 0.2 N HCl.

obtained as given by curve **d** or **e** (Fig. 3) or **d** superimposed upon **e**.

In H_2O, the electron transfer (Eq. 7, X = OH) is a slow reaction in comparison with the equilibrium (Eq. 15) and the decompositions of **9**. The rate constant for the electron transfer was found to be 935 $M^{-1}min^{-1}$ at pH 3.52.

When the equilibrium (Eq. 15) was shifted either to the right or the left, one of the reactants for Eq. 7 would be present in a trace amount leading to a very slow radical formation. Consequently, an increase of the reaction rate, with a maximum at the apparent pK'_a, may be expected if the pH of the reaction solution was adjusted to be closer to the pK'_a.

On defining : $R = [5\text{-EtFl-4a-OH}] / [5\text{-EtFl}^+_{ox}]$ and

$$[Fl_{tot}] = [5\text{-EtFl}^+_{ox}] + [5\text{-EtFl-4a-OH}] \text{ at time t,}$$

the rate of the disappearance of 5-EtFl^+_{ox} (Eq. 16) can be represented by Eq. 17:

$$-d[5\text{-EtFl}^+_{ox}] / dt = k \cdot [5\text{-EtFl}^+_{ox}] \cdot [5\text{-EtFl-4a-OH}] \qquad (16)$$

$$-d[5\text{-EtFl}^+_{ox}] / dt = k \cdot [Fl_{tot}]^2 \cdot [R / (1 + R)^2] \qquad (17)$$

The rate of the disappearance of 5-EtFl^+_{ox} is proportional to $R / (1 + R)^2$. Starting from the equilibrium constant $K' = [H^+] \cdot [5\text{-EtFl-4a-OH}] / [5\text{-EtFl}^+_{ox}]$ (Eq. 15) and using the value determined for the pK'_a (= 4.15) in 0.1 M citrate buffer, a pH-profile was calculated (Fig.4) showing the relative initial rate ($V_{pH} / V_{pK'a}$) versus the pH with a maximum ratio of

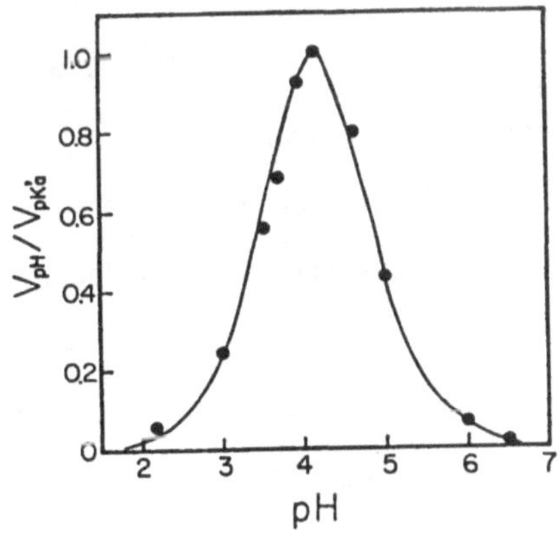

Fig. 4. One-electron transfer (Eq. 7, X = OH; Scheme 2) in aqueous solution. The relative initial rate ($V_{pH} / V_{pK'a}$) versus the pH. (Theoretical profile, ——— ; observed values, ●).

1 at pH = pK'$_a$. The excellent agreement of the observed values with the theoretical pH profile (Fig. 4) provided unambiguous evidence for the mechanism of the process (Scheme 2) which, so far, has only been postulated on account of qualitative observations.

A minor product, 3-methyllumiflavin (Fl_{ox}, mostly 2-5%), was found to result from the disproportionation of **9**. Considering this and the very high yields of 5-EtFl• (up to 98%), our findings excluded the possibility that a solvolysis of the starting compound (Eq. 18) could have played an essential role as proposed for the conversion of 5-methylated flavins [7] which would give a theoretical yield of 33.3% for Fl_{ox} and 66.7% for 5-EtFl•.

$$\text{5-EtFl}^+_{ox} \xrightarrow[H^+]{H_2O} \text{5-(CH}_3\text{CHOH)-Fl-1-H} \xrightarrow[CH_3CHO]{} FlH_2$$

$$\xrightarrow{2\ \text{5-EtFl}^+_{ox}} Fl_{ox} + 2\,H^+ + 2\ \text{5-EtFl}\bullet \qquad (18)$$

3. THE DEVELOPMENT OF FLAVIN LUMINESCENT SYSTEMS

After having established the existence of the 4a- and 10a-substituted flavin radicals **9** and **13**, we have aimed to show that the transfer of an electron to these radicals may generate excited species. In particular, flavin radicals of type **9** are closely related to **2**, postulated as a key intermediate in bacterial bioluminescence. However, 5-alkyl-4a-substituted flavins are known to be non-fluorescent, which makes any luminescence found very likely to have arisen from energy transfer. The primary product(s) of excitation may not be easy to determine. In contrast, since 10a-flavin adducts are fluorescent, the excited species in this series may be less difficult to identify. This was already realized by McCapra [8] who succeeded in constructing a flavin luminescent system and identifying the 10a-hydroxyflavin pseudobase as the excited species. Starting from the 10a-peroxyflavin adduct **15**, a decomposition to the 10a-hydroxyflavin and a perester was assumed. The perester was proposed to be reduced by the flavin pseudobase following an *intermolecular* CIEEL mechanism.

McCapra concluded that it is difficult to interpret these results in other terms. On the other hand, from studies on related 1-methylquinoxalines, McCapra found that electron-donating substituents did not enhance the quantum yield. He then considered a mechanism involving excited species produced from the perester alone, with the quinoxaline pseudobases as fluorescent acceptors but not active participants.

We want to suggest that an alternative *intramolecular* CIEEL mechanism (Scheme 3), similar to the formulation in Scheme 1, should also be considered. The nature of the electron donor is apparently important in the excitation step of the CIEEL mechanism. We observed that the one-electron transfer from superoxide radicals in Eq. 14 only resulted in a dark $10,10^{a}$-ring opening.

Chemiluminescence accompanying the decomposition of 5-ethylflavin-4^{a}-alkylperoxides, closely related to **1** (Scheme 1), was reported by Kemal and Bruice [9]. These authors concluded that the luminescent reaction was a minor pathway. We have also reached the same conclusion based on our analysis of the decomposition products of 5-EtFl-4^{a}-OOH [10]. The λmax of emission at ~530 nm proved that the flavin pseudobase was not the emitter. We believe that energy was probably transferred from the yet unknown primary excited species to 3-methyllumiflavin, a minor product of N^{5}-dealkylation.

In current studies we have tried to effect luminescence by first accumulating $(5\text{-EtFl-}4^{a}\text{-X})^{\dot{+}}$ (X = OH or A; cf. Fig. 1) and subsequently adding an electron donor. Indeed we have

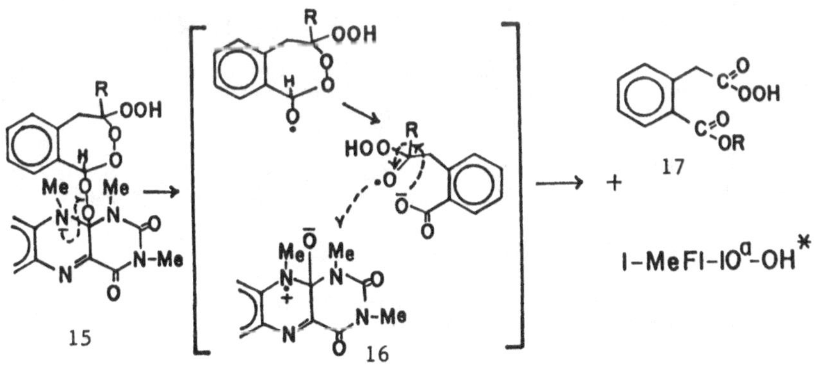

Scheme 3

obtained chemiluminescence in benzene with the recovery of 4a-flavin adducts using dibenzoylperoxide as the electron donor. This chemiluminescence was nullified by the addition of H_2O which is consistent with the recent finding on the instability of (5-EtFl-4a-OH)$^{\dot{+}}$ in water. This instability has led to a search for more stable flavin radicals to be used in the further development of luminescent systems. In an accompanying paper [11] we have described some new adducts obtained from addition reactions with dithionite and sulfoxylate radical ions which may be relevant to the mechanism of a H_2O_2-induced bacterial bioluminescence.

References

1. Schuster, G. B. 1979. Acc. Chem. Res. 12, 366-373.

2. Mager, H. I. X., R. Addink. 1984. In: Flavins and Flavoproteins (R. C. Bray, P. C. Engel, and S. G. Mayhew, eds.). W. de Gruyter, Berlin, pp. 37-40.

3. Mager, H. I. X., R. Addink: 1982. In: Flavins and Flavoproteins (V. Massey and C. H. Williams, Jr., eds.). Elsevier, Amsterdam, pp. 284-293; 1983. Tetrahedron 39, 3359-3366; 1984. In: Flavins and Flavoproteins (R. C. Bray, P. C. Engel, and S. G. Mayhew, eds.). W. de Gruyter, Berlin, pp. 29-32.

4. Mager, H. I. X., R. Addink. 1985. Tetrahedron 41, 183-190.

5. Mager, H. I. X., S.-C. Tu. unpublished results.

6. Ghisla, S., U. Hartmann, P. Hemmerich, F. Müller. 1973. Ann. Chem. 1388-1415.

7. Kemal, C., T. C. Bruice. 1976. J. Am. Chem. Soc. 98, 3955-3964.

8. McCapra, F., P. Leeson: 1976. J. Chem. Soc. Chem. Comm. 1037-1038; 1979. J. Chem. Soc. Chem. Comm. 114-117; 1985. In: Chemi- and Bioluminescence (J. G. Burr, ed.) Marcel Dekker, New York, pp. 283-286.

9. Kemal, C., T. C. Bruice: 1976. Proc. Nat. Acad. Sci. USA 73, 995-999; 1977. Proc. Nat. Acad. Sci. USA 74, 405-409; 1977. J. Am. Chem. Soc. 99, 7064-7076.

10. Mager, H. I. X. 1979. Tetrahedron Lett. 3549-3552.

11. Mager, H. I. X., S.-C. Tu. "Characterization of New 'Bleached' Products Obtained on Treating Flavins and Flavoproteins with Dithionite " This volume.

CHARACTERIZATION OF NEW "BLEACHED" PRODUCTS OBTAINED ON TREATING FLAVINS AND FLAVOPROTEINS WITH DITHIONITE*

Humphrey I. X. Mager§ and Shiao-Chun Tu

Department of Biochemical and Biophysical Sciences, University of Houston-University Park, Houston, Texas 77004

In our search for new 4a-substituted flavin radicals, we have examined the feasibility of the following addition reaction:

$$5\text{-EtFl}^{+}_{ox} + X^{\cdot-} \rightarrow 5\text{-EtFl-4a-X}\cdot \qquad (1)$$

For $X^{\cdot-}$, attention has been given to the sulfoxylate radical ion which is in equilibrium with the dithionite ion:

$$S_2O_4^{=} \rightleftharpoons 2\ SO_2^{\cdot-} \qquad (2)$$

For the starting flavin species, same results were obtained with either the 5-ethyl-3-methyl-lumiflavinium cation (5-EtFl^{+}_{ox}) or the pseudobase (5-EtFl-4a-OH) because of the rapid equilibrium [1]:

$$5\text{-EtFl}^{+}_{ox} + H_2O \rightleftharpoons 5\text{-EtFl-4a-OH} + H^{+} \qquad (3)$$

We have now found that "bleaching" could be ascribed to the formation of several new covalent adducts absorbing in the 300-400 nm region.

1) Conversions to a 4a-sulfite adduct (Eq. 4) and a 4a-dithionite adduct (Eq.5) have been obtained, as indicated by the spectral change from curves **a/d** to curves **b** and **c**,

$$5\text{-EtFl-4a-OH} \underset{H_2O}{\overset{H^{+}}{\rightleftharpoons}} 5\text{-EtFl}^{+}_{ox}$$

$$5\text{-EtFl}^{+}_{ox} \xrightarrow{SO_3^{=}} 5\text{-EtFl-4a-}SO_3^{-} \qquad (4)$$

$$5\text{-EtFl}^{+}_{ox} \xrightarrow{S_2O_4^{=}} 5\text{-EtFl-4a-}S_2O_4^{-} \qquad (5)$$

$$5\text{-EtFl}^{+}_{ox} \xrightarrow{SO_2^{\cdot-}} 5\text{-EtFl-4a-}SO_2\cdot \rightleftharpoons 5\text{-EtFl}\cdot + SO_2 \qquad (6)$$

---

*This work was supported by a grant from the Texas Advanced Technology Research Program to S.C.T. through the University of Houston, and by grants E-1030 from the Robert A. Welch Foundation and GM25953 from NIH.

§On sabbatical leave from September 1986 until September 1987. Permanent address: 34 de Ruyterstraat, 2266 KZ LEIDSCHENDAM, The Netherlands.

Flavins and Flavoproteins 1987

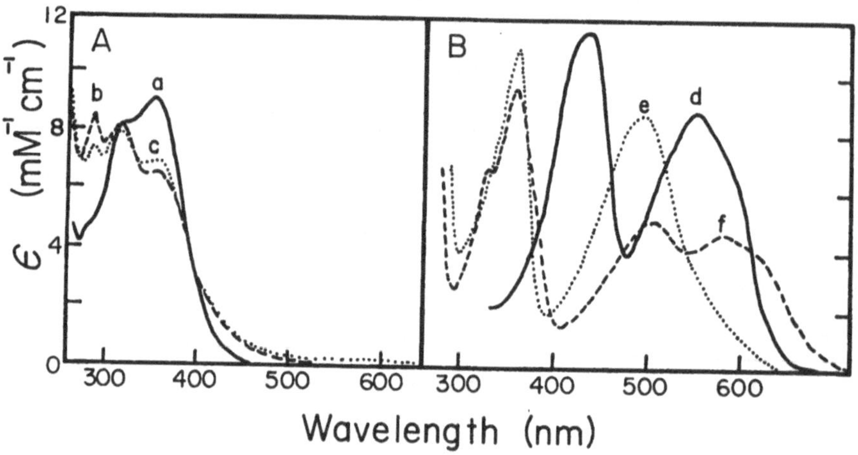

Fig. 1A: Curves **a**, **b**, and **c** are 5-EtFl-4^{a}-OH, 5-EtFl-4^{a}-SO_3^-, and 5-EtFl-4^{a}-$S_2O_4^-$, respectively, all in 0.1 M phosphate, pH 7.00. Fig. 1B: Curves **d** and **e** are 5-EtFl$^+_{ox}$ and 5-EtFlH$^{\dot{+}}$, respectively, in 0.2 N HCl. Curve **f** is 5-EtFl• in 0.1 M phosphate, pH 7.00.

respectively (Fig. 1A, B). These two new adducts were spectrally similar but could be distinguished on acidification: the 4^{a}-sulfite adduct produced the 5-EtFl$^+_{ox}$ (curve **d**) whereas the 4^{a}-dithionite adduct yielded the flavosemiquinone (curves **e**, **f**).

2) The Eq. 5 reaction had to compete with pathways providing 5-EtFl• via a very reactive adduct considered to be the 4^{a}-sulfoxylate-flavin radical (Eqs. 6, 7). For example, under strictly anaerobic conditions at pH 7.00 - 8.00 and with $S_2O_4^=$ / Fl = 0.75, the competitive pathways gave rise to a final superimposed spectrum of curves **c** + **f** in about a 1:1 ratio. 5-EtFl-4^{a}-SO_2• can also be obtained by first accumulating 5-EtFl•, followed by adding $SO_3^=$ (Eqs. 6, 7). In 0.1 M phosphate, pH 7.00, the apparent

K = [5-EtFl•] [$SO_3^=$] / [5-EtFl-4^{a}-SO_2•] = 1.1 x 10^{-2} M.

5-EtFl-4^{a}-SO_2• reacted with NH_2OH to give another 4^{a}-flavin adduct which also gave flavosemiquinone on acidification. *Interestingly, 5-EtFl-4^{a}-SO_2• is considerably more sensitive toward O_2 than 5-EtFl•. Consequently the formation of 5-EtFl• following pathways 6, 7 could be effectively suppressed by the removal of 5-EtFl-4^{a}-SO_2• with O_2!*

This gave an elegant way to obtain quantitative formation of 5-EtFl-4a-$S_2O_4^-$ (curve **c**). Using a slight excess of dithionite (e.g. $S_2O_4^{=}$ / Fl = 1.2) and in the initial presence of O_2, 5-EtFl-4a-SO_2• was rapidly oxidized to 5-EtFl-4a-SO_3^- which , by subsequent nucleophilic displacement by $S_2O_4^{=}$, recycled to the dithionite adduct. The ratio of the nucleophiles present and the prompt removal of O_2 by N_2 bubbling after the dithionite addition are important experimental factors. After the accumulation of the dithionite adduct and on adding more $S_2O_4^{=}$ ($\rightleftharpoons$ 2 $SO_2^{\dot{-}}$) under N_2, the dithionite adduct also appeared to undergo a nucleophilic displacement:

$$\text{5-EtFl-4a-}S_2O_4^- + SO_2^{\dot{-}} \xrightarrow[S_2O_4^{=}]{} \text{5-EtFl-4a-}SO_2\cdot \rightleftharpoons \text{5-EtFl}\cdot + SO_2 \qquad (7)$$

3) On increasing the initial $S_2O_4^{=}$ / flavin ratio or on first accumulating 5-EtFl• and subsequently adding $S_2O_4^{=}$, another adduct was formed, which absorbed at about 340 nm and yielded the 1,5-dihydroflavin (5-EtFlH) on acidification. These findings suggested that this third adduct is the 4a–sulfoxylateflavin anion:

$$\text{5-EtFl}\cdot + SO_2^{\dot{-}} \longrightarrow \text{5-EtFl-4a-}SO_2^- \xrightarrow{H^+} \text{5-EtFlH} + SO_2 \qquad (8)$$

A mixture of the sulfoxylateflavin anion, the sulfoxylateflavin radical, and the flavosemiquinone was formed during a slow, spontaneous decomposition of the dithionite-flavin adduct under N_2 (Eqs. 9, 10).

$$\text{5-EtFl-4a-}S_2O_4^- \longrightarrow \text{5-EtFl-4a-}SO_2^- + SO_2 \qquad (9)$$

$$\text{5-EtFl-4a-}S_2O_4^- \xrightarrow[SO_2^{\dot{-}}]{} \text{5-EtFl-4a-}SO_2\cdot \rightleftharpoons \text{5-EtFl}\cdot + SO_2 \qquad (10)$$

It is concluded that on treating 5-EtFl$^+_{ox}$ with $Na_2S_2O_4$ in aqueous solution the occurence of several parallel and sequential reactions (Eqs. 4-10) has to be considered.

Bleaching obtained on treating flavins with dithionite, as judged by the disappearance of absorption in the visible region, has been generally used as a criterion that the reduction to the 1,5-dihydroflavin has taken place. This criterion has now appeared to be unreliable.

Particularly, on applying stoichiometric amounts of dithionite, it could sometimes lead to erroneous conclusions.

The sodium dithionite treatment of flavodoxin from *Peptostreptococcus elsdenii* is one literature example [2] in which some conclusions should be reevaluated. The properties found for the bleached product do not fit those of a 1,5-dihydroflavin, as concluded originally, but are in complete agreement with those of 5-EtFl-4^{a}-$S_2O_4^-$ and/or 5-EtFl-4^{a}-$SO_2\cdot$ described in this report.

Another example of great relevance to our studies is a recent paper describing a bacterial bioluminescence enhanced by H_2O_2 and $FMNH_2$ [3]. We found that this luminescence only occurred when FMN was treated with dithionite, but not when it was fully reduced photochemically or catalytically by H_2. This suggests that a new enzyme-bound flavin adduct, arisen from FMN and $S_2O_4^=$ and/or $SO_2^{\dot{-}}$, could have played an essential role in a light producing reaction with H_2O_2. In particular, further studies of this bioluminescent system might answer the question whether the light emission has followed a CIEEL mechanism.

References

1. Mager, H. I. X., S.-C. Tu. "One-Electron Transfers in Flavin Systems: Relevance to the Postulated CIEEL Mechanism in Bacterial Bioluminescence." This volume.

2. Mayhew, S. G., E. J. F. van Arem, M. J. J. Strating, J. H. Wassink. 1976. In: Flavins and Flavoproteins (T. P. Singer, ed.). Elsevier, Amsterdam, 1976, pp. 411-421.

3. Watanabe, H., J. W. Hastings. 1987. J. Biochem. (Japan) 101, 279-282.

FLAVIN-MEDIATED HYDROGEN PEROXIDE PRODUCTION BY BIOLOGICAL AND CHEMICAL PHOTOSYSTEMS

M.A. De la Rosa

Department of Biochemistry, Faculty of Biology and C.S.I.C., University of Seville, Apdo. 1095, 41080-Sevilla, Spain

P.F. Heelis

The North E Wales Institute, Research Division, Deeside, Clwyd CH5 4BR, UK

K.K. Rao and D.O. Hall

Department of Biology, King's College London, Campden Hill Road, London W8 7AH, UK

Introduction

Solar energy conversion technologies aimed towards the photoproduction of fuels and chemicals have aroused serious interest over the last 15 years or so. In this context, we have designed several biological and chemical photosystems to produce hydrogen peroxide -this is considered to be not only an useful source of oxygen but also a powerful source of energy-, which are based on the well-known reduction of O_2 catalyzed by flavins (1-3). In biological photosystems (thylakoids) the flavin acts as a simple redox catalyst transferring electrons from the terminal acceptor of PSI to O_2, the light-transducing system being the natural photosynthetic apparatus itself (3). In artificial model systems, by contrast, the flavin acts as a highly efficient redox photosensitizer accepting electrons when photoexcited in its triplet state from a sacrificial electron donor and reducing, in the dark now, O_2 to H_2O_2 (2,4) (see Fig. 1).

Flavins and Flavoproteins 1987

Results and Discussion

Spinach thylakoids are able to carry out O_2 photoreduction with the concomitant formation of H_2O_2 in a process that is drastically accelerated by riboflavin. Under optimal conditions, the initial rate of H_2O_2 production ranged between 200 and 300 µmol H_2O_2 mg^{-1}(Chl) h^{-1}. In studies with thylakoids immobilized in several supports, only agar-immobilized thylakoids provided a final peroxide yield similar to or even higher than that obtained from free thylakoids (3).

As an alternative to biological systems, several artificial flavin photosystems based on the very peculiar photochemical properties of flavins were designed. Actually, flavins are energy-transducing redox photosystems of the energized--oxidized form type and can therefore exist in two redox pairs that share the same reduced form: the low potential or

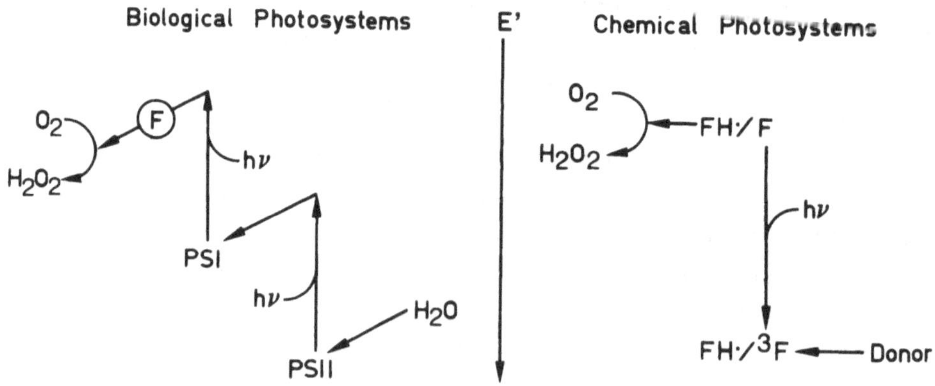

Figure 1. Schematic drawing of the operating mechanism of flavin-mediated hydrogen peroxide photoproduction by biological (left) and chemical (right) systems. Please note that flavin acts as a simple redox catalyst in biological photosystems and as an energy-transducing redox photosensitizer in artificial model systems.

basal pair in which the oxidized form is unenergized and stable (E'_o (FH./F), pH 7 = -0.22 V) and the high potential pair in which the oxidized form is photoexcited and unstable (E'_o (FH./^{3}F), pH 7 = +1.85 V). The flavin photosystems can thus promote the light-driven uphill transfer of electrons from an electron donor of high potential to an electron acceptor of low potential (4) (Fig. 1).

H_2O_2 photoproduction by flavin photosystems was found to be markedly dependent on the type of flavin, lumiflavin appearing to be the most efficient photosensitizer (1). The efficiency of several compounds as electron donors for the photochemical reduction of lumiflavin was afterwards investigated by laser flash photolysis (2), the bimolecular rate constant of quenching of the lumiflavin triplet state (K_q) being determined for each substrate (Table 1).

Observation of flavosemiquinone formation after the triplet decay in the presence of all the substrates studied suggested that the mechanism of lumiflavin photoreduction involves an

Table 1. Second-Order Rate Constants (K_q) of Lumiflavin Triplet Quenching and Efficiency of Conversion (ß) of the Semireduced Species to Fully Reduced Flavin.

Substrate	$K_q/M^{-1}\ s^{-1}$	ß/%
Dithioerythritol	1.3×10^8	25
EDTA	3.1×10^8	75
Hydroxylamine	1.0×10^7	70
Semicarbazide	4.8×10^8	85
Tricine	3.0×10^7	70
Triethanolamine	5.1×10^8	30

initial one-electron equivalent transfer from the substrate to the flavin triplet state, with the concomitant formation of flavosemiquinone radicals. The efficiency of conversion of flavosemiquinone into fully reduced flavin (ß) was therefore determined (Table 1). For semicarbazide ß reached its higest value (85%), what indicates that flavosemiquinone disproportionation and second one-electron reduction are the dominat processes (2).

Semicarbazide was actually shown to be the most efficient electron donor for lumiflavin-sensitized H_2O_2 photoproduction, initial peroxide formation rates of ca. 125 µmol H_2O_2 l^{-1} min^{-1} being obtained. This rate was almost three times higher than that obtained with EDTA, a compound that has been widely used for flavin photoreduction.

We would like to conclude by saying that flavin-mediated H_2O_2 photoproduction by either biological or chemical systems may be, for its simplicity and requirements, most suitable for the conversion and storage of solar energy.

References

1. De la Rosa, M.A., J.A. Navarro, F.F. De la Rosa and M. Losada. 1983. Stabilization by high pH of hydrogen peroxide production with flavin photosystems. Photobiochem. Photobiophys. 5, 93-103.

2. Heelis, P.F., M.A. De la Rosa and G.O. Phillips. 1985. A laser flash photolysis study of the photoreduction of the lumiflavin triplet state. Photobiochem. Photobiophys. 9, 57-63.

3. De la Rosa, M.A., K.K. Rao and D.O. Hall. 1986. Hydrogen peroxide photoproduction by free and immobilized spinach thylakoids. Photobiochem. Photobiophys. 11, 173-187.

4. Losada, M., M. Hervás, M.A. De la Rosa and F.F. De la Rosa. 1983. Energy transduction by bioelectrochemical systems. Bioelectrochem. Bioenerg. 11, 193-230.

ON THE CHEMICAL MECHANISMS OF BACTERIAL LUCIFERASE, AN UPDATING

Sandro Ghisla, Jens Eckstein and Peter Macheroux
Fakultät Biologie der Universität, D-7750 Konstanz, FRG

The purpose of this review is to focus on the chemical and physicochemical problems inherent to the mechanism of generation of excited states in the bacterial luminescence reaction. We refer to some excellent reviews by others for the discussion of biochemical/enzymological [1,2,3], bacteriological [4] and general problems including the different chemistry of bioluminescence in other organisms [5-8].
The net reaction can be reduced to the following equation:

$$2\ NAD(P)H + ATP + O_2 + 2\ H^+ \Longleftrightarrow 2\ NAD(P)^+ + ADP + 2\ H_2O + h\nu$$

Two molecules play particular roles in the process: Long chain aldehyde (C_{14}, myristic aldehyde) and FMN are cosubstrates of the enzyme luciferase. The former is oxidized to the corresponding carboxylic acid and then reconverted to the aldehyde by a reductase using NADPH and ATP (2 ATP may be required). FMN is reduced by FMN reductase, serves in the activation of O_2 and can play the role of the emitter:

$$FMNH_2 + RCH{=}O + O_2 \longrightarrow FMN + R\text{-}COOH + H_2O + h\nu\ (0.1 - 0.2)$$

The following properties, steps and intermediates, which have been disputed at one time or another, are now accepted in their essence, although discrepancies might still exist about details:

- Substrates [9-12] and stoichiometry of the reaction [9,10]
- Incorporation of 1/2 $^*O_2$ in the products R-C*OOH and $H_2{}^*O$ [13]
- Physical characteristics of emitted radiation and quantum yield [1,2,8]
- Involvment of flavin in the emission [14]

Flavins and Flavoproteins 1987

- Turnover mechanism [15], and details of kinetic steps [3,16,17]
- Formation of flavin-4a-hydroperoxide prior to reaction with R-CHO [18-20].
- Formation of flavin-4a-hydroxide subsequent to light emisssion [21,22]
- Exclusion of N(1) protonated flavin (cation) as emitter [23]
- Flavin-4a-hydroxide as emitter [21]
- Blue- and Yellow-fluorescent proteins (BFP [8,24], and YFP [25-27]) as alternative emitters

In contrast to the above, the following questions are still unsolved:

- Mode of oxidation of aldehyde by the flavin-4a-hydroperoxide [28-31].
- Isotope effects on light emission using C(1)-H/D aldehydes [31,32]
- Mode of population of the excited state [33].
- Chemical structure of the primary (excited) state.
- Existence of more than one excited state (or emitter)
- Mechanism of excitation of, or mode of transfer of energy to "alternative emitters" such as blue- or yellow-fluorescent proteins.
- The question of light emission in the absence of aldehyde, i.e. the mechanism of the so called "endogenous light" [34]
- The nature of the product and of the fluorescence increase observed upon photoirradiation of luciferase flavin-4a-hydroperoxide [35].

The chemical reactivity of luciferase bound flavin-4a-hydroperoxide

For an elucidation of the mechanism of oxidation of the substrate aldehyde, an understanding of the chemical reactivity of luciferase flavin-4a-hydroperoxide is important. Bruice [28] has pointed out that the properties of this peroxide are quite different from those of aliphatic hydroperoxides, as deduced e.g. from the low pK of ~9.2 of the corresponding flavin-4a-hydroxide (dissociation 4a-OH $\Longleftrightarrow$ 4a-O^- + H^+ [29]). The reaction of flavin-4a-hydroperoxide with thioxane is 10^5 faster than that of t-butyl hydroperoxide [30]. Enzyme bound flavin-4a-hydroperoxides exhibit typical reactivities like the oxygenation of sulfur-containing substrate analogs by the aromatic hydroxylases [36], by N,S-monooxygenase [37], and by cyclohexanone monooxygenase [38]. The latter performs catalysis by a Baeyer-Villiger mechanism including the monoxygenation of boronic acids to the corresponding alcohols. The study of the reactions of luciferase flavin-4a-hydroperoxide with these two analogs would thus be of particular interest. The expected reactions would be the following:

$$L\text{-FlH-OOH} + R\text{-}CH_2\text{-SH} \longrightarrow L\text{-}Fl_{ox} + R\text{-}CH_2\text{-SOH} + H_2O + h\nu \; ?$$

$$L\text{-FlH-OOH} + R\text{-}CH_2\text{-}B(OH)_2 \longrightarrow L\text{-}Fl_{ox} + R\text{-}CH_2\text{-OH} + B(OH)_3 + h\nu \; ?$$

Finding of oxidation with concomitant light emission with either analog would exclude the requirement of aldehyde C-H bond breaking, and also the involvment of an excited carboxylate in the light emission process [1]. The latter has been considered as a possible "primary" excited state with potential for transferring its excitation energy to different possible emitters, thus a "branched" type of pathway [1,39]. We found that 1-decanethiol actually stabilizes luciferase flavin-4a-hydroperoxide [40] in a way similar to that observed with various compounds including long chain alcohols [34,41]. Thus in the presence of a 5-fold excess of this analog, the decay rate of the hydroperoxide at 0° was decreased by -50%, and no further effect on the light emission was observed [40]. Although an analysis of the possible products was not done, we conclude that, in contrast to other peroxides, luciferase-flavin-4a-hydroperoxide is not particularly reactive with thiols.
The second analog, long chain boronic acid, has been thought to induce light emission and thus to be reactive [39]. We have reinvestigated this reaction paying particular attention to the purification of nonylboronic acid, and have found, that most preparations contain trace impurities, which induce light emission with luciferase [42]. They could be long chain aldehydes arising from autoxidation of the boronic acid. With the purest preparations no light emission above the level of "endogenous light" was obtained. Actually, nonylboronic acid is an inhibitor competitive with aldehyde in the normal luciferase reaction: Added to purified luciferase flavin-4a-hydroperoxide in the absence of aldehyde it slightly enhances the rate of decay (~ twofold effect on decay rate at 0°, and at ~20 μM, equimolar concentrations) [42]. No significant amounts of nonanol and of boric acid, the two expected products, could be detected in the incubation mixtures. Since boronic acids typically react according to a Baeyer-Villiger mechanism, our results suggest that bacterial luciferase is not set up to promote this type of reaction.

On the nature of the emitter(s)

This has been a major point of controversy due to the lack of coincidence of the fluorescence emission spectrum of Fl_{ox} (λ_{max} ~520 nm), the product found in the spent luciferase reaction, with the luciferase luminescence (λ_{max}~490 nm). Although earlier porposals envisaged different emitters, the involvment of flavin was clearly documented by the use of flavin analogs [14]. Among the proposals one was $1H\text{-}Fl_{ox}^{+}$ (1) [43], which, in spite of clear cut evidence to the contrary [23], is still encountered in newer text books and reviews. 1-Deaza-FMN (2) is competent in light emission, the quantum yield being ~15 % that of FMN, it forms a hydroperoxide which has a fluorescence emission λ_{max}~485 nm. This FMN analog cannot be protonated at position 1; its cationic form (protonation occurs probably at N(5) with a pK < 0), has an absorption λ_{max} ~510 nm, which virtually excludes an emission below 500 nm [23].

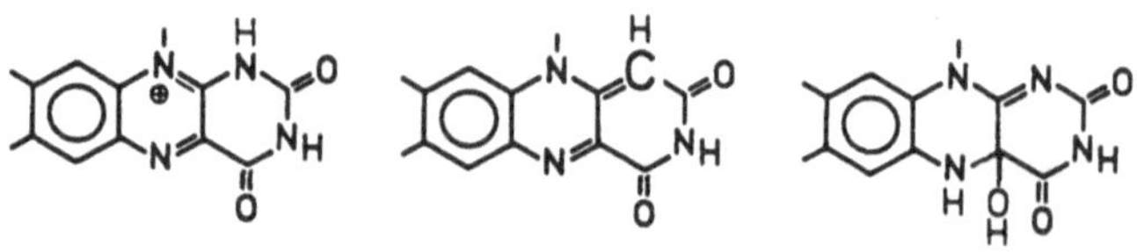

1 2 3

Very recently we succeeded in isolating and characterizing luciferase-flavin-4a-hydroxide (3). Its formation parallels emission of light, and its fluorescence emission spectrum is essentially identical to that of luciferase luminescence emission under the same conditions [21,22]. This, and the logic of the chemical reaction strongly suggest, that 3 is the emitter, at least under the specific experimental conditions used. In our opinion Lee's fluorescent transient [17], proposed as one of the emitters, is identical with 3. In contrast to this, the role as emitters of the so called sensitizer protein BFP and YFP, which contain 8-ribityllumazine [8,24] and FMN [25,27] as chromophores, is undisputed. Therefore the discussion shall be restricted to mechanistic aspects.

On the mechanism of generation of the excited state

Recently, based on the original proposal of Schuster and coworkers [44], a so called CIEEL Mechanism (Chemically Initiated Electron Exchange

Luminescence) has been put forward [2,39] as an alternative to previous proposals such as the Baeyer-Villiger fragmentation [45]. The shortcomings of the latter and of several other mechanisms, were the lack of a satisfactory explanations for the generation of the excited state, the postulation of the "wrong" emitter (cf. above), or the requirement of several chemical transformations involving excited molecules. The original CIEEL scheme can be adapted for an internal activation as shown in Scheme 1 [33,45].

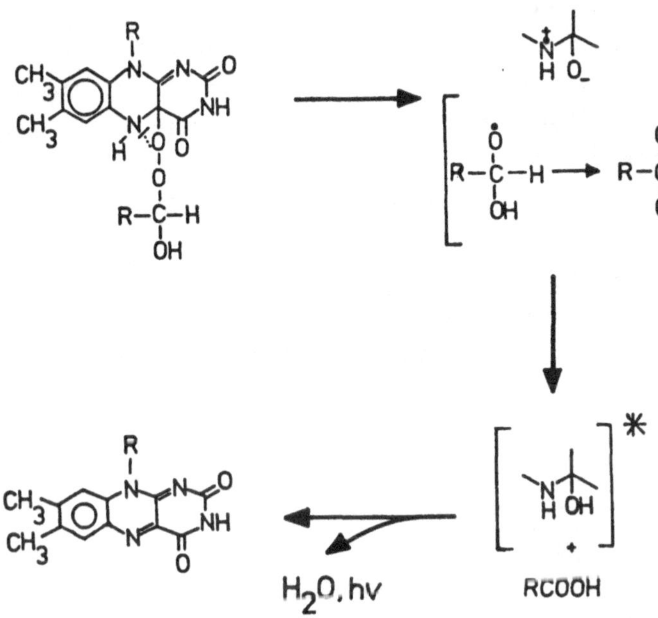

Scheme 1) Adaptation of the CIEEL mechanism to the bacterial luciferase reaction. In this case the 4a,5-dihydroflavin moiety acts as a donor and induces breaking of the peroxide bond. In the crucial step the aldehyde C(1)-H bond is broken via a radical mechanism and the resulting fragment back-donates $1e^-$ to the flavin to generate the excited state. Upon relaxation the flavin-4a-hydroxide eliminates water to reform FMN_{ox}

Here the activator is the o-phenylenediamine moiety of the 4a,5-dihydro-4a-hydroperoxy-flavin itself, and it would initiate the reaction by transiently transferring an electron to the peroxide of the peroxyhemiacetal, and resulting in the breakage of the O-O bond, as shown in Scheme 1. The occurrence of the peroxyhemiacetal is perhaps somewhat speculative, although kinetic evidence obtained with luciferase [32], as well as with flavin models [46,47] is compatible with its formation prior to the light generating step. The (co)involvement of a sulfhydryl residue analogous to the proposal of Kosower [48] can also be formulated. In order to test this CIEEL hypothesis, we have attempted the correlation of the rate of k_{obs} (light emission rise or decay) to the redox poten- tial or Hammett parameters of a series of 8-substituted flavin analogs. The results [49] are clearly compatible with Schuster's mechanism and with a radical type reaction, but they argue against a Baeyer-Villiger formu- lation. A few considerations might add weight to this interpretation: Aliphatic aldehydes are known to be prone to radical oxidation by oxygen/peroxides [50]. The flavin-4a-hydroperoxide is a molecule containing an oxidizing (peroxide) combined with a reducing moiety (phenylene diamine). That the latter type is a good donor and easily autoxidable does not need particular documentation. A further example of excellent $1e^-$ oxidability is the close relative 1,5-dihydro-flavin. A CIEEL formulation is also consistent with model work from Bruice's lab, who studied the luminescence accompanying the decay of N(5)-blocked flavin-4a-peroxihemiacetals [46,47].

On the mechanismus of population of excited states of "alternative emitters"

A central issue is whether the so called "fluorescent proteins" are excited directly as a consequence of the energizing chemical step, or whether they accept the excitation from a higher energy "primary" excited chromophore. Emission by a number of alternative emitters demands that there be branching; an unsolved question is whether this branching occurs before or after the formation of an electronically excited state, i.e. whether excitation is direct or not [2,8,39]. These differences are are summarized in Scheme 2. A strict requirement of the indirect mechanism is the occurrence of an excited chromophore of still unknown nature [L-?]*, scheme 2, which possesses enough energy for efficient transfer to all emitters, in particular to BFP. This excited species must precede the flavin-4a-

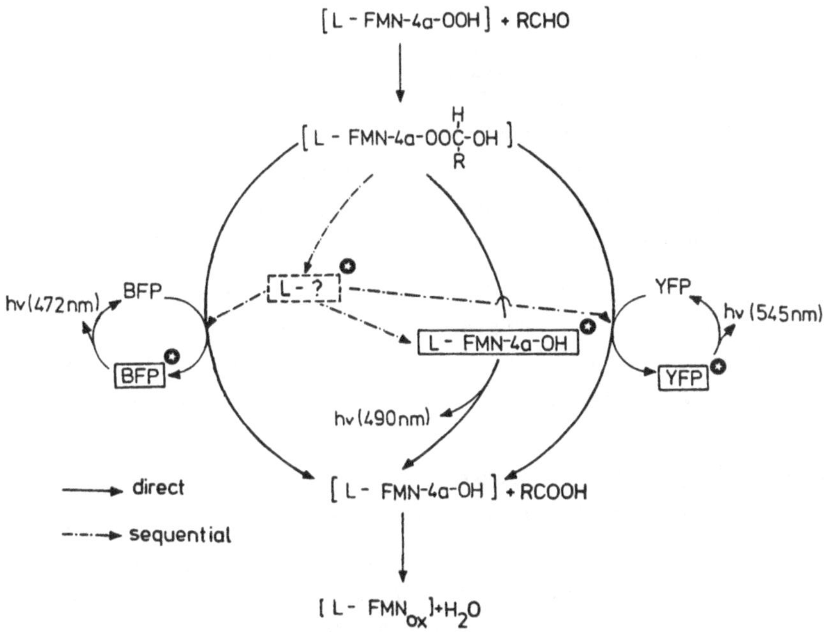

Scheme 2) Different modes for the population of the emitter excited states in the bacterial luciferase reaction. In the so-called "direct" mechanism (——) the external emitters compete with the 4a,5-dihydroflavin itself in the activation of the peroxide fragmentation e.g. via a CIEEL mechanism. The "indirect" mechanism foresees primary formation of an excited state of still unknown chemical constitution [L-?], the energy of which must be sufficient for transfer to the secondary emitters (BFP or YFP) e.g. via a Förster transfer (-.-.-). Alternatively [L-?] relaxes to excited flavin-4a-hydroxide.

hydroxide as excited state. It has been suggested that this could be the excited carboxylic acid product [1]. Its formation by a CIEEL mechanism is, however, difficult to envision. Another candidate might be a flavin-4a-hydroxide in a higher energy conformation. We do not believe in the existence of mysterious chromophores, which might be present in luciferase. Since this enzyme can be prepared in gram quantities (0.1 mMoles), if such chromophores would exist it is unlikely they could have escaped detection with modern analytic techniques. The choice of such a putative chromophore should thus be restricted to different forms of the flavin itself, to species derived from the substrate or to amino acid residues present at the enzyme active center. In this context it is tempting to speculate about an alternative, which has not yet been considered: 4a,5-Dihydroflavins are nonfluorescent in solution, but can be highly fluorescent when bound to proteins (examples: luciferase flavin-4a-hydroxide [21] and luciferase flavin-4a-hydroperoxide [35]) or in frozen solution [51]. The stokes shift in the case of the 4a,5-dihydroflavin chromophore is unusually large suggesting the presence of energetically different conformers. The absorption spectra of 4a,5-dihydroflavins have maxima ranging from 330 to 390 nm depending on the substituents at positions N(5) or C(4a), and are also strongly dependent on the temperature [51,52]. This has been interpreted in terms of a different degree of bending of the flavin depending on (sterical interactions of) the substituents. NMR investigations of 1,5-dihydro-FMN bound to luciferase indicate a planar structure [52], while the indroduction of a substituent at C(4a) induces bending to some degree. From this we think it is worth considering that conformers of 4a,5-dihydroflavin are stabilized by the enzyme, and play the role of the primary excited state.

In the indirect mechanism any primary excited state is required to be much less fluorescent than 4a-hydroxy-flavin, but to have a life time sufficiently long for efficient transfer of energy to a secondary emitter. It is interesting that Shepherd and Bruice have postulated a primary, nonfluorescent excited state X* to be formed from the decomposition of 4a-peroxyhemiacetals of N(5) substituted flavins, which transfers excitation to several (secondary) emitters [47]. Clearly also other species which can be formed from a flavin-4a-hydroperoxide and which can be converted to flavin-4a-hydroxide could play the same role.

In contrast to the above, the so called "direct" mechanism [2,39] envisages several external chromophores as well as the 4a,5-dihydroflavin nucleus itself to act as alternative reactants, e.g. according to the CIEEL mechanism. Preliminary experiments from our laboratory designed to give an answer to this problem, involved the use of purified luciferase flavin-4a-hydroperoxides formed using different flavin analogs (iso-FMN, FMN, 2-thio-FMN), which have emission spectra ranging from 478 to 540 nm, and YFP. The results [53,54] are compatible with a direct mechanism; they cannot, however exclude an indirect one (Scheme 2). We will pursue this lead, expanding the range of FMN analogs used, in particular varying the type of FMN substitution when bound to YFP. A prediction for this mechanism is that a correlation similar to that presented in the accompanying paper

[49] for luciferase itself, should be found when k_{obs} is plotted against the redox potential of the FMN bound to YFP (of BFP). Note that in this case an "oxidized" species (FMN, and 8-ribityllumazine in the case of YFP and BFP respectively) must serve as donor in spite of the fact that it is considered an "acceptor", and that the redox potential would be that for the oxidation of the species. Since information of the last point is not available an assessment of the role of these molecules as "donors" in a putative CIEEL mechanism is difficult.

Conclusions

From the experiments discussed in this and in the accompanying paper, and from the chemical considerations discussed above the following points emerge:

- A Baeyer-Villiger mechanism appears improbable.
- No experimental result contradicts a radical (or a CIEEL) mechanism. This also would be in agreement with most recent findings from Lee's group [55], showing production of light from luciferase-flavin-4a-hydroperoxide and free radicals.
- A differentiation between a "direct" or an "indirect" type mechanism (Scheme 2) is not (yet) possible.
- The chromophore of a putative primary excited state has to be found among the species and intermediates directly involved in the reaction.

Finally we would like to stress the point, that some ideas and speculations put forward above should be taken as such; they are intended to promote discussion and hopefully to help find a solution to the fascinating puzzle of flavin mediated luminescence.

This work was supported by the Deutsche Forschungsgemeinschaft (Gh 2/4-5). We would like to thank Drs. J.W. Hastings and T. Wilson for many valuable discussions and suggestions and for reviewing this manuscript.

References

1. Ziegler, M.M., and Baldwin, T.O., (1981) in "Current Topics in Bioenergetics", (Sanadi, D.R., ed.), Vol. 12, Academic Press, New York, pp. 65-113.

2. Hastings, J.W., Potrikus, C.J., Gupta, S.C., Kurfürst, M., and Makemson, J.C., (1985) Adv. Microb. Physiology 26, 235-291

3. Hastings, J.W., and Nealson, K.H., (1977) Ann.Rev. Microbiol. 31, 549-595

4. Hastings, J.W., and Nealson, (1981) in "The Prokaryotes" (Mortimer, P.S., Stolp, H., Trüper, H.G., Balows, A., and Schlegel, H.G., eds) Springer, Berlin, Heidelberg, pp. 1332-1345

5. Shimomura, O., (1982) in "Chemical and Biological Generation of Excited States" (Adam, W., and Cilento, G., eds.) Academic Press, New York, pp. 249-276

6. Hastings, J.W., (1983) J.Mol.Evol. 19, 309-321

7. Hastings, J.W., (1978) Methods in Enzymol. LVII, 125-135

8. Lee, J., (1985) in "Chemi- and Bioluminescence" (Burr, J.G., ed) Plenum Press pp. 401-437

9. Hastings, J.W., and Balny, C., (1975) J. Biol. Chem. 250, 7288-7292

10. Becvar, J., and Hastings, J.W., (1975) Proc. Natl. Acad. Sci. USA 72, 3374-3376

11. McElroy, W.D., Hastings, J.W., Sonnenfeld, V., and Coulombre, J., (1953) Science 118, 385-386

12. Strehler, B., and Cormier, M.J., (1954) J. Biol. Chem. 211, 213-225

13. Suzuki, K., Kaidoh, T., Katagiri, M., and Tsuchiya, T., (1983) Biochim. Biophys. Acta 722, 297-301

14. Mitchell, G., and Hastings, J.W., (1969) J. Biol. Chem. 10, 2572-2576

15. Hastings, J.W., and Gibson, Q.H., (1963) J. Biol. Chem. 238 2537-2554

16. Gibson, Q.H., Hastings, J.W., Weber, G., Duane, W., and Massa, J., (1966) in "Flavins and Flavoproteins" (Slater, E.C., ed) Elsevier, Amsterdam, pp. 341-359

17. Matheson, I.B.C., and Lee, J., (1983) Photochem. Photobiol. 38, 231-240

18. Hastings, J.W., Balny, C., LePeuch, C., and Douzou, P., (1973) Proc. Natl.Acad.Sci.USA 70, 3468-3472

19. Ghisla, S., Hastings, J.W, Favaudon, V., and Lhoste, J.-M., (1978) Proc. Natl. Acad. Sci. USA 75, 5860-5863

20. Vervoort, J., Müller, F., Lee, J., vandenBerg, W.A.M., and Moonen, C.T.W., (1986) Biochemistry 25, 8062-8067

21. Kurfürst, M., Macheroux, P., Ghisla, S., and Hastings, J.W., (1987) Biochim. Biophys. Acta 924, 104-110

22. Kurfürst, M., Ghisla, S., and Hastings, J.W., (1984) Proc. Natl. Acad. Sci. USA 81, 2990-2994

23. Kurfürst, M., Ghisla, S., and Hastings, J.W., (1982) in "Flavins and Flavoproteins" (Massey, V., and Williams, C.H., eds) Elsevier, North-Holland, N.Y., pp. 353-358

24. Gast R., and Lee, J., (1978) Proc. Natl. Acad. Sci. USA 75 833-837

25. Leisman, G., and Nealson, K.H., (1982) in "Flavins and Flavoproteins" (Massey, V., and Williams, C.H., eds) Elsevier, New York, pp. 382-386

26. Ruby, E.G., and Nealson, K.H., (1977) Science 196, 432-434

27. Macheroux, P., Schmidt, K.U., Steinerstauch, P., Buntic, G., Colepicoli, P., Hastings, J.W., and Ghisla, S., (1987) Biochem. Biophys. Res. Commun. in press

28. Bruice, T.C., Noar, J.B., Ball, S.S., and Venkataram, U.V., (1983) J.Amer. Chem. Soc. 105, 2452-2463

29. Bruice, T.C., (1984) in "Flavins and Flavoproteins" (Bray, R., Engel, P., and Mayhew, S.G., eds.) DeGruyter, Berlin, N.Y., pp 45-55

30. Kemal, C., Chan, T.W., and Bruice, T.C., (1977) Proc. Natl. Acad. Sci. USA 74, 405-409

31. Shannon, P., Presswood, R.P., Spencer, R., Becvar, J.E., Hastings, J.W., and Walsh, C., (1978) in "Mechanisms of Oxidizing Enzymes" (Singer, T.P., and Ondarza, R.N., eds) Elsevier, pp. 69-78

32. Macheroux, P., Boschert, W., Ghisla, S., and Hastings, J.W., (1987) in preparation

33. Wilson, T., (1985) in "Singlet Oxygen", vol. II (Frimer, A.A., ed) CRC Press, pp. 37-65

34. Hastings, J.W., Gibson, Q.H., Friedland, J., and Spudich, J. (1966) in "Bioluminescence in Progress" (Johnson, F.H., and Haneda, Y., eds.) Princeton Univ. Press, Princeton, pp. 151-186

35. Balny,C., and Hastings, J.W., (1975) Biochemistry 14, 4719-4723

36. Ballou, D.P. (1984) in "Flavins and Flavoproteins" (Bray, R., Engel, P., and Mayhew, S.G., eds) DeGruyter, Berlin, N.Y., pp. 605-618

37. Jones, K., and Ballou, D.P., ibid, pp. 619-622

38. Branchaud, B.P., and Walsh, C., (1985) J.Amer. Chem. Soc. 107 2153-2161

39. Macheroux, P., and Ghisla, S., (1985) Nachr. Chem. Tech. Lab. 33 785-790

40. Kurfürst, M., (1983) Ph.D. Thesis, Univ. of Konstanz

41. Tu, S.C., (1982) J. Biol. Chem. 257, 3719-3725

42. Macheroux, P., Ahrens, M., Ghisla, S., Branchaud, B.P., Eberhard, A., and Hastings, J.W., (1987) in preparation

43. Eley, M., Lee, J., Lhoste, J.M., Cormier, M.J., and Hemmerich, P., (1970) Biochemistry 9, 2902-2908

44. Schuster, G.B., (1979) Acc. Chem. Res. 12, 366-373

45. Eberhard, A., and Hastings, J.W., (1972) Biochem. Biophys. Res. Commun. 47, 348-353

46. Kemal, C., and Bruice, T.C., (1977) J. Amer. Chem. Soc. 99, 7064-7067

47. Shepherd P.T., and Bruice, T.C., (1980) J.Am.Chem.Soc. 102, 7776-7778

48. Kosower, E.M., (1980) Biochem. Biophys. Res. Commun. 92, 356-364

49. Macheroux, P., Eckstein, J., and Ghisla, S., paper in this volume

50. Cooper, H.R., and Melville, H.W., (1951) J. Chem. Soc,. 1984-2002

51. Ghisla, S., Massey, V., Lhoste, J.M., ans Mayhew, S.G., (1974) Biochemistry 13, 589-597

52. Ghisla, S., Hartmann, U., Hemmerich, P., and Müller, F., (1973) Liebigs Ann. Chem. 1388-1415

53. Macheroux, P., (1986) Ph.D. Thesis, Univ. of Konstanz (to be published)

54. Buntic, R., (1986) BS Thesis, Biochemical Sciences, Harvard University

STUDIES ON THE MECHANISM OF BACTERIAL BIOLUMINESCENCE. EVIDENCE COMPATIBLE WITH A ONE ELECTRON TRANSFER PROCESS AND A CIEEL MECHANISM IN THE LUCIFERASE REACTION

Peter MACHEROUX, Jens ECKSTEIN and Sandro GHISLA

Faculty of Biology, University of Konstanz,
D-7750 Konstanz, FRG

Introduction and Premises

In the key step of bacterial bioluminescence a long chain aldehyde is oxidized to the corresponding fatty acid, with the concomitant generation of an excited state. This oxidation is carried out by a flavin-C(4a)-hydroperoxide, possibly over a flavin-C(4a)-peroxyhemiacetal [1,2]. The mechanism of the key step, the generation of the excited emitter, is still unclear. None of the several previous proposals, including a "Baeyer-Villiger" oxidation, or a dissociative electron transfer mediated by a cysteine sulfhydril, were satisfactory [3,4]. As introduced in the preceding review article [5], the recently proposed CIEEL mechanism (Chemically Initiated Electron Exchange Luminescence) [6] merits careful consideration. In preliminary experiments [7,8] aiming at its verification, we attempted a correlation of the rate of emission decay with the redox potential of flavins, substituted at various ring positions. The rather poor correlation obtained was, however, attributed to specific effects induced by the flavin modification, e.g. alteration of protein-flavin interactions. It has been shown, using FMN analogues carrying various substituents at position C(8), that these are recognized by luciferase, and are competent in bioluminescence, suggesting that this position is "free",

Flavins and Flavoproteins 1987

i.e. exposed to solvents [9,10]. Based on this we reasoned, that modifications restricted to this position would conserve intact flavin-protein interactions, and not prejudice the catalytic event.

An important assumption needs discussion: Verification of the CIEEL mechanism according to Schuster [6] bases on a correlation with the redox potential of the donor involved, in our case this would be N(5) of 4a,5-dihydroflavin [5]. Concrete values on the redox potential of a couple Fl_{ox} – 4a,5-dihydro-Fl_{red} do not exist. 4a,5-dihydro-flavins are, however, simply tautomeric forms of 1,5-dihydro-flavins,

the second beeing clearly thermodynamically favored. It is reasonable to assume, that the free energy difference between the two forms is not very large: Alkyl substituents have been found to migrate reversibly between the two positions [11,12]. Slight variations of the alkylation conditions lead to N(5) or C(4a) derivatives of $Fl_{red}H^-$. Adduct formation of some flavins with sulfite has been found to yield an equilibrium mixture of 5- and 4a-adducts [13]. The difference in free energy between the 4a,5- and the 1,5-tautomer thus probably is much smaller than that between Fl_{ox} and Fl_{red}. Consequently substituents affecting the Fl_{ox}/1,5-H_2-Fl_{red} redox potential should affect that of Fl_{ox}/4a,5-H_2-Fl_{red} in an analogous way. As a further test, we resorted to the classical Hammett correlation, since the key "CIEEL - step" can be viewed formally as an (internal) activation of a hydroperoxihemiacetal by a p-X-substituted o-phenylene (di)amine, in which "only" X is varied:

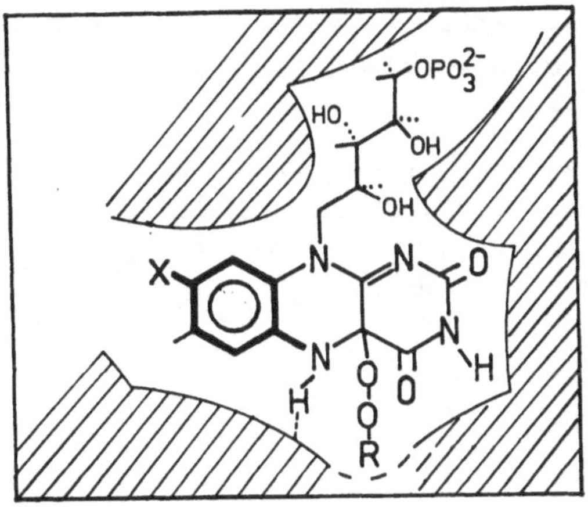

By introducing a reference system (i.e. nonsubstituted FMN) the Hammett equation can be formulated as follows:

$$\log k/k_o = r * \sigma`,$$

where k_o is the rate constant of the reference system, r (rho) is the "reaction parameter" and σ` is the "substituent parameter". We have estimated σ` and r for a series of 8-substituted flavins, and used the values for the plot of Fig. 2 described below.

Results and Discussion

We have expanded the previous measurements [7] using the 8 flavins shown in the figures of redox potential in the range -50 to -320 mV. The correlation of log k (rate constant of the light emission decay) with the redox potential plotted according to Schuster [6] yields a slope of -14 V^{-1}. This means a value of 0.8 for the so called "diagnostic constant" or "Brönsted slope" α, which has been suggested to reflect the extent of one-electron-transfer occuring during a reaction [6,14,17]. α can have values between 0 and 1, the second extreme reflecting complete charge separation. According to Schuster values of α<1, can also be compatible with transfer of one electron, when changes in bonding geometry are taken into account.

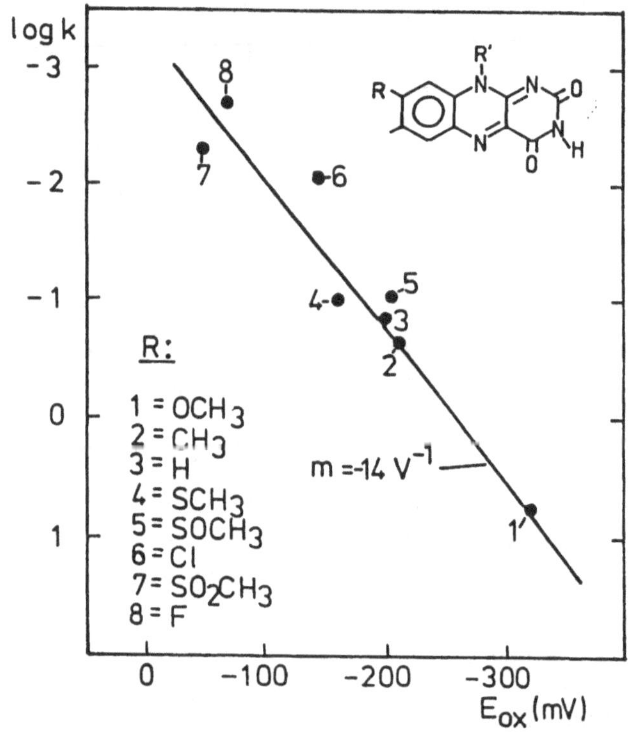

Fig. 1)
Dependence of luminescent decay rate on the redox potential of 8-substituted FMN derivates: The luminescence emission was measured in a dithionite assay [17]. The aldehyde was decanal at a final concentration of ≈4×10^{-3} M. Luciferase and FMN were 6.7×10^{-7} M and 1×10^{-6} M, respectively. The data shown are the average of at least 10 measurements. The FMN derivative #3 is "iso-FMN" (6,7-dimethyl-8-demthyl-FMN). The slope is -14 V^{-1}, it can also be expressed as $-\alpha/2.3RT$, yielding a value of 0.8 for α.

The Hammett plot shown in Fig. 2 yields a reaction parameter of -0.8. This value reflects a typical value for reactions, where radicals are involved [15,16]. It should be pointed out, that the σ` values were calculated by substituting the equilibrium constant (K) in the classical Hammett equation with the redox potential according to the method of Kochi et.al.[17] using unsubstituted FMN as reference system.

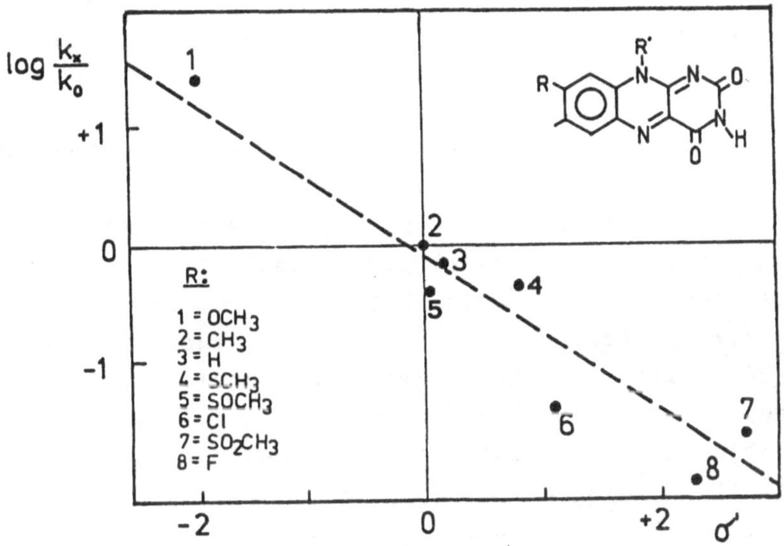

Fig. 2)
Correlation of log k/k₀ with the substituent parameters σ` in a Hammett-type plot: The FMN derivate #3 is "iso-FMN" (6,7-dimethyl-8-demthyl-FMN). The slope corresponds to a value of -0.8 for the rate parameter r.

Conclusions

Linear free energy relationships have been extensively studied for electron transfer and related reactions in both inorganic and organic systems, they allow reliable mechanistic conclusions [17]. Our value of $\alpha \approx 0.8$ clearly is compatible with Schuster's CIEEL mechanistic interpretation of luminescent reactions. In the examples reported by his group α had values ≈ 0.3. According to Kochi et.al. [17] α Brönsted values of $0.5 < \alpha < 1$ can be interpreted with confidence to reflect the "intrinsic barrier" of a system for a one electron transfer process. The data were plotted according to Hammett also using classic σ parameters. In this case the correlation is generally poorer and depends on the type of effect (M- or I-effect) the parameters were derived for. When the M- and I-effected parameters are averaged (σ^+) a value $r \approx -1.7$ can be estimated. According to the data reviewed by Jaffé [15] values of $r > 3$ or $r < -3$ are typical for processes involving ionic transition states. Wells [16] points out that radical mechanisms yield r values 0 ± 1, and indeed a value $r = -0.76$ was found for the photohalogenation of substituted toluenes, a typical radical process. It should be pointed out that the coincidence of our value $r \approx -0.8$ with that of typical radical mechanism is remarkable. We would like to stress the point that this, of course, might be coincidental: Since reliable data on chemical systems which might be taken as models for our reaction are not available, this cannot be taken as proof for a radical mechanism. On the other hand a Baeyer-Villiger mechanism would predict the reaction rate to increase with increasing redox potential of the flavin. This mechanism should be reflected by a Hammett dependence with a positive slope. As a minimal interpretation our results are in disaccord with a Baeyer-Villiger type mechanism.

References

1. Hastings, J.W., C. Balny, C. Le Peuch P. and Douzou 1973. Proc.Natl.Acad.Sci.USA, 70, 3468-3489.

2. Ghisla, S., J.W. Hastings, W. Favaudon and J.M.Lhoste 1978. Proc.Natl.Acad.Sci.USA, 75, 5860-5863.

3. Hastings, J.W. and K.H. Nealson 1977. Annu.Rev. Microbiol., 31, 549-595.

4. Kosower, E.M. 1980. Biochem.Biophys.Res.Commun., 92, 356-364.

5. Ghisla, S., J. Eckstein and P. Macheroux this volume.

6. a.) Koo, J.Y. and G.B. Schuster 1977. J.Am.Chem.Soc., 99/18, 6107-6109.
   b.) Schuster G.B. 1979. Acc.Chem. Res., 12, 366-373.

7. Macheroux, P. 1986. Ph.D.Thesis Univers.Konstanz, FRG.

8. Macheroux, P., S. Ghisla, J.W. Hastings and M. Kurfürst 1984. In: Flavins and Flavoproteins (Bray R., P. Engel and S.G. Mayhew eds.), de Gruyter, Berlin, pp. 669-672.

9. Watanabe,T., K. Matsui, S. Kasai and T. Nakamura 1978. J.Biochem, 84, 1441-1446.

10. Chen, L. and T.O. Baldwin 1984. In: Flavins and Flavoproteins (Bray R., P. Engel and S.G. Mayhew eds.), de Gruyter, Berlin, pp. 785-788.

11. Ghisla S., U. Hartmann, P. Hemmerich and F. Müller 1973. Liebigs Ann.Chem., 1388-1415.

12. Haas, W. and P. Hemmerich 1975. In: Reactivity of Flavins (Jagi, K. ed.), Univ. of Tokyo Press, pp. 1-13.

13. Bruice, T.C., L. Hevesi and S. Shinkai 1973. Biochem., 12, 2083-2089.

14. Wilson, T. 1985. In: Singlet O_2 (Frimer, A. ed.), CRC Press Inc., II, pp.37-65.

15. Jaffé, H.H. 1953. Chem.Rev., 53, 191-260.

16. Wells, P.R. 1963. Chem.Rev., 63, 171-219.

17. Fukuzumi, S. and J.K. Kochi 1983. Bull.Chem.Soc.Jpn., 56, 969-979.

18. Hastings, J.W., T.O. Baldwin and M.Z. Nicoli 1978. Methods in Enzymology, 57, 135-152.

STRUCTURAL ANALYSIS OF BACTERIAL LUCIFERASE

T. O. Baldwin, L. H. Chen, L. J. Chlumsky, J. H. Devine, T. C. Johnston, J.-W. Lin, J. Sugihara, J. J. Waddle & M. M. Ziegler

Department of Biochemistry and Biophysics, Texas A&M University, and the Texas Agricultural Experiment Station, College Station, Texas 77843

Introduction

Approximately 1-2% of all marine bacteria emit light when grown on a complex medium, making bioluminescence one of the more common phenotypic traits among marine bacteria (1). Expression of bioluminescence is the result of a regulon consisting of two operons and at least seven structural genes (2). The enzyme responsible for the light-emitting reaction, bacterial luciferase, is a flavin hydroxylase. The luciferase is a heterodimer ($\alpha\beta$) with a single active center that resides primarily on the α subunit (3). The luciferase subunits are encoded by the *luxAB* genes. The aldehyde substrate is supplied *in vivo* by a total of three proteins, encoded by the *luxC*, *luxD* and *luxE* genes. The product of the *luxD* gene is a fatty acyltransferase that apparently supplies tetradecanoic acid to the fatty acid reductase complex, which consists of four copies each of the products of the *luxC* and *luxE* genes (4). In a reaction requiring ATP and NADPH, the fatty acid reductase complex reduces tetradecanoic acid, produced by the bioluminescence reaction or supplied by the fatty acyltransferase, to tetradecanal (5). This model explains the observation that there is no pool of aldehyde substrate. Apparently, the complex produces aldehyde at a level required to supply the luciferase.

The enzyme binds $FMNH_2$ from solution to form an enzyme-reduced flavin complex which reacts rapidly with O_2 to form the enzyme-bound flavin C4a peroxydihydroflavin, intermediate II (3). In the absence of the aldehyde substrate, the intermediate II decays in a dark reaction to yield hydrogen peroxide. Intermediate II binds a long-chain aldehyde to form intermediate IIA, which decays through several intermediates to yield FMN, the carboxylic acid and blue-green light. It has been proposed that the primary excited state formed in the reaction is the C4a hydroxy flavin (6). While the spectral data available are consistent with the proposal that the pseudobase is the emitter in the reaction *in vitro*, there is little solid data to support the pseudobase as the primary excited state. The work of Lee and his associates suggests that, at least for some species of luminous bacteria, light emission *in vivo* occurs from a secondary emitter protein, the lumazine protein, or LumP (7). Since the reaction *in vitro* with pure luciferase is clearly luminescent, LumP must not be essential for bioluminescence. Work from the laboratory of K. Nealson (8) and recently from our laboratory (9) demonstrates that addition of the yellow fluorescence protein (YFP) purified from *V. fischeri* strain Y-1 results in a dramatic alteration in the bioluminescence emission spectrum with the generation of a new peak of emission centered at about 540 nm, in addition to the 485 nm peak. YFP carries FMN and is a homodimer of 22,000 M_r subunits (9). The precise mechanism for any of these possible energy transfer systems is not known. The reaction pathway for purified luciferase *in vitro* as it is usually depicted is presented in Figure 1.

Flavins and Flavoproteins 1987

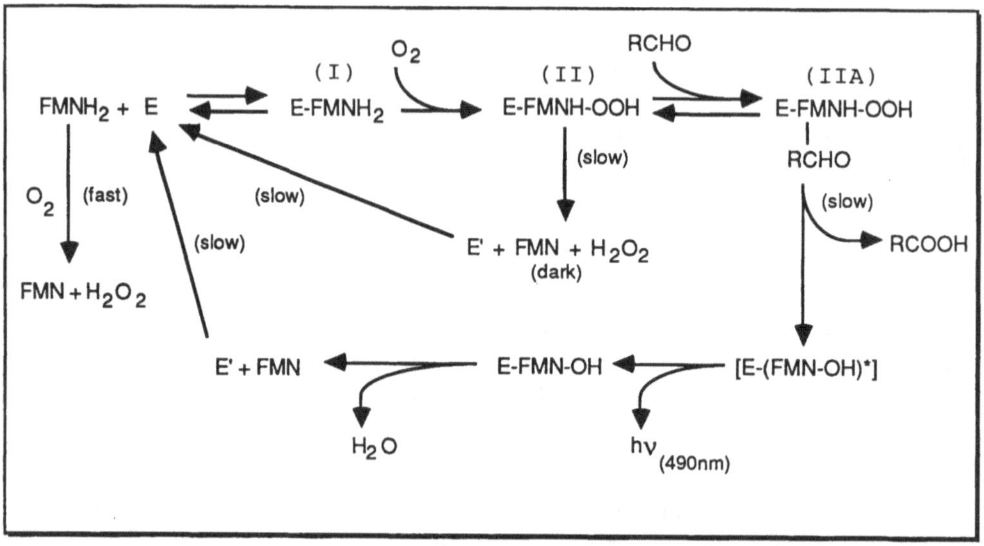

Figure 1. Reaction pathway in the bacterial bioluminescence reaction. The pathway as shown is that followed in the dithionite assay, in which the enzyme is incubated with $FMNH_2$ in the presence of dithionite; the reaction is initiated by injection of an air-equilibrated aqueous suspension of aldehyde.

The active center of bacterial luciferase must accommodate $FMNH_2$, O_2, and the aliphatic aldehyde. At this time, we do not have a crystal structure of the luciferase to assist in our understanding of the interactions between the enzyme and the substrates that result in the stabilization of the C4a peroxydihydroflavin or prevent quenching of the excited state. Our approach has been to rely on analysis of mutant forms of the enzyme in conjunction with chemical modification and limited proteolysis. About 6 years ago, we and others cloned the genes, *luxAB*, encoding the subunits of luciferase from *Vibrio harveyi* (10,11), and since then, we have determined the sequences of the genes, and therefore the amino acid sequences of the luciferase subunits, from *V. harveyi* (12,13), *V. fischeri*, *Photobacterium leiognathi*, and *P. phosphoreum*. We have cloned and sequenced the *luxAB* genes for several luciferase mutants of *V. harveyi* and have employed the techniques of oligonucleotide-directed site-specific mutagenesis and deletion mutagenesis to study the structure of the active center of the luciferase. This paper presents an overview of our current understanding of the structure of the active center and the role of other portions of the protein in achieving and maintaining the active structure of the enzyme.

Amino Acid Sequences of the α and β Subunits of Luciferases from *V. harveyi* and *V. fischeri*

The sequences of amino acids comprising the α and β subunits of the luciferases from *V. harveyi* and *V. fischeri* are presented in Fig. 2. The first and most striking observation to be made from this comparison is the homology, not only between *V. harveyi* and *V. fischeri*, but between the α and β subunits. As was suggested earlier on the basis of limited amino acid sequence data (14), it is clear that the two subunits arose by gene duplication. The alignment presented in Fig. 2 was adjusted to

```
Vfa-MET Lys Phe Gly Asn Ile Cys Phe Ser Tyr Gln Pro Pro Gly Glu Thr His Lys Gln Val Met Asp Arg Phe Val Arg Leu Gly Ile Ala Ser Glu Glu Val Gly         Phe Asp Thr
Vha-MET Lys Phe Gly Asn Phe Leu Leu Thr Tyr Gln Pro Pro Glu Leu Ser Gln Thr Glu Val Met Lys Arg Leu Val Asn Leu Gly Lys Ala Ser Glu Gly Cys Gly         Phe Asp Thr
Vhb-MET Lys Phe Gly Leu Phe Phe Leu Asn Phe Met Asn Ser Lys Arg Ser Ser Asp Gln Val Ile Glu Glu Met Leu Asp Thr Ala His Tyr Val Asp Gln Leu Lys         Phe Asp Thr
Vfb-MET Lys Phe Gly Leu Phe Phe Leu Asn Phe Gln Lys Asp Gly Ile Thr Ser Glu Glu Thr Leu Asp Asn Met Val Lys Thr Val Thr Leu Ile Asp Ser Thr Lys Tyr His Phe Asn Thr

Vfa-Tyr Trp Thr Leu Glu His His Phe Thr Glu Phe Gly Leu Thr Gly Asn Leu Phe Val Ala Ala Ala Gly Leu Leu Gly Arg Thr Lys Thr Leu Asn Val Gly Thr Met Gly Val Val Ile
Vha-Val Trp Leu Leu Glu His His Phe Thr Glu Phe Gly Leu Leu Gly Asn Pro Tyr Val Ala Ala Ala His Leu Leu Gly Ala Thr Glu Thr Leu Asn Val Gly Thr Ala Ala Ile Val Leu
Vhb-Leu Ala Val Tyr Glu Asn His Phe Ser Asn Asn Gly Val Val Gly Ala Pro Leu Thr Val Ala Gly Phe Leu Leu Gly Met Thr Lys Asn Ala Lys Val Ala Ser Leu Asn His Val Ile
Vfb-Ala Phe Val Asn Glu His His Phe Ser Lys Asn Gly Ile Val Gly Ala Pro Ile     Ala Ala Gly Phe Leu Leu Gly Leu Thr Asn Lys Leu His Ile Gly Ser Leu Asn Gln Val Ile

Vfa-Pro Thr Ala His Pro Val Arg Gln Leu Glu Asp Val Leu Leu Leu Asp Gln Met Ser Lys Gly Arg Phe Asn Phe Gly Thr Val Arg Gly Leu Tyr His Lys Asp Phe Arg Val Phe Gly
Vha-Pro Thr Ala His Pro Val Arg Gln Ala Glu Asp Val Asn Leu Leu Asp Gln Met Ser Lys Gly Arg Phe Arg Phe Gly Ile Cys Arg Gly Leu Tyr Asp Lys Asp Phe Arg Val Phe Gly
Vhb-Thr Thr His His Pro Val Arg Val Ala Glu Glu Ala Cys Leu Leu Asp Gln Met Ser Glu Gly Arg Phe Ala Phe Gly Phe Ser Asp Cys Glu Lys Ser Ala Asp Met Arg Phe Phe Asn
Vfb-Thr Thr His His Pro Val Arg Val Ala Glu Glu Ala Ser Leu Leu Asp Gln Met Ser Glu Gly Arg Phe Ile Leu Gly Phe Ser Asp Cys Glu Ser Asp Phe Glu Met Glu Phe Phe Lys

Vfa-Val Asp Met Glu Glu Ser Arg Ala Ile Thr Gln Asn Phe Tyr Gln Met Ile Met Glu Ser Leu Gln Thr Gly Thr Val Ser Ser Asp Ser Asp Tyr Ile Gln Phe Pro Asn Val Asp Val
Vha-Thr Asp Met Asp Asn Ser Arg Ala Leu Met Asp Cys Trp Tyr Asp Leu Met Lys Glu Gly Phe Asn Glu Gly Tyr Ile Ala Ala Asp Asn Glu His Ile Lys Phe Pro Lys Ile Gln Leu
Vhb-Arg Pro Thr Asp Ser Gln Phe Gln Leu Phe Ser Glu Cys His Lys Ile Ile Asn Asp Ala Phe Thr Thr Gly Tyr Cys His Pro Asn Asn Asp Phe Tyr Ser Phe Pro Lys Ile Ser Val
Vfb-Arg His Ile Pro Ser Arg Gln Gln Gln Phe Glu Ala Cys Tyr Glu Ile Ile Asn Asp Ala Leu Thr Thr Gly Tyr Cys His Pro Gln Asn Asp Phe Tyr Ile Phe Pro Lys Val Ser Ile

Vfa-Tyr Pro Lys Val Tyr Ser Lys Asn     Val Pro Thr Cys Met Thr Ala Glu Ser Ala Ser Thr Thr Glu Trp Leu Ala Ile Gln Gly Leu Pro Met Val Leu Ser Trp Ile Ile Gly Thr
Vha-Asn Pro Ser Ala Tyr Thr Gln Gly Gly Ala Pro Val Tyr Val Val Ala Glu Ser Ala Ser Thr Thr Glu Trp Ala Ala Glu Arg Gly Leu Pro Met Ile Leu Ser Trp Ile Ile Asn Thr
Vhb-Asn Pro His Ala Phe Thr Glu Gly Gly Pro Ala Gln Phe Val Asn Ala Thr Ser Lys Glu Val Val Glu Trp Ala Ala Lys Leu Gly Leu Pro Leu Val Phe Arg Trp Asp Asp Ser Asn
Vfb-Asn Pro His Cys Tyr Ser Asp Asn Gly Pro Lys Gln Tyr Val Ser Ala Thr Ser Lys Glu Val Val Met Trp Ala Ala Lys Lys Ala Ser Pro Leu Thr Phe Lys Trp Glu Asp     Asn

Vfa-Asn Glu Lys Lys Ala     Gln Met Glu Leu Tyr Asn Glu Ile Ala Thr Glu Tyr Gly His Asp Ile Ser Lys Ile Asp His Cys Met Thr Tyr Ile Cys Ser Val Asp Asp Asp Ala Gln
Vha-His Glu Lys Lys Ala     Gln Leu Asp Leu Tyr Asn Glu Val Ala Thr Glu His Gly Tyr Asp Val Thr Lys Ile Asp His Cys Leu Ser Tyr Ile Thr Ser Val Asp His Asp Ser Asn
Vhb-Ala Gln Arg Lys Glu     Tyr Ala Gly Leu Tyr His Glu Val Ala Gln Ala His Gly Val Asp Val Ser Gln Val Arg His Lys Leu Thr Leu Leu Val Asn Gln Asn Val Asp Gly Glu
Vfb-Leu Glu Thr Lys Glu Arg Tyr Ala Ile Leu Tyr Asn Lys Thr Ala Gln Gln Tyr Gly Val Asp Ile Ser Asp Val Asp His Gln Leu Thr Val Ile Ala Asn Leu Asn Ser Asp Arg Ser

Vfa-Lys Ala Gln Asp Val Cys Arg Glu Phe Leu Lys Asn Trp Tyr Asp Ser Tyr Val Asn Ala Thr Asn Ile Phe Asn Asp Ser Lys Gln Thr Arg Gly Tyr Asp Tyr His Lys Ala Gln Trp
Vha-Arg Ala Lys Asp Ile Cys Arg Asn Phe Leu Gly His Trp Tyr Asp Ser Tyr Val Asn Ala Thr Lys Ile Phe Asp Asp Ser Asp Gln Thr Lys Gly Tyr Asp Phe Asn Lys Gly Gln Trp
Vhb-Ala Ala Arg Ala Glu Ala Arg Val Tyr Leu Glu Glu Phe Val Arg Glu Ser Tyr Ser Asn Thr Asp Phe Glu Gln Lys Met Gly Glu Leu Leu Ser Glu Asn Ala
Vfb-Thr Ala Gln Glu Glu Val Arg Glu Tyr Leu Lys Asp Tyr Ile Thr Glu Thr Tyr Pro Gln Met Asp Arg Asp Glu Lys Ile Asn Cys Ile Ile Glu Glu Asn Ala

Vfa-Arg Asp Phe Val Leu Gln Gly His Thr Asn Thr Asn Arg Arg Val Asp Tyr Ser Asn Gly Ile Asn Pro Val Gly Thr Pro Glu Gln Cys Ile Glu Ile Ile Gln Arg Asp Ile Asp Ala
Vha-Arg Asp Phe Val Leu Lys Gly His Lys Asp Thr Asn Arg Arg Ile Asp Tyr Ser Tyr Glu Ile Asn Pro Val Gly Thr Pro Glu Glu Cys Ile Ala Ile Ile Gln Gln Asp Ile Asp Ala
Vhb-                                                                                            Ile Gly Thr Tyr Glu Glu Ser Thr Gln Ala Ala Arg Val Ala Ile Glu Cys
Vfb-                                                                                            Val Gly Ser His Asp Asp Tyr Tyr Glu Ser Ile Lys Leu Ala Val Glu Lys

Vfa-Thr Gly Ile Thr Asn Ile Thr Cys Gly Phe Glu Ala Asn Gly Thr Glu Asp Glu Ile Ile Ala Ser Met Arg Arg Phe Met Thr Gln Val Ala Pro Phe Leu Lys Glu Pro Lys•
Vha-Thr Gly Ile Asp Asn Ile Cys Cys Gly Phe Glu Ala Asn Gly Ser Glu Glu Glu Ile Ile Ala Ser Met Lys Leu Phe Gln Ser Asp Val Met Pro Tyr Leu Lys Glu Lys Gln•
Vhb-Cys Gly Ala Ala Asp Leu Leu Met Ser Phe Glu Ser Met Glu Asp Lys Ala Gln Gln Arg Ala Val Ile Asp Val Val Asn Ala Asn Ile Val Lys Tyr His Ser•
Vfb-Thr Gly Ser Lys Asn Ile Leu Leu Ser Phe Glu Ser Met Ala Asp Phe Lys Gly Val Lys Glu Ile Ile Asp Met Leu Asn Gln Lys Ile Glu Lys Asn Leu Pro•
```

Figure 2. Alignment of the amino acid sequences of the α (a) and β (b) subunits of the luciferases from Vibrio harveyi *(Vh) and* Vibrio fischeri *(Vf).*

maximize conservative alignments and to minimize the number of gaps required in the alignment. A gap was required, since the β subunits are shorter by about 30 residues than the α subunits, and to maximize conservative alignments, it was apparent that alignment of the C-terminal sequences of the α and β subunits was necessary.

The most highly conserved regions of multimeric enzymes are the regions comprising the substrate and effector binding sites and the intersubunit contact regions. We have compared the alignments presented in Fig. 2 and find three major regions within the α subunit that appear to be highly conserved, the N-terminal 5 residues, residues 40-120, and residues 175-340. There are two highly basic regions within the α subunits, residues 98-120 and 274-291, that contain the protease labile regions of the luciferase (15) and reside within the highly conserved regions. The reactive thiol of the luciferase from *V. harveyi* resides at position 106 (16); the analogous residue in the luciferase from V. *fischeri* is valine.

The β subunits also have long stretches of highly conserved amino acid sequence. These are the N-terminal 10 residues, residues 40-120 and residues 134-225. The C-terminal sequences of the β subunits appear to be quite different, in comparison to the central regions of the subunits. This point will be discussed later in this paper in the context of a series of β subunit C-terminal deletion mutants that we have constructed. It is also interesting to note that there are long stretches of sequence in the subunits that are identical, or nearly so, between all four polypeptides, especially the regions from residues 40-65 and residues 77-104.

The percent identity between the various pairwise combinations of subunits is presented in Table 1. The two α subunits are identical at 64% of the residues, while the β subunits are identical at 52% of the residues. All of the α/β subunit comparisons had about 30% identity. For reference, the α and β subunits of hemoglobins, which have virtually identical folding patterns, are identical at roughly 30% of the positions (17). An earlier comparison of the N-terminal 200 residues of the α and β subunits of the luciferase from *V. harveyi* showed that through this region preceding the apparent deletion in the β subunit, 36% of the positions were identical (18).

Analysis of Randomly Generated Mutants of the *V. harveyi* Luciferase

Cline and Hastings selected and characterized a large set of mutants having lesions in the *V. harveyi* luciferase structural genes resulting in altered bioluminescence decay kinetics, the altered kinetics (AK) mutants; all had lesions in the α subunits (19). We have taken advantage of these mutants and for our initial investigations of the structure of the active center, we have chosen two mutants, AK-6

Table 1. Percent Identity Between the Various Subunits Presented in Figure 2.

	V. harveyi α	*V. fischeri* α	*V. harveyi* β	*V. fischeri* β
V. harveyi α	100%			
V. fischeri α	64%	100%		
V. harveyi β	30%	26%	100%	
V. fischeri β	28%	30%	52%	100%

and AK-20. These mutants were selected on the basis of their dramatically altered enzymatic properties and their normal thermal stability, suggesting that the gross conformation of the proteins had not changed as a result of the lesions. The mutant AK-6 luciferase was reported to have normal aldehyde binding but very weak flavin binding, while the mutant AK-20 luciferase was reported to have normal flavin binding properties but dramatically altered aldehyde binding and aldehyde chainlength dependence on the intermediate IIA decay rate (19). By analysis of these two mutants, we hoped to gain information regarding both aspects of the active center, the flavin binding site and the aldehyde binding site.

Thermal denaturation studies with the luciferase from AK-6 have suggested that the gross conformation of the enzyme is not dramatically different from that of the wild-type enzyme. The quantum yield is, however, greatly reduced (19). The dissociation constant for the enzyme:$FMNH_2$ complex is about 1000 times the wild-type value, but the C4a peroxydihydroflavin is much more stable on the AK-6 enzyme than wild-type enzyme, especially at elevated values of pH. At pH 7.3, 25° with *n*-decanal as substrate, the intermediate IIA half-life was 69 sec for AK-6 and 3 sec for the wild-type enzyme (20). The bioluminescence emission spectrum for the AK-6 enzyme was red-shifted both *in vivo* and *in vitro* about 12 nm (21). We have cloned the *luxAB* genes from AK-6, and from the sequence of the *luxA* gene, we have shown that the lesion of AK-6 caused a substitution of Asn for Asp at position α113. By site-directed mutagenesis, we reverted the lesion of the AK-6 *luxA* gene to the wild-type sequence and regained wild-type luciferase, demonstrating that the Asp→Asn change was responsible for the unique properties of the AK-6 luciferase.

The Asp at position α113 is conserved in all luciferases that we have sequenced. To determine whether the Asp plays a specific role that is not fulfilled by the Asn, or the Asn has specific features that render the enzyme inactive, we incorporated a series of other amino acids into the α113 position by site-directed mutagenesis. The residues incorporated, Glu, Ser, Thr, and Lys, are listed in Table 2. None of the other residues functioned as well as the Asp; the quantum yield of the reaction catalyzed by the Glu substitution, ca. 10% of the wild-type enzyme, was the least affected of the series. The decay rates of the reactions catalyzed by the enzymes are presented in Table 2. The Lys mutant, αD113K, had such a low quantum yield that we have thus far been unable to determine with confidence the decay rates for the various aldehyde substrates. From the results of these experiments, we conclude that the Asp at α113 resides at or very near to the flavin binding pocket.

Table 2. Intermediate IIA Decay Rates of the Wild-Type Luciferase and α113 Mutants[a].

Enzyme [b]	*n*-octanal	*n*-decanal	*n*-dodecanal
Wild-Type	0.058	0.386	0.053
αD113E	0.057	0.175	0.035
αD113S	0.042	0.050	0.022
αD113T	0.012	0.031	0.006
αD113N	0.022	0.052	0.010
αD113K	N.D.	N.D.	N.D.

[a]*Decay rates are given in sec^{-1} and were determined at 25° in 50 mM phosphate, 0.2% BSA, pH 7.0 with 10μl of a 0.1% sonicated suspension of the indicated aldehyde.*
[b]*Enzymes were from* E. coli *carrying pUC-derived plasmids into which the* luxAB *genes had been cloned. The mutants were Glu (E), Ser (S), Thr (T), Asn (N; AK-6), and Lys (K). N.D.=not determined.*

The AK-6 enzyme appears to suffer a low quantum yield, not due to the specific properties of the Asn residue, but rather due to a specific role for the carboxylate of Asp that is not satisfied by the Asn and is only partially satisfied by the Glu. Whether the effects of these substitutions are due to direct interactions with the flavin or alterations in interactions that result from secondary level structural changes, such as loss of a stabilizing salt interaction, cannot be determined without more experimentation. The other mutant from the collection of Cline and Hastings that we have studied is AK-20. The luciferase from AK-20 has a dramatically altered aldehyde chainlength effect on the intermediate IIA decay rate. With the wild-type luciferase at pH 7.0, 25°, the decay rates are 0.068 sec^{-1}, 0.335 sec^{-1}, and 0.053 sec^{-1} for *n*-octanal, *n*-decanal and *n*-dodecanal, while for the same aldehydes and experimental conditions, the decay rates for the AK-20 enzyme are 0.200 sec^{-1}, 0.069 sec^{-1}, and 0.111 sec^{-1}. AK-20 has been shown to possess a lesion in the α subunit of the heterodimeric luciferase, resulting in a reduced bioluminescence activity, a slight enhancement of the $FMNH_2$ binding and decreased affinity for the aldehyde substrate (19). By a different approach, we have shown that the affinity of the AK-20 luciferase for aldehyde substrate is not greatly altered, but that the affinity for the second (inhibitory) aldehyde site is much less than that of the wild-type enzyme. We found that the $FMNH_2$ binding affinity of the AK-20 enzyme depended on the aldehyde chain length, consistent with the proposal that substrate binding may proceed by a random mechanism (22) rather than the commonly depicted ordered sequential reaction mechanism (Fig. 1). Chemical modification of the reactive sulfhydryl group indicated that although the substrate binding sites have been perturbed by the lesion in the α subunit, the reactivity and accessibility of Cys-106 were not greatly altered. The AK-20 luciferase was more sensitive than the wild type to inactivation by trypsin or chymotrypsin in Bis-Tris, and the extent of protection by phosphate was less for the AK-20 luciferase than for the wild type. Despite the increased susceptibility to proteolytic inactivation, the AK-20 luciferase was more stable (E_{act} = 77.1 kcal/mol) than the wild type (E_{act} = 63.0 kcal/mol) to thermal inactivation.

We have cloned and determined the sequence of the *luxAB* genes from AK-20 and found the only lesion to be a substitution of Phe for Ser at α227. As with the AK-6 mutant, we have confirmed the lesion by reversion to the wild-type sequence. To further investigate the functional significance of the AK-20 lesion, we constructed variants with the following substitutions at α227: Cys, Leu, Tyr, Asn, Gly and Asp. All of these variants show the normal "slow-fast-slow" kinetic pattern with the 8, 10 and 12 carbon aldehyde substrates, and all have quantum yields close to that of the wild-type enzyme. For example, the Asn substitution has intermediate IIA decay rates of 0.078 sec^{-1}, 0.395 sec^{-1}, and 0.057 sec^{-1} for *n*-octanal, *n*-decanal and *n*-dodecanal, determined under the same conditions described in Table 2. The quantum yield of the reaction catalyzed by the Asn variant with *n*-octanal as substrate was 4 times the quantum yield of the wild-type enzyme assayed under identical conditions (23).

We conclude from these experiments that while position α227 is close to or within the aldehyde binding pocket, and therefore close to the flavin binding pocket, the effect of the Ser→Phe is due more to the properties of the Phe than to loss of any critical features of the Ser. The other luciferases that we have sequenced have Thr at that position, and even Tyr at α227 had intermediate IIA decay rates virtually indistinguishable from the wild type. The specific effects of Phe on the binding of the second (inhibitory) molecule of aldehyde to the enzyme, as well as the effect of the other substitutions on the aldehyde substrate inhibition phenomenon, are currently under investigation.

There is a conserved His at α224, in close proximity to the α227 site. This residue was of interest due to its proximity to the α227 and the observation by Cousineau and Meighen that the enzyme is inactivated by ethoxyformic anhydride in a fashion suggestive of an essential histidinyl residue on the α subunit in or near the active center (24). We have changed the His codon to a Leu codon at α224; the resulting enzyme appears to be unaltered, suggesting that the His at α224 was not the target of the experiments of Cousineau and Meighen.

The most often modified residue in bacterial luciferase has been the highly reactive cysteinyl residue at position α106 of the *V. harveyi* enzyme (15,25-28). In spite of extensive efforts to demonstrate that the thiol was not required in the bioluminescence reaction, we have been unable to modify the enzyme, even by incorporation of an $-SCH_3$ group, without completely inactivating the enzyme (16). On chemical model grounds, it has been proposed that the sulfhydryl plays a catalytic role (29). That proposal was put into serious question when we found that all the other luciferases have Val at α106. The luciferase from *V. fischeri*, even though it lacks the highly reactive thiol at α106, is nonetheless inactivated by thiol directed reagents. The inactivation of the luciferase from *V. fischeri* by N,*n*-alkylmaleimides showed the same chainlength dependence as did the enzyme from *V. harveyi*, even though the absolute rates of inactivation were much slower for the *V. fischeri* enzyme, suggesting that the *V. fischeri* enzyme also has a cysteinyl residue residing in a hydrophobic cleft that is near the active center (30).

To more thoroughly understand the role of the cysteinyl residue at position α106 in the structure of the active center, we constructed a variant with Ser at position α106. The inactivation of the variant by reaction with N-ethylmaleimide was much slower than for the wild-type enzyme, although the rate was still faster than for the enzyme from *V. fischeri* (Fig. 3). The results of the experiment presented

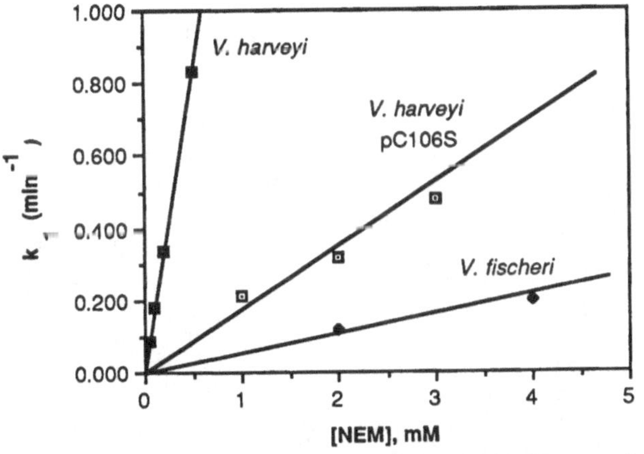

Figure 3. Pseudo-first order rate constants for inactivation of luciferases from wild-type V. harveyi *(26), pC106S mutant, and* V. fischeri *(30) as a function of N-ethylmaleimide concentration. Luciferase from the pC106S mutant (final concentration 0.27 μM) was mixed with different concentrations of N-ethylmaleimide in 20 mM phosphate, pH 7.08, at 25°. Aliquots were removed at various times and diluted 1:200 into assay buffer to stop the reaction; activity remaining was determined by the standard $FMNH_2$ injection assay , and pseudo-first order rate constants for inactivation were determined from plots of % activity remaining vs. time.*

Table 3. Decay Rates of *V. harveyi* Wild-Type and pC106S Mutant Intermediates IIA[a].

		k_1 (sec^{-1})	
Enzyme	*n*-octanal	*n*-decanal	*n*-dodecanal
wild-type	0.0432	0.304	0.0416
pC106S	0.0958	0.389	0.0604

[a]*Decay rates were determined in the standard $FMNH_2$ injection assay (with BSA), using aldehyde concentrations giving maximal I_o for each enzyme with each aldehyde.*

in Fig. 3 confirm that the cysteine at α106 is the reactive cysteine in the *V. harveyi* enzyme. This experiment further demonstrates the existence of at least one other cysteinyl residue that is reactive and, if modified by reaction with N-ethylmaleimide, results in inactivation of the enzyme. This observation is very interesting, since there are several reports in the literature of attachment of probes to the reactive cysteine for spectroscopic studies (31,32). The observation of other reactive cysteinyl residues on the luciferase brings into question whether these various probes were all attached to the cysteinyl residue at α106.

To further investigate the properties of the α106S mutant, we have studied the kinetics of the intermediate IIA decay and measured the substrate binding constants. The intermediate IIA decay rates were measured for the mutant and for the wild type under identical conditions and are presented in Table 3. The α106S mutant enzyme has a faster rate of decay with all the aldehydes tested. In the process of making these measurements, we found that not only was the α106S mutant active, but it appeared to be less sensitive to aldehyde substrate inhibition than the wild type, having a higher quantum yield than the wild type with optimal aldehyde concentrations. The aldehyde binding profiles for both the wild type and the mutant are presented in Fig. 4. The K_m for *n*-decanal with the α106 serine mutant was calculated from the data in Fig. 4 to be 7.5 μM, somewhat

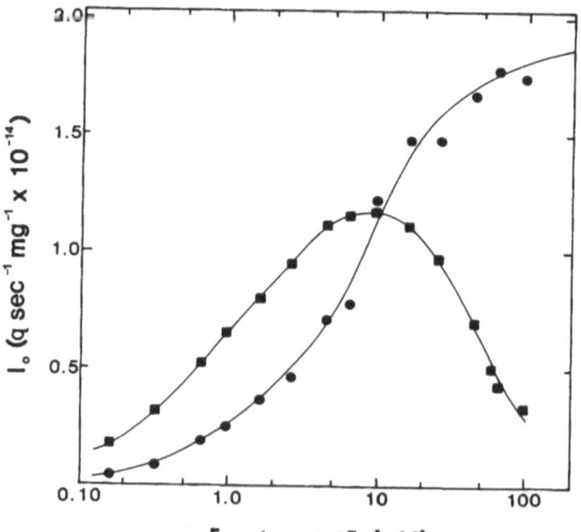

Figure 4. Initial light intensity (activity) of V. harveyi wild-type (■) and the αC106S mutant (pC106S, ●) luciferases as a function of n*-decanal concentration. Standard $FMNH_2$ injection assays were performed in 200 mM phosphate, pH 7.0 (no BSA) following ~15 s incubation of the enzyme (0.14 μM) at the specified aldehyde concentration. Stock solutions of* n*-decanal were prepared in N_2-saturated 200 mM phosphate.*

higher than the K_d calculated for the first molecule of *n*-decanal bound to the wild-type enzyme under these conditions, 4.3 µM, after correction for substrate inhibition (22). However, the most striking difference between the wild-type and the mutant enzyme apparent from Fig. 4 is that the mutant is not subject to aldehyde inhibition. It is noteworthy in this regard that the luciferases from *V. fischeri*, *P. phosphoreum*, and *P. leiognathi*, which all have valine at α106, are similarly insensitive to aldehyde substrate inhibition, suggesting that the cysteine at α106 of wild-type *V. harveyi* may be intimately involved in the substrate inhibition process.

The α106S mutant has several other very interesting properties. It has normal thermal stability, and decreased affinity for the flavin substrate, as well as for the aldehyde substrate. The K_d for $FMNH_2$, determined by the dithionite assay technique, was 23 µM, compared with 0.3 µM for the wild-type enzyme determined in 50 mM phosphate, pH 7.0, 25°.

The inactivation of the *V. harveyi* luciferase by alkylating reagents appears to be due to steric blocking of the flavin binding. The $-SCH_3$ mixed disulfide formed by reaction of the α106 cysteinyl residue with methyl methanethiolsulfonate is larger than the Val side chain found at α106 in the luciferases from other species. The aldehyde substrate inhibition appears to be due to nonproductive binding of aldehyde in the flavin binding pocket, since the enzyme:$FMNH_2$ complex is insensitive to aldehyde substrate inhibition (33). If by substitution of the Ser for Cys at α106, we have reduced the affinity of the aldehyde for the nonproductive binding site in the flavin pocket, the result would be to relieve the enzyme of the inhibitory effect of the aldehyde substrate. The effect of the Ser residue on the affinity of the enzyme for the flavin substrate is likely due to the hydrogen bonding potential of the serine hydroxyl. By comparison, sulfhydryl groups are hydrophobic and do not tend to donate or accept H-bonds. The sulfhydryl group is in a hydrophobic pocket (26,31), so it is more likely that the Ser hydrogen bonds to the peptide backbone than to functional groups within the hydrophobic pocket.

Structural Features Affecting Dimer Formation

We have recently constructed a series of C-terminal deletions in the β subunit of the luciferase from *V. harveyi*. The purpose of these experiments was to determine the role of the C-terminal region of the β subunit, which is one of the least conserved regions of the β subunit (Fig. 2), in stabilizing the heterodimeric structure of the luciferase. The β subunit C-terminal amino acid sequences of the various constructions are given in Fig. 5.

Luciferase produced from plasmid pJS23 was indistinguishable from wild-type luciferase, indicating that the C-terminal sequence of the β subunit was unimportant. Cells carrying plasmid pJS12 produced large amounts of active luciferase when grown at 20°, but very little at 37°, suggestive of thermal instability. However, when the luciferase was purified from cells carrying pJS12 grown at 20°, the luciferase was found to have normal thermal stability, protease lability, flavin binding affinity,

										•320						
Wild-Type	Val	Ile	Asp	Val	Val	Asn	Ala	Asn	Ile	Val	Lys	Tyr	His	Ser	•	
pJS 23	Val	Ile	Asp	Phe	Cys	Ala	Arg	Cys	Asp	Val	Phe	Leu	Leu	Asn	Arg	•
pJS 12	Val	Ile	Ala	Met	Trp	•										
pJS 2	Val	Ile	Gly	•												

Figure 5. C-terminal variants of the β subunit of V. harveyi *luciferase. All variants were constructed by deletion of sequences at the 3' end of the* luxB *gene.*

intermediate IIA decay rate and quantum yield. The only apparent deficiency in the enzyme is that, while it refolds from urea as well as the wild-type enzyme at 20°, it is deficient in refolding from urea at 37°. That is, the luciferase encoded by pJS12 appears to be a temperature dependent protein folding mutant. Cells carrying pJS2 produce only about 2% as much luciferase at 20° as cells carrying the wild-type plasmid. The luciferase produced from pJS2 has normal properties, like the pJS12-encoded enzyme, except that the pJS2 enzyme appears to be folding deficient at 20° as well as at 37°. Thus it appears that the C-terminal sequence of the β subunit has little to do with the stability of the final αβ dimeric structure of the luciferase. The length of the protein appears to be critical in the ability of the subunit to dimerize with the α subunit, but once the dimer has formed, the C-terminus is not important.

In a related series of experiments studying the formation of active luciferase *in vivo* from two separate plasmids, one carrying the α subunit gene and the other carrying the β subunit gene, we have found that the initial encounter complex between the two subunits must be as partially folded intermediates (34). Subunits produced and folded in separate cells do not dimerize unless they are first denatured with urea. Only if the subunits are produced in the same cell, such that they can interact during the folding process, will they dimerize properly to form active luciferase.

Summary

Since the initial cloning of the bacterial luciferase genes, much has been learned of the structure of the enzyme. The luciferases from four different species have been sequenced and numerous mutants have been constructed and partially characterized. Access to the cloned genes has even allowed the initial inspection of the processes of protein folding and dimerization *in vivo*. A high resolution crystal structure is of obvious importance to our ultimate understanding of the properties of this enzyme. Solution of the structure of the enzyme is currently underway in the laboratory of Prof. Brian Matthews.

Acknowledgments

This work was supported in part by grants from the National Science Foundation (NSF-DMB 85-10784), the Robert A. Welch Foundation (A865), and the Texas Advanced Technology Research Program.

References

1. Nealson, K. H. (personal communication).
2. Engebrecht, J., K. Nealson and M. Silverman. 1983. Cell 32, 773-781.
3. Ziegler, M. M. and T. O. Baldwin. 1981. Curr. Top. Bioenerg. 12, 65-113.
4. Wall, L. and E. A. Meighen. 1986. Biochemistry 25, 4315-4321.
5. Rodriguez, A., L. Wall, D. Riendeau and E. Meighen. 1983. Biochemistry 22, 5604-5611.
6. Kürfurst, M., S. Ghisla and J. W. Hastings. 1984. Proc. Natl. Acad. Sci. U.S.A. 81, 2990-2994.
7. O'Kane, D. J., V. A. Karle and J. Lee. 1985. Biochemistry 24, 1461-1467.

8. Leisman, G. and K. H. Nealson. 1982. In: Flavins and Flavoproteins (V. Massey and C. Williams, eds.). Elsevier, Amsterdam. pp. 383-386.

9. Daubner, S. C., A. M. Astorga, G. Leisman and T. O. Baldwin. 1988. This volume.

10. Baldwin, T. O., T. Berends, T. A. Bunch, T. F. Holzman, S. K. Rausch, L. Shamansky, M. L. Treat and M. M. Ziegler. 1984. Biochemistry 23, 3663-3667.

11. Belas, R., A. Mileham, D. Cohn, M. Hilmen, M. Simon and M. Silverman. 1982. Science 218, 791-793.

12. Cohn, D. H., A. J. Mileham, M. I. Simon, K. H. Nealson, S. K. Rausch, D. Bonam and T. O. Baldwin. 1985. J. Biol. Chem. 260, 6139-6146.

13. Johnston, T. C., R. B. Thompson and T. O. Baldwin. 1986. J. Biol. Chem. 261, 4805-4811.

14. Baldwin, T. O., M. M. Ziegler and D. A. Powers. 1979. Proc. Natl. Acad. Sci. U.S.A. 76, 4887-4889.

15. Rausch, S. K. and T. O. Baldwin. 1988. Submitted for publication.

16. Ziegler, M. M., M. V. Merritt, S. K. Rausch, J. J. Dougherty, Jr. and T. O. Baldwin. 1988. Submitted for publication.

17. Dayhoff, M. O. 1972. Atlas of Protein Sequence and Structure, Vol. 5. National Biomedical Research Foundation, Silver Spring, MD.

18. Baldwin, T. O., T. C. Johnston and R. F. Swanson. 1984. In: Flavins and Flavoproteins (R. C. Bray, P. C. Engel and S. G. Mayhew, eds.). Walter de Gruyter, Berlin. pp. 345-358.

19. Cline, T. W. and J. W. Hastings. 1972. Biochemistry 11, 3359-3370.

20. Cline, T. W. 1973. Ph.D. thesis, Harvard University, Cambridge, MA.

21. Cline, T. W. and J. W. Hastings. 1974. J. Biol. Chem. 249, 4668-4669.

22. Holzman, T. F. and T. O. Baldwin. 1983. Biochemistry 22, 2838-2846.

23. Baldwin, T. O., J.-W. Lin, T. C. Johnston, L. H. Chen, L. J. Chlumsky and M. M. Ziegler. 1987. In: Bioluminescence and Chemiluminescence (J. Schölmerich, ed.). John Wiley & Sons, Berlin New York. in press.

24. Cousineau, J. and E. A. Meighen. 1976. Biochemistry 15, 4993-5000.

25. Ziegler-Nicoli, M., E. A. Meighen and J. W. Hastings. 1974. J. Biol. Chem. 249, 2385-2392.

26. Ziegler-Nicoli, M. and J. W. Hastings. 1974. J. Biol. Chem. 249, 2393-2396.

27. Tu, S.-C. and J. Henkin. 1983. Biochemistry 22, 519-523.

28. Fried, A. and S.-C. Tu. 1984. J. Biol. Chem. 259, 10754-10759.

29. Kosower, E. M. 1980. Biochem. Biophys. Res. Commun. 92, 356-364.

30. Currie, M. S. 1974. Undergraduate thesis, Harvard University, Cambridge, MA.

31. Merritt, M. V. and T. O. Baldwin. 1980. Arch. Biochem. Biophys. 202, 499-506.

32. Tu, S.-C., C.-W. Wu and J. W. Hastings. 1978. Biochemistry 17, 987-993.

33. Meighen, E. A. and R. E. MacKenzie. 1973. Biochemistry 12, 1482-1491.

34. Waddle, J. J., T. C. Johnston and T. O. Baldwin. 1987. Biochemistry (in press).

Sequence Similarity between the α Subunit of Luciferase from *Vibrio harveyi* and Clostridial Flavodoxin

M. Murata and John Lee

Department of Biochemistry, School of Chemical Sciences
University of Georgia, Athens, Ga 30602

Introduction

The complete amino acid sequences of the α (1) and β (2) subunits of the luciferase from *Vibrio harveyi* have been reported. Although sequence homology between the N-terminal regions of the subunits was suggested previously (3), no comparison has been made on the similarity, or the lack of it, of the entire subunit sequences. Here we present the comparison of the entire luciferase sequences. To find the flavin binding site, the subunit sequences were compared with those from other flavin binding proteins whose three dimensional structures were known.

Results

Fig. 1 shows the alignment of the α and β subunit sequences of luciferase from *Vibrio harveyi*. Of the 324 residues aligned, 102 residues (31%) are matched by identical amino acids. The standardized score for this alignment is 24.48 indicating that the similarity between the sequences is statistically highly significant. The alignment in Fig. 1 produces in the β subunit a single gap of 29 residues, between Thr 258 and Asp 259. The high degree of sequence similarity observed undoubtedly reflects the overall structural similarity between the two subunits, with the exception of the region corresponding to the particularly long deletion in the β subunit. It is generally

Flavins and Flavoproteins 1987

```
           10        20        30        40        50        60        70
            +         +         +         +         +         +         +
Lα MKFGNFLLTYQPPELSQTEVMKRLVNLGKASEGCGFDTVWLLEHHFTEFGLLGNPYVAAAHLLGATETLNVG
Lß MKFGLFFLNFMNSKRSSDQVIEEMLDTAHYVDQLKFDTLAVYENHFSNNGVVGAPLTVAGFLLGMTKNAKVA
   .... . .       .   .               ...    . ..   .  . .   .  ...      .

         80        90       100       110       120       130       140
          +         +         +         +         +         +         +
Lα TAAIVLPTAHPVRQAEDVNLLDQMSKGRFRFGICRGLYDKDFRVFGTDMDNSRALMDCWYDLMKEGFNEGYI
Lß SLNHVITTHHPVRVAEEACLLDQMSEGRFAFGFSDCEKSADMRFFNRPTDSQFQLFSECHKIINDAFTTGYC
       .  . .... ..   ...... ... ..        . .  .    .    .           .  ..

      150       160       170       180       190       200       210
        +         +         +         +         +         +         +
Lα AADNEHIKFPKIQLNPSAYTQGGAPVYVVAESASTTEWAAERGLPMILSWIINTHEKKAQLDLYNEVATEHG
Lß HPNNDFYSFPKISVNPHAFTEGGPAQFVNATSKEVVEWAAKLGLPLVFRWDDSNAQRKEYAGLYHEVAQAHG
      .    ....  .. . . ..    . . .    ....  ...    .       .    .. ...  ..

    220       230       240       250       260       270       280
      +         +         +         +         +         +         +
Lα YDVTKIDHCLSYITSVDHDSNRAKDICRNFLGHWY-DSYVNATKIFDDSDQTKGYDFNKGQWRDFVLKGHKD
Lß VDVSQVRHKLTLLVNQNVDGEAARAEARVYLEEFVRESYSNT-----------------------------D
    ..    . .        .   .    .  .      .. .                              .

  290       300       310       320       330       340       350
    +         +         +         +         +         +         +
Lα TNRRIDYSYEINPVGTPEECIAIIQQDIDATGIDNICCGFEANGSEEEIIASMKLFQSDVMPYLKEKQ
Lß FEQKMGELLSENAIGTYEESTQAARVAIECCGAADLLMSFESMEDKAQQRAVIDVVNANIVKYHS
              .  .. ..        .   .       ..         .           .
```

Fig. 1: Sequence alignment of the α and ß subunits of luciferase. Sequence comparisons were made on a VAX-11/750 using the programs based on the algorithm of Needleman and Wunsch (4). The weighting system of the mutation data matrix by Dayhoff et al. (5) was used. The standardized alignment score was calculated as the number of standard deviations by which the score of the real sequences differs from the mean score of 200 scrambled sequences.

accepted that the active site of luciferase resides exclusively on the α subunit, although the presence of the ß subunit is required for the enzymatic reaction (6). If this functional distinction of the two subunits is correlated to the structural difference, the alignment shown in Fig. 1 suggests that the region of the α subunit corresponding to the large gap in the ß subunit is involved in the catalytic activity of the enzyme. It is noted that one of the cleavage sites which produce the

the proteolytic labile fragments has been located within this region at residue 280 (Phe) (1).

The subunit sequences of luciferase were compared with those from other flavin binding proteins including glutathione reductase, p-hydroxybenzoate hydroxylase, D-amino acid oxidase and four flavodoxins; no sequence similarity was found. However, the sequence of Clostridial flavodoxin (7) was found to be statistically (i.e. standardized score > 3) similar to a segment of the α subunit of luciferase. The segment corresponds to residues 190-318. The alignment of this segment with the flavodoxin is shown in Fig. 2. The standardized score obtained was 5.69. Of the 126 aligned residues, 28 residues were matched (22%). Note that the region of the α subunit (residues 258-286, corresponding to the gap in the β subunit) whose possible involvement in the catalytic activity was mentioned above is enclosed in this part of the aligned sequences. The sequence similarity observed here is particularly noteworthy in view of the observed similarity of the Raman spectrum of FMN bound to these two proteins (8).

In the crystal structure of clostridial flavodoxin (9), the

```
   190        200       210       220       230       240           250
    +          +         +         +         +         +             +
Lα  M-ILSWIINTHEKKAQLDLYNEVATEHGYDVTKIDHCLSYITSVDHDSNRAKDICRNF----LGHW-YDSY
    . .  .       .    .      . . ..  .      .             .       .
FD  MKIVYWSGTGNTEKMA-ELIAKGIIESGKDVNTINVSDVNIDELLNEDILILG-CSAMGDEVLEESEFEPF
             +          +         +         +         +          +
             10         20        30        40        50         60

          260        270        280         290       300       310
           +          +          +           +         +         +
Lα  VN--ATKIFDDSDQTKG-YDFNKGQW-RDFV--LKGHKDTNRRIDYSYEINPVGTPEECIAIIQQDIDAT
         ...        . .    . . ...     .               .      ..      .
FD  IEEISTKISGKKVALFGSYGWGDGKWMRDFEERMNGYCVVVETPLIVQNEPDEAEQDCIEF-GKKIANI
    +         +         +         +         +         +           +
   70        80        90        100       110       120         130
```

Fig. 2: Sequence alignment of the α subunit of luciferase (Lα) and flavodoxin from Clostridium MP (FD).

flavin binding site is located at one end of the molecule surrounded by the two loops formed by residues 56-59 and residues 89-91. The isoalloxazine ring is sandwiched by Met 56 and Trp 96. The alignment shown in Fig. 2 places phenylalanyl residues at the corresponding residues in the luciferase sequence. Two tryptophan residues, Trp 6 and Trp 95 of the flavodoxin, which are near the flavin binding site are also seen in the α subunit. The alignment shows a concentration of matched residues (residues 86-98) around one of the flavin binding loops in flavodoxin.

Acknowledgments

Supported by grants from NSF DMB85-12361 and NIH GM28139

References

1. Cohn, D.H., A.J. Mileham, M.I. Simon, K.H. Nealson, S.K. Rausch, D. Bonam, T.O. Baldwin. 1985. J. Biol. Chem. 260, 6139.

2. Johnston, T.C., R.B. Thompson, T.O. Baldwin. 1986. J. Biol. Chem. 261, 4805.

3. Baldwin, T.O., M.M. Ziegler, D.A. Powers. 1979. Proc. Natl. Acad. Sci. (USA) 76, 4887.

4. Needleman, S.B., C.D. Wunsch. 1970. J. Mol.Biol. 48,443.

5. Dayhoff, M.O., W.C. Baker, L.H. Hunt. 1983. Meth. Enzymol. 91, 524.

6. Ziegler, M.M., T.O. Baldwin. 1981. Curr. Top. Bioenerg. 12, 65.

7. Tanaka, M., M. Haniu, K.T. Yashunobu, S.G. Mayhew. 1974. J. Biol. Chem. 249, 4393.

8. Visser, A.J.W.G., J. Vervoort, D.J. O'Kane, J. Lee, L.A. Carreira. 1983. Eur. J. Biochem. 131, 639.

9. Burnett, R.M. G.D. Darling, D.S. Kendall, M.E. LeQuesne, S.G. Mayhew, W.W. Smith, M.L. Ludwig. 1974. J. Biol. Chem. 249, 4383.

ISOLATION AND CHARACTERIZATION OF A YELLOW FLUORESCENT PROTEIN FROM *VIBRIO FISCHERI* STRAIN Y-1

Susan Colette Daubner, Adrienne M. Astorga, Gary Leisman, and Thomas O. Baldwin

Department of Biochemistry and Biophysics, Texas A&M University, and the Texas Agricultural Experiment Station, College Station, Texas 77843

Introduction

A decade ago, a yellow light-emitting bacterium, *Vibrio fischeri* strain Y-1, was isolated from seawater off the California coast. This isolate was discovered to possess a bimodal emission spectrum with maxima of 545 nm and 500 nm *in vivo* (1). The emission of yellow light, unique to Y-1, was later determined to be due to a secondary emitter, the yellow fluorescent protein (YFP) (2). When added to luciferase *in vitro* YFP shifts the emission maximum from 485 to 535 nm, with a shoulder at 485 nm. Leisman and Nealson (2) found that the interaction between YFP and luciferase is temperature-sensitive both *in vitro* and *in vivo*, and is affected by pH and aldehyde chain length. Here we present a purification procedure for the yellow fluorescent protein in its active form.

Methods

Absorbance spectra were recorded on a Kontron Uvikon 810 spectrophotometer. Fluorescence and bioluminescence spectra were recorded on a SLM fluorometer. Luciferase assays coupled to FMN oxidoreductase were performed as described by Drueckhammer *et al.* (3). Protein concentrations were determined with the BioRad protein assay using carbonic anhydrase as a standard, and by absorbance at 280 nm. Protein determinations with the BioRad assay were found to agree well with values obtained with the Lowry method (4). Electrophoretic analysis of protein samples employed slabs of 12 % polyacrylamide (5). Gels were stained in 10% acetic acid, 10% isopropyl alcohol, 0.05% Coomassie Blue, and destained with the same solution omitting the blue dye. Calibration curves were constructed using bovine serum albumin, ovalbumin, glyceraldehyde-3-P-dehydrogenase, carbonic anhydrase, trypsinogen, soybean trypsin inhibitor, and α-lactalbumin.

Strain Y-1 was grown in 1.5 liters sodium chloride complete medium at 18 °C in 2.8 liter flasks or in a 16 liter fermentor in the same medium. The cells were harvested at late log phase and stored as pellets at -20 °C. Purification steps were conducted at 4 °C. Frozen cells were thawed and lysed by osmotic shock in 10 mM EDTA, 1 mM DTT, pH 7.0. Cell debris was removed by centrifugation. The fraction precipitating between 40 and 70% saturated ammonium sulfate was resuspended in and dialyzed against 30 mM sodium phosphate, 1 mM EDTA, 1 mM DTT, pH 7.0. DE 52 (DEAE-cellulose) was

Flavins and Flavoproteins 1987

added to the ammonium sulfate precipitate and allowed to equilibrate. The unbound material was applied to an Ultrogel AcA 54 column and eluted with 30 mM sodium phosphate, 1 mM EDTA, 1 mM DTT, 2% glycerol, pH 7.0. The YFP fraction was identified both by flourescence and by molecular weight on 12 % polyacrylamide-SDS gels, and was concentrated against sucrose. The sample was dialyzed *vs* 30 mM sodium phosphate, 1 mM EDTA, 1 mM DTT, 10% glycerol, pH 7.0, and applied to a column of DEAE-Sephacel. Pure YFP was eluted with a 30-250 mM sodium phosphate gradient.

Results and Discussion

Figures 1a and 1b show absorbance spectra of the pooled yellow fractions resulting from chromatography of Y-1 lysate on Ultrogel and DEAE-Sephacel respectively. On SDS-PAGE YFP appears to be about 90% pure after Ultrogel and homogeneous after DEAE-Sephacel. The spectrum in Figure 1a is obscured by the presence of a cytochrome (to be discussed below), but the spectrum in Figure 1b is that of a typical flavoprotein, with peaks at 384 and 454 nm.

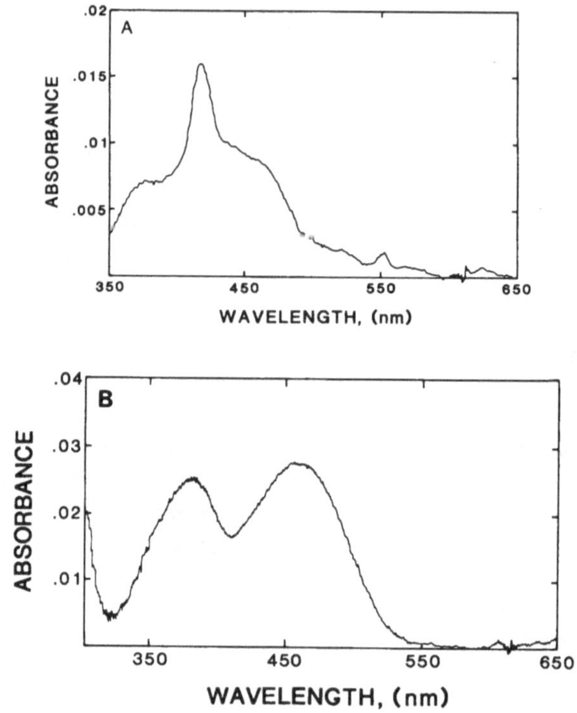

Figure 1a. Absorbance spectrum of a typical elution fraction from an Ultrogel AcA 54 column. The sample is in 30 mM sodium phosphate, 1 mM EDTA, 1 mM DTT, 2 % glycerol, pH 7.0.
Figure 1b. Absorbance spectrum of fluorescent fractions eluting off a DEAE-Sephacel column. The sample is in about 50 mM sodium phosphate, 1 mM EDTA, 1 mM DTT, 10 % glycerol, pH 7.0.

The presence of YFP alters the emission spectrum of luciferase. Luciferase was mixed with the FMN oxidoreductase reaction mixture (FMN, NADH, enzyme), and aldehyde, and showed a typical bioluminescence emission spectrum, with a peak at 485 nm (Figure 2a). When YFP was added to the same mixture, a new peak developed at 535 nm (Figure 2b). Strain Y-1 contains another yellow fluorescent protein, which elutes in the void volume of the Ultrogel AcA 54 column, but does not alter the bioluminescence emission spectrum of luciferase. The other yellow protein does not co-elute with luciferase.

YFP elutes from Ultrogel AcA 54 with an apparent molecular weight of 44,000. On SDS gels it has an apparent molecular weight of 22,000, suggesting a dimeric structure for the holoprotein.

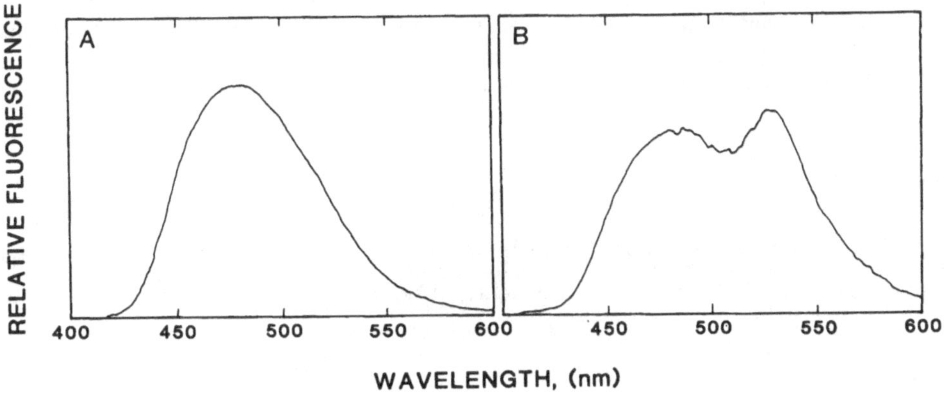

Figure 2. Bioluminescence emission spectra of luciferase in the presence and absence of YFP. Luciferase assay was coupled to FMN oxidoreductase and carried out in the fluorimeter with the excitation lamp turned off. The spectrum in panel A is the control, 1.5 μM luciferase, and the spectrum in panel B shows the result of adding 6 μM YFP. Total light in panel B is about 75% that in panel A. The assay was carried out at pH 7.2 in 100 mM phosphate, at 18 °C.

Table 1. Purification Parameters for YFP Isolation

sample	415/445	280/445	mg/ml/445
Fernbach-grown cells			
NH_4SO_4 ppt	1.7	120.4	26.1
DEAE filtrate	1.8	37.0	16.3
Ultrogel eluate	2.1	13.6	5.9
DEAE-Sephacel eluate	0.67	6.4	2.7
Fermentor-grown cells			
NH_4SO_4 ppt	2.1	213.0	25.0
DEAE filtrate	2.6	49.4	40.0
Ultrogel eluate	5.3	33.2	15.8

Strain Y-1 contains a cytochrome c which is difficult to remove from YFP; the molecular weight of the cytochrome is about 46,000, so it chromatographs closely on Ultrogel and can be seen in Figure 1a as a large peak at 418 nm and a small peak at 550 nm. When Y-1 is grown in Fernbach flasks with about 170 rpm of agitation the cell lysate contains much less of the cytochrome than if the cells are grown in the fermentor (60-180 rpm, 15-45 l/min aeration). Table 1 displays the spectral parameters used in assessing the separation of YFP from the cytochrome (A_{415}/A_{445}) and from other proteins (A_{280}/A_{445} and mg per ml/A_{445}). A_{415}/A_{445} is lower at all intermediate steps for Y-1 grown in flasks, reflecting the lower level of cytochrome in these cells.

Purified YFP was subjected to Edman sequence analysis. The N-terminal sequence was determined to be the following: Met-Phe-Lys-Gly-Ile-Val-Glu-Gly-Ile-Gly-Ile-Ile-Glu-Lys-Ile. Using the N-terminal sequence we have constructed an oligonucleotide for the cloning of YFP. Once YFP has been cloned into *E. coli* larger amounts of protein will be available for the study of the mechanism by which YFP and luciferase interact in the emission of yellow light.

Acknowledgements

This work was supported in part by grants from the National Science Foundation (NSF DMB 85-10784), the Robert A. Welch Foundation (A865), and the Texas Advanced Technology Research Program. Special thanks to Dr. Finn Wold, University of Texas Health Science Center in Houston, for the N-terminal sequence analysis.

References

1. Ruby, E. G., and Nealson, K. H. 1977. Science 196, 432.

2. Leisman, G. and Nealson, K. H. 1982. In: Flavins and Flavoproteins, (V. Massey and C.H. Williams, Jr., eds.)Elsevier/North Holland, Amsterdam, pp. 383.

3. Drueckhammer, D. G., Riddle, V. W., and Wong, C.-H. 1985. J. Org. Chem. 50, 5387.

4. Lowry, O. H., Rosebrough, N. J., Farr, A. L., and Randall, R. J. 1951. J. Biol. Chem. 193, 265.

5. Laemmli, U. K. 1970. Nature (London), 227, 680.

PURIFICATION AND CHARACTERIZATION OF AN UNUSUAL NON-FLUORESCENT FLAVOPROTEIN FROM PHOTOBACTERIUM LEIOGNATHI

D.J. O'KANE*, J. VERVOORT**, F. MULLER** AND J. LEE*
*Department of Biochemistry, University of Georgia, Athens, GA 30602; **Department of Biochemistry, Agricultural University, De Dreijen 11, 6703 BC, Wageningen, The Netherlands.

Introduction

An unusual non-fluorescent flavoprotein (NFP) from *Photobacterium leiognathi* was identified by its distinctive coherent anti-Stokes Raman spectrum (1). NFP can be purified easily in large amounts and is a major flavoprotein of *P. leiognathi*. The flavin chromophore of NFP is relatively hydrophobic and resembles the unusual "p-flavin" (2) previously obtained from *P. phosphoreum* luciferase (Lase) (3).

Results

NFP is usually obtained as side fractions during the purification of Lase and lumazine protein from *Photobacterium* (4,5). NFP can also be purified to homogeneity by a four-column procedure involving preparative HPLC that will be described in detail elsewhere. This purification procedure is summarized in Table 1. The final preparation migrates as a single band in SDS-PAGE. NFP has absorbance maxima at 271, 387 and 442 nm with shoulders at 283 and 462 nm (Fig. 1). There is a clear absence of a Soret band in the 400 to 420 nm region indicating an absence of heme. Plasma emission spectroscopy also reveals an absence of metal ligands. However, phosphorus is found in stoichiometric amounts (1.03 mole/mole flavin, assuming ε_{442} = 12000 M^{-1} cm^{-1}).

Flavins and Flavoproteins 1987

TABLE 1. Summary of NFP Purification from P. leiognathi[a].

STAGE	A280	A442	A280/A442	F520[b]	LASE[c]
CELL-FREE EXTRACT	215000[d]	4360[d]	49.3	36000	3.46×10^6
PHENYL-AGAROSE	25250	787	32.0	15540	2.18×10^5
TSK PHENYL-5-PW	3380	284	11.9	415	1276
TSK-G3000SWG (SEC)	1530	262	5.8	275	ND (<70)
TSK-DEAE 3SW (ANION)	787	247	3.2	45	ND (<100)

[a] Based on 1 kg (wet weight) of frozen cell paste.

[b] Relative fluorescence ($Ex_{470} \rightarrow Em_{520}$).

[c] Units are 10^{12} photons.s^{-1} using tetradecanal in a one ml assay volume, performed and calibrated as described in (4).

[d] The sample is turbid and demonstrates considerable light scattering that is recorded as absorbance.

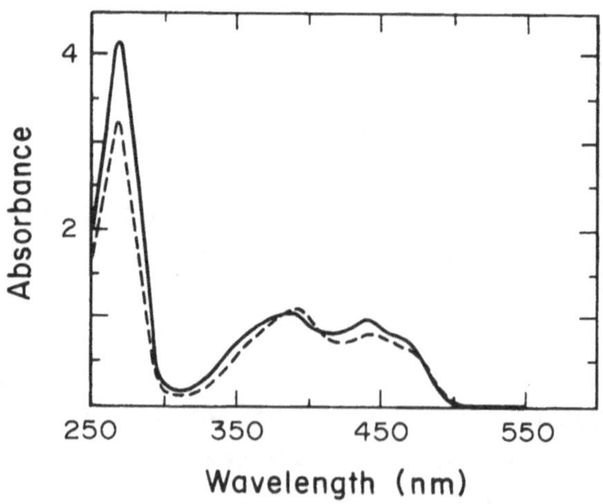

Figure 1. Absorbance spectrum of P. leiognathi NFP (———) and the isolated flavin chromophore (------) in 50 mM phosphate, pH7.0.

TABLE 2. Fluorescence Enhancement and Liberation of the Flavin Chromophore from P. LEIOGNATHI NFP.

TREATMENT	RELATIVE FLUORESCENCE SOLUTION	ULTRAFILTRATE
NONE	1.0	---
$KBr/(NH_4)_2SO_4$	6.3	4.7
8M UREA	3.7	2.1
6M Gdn·HCl	9.7	1.5
SUBTILISIN 6M Gdn·HCl$^+$	14.9	11.0

Native NFP is weakly fluorescent (Q_f ca. 5 X 10^{-4}) with a corrected emission maximum at 522 nm. The flavin is immobile: the steady state fundamental fluorescence anisotropy of NFP is 0.33 from a Perrin plot. The flavin is tightly, but non-covalently bound and cannot be removed by usual procedures (Table 2). The flavin chromophore is extractable into organic solvents in high yields after digestion with subtilisin in 6 M guanidinium·HCl. The fluorescence maximum of the free flavin chromophore shifts to 543 nm and Q_f increases two orders of magnitude. A second fluorescent compound (λ max = 489 nm) is detected in the guanidine solutions and isobutanol extracts. This may be a lumichrome-type degradation product of the flavin chromophore. The flavin derivative fails to co-migrate with riboflavin, FMN or FAD during TLC on cellulose. The flavin chromophore, however, can serve as a substrate in bacterial bioluminescence (Figure 2). Undialyzed solutions containing apo-NFP and liberated flavin are also active in bioluminescence, but freshly dialyzed native NFP is inactive. NFP has no intrinsic bioluminescence in the absence of added Lase. Physiochemical properties (Table 3) also distinguish NFP from Lase. Native NFP has a smaller molecular weight, sedimentation coeffecient and Stokes radius than Lase. Furthermore, the ratio of flavin derivative to protein indicated two moles of flavin per 45000 M_r holoprotein while Lase from P. phosphoreum reportedly has 0.2 mole "p-flavin"

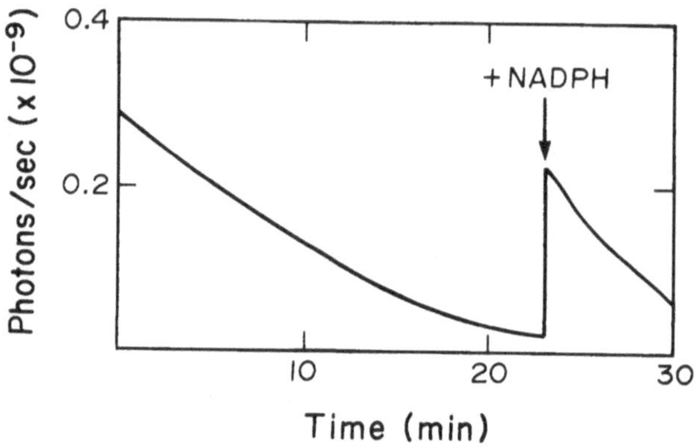

Figure 2. Bioluminescence obtained with NFP chromophore in a Lase-NADPH coupled assay. Additional NADPH added where indicated.

Table 3. Physical Properties of P. leiognathi NFP

Stokes radius, (Å)	27
Sedimentation coeffecient, $s^{o}_{20,w}$ (S)	4.05
Apparent partial specific volume, $\bar{v}_2$(cm^3 g^{-1})	0.725
Molecular weight	
Low speed sedimentation equilibrium	455000 $\pm$ 1500
$s^{o}_{20,w}$ and Stokes radius	45000
Size-exclusion chromatography	41000
SDS-PAGE	27500
Minimum M_r of apo-NFP from Biuret and flavin	22000
Mininum M_r of apo-NFP from P content and flavin	21800
Molar volume, V (cm^3 mole^{-1})	53200
Hydration, (g H_2O/g protein):	
from the amino acid composition	0.45
V and M_r	0.43
Fluorescence Lifetime, τ (nsec):	
of native NFP	12.5
of isolated chromophore	3.6

bound per mole Lase (3). The NFP chromophore differs from "p-flavin" because it is not extractable by chloroform. A similar NFP from P. phosphoreum has recently been obtained free of Lase activity. Work is in progress to elucidate the structure of the NFP chromophore and to discern the function of NFP.

Acknowledgement

Supported in part by grant DMB-85 12361 from the National Science Foundation.

References

1. Visser, A.J.W.G., J. Vervoort, D.J. O'Kane, J. Lee, and L.A. Carreira. 1983. Eur. J. Biochem. 131, 639.

2. Kasai, S., K. Mastui, and T. Nakamura. 1982. In: Flavins and Flavoproteins (V. Massey and C.H. Williams, eds.). Elsevier/North Holland, p. 459.

3. Matsuda, K. and T. Nakamura. 1972. J. Biochem. (Tokyo) 72, 951.

4. O'Kane, D.J., M. Ahmad, I.B.C. Matheson, and J. Lee. 1986. Meth. Enzymol. 133, 109.

5. O'Kane, D.J. and J. Lee. 1987. Meth Enzymol. 133, 149.

PURIFICATION AND SOME PROPERTIES OF FP_{390} FROM P. PHOSPHOREUM

Sabu Kasai, Kunio Matsui* and Takao Nakamura**

Research Institute for Atomic Energy, Osaka City University, Sumiyoshi-ku, Osaka 558, Japan, *Teikoku Women's University, Department of Home Economics, Moriguchi-shi, Osaka 570, Japan, and **Faculty of Science, Osaka University, Toyonaka-shi, Osaka 560, Japan

Introduction

Luciferase prepared from luminescent cells contains a yellow or brown pigment called p-flavin (PF, from Photobacterium phosphoreum) or b-flavin (from Vibrio harveyi)(1,2), whereas luciferase from non-luminescent cells (e.g. aldehyde deficient mutants) is colorless. We tried to determine the chemical structure of PF and obtained evidence that PF consists of FMN and myristic acid (3). However, the structure is not yet established, as the sample was a mixture of two similar flavins. PF seems to become partially modified during extraction. On the other hand, during purification of yellow luciferase, we found that luminescent cells contain large amounts of another yellow protein, the prosthetic group of which seems to be a flavin having a similar structure as PF. The protein is named FP_{390} as its absorption maximum is at 390 nm. Though the physiological function of FP_{390} is not known, this protein seems to be involved in the luminous reaction. Therefore we started to purify and to investigate this protein.

Materials and Methods

A brightly luminescent strain of P. phosphoreum (IFO 13896) was

Flavins and Flavoproteins 1987

cultured in medium (4) containing 3% of NaCl. Frozen cells were extracted with 50 mM buffer (all buffers were phosphate buffer pH 7.0 containing 1 mM EDTA and 5 mM mercaptoethanol). FP_{390} was precipitated with $(NH_4)_2SO_4$ (fraction between 40 and 75% saturation), dialyzed and chromatographed on DEAE Sephadex in 0.35 M buffer. The yellow fractions were pooled and rechromatographed on Sephadex G-75 (50 mM buffer). In the final stage, the sample was applied to a butyl-cellulose column (prepared by coupling n-butylamine to epoxy-activated cellulose as described in (5)) equilibrated with 50 mM buffer containing 1.5 M $(NH_4)_2SO_4$. Elution was achieved by decreasing the concentration of $(NH_4)_2SO_4$ to 1 M. The purified protein was crystallized in a sitting drop by vapour diffusion.

Results and Discussion

The possible involvement of FP_{390} in bioluminescence was studied. The ability of cells to luminesce can be suppressed by mutation or by lowering the NaCl concentration in the culture medium (6,7). Luminous bacteria grow in media with less than 1% NaCl. However, light emission decreases at NaCl concentrations less than about 1.5%, and at 0.5% NaCl the cultures are almost dark. As shown in Fig. 1, cells grown at 2 and 3% NaCl contain several proteins in larger amounts than those grown at lower NaCl concentrations: besides the prominent luciferase, FP_{390} is one of them. At 0.5 or 1% NaCl concentration, the FP_{390} band becomes faint concomitant with the bands of the luciferase subunits. These results suggest a connection between FP_{390} and light emission.

FP_{390} was purified to complete homogenity and was found to consist of a single polypeptide chain of appr. 23,500 dalton as judged by SDS PAGE. Table 1 shows the amino acid composition of FP_{390}. The content in Asp/Asn is remarkably high, similar to the luciferases from P.phosphoreum (8) and V.harveyi (9). At present, the partial amino acid sequence is being deter-

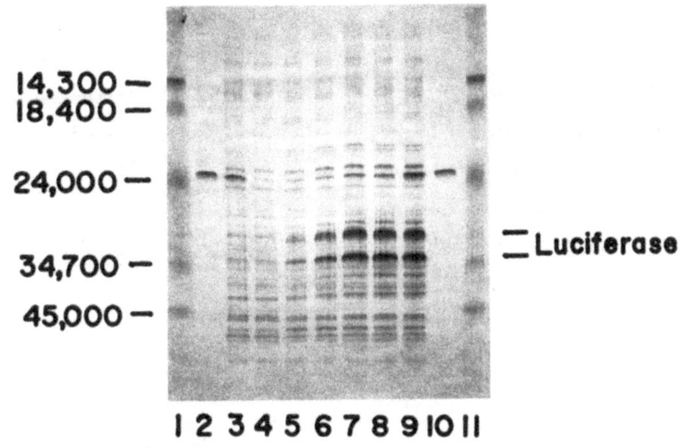

Figure 1. SDS PAGE of FP_{390} and extracts of bacterial cells cultured at various NaCl concentrations and harvested in the logarithmic phase. 1 g of frozen cells were each extracted with 2 ml of 50 mM buffer, and 2 µl of each extract were charged on a 12.5% gel. Lanes 1 and 11, molecular weight markers, lysozyme (14,300), β-lactoglobulin (18,400), trypsinogen (24,000), pepsin (34,700), egg albumin (45,000). Lanes 2 and 10, FP_{390}. Lanes 3 and 9, FP_{390} plus extract of cells cultured at 0.5 and 5% NaCl, respectively. Lanes 4-8, extracts of cells cultured in 0.5, 1, 1.5, 2 and 3%, respectively.

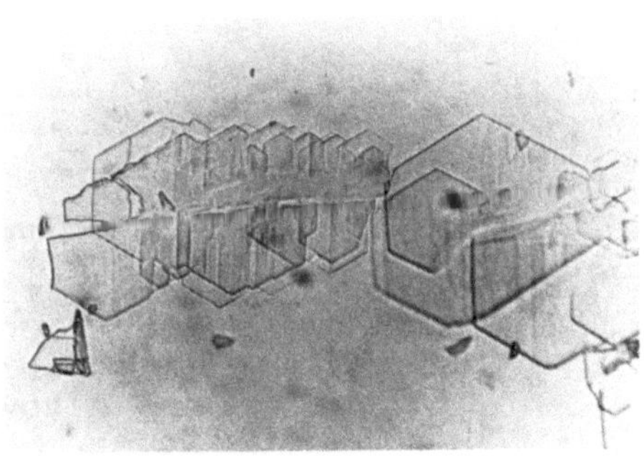

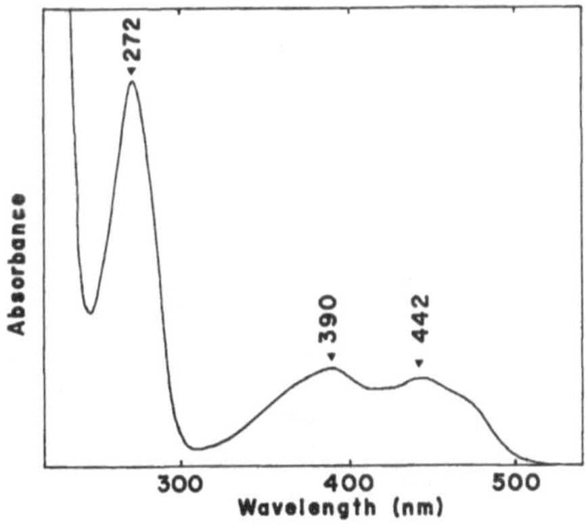

Figure 2. Crystals of FP_{390}. Figure 3. Absorption spectrum of FP_{390}.

mined in order to determine if there is a structural relationship between luciferase(s) and FP_{390}. In addition, FP_{390} was crystallized (Fig. 2), but no crystals suitable for

X-ray analysis could yet be obtained. Fig. 3 shows the absorption spectrum of FP_{390} in 50 mM buffer. No sigificant spectral change was observed in the visible region upon adding urea (8 M final; incubated overnight at 4°C). When luciferase was treated in the same way, the absorption spectrum changed, but the resulting spectrum was different from the one of the FP_{390} sample. On the other hand, if both yellow luciferase and FP_{390} were extracted with hot butanol and the extracts purified by TLC, they showed a similar spectrum. It therefore appears that the two proteins contain structurally related, but slightly differing chromophores.

Table 1. Amino acid composition of FP_{390}, assuming a molecular weight of 23,500 dalton.

Amino acid	moles
Lysine	18.5
Histidine	2.7
Arginine	3.6
Aspartic acid	38.0
Threonine	8.6
Serine	13.0
Glutamic acid	20.6
Proline	3.2
Glycine	5.7
Alanine	11.4
Half-cystine	3.1
Valine	14.2
Methionine	4.4
Isoleucine	12.4
Leucine	20.1
Tyrosine	10.6
Phenylalanine	9.8
Tryptophan	2.0

The authors thank to Mr. Sigehiro Kamitori and Dr. Ken Hirotsu for the crystallization of FP_{390}, and Dr. Saburo Hara for the amino acid analysis.

References

1. Matsuda, K., Nakamura, T. 1972. J. Biochem. 72, 951.
2. Mitchell, G. W., Hastings, J. W. 1970. Biochemistry 9, 2699.
3. Kasai, S., Matsui, K., Nakamura, T. 1982. In: Flavins and Flavoproteins (V. Massey and C. H. Williams, eds.). Elsevier North Holland, p. 459.
4. Nakamura, T., Matsuda, K. 1971. J. Biochem. 70, 35.
5. Sundberg, L., Porath, J. 1974. J. Chromatogr. 90, 87.
6. Dunlap, P. 1985. Arch. Microbiol. 141, 44.
7. Watanabe, H., Hastings, J. W. 1986. Arch. Microbiol. 145, 342.
8. Yoshida, K., Nakamura, T. 1973. J. Biochem. 74, 915.
9. Johnston T. C., Thompson R. B., Baldwin T. O. 1986. J. Biol. Chem. 261, 4805.

Structure, Mechanism, and Biosynthesis of Flavoenzymes Containing Covalently-Bound Flavins

8α-IMIDAZOLYLFLAVINS: APPLICATION OF THEIR REDOX AND SPECTRAL PROPERTIES TO FLAVOENZYME SYSTEMS

D.E. Edmondson, R. De Francesco
Department of Biochemistry, Emory University, Atlanta, GA. 30322

Introduction

The number of known flavoenzymes which contain 8α-substituted flavins has increased substantially over the past 20 years. In all known cases, with the exception of 6-substituted flavins in trimethylamine dehydrogenase and in dimethylamine dehydrogenase, covalent attachment of the flavin to the protein involves a linkage between the the 8α-methylene group of the isoalloxazine ring and the following nucleophilic amino acids: N(1) or N(3) histidyl, O-tyrosyl, or S-cysteinyl (Fig.1) (see (1) for a recent review).

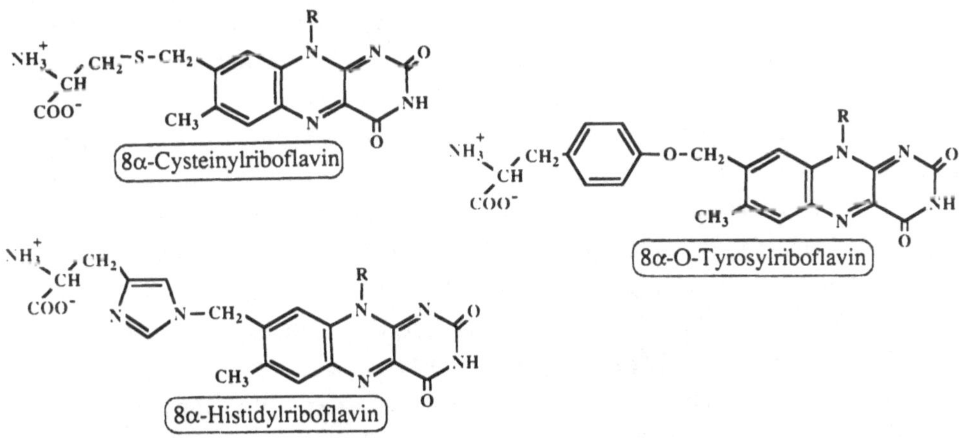

Fig. 1 - Structure of 8α-substituted flavins naturally occurring in some flavoenzymes. R denotes the rest of the FAD molecule.

Flavins and Flavoproteins 1987

To date, little is known regarding the question as to the reason why some flavoenzymes require a covalently bound flavin while in the majority of cases, the flavin coenzyme is bound to the apoenzyme via non-covalent interactions. Other redox enzymes have cofactors which may also be bound by either covalent or non-covalent interactions. The hemoproteins constitute one well-studied example while the pyrroloquinoline (PQQ)-dependent enzymes constitute another, less well-characterized system. This paper will summarize both published and unpublished studies from our laboratory which attempt, as a long range goal, to answer the question as to why certain flavoenzymes require a covalent flavin. The 8α-imidazolylflavins will be the subject of discussion as their properties are amenable to a number of experimental approaches.

Influence of Flavin Oxidation Level on the Ionization Properties of the 8α-Imidazole Ring.

It has been known since early studies on the 8α-N(3)-histidylflavin of succinate dehydrogenase (2,3) that ionization of the 8α-imidazole ring leads to pronounced absorption and fluorescence alterations of the flavin ring in its oxidized form. The pK_a of the imidazole ring in 8α-N(3)-histidylriboflavin is 4.7 while that of the N(1) isomer is 5.2 (4). Recent studies on a model 8α-imidazolylriboflavin analog showed the imidazole pK_a to be 6.0 using a variety of spectral and potentiometric techniques (5,6). Thus, the lower pK_a values observed for the synthetic or natural 8α-histidylflavins is a reflection of alterations induced by the presence of the carboxyl and α-amino groups present on the amino acid side chain. The coincidence of pK_a values determined by either absorption or fluorescence spectral studies shows the pH-dependence of flavin fluorescence quantum yield to be due to a ground-state process rather than to an excited state process.

Recent redox potential data (5) showed the pK_a of the imidazole ring of 8α-imidazolylriboflavin to increase from 6.0 to 6.5 on semiquinone formation and to further increase to 7.0 on hydroquinone formation. ^1H NMR data on the pH-dependence of chemical shifts of the imidazole protons also showed the pK_a of the 8α-imidazole substituent to be 7.0 on two-electron reduction of the flavin. These data demonstrate that the redox level of the flavin can modulate the pK_a of the 8α-imidazole ring in model flavin systems. This behavior is also expected to occur with 8α-histidylflavins in flavoenzyme systems. Previous flash photolysis data (7) showed that the pK_a of the N(5) position of the 8α-histidylflavin semiquinone is 7.2 which is approximately 1 pK unit lower than that for riboflavin. Redox potential data (5) show the pK_a of the N(1) position in 8α-imidazolylriboflavin hydroquinone to be also lower than that of riboflavin. Fig.2 shows the relevant pK_a values of the various positions of 8α-imidazolylriboflavin at the various redox levels that would be relevant in the physiological pH range.

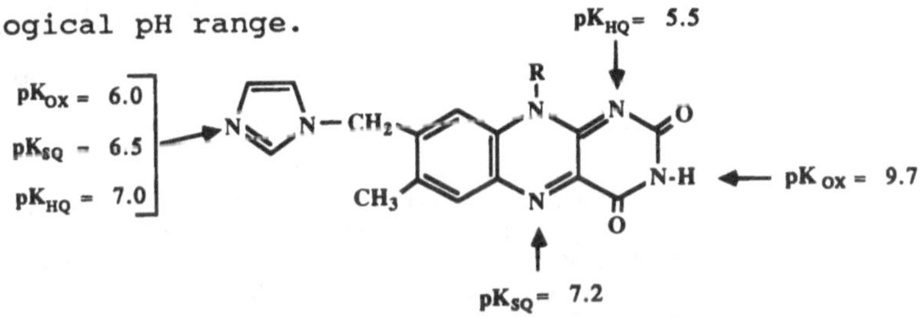

Fig. 2 - pK_a values of the various positions of the 8α-imidazolylriboflavin at different redox level. Only pK_a values that are relevant in the physiological range are shown.

As a first step towards investigating the effect of imidazole ionization on the redox properties of a protein-bound 8α-histidylflavin, we have studied 8α-imidazole FMN bound to apoflavodoxins from A. vinelandii and from C. pasteurianum. Previous work (8) has shown 8α-substituted flavins and flavin peptides to be tightly bound to apoflavodoxin. For this synthetic FMN analog, the K_d values determined for binding to these two apoflavodoxins is ≤1 nM. In both cases, the 8 position is exposed to solvent so that only minor perturbations of the pK_a of the imidazole ring should be observed. The pH-dependence of the absorption spectrum is shown as difference spectra in Fig.3 for the reconstituted <u>Azotobacter</u> apoflavodoxin complex. The spectral perturbations are similar to those observed for free 8α-imidazolylriboflavin in aqueous solution.

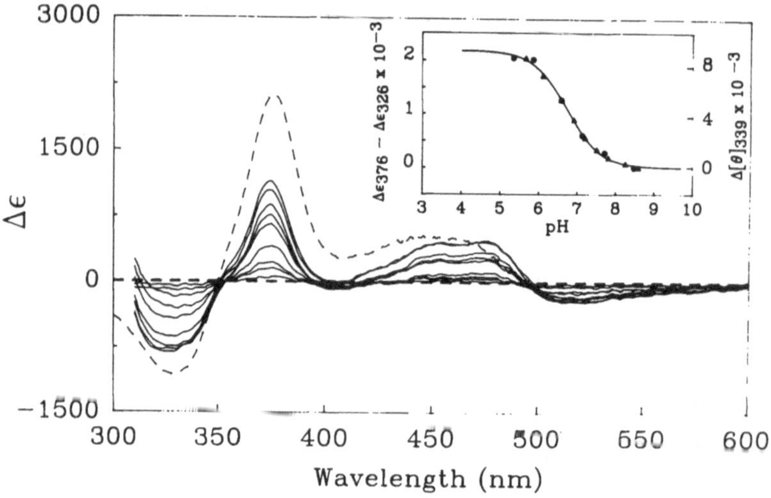

Fig. 3 - Set of difference spectra obtained during a pH titration of 8α-imidazolyl-FMN <u>Azotobacter</u> flavodoxin. Spectra obtained at the different pH values were subtracted from the starting spectrum taken at pH 8.2. DOTTED LINE: difference spectrum (deprotonated minus protonated) for 8α-imidazolylriboflavin. INSET: Best fit of the experimental points to a single proton-dissociation curve with a pK_a of 6.7.

▲ - Difference spectral data.
● - CD spectral titraton

Analysis of the data show a single proton dependence and pK_a values of 6.7 and 6.8 for the 8α-imidazole ring in complexes with apoflavodoxins from <u>A. vinelandii</u> and <u>C. pasteurianum</u> respectively. The pK_a values for the imidazole ring in neutral flavin semiquinone complexes of 8α-imidazolylFMN complexes of the above apoflavodoxins have been determined by measuring the pH dependence of the rate of electron transfer to cytochrome c (9). Both flavodoxin complexes show a pK_a of 7.0 which can be shown to be due to the ionization of the 8α-imidazole ring.

These pH-dependent spectral studies were extended to the cholesterol oxidases from Schizophylum commune (S.c.) and from Gloeocystidium chrysocreas (G.c.). The former enzyme is known to contain an 8α-N(1)-histidylFAD (10) while the latter contains an 8α-histidylFAD (whether N(1) or N(3) still remains to be determined). Difference spectra at various pH values of either cholesterol oxidase are similar to those observed with free 8α-imidazolylflavins and are completely reversible. The experimental data in Fig. 4 for G.c. cholesterol oxidase demonstrate that the pH dependence of absorption spectra are also in agreement with similar fluorescence and circular dichroism data (11).

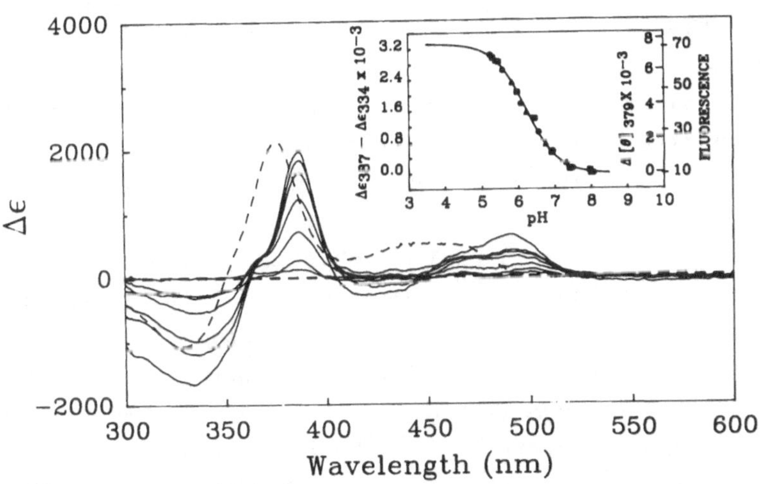

Fig. 4 - Set of difference spectra obtained during a pH titration of cholesterol oxidase from Gloeocystidium chrysocreas. Experimental details as in Fig. 3.
DOTTED LINE: difference spectrum (deprotonated minus protonated) for 8α-imidazolylriboflavin.
INSET: Best fit of the experimental points to a single proton-dissociation curve with a pK_a of 6.2.
● - Difference spectral data
▲ - CD spectral titration
■ - Fluorescence titration

Analysis of the data show the pK_a of the 8α-imidazole ring for the oxidized flavin is 5.4 for S.c. cholesterol oxidase and 6.1 for the G.c. enzyme. In both enzymes, the 8α-imidazole ring appears to be accessible to the solvent since the pK values are either the same or slightly lower than that observed with the free model flavin in solution. It remains for future work to determin the effect of flavin redox level on the pK_a values of the 8α-imidazole ring in these enzymes.

Does Any Spin Density Reside on the Imidazole Ring in 8α-Imidazolylflavin Semiquinones?

ESR and ENDOR studies (see (12) for a recent review) have demonstrated that the 8 position of the isoalloxazine ring has appreciable spin density both in free and protein-bound flavin semiquinones. It is of interest to determine whether any appreciable spin density could be distributed on the 8α-imidazole ring since enzymes containing the flavins undergo 1 e^- redox reactions and such information would be important in understanding whether the imidazole ring might serve as a reactive site in such reactions. ESR spectra of the semiquinone forms of flavoenzymes containing 8α-substituted flavins (13) suggest no major changes in spin distribution as compared with data on non-covalent flavoproteins. One approach to address this question with free 8α-imidazoylflavins is to perform NMR experiments on partially-reduced flavin solutions. It is well-known that an equilibrium concentration of flavin semiquinone is present in partially-reduced solutions in which the rates in the forward and reverse directions are rapid on the NMR time scale. ^{1}H Resonances in ring systems will be extensively broadened to a level beyond detection while those resonances in portions of the molecule containing little or no spin density will still be observable or broadened only slightly due to dipolar interactions. Both ^{1}H and ^{13}C NMR experiments on partially-reduced 8α-imidazolylflavin solutions show the isoalloxazine resonances to be broadened beyond detection while the imidazole resonances are still detectable (14). These data show the free electron to be localized on the isoalloxazine ring and little or no spin density to be situated on the imidazole ring. Whether this situation also applies to enzyme-bound 8α-flavins remains for future work.

Considerations of Oxidation-Reduction Potential Data on Flavoenzymes Containing Covalent Flavins

Although detailed pH-dependent redox potential measurements have not been done on flavoenzymes containing 8α-histidylflavins, the potentials of several enzymes have been measured at single pH values. Thiamine oxidase (which contains 8α-histidylFAD (15)) has an E_m value of +55mV, $E_{ox/sq}$= +80mV, and $E_{sq/hq}$ = +30 mV (16). Unpublished work in our laboratory on S.c. cholesterol oxidase shows an E_m = + 21 mV, $E_{ox/sq}$ = + 30 mV, and $E_{sq/hq}$ = +12 mV (17). These values are quite positive with respect to the corresponding potentials of model 8α-imidazolylflavins (5). It has been a traditional practice in flavin enzymology to treat differences in redox potential between free and protein-bound flavins in terms of differential binding affinities of the apoprotein for the oxidized, semiquinone, and hydroquinone forms of the flavin. This approach is illustrated in terms of a Born cycle in Scheme 1. For flavoenzymes exhibiting a potential higher than that of the free flavin, binding of the flavin hydroquinone to the apoprotein is calculated to be tighter than for the oxidized form. We have confirmed these calculations experimentally by determining the binding constant of reduced FMN to *Azotobacter* apoflavodoxin. Binding constants determined by direct titrations are in good agreement with those calculated from redox potential data.

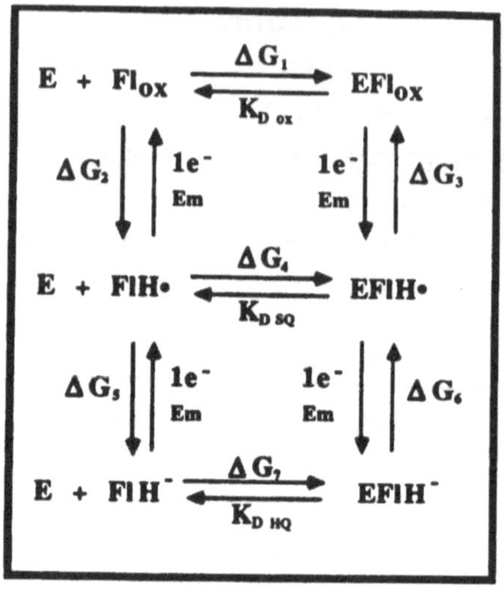

Scheme 1 - Born cycle for a flavoenzyme containing a noncovalent-bound flavin coenzyme.

For flavoenzymes containing covalently-bound flavins, the above considerations are invalid inasmuch as the flavin is not dissociable from the apoprotein. Thus, alternative considerations must be used in analysis of redox potentials of these enzymes. It is instructive to compare similar data in comparison of the redox potential of the heme peptide of cytochrome c with that of the covalently-bound heme in the intact protein. The difference of free energy is 7 Kcal/mole which is attributed to a destabilization of the oxidized heme due to the effective dielectric mileau of the protein environment relative to that of aqueous solution (18). It is of interest that the favorable free energy change observed for thiamine oxidase (9.6 Kcal/mole for the 2-electron process) is of the same order of magnitude as the 1-electron process for cytochrome c (Scheme 2).

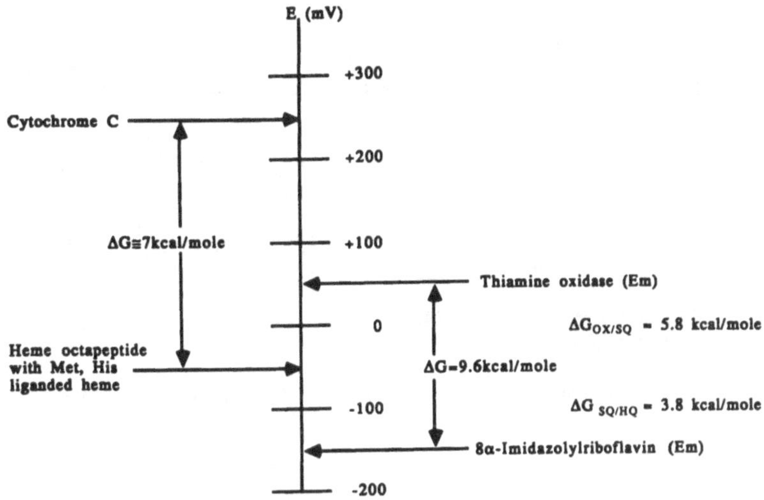

Scheme 2 - Free energy diagraham comparing the free energy of cytochrome c reduction with the one-electron steps of thiamine oxidase reduction.

The destabilization of the oxidized covalent flavin by the protein environment could be a major determinant in redox potential values. These values could be modulated by the pK_a values of both the isoalloxazine and imidazole rings in the three redox levels of the flavin and further modulated by substrate binding to the active site. Further information on these ideas will require more detailed structural and redox potential information.

Acknowledgement

This work has been supported by a grant from the National Institutes of Health (GM-29433).

References

1. Singer, T.P., W.S. McIntire. 1984. Meth. Enzymol. 106, 369.

2. Kearney, E.B., T.P. Singer. 1955. Arch. Biochem. Biophys. 60, 225.

3. Walker, W.H., T.P. Singer, S.Ghisla, P. Hemmerich. 1972. Eur. J. Biochem. 26, 279.

4. Edmondson, D.E., W.C. Kenney, T.P. Singer. 1976. Biochemistry 15, 2937.

5. Williamson, G., D.E. Edmondson. 1985. Biochemistry 24, 7790.

6. Williamson, G., D.E. Edmondson. 1985. Biochemistry 24, 7918.

7. Edmondson, D.E., F. Rizzuto, G. Tollin. 1977. Photochem. Photobiol. 25, 445.

8. Oestreicher, G., D.E. Edmondson, T.P. Singer. 1976. In: Flavins and Flavoproteins (T.P. Singer, ed.) Elsevier Pub. Co., p.447.

9. De Francesco, R., G. Tollin, D.E. Edmondson. 1987. Biochemistry, in press.

10. Kenney, W.C., T.P. Singer, M. Kukyama, Y. Miyake. 1979. J. Biol. Chem. 254, 4689.

11. De Francesco, R., D.E. Edmondson. manuscript in preparation.

12. Edmondson, D.E. 1985. Biochem. Soc. Trans. 13, 593.

13. Edmondson, D.E., B.A.C. Ackrell, E.B. Kearney. 1981. Arch. Biochem. Biophys. 208, 69.

14. VerVoort, J., G. Williamson, F. Muller, D.E. Edmondson. manuscript in preparation.

15. Kenney, W.C., D.E. Edmondson, R.L. Seng. 1976. J. Biol. Chem. 251, 5386.

16. Gomez-Moreno, C., M. Choy, D.E. Edmondson. 1979. J. Biol. Chem. 254, 7630.

17. Williamson, G., D.E. Edmondson. unpublished observations.

18. Churg, A.K., A. Warshel. 1986. Biochemistry 25, 1675.

X-RAY STUDY OF FLAVOENZYMES CONTAINING COVALENTLY BOUND FLAVINS

F.S. Mathews and L.W. Lim

Departments of Cell Biology and Physiology and of Biological Chemistry
Washington University School of Medicine, St. Louis, MO 63110, USA

Introduction

In a number of flavoproteins the flavin is covalently attached to the polypeptide chain, usually through the 8α-methyl position of the isoalloxazine ring to a cysteine or histidine side chain. Neither the functional role of the covalent binding nor the mechanism of attachment of the flavin to the protein is well understood. As part of our continued interest in the structures of electron transport proteins, in particular complex flavoenzymes, we have undertaken the study of two enzymes containing covalently bound flavin, trimethylamine dehydrogenase (TMADH) and *p*-cresol methylhydroxylase (PCMH). In each case the mode of attachment of the flavin to the protein is unusual. In the former, FMN is attached through the 6 position to a cysteine sulfur (1). In the later, FAD is attached through the 8α-methyl position to a tyrosine hydroxyl (2). The structure of PCMH has been analyzed at 6.0 Å resolution (3) and extended to 3.0 Å resolution (4). We report here the current status of the TMADH investigations at 2.4 Å resolution.

Structure of TMADH

TMADH is a dimer of subunit M_r 83,000 containing one $[4Fe{-}4S]^{2+}$ cluster and one covalently bound FMN per subunit (5,6). It is found in methylotrophic bacteria, such as W3A1, where it catalyzes the oxidative demethylation of trimethylamine to form dimethylamine and formaldehyde. Subsequently electrons are passed to an electron transfer flavoprotein *in vivo* (7). The amino acid sequence of TMADH is not known.

Flavins and Flavoproteins 1987

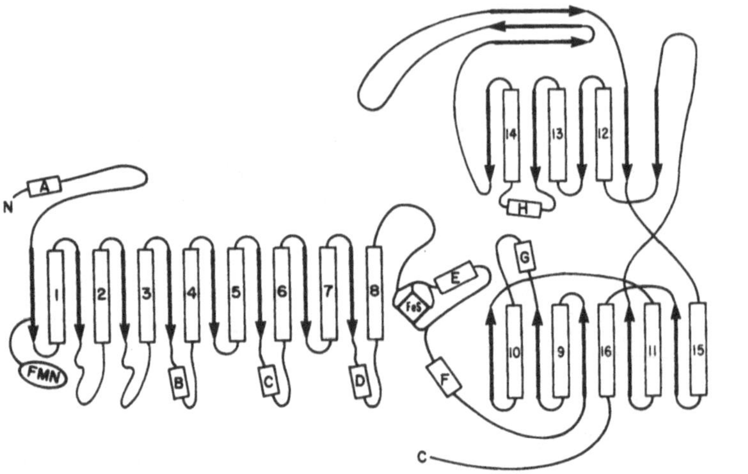

Fig. 1. Topology and secondary structure of TMADH. Arrows represent β-strands and rectangles represent α-helices. Helices within elements of supersecondary structure are numbered while other helices are lettered. Covalent attachment of FMN and $[4Fe\text{-}4S]^{2+}$ are indicated.

The structure of TMADH was solved initially at 2.4 Å (8) resolution by tracing an idealized α-carbon chain through the electron density displayed on a molecular graphics system. The electron density map was obtained from data collected with an area detector system and was averaged about the non-crystallographic 2-fold axis.

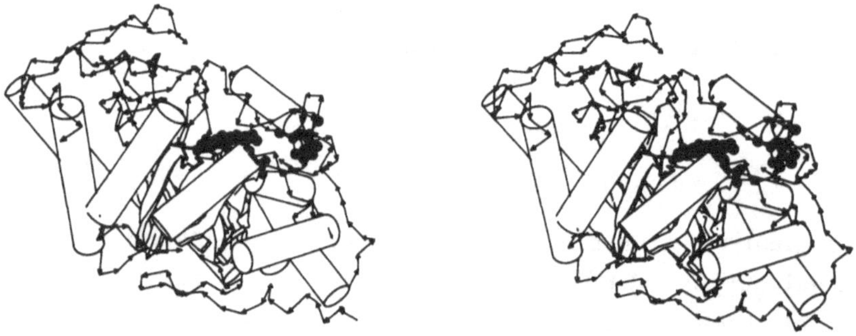

Fig. 2. Large domain of TMADH. The β-sheet and α-helices of the $\beta_8\alpha_8$ supersecondary structure are indicated by arrows and cylinders, respectively. The atoms of the FMN and $(4FE\text{-}4S)^{2+}$ are drawn as shaded circles. Picture produced by a computer program written by A. M. Lesk and K. D. Hardman (16,17).

Each subunit of TMADH is folded into 3 domains (Figure 1). The largest domain is made up of approximately the first 380 residues from the N-terminus. It is folded into a parallel $\beta_8\alpha_8$ barrel and contains both the FMN and the $[4Fe\text{-}4S]^{2+}$ cofactors (Figure 2). The FMN is located at the C-terminal end of the β barrel, with atom C-6 of the flavin ring bound to a cysteine sulfur atom at the end of the first β strand and 0-2 nearly coincident with the barrel axis. The $[4Fe\text{-}4S]^{2+}$ is bound to an α/β hairpin connected to the C-terminus of the eighth helix of the barrel. This hairpin lies on the external surface of the $\beta_8\alpha_8$ barrel and partially covers the benzenoid ring of the flavin. The 8-α methyl group of the flavin is approximately in van der Waals contact with one of the cysteine sulfur ligands of the [4Fe-4S] cluster. Both the cluster and the flavin are deeply buried in the protein.

The second (medium) and third (small) domains each contain 5-stranded parallel beta sheets (Figure 3). In the medium domain the central β-sheet is flanked by two or three α-helices on each side. This domain resembles the nucleotide-binding domain of lactate dehydrogenase in its folding topology. In the small domain, the β-sheet is flanked by three helices on one side and by 3-stranded antiparallel β-sheet on the other. Its folding topology closely resembles that of the FAD-binding domain of glutathione reductase (9).

The medium and small domains are connected at two places. The polypeptide chain, after leaving the large domain, forms about 3/4 of the medium domain, goes on to form the small domain and finally reenters and completes the medium domain. The topological connectivity of the medium and small domains together closely resembles that of the NADPH and FAD binding domains of glutathione reductase (Figure 4). When the α-carbon positions of both pairs of domains are compared, a large part of the secondary structural elements are found to nearly superimpose. The central β-strands, the strands connecting the pairs of domains and the antiparallel β-sheet of the small domain of TMADH match quite closely the corresponding elements in glutathione reductase.

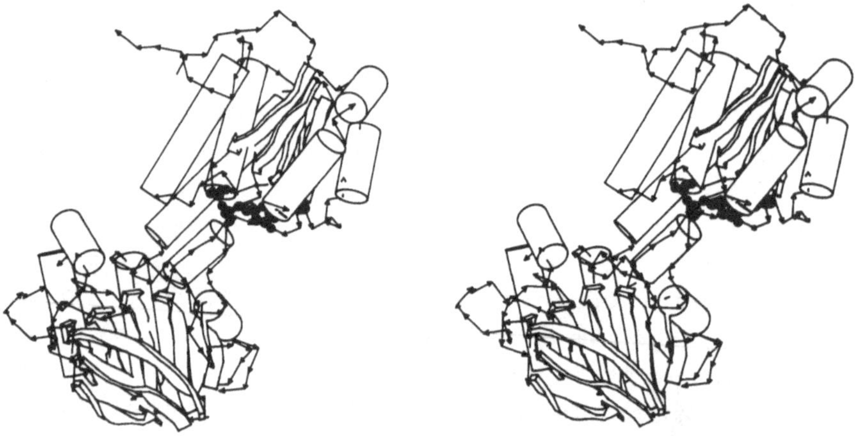

Fig. 3. Medium (upper) and small (lower) domains of TMADH. The shaded circles represent the atoms of the newly-discovered ADP cofactor (see text). Picture produced by a computer program written by A. M. Lesk and K. D. Hardman (16,17).

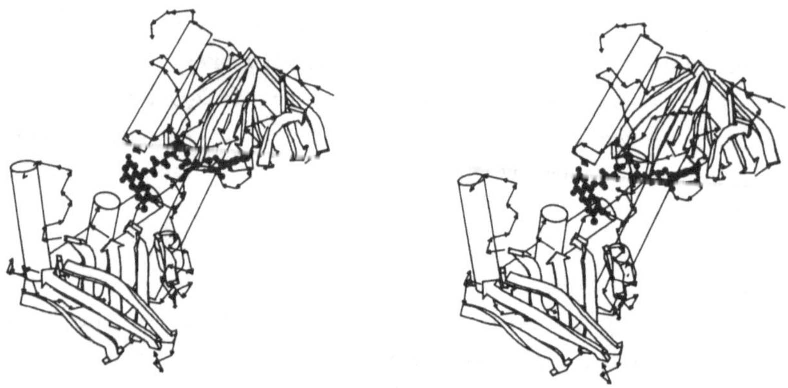

Fig. 4. Residues 18 to 360 of glutathione reductase comprising the FAD-containing (upper) and NADPH-containing (lower) domains. The atoms of the FAD moiety are indicated by shaded circles. Picture produced by a computer program written by A. M. Lesk and K. D. Hardman (16,17).

Refitting the Electron Density Map of TMADH

We have recently improved the quality of the electron density map by solvent leveling procedures (10). The purpose in doing this was to define the geometry of the prosthetic groups more clearly, fit the full polypeptide backbone chain and attempt to deduce the amino acid sequence from the shapes of the side chain density. The crystals contain 54% solvent. As a precaution against artificially truncating external side chains, the solvent content parameter was set to 51%. The solvent-leveled map was then averaged about the molecular 2-fold axis prior to interpretation.

An hypothetical "x-ray" sequence has been fitted to the solvent-leveled and averaged 2.4 Å resolution electron density. The fitting was carried out on the MMS-X molecular graphics system using the graphics program NEWNIP (11). Individual amino acids were added one at a time to the growing polypeptide chain and adjusted to fit the side chain density. Often several different amino acids were tested to achieve the closest match to the density. A total of 729 amino acids were identified and fitted to the density, compared to 733 residues previously estimated from the α-carbon fitting. The amino acid composition derived from the "x-ray" sequence is compared in Table 1 with that determined chemically (6). In well defined regions of the map almost all of the amino acids could be

TABLE 1

Comparison of Chemical and "X-ray" Amino Acid Compositions of Trimethylamine Dehydrogenase

Amino Acid	"X-ray" Composition	Chemical Composition
ASX	86	73
THR	39	40
SER	49	39
GLX	68	82
PRO	42	47
GLY	73	69
ALA	72	65
VAL	44	39
MET	16	16
ILE	37	39
LEU	37	49
TYR	30	31
PHE	18	20
HIS	20	22
LYS	35	41
TRP	16	18
ARG	39	46
CYS	8	9
TOTAL	729	745

easily identified by shape. In other regions, especially on surface loops, only the larger aromatic amino acids were unambiguous.

An anomalous difference Fourier map was computed from the native x-ray data which had peaks at the known iron, phosphate and cysteine positions and verified most of the methionine residues chosen. All of the eight cysteine residues occur at chemically identified locations, i.e., the five covalent linkages to the FMN and $[4Fe-4S]^{2+}$ cofactors and the three mercury binding sites used for phase determination.

Discussion

The comparison between the chemical and "x-ray" amino acid sequences indicates the difficulties of determining sequence from electron density. The number of short side chains, such as Ala, Asx, Gly, Ser, etc. are over estimated while that of longer side chains, such as Lys, Arg, Glx etc. are under estimated. This effect can arise if surface side chains undergo large thermal motions or are disordered.

The known FMN-binding peptide (1) was easily fit to the electron density map. In addition, two short tryptophan-containing peptides of known sequence (Steenkamp and Smith, personal communication) were located and

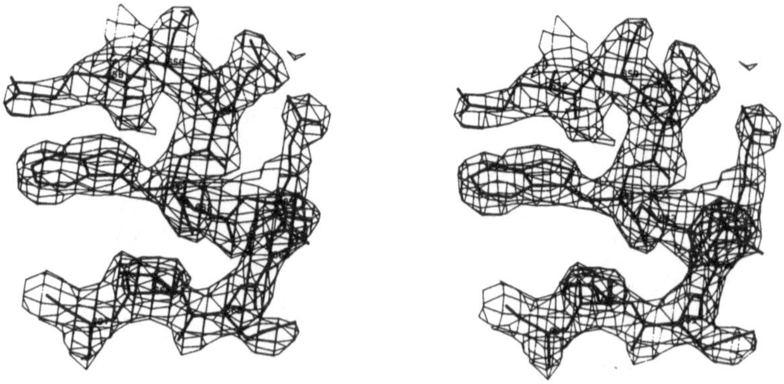

Fig. 5. Residues 658-667 of TMADH fitted to the solvent-leveled electron density map. The known sequence of amino acids, starting at position 658 at the top is Glu, Ser, Glu, Trp, Ala, Glu, Asp, Asp, Ile, Lys. In the figure the last residue is fitted with Ser.

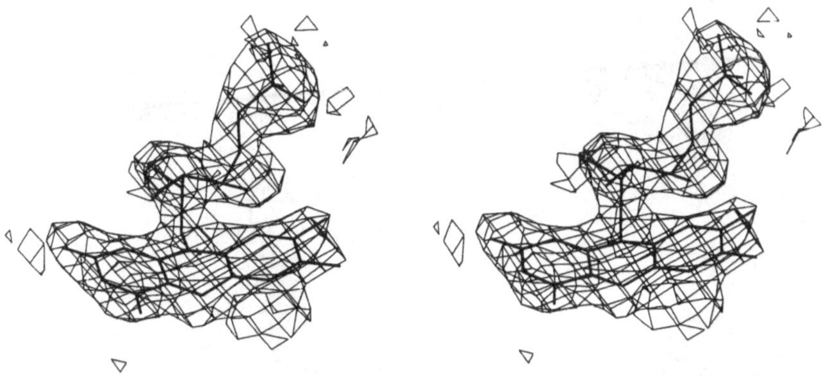

Fig. 6. FMN cofactor of TMADH fitted to the solvent-leveled 2.4 Å resolution electron density map. The bend angle of the flavin ring is about 20°. The open density at the lower right edge of the isoalloxizine ring represents the sulfur atom of Cys 30 covalelently attached to position 6 of the ring.

identified. These occur at positions 77-84 and 658-667 in the "x-ray" sequence. The latter segment, superimposed on the electron density, is shown in figure 5.

The electron density around the FMN is shown in Figure 6. The density is more clearly defined than previously, enabling the atoms of the ribityl and phosphate groups to be placed with greater confidence. Furthermore, the flavin ring is clearly bent. The density of the flavin ring can be most easily matched by introducing a 20° "butterfly" bend about an axis defined by the N-5 to N-10 bond. This bend angle is considerably larger than observed in other flavoproteins and approaches that observed in crystal structures of model flavin compounds (12).

During the fitting of the hypothetical "x-ray" sequence to the electron density, an unconnected piece of density was discovered which previously had been interpreted as part of the polypeptide chain. This density corresponded in position to the ADP moiety of FAD in the analogous position in glutathione reductase. This density, with ADP superimposed, is shown in Figure 7. The location of ADP in the medium and small domains

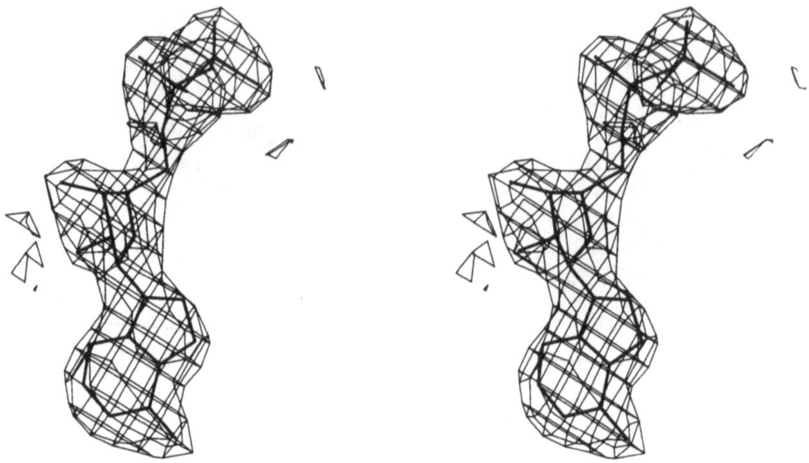

Fig. 7. ADP fitted to the solvent-leveled electron density map of TMADH.

is also included in Figure 3 and can be compared with the position of FAD in glutathione reductase, figure 4. The anomalous difference Fourier also contained peaks of high density at the diphosphate position of magnitude comparable to the single phosphate of FMN. ADP has also been recently identified in the supernatant of denatured TMADH (Steenkamp, personal communication). Thus, the structural similarity between TMADH and glutathione reductase now includes the presence of the ADP fragment of FAD.

The functional significance of ADP in TMADH is unknown. There is no electron density beyond the diphosphate portion of the ADP which would correspond to the riboflavin portion of FAD. In fact, that region of the map corresponds to a solvent-filled cavity near the surface of the protein. One possibility is that ADP is a vestigial fragment from a glutathione reductase-like precursor of TMADH. Another possibility is that FAD was originally present in TMADH but was cleaved during isolation. Modeling studies suggest that if FAD were present, it could lie quite close to the $[4Fe-4S]^{2+}$ and serve as a conduit for electrons to the electron transfer flavoprotein.

The mode of binding of FMN in TMADH is remarkably similar to that in flavocytochrome b_2 (13) and glycolate oxidase (14). The flavin-binding domain in each of these enzymes is made up of a parallel $\beta_8\alpha_8$ barrel structure. In all three structures, the FMN cofactors nearly coincide when the β barrels are optimally aligned. Thus, the covalent nature of the binding of FMN to TMADH does not result in a mode of binding significantly different from that in other, non-covalently bound FMN containing enzymes. The identification of more subtle differences in FMN binding between the covalently and non-covalently bound cases, such as alteration in the degree of bending or in the electronic structures of the flavin ring requires continued investigation.

Acknowledgement

This research has been supported by NIH grant No. GM-31611

References

1 Steenkamp, D. J., McIntire, W. and Kenney, W. C. (1978) J. Biol. Chem. 253, 2818-2824

2 McIntire, W. S., Edmundson, D. E., Hopper, D. J. and Singer, T. P. (1981) Biochemistry 20, 3068-3075.

3 Shamala, N., Lim, L. W., Mathews, F. S., McIntire, W., Singer, T. P. and Hopper, D. J. (1985) J. Mol. Biol. 183, 517-518.

4 Bellamy, H. D., Mathews, F. S., McIntire, W. and Singer, T. P. (1987) This volume.

5 Lim, L.W., Mathews, F. S. and Steenkamp, D. J. (1982) J. Mol. Biol. 162, 869-876.

6 Kasprzak, A. A., Papas, E. J. and Steenkamp, D. J. (1983) Biochem. J. 211, 535-541.

7 Steenkamp, D. J., and Mallinson, J. (1976) Biochim. Biophys. Acta 429, 705-719.

8 Lim, L. W., Shamala, N., Mathews, F. S., Steenkamp, D. J., Hamlin, R., and Xuong, Ng.h. (1986) J. Biol. Chem. 261, 15140-15146.

9 Thieme, R., Pai, E. F., Schirmer, R. H. and Schulz, G. E. (1981) J. Mol. Biol. 152, 763-782.

10 Wang, B. C. (1982) Methods Enzymol. 115, 90-112.

11 Lederer, F., Glatigny, A., Bethge. P. H., Bellamy, H. D. and Mathews, F. S. (1981) J. Mol. Biol. 148, 427-448.

12 Glehn, M., Stensland, B. and Gärtner, B. (1977) Acta Cryst. B33, 2388 2391.

13 Xia, Z.-x., Shamala, N., Bethge, P. H., Lim, L. W., Bellamy, H. D.,

14 Xuong, Ng.-h., Lederer, F. and Mathews, F. S. (1987) Proc. Natl. Acad. Sci. USA 84, 2629-2633.

15 Lindqvist, Y., and Brändén, C.-I. (1985) Proc. Natl. Acad. Sci. USA 82, 6855-6859.

16 Lesk, A. M. and Hardman, K. D. (1982) Science 216, 539-540.

17. Lesk, A. M. and Hardman, K. D. (1985) Methods Enzymol. 115, 381-390.

STRUCTURAL STUDIES OF P-CRESOL METHYLHYDROXYLASE

H. D. Bellamy and F. S. Mathews

Department of Cell Biology and Physiology, Washington University School of Medicine, St. Louis, MO 63110, USA

W. S. McIntire

Molecular Biology Division, Veterans Administration Medical Center, San Francisco, CA 94121, USA

T. P. Singer

Molecular Biology Division, Veterans Administration Medical Center, San Francisco, CA 94121, USA, and Departments of Biochemistry-Biophysics and Pharmaceutical Chemistry, University of California, San Francisco, CA 94143

Introduction

PCMH is an $\alpha_2\beta_2$ dimer of M_r about 116,000 found in a number of soil bacteria (1). It catalyzes the hydroxylation of p-cresol to first form p-hydroxybenzyl alcohol and then p-hydroxybenzaldehyde. One type of subunit, of M_r 50,000, contains FAD bound covalently to a tyrosine residue through the 8α methyl group of the flavin ring (2). The other, of M_r 8,000 (3) is a c-type cytochrome. The tetramer is very stable with respect to dissociation; however the subunits can be isolated by isoelectric focusing. The isolated flavoprotein subunit has only 2% of the catalytic activity of the intact enzyme (4). The amino acid sequence of the cytochrome subunit has been determined as has a portion of that of the flavoprotein subunit (3). A 6.0 Å resolution study of the structure showed that the flavoprotein subunits are tightly associated across a non-crystallographic 2-fold axis (5). The locations of the iron atoms are firmly established from an anomalous scattering difference Fourier map. The cytochrome subunits lie on the periphery of the dimer of flavoprotein subunits, each associated with a single subunit and lying on the dimer surface.

Flavins and Flavoproteins 1987

Results

Data have been collected to 2.8 Å resolution at the area detector facility at the University of California at San Diego (5). Crystals of the native enzyme and the two heavy atom derivatives utilized in the low resolution study were used. The crystals exhibit pronounced anisotropy in scattering. Scattering extends to at least 2.8 Å resolution in the _a_ and _c_ directions but only to about 3.5 Å resolution in the _b_ direction. The heavy-atom positions have been refined with anisotropic temperature factors and a 3.0 Å resolution map calculated. The map has been subjected to 3 cycles of solvent leveling (7), each with a new molecular envelope. The solvent content of the crystals is about 55%. The figure of merit improved from 0.27 to 0.68.

The solvent leveled map was averaged about the non-crystallographic 2-fold axis (8). The electron density corresponding to the cytochrome subunit was displayed on a molecular graphics system and the course of the polypeptide chain traced (figure 1). The heme was also placed. This procedure was facilitated by knowledge of the iron position and the approximate orientation of the cytochrome skeleton obtained at low resolution (5). The amino-acid sequence is presently being fitted to the density. As predicted from the low resolution study and the sequence (3)

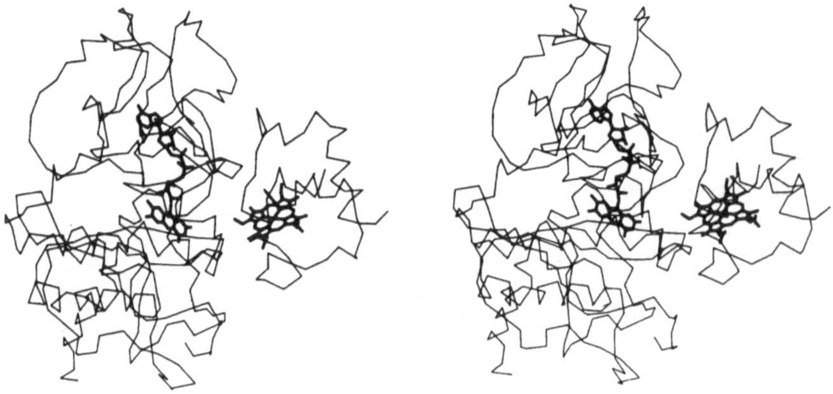

Fig. 1 Stereograph showing the flavoprotein and cytochrome subunits of p-cresol Methylhydroxylase. The polypeptide chain tracing shows the C_α positions. The heme and FAD are also shown. The other monomer binds to the surface of the flavoprotein at the left.

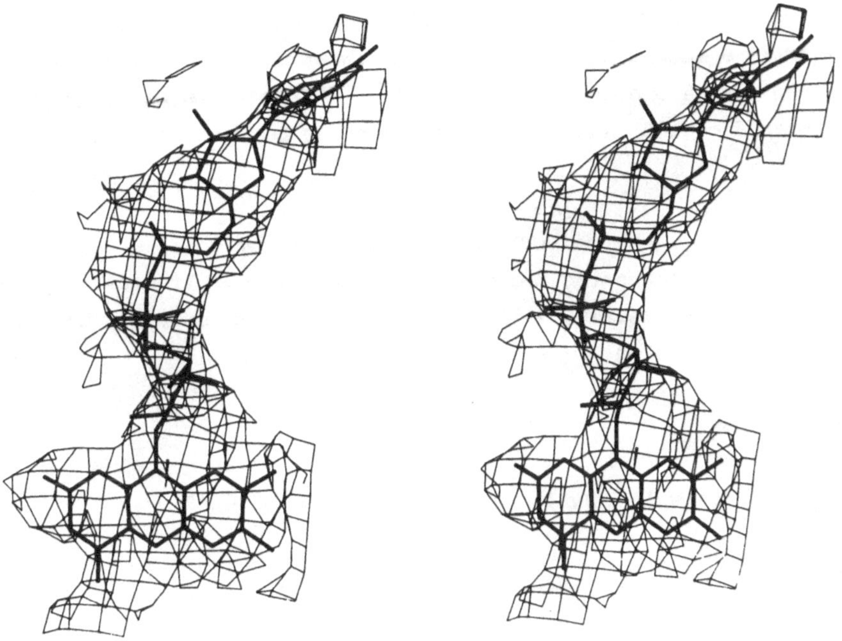

Fig. 2 Stereograph of the FAD and the density around it. In this orientation the heme (not shown) would be to the left with the iron 13 Å from atom O2 of the flavin ring.

the structure of the cytochrome subunit resembles that of cytochrome $\underline{c}_5$ (9) more closely than that of other cytochromes of known structure.

In order to analyze the flavoprotein subunit, contours of the sectioned electron density map were transferred to mylar sheets and stacked to form a minimap. The density map is very noisy in places and difficult to interpret. A tentative tracing of the polypeptide chain in the flavoprotein subunit has been obtained (figure 1). One domain near the N-terminal helix appears to contain a central mixed β-sheet arrangement connected in part by α-helices.

Density corresponding to the FAD has also been tentatively identified and fitted on the graphic system (figure 2). The nearest part of the flavin ring is about 13 Å from the known position of the iron atom. The relationship between the FAD and the heme is shown in figure 1. The flavin ring is close to the mixed sheet domain whereas the adenine portion is located in a second domain. The FAD is shown with the tyrosine-8α methyl bond on the opposite side from the heme but it is possible that the FAD is rotated by 180° and that the tyrosine is between the FAD and the heme.

Attempts to improve the quality of the map either by various phase refinement and/or extension techniques will continue. If necessary, data from additional derivatives will be recorded and from some of the same derivatives recollected.

Acknowledgement

This research was supported by NSF grant No. DMB-8518583.

References

1 Keat, M. H. and Hopper, D. J. (1978) Biochem. J. 175, 649-658.

2 McIntire, W. S., Edmundson, D. E., Hopper, D. J. and Singer, T. P. (1981) Biochemistry 20, 3068-3075.

3 McIntire, W., Singer, T. P., Smith, A. J. and Mathews, F. S. (1986) Biochemistry 25, 5975-5981.

4 McIntire, W., Hopper, D. J., and Singer, T. P. (1985) Biochem. J. 228, 325-335.

5 Shamala, N., Lim, L. W., Mathews, F. S., McIntire, W., Singer, T. P. and Hopper, D. J. (1986) Proc. Natl. Acad. Sci. U. S. A., 83, 4626-4630.

6 Xuong, Ng.h., Sullivan, D., Nielsen, C. and Hamlin, R. (1985) Acta Cryst. B41, 267-269.

7 Wang, B. C. (1982) Methods Enzymol. 115, 90-112.

8 Bricogne, G. (1976) Acta Crystallographica A32, 832-847.

9 Ghosh, D., O'Donnell, S., Furey,S., Jr., Robbins, A. H. and Stout, C. D. (1982) J. Mol. Biol. 158, 73-109.

THE CHEMICAL AND STEREOCHEMICAL COURSE OF OXIDATION OF 4-ETHYLPHENOL AND OTHER 4-ALKYLPHENOLS BY *p*-CRESOL METHYLHYDROXYLASE

William S. McIntire

Molecular Biology Division, Veterans Administration Medical Center, San Francisco, CA 94121.

Craig Bohmont

Department of Biochemistry and Biophysics, University of California, San Francisco, CA 94143.

Introduction

Previous work on the flavocytochrome *p*-cresol methylhydroxylase (PCMH) from *Pseudomonas putida* (N.C.I.B. 9869) had shown that this enzyme is able to oxidize a wide variety of 4-alkylphenols to the corresponding α-carbinols, some of which can be further oxidized to the α-carbonyl compounds (1,2). In order to describe the mechanism in more detail, an analysis of the stereochemistry of the oxidation of 4-ethylphenol was undertaken. A preliminary study indicated that this substrate was converted to 1-(4'-hydroxyphenyl)ethanol with a *R*:*S*-isomer ratio of 1:2 (3). The reaction was carried out with phenazine methosulfate (PMS) as the reoxidant for the enzyme at pH 9.5. PMS is known to be deleterious to enzymes, and can undergo photo- and alkaline degradation (4 and ref. therein). During isolation of the alcohol extensive purification was necessary to remove PMS and its breakdown products. Also, the high pH of the medium used for the enzymic conversion of 4-ethylphenol is not optimal (2). Finally, no account was taken of the effect the further oxidation to 4-hydroxyacetophenone would have, i.e., the *R*- and *S*-alcohols could have been oxidized at different rates or one not at all. All these factors could have affected the stereochemical outcome of the experiment. With this in mind, a new method was developed using cytochrome *c*, coupled to beef liver mitochondrial cytochrome oxidase (and ultimately O_2), as the reoxidant for PCMH

Flavins and Flavoproteins 1987

at pH 7.6. The new procedure allowed rapid product and stereochemical analysis. Surprisingly, in addition to 1-(4'-hydroxyphenyl)ethanol and 4-hydroxyacetophenone, 4-vinylphenol was a major product. In order to test the generality of the reaction, four other 4-alkylphenols were tested. This paper presents the results obtained using the new method.

Experimental procedures

The reaction mixtures contained 3 to 4 mM substrate, 50 µM horse heart cytochrome *c*, 0.4 mg/ml of cytochrome oxidase (purified to the red-green split stage, ref. 5) and 0.1 to 0.4 µM PCMH in 0.01 M Tris-HCl, I = 0.01 and pH 7.6 at room temperature. The mixture was well stirred to allow oxygen from the air to continuously dissolve. Aliquots of the suspension (1 mL) were quenched at various times with 0.5 mL of saturated $(NH_4)_2SO_4$ on ice. For the reactions involving 4-*n*-propylphenol and 4-isopropylphenol, 3 mL aliquots were added to 1.5 mL of the $(NH_4)_2SO_4$ solution on ice. Each sample was extracted 4 x 250 mL with ether and the final volume brought to exactly 1 mL. For HPLC analysis, 5 µL of the ether solutions were injected directly onto a 0.4 x 7.5 cm, 3 µ octadecylsilyl-silica gel column ($MeOH/H_2O$, 55:45; 1 mL/min; 280 nm). The reaction was also scaled to accommodate 250 mg of 4-ethylphenol. After quenching and ether extraction, the 1-(4'-hydroxphenyl)ethanol was purified by "flash chromatography" (6) on a 40 µ silica gel column (1 x 28 cm) with $CHCl_3$/ether, 3:2, in a single step. The alcohol was treated with acetic anhydride and triethylamine, 1:1 (v/v), and the optical purity of the diacetylated material was analyzed on a 0.4 x 25 cm "Pirkle column" (7).

For the reaction of *R,S*-1-(4'-hydroxyphenyl)ethanol with the enzyme, the ether extracts for each time point were carefully dried in a stream of Ar, taken up in 60:40 $MeOH/H_2O$, and the remaining alcohol purified on a 0.4 x 25 cm, 5 µ octadecylsilyl-silica gel column by HPLC using this solvent system. The stereochemistry of the alcohol was determined by a method similar to the one described earlier.

Rapid scan spectra were obtained during some HPLC runs using a 8 µL cuvette style flow cell (Hellma Cells, Inc.) placed in the cell holder of a Hewlett-Packard 8451A Diode Array Spectrophotometer.

Results and Discussion

The mechanism for the oxidation of 4-ethylphenol by PCMH is given by:

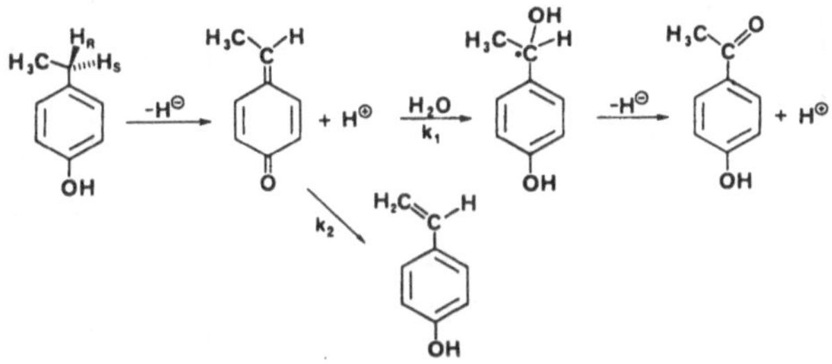

Mechanism I

There are three distinct and potentially stereoselective steps: 1) The removal of a prochiral hydrogen of 4-ethylphenol; 2) The addition of H_2O to the *re* or *si* side of the quinone methide; 3) The preferential oxidation of the *R*- or S-alcohol. The HPLC analysis for this reaction is given in Fig. 1. It is clear from this figure that a third product was formed, which was identified as 4-vinylphenol by comparison of its HPLC and spectral properties with those of 4-vinylanisole after methylation with CH_2N_2. It

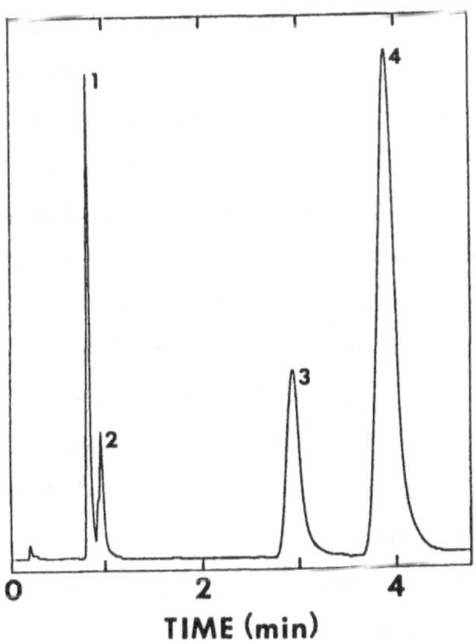

Fig. 1: HPLC run for the 4-ethylphenol reaction. 1) 1-(4'-hydroxyphenyl)-ethanol; 2) 4-hydroxyacetophenone; 3) 4-vinylphenol; 4) 4-ethylphenol.

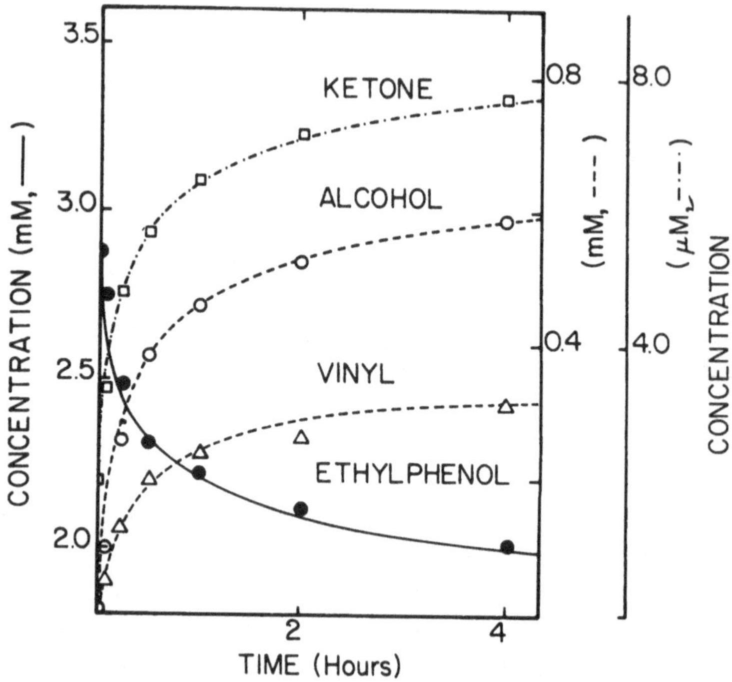

Fig.2: Time course for the reaction of 4-ethylphenol with PCMH.

is postulated that it forms by rearrangement of the quinone methide. Fig. 2 shows the time course for the reaction. Notice that only 1/3 of the 4-ethylphenol is oxidized. Preliminary studies indicate that the accumulating 4-vinylphenol is competitive inhibitor for this reaction.

A large amount of the enzymically produced 1-(4'-hydroxyphenyl)-ethanol was quantitatively isolated from the reaction mix and purified in a single step. This material was found to be > 97% S-isomer, which is in sharp contrast to the 67% S-isomer found earlier (3). As mentioned in the "Introduction", any number of factors can account for this discrepancy.

A reaction of *R,S*-1-(4'-hydroxyphenyl)ethanol with the enzyme was also monitored (Fig. 3). Evidently, the more rapid oxidation of the S-isomer progressively enriches the alcohol in the R-isomer; however, a large amount of ketone would need to be formed for this to contribute significantly to the optical purity of the alcohol. As seen in Fig. 2, the ketone is formed at about 1% of the concentration of the alcohol when PCMH and 4-ethylphenol are incubated.

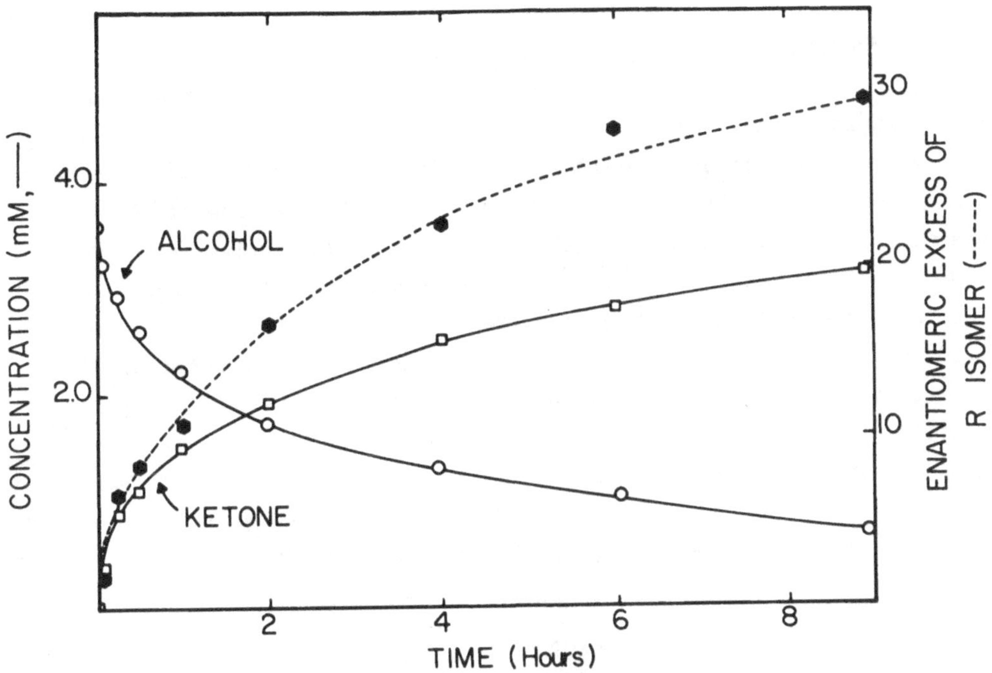

Fig. 3: The analysis of the reaction of *R,S*-1-(4'-hydroxyphenyl)ethanol with PCMH.

In order to explain the formation of both 4-vinylphenol and 1-(4'-hydroxyphenyl)ethanol from 4-ethylphenol, two mechanisms were considered. The common intermediate pathway given by Mechanism I, and the parallel pathway given by Mechanism II.

enzyme + 4-ethylphenol → → → alcohol [P_1]
enzyme + 4-ethylphenol → → → 4-vinylphenol [P_2]

Mechanism II

For I the ratio $[P_1]/[P_2] = k_1/k_2$ and for II it is equal to $v_1/v_2 = V_1/K_1[S_1]/(V_2/K_2[S_2])$ where $S_1 = S_2 =$ 4-ethylphenol (8). The value for $[P_1]/[P_2]$ was found to be invariant from 0.1 to 3.4 mM 4-ethylphenol as predicted by either mechanism. A value of 2.11 ± 0.08 was measured for this ratio. When 4-ethylphenol-α,α-d_2 and 4-ethylphenol-β,β,β,-d_3 were substrates, the ratios were 2.53 ± 0.05 and 4.05 ± 0.13, respectively. Steady-state and stopped-flow analysis (4) indicated that the V/K ratio is not sensitive to any process following the removal of the α-hydrogen. Therefore, if the chemistries of hydrogen abstraction for the two branches of Mechanism II are similar, the $[P_1]/[P_2]$ ratio should be little affected by the deuterium labeling substrates. In contrast, k_1 and k_2 in Mechanism I should be differentially sensitive to the isotopic substitution and the ratio should change. The results offer evidence in favor of Mechanism I.

A final test involved incubation of either 1-(4'-hydroxyphenyl)ethanol or 4-vinylphenol with reduced PCMH under Ar. A small amount of 4-vinylphenol and a large amount of the alcohol, respectively, were formed. Reactions did not occur when PCMH was absent or when 4-vinylphenol was incubated with oxidized PCMH in the presence of cytochrome *c*, cytochrome oxidase, and O_2. These observations are consistent with Mechanism II only if the conversion of 4-ethylphenol to these products is reversible. However, formation of equimolar amounts of 4-ethylphenol, and 4-hydroxyacetophenone are expected for the anaerobic experiment involving the alcohol, since reversal would produce 4-ethylphenol and oxidized enzyme, which would again be reduced by the large excess of the alcohol to form the

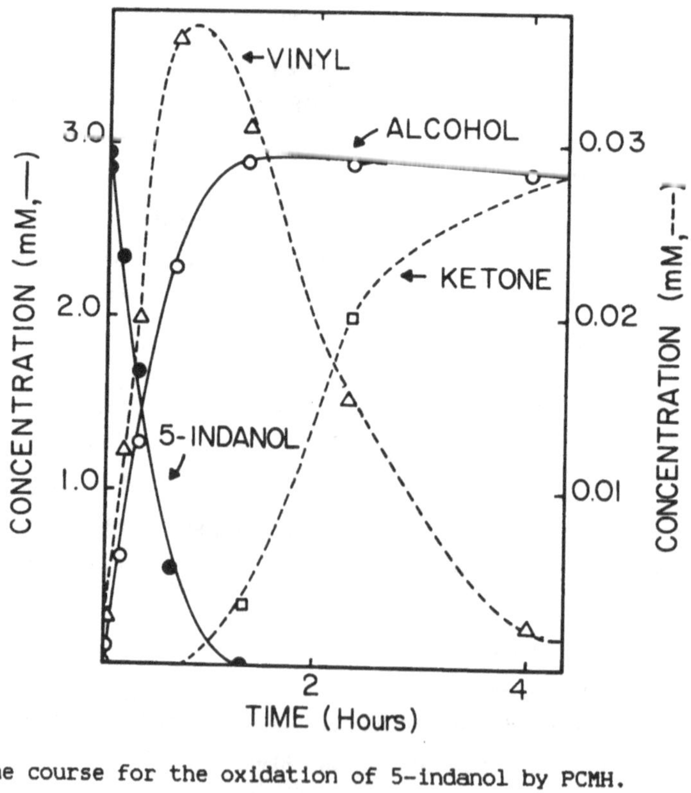

Fig. 4: Time course for the oxidation of 5-indanol by PCMH.

ketone. Neither 4-ethylphenol nor the acetophenone were observed. On the other hand, Mechanism I allows interconversion of the styrene and the alcohol through the intermediate quinone methide. In conclusion, all the evidence is compatible with Mechanism I and not Mechanism II.

The generality of the reaction was investigated with 4-*n*-propylphenol, 4-isopropylphenol, 6-hydroxytetralin (1), and 5-indanol (2) as substrates.

The products were identified by comparing chromatograms and rapid scan UV spectra obtained during HPLC runs on the ether extracts of the reaction mixtures. Fig. 4-7 present the kinetic data for these reactions. The corresponding alcohols and styrenes were produced from the propylphenols, 6-hydroxytetralin, and 5-indanol, but the ketone for the latter only. The alcohols formed from 4-*n*-propylphenol and 6-hydroxytetralin are probably

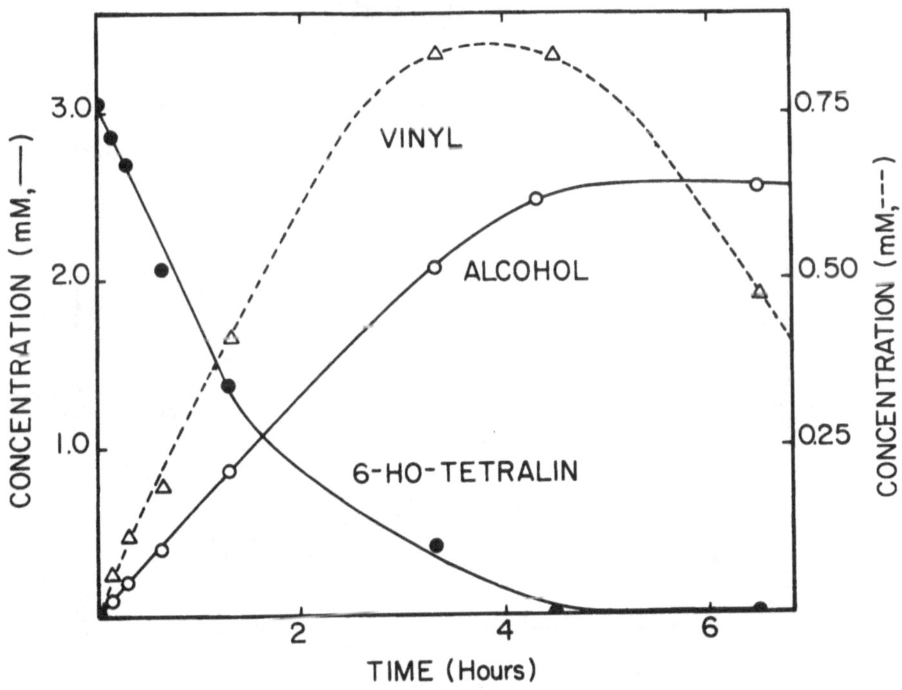

Fig. 5: Time course for the oxidation of 6-hydroxtetralin by PCMH.

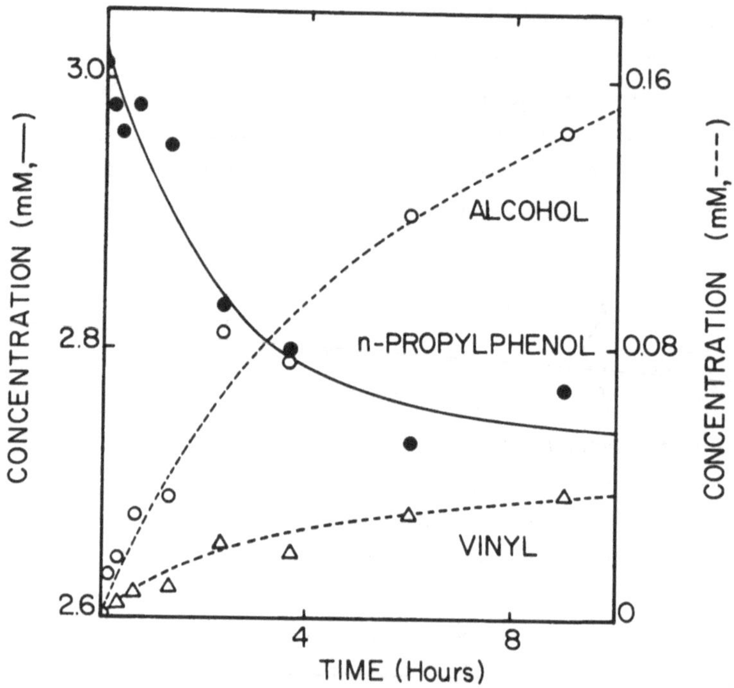

Fig. 6: Time course for the oxidation of *n*-propylphenol by PCMH.

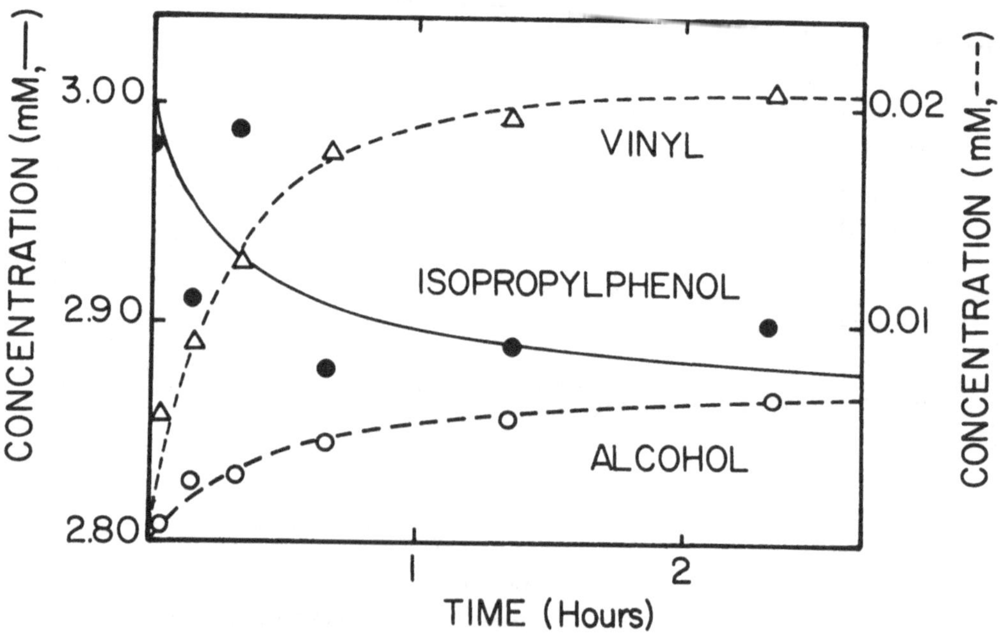

Fig. 7: Time course for the oxidation of isopropylphenol by PCMH.

poor substrates for PCMH. Interestingly, the two bicyclic alkylphenols were completely oxidized whereas only a small percentage of the propylphenols were oxidized. Since the yield was so poor, there was insufficient alcohol from the 4-*n*-propylphenol reaction for stereochemical analysis. Attempts to isolate the alcohols formed from the bicyclic alkylphenols failed, since the alcohols and/or styrenes seemed to be very prone to polymerization.

Pure *R*- and S-4-ethylphenol-α-d_1, *R*- and S-1-(4'-hydroxyphenyl)-ethanol, and *R*- and S-1-(4'-hydroxyphenyl)ethanol-α-d_1 have been synthesized in the laboratory of Dr. John C. Craig at the University of California in San Francisco. As a collaborative effort with his group, we are carrying out experiments with these compounds which will define the stereochemistry for all the important transformations, and which will also offer insight into the chemical mechanism by yielding pure primary and secondary deuterium isotope effects.

Acknowledgements

This work was supported by Program Project Grant HL 16251 from the National Institutes of Health, by Grant DMB 8416967 from the National Science Foundation, and by the Veterans Administration.

References

1. Keat, M.J., and Hopper, D.J. (1978) *Biochem. J.* **175**, 649-658.

2. McIntire, W.S., Hopper, D.J., and Singer, T.P. (1985) *Biochem. J.* **228**, 325-335.

3. McIntire, W.S., Hopper, D.J., Craig, J.C., Everhart, E.T., Webster, R.V., Causer, M.J., and Singer, T.P. (1984) *Biochem. J.* **224**, 617-621.

4. McIntire, W.S., Hopper, D.J., and Singer, T.P. (1987) *Biochemistry* **26**, in press.

5. Hatefi, Y., and Reiske, J.S. (1967) *Methods in Enzymol.* **10**, 225-231.

6. Still, W.C., Kahn, M., and Mitra, A. (1978) *J. Org. Chem.* **43**, 2923-2925.

7. Kasai, M., Froussio, C., and Ziffer, H. (1983) *J. Org. Chem.* **48**, 459-464.

8. Fersht, A. (1985) *Enzyme Structure and Function*. 2nd ed., W.H. Freeman and Co., New York, Chapter 3, pp. 98-120.

REDOX PROPERTIES OF TRIMETHYLAMINE DEHYDROGENASE

M.T. Stankovich
Department of Chemistry, University of Minnesota, 207 Pleasant St., S.E., Minneapolis, MN 55455

D. J. Steenkamp
Department of Chemical Pathology, Medical School, Observatory 7925, Cape Town, South Africa

Introduction

Trimethylamine dehydrogenase (TMADH) from bacterium W3A1 is composed of two identical subunits, each containing one iron-sulfur cluster (Fe-S) and one flavin mononucleotide which is covalently attached through a 6-S-cysteinyl linkage (1,2). Reaction begins with transfer of two electrons from trimethylamine to the flavin. These electrons are redistributed to form flavin radical and reduced Fe-S (3,4,5). We report here the results of spectroelectrochemical studies which reveal the thermodynamic basis for electron distribution in TMADH.

Results

First, the redox properties for the Fe-S cluster alone were determined by using TMADH which had been treated with phenyl hydrazine. This treatment produces a covalent flavin adduct which is redox inactive (6). Surprisingly, the two Fe-S in the inactive TMADH dimer differ in potential; one has $E^{0\prime}$ = 0.000 V and the other has $E^{0\prime}$ = -0.066 V versus the standard hydrogen electrode (SHE). This difference in redox potential

Flavins and Flavoproteins 1987

is suggested by the change in form of difference spectra during the titration.

Next, active TMADH was titrated. The flavins behave identically in native TMADH. The potentials of the individual electrons are $E_1^{0'}$ = -0.008 V and $E_2^{0'}$ = -0.029 V versus SHE for, respectively, the first and second electron transfers to the flavin. The maximum amount of flavin radical generated was 44%. In native TMADH, electrons are distributed between flavin and Fe-S, producing a substantial amount of flavin radical and reduced Fe-S. However, the two Fe-S again have different potentials ($E^{0'}$ = +0.070 V and $E^{0'}$ = -0.043 V) as shown in Figure 1. Interestingly, when comparing the potentials of native and inactive TMADH, the potentials of the Fe-S appear to be affected by the oxidation state of the flavin; the $E^{0'}$ of the more positive Fe-S is shifted considerably negative in inactive TMADH where flavin is formally reduced.

We then titrated TMADH in the presence of the substrate analog and inhibitor, tetramethylammonium chloride (TAC). Under this condition, flavin $E_1^{0'}$ undergoes a marked positive shift to +0.134 V and $E_2^{0'}$ shifts more negative than -0.4 V (reduced methyl viologen could not transfer the second electron). As shown in Figure 1, the Fe-S clusters are isopotential ($E^{0'}$ = +0.030 V) near their former midpoint potential. The 0.074 V Nernst slope is nearly ideal. The large changes in flavin potentials are consistent with inhibitor binding the flavin site which is also shown by x-ray crystallography (7). The large separation of the potentials of the flavin electrons produces essentially complete stabilization of flavin radical for inhibited TMADH. Of greater significance, flavin $E_1^{0'}$ is much more positive than $E^{0'}$ for Fe-S. Since electron transfer

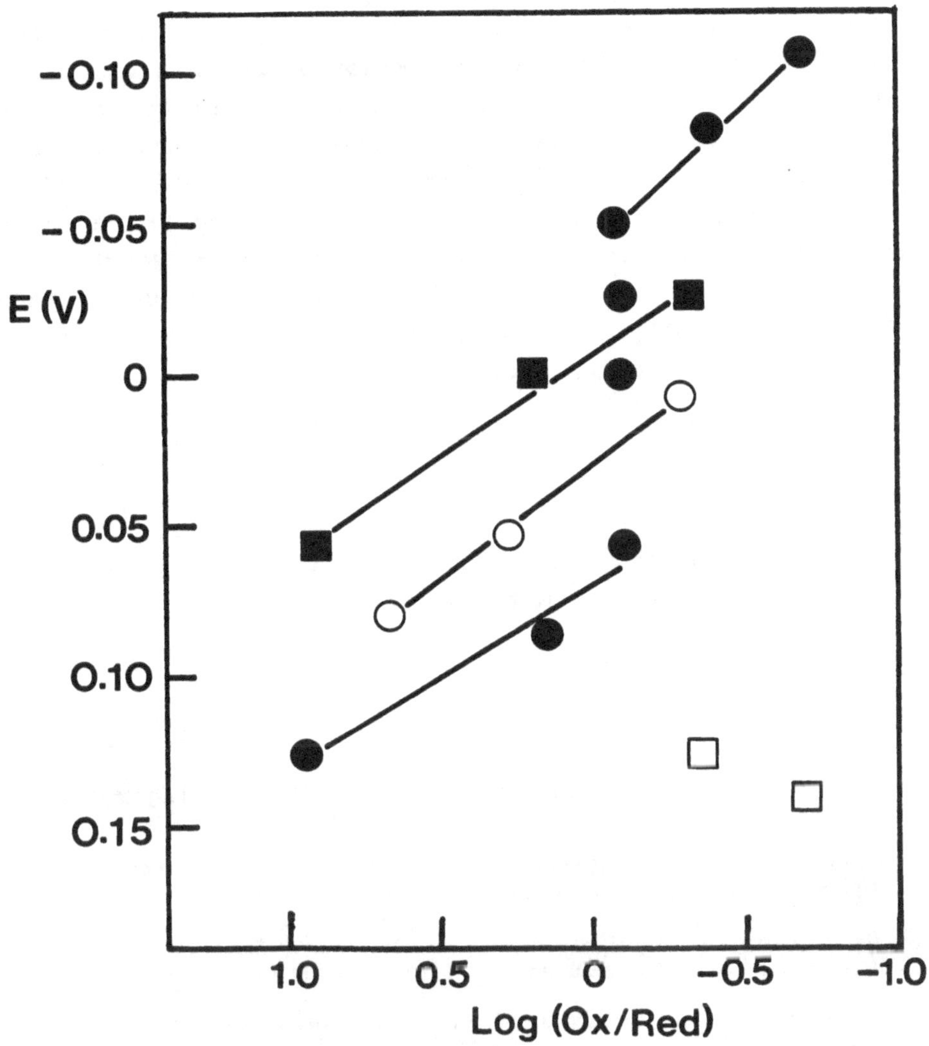

FIGURE 1. Nernst plot for TMADH. Potentiometric titration of 3-5 μM TMADH dimer was performed at 10 C in 0.05 M PPi, pH 8.5 with 0.10 mM methyl viologen as mediator of electron transfer (8) and with 2 μM of each of the poising dyes, methylene blue, pyocyanine, indigo disulfonate and phenosafranine. Tetramethylammonium chloride (0.2 mM) was also present for titration of inhibited TMADH. Values are plotted for reduction of the iron-sulfur cluster (circles) or for reduction of the oxidized flavin cofactor to flavin radical (squares) in native TMADH (closed symbols) or inhibited TMADH (open symbols).

to the natural electron acceptor, electron transferring flavoprotein, is believed to occur through the Fe-S (5), TAC-bound TMADH is trapped in a partially reduced state in which it cannot accept more electrons or transfer the one it has. From the shifts of flavin $E_1^{0'}$ and $E_2^{0'}$, the relative binding constants for TAC may be calculated. TAC binds TMADH which contains flavin radical 10^{+3} times tighter than TMADH which contains oxidized flavin and 10^{+7} times tighter than TMADH which contains fully reduced flavin. This is the thermodynamic basis for TAC inhibition.

Acknowledgement

This research was supported by the National Institute of General Medical Sciences (GM29344).

References

1. Hill, C.L., Steenkamp, D.J., Holm, R.H., Singer, T.P. 1977. Proc. Natl. Acad. Sci. USA 74, 547-551.

2. Steenkamp, D.J., McIntire, W., Kenney, W. 1978. J. Biol. Chem. 253, 2818-2824.

3. Steenkamp, D.J., Singer, T.P., Beinert, H. 1978. Biochem. J. 169, 361-369.

4. Steenkamp, D.J., Beinert, H. 1982. Biochem. J. 207, 233-239.

5. Steenkamp, D.J., Beinert, H. 1982. Biochem. J. 207, 241-252.

6. Nagy, J., Kenney, W.C., Singer, T.P. 1979. J. Biol. Chem. 254, 2684-2688.

7. Bellamy, H.D. 1985. Ph.D. Thesis, Washington University, St. Louis, MO.

8. Stankovich, M.T. 1980. Anal. Biochem. 109, 295-308.

REDOX PROPERTIES OF *ESCHERICHIA COLI* FUMARATE REDUCTASE

Brian A.C. Ackrell, Bruce Cochran, Edna B. Kearney, and Gary Cecchini

Department of Biochemistry and Biophysics, University of California, San Francisco, California 94143, and Molecular Biology Division, Veterans Administration Medical Center, San Francisco, California 94121.

Introduction

Fumarate reductase (FRD) of *Escherichia coli* is a membrane-bound flavoprotein produced during anaerobic growth when fumarate is the terminal electron acceptor. The catalytic properties of this enzyme bear a striking resemblance to those of cardiac succinate dehydrogenase (SDH), including that of being converted by tightly-bound oxaloacetate (OAA) to an inactive form that can be activated by reduction (1). There are indications from kinetic studies, however, that the OAA-FRD complex is less stable than the OAA-SDH complex. Both enzymes contain an 8α-[N(3)]-histidyl FAD moiety and the same types of Fe-S clusters: Center 1 (2Fe-2S), Center 2 (4Fe-4S), and Center 3 (3Fe-4S) (2,3). The redox potentials (Em) of these components, however, are such that SDH is predisposed for reduction of ubiquinone (Em = +77 mV) and FRD for reoxidation of reduced menaquinone (Em = -74 mV) generated during anaerobic growth of *E. coli*. Thus, the FAD/$FADH_2$ couple and Centers 1,2, and 3 have Em values of -90 (4), -5, -400, and +60 mV (5), respectively, in cardiac SDH. In FRD, the flavin potential has not been determined; the Em values for Centers 1,2, and 3 are -20, -330, and -70 mV, respectively (6). The role of the very low potential tetranuclear clusters (Centers 2) in these enzymatic functions is not yet understood.

Results and Discusssion

To complete the comparative picture, we determined the Em for the FAD/$FADH_2$ couple in FRD to see if it differs from that in SDH, and if

Flavins and Flavoproteins 1987

so, whether reduction of the FAD moiety will cause dissociation of bound OAA with restoration of activity, as it does in the cardiac enzyme. These aims were achieved by studying the reductive activation of OAA-bound FRD in redox titrations using mediator dyes and dithionite and ferricyanide as titrants, with a known concentration of OAA present to achieve ready equilibration between bound and unbound forms of the enzyme. Activation was monitored in the succinate-PMS assay (15°) (7), for which aliquots of enzyme were withdrawn anaerobically from the redox cell and delivered directly into the assay mixture, which contained 20 mM succinate; the latter also served to displace loosely-bound OAA from the enzyme and prevent its reversion to the deactivated form. Activity was shown to be a sigmoidal function of the oxidation-reduction potential (E_h). Titration data obtained at pH 7.8 with a preparation of

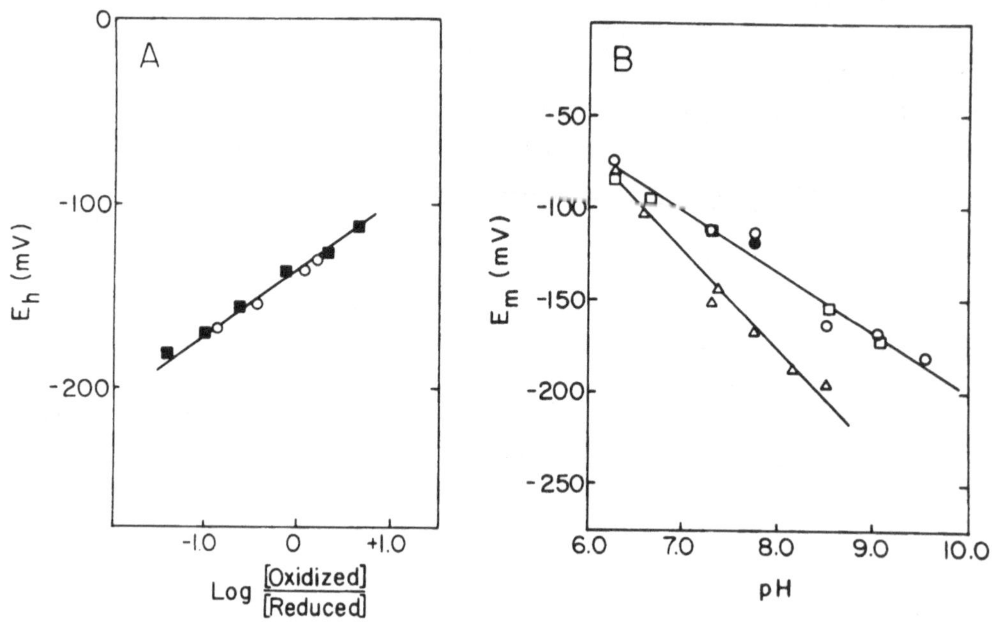

Fig. 1(A) Nernst plot for activation of FRD complex at pH 7.8 (titrants: dithionite (■) and ferricyanide (o)). (B) pH dependence of the midpoint potential (Em) for activation of *E. coli* FRD and SDH [(o), FRD complex; (□), *E. coli* membranes; (Δ), cardiac Complex II; (●), titration conducted at half concentrations of mediator dyes].

membrane-bound fumarate reductase are presented in Fig. 1A as a plot of E_h against log (oxidized components)/(reduced components) according to the Nernst equation, where oxidized component is equated with the deactivated fraction of the enzyme and reduced component with the active fraction of enzyme (4). The fact that the data points fit a single straight line having a slope of -30 mv implies that destabilization of the oxaloacetate bond is the result of reduction of only one of the several components of the enzyme and is associated with the uptake of $2e^-$. In titrations of membrane vesicles and isolated fumarate reductase complex, which gave identical results, it was further shown that the Em for activation, and thus of the critical component, was linearly dependent on pH in the range pH 6.3 to 9.6 and decreased 0.03 V/pH unit increase (Fig. 1B). Hence, as with SDH, reduction of the FAD moiety was responsible for the dissociation of bound OAA, since reduction of Center 1, 2, or 3 would require $1e^-$ and is pH-independent. Em values obtained for the FAD/$FADH_2$ couples of cardiac SDH and FRD were -90 and -50 mV (pH 7.0), respectively.* The data also indicate that activation involves a single H^+, consistent with reduction of FAD to its anionic hydroquinone form. Fig. 1B also shows that the Em for activation of SDH decreases 0.06V/pH unit increase, a 2 H^+ change. A major difference is thus implied in the processes for the two enzymes, which is not presently understood. Model studies have shown, however, that ionization of the imidazole substituent (pKr = 7.0) markedly influences the redox properties of 8α-N-imidazolylriboflavin and changes the slope of the Em versus pH plot from -60 to -30 mV (9), which is the slope observed for FRD. The possibility arises, therefore, that the pKr of the FAD-linked imidazole in SDH is significantly different from that in FRD so that the inflection point occurs outside the pH range studied or is too close to the pK of an ionizable group in the oxidized flavin to be detectable.

*Calculations include an adjustment for the presence of OAA in the titration, since this ligand preferentially binds to the oxidized forms of the enzymes. The binding affinities of OAA for the oxidized and reduced forms of FRD, determined as part of this titration, were 0.1 and 7 μM, respectively, about 10-fold higher than the corresponding values with cardiac SDH (8).

The tight binding of OAA to cardiac SDH has been attributed to thiohemiacetal bond formation with the active site thiol, which is stabilized by the conformation of the protein (10). This mechanism may also pertain in FRD, since both OAA and substrate afford protection against irreversible inhibition by the sulfhydryl reagent N-ethylmaleimide (unpublished data). The looser binding of OAA to FRD than SDH is nevertheless indicative of a difference in protein structure at the active site, as is also evidenced by the different E_m values for the $FAD/FADH_2$ couples.

Acknowledgment:

This work was supported by the Veterans Administration and National Institutes of Health Program Project grant HL-16251.

References

1. Cecchini, G., B.A.C. Ackrell, J.O. Deshler, R.P. Gunsalus. 1986. *J. Biol. Chem.* 261, 1808.

2. Weiner, J.H., P. Dickie. 1979. *J. Biol. Chem.* 254, 8590.

3. Morningstar, J.E., M.K. Johnson, G. Cecchini, B.A.C. Ackrell, E.B. Kearney. 1985. *J. Biol. Chem.* 260, 13631.

4. Ackrell, B.A.C., E.B. Kearney, D. Edmondson. 1975. *J. Biol. Chem.* 250, 7114.

5. Ohnishi, T. 1981. In: Mitochondria and Microsomes (C.P. Lee, G. Schatz, and G. Dallner, eds.). Addison-Wesley, p. 191.

6. Cammack, R., D.S. Patil, C. Condon, P. Owen, S.T. Cole, J.H. Weiner. 1984. In: Flavins and Flavoproteins (R.C. Bray, P.C. Engel, and S.G. Mayhew, eds.). Walter de Gruyter, p. 551.

7. Singer, T.P. 1974. *Methods Biochem. Anal.* 22, 123.

8. Kotlyar, A.B., A.D. Vinogradov. 1984. *Biochem. Biophys. Acta* 784, 24.

9. Williamson, G., D.E. Edmondson. 1985. *Biochemistry* 24, 7790.

10. Vinogradov, A.D., D. Winter, T.E. King. 1972. *Biochem. Biophys. Res. Commun.* 49, 441.

REACTIONS OF LUMIFLAVIN WITH AMINYL AND α-AMINO RADICALS OF PIPERAZINE

P.S. Surdhar and D.A. Armstrong
Chemistry Department, University of Calgary, Calgary, Alberta T2N 1N4

Introduction

Earlier studies (1,2) have shown that amine radicals react with lumiflavin (Fl) to produce •FlH which eventually decays by reaction [1] to Fl + FlH_2 (3,4).

$$2\ \bullet FlH \longrightarrow Fl + FlH_2 \qquad [1]$$

The radicals were produced by one or the other of the following two reactions:

$$AH_2 + \bullet OH \longrightarrow AH\bullet + H_2O \quad or \quad \bullet AH_2^{+} + OH^{-} \qquad [2]$$

$$AH_2 + Fl^{3} \longrightarrow AH\bullet + \bullet FlH \quad or \quad AH_2^{+} + Fl^{\bullet -} \qquad [3]$$

Here AH_2 represents the amine, Fl the lumiflavin and Fl^3 the triplet excited state.

Recent studies have demonstrated that amines give rise to nitrogen-centered oxidising radicals ($=N^{\bullet\pm}$) as well as the reducing α-amino radicals (-NH-$\dot{C}$H-) (5,6). In the light of this finding we have chosen to reexamine the reactions of piperazine (Pip, for structure see end of next section) radicals with Fl by the technique of pulse radiolysis. Details will be published in reference (6).

Results and Discussions:

Piperazine Radicals: Figure 1 shows the spectrum of the -NH-$\dot{C}$H- radical which can be formed preferentially in reaction [4] (7), and the combined spectra of this radical and the

Flavins and Flavoproteins 1987

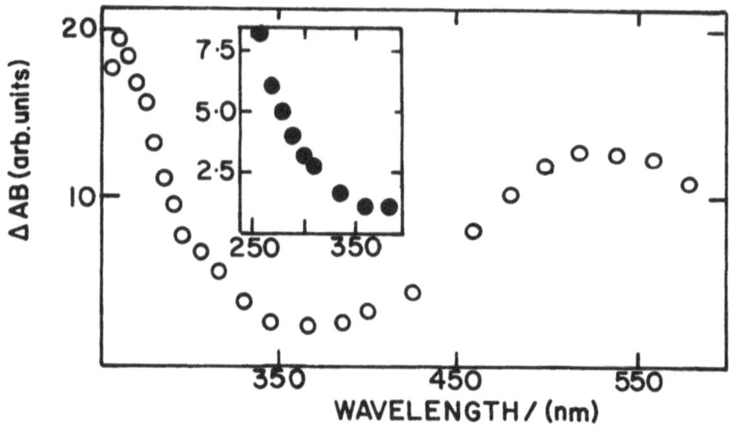

Figure 1. Combined spectra at 20 μs after pulsing 500 μM Pip in N_2O-saturated solutions. Inset: -NH-ĊH- in argon-saturated 500 μM Pip solutions containing 5 mM t-ButOOt-But and 2 M But-OH.

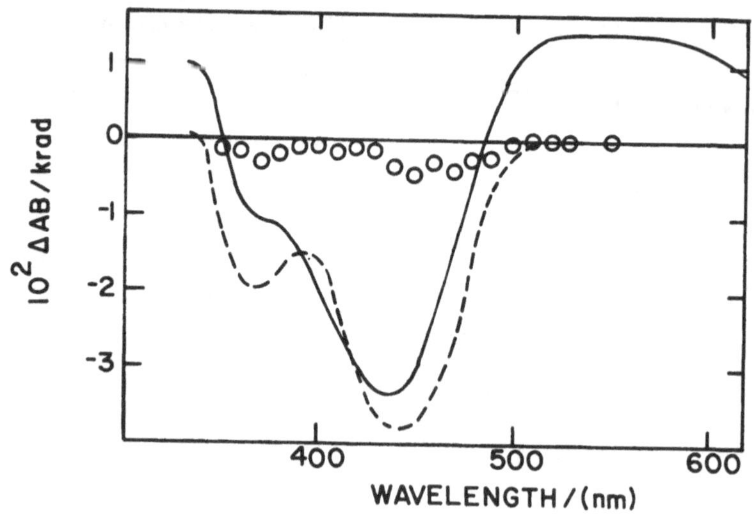

Figure 2. Difference spectra after pulsing N_2O-saturated Fl solutions: containing 0.01 M Pip at 20 μs corrected for N.•.N^+ absorbance (——), and after decay at 4 ms (0 0); after decay at 4 ms of •FlH in 20 mM HCO_2^- (---).

t-ButO• + Pip $\longrightarrow$ t-But-OH + Pip(-NH-ĊH-) [4]

nitrogen-centered species, which at pH 7 constitute 40 and 60% respectively of the total •AH radicals produced in reaction [2] (6). At piperazine concentrations > 5 mM the initially formed $=N^{\cdot\pm}$ species is complexed with piperazine to form an $N{\cdot}{\cdot}N^+$ species. It possesses a broad absorbance with a maximum at 520 nm. The -NH-ĊH- has a peak at 265 nm and negligible absorbance above 360 nm. This quite closely resembles the spectrum for the corresponding trimethylamine species (5).

Reactions with Fl: The solid line in figure 2 shows the difference spectrum produced at 20 µs after pulsing N_2O or argon-saturated solutions containing 0.1 M Pip. The spectrum when normalised at 550 nm to that produced by the $\cdot CO_2^-$ radical (3) showed an excellent fit and must be attributed to •FlH formed by the α-amino radicals of Pip in reaction [5]:

-NH-ĊH- + Fl $\longrightarrow$ -N=CH- + •FlH [5]

This process was shown to have a second order rate constant of $3.3 \pm 0.3 \times 10^9\ M^{-1}\ s^{-1}$.

In the pulse radiolysis of N_2O-saturated solutions the -NH-ĊH- and $N{\cdot}{\cdot}N^+$ radical species are formed in roughly equal proportions. The latter are inert to Fl. This means that 20 µs after pulse there are roughly equal amounts of •FlH and $N{\cdot}{\cdot}N^+$, both of which contribute to the absorbance at 550 nm. This absorbance decays by a second order process which must be attributed to a combination of reactions [1], [6] and [7].

•FlH + $N{\cdot}{\cdot}N^+$ $\longrightarrow$ Fl + 2 Pip + H^+ [6]

2 $N{\cdot}{\cdot}N^+$ $\longrightarrow$ Products [7]

The open circles show the spectrum at 4 ms after the pulse, when the decay of the radical species was complete. The dashed line shows for comparison the spectrum after the decay of •FlH formed in reaction [8]. Here reaction [1] is the only

$\cdot CO_2^-$ + Fl $(+H^+)$ $\longrightarrow$ CO_2 + •FlH [8]

decay process. The shapes of the two long time spectra are identical, showing that Fl + FlH_2 are the only products of the

Pip radicals which absorb in this region. The very small yield of FlH_2 produced by the decay of the •FlH + N.•.N$^+$ mixture, the total concentration of which was equal to that of the •FlH formed by •CO_2^-, indicates that reaction [6] is the dominant process. A second order rate constant of 2.7 x 10^9 M^{-1} s^{-1} was estimated for this reaction.
Similar low yields of FlH_2 were found from •AH radicals formed in reaction [3]. Thus the oxidising character of the aminyl radical form must be taken into account in flavin-amine interactions involving flavin photolysis or free radical formation by other means.

Piperazine at pH 7:

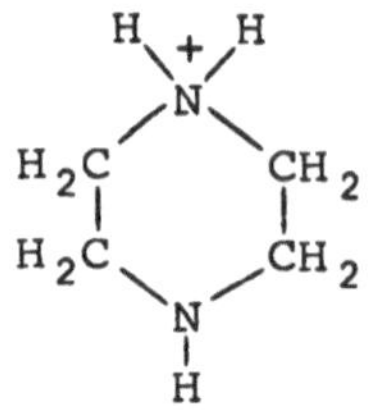

References

1. Surdhar, P.S., Bader, D.E., Armstrong, D.A.: Can. J. Chem. 63, 1357 (1985).

2. Heelis, P.F.: Chem. Soc. Rev. 11, 15 (1982).

3. Surdhar, P.S., Armstrong, D.A., Schmidt, K.H., Mulac, W.: Int. J. Rad. Phys. Chem. (in press).

4. Traber, R., Kramer, H.E.A., Hemmerich, P.: Biochemistry, 21, 1687 (1982).

5. Das, S., Sonntag, von. C.: Z. Naturforsch. 41b, 505 (1986).

6. Surdhar, P.S., Armstrong, D.A.: Submitted to Can. J. Chem. for publication.

7. Nazran, A.S., Griller, D.: J. Am. Chem. Soc. 105, 1970 (1983).

KINETIC PROBES OF THE PROPOSED RADICAL MECHANISM OF MONOAMINE OXIDASE

Mark C. Walker and Dale E. Edmondson
Dept. of Biochemistry, Emory University, Atlanta, Ga. 30322

Introduction

The neuropharmacological importance of Monoamine Oxidase (MAO) makes elucidation of its catalytic mechanism of particular interest. Studies by Silverman et.al (1,2) have suggested that the catalytic mechanism of oxidative deamination proceeds by a radical mechanism. In this mechanism, C-H bond cleavage at the α-position occurs subsequent to the first $1e^-$ transfer to the flavin. Formation of a substrate radical intermediate is suggested by product analysis of the action of MAO on a number of N,N-dialkyl amines. Previous steady state and stopped-flow kinetic experiments on MAO-B (3) show large deuterium kinetic isotope effects on benzylamine turnover ($^{D}V_{max}$= 6.7) and during anaerobic reduction ($^{D}k_{red}$= 8.8). Depending upon the relative equilibria and rate constants involved, one would expect the accumulation of a flavin semiquinone intermediate under conditions where C-H bond cleavage is rate limiting, however, none was observed. We have investigated both steady state and reductive-half-reaction kinetic properties of a series of para and meta substituted benzylamines (both α,α-^{1}H and α,α-^{2}H) as substrates for MAO-B. Previous studies of substituent effects on the rate of oxidation of benzylamine analogues by MAO show V_{max} values at air saturation are significantly altered by the nature of the substituent (4,5). As reported here, no evidence supporting the radical mechanism is found which suggests other mechanistic routes may be operative.

Flavins and Flavoproteins 1987

Results and Discussion

The results in Table 1 are a compilation of steady state and stopped-flow reductive-half-reaction kinetic data on the interaction of MAO-B with a series of para and meta substituted benzylamines. Both turnover and k_{red} values are

Table 1: Steady-state and Pre-steady-state Kinetic Parameters For Bovine Liver Mitochondrial Monoamine Oxidase

SUBSTRATE	V_{max} (min-1)	$^{D}(V_{max})$	$^{D}\frac{V_{max}}{K_m}$	$*k_{red}$ (min^{-1})	$^{D}(k_{red})$
Benzylamine	505	7.9	4.9	780	10
p-MeO-BA	244	8.9	8.1	340	11
p-Me-BA	62.8	9.1	7.0	60	8.7
p-F-BA	398	7.8	5.2	600	10
p-Cl-BA	181	7.0	5.0	304	8.0
p-Br-BA	78.1	7.2	7.2	120	7.1
p-CF_3-BA	12.9			10.8	9
p-NO_2-BA				3.6	
p-$(CH_3)_2$N-BA				161	12
m-MeO-BA	720	6.7	3.4	1690	11
m-Me-BA	770	6.0	3.8	1860	12
m-Cl-BA	667	5.5	3.2	2280	14
m-CF_3-BA	944	7.3	7.5	2050	13

* k_{red} are obtained as described in ref. (6)

altered by factors of 100 and 500, respectively, on alteration of the substituent. Reasonably large kinetic isotope effects are apparent for all substrates investigated. Although space does not permit a detailed discussion of these data, it is clear that carbon-hydrogen bond cleavage contributes substantially to rate limitation in catalytic turnover and in reductive-half-reaction experiments for all analogues tested. No spectral evidence was observed for any anionic or neutral flavin semiquinone formation with any of the proteo or deutero analogues tested in stopped-flow experiments. Fig. 1 shows representative traces for the progress of enzyme reduction by three different substrate analogues. Absorbance traces at 375nm followed an identical time course as the bleaching of 450nm absorbance.

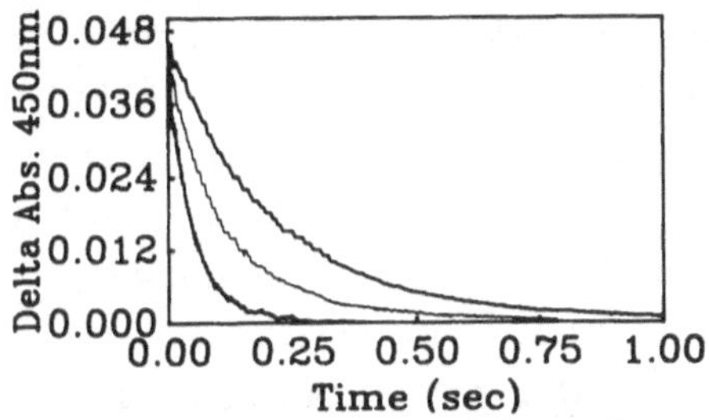

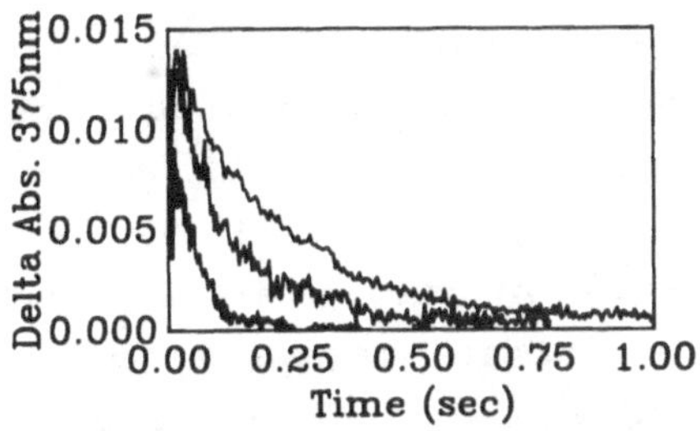

Fig.1 Flavin Bleaching due to reduction with substrate as monitored at 450 and 375nm in 50mM HEPES pH 7.50 with 5.0mg/ml RTX-100 (a catalytically hydrogenated form of Triton X-100) at 25 C. Substrate and enzyme concentrations are 2.0mM and 3μM respectively.

Slow: p-Methoxy-benzylamine (k_{obs}= 288 min^{-1})
Moderate: p-Fluoro-benzylamine (k_{obs}= 545 min^{-1})
Fast: m-Methoxy-benzylamine (k_{obs}= 1210 min^{-1})

If anionic flavin semiquinone were formed in any quantities at all, one would expect the bleaching of 450nm to precede any bleaching of 375nm absorbance since the difference in 450nm absorption between the anionic flavin semiquinone and flavin hydroquinone are minimal.

One substrate analogue (p-dimethylamino-benzylamine) is of interest in that its imine product can exist in several tautomeric and protonated forms. Several spectral intermediates are observed on following substrate conversion to the aldehyde product during catalytic turnover (Fig. 2).

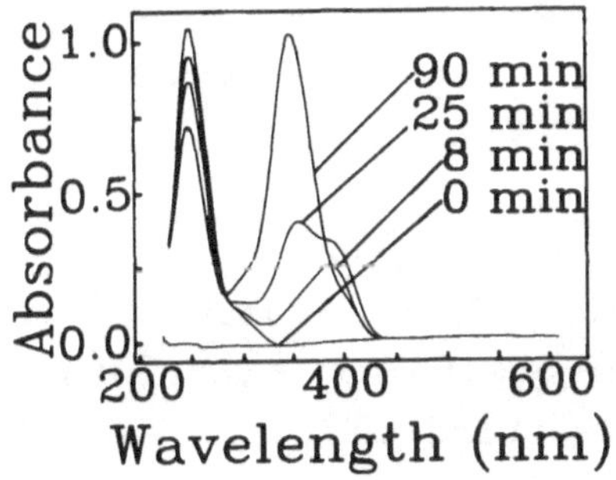

Fig.2 Accumulation of a reaction intermediate (395nm) prior to the appearance of the final product aldehyde (350nm) during catalytic turnover with 83μM p-dimethylamino-benzylamine. The reaction was performed in a 3ml cuvette at 25 C and 250μM O_2 using a catalytic amount of enzyme.

Anaerobic stopped-flow studies show a spectral intermediate (Fig. 3) with a λ_{max} at 375nm and an extinction of approximately $32{,}500M^{-1}cm^{-1}$, assuming a 1:1 stoichiometry with enzyme-flavin concentration. This intermediate is formed at a rate (88 min^{-1}) similar to catalytic turnover but slower than flavin 450nm bleaching (161 min^{-1}). Both steps exhibit an isotope effect of 12 which shows C-H bond cleavage precedes both steps. To test whether this intermediate was in fact the flavin anionic semiquinone intermediate, rapid freeze-quench EPR kinetic studies were performed under anaerobic conditions identical to those used in stopped-flow experiments. No paramagnetic intermediates were detected although the sensitivity of the experiment was designed such that 10% flavin radical (based on total enzyme) could have readily been detected.

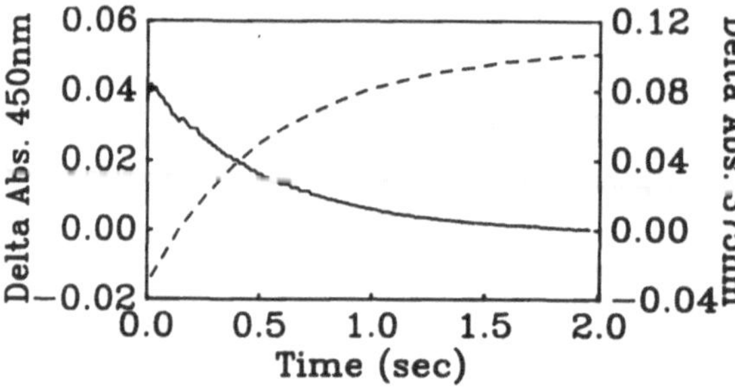

Fig.3 Absorbance changes during anaerobic reduction of MAO with p-dimethylamino-benzylamine. Conditions are identical to those in Fig.1. Solid; 450nm, k_{obs}= 112 min^{-1}. Dashed; 375nm, k_{obs}= 69 min^{-1}.

Our preliminary assignment is that the 375nm intermediate is the imine product of the p-dimethylamino- substituted substrate. This assignment is based on literature spectral values for similar substituted hydrazones (7). Subsequent decay of the 375nm species occurs at rates slower than catalysis which are due to nonenzymatic processes in conversion of the imine to the aldehyde.

In conclusion, the above data do not support the proposed radical mechanism for MAO catalysis. Further work is needed to verify the identity of the 375nm absorbing intermediate as the imine product. If so, the bleaching of oxidized flavin absorbance at 450nm would precede imine formation in an isotopically sensitive step. Whether this is due to flavin hydroquinone formation or to substrate-adduct formation remains for future investigations.

Acknowledgement

This work was supported by NIH Grant GM-29433.

References

1. Silverman, R.B., S.J. Hoffman, W.B. Catus. 1972. In: Flavins and Flavoproteins (V. Massey and C.H. Williams ed.) Elsevier North Holland. pp 213-216.

2. Silverman, R.B., P.A. Zieske. 1986. Biochemistry 25, 341-346.

3. Husain, M., D.E. Edmondson, T.P. Singer. 1982. Biochemistry 21, 595-600.

4. Hellerman, L., V.G. Erwin. 1986. J. Biol. Chem. 243, 5234-5243.

5. Zeller, E.A., M. Hsu, P.K. Li, J. Ohlsson, K. Subbarao. 1971. Chimia, v25, pp 403-405.

6. Simondsen, R.P., P.C. Weber, F.R. Salemme, G. Tollin. 1982. Biochemistry 21, 6366-6375.

7. Zvrerv, V.V., T.N. Pylaeva, A.P. Stolyarov, V.I. Savin, Yu.P. Kitaev. 1975. Izv. Akad. Nauk SSSR, Ser. Khimich. 24, 2699-2703.

STOPPED-FLOW STUDIES ON THE MECHANISM OF THE OXIDATION OF N-METHYL-4-PHENYL-1,2,3,6-TETRAHYDROPYRIDINE (MPTP) BY MONOAMINE OXIDASE

Rona R. Ramsay, Steven C. Koerber, and Thomas P. Singer

Departments of Biochemistry/Biophysics and Pharmaceutical Chemistry, University of California, San Francisco, and Molecular Biology Division, Veterans Administration Medical Center, San Francisco, California 94121.

INTRODUCTION

Monoamine oxidase (MAO) plays a crucial role in synaptic transmission and in the detoxification of amines in peripheral tissues. MAO exists in two forms (A and B) which differ in amino acid sequence in a limited region which confers specificity toward substrates and inhibitors (1). Steady state (2,3) and pre-steady state (3) experiments on MAO-B showed that phenylethylamine was oxidized by a binary mechanism but that with benzylamine a ternary complex was involved.

The unexpected observation that MAO (both A and B) oxidized the ternary amine, MPTP (4,5), provided an opportunity to extend these studies. The product, MPDP, is sufficiently stable to allow examination of the oxidative half-reaction in the presence of product. We outline here experiments on MAO-B which show that there are three competing pathways for the reoxidation of reduced enzyme (Scheme 1). Results from similar experiments using the other form of the enzyme are consistent with a binary mechanism for MAO-A.

METHODS

Purified preparations of MAO-A (6) and MAO-B (7) were used. Anaerobic stopped-flow spectrophotometry and the data analysis will be described (8). In studies of the oxidative half-reaction, reduced MAO was prepared by anaerobic titration with the substrate.

Flavins and Flavoproteins 1987

TABLE 1 : Kinetic Parameters for Monoamine Oxidase from Steady-State and Stopped-flow Half-reaction Experiments

	Steady-State		Reductive Half-Reaction		Oxidative Half-Reaction	
	turnover number s^{-1}	K_m mM	k_{obs} s^{-1}	K_D mM	app. 2nd order rate constant $M^{-1}s^{-1}$	app. 1st order rate const. at air saturation s^{-1}
A. MAO-A						
Kynuramine	1.9	0.11	1.0	0.12	6,500	1.7
MPTP	0.16	0.04	0.12	0.06	6,420	1.7
B. MAO-B						
Benzylamine	8.8	0.38	10.9	0.14	29,400	7.6
MPTP	3.3	0.30	3.7	0.037	23,300	6.0
MPDP	0.28	0.48	n.d.	n.d.	21,800	5.6
dithionite	--	--	n.d.	n.d.	6,000	1.6

SCHEME 1.

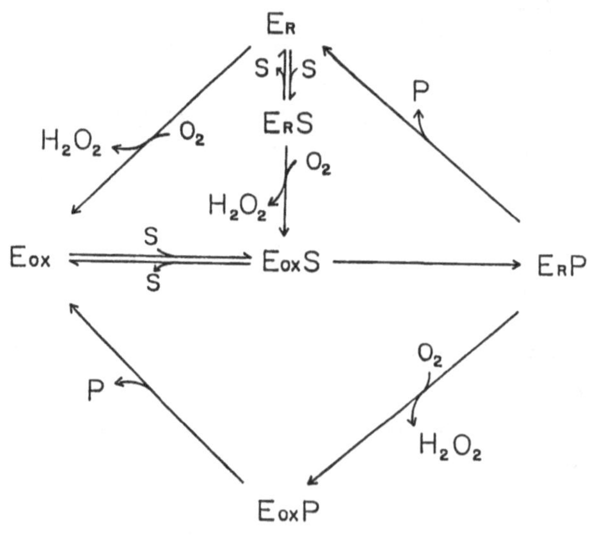

RESULTS AND DISCUSSION

Rate constants for the reductive and oxidative half reactions of MAO-A and MAO-B are compared with the steady-state parameters in Table 1. Considering first MAO-B, the observed rate constant for the reductive half-reaction with MPTP is the same as the turnover number in the steady state reaction and with benzylamine is marginally higher. The apparent second order rate constants for the oxidative half-reaction are not independent of the substrate concentration as would be predicted for a ping-pong mechanism., Rather, the presence of product or of substrate increases the observed rates from 6 X 10^3 M^{-1} s^{-1} for reoxidation in the absence of any ligand (3) to four times that in the presence of substrate or product. Benzylamine induced the fastest rate (Table 1), followed by MPTP and then $MPDP^+$. The dissociation constant for the enzyme $MPDP^+$ complex was less than 20 µM because the second order reaction rate was the same at all concentrations used (0.1-2 mM). In the presence of substrates, the data obtained was more complicated, in that both the slopes and the intercepts of the K_{obs} versus $[O_2]$ plots increased. The increase in the slope with increasing substrate concentration is evidence for a specific reduced enzyme-substrate complex which reacts faster with O_2 than does reduced enzyme alone. Saturation was not observed, so the dissociation constants must be larger than 2 mM.

The data suggest that the mechanism of MAO-B is not exclusively binary or ternary. Instead, we propose that the mechanism for a given substrate is determined by competition between the three alternate pathways shown in Scheme 1. The flux through each pathway is determined by the relative rate constants for the reoxidation of the free, reduced enzyme, the reduced enzyme-substrate complex and the reduced enzyme-product complex, and by their relative dissociation constants. The possibility that other flavin oxidases follow a similar scheme and are not limited to one or other mechanisms as previously suggested (3,9), must be considered. For MPTP, and presumably for MPDP also, the steady-state turnover is clearly dominated by the reductive half-reaction (Table 1). For benzylamine the values for turnover, reduction and oxidation at a saturation are numerically close, suggesting that the steady-state rate is determined by a function of both the reductive and oxidative half-reactions.

In contrast, for MAO-A, there is no evidence for any ternary complex formation. The oxidative half-reaction is second order and independent of added kynuramine or MPTP. However, with kynuramine (as for benzylamine with MAO-B), the rates of reduction and oxidation are numerically close (Table 1). With MPTP, reduction is limiting, Reoxidation of MAO-A was unaffected by the presence of low concentrations of $MPDP^+$, in contrast to the B-enzyme. The data are consistent with the operation of only the upper part of the loop ($E_RP \rightarrow E_R \rightarrow E_{OX}$) in Scheme 1 in MAO-A.

Acknowledgments: This work was supported by grants HL 16251 and NS 23066 from the National Institutes of Health, by DMB 8416967 from the National Science Foundation, and by the Veterans Administration.

REFERENCES

1. Singer, T.P. 1985. In: Biochemical Pharmacology and Toxicology, Vol. 1 (D. Zakim and D.A. Vessey, eds.). John Wiley & Sons, Inc., p. 229.

2. Pearce, L.B. and J. A. Roth. 1985. *Biochemistry* **24**, 1821.

3. Husain, M., D.E. Edmondson, and T.P. Singer. 1982. *Biochemistry* **21**, 595.

4. Chiba, K., A. Trevor, and N. Castagnoli, Jr. 1984. *Biochem. Biophys. Res. Commun.* **120**, 574.

5. Singer, T.P., J.I. Salach, N. Castagnoli, Jr., and A. Trevor. 1986. *Biochem. J.* **235**, 785.

6. Weyler, W. and J.I. Salach. 1985. *J. Biol. Chem.* **260**, 13199.

7. Salach, J.I. 1979. *Arch. Biochem. Biophys.* **192**, 128.

8. Ramsay, R.R., S.C. Koerber, and T.P. Singer. 1987. *Biochemistry* (in press).

9. Bright, H.J. and D.J.T. Porter. 1975. *Enzymes (3rd ed.)* **12**, 421-505.

METHOXATIN CONTAINING METHYLAMINE DEHYDROGENASE; ENZYME REDUCTION BY PRIMARY AMINES.

Robert B. McWhirter and Michael H. Klapper
Department of Chemistry, The Ohio State University, Columbus OH, 43210

There is a group of bacterial (1) and mammalian (2) dehydrogenases and oxidases that contain the orthoquinone methoxatin. Although a detailed mechanism has been proposed for amine oxidation by methoxatin analogs (3), there is little information on the mechanism of any methoxatoenzyme (4).

The methoxatoenzyme methylamine dehydrogenase (MADH) from the unclassified bacterium W3A1 catalyzes the reaction:

$$H\text{-}(CH_2)_nCH_2NH_3^+ + 2A + H_2O \longrightarrow H\text{-}(CH_2)_nCHO + 2A^- + NH_4^+ + 2H^+$$

with little specificity for the 1-electron acceptor A (5). Since the protein stabilizes the 1-electron reduced semiquinone (QH•) of methoxatin (6), we have been investigating the possible function of QH• in the enzymatic mechanism.

We find that, in the absence of an electron acceptor, primary amines reduce oxidized MADH (Q) to the fully reduced, hydroquinone species (QH_2) with a stoichiometry of 0.94±0.09 QH_2 formed per amine added (Figure 1). The isosbestic point at 356 nm indicates no intermediate QH•, and the pseudo-1st order reduction rate constant, k_{app}, may be hyperbolicly dependent on amine concentration. On the assumption of a rapidly formed preequilibrium intermediate in which there is a slower reduction, we calculated the equilibrium constant K, the 1st-order reduction rate constant k_r, and the 2nd order rate constant k_r/K (Table I). The intermediate assumed cannot be the MADH semiquinone, QH•, for reasons presented below.

Flavins and Flavoproteins 1987

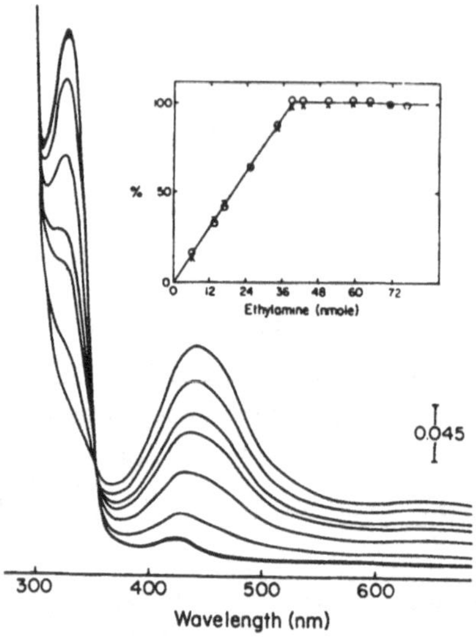

Figure 1. Reduction of MADH by ethylamine
Spectra were obtained within 1 minute after ethylamine addition to MADH (42 nmoles methoxatin) in 2.0 ml 0.1 M phosphate, pH 7.5 at 23 °C. Reduction proceeds from the fully oxidized MADH with λ_{max} at 444 nm to the fully reduced MADH with λ_{max} at 325 nm.
Inset - the percent absorbance changes at 440 and 325 nm with the best fit equivalence point at 39.5±1.0 nmoles.

Table I — Stopped-flow and steady-state kinetic constants

CH_3NH_2 Reduction of MADH			CH_3NH_2 steady-state oxidation		
k_r	s^{-1}	195(6)	k_c	s^{-1}	12.6(0.3)
K	µM	95(8)	K_m	µM	13(2)
k_r/K	$s^{-1}\mu M^{-1}$	2.1(0.2)	k_c/K_m	$s^{-1}\mu M^{-1}$	1.0(0.1)

Reactions conditions were those of figure 1. The rapid reduction with no electron acceptor added was monitored with by stopped-flow at 440 nm. The steady-state experiments were based on the protocol of Eady and Large (8) with phenazine ethosulfate substituted as electron acceptor.

Comparison of the constants in Table I suggest that, for methylamine at least, 2-electron methoxatin reduction is on the catalytic path.

With ethylamine as reductant and no electron acceptor present, acetaldehyde is produced within the manual mixing time. The reaction stoichiometry of 1.05±0.11, aldehyde released to QH_2 formed, was measured with the yeast alcohol dehydrogenase catalyzed NADH reduction of acetaldehyde. Therefore, aldehyde is released rapidly and before QH_2 reoxidation.

Kenney and McIntire (7) reported QH• formation upon addition of methylamine to MADH. We also observe the semiquinone, but only after QH_2 formation is complete. For example, 5 minutes after addition of butylamine to an anaerobic MADH solution most of the methoxatin is fully reduced, as seen by a peak at 325 nm (Figure 2b). Since QH• formation is very slow (seen by the simultaneous absorbance increase at 425 nm and decrease at 325 nm; Figure 2 inset), there can be no catalytic significance to direct amine reduction of Q to QH•.

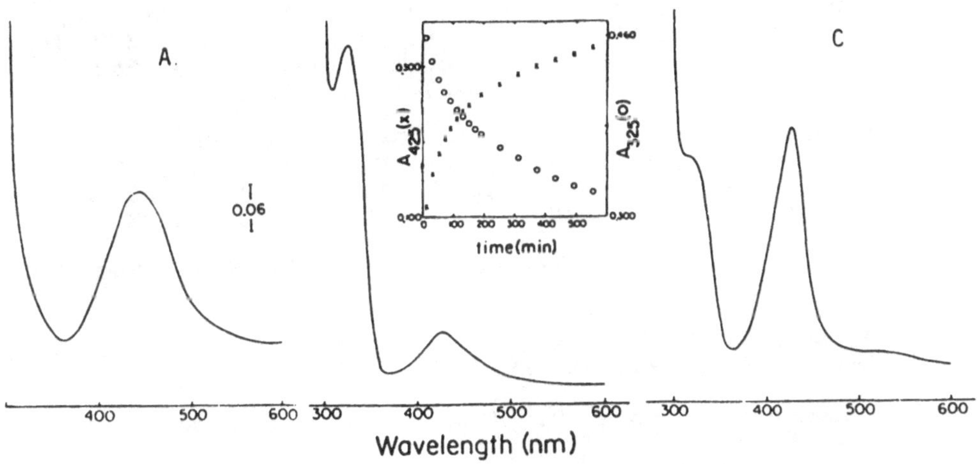

Figure 2. Slow formation of the MADH semiquinone
Spectra were obtained under anaerobic conditions in 0.1 M phosphate, pH 7.5 and 23 °C.
A - Spectrum of oxidized MADH (62 nmole of methoxatin)
B - Spectrum 5 min after addition of 44 nmole n-butylamine. Cerca 70% (expected 71%) of the methoxatin has been reduced based on absorbance changes at 324 and 440 nm.
Inset - Subsequent slow formation of QH•
x - A_{425}; o - A_{325}
C - Spectrum cerca 24 hr after addition of n-butylamine

We propose that the MADH catalyzed dehydrogenation of primary amines is a ping-pong reaction as outlined by the following, still incomplete scheme.

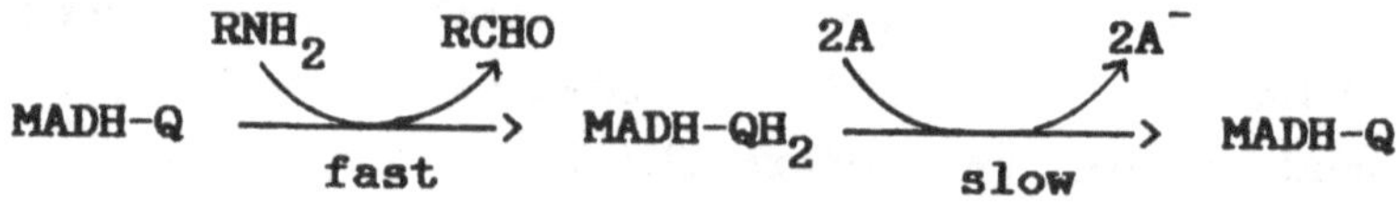

A similar proposal has been made with regard to methanol dehydrogenase (9). We are now attempting to find the NH_3 that must also be released and to determine whether the methoxatin semiquinone is involved in the slower QH_2 reoxidation.

This research was supported partially by NIH grant no. GM35718

References

1. J. Westerling, et al. (1979) Biochem. Biophys. Commun. 87 719; J. A. Duine, et al. (1979) FEBS Lett. 108, 443; R. DeBeer, et al. (1980) Biochim. Biophys. Acta 622, 370; M. Iwake, et al. (1983) Arch. Biochem. Biophys. 220, 253; F. Kawai, et al. (1985) Agric. Biol. Chem. 49, 1071

2. M. Ameyama, et al. (1984) Agric. Biol. Chem. 48, 561; C. L. Lobenstein-Verbeek, et al. (1984) FEBS Lett., 170, 305; M. Ameyama, et al., (1985) Agric. Biol. Chem. 49, 3623; P. R. Williamson, et al. (1986) J. Biol. Chem. 261, 16302

3. T. S. Eckert and T. C. Bruice (1983) J. Am. Chem. Soc. 105, 4431

4. J. A. Duine and J. Frank (1980) Biochem. J. 187, 213; C. Parkes and R. H. Abeles (1984) 23, 6355

5. R. Chandrasekar and M. H. Klapper (1986) J. Biol. Chem. 261, 3616

6. W. C. Kenney and W. McIntire (1983) Biochemistry 22, 3858

7. W. C. Kenney and W. McIntire (1984) in Proc. Int. Symp. Microb. Growth C1 Compd, ed., R. L. Crawford and R. S. Hanson (Am Soc. Microbiol., Washington, D. C.) 165

8. R. R. Eady and P. J. Large (1968) Biochem. J. 106, 245

9. R. de Beer, et al. (1983) Eur. J. Biochem 130, 105

INTERACTION OF BACTERIAL SARCOSINE OXIDASE WITH FOLATE DERIVATIVES

Marilyn Schuman Jorns and Kalla Kvalnes-Krick

Department of Biological Chemistry, Hahnemann University School of Medicine, Philadelphia, PA 19102

INTRODUCTION

Sarcosine oxidase from Corynebacterium sp. P-1 contains both covalently bound FAD [8α-(N^3-histidyl)-FAD] and noncovalently bound FAD. The enzyme is composed of 4 nonidentical subunits (M_r, 100,000, 42,000, 20,000 and 6,000). This unusual subunit composition is a genuine property of the enzyme and not a proteolytic artifact (1). A similar sarcosine oxidase has been isolated from Corynebacterium sp. U-96 (1-3). The presence of both covalent and noncovalent flavin distinguishes these enzymes from all other known flavoproteins. Their complex quaternary structure is also unique as compared with other sarcosine oxidizing enzymes. Mammalian sarcosine dehydrogenase is a monomeric protein, containing only covalently bound flavin (4-6). That the mammalian enzyme has a catalytically significant binding site for tetrahydrofolate (H_4folate) (7,8) prompted studies (9) to determine whether a similar property might be observed with a more complex bacterial enzyme.

RESULTS AND DISCUSSION

Formaldehyde was not detectable during growth of Corynebacterium sp. P-1 with sarcosine as carbon and nitrogen source. Oxidation of formaldehyde to formic acid was excluded by the absence of formaldehyde dehydrogenase activity in crude cell extracts. That the further metabolism of formaldehyde might involve entry into the one-carbon pool prompted studies to determine whether the organism's sarcosine oxidase might

Flavins and Flavoproteins 1987

possess a binding site for H_4folate. We found that the enzyme bound 2.19 mol H_4folate/mol enzyme, exhibiting the same affinity (K_D = 8.6 uM) for both ligands. Intracellularly, folate derivatives are present mainly as polyglutamate derivatives. Folate tetraglutamate derivatives predominate in corynebacteria (10). Binding parameters observed with tetrahydropteroyltetraglutamate ($H_4PteGlu_4$) (2.15 mol $H_4PteGlu_4$/mol enzyme, K_D = 15.4 uM) were similar to that observed with H_4folate. β-Mercaptoethanol and anerobic conditions were used during binding measurements to prevent oxidation of the reduced folate derivatives. Since β-mercaptoethanol reduced the flavins in sarcosine oxidase, complex formation involved fully reduced enzyme. However, similar binding was observed with oxidized and reduced enzyme in studies with 5-formyltetrahydrofolate (5-HCO-H_4folate), an air-stable derivative. The oxidized enzyme could also be specifically adsorbed onto an affinity matrix prepared by coupling 5-HCO-H_4folate to AH-Sepharose.

The oxidation of sarcosine did not depend on the presence of H_4folate nor did H_4folate affect the rate of sarcosine oxidation. However, H_4folate did affect formaldehyde production. In the absence of H_4folate, reaction of the enzyme with sarcosine under anaerobic conditions resulted in the formation of a stoichiometric amount of formaldehyde (0.94 mol/mol flavin). No formaldehyde was detected when the experiment was repeated in the presence of a H_4folate (Figure IA). 5,10-Methylenetetrahydrofolate (5,10-CH_2-H_4folate) was formed instead of formaldehyde (1.2 mol 5,10-CH_2-H_4folate/mol flavin), as judged by NADPH formation in a coupled assay with 5,10-CH_2-H_4folate dehydrogenase (Figure 1B). 5,10-CH_2-H_4Folate formation was catalyzed by sarcosine oxidase and not due to a nonenzymic reaction between free H_4folate and formaldehyde since: 1. The same rate of NADPH formation was observed in the coupled assay with an equivalent amount of authentic 5,10-CH_2-H_4folate; 2. A much slower rate was observed when the coupled assay was initiated by adding formaldehyde in the presence of H_4folate (Figure 1B). Formation of 5,10-CH_2-H_4folate was not

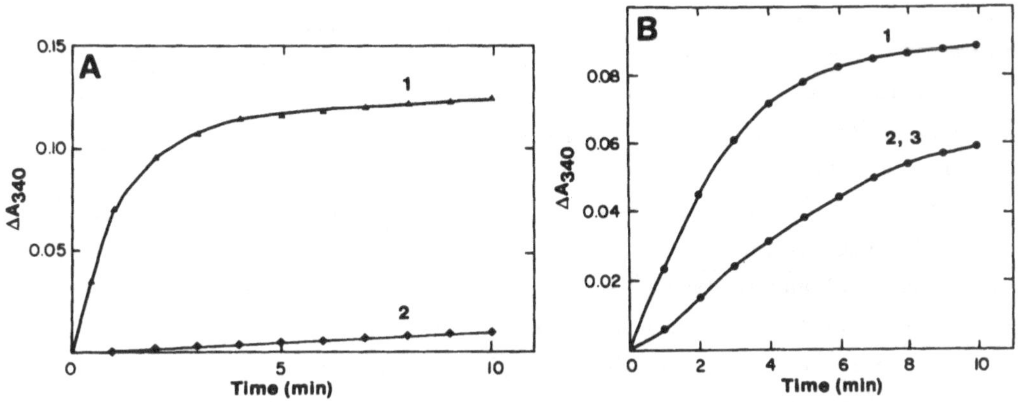

Figure 1: Effect of H4folate on catalysis. PANEL A: Formaldehyde was measured by the increase in A_{340} due to NADH formation in a coupled assay with formaldehyde dehydrogenase. Reactions were initiated in the absence (curve 1) or presence (curve 2) of [6R,S]-H_4folate (0.12 mM) by adding 4.0 mM sarcosine to an anaerobic solution of sarcosine oxidase (21.4 uM with respect to flavin). PANEL B: 5,10-CH_2-H_4Folate was measured by the increase in A_{340} nm due to NADPH formation in a coupled assay with 5,10-CH_2-H_4folate dehydrogenase. Reactions were initiated by adding 4.0 mM sarcosine (curve 1) or 20 uM formaldehyde (curve 2) to an anaerobic solution of 20.7 uM sarcosine oxidase plus 0.10 mM [6R,S]-H_4folate. The reaction shown in curve 2 was also conducted in the absence of sarcosine oxidase (curve 3).

catalyzed by sarcosine oxidase when exogenous formaldehyde was added to solutions containing sarcosine oxidase plus H_4folate in the absence (Figure 1B) or the presence of glycine (2.0 mM). The latter was tested since, in the catalytic reaction, H_4folate might react with the imine formed during sarcosine oxidation and the same imine should exist at equilibrium in solutions containing glycine plus formaldehyde. The results indicate that the formaldehyde (or a precursor), formed during sarcosine oxidation, reacts faster with enzyme-bound H_4folate as compared to the rate of its release into solution. This suggests that H_4folate serves primarily to trap formaldehyde as it is formed at the active site.

That a catalytically significant binding site for H4folate is present in both bacterial sarcosine oxidase and mammalian sarcosine dehydrogenase suggests that this is a general property retained by sarcosine oxidizing enzymes despite major differences in quaternary structure and coenzyme composition. That H_4folate and H_4PteGlu$_4$ bind to bacterial sarcosine oxidase with a 2:1 stoichiometry is noteworthy since only one of the 2 nonequivalent flavins in the enzyme acts as an input site for electrons (1,2) and the enzyme appears to have only a single sarcosine binding site (11). It is conceivable that only one of the 2 folate sites is functional. On the other hand, the presence of 2 functional folate sites, flanking the sarcosine site, would eliminate the need to displace 5,10-CH_2-H_4folate in the half-reduced enzyme with fresh H_4folate. This might enhance the efficiency of formaldehyde trapping during reduction of the half-reduced enzyme.

References

1. Kvalnes-Krick, K., Jorns, M.S.: Biochemistry 25, 6061-6069 (1986).
2. Jorns, M.S.: Biochemistry 24, 3189-3194 (1985).
3. Suzuki, M.: J. Biochem. (Tokyo) 89, 599-607 (1981).
4. Sato, M., Ohnishi, N., Yagi, K.: Biochem. Int. 3, 89-92 (1981).
5. Patek, D.R., Frisell, W.R.: Arch. Biochem. Biophys. 150, 347-354 (1972).
6. Cook, R.J., Misono, K.S., Wagner, C.: J. Biol. Chem. 259, 12475-12480 (1984).
7. Wittwer, A.J., Wagner, C.: J. Biol. Chem. 256, 4102-4108 (1981).
8. Steenkamp, D.J., Husain, M.: Biochem. J. 203, 707-715 (1982).
9. Kvalnes-Krick, K., Jorns, M.S.: submitted.
10. Foo, S.K., Cichowicz, D.S., Shane, B. Anal. Biochem. 107, 109-115 (1980).
11. Hayashi, S., Suzuki, M., Nakamura, S.: J. Biochem. (Tokyo) 94, 551-558 (1983).

SULFHYDRYL GROUPS AND THE BINDING SITE OF NONCOVALENTLY BOUND FAD OF *Corynebacterium* SARCOSINE OXIDASE

Haruo Suzuki, Yasuko Kawamura-Konishi

Department of Biophysical Chemistry, Kitasato University School of Medicine, Sagamihara, Kanagawa 228, JAPAN

Protein-flavin interaction is a key determinant of the function of flavoprotein. A sarcosine oxidase from *Corynebacterium* sp. U-96 is a unique flavoprotein containing both covalently and noncovalently bound FAD(1). The enzyme consists of four non-identical subunits(A,Mr 110,000;B, Mr 44,000;C, Mr 21,000;D, 10,000) and the covalently bound flavin is attached to subunit B(1). Hayashi *et al.*(2) reported that the enzyme contains 8 sulfhydryl(SH) groups per mol of enzyme and is completely inactivated by the modification of 2 SH groups by iodoacetamide(IAA). Moreover, the carboxyamidation of the SH groups was prevented by sodium acetate, a competitive inhibitor with respect to sarcosine(2). This suggests the presence of the SH groups at a sarcosine-binding site of the enzyme. On the other hand, Jorns(3) reported that sarcosine reacts with noncovalent flavin and oxygen with covalent one. These facts make us to think that IAA-reactive SH groups must be present near noncovalent FAD and to work on the reactivity of SH reagents with holo- and semiapoenzymes.

Firstly, we incubated holoenzyme with ^{14}C-IAA for given times at 30 °C and the reaction mixture was analyzed by sodium dodecyl sulphate polyacrylamide gel electrophoresis(SDS/PAGE) (Fig.1). ^{14}C-IAA was incorporated selectively into subunit B and the ^{14}C-incorporation was inhibited by sodium acetate. Since acetate is a competitive inhibitor of enzyme(2,4), these SH groups might be located near acetate-binding site, that is, sarcosine-binding site of enzyme. Moreover, Hayashi(5) suggested that the covalent and noncovalent FAD have different

Flavins and Flavoproteins 1987

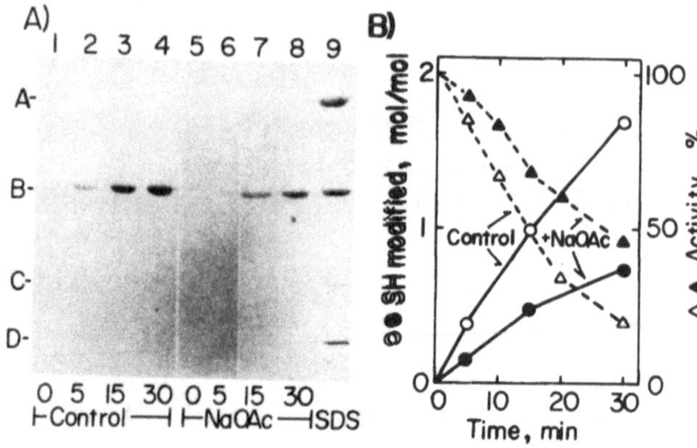

Fig.1 A) The enzyme(2.1 μM) was incubated with 1 mM ^{14}C-IAA in the presence(lanes 5-8) and absence(lanes 1-4) of sodium acetate in 50 μℓ of 20 mM pyrophosphate buffer(pH 8.0) for given times at 30 °C, then 1 μℓ of 1 M mercaptoethanol and the 15 μℓ of buffer(0.31 M Tris-HCl,pH 6.8, 10% SDS,50% glycerol, 25% mercaptoethanol) were added. After boiling the mixture,20 μℓ aliquot of the mixture was analyzed by SDS/PAGE(12.5% acrylamide, 0.33% bisacrylamide). Lane 9, the incubation for 5 min at 60 °C in the presence of 0.2% SDS. Autoradiograph of the dried gel is shown. From the relative intensities of the exposed film, mol SH in A,B,C and D subunits were calculated to be 3.1±0.2(3), 2.5±0.2(2-3), 1±0.1(1) and 1.2±0.3(1-2), respectively. B) The residual activity of enzyme(7.4 μM) after incubation with 1 mM IAA in the presence(▲) and absence(△) of 50 mM sodium acetate. Band intensities of the exposed film shown in Fig.1A are also plotted agaist the incubation time in the presence(●) and absence(o) of sodium acetate. The total number of SH groups per mol enzyme was determined to be 7.7 by the use of DTNB.

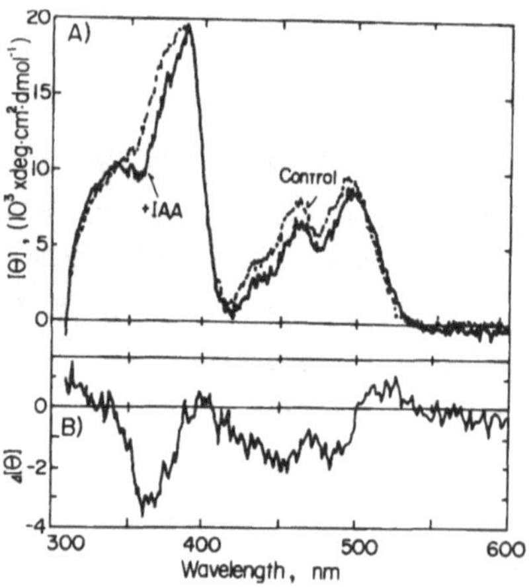

Fig.2
A) CD spectrum of the oxidized form of enzyme (82.4 μM) after incubation for 30 min at room temperature with 1 mM IAA (solid line). The dashed line shows a CD spectrum of the oxidized form of enzyme.
B) The difference CD spectrum of the oxidized form of enzyme(solid line minus dashed line in Fig.2A)

roles in the catalysis, and Jorns(3) also suggested that noncovalent FAD reacts with sarcosine and covalent FAD with oxygen. These facts indicate that the IAA-reactive SH groups of subunit B are located near noncovalent bound FAD. It is therefore possible that the modification of enzyme with IAA may change the conformation around the flavin moiety. To check this, we measured the absorption and CD spectra after incubation of enzyme with IAA. In the visible wavelength region, no change in the absorption spectrum was observed(data not shown), but significant change in CD spectrum was observed (Fig.2). On the other hand, the far-UV CD was not changed after incubation with IAA(data not shown). These data suggest that the IAA-treatment of enzyme did not cause any detectable change in the secondary structure, but a local conformation around the flavin moiety was changed by the IAA-treatment of enzyme.

The amounts of 5,5'-dithiobis(2-nitrobenzoic acid)(DTNB) -reactive SH groups of the holoenzyme was less than one mol per mol of emzyme, but the removal of noncovalent FAD exposed 2 SH

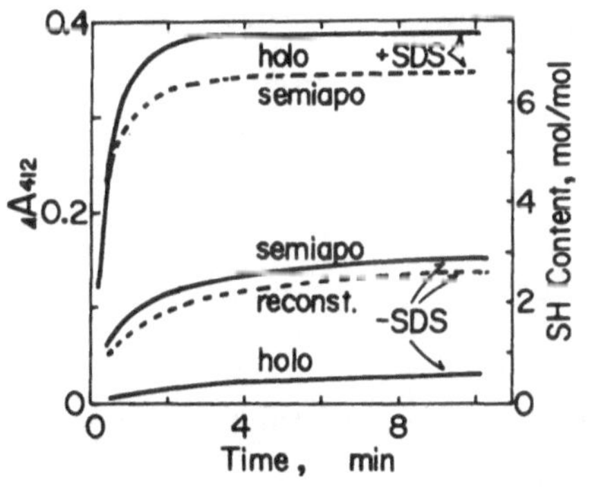

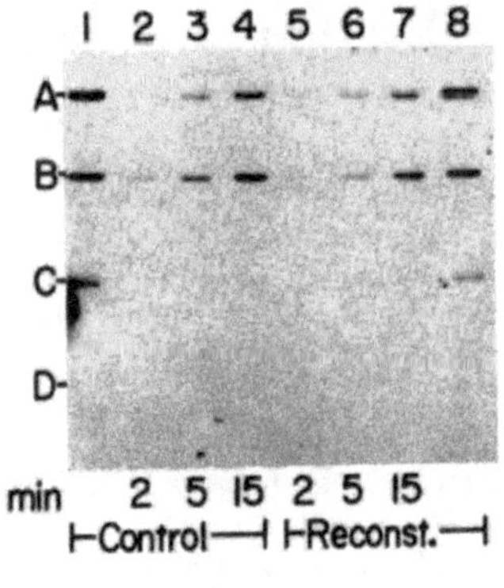

Fig.3(Left) Effect of time on the reaction of enzyme with DTNB. The enzyme(3.8 μM) was incubated with 0.6 mM DTNB in the presence or absence of SDS at room temperature. The absorbance change at 412 nm and the calculated mol SH reacted per mol of enzyme is shown.
Fig.4(Right) Effect of time on the incorporation of ^{14}C-IAA into subunits of semiapoenzyme(lanes 1-4) and reconstituted enzyme(lanes 5-8). The enzyme(1.9 μM) was incubated for given times with IAA in 50 μℓ incubation mixture and the mixture was processed as in Fig.1A. The autoradiograph of the dried gel is shown. Lanes 1 and 8 show the incubation of enzyme with SDS.

groups of enzyme(Fig.3). Experiments similar to those in Fig.1 showed that ^{14}C-IAA mainly incorporated into subunit B and the rate of ^{14}C-incorporation into subunit B seemed to be greater for semiapoenzyme than holoenzyme(compare Fig.1 with Fig.4). Removal of noncovalent FAD does not change the secondary structure of enzyme(6), and the reconstituted enzyme(12% active) had lower reactivity with ^{14}C-IAA(Fig.4), so noncovalent FAD may be located near the SH groups of subunit B.

These results make us to propose that both covalent and noncovalent FAD are localized in subunit B to facilitate the electron transfer between these two FAD groups(7). This proposal can be clarified directly by the binding experiments of FAD analogs with semiapoenzyme when we get highly active semiapoenzyme to be reconstituted.

Acknowledgements

We are grateful to Dr.M.Suzuki, Noda Institute for Scientific Research, for his kind supply of sarcosine oxidase, and Prof.S.Horie, Dept. of Biochem., and Dr.S.Hayashi-Taguchi for the valuable discussions during the course of this work.

References

1) Suzuki,M. 1981. J.Biochem., 89, 599

2) Hayashi,S., M.Suzuki, S.Nakamura. 1983. J.Biochem., 94,551

3) Jorns,M.S. 1985. Biochemistry, 24, 3189

4) Hayashi,S., M.Suzuki, S.Nakamura. 1983. Biochim.Biophys.Acta, 742, 630

5) Hayashi,S. 1984. J.Biochem., 95, 1201

6) Hayashi,S., S.Nakamura, M.Suzuki, 1980. Biochem.Biophys. Res.Commun., 96,924

7) Kawamura-Konishi,Y. and H.Suzuki, 1986. Seikagaku, 58,743

DIMETHYLGLYCINE DEHYDROGENASE FROM PIG LIVER MITOCHONDRIA: PURIFICATION AND PROPERTIES

Chikako Ichihara, Nobuko Ohishi, Kunio Yagi
Institute of Applied Biochemistry, Yagi Memorial Park, Mitake, Gifu 505-01, Japan

Introduction

Sarcosine and dimethylglycine dehydrogenases (EC 1.5.99.1 and EC 1.5.99.2) catalyze the oxidative N-demethylation of sarcosine and dimethylglycine (DMG), respectively, at the last stage of choline degradation. These two flavoproteins were purified from rat liver and pig liver mitochondria (1-3), and were found to have folate-binding capacity (2). Dimethylglycine dehydrogenase can demethylate, in addition to DMG, sarcosine, N-methylalanine, and ε-N-methyllysine (4). The kinetic analysis of the enzyme reaction showed that the reaction rate is not saturable at high concentrations of the substrate, DMG, indicating that this enzyme has a second binding site for DMG, which enhances the reaction rate (5). This result suggests the possibility that the reaction is regulated through the second binding site. To clarify this point, we further purified this enzyme from pig liver mitochondria to homogeneity and tried to characterize the site responsible for this phenomenon.

Results and Discussion

Dimethylglycine dehydrogenase has been purified from pig liver mitochondria by ammonium sulfate fractionation, gel filtration, and DEAE-Sephadex and affinity chromatographies.

Flavins and Flavoproteins 1987

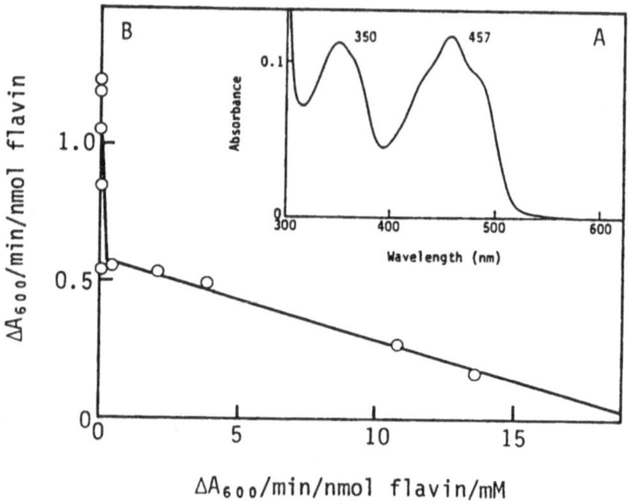

Fig. 1. Absorption spectrum of dimethylglycine dehydrogenase purified from pig liver mitochondria and Eadie-Hofstee plot of its activity toward dimethylglycine. A, Sample contained 0.77 mg/ml of the protein in 10 mM Tris-HCl buffer (pH 7.4) with 0.05 M NaCl. B, The activity (the enzyme, 0.1 nmol) was determined by measuring the reduction of DCIP through PMS. The decrease in absorbance at 600 nm was assayed with various concentrations of dimethylglycine at 37°C and pH 7.4.

The purified enzyme migrated as a single band in SDS-polyacrylamide gel electrophoresis, and its molecular weight was estimated to be approximately 90,000. It contained 1 mole of covalently bound flavin/mole of protein, and its absorption spectrum had maxima at 350 and 457 nm at pH 7.4 (Fig. 1A).

The purified enzyme showed dehydrogenase activities toward both DMG and sarcosine. The K_m values for DMG and sarcosine were 26 μM and 29.4 mM, respectively. Under anaerobic conditions, approximately 30 moles of sarcosine per mole of flavin were needed to bleach the 457 nm absorption, whereas one DMG per flavin was enough to bleach it, indicating that sarcosine is a poor substrate for dimethylglycine dehydrogenase.

The rate of the dehydrogenase reaction at high concentrations of DMG showed a similar non-saturable kinetics as reported for dimethylglycine dehydrogenase from rat liver (5)

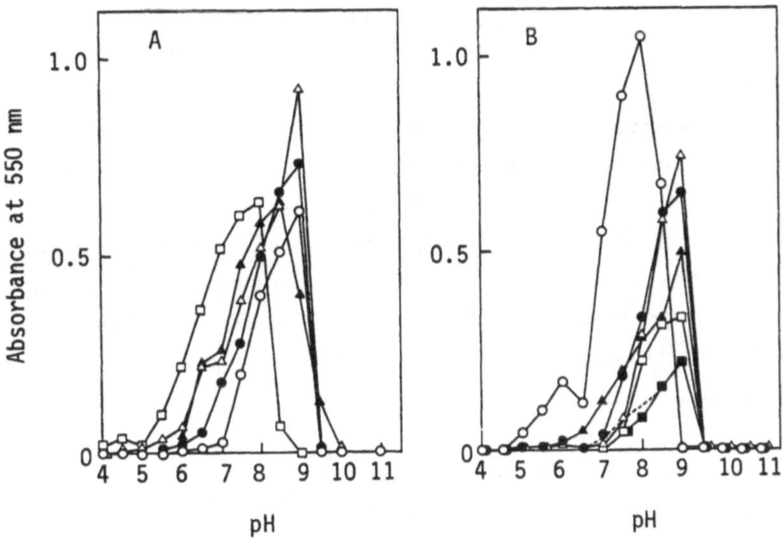

Fig. 2. The pH profile of dimethylglycine dehydrogenase reaction at various concentrations of dimethylglycine and effects of accelerators. The activity was estimated by measuring the reduction of NBT through PMS at 550 nm in 0.1 N HCl. A, The enzyme (0.1 nmol) was incubated with 0.2 (○), 2 (●), 20 (△), 100 (▲), and 200 (□) mM dimethylglycine at 37°C for 30 min at various pH's. B, The enzyme (0.05 nmol), NBT, PMS, and 2 mM substrate dimethylglycine were incubated with 0.1 M acetoacetic acid (○), 10 mM methoxyacetic acid (●), 0.1 M acetic acid (△), 0.1 M sarcosine (▲), or 1.3 M formic acid (□), or without accelerator (■). Dotted line shows the activity toward 0.1 M sarcosine.

(see Fig. 1B), thus indicating that this enzyme also has a second binding site for DMG which increases the enzymatic reaction rate. This enhancement was also observed by the addition of some carboxylic acid derivatives, such as acetoacetic acid, methoxyacetic acid, acetic acid, sarcosine, and formic acid in the presence of more than 200 μM DMG. At low concentrations of DMG, however, methoxyacetic acid and acetic acid (or sodium acetate) inhibited the reaction competitively.

Figure 2A shows the pH profile at various concentrations of DMG. At less than 20 mM DMG, the pH optimum was about 9 and the activity in the neutral region was low, whereas at 100 or

200 mM DMG, pH optimum moved to 7.5-8.5 and the activity in the neutral region became high. Figure 2B shows the pH profile of the oxidative reaction toward 2 mM DMG in the presence of the above-mentioned accelerators. Methoxyacetic acid, acetic acid, and formic acid accelerated the activity without changing the pH optimum, while acetoacetic acid altered it from 9 to 8 and at the latter pH the activity became much higher. Sarcosine also enhanced the activity at neutral pH, but its rate was much lower than that at pH 9. These results clearly support our view that dimethylglycine dehydrogenase has a second binding site other than the first substrate binding site.

When DMG binds to the second binding site of the enzyme, the conformation of the enzyme might be changed to accelerate the enzyme reaction at neutral pH; and one of the accelerators, acetoacetic acid, provokes the same conformational change by combining with the second binding site. For this enhancement by substances other than the substrate DMG, not only the presence of a carboxyl residue but also the same molecular size as DMG seems to be requisite.

References

1. Sato, M., N. Ohishi, K. Yagi. 1979. Biochem. Biophys. Res. Commun. 87, 706.
2. Wittwer, A. J., C. Wagner. 1980. Proc. Natl. Acad. Sci. USA 77, 4484.
3. Steenkamp, D. J., M. Husain. 1982. Biochem. J. 203, 707.
4. Wittwer, A. J., C. Wagner. 1981. J. Biol. Chem. 256, 4109.
5. Porter, D. H., R. J. Cook, C. Wagner. 1985. Arch. Biochem. Biophys. 243, 396.

COMPARISON OF PARTIAL AMINO ACID SEQUENCES DEDUCED FROM THE NUCLEOTIDE SEQUENCE OF A BOVINE ADRENAL MONOAMINE OXIDASE cDNA CLONE TO AMINO ACID SEQUENCES OBTAINED FROM BOVINE LIVER MONOAMINE OXIDASE TYPE B.

Walter Weyler[1], John Powell[2], Yun-Pung P. Hsu[3,4], Shiuan Chen[5], David P. Corey[6,4], Margot Utterback[3,4], Christian Titlow[3,4], Xandra O. Breakefield[3,4]

[1]Veterans Administration Medical Center, Molecular Biology Division, San Francisco; [2]Clinical Research Center, Harrow, Middlesex, England; [3]E.K. Shriver Center, Waltham, Mass.; [4]Neuroscience Program (Neurology) Harvard Medical School, Boston, Mass; [5]City of Hope, Division of Immunology, Duarte CA, [6]Hughes Institute, Massachusetts General Hospital, Boston, Mass.

Monoamine oxidase (MAO) [monoamine: O_2 oxidoreductase (deaminating), EC 1.4.3.4], a covalent FAD flavoenzyme, is of considerable pharmacologic interest because of its central position in the metabolism of neurotransmitter amines. Isoactivities were noted early, and categorized as type A and type B. Although each form of the enzyme activity has been isolated in pure form (1) this has not been accomplished from the same species, thereby, precluding direct comparison. Compelling but indirect evidence suggests that distinct protein species are responsible for the unique activities(2). The enzyme is thought to be composed of two equal-sized subunits of similar amino acid sequence (3), only one of which carries the covalent FAD. Separation of subunits has not yet been achieved.

We have sequenced a MAO cDNA clone from a bovine adrenal cDNA library and compared the deduced amino acid sequence to amino acid sequences of several peptides from bovine liver MAO type B. These studies provided evidence that there are multiple mRNA species coding for bovine MAO. We also discuss local homologies of two regions of the deduced amino acid sequence with other flavoproteins. Methods will be presented elsewhere.

RESULTS and DISCUSSION

So far a span of 492 amino acid residues have been deduced from the

Flavins and Flavoproteins 1987

A

cDNA	NWCQEQYSGGCYTAYF
FAD peptide	NWC<u>E</u>EQYSGGCYTAYF
	*

B

	*
cDNA	YKEAFWKKKDYCGCMII
PEP 1&2	YKEPFWKNMDYCGSMII
PEP 3	YKEPFWIPPPM

Fig. 1. **Comparison of peptide sequence from bovine liver MAO B to sequenc fragments deduced from bovine adrenal MAO cDNA.** A. Comparison of FAD peptide. Position of FAD is indicated by asterisk. Deviation from deduced sequence is underlined. B. Comparison of three peptide seqences to a segment of deduced sequence. Asterisk indicates start of peptide 2.

nucleotide sequence of the adrenal MAO cDNA clone. We have also sequenced 12 peptides of bovine liver MAO type B containing 153 amino acid residues. Of these 72.5% are identical to the deduced cDNA sequence. The flavin containing peptide and three other peptides are of particular interest. As is shown in Figure 1A, 15 of 16 amino acid residues flanking the covalent flavin binding site are identical to the deduced sequence with a gln changed to glu. Figure 1B shows the amino acid sequence of three peptides. Peptide one and two are adjacent on the deduced sequence. The third peptide is homologous with the first peptide but then deviates from the contiguous second peptide. This result indicates that the cloned cDNA is not the product of mRNA for MAO B but is most likely derived from mRNA coding for one subunit of MAO A. To date, this is the most direct demonstration that there are multiple mRNA species coding for MAO. For this conclusion it is assumed that there are no internal sequence repeats in MAO B; however, none are found in the deduced sequence reported on here. In addition, these findings also support the two hypotheses: 1. that the bovine liver MAO B has two subunits and 2. that these are similar in primary sequence but not identical. Therefore, although the flavin peptide is 94% identical to the deduced amino acid sequence it is possible that the deduced sequence represents the non-flavin subunit.

Comparison of the deduced amino acid sequence to the protein database, using the Dayhoff (11) mutational data matrix, revealed a number of homologies. Near the N-terminus of the deduced sequence MAO is homologous with the N-terminus of a number of FAD flavoproteins (Figure 2). This region forms the well known nucleotide binding fold which is the binding site for the diphosphate ester and the adenine ring system of FAD. In addition to other similarities a highly conserved triad of glycine residues, thought to

Enzyme[a]	Start Position	-------- Sequence --------					Ref.
			* * *				
MAO	20	AGGMFDVVVI	GGGISGLSAA	KLLAEHEVNV	LVLEARERVGG	R	
MerA	100	QVAVI	GSGGAAMAAA	LKAVEQGAQV	TLIERG		10
GR	17	AVASYEYLVI	GGGSGGLASA	RRAAELGARA	AVVESHKLGGT	CV	8
AAO (pig)	2	MRVVVI	GAGVIGLSTA	LCIHERYHSV	LQPLDVKVYAD	R	11
LDH (E.coli)	3	TEIKTGVVVL	GAGPAGYSAA	FRCADLGLET	VIVERYNTLGG	V	9
FRDA (E.coli)	2	QTFQADLAIV	GAGGAGLRAA	IAAAQANPNA	KIALISKVYPM	RSH	7
SDHA (B.subtilis)	2	SQSSIIVV	GGGLAGLMAT	IKAAESGMAV	KLFSIVPVKRS	HSV	6
AlO (yeast)	3	IPEEFDILVL	GGGSSGGSCI	AGRLANLDHS	LK		12
pHBH (Pseud.fluo.)		MKTQVAII	GAGPSGLLLG	QKKHKAGIDN	VILERQTPDYV	L	5

Figure 2. **Comparison of N-terminus of a number of FAD proteins with deduced sequence of MAO.** Homologies with MAO are underlined. Only a representative member of each enzyme is shown. (*), marks conserved glycine triad. [a]Abreviations used are GR, glutatione reductase; AAO, amino acid oxidase; LDH, lippoamide dehyd-rogenase; FRDA, fumarate reductase subunit A; SDHA, succinate dehydrogenase subunit A; AlO, alcohol oxidase; pHBH, para-hydroxy benzoate hydroxylase.

Source	Start Position	----Sequence------			References
MAO B	438	IYFA	GTETA	T	
mercuric reductase	16	PVWA	ATQTV	TLSVPGM	10
Cyt P450 reductase	82	VFY-	GSQTG	TAEEFAN	13
Azotobacter vinelandii	5	LFF-	GSNTG	KTRKVAK	13
Desulfovibrio vulgaris	6	IVY-	GSTTG	NTEYTAE	13
Megasphaera elsdenii	4	IVY-	WSGTG	NTEAMAN	13
Clostridium MP	3	IVY-	WSGTG	NTEKMAE	13

Figure 3. **Comparison of MAO sequence segment to cytochrome P450 reductase, mercuric reductase and four bacterial flavodoxins.** This figure is modified from Porter and Kasper(13) with mercuric reductase and MAO added. Features thought to be important for bonding to the phosphate of FMN are underlined. (-), indicates space introduced to give better fit.

be an important structural element in the formation of a Rossmann-fold, is present. These homologies suggest that the adenine moiety and the diphosphate ester of the covalent FAD in MAO, fumarate reductase and succinate dehydrogenase make similar contacts to the protein supersecondary structure as FAD in noncovalent flavoenzymes. In marked contrast to the latter two covalent flavoenzymes, in which the FAD binds to a histidyl residue near position 40, in MAO the covalent FAD bond is to a cysteinyl residue at position 415. If the noncovalent bonding of FAD in MAO is similar to other FAD proteins then the obvious implication for MAO would be that the C-terminal end of the enzyme loops back toward the N-terminal end.

An additional short segment of 11 amino acids, starting with residue 438, is also homologous with short regions of other proteins including the flavoproteins in Figure 3. This homology between cyt P450 reductase and the flavodoxins has previously been discussed in terms of these residues making contacts with the phosphate group of FMN (13). Including this last homologous peptide, therefore, at least three possible sites may be considered as being involved in a hypothetical noncovalent FAD binding site in MAO.

Aknowledgements: This work was supported by the Veterans Administration; NIH grants MH42462 (S.C.) and NS 21921 (X.O.B.); and the Scottish Rite Foundation. One of us (W.W.) thanks John W. Winslow for helpful discussions.

REFERENCES

1. Weyler, W., Salach, J.I. (1985) J.Biol.Chem. **260**, 13199 & ref. therein
2. Cawthon, R.M., Breakefield, X.O. (1983) Biochem. Pharmacol. **32** 441 and references therein
3. Minamiura, N., Yasunobu, K.T. (1978) Arch. Biochem. Biophys. **189**, 481
4. Dayhoff, M.O. Baker, W.C., Hunt, L.T. (1983) Meth. Enzymol. **91**, 534
5. Vereijken, J.M., Hofsteenge, J., Bak, H.J., Beintema, J.J. (1980) Eur. J. Biochem. **113**, 151
6. Phillips, M.K., Hederestedt, L., Hasnain, S., Rutberg, L., Guest, J.R. (1987) J. Bacteriol. **169**, 864
7. Cole, S.T. (1982) Eur. J. Biochem. **122**, 479
8. Kraut-Siegel, R.L., Blatterspiel, R., Saleh, M., Schiltz, E., Shirmer, R.H., Untucht-Grau, R. (1982) Eur. J. Biochem. **121** 259
9. Rice, D.W., Schulz, G.E., and Guest, J.R.(1984) J. Mol. Biol. **174**, 483
10. Nigel L. Brown, N.L., Dean Goddette, D. (1984) Flavins and Flavoproteins. R.C. Bray, P.C. Engel, S.G. Mayhew eds. p 165. Walter de Gryter.
11. Ronchi, S., Minchiotti, L., Falliano, M., Curti, B., Swenson, R.P., Williams, C.H., Massey, V. (1982) J. Biol. Chem. **257**, 8824
12. Ellis, S.B., Brust, P.F., Koutz, P.A., Waters, A.F., Harpold, M.M., Gingeras, T.R. (1985) Mol. Cell. Biol. **5**, 1111
13. Porter, T.D., Kasper, C.B. (1986) Biochemistry **25** 1682 & ref. therein.

COVALENT BINDING OF FAD TO *BACILLUS SUBTILIS* SUCCINATE DEHYDROGENASE

L. Hederstedt
Department of Microbiology, University of Lund, Sölvegatan 21, S-223 62 Lund, Sweden

About 25 different enzymes have been found to contain flavin covalently bound to a specific His, Tyr or Cys residue, depending on the protein (1, 2). The functional significance of the bond is not understood. Does it alter the redox properties of the flavin, is it important for protein structure or is it there just to keep the flavin in place? Other unsolved problems are exactly *when* and *how* during biosynthesis of the flavoprotein the modification is accomplished and if specific cell-factors/enzymes are required. We address these questions by using the Krebs' cycle enzyme succinate dehydrogenase (SDH) in the bacterium *Bacillus subtilis* as a model system (3). Our experimental approaches are (i) amino acid sequence comparisons, (ii) mutant studies, and (iii) expression of the flavoprotein in heterologous cells.

SDH

Eucaryotic and procaryotic SDH are very similar (4,5). They consist of two unequal protein subunits with one FAD linked 8α-N(3)His to the larger, Fp, subunit (6). Three different iron-sulfur (FeS) centers, detectable by EPR and designated S-1, S-2 and S-3, are present in the enzyme. All the centers seem ligated to the smaller, Ip, subunit. S-1 is a $[2Fe-2S]^{2+,1+}$, S-2 a $[4Fe-4S]^{2+,1+}$, and S-3 a $[3Fe-xS]$

Flavins and Flavoproteins 1987

(x=3 or 4) center (7, 8). Centers S-1 and S-3 can be reduced by succinate, whereas center S-2 can only be reduced by stronger reductants.

Structure of Fp

The nucleotide sequence of the structural genes for *B. subtilis* Fp and Ip have been determined and confirmed by N-terminal amino acid sequencing (9, 10). Fp contains 585 amino acid residues. The His-40 residue in Fp to which FAD is covalently bound has been identified from the strong homology of the predicted amino acid sequence to the sequence of a flavin-peptide isolated from bovine heart SDH (11) (Fig. 1). The primary sequence around the His is conserved in different succinate oxidoreductases (12), but not in all proteins with 8α-N(3)His bound FAD (1). This suggests that the conserved sequence is more important for enzyme function than for binding of flavin.

Figure 1. Conserved amino acid residues in the 8α-[N(3)His]-FAD region of *B. subtilis*, *E. coli* and bovine heart mitochondrial Fp subunit of SDH.

Amino acid sequence comparisons with *E. coli* SDH and fumarate reductase indicate that in the folded Fp polypeptide both the N-terminal region and an internal segment around residue 360 contact the adenine phosphate part of FAD (9). Recent experiments done in our laboratory, in which the *B. subtilis* SDH has been redesigned to more closely resemble *E. coli* SDH, have demonstrated that the conserved segment at residues 250 to 257 forms part of the substrate binding site (13). Conserved His residues at positions 235 and 381 may be directly involved in catalysis and located close to the isoalloxazine ring.

Flavin mutants

Several mutants of *B. subtilis* lacking covalently bound flavin in Fp have been isolated (14). These mutants totally lack SDH activity. In Table 1 some properties of SDH in one such mutant are compared with those of SDH from wild type bacteria and a mutant SDH containing FAD. Surprisingly, membrane binding of SDH, assembly of FeS centers and electron transfer from the respiratory chain to centers S-1 and S-3 are not affected by the lack of the relatively large FAD molecule. Furthermore, the EPR signals and the midpoint redox potentials of the FeS centers and the spin-interactions between S-1 and S-2 and between S-1 and S-3 are identical in flavin deficient and wild type SDH (12, 15). The primary defect is in the Fp subunit in both these mutant SDH as determined by cloning and sequencing mutated genes (12, 16). The respective mutations are amino acid substitutions (see Table 1). The replacement of Gly-46 with Asp results in a flavin deficient SDH. The approximative location of other mapped, but not yet sequenced, missence mutations in the Fp which also result in lack of covalently bound flavin are shown in Figure 2. The data from the mutants demonstrate that amino acid substitutions far downstream from the His-40 region can effect flavinylation of *B. subtilis* Fp.

Table 1. Properties of Membrane Bound SDH in *B. subtilis* Wild Type Cells and Two Mutant Strains.

Property	Wild type	Strain KA97124	KA97101
SDH activity	+	−	−
MW of protein subunits			
Fp	65 186	65 243	65 257
Ip	28 285	28 285	28 285
FAD_{cov} present	+	−	+
FeS centers S-1,S-2, and S-3 present	+	+	+
Mutation in Fp		G-46 to D	G-116 to E

Flavinylation

Flavin is known to be covalently bound to the Fp polypeptide independently of the Ip polypeptide and before the subunits assemble into SDH on the membrane (14). Mutations which interfere with the covalent binding of FAD are spread over the Fp polypeptide (Fig. 2). The NH_2-terminal end of the protein, encompassing the first 150 residues with the His-40, is thus not sufficient for FAD to be covalently bound. These findings strongly suggest that the newly translated protein has to be folded into a structure exposing a flavin "binding pocket" before FAD (activated at the 8α-methyl?) can be bound. The subsequent covalent bond formation between the methyl group of the isoalloxazine ring and the tele-nitrogen of the imidazole side-group in His-40 is possibly catalyzed by the Fp polypeptide itself (2).

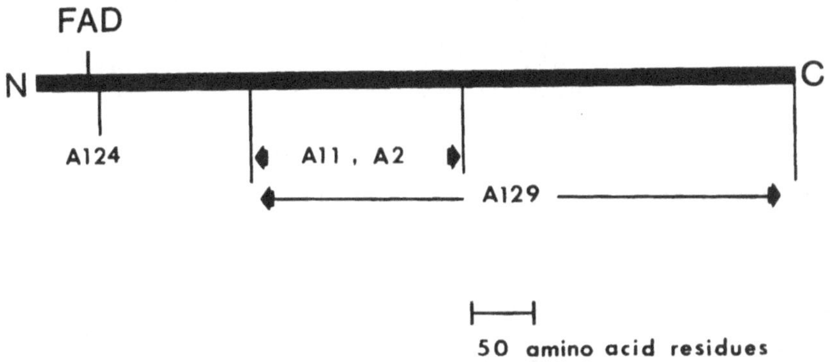

Figure 2. Map of the *B. subtilis* Fp polypeptide. Indicated are the locations of the FAD attachment site (His-40) and different amino acid substitutions which result in a not flavinylated SDH.

There is no genetic evidence for an enzyme in *B. subtilis* with the only function to flavinylate Fp. Fp is the *only* protein containing a covalently bound flavin in *B. subtilis* cells (14) and the 35 mutants defective in SDH which have been analysed are all mutated in the *sdh* genes (3). On the other hand, *B. subtilis* Fp polypeptide expressed from the cloned structural gene in *E. coli* does not contain covalently bound flavin (10). This lack of flavin is not a result of a mutational change in the cloned gene since the isolated gene codes for a functional Fp when reintroduced into *B. subtilis* (17). The explanation(s) to why FAD is not covalently bound to *B. subtilis* Fp in *E. coli* must therefore be found at the post-translational level. The N-terminus of the heterologous Fp is processed normally in *E. coli*. Furthermore, *E. coli* proteins are flavinylated also in cells expressing the *B. subtilis* SDH polypeptides. These results show that *B. subtilis* cell specific factors directly or indirectly are required for flavinylation or that factors in *E. coli* prevent the covalent binding of flavin to *B. subtilis* Fp.

Acknowledgement

This work was supported by grants from the Swedish Medical Research Council and Emil och Wera Cornells Stiftelse.

References

1. Singer, T.P. and McIntire, W.S. 1984. Methods Enzymol. *106*, 369-378.

2. Decker, K. 1982. In: Flavins and Flavoproteins (Massey and Williams, eds.), Elsevier North Holland Inc., pp. 465-472.

3. Hederstedt, L. 1986. Methods Enzymol. *126*, 399-414.

4. Hederstedt, L. and Rutberg, L. 1981. Microbiol. Rev. *45*, 542-555.

5. Hatefi, Y. 1985. Ann. Rev. Biochem. *54*, 1015-1069.

6. Salach, J., Walker, W.H., Singer, T.P., Ehrenberg, A., Hemmerich, P., Ghisla, S. and Hartman, U. 1972. Eur. J. Biochem. *26*, 267-278.

7. Singer, T.P. and Johnson, M.K. 1985. FEBS Lett. *190*, 189-198.

8. Ohnishi, T. 1987. In: Current Topics in Bioenergetics (Lee ed.) Academic Press, New York.

9. Phillips, M.K., Hederstedt, L., Hasnain, S., Rutberg, L. and Guest, J.R. 1987. J. Bacteriol. *169*, 864-873.

10. Hederstedt, L., Bergman, T. and Jörnvall, H. 1987. FEBS Lett. *213*, 385-390.

11. Kenney, W.C., Walker, W.H., and Singer, T.P. 1972. J. Biol. Chem. *247*, 4510-4513.

12. Maguire, J.J., Magnusson, K. and Hederstedt, L. 1986. Biochemistry, USA. *25*, 5202-5208.

13. Hederstedt, L. and Hedén, L.-O. 1987. Manuscript in preparation.

14. Hederstedt, L. 1983. Eur. J. Biochem. *132*, 589-593.

15. Hederstedt, L., Maguire, J.J., Waring, A.J. and Ohnishi, T. 1985. J. Biol. Chem. *260*, 5554-5562.

16. Fridén, H. and Hederstedt, L. 1986. Unpublished.

17. Fridén, H., Hederstedt, L. and Rutberg, L. 1987. FEMS Microbiol. Lett. *41*, 203-206.

6-HYDROXY-D-NICOTINE OXIDASE: EXPRESSION OF THE FLAVOENZYME AS A COVALENTLY FLAVINYLATED POLYPEPTIDE

Roderich Brandsch

Biochemisches Institut der Universität, D-7800 Freiburg i. Br., FRG

Introduction

In flavoproteins (FPs) the prosthetic group is tightly, but in most cases noncovalently bound to the polypeptide chain. There is, however, a significant number of FPs known with covalently attached flavin (cFPs). The amino acid residues of the polypeptide chain involved in this type of binding were identified as histidine, cysteine and tyrosine. The occurence of cFPs and the type of linkage between the flavin and the polypeptide moiety has been reviewed extensively (1,2,3,4). From these studies no obvious physiological advantage of this type of cofactor binding could be deduced. A rational explanation for the need of covalent flavin attachment to cFPs is still lacking. Another open question is how the covalent linkage is accomplished. In other known situations of covalently bound prosthetic groups, like biotin, the reaction is enzymatically catalyzed (5). Whether an enzyme is involved in covalent flavinylation is still unknown.

We are investigating the structure and regulation of the nicotine regulon, a plasmid-encoded catabolic pathway in *Arthrobacter oxidans*. It comprises several flavoenzymes (6,7,8). One of these enzymes, 6-hydroxy-D-nicotine oxidase (6-HDNO) possesses an 8α-riboflavin-N_3-histidyl linkage (9).

Flavins and Flavoproteins 1987

The enantiozyme, 6-hydroxy-L-nicotine oxidase (6-HLNO) with noncovalently bound FAD is coinduced by nicotine in this organism. The expression of the two enzymes, however, is differently regulated. 6-HLNO is synthesized already during logarithmic growth, while 6-HDNO starts to appear in the stationary phase. It is not known if the difference in FAD-binding observed between the two enantiozymes (6) is due to the coinduction of a flavinylating enzyme together with 6-HDNO. Since the bacterial cells also undergo in the stationary phase a morphological differentiation from rod-shaped bacteria to coccoid forms, a switch in the expression of certain bacterial gens seems not surprising.
We were interested to elucidate two questions: 1. at what stage during the synthesis of 6-HDNO is the cofactor bound, e.g. is it a cotranslational or a posttranslational process, and 2. can we identify a flavinylating enzyme (a FAD-hologenzyme synthetase) for the 6-HDNO?
The availability of the cloned 6-HDNO gene (10) allowed us to extend previously performed studies (11) and to analyze these questions in greater detail.

Results

Covalent flavinylation of 6-HDNO expressed in *E.coli*. The 6-HDNO gene cloned into *E.coli* allowed the synthesis of a polypeptide which is undistinguishable enzymatically and immunologically from authentic 6-HDNO (10). Using an in vitro coupled transcription-translation system (12) primed with plasmid DNA carrying the 6-HDNO gene we were able to show that the polypeptide synthesized in *E.coli* is also covalently flavinylated (13). That was proof that it is possible to express in a heterologous system the same covalently flavinylated enzyme as in the organism of origin. Assuming the participation of a flavinylating enzyme in this process, this enzyme must be species unspecific.

In vitro synthesis of 6-HDNO apoenzyme. In a riboflavin-requiring mutant of *Arthrobacter oxidans*, the synthesis of 6-HDNO was riboflavin dependent. In the presence of limiting riboflavin concentrations no 6-HDNO apoenzyme could be identified (11). In order to make 6-HDNO apoenzyme available as substrate for an enzyme-catalyzed flavinylation reaction we used the *E.coli*-derived in vitro transcription-translation system. When FAD was removed from the in vitro system an enzymatically inactive polypeptide, corresponding to apo-6-HDNO, was synthesized. However, addition of FAD to the in vitro system containing apo-6-HDNO did not result in the regeneration of an enzymatically active holoenzyme (14). Also no TCA-precipitable $[^{14}C]$FAD could be demonstrated.
From these results we concluded that the conformation of the polypeptide chain synthesized in the absence of the cofactor seems not to allow the covalent binding of FAD. It also follows that attachment of FAD occurred during the synthesis of the polypeptide chain, that is cotranslationally.

Cotranslational flavinylation. We investigated the possibility of cotranslational flavinylation by using various partly delected 6-HDNO genes inserted into appropriate plasmid vectors (15). The polypeptides expressed from these constructs were analyzed in vitro in the *E.coli* transcription-translation system.
Two sets of deletions were performed. Carboxy-terminal deletions resulted in truncated 6-HDNO-polypeptides. Deletions on the amino-terminal end were constructed by in-frame-fusions to the P_L-promoter carried on an expression plasmid vector. No 6-HDNO activity could be detected in extracts obtained from *E.coli* cells transformed with the corresponding plasmid constructs.

The polypeptides expressed in vitro in the E.coli coupled transcription-translation system were analyzed for covalent [^{14}C]FAD incorporation by measuring the amount of TCA-precipitable radioactivity. The data show that removal of approximately 40% from the carboxy-terminal part of the 6-HDNO polypeptide did not inhibit covalent flavinylation of the truncated protein. If a fusion protein was constructed consisting of a truncated 6-HDNO polypeptide and the ß-lactamase of pBR322 the resulting protein still was covalently flavinylated. From these findings we conclude that flavinylation takes place cotranslationally during the synthesis of 6-HDNO.

Amino-terminal deletions showed that the resulting polypeptides were flavinylated if the first 30 amino acids were removed. The histidine involved in FAD-binding was shown to be situated at position #70 in the 6-HDNO amino acid chain. If the sequence upstream of the histidine, comprising the glycine residues was replaced by seven amino acid residues derived from the cloning vector, flavinylation could no longer be detected.

Search for flavinylating activity in the E.coli coupled transcription-translation system. Since the in vitro synthesized apoenzyme does not serve as substrate in a flavinylation reaction, we tried to investigate the involvement of an enzyme in the covalent binding of FAD to 6-HDNO by the use of an in vitro transcription-translation system consisting of well defined components only. Since not all the required factors can be obtained in pure form, we used in a first approach a semi-defined cell-free transcription-translation system (16). It consists of sedimented ribosomes and a post-ribosomal S-150 supernatant. The S-150 supernatant was fractionated on a 10% to 30% sucrose gradient and 19 fractions were collected. One of these fractions containing the tRNA-amino acyl synthetases supported optimal protein synthesis;

it was used in the reconstituted cell-free system supplemented with RNA polymerase, phosphoenolpyruvate and tRNA. It was hoped that the hypothetical holoenzyme-synthetase would be removed from the fraction of the S-150 supernatant supporting cell free synthesis. In this case, the apo-6-HDNO synthesized in this in vitro system should show no enzyme activity. Control assays performed in the presence of [^{35}S]methionine revealed that the 6-HDNO polypeptide has been synthesized. Enzyme activity measured in this reconstituted transcription-translation system is shown in Fig. 1.

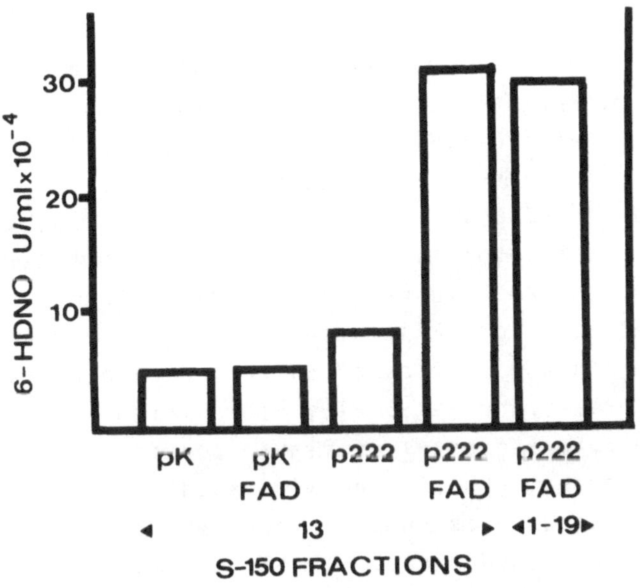

Fig. 1. In vitro synthesized 6-HDNO activity. A S-150 cell free *E.coli* extract was fractionated on a 10-30% sucrose gradient. Fraction #13 was used to support transcription-translation from plasmid DNA in the absence or presence of FAD. pK: plasmid vector pBR322; p222: recombinant plasmid carrying the 6-HDNO gene. Addition of sucrose gradient fractions #1-19 does not result in an increase in 6-HDNO activity.

In the absence of FAD enzyme activity could not be detected. Addition of FAD, however, resulted in the synthesis of enzymatically active 6-HDNO. If different fractions of the sucrose gradient were added to this in vitro system, no stimulating effect on enzyme activity was found. Thus, either the flavinylating enzyme was still present in the sucrose fraction or the ribosomal pellet used in the reconstituted system, or no enzyme is required for FAD-binding.
Although we favour the second possibility, we are still trying to obtain a completely defined in vitro system to clarify this aspect.

Comparison of the amino acid sequence of flavin-peptides from cFPs. Succinate dehydrogenase (SDH) and fumarate reductase were the first enzymes with covalently bound FAD for which the primary structure became known (17,18), the complete sequence of 6-HDNO (19) is now also established. Two partial amino acid sequences became available for sarcosine dehydrogenase (SADH) and dimethylglycine dehydrogenase (DMGDH) from rat liver mitochondria (20). Until now a comparison of the amino acid sequences derived from the flavin-binding peptides of cFPs failed to reveal common features (3,4). The characteristics of SDH and FR are compared to 6-HDNO elsewhere (Brandsch et al., this volume). Fig. 2 lists known amino acid sequences of flavin peptides taking recent data into account (20). The amino acid sequence of the 6-HDNO flavin peptide is shown as derived from the nucleotide sequence of the 6-HDNO gene. In the enzymes with extensive amino acid sequences available, the flavinylated histidine is situated in a stretch of amino acids flanked by glycine residues (Fig.2, DMGDH, SADH, 6-HDNO). This sequence reminds one on the phosphate binding loop discussed for mononucleotide-containing proteins (21). If it reflects indeed a related secondary structure in the FAD-binding domain of cFPs is to early to tell.

DMGDH (Cook et al.,85)	S-E-L-T-A-(G)-S-T-W-H-A-A-(G)-L-T-T
SADH (Cook et al.,85)	L-T-S-(G)-T-T-W-H-T-A-(G)-L-G-R
6-HDNO (Brandsch et al.,86)	R-S(G)-G-H-N-P-N-(G)-Y-A-T-N
SADH (Pinto and Fissel,75)	K-S-E-(G)-H-(T-A-D)
MAO (Walker et al.,71)	S-(G)-G-C-Y

Fig. 2 Comparison of flavin-peptides from cFPs.
The flavinylated amino acid residue is underlined; flanking glycine residues are set in brackets.

Such an arrangement, however, would help to bring the dimethyl-isoalloxazinring and the side-chain of the amino acid residue into proper orientation to allow the formation of a covalent bond.

Acknowledgement

I thank Veronika Bichler for excellent technical assistance and Prof. Dr. Karl Decker for stimulating discussions. This work was supported by a grant from the Deutsche Forschungsgemeinschaft, Bonn-Bad Godesberg (SFB 206).

References

1. Kenney, W.C., W.H. Walker. 1972. FEBS-Lett. 20, 297.
2. Singer, T.P., D.E. Edmondson. 1974. FEBS-Lett. 42, 1.
3. Edmondson, D.E., T.P. Singer. 1976. FEBS-Lett. 64, 255.
4. Decker, K. 1982. In: Flavins and Flavoproteins (V. Massey and C.H. Williams, eds.) Elsevier North Holland, p. 465.
5. Lane, D.M., D.L. Young, F. Lynen. 1964. J. Biol. Chem. 239, 2858.
6. Brühmüller, M., A. Schimz, L. Messmer, K. Decker. 1975. J. Biol. Chem. 259, 7747.
7. Hochstein, L.I., B.P. Dalton. 1967. Biochim. Biophys. Acta 139, 56.
8. Holmes, P.E., S.C. Rittenberg. 1972. J. Biol. Chem. 247, 7628.
9. Singer, T.P., E.B. Kearney, V. Massey. 1955. Arch. Biochem. Biophys. 60, 255.
10. Brandsch, R., W. Faller, K. Schneider. 1986. Mol. Gen. Genet. 202, 96.
11. Hamm, H.-H., K. Decker. 1980. Eur. J. Biochem. 104, 391.
12. Müller, M., G. Blobel. 1984. Proc. Natl. Acad. Sci. USA 81, 7421.
13. Brandsch, R., V. Bichler. 1985. FEBS-Lett. 192, 204.
14. Brandsch, R., V. Bichler. 1986. Eur. J. Biochem. 160, 285.
15. Brandsch, R., V. Bichler, H. Nagursky. 1987. Eur. J. Biochem. (in press)
16. Müller, M., G. Blobel. 1984. Proc. Natl. Acad. Sci. USA 81, 7737.
17. Cole, S.T. 1982. Eur. J. Biochem. 122 479.
18. Wood, D., M.G. Darlison, R.J. Wildw, J.R. Guest. 1984. Biochem. J. 222, 519.

19. Brandsch, R., A.E. Hinkkanen, L. Mauch, H. Nagursky, K. Decker: Eur. J. Biochem. (in press)

20. Cook, R.J., K.S. Misono, C. Wagner. 1985. J. Biol. Chem. 260, 12998.

21. Möller, W., R. Amons. 1985. FEBS-Lett. 186, 1.

22. Pinto, J.T., W.R. Frisell. 1975. Arch. Biochem. Biophys. 169, 483.

23. Walker, W.H., E.B. Kearney, R.L. Seng, T.P. Singer. 1971. Eur. J. Biochem. 24, 328.

THE FLAVIN-BINDING SITES OF 6-HYDROXY-D- AND 6-HYDROXY-L-NICOTINE OXIDASE

Roderich Brandsch, Ari E. Hinkkanen, Ludwig Mauch,
Heiner Nagursky and Karl Decker
Biochemisches Institut der Universität, Freiburg i.Br., FRG

Introduction

In the presence of nicotine a metabolic pathway consisting of a series of flavoenzymes is induced in *Arthrobacter oxidans*. Two of these enzymes, 6-hydroxy-D-nicotine oxidase (6-HDNO) and 6-hydroxy-L-nicotine oxidase (6-HLNO) contain FAD; noncovalently bound in 6-HLNO and covalently bound via an 8α(N3-histidyl)-riboflavin linkage in 6-HDNO. It is the same type of linkage also found in succinate dehydrogenase (SDH) and fumarate reductase (FR) (2,3). The amino-terminal part of the flavinylated subunit exhibits in both enzymes an AMP-binding site characteristic for many FAD and NAD(P)-dependent enzymes (3). This site is followed seven to eight amino acid residues further downstream by the histidine residue involved in covalent flavinylation.
We asked the question whether this particular arrangement is also present in 6-HDNO and how the FAD-binding sites of the enantiozymes 6-HDNO and 6-HLNO compare with each other.

Results and Discussion

We determined the primary structure of the 6-HDNO by sequencing the cloned gene (4). The amino acid sequence of the 6-HDNO FAD peptide was previously determined (5) and was used to establish the correct reading frame. The 6-HLNO polypep-

Flavins and Flavoproteins 1987

tide can readily be purified (6) and the amino acid composition as well as the amino acid sequence of the FAD peptide have been determined. In Tab. 1 the amino acid composition of the 6-HLNO is compared with the amino acid composition derived from the nucleotide sequence of the 6-HDNO gene.

Table 1 Comparison of the Amino Acid Composition of 6-HDNO with the Composition of 6-HLNO.
The amino acid composition of 6-HDNO was derived from the DNA-sequence; the composition of 6-HLNO was obtained from the purified protein. The content of each amino acid is expressed as a percentage of the total number of amino acids.

amino acid	6-HDNO	6-HLNO
Ala	10.3	9.9
Arg	7.2	5.6
Aps/Asn	10.8	9.4
Cys	1.1	1.2
Glu/Gln	8.1	10.3
Gly	9.2	9.2
His	1.5	3.5
Ile	4.9	5.2
Leu	10.5	8.8
Lys	2.7	2.9
Met	1.6	1.0
Phe	3.1	3.7
Pro	5.4	5.9
Ser	7.4	5.7
Thr	4.0	4.4
Trp	2.2	0.8
Tyr	2.2	2.7
Val	7.4	8.8
Total M_r	48,261	53,300

There are marked differences in the content of certain amino acids (e.g. Arg, Asp/Asn, Glu/Gln, His, Leu, Ser, Trp, Val) which argue against a close genetic relationship between the two enzymes.
Direct evidence that the polypeptides of the enantiozymes do not share any common feature, at least in their amino-terminal part, came from the comparison of the amino acid stretch of 35 amino acid residues could be determined. A comparison with the corresponding DNA-derived amino acid

sequence of 6-HDNO did not reveal any homologies (Fig. 1).

```
6-HDNO      MVSSKLATPLSIQGEVIYPDDSGFDPIANIWDGRH..
                                ..(33)SGGH71NPNGYA
               ß         a          ß
6-HLNO      MYDAIVVGGGFSGLKAARDLTEAGCEVLLPEGGER
FR         (6)DLAIVGAGGAGLRAAIAAAQANPNAKIALISK...
                                ...(3)MRSH44TVAAEG  (Cole,82)
SDHA       (8)DAVVIGAGGAGIARLAQISQSGQT  CALLSK...
                                ...(3)TRSH45TVSAQG  (Wood et al,84)
```

Fig. 1 H_3N-terminal amino acid sequences of 6-HDNO and 6-HLNO; comparison with the FAD-binding sites of the covalently flavinylated polypeptides of FR and SDH. Overlined are the secondary structure elements of the AMP-binding domain; highly conserved amino acid residues are underlined; the position of the histidine involved in covalent flavinylation is indicated.

The amino acid sequence of 6-HLNO exhibits a typical AMP-binding site which is absent from the corresponding 6-HDNO sequence. These data taken together show that the two enantiozymes belong to different structural classes of proteins. 6-HLNO apparently resembles other known dinucleotide binding proteins but 6-HDNO not.

6-HDNO is, to our knowledge, the first enzyme, besides FR and SDH, with covalently bound FAD of which the primary structure has been determined. In FR and SDH we have the peculiar situation that an AMP-binding site, as found in enzymes with noncovalently bound dinucleotides, coexists with a covalently bound FAD. This situation induced us to hypothesize (7) that such an arrangement will be found in other covalently flavinylated enzymes, too. In a first step, FAD will be retained by the AMP site of the polypeptide chain, followed in a second step by the formation of a covalent bond between the isoalloxazine ring and a particular amino acid residue (His, Tyr, Cys). Our results show, however, that this is not the case with 6-HDNO. They do not exclude the possibility that the situation characteristic for FR and SDH will be found in other covalently flavinylated enzymes, but they indicate that this particular arrangement is not required for covalent flavinylation. Secondary structure elements of the polypeptide chain are known to be involved in this process (8,9); thus, a two-step mechanism similar to the one outlined above can still hold true.

Acknowledgements

This work was supported by a grant from the Deutsche Forschungsgemeinschaft, Bonn-Bad Godesberg (SFB 206).

References

1. Decker, K., H. Bleeg. 1965. Biochem. Biophys. Acta 105, 313.
2. Cole, S.T. 1982. Eur. J. Biochem. 122, 479.
3. Wood, D., M.G. Darlison, R.J. Wilde, J.R. Guest. 1984. Biochem. J. 222, 519.
4. Brandsch, R., W. Faller, K. Schneider. 1986. Mol. Gen. Genet. 202, 96.
5. Brühmüller, M., K. Decker. 1973. Eur. J. Biochem. 37, 256.
6. Hinkkanen, A., E.-M. Lilius, J. Nowack, R. Maas, K. Decker. 1983. Hoppe-Seyler's Z. Physiol. Chem. 364, 801.
7. Brandsch, R., V. Bichler. 1985. FEBS-Lett. 192, 204.
8. Marguire, J., K. Magnusson, L. Hederstedt. 1986. Biochem. 25, 5202.
9. Brandsch, R., V. Bichler, H. Nagursky. 1987. Eur. J. Biochem. in press

List of Participants

CANADA

SURDHAR, P.S.

ENGLAND

ANDERSON, R.F.

BERRY, A.

ELLISON, P.

ENGEL, P.C.

SCRUTTON, N.J.

FRANCE

LEDERER, F.

SAINT-MARTIN, P.

TEGONI, M.

URBAN, P.

GERMANY

BACHER, A.

BECKER, K.

BRANDSCH, R.

DECKER, K.

ECKSTEIN, J.

GHISLA, S.

KARPLUS, P.A.

KRAUTH-SIEGEL, R.L.

MANSTEIN, D.J.

PAI, E.F.

SCHIRMER, R.H.

SCHÖLLHAMMER, T.

IRELAND

MAYHEW, S.G.

ITALY

CURTI, B.

MALCOVATI, M.

SIMONETTA, M.

VANONI, M.

ZANETTI, G.

JAPAN

FUKUI, K.

HORIIKE, K.

KAKINUMA, K.

KOYAMA, H.

MATSUBARA, H.

MATSUI, J.

MIURA, R.

MIYAKE, Y.

NAKAMURA, T.

NISHINO, T.

NISHINO, T.

NISHIMOTO, Y.

SAITO, T.

SHIGA, K.

SUZUKI, H.

JAPAN (continued)

TSUSHIMA, K.

YAGI, K.

YAMAZAKI, K.

YOSHIKAWA, S.

POLAND

KOZIOL, J.

SPAIN

FILLAT, M.

GOMEZ-MORENO, C.

SWEDEN

HEDERSTEDT, L.

LINDSKOG, S.

NEUJAHR, H.

SANDSTRÖM, A.

SEJLITZ, C.T.

THE NETHERLANDS

BENEN, J.A.E.

BONANTS, P.

DE KOK, A.

MAGER, H.I.X.

MÜLLER, F.

SCHREUDER, H.

VAN BERKEL, W.J.H.

VERVOORT, J.

VISSER, J.W.G.

UNITED STATES

ACKRELL, B.A.C.

AHMED, A.

ARSCOTT, L.D.

BALDWIN, T.O.

BALLOU, D.P.

BARBER, M.J.

BELLAMY, H.D.

BOWERS-KOMRO, D.

BOWMAN, B.

BRISSETTE, P.

BROWN, K.

BRYANT, C.

BYRON, C.M.

CHASTAIN, J.

CLAIBORNE, A.

COOK, R.

CUSANOVICH. M.A.

D'ARDENNE, S.

DAUBNER, C.

DE FRANCESCO, R.

DERVARTANIAN, D.

DISTEFANO, M.

EDMONDSON, D.E.

EINARSDOTTIR, G.H.

EREN, M.

UNITED STATES (continued)

FISH, K.M.

FITZPATRICK, P.

GIEGEL, D.A.

GODDETTE, D.

GREEN, J.M.

GUSTAFSON, W.G.

HALL, C.L.

HILLE, R.

HOPKINS, T.R.

HUYNH, B.

HUSAIN, M.

IACOBUCCI, G.A.

JONES, D.P.

JORNS, M.S.

KAMIN, H.

KIM, J.

KLAPPER, M.H.

KREY, G.D.

LEE, J.

LEE, J.S.

LE GALL, J.

LIVE, D.

LUDWIG, M.L.

MARKLEY, J.L.

MASSEY, V.

MATHEWS, F.S.

MATTHEWS, R.G.

MCCORMICK, D.B.

MCINTIRE, W.

MCWHIRTER, R.B.

MERRILL, A.H. JR.

MEYER, T.E.

MILLER, S.M.

MOORE, M.J.

MOOS, W.H.

MURATA, M.

NEGRI, A.

O'KANE, D.

PATEL, D.

PELEATO, M.

POOLE, L.B.

POULSEN, L.

POWELL, P.J.

POWLOWSKI, J.

PRONGAY, A.

RAMSEY, R.

RIVLIN, R.S.

SAHLMAN, L.

SCHOPFER, L.M.

SINGER, T.P.

STANKOVICH, M.

STEWART, D.

SUDDATH, B.

UNITED STATES (continued)

SWENSON, R.P.

TAYLOR, M.G.

TITLOW, C.C.

TOLLIN, G.

TU, S-C.

WALKER, M.C.

WALSH, C.

WAMPLER, J.

WEYLER, W.W.

WILLIAMS, C.H. JR.

ZELLER, H-D.

ZHOU, J-Z.

ZIEGLER, M.M.

Author Index

Subject Index

Kolibri 360 GmbH
Ferdinand-Jühlke-Straße 7
99095 Erfurt, Deutschland
produktsicherheit@kolibri360.de